Spon's Civil Engineering and Highway Works Price Book

2007

Spon's Civil Engineering and Highway Works Price Book

Edited by

DAVIS LANGDON

2007

Twenty-first edition

First edition 1984
Twenty-first edition published 2007
by Taylor & Francis
2 Park Square, Milton Park, Abingdon, Oxon OX14 4RN

Simultaneously published in the USA and Canada
by Taylor & Francis
270 Madison Avenue, New York, NY 10016

Taylor & Francis is an imprint of the Taylor & Francis Group, an informa business

Printed and bound in Great Britain by
TJ International Ltd, Padstow, Cornwall

Publisher's note
This book has been produced from camera-ready copy supplied by the authors.

British Library Cataloguing in Publication Data
A catalogue record for this book is available from the British Library

ISBN10: 0-415-39381-7
ISBN13: 978-0-415-39381-2
ISSN: 0957-171X

Contents

PART 4: UNIT COSTS (CIVIL ENGINEERING WORKS) – continued

PART 5: UNIT COSTS (HIGHWAY WORKS)

PART 6: LAND REMEDIATION

PART 7: UNIT COSTS (ANCILLARY BUILDING WORKS)

PART 8: ONCOSTS AND PROFIT

PART 9: COSTS AND TENDER PRICES INDICES

PART 10: DAYWORK

PART 11: PROFESSIONAL FEES

PART 12: APPROXIMATE ESTIMATES

PART 13: OUTPUTS

PART 14: TABLES AND MEMORANDA

Preface

Following the major price increases in steel products throughout 2005 there are now signs that prices are steadying in the first half of 2006. Analysts are still uncertain as to the direction oil prices are moving, with some believing that prices could go even higher; certainly for the foreseeable future. Oil prices continue to impact on manufacturing, energy and transportation costs, this is reflected in numerous materials price increases since the beginning of 2006.

For the 2007 edition, we have carried out a general revision of all prices up to May/June 2006 in consultation with leading manufacturers, suppliers and specialist contractors. Our efforts have been directed at reviewing, revising and expanding the scope, range and detail of information to help enable the reader to compare or adjust any unit costs with reference to allocated resources or outputs.

The rates, prices and outputs included in the Resources and Unit Cost calculations, including allowances for wastage, normal productivity and efficiency, are based on medium sized Civil Engineering schemes of about £8 million in value, with no acute access or ground condition problems. However, they are equally applicable, with little or no adjustment, to a wide range of construction projects from £2 million to £50 million. Where suitable, tables of multipliers have been given to enable easy adjustment to outputs or costs for varying work conditions.

As with all attempts to provide price guidance on a general basis, this must be loaded with caveats. In applying the rates to any specific project, the user must take into account the general nature of the project, i.e. matters such as scale, site difficulties, locale, tender climate etc. This book aims at providing as much information as possible about the nature of the rate so as to assist the user to adapt it if necessary.

This edition continues to provide the reader with cost guidance at a number of levels, varying from the more general functional costs shown in Part 7 and Part 12, through the detailed unit costs in Parts 4 and 5 which relate respectively to the CESMM3 and the Highways Method of Measurement bills of quantities formats, down to the detailed resource costings given in Part 3 supplemented by the further advice on output factors in Part 13.

The Unit Costs sections (Parts 4, 5 and 7) cover a wide range of items and, where appropriate, notes are included detailing assumptions on composition of labour gang, plant resources and materials waste factors. Part 3 (Resources) includes detailed analysis of labour and plant costs, allowing the user to adjust unit costs to individual requirements by the substitution of alternative labour, materials or plant costs. Unit Costs are obviously dependent upon the outputs or man-hours used to calculate them. The outputs used in this work have been compiled in detail from the editors' wide ranging experience and are based almost exclusively on time and motion studies and records derived from a large number of recent Civil Engineering schemes. This information is constantly being re-appraised to ensure consistency with current practice. A number of prices and outputs are based upon detailed specialist advice and acknowledgements to the main contributors are included on page xiii.

The market could change very rapidly and to monitor this and maintain accuracy levels readers should use the free price book update service, which advises of any significant changes to the published information over the period covered by this edition. The Update is circulated free every three months, until the publication of the next annual Price Book, to those readers who have registered with the publishers; in order to do so readers should complete and return the coloured card bound within this volume.

Whilst all efforts are made to ensure the accuracy of the data and information used in the book, neither the editors nor the publishers can in any way accept liability for loss of any kind resulting from the use made by any person of such information.

DAVIS LANGDON
MidCity Place
71 High Holborn
London
WC1V 6QS

Abbreviations

BS	British Standard
BSS	British Standard Specification
cwt	hundredweight
cu	cubic
DL	Davis Langdon LLP
DERV	diesel engine road vehicle
Defra	Department for Environment, Food and Rural Affairs
DfT	Department for Transport
ft	foot
ft^3	cubic foot
ft^3/min	cubic foot per minute
ha	hectare
HMSO	Her Majesty's Stationery Office
hp	horsepower
hr	hour
in	inch
kg	kilogramme
km	kilometre
kVA	kilovolt ampere
kw	kilowatt
m	metre
m^2	square metre
m^3	cubic metre
μu	micron (10^{-3} millimetre)
mm	millimetre
mm^2	square millimetre
N	newton
ne	not exceeding
nr	number
pa	per annum
PC	Prime Cost
sq	square
t	tonne
wk	week
yd	yard
yd^3	cubic yard

Acknowledgements

The Editors wish to record their appreciation of the assistance given by many individuals and organisations in the compilation of this edition.

Materials suppliers and sub-contractors who have contributed this year include:

Abacus Lighting Ltd
Oddicroft Lane
Sutton-in-Ashfield
Notts NG17 5FT
Lighting and street furniture
Tel: 01623 511111
Fax: 01623 552133
Website: www.abacuslighting.com
e-mail: sales@abacuslighting.com

ACO Technologies Plc
Hitchin Road
Shefford
Beds SG17 5TE
Drainage systems
Tel: 01462 816666
Fax: 01462 815895
Website: www.aco.co.uk

Aggregate Industries (Charcon)
Hulland Ward
Ashbourne
Derby DE6 3ET
Kerbs, edgings and pavings
Tel: 01335 372222
Fax: 01335 370074
Website: www.charcon.com

Akzo-Nobel Decorative Coatings Ltd
PO Box 37
Crown House
Hollins Road
Darwen
Lancs BB3 0BG
Paints/finishes
Tel: 01254 704951
Fax: 01254 774414
Website: www.crownpaint.co.uk

Akzo-Nobel Wood Treatment
Meadow Lane
St Ives
Cambs PE27 4UY
Timber treatment
Tel: 01480 496868
Fax: 01480 496801
Website: www.woodprotection.co.uk
e-mail: woodcare@sis.akzonobel.com

AMEC Specialist Businesses
Cold Meece
Swynnerton
Stone
Staffordshire ST15 0QX
Tunnelling
Tel: 01785 760022
Fax: 01785 760762
Website: www.amec.com

Andrew Sykes PLC
Premier House
Darlington Street
Wolverhampton
West Midlands WV2 4JJ
Groundwater controls
Tel: 01902 328700
Fax: 01902 422466
Website: www.andrews-sykes.com
e-mail: info@andrews-sykes.com

Arminox UK Ltd
Oundle
Peterborough
Cambs PE8 4JT
Stainless steel reinforcement
Tel: 01832 272109
Fax: 01832 275759
Website: www.arminox.com

Artex Rawlplug Company Ltd
Skibo Drive
Thornlie Bank Industrial Estate
Glasgow G46 8JR
Fixings
Tel: 0141 638 7961
Fax: 0141 638 7397
Website: www.rawlplug.co.uk

Arthur Fischer (UK) Ltd
Hithercroft Road
Wallingford
Oxon OX10 9AT
Fixings
Tel: 01491 833000
Fax: 01491 827953
Website: www.fischer.co.uk
e-mail: sales@fischer.co.uk

Asset International Ltd
Stephenson Street
Newport
South Wales NP19 0XH
Steel culverts
Tel: 01633 271906
Fax: 01633 290519
Website: www.assetint.co.uk

Bachy Solertanche Ltd
Foundation Court
Godalming Business Centre
Catteshall Lane
Godalming
Surrey GU7 1XW
Diaphragm walling
Tel: 01483 427311
Fax: 01483 417021
Website: www.bacsol.co.uk

Barnard Group Ltd
Lower Tower Street
Birmingham B19 3NL
Coated steel pipes and fittings
Tel: 0121 359 5531
Fax: 0121 359 8350
Website: www.barnard.co.uk
e-mail: admin@barnard.co.uk

Beaver 84 Ltd
Beaver House
Crompton Close
Basildon
Essex SS14 3AY
Scaffolding supply and hire
Tel: 01268 530888
Fax: 01268 531888
Website: www.beaver84.co.uk
e-mail: sales@beaver84.co.uk

Bell & Webster Concrete Ltd
Alma Park Road
Grantham
Lincs NG31 9SE
Precast concrete retaining walls
Tel: 01476 562277
Fax: 01476 562944
Website: www.eleco.com/bellandwebster
e-mail: bellandwebster@eleco.com

Blockleys Plc
Sommerfield Road
Trench Lock
Telford
Shropshire TF1 5RY
Bricks and pavers
Tel: 01952 251933
Fax: 01952 265377
Website: www.blockleys.com

BLüCHER UK LTD
Station Road Industrial Estate
Tadcaster
N Yorkshire LS24 9SG
Stainless steel pipes and fittings
Tel: 01937 838000
Fax: 01937 832454
Website: www.blucher.co.uk
e-mail: mail@blucher.co.uk

BRC Special Products
Astonfield Industrial Estate
Stafford ST16 3BP
Masonry reinforcement and accessories
Tel: 01785 222288
Fax: 01785 240029
Website: www.brc-special-products.co.uk

BSS Group PLC
Fleet House
Lee Circle
Leicester LE1 3QQ
Steel pipes and fittings
Tel: 0116 2623232
Fax: 0116 2531343
Website: www.bssuk.co.uk

Capital Demolition
Capital House
Woodham Park Road
Addlestone
Surrey KT15 3TG
Demolition and clearance
Tel: 01932 346222
Fax: 01932 340244
Website: www.capitaldemolition.co.uk

Caradon Parker Ltd
Units 1-2 Longwall Avenue
Queen's Drive Industrial Estate
Nottingham NG2 1NA
Geotextiles
Tel: 0115 986 2121

Castle Cement
Park Square
3160 Solihull Parkway
Birmingham Business Park
Birmingham B37 7YN
Cement products
Tel: 0121 606 4000
Fax: 0121 606 1412
Website: www.castlecement.co.uk
email: information@castlecement.co.uk

CCL Stressing Systems
Unit 4
Park 2000
Millennium Way
Westland Road
Leeds LS11 5AL
Bridge Bearings
Tel: 0113 270 1221
Fax: 0113 271 4184
Website: www.ccclstressing.com
e-mail: sales@cclstressing.com

Celsa Steel (UK) Ltd
Building 58
East Moors Road
Cardiff CF24 5NN
Bar reinforcement
Tel: 0292 035 1800
Fax: 0292 035 1801
Website: www.celsauk.com

Cementation Foundations Skanska Ltd
Maple Cross House
Denham Way
Rickmansworth
Herts WD3 9AS
Piling
Tel: 01923 423 100
Fax: 01923 777 834
Website: www.cementationfoundations.
skanska.co.uk
e-mail: cementation.foundations@
skanska.co.uk

Corus Construction & Industrial
PO Box 1
Brigg Road
Scunthorpe
North Lincoln DN16 1BP
Structural steel sections
Tel: 01724 404040
Fax: 0870 902 3133
Website: www.corusgroup.com
www.corusconstruction.com
e-mail: corusconstruction@corusgroup.com

CU Phosco Lighting Ltd
Charles House
Great Amwell
Ware
Herts SG12 9TA
Luminaires, bollards, street furniture
Tel: 01920 462272
Fax: 01920 860635
Website: www.cuphosco.co.uk

Darfen Durafencing
Unit B1
Castle Road
Sittingbourne
Kent ME10 3RL
Fencing Systems and Gates
Tel: 01795 414180
Fax: 01795 414190
Website: www.darfen.co.uk

Don & Low Ltd
Newford Park House
Glamis Road
Forfer
Angus DD8 1FR
Geotextiles
Tel: 01307 452200
Fax: 01307 452422
Website: www.donlow.co.uk

Entec UK Ltd (Consulting Engineers)
160-162 Abbey Foregate
Shrewsbury
Shropshire SY2 6BZ
Land Remediation
Tel: 01743 342000
Fax: 01743 342010
Website: www.entecuk.com

Expanded Metal Company Ltd
PO Box 14
Longhill Industrial Estate (North)
Hartlepool
Cleveland TS25 1PR
Expanded metal products
Tel: 01429 867388
Fax: 01429 866795
Website: www.expamet.co.uk
e-mail: sales@expamet.co.uk

Finning UK Ltd
Watling Street
Cannock
Staffs WS11 3LL
Plant agents (Caterpillar)
Tel: 01543 462551
Fax: 01543 416700
Website: www.finning.co.uk
e-mail: mailbox@finning.co.uk

Fosroc Ltd
Coleshill Road
Tamworth
Staffordshire B78 3TL
Waterproofing and expansion joints
Tel: 01827 262222
Fax: 01827 262444
Website: www.fosrocuk.com
e-mail: sales@fosrocuk.com

Freyssinet Ltd
Hollinswood Court
Stafford Park
Telford
Shropshire TF3 3DE
Reinforced earth systems
Tel: 01952 201901
Fax: 01952 201753
Website: www.freyssinet.co.uk

GEM Professional (Joseph Metcalfe Ltd)
Brookside Lane
Oswaldtwistle
Accrington
Lancs BB5 3NY
Landscaping materials
Tel: 01254 356600
Fax: 01254 356677
Website: www.gemgardening.co.uk
e-mail: sales@gemgardening.co.uk

Geberit UK Ltd
New Hythe Business Park
Aylesford
Kent ME20 7PG
UPVC pipes
Tel: 01622 717811
Fax: 01622 716920
Website: www.geberit.co.uk

Grace Construction Products Ltd
Ajax Avenue
Slough
Berks SL1 4BH
Waterproofing/expansion joints
Tel: 01753 692929
Fax: 01753 691623
Website: www.graceconstruction.com

Greenham Trading Ltd
Greenham House
671 London Road
Isleworth
Middx TW7 4EX
Contractors site equipment
Tel: 020 8560 1244
Fax: 020 8568 8423
Website: www.greenham.com
e-mail: isleworth.sales@greenham.com

Halfen Ltd.
Humphrys Road
Woodside Estate
Dunstable
Beds LU5 4TP
Concrete Inserts
Tel: 01582 470300
Fax: 08705 316304
Website: www.halfen.co.uk
e-mail: info@halfen.co.uk

Hanson Building Products
Stewartby
Bedford
Beds MK43 9LZ
Bricks
Tel: 08705 258258
Fax: 01234 762040
e-mail: info@hansonbrick.com

Hanson Concrete Products
PO Box 14
Appleford Road
Sutton Courtney
Abingdon
Oxfordshire OX14 4UB
Blockwork and precast concrete products
Tel: 01235 848808
Fax: 01235 846613
Website: www.hanson-europe.com

Hepworth Building Products Ltd
Hazlehead
Crow Edge
Sheffield S36 4HG
Clayware Pipes
Tel: 01226 763561
Fax: 01226 764827
Website: www.hepworthdrainage.co.uk
e-mail: info@hepworthdrainage.co.uk

Hepworth Concrete Pipes Ltd
White Hill Road
Ellistown
Leicester LE67 1ET
Concrete pipes
Tel: 01530 240024
Fax: 01530 240025
Website: www.hepworthconcrete.co.uk
e-mail: info@hepworthconcrete.co.uk

H E Services (Plant Hire) Ltd
Whitewall Road
Strood
Kent ME2 4DZ
Plant Hire
Tel: 01634 291290
Fax: 01634 295626
Website: www.heservices.co.uk
e-mail: strood@heservices.co.uk

HM Plant Ltd
38 Castlefields Industrial Estate
Bridgewater
Somerset TA6 4ZE
Plant agents (Hitachi Excavators)
Tel: 01278 425533
Fax: 01278 452511
Website: www.hmplant.ltd.uk
e-mail: info@hmplant.ltd.uk

Hughes Concrete Ltd
Barnfields
Leek
Staffs ST13 5QG
Precast concrete manhole units
Tel: 01538 380500
Fax: 01538 380510
Website: www.hughesconcrete.co.uk

Inform UK Ltd
Industrial Park
Ely Road
Waterbeach
Cambridge CB5 9PG
Formwork and accessories
Tel: 01223 862230
Fax: 01223 440246
Website: www.inform.co.uk
e-mail: general@inform.co.uk

James Latham Plc
Unit 3
Finway Road
Hemel Hempstead HP2 7QU
Timber merchants
Tel: 01442 849100
Fax: 01442 267241
Website: www.lathamtimber.co.uk
e-mail: plc@lathams.co.uk

JC Banford Excavators Ltd
Rocester
Staffs ST14 5JP
Plant manufacturers
Tel: 01889 590312
Fax: 01889 591507
Website: www.jcb.co.uk

Klargester Environmental Products Ltd
College Road
Aston Clinton
Aylesbury
Bucks HP22 5EW
Cesspools, septic tanks and interceptors
Tel: 01296 633000
Fax: 01296 633001
Website: www.klargester.co.uk
e-mail: uksales@klargester.co.uk

Leighs Paints
Tower Works
Kestor Street
Bolton
Lancs BL2 2AL
Paints/finishes
Tel: 01204 521771
Fax: 01204 382115
Website: www.leighspaints.co.uk

Maccaferri Ltd
The Quorum
Oxford Business Park Nr
Garsington Road
Oxford OX4 2JZ
River and sea gabions
Tel: 01865 770555
Fax: 01865 774550
Website: www.maccaferri.co.uk
e-mail: oxford@maccaferri.co.uk

Marley Plumbing & Drainage Ltd
Dickley Lane
Lenham
Maidstone
Kent ME17 2DE
UPVC drainage and rainwater systems
Tel: 01622 858888
Fax: 01622 858725
Website: www.marley.co.uk
e-mail: marketing@marley.co.uk

Marshalls Mono Ltd
Southowram
Halifax
West Yorkshire HX3 9SY
Kerbs, edgings and pavings
Tel: 01422 312000
Fax: 01422 330185
Website: www.marshalls.co.uk

Melcourt Industries
Boldridge Brake
Long Newnton
Tetbury
Glos GL8 8JG
Horticultural products
Tel: 01666 502711
Fax: 01666 504398
Website: www.melcourt.co.uk
e-mail: mail@melcourt.co.uk

Naylor Industries Ltd
Clough Green
Cawthorne
Barnsley
South Yorkshire S75 4AD
Clayware pipes and fittings
Tel: 01226 790591
Fax: 01226 790531
Website: www.naylor.co.uk
e-mail: sales@naylor.co.uk

Netlon Group (The)
New Wellington Street
Mill Hill
Blackburn BB2 4PJ
Geotextiles
Tel: 01254 262431
Fax: 01254 266867
Website: www.netlon.com
www.tensar-international.com
e-mail: sales@netlon.co.uk
sales@tensar.co.uk

Parker Merchanting Ltd
The Orbital Centre
Southend Road
Woodford Green
Essex IG8 8HF
Geotextiles, Consumable stores
Tel: 0208 709 7600
Fax: 0208 709 7636
Website: www.parker-merchanting.com

P D Edenhall Ltd
Danygraig Road
Risca
Newport NP11 6DP
Bricks, concrete products
Tel: 01633 612671
Fax: 01633 601280
Website: www.pd-edenhall.co.uk
e-mail: enquiries@pd-edenhall.co.uk

PHI Group Ltd
Harcourt House, Royal Crescent
Cheltenham
Glos GL50 3DA
Retaining walls, gabions
Tel: 0870 3334120
Fax: 0870 3334121
Website: www.phigroup.co.uk
e-mail: info@phigroup.co.uk

Platipus Anchors Ltd
Kingsfield Business Centre
Philanthropic Road
Redhill
Surrey RH1 4DP
Earth anchors
Tel: 01737 762300
Fax: 01737 773395
Website: www.platipus-anchors.com
e-mail: info@platipus- anchors.com

Polyfelt Geosynthetics (UK) Ltd
Unit B2
Haybrook Industrial Estate
Halesfield 9
Telford
Shropshire TF7 4QW
Geotextiles
Tel: 01952 588066
Fax: 01952 588466
Website: www.polyfelt.com
e-mail: fraser@polyfelt.co.uk

Protim Solignum Ltd
Field House Lane
Marlow
Buckinghamshire SL7 1DA
Protective Coatings
Tel: 01628 486644
Fax: 01628 481276
Website: www.osmose.co.uk
e-mail: info@osmose.co.uk

Rigidal Industries
Unit 62
Blackpole Trading Estate West
Worcester WR3 8ZJ
Cladding
Tel: 01905 754030
Fax: 01905 750555
Website: www.rigidal-industries.com
e-mail: sales@rigidal.co.uk

RIW Ltd
Arc House
Terrace Road South
Binfield, Bracknel
Berks RG42 4PZ
Waterproofing systems
Tel: 01344 861988
Fax: 01344 862010
Website: www.riw.co.uk
e-mail:enquiries@riw.co.uk

RMD Kwikform
Brickyard
Aldridge
Walsall WS9 8BW
Temporary works and access equipment
Tel: 01922 743743
Fax: 01922 743400
Website: www.rmdquickform.com

ROM Ltd
710 Rightside lane
Sheffield
South Yorkshire S9 2BR
Formwork and accessories
Tel: 0114 231 7900
Fax: 0114 231 7905
Website: www.rom.co.uk
e-mail: sales@rom.co.uk

ROM (NE) Ltd
Unit 10
Blaydon Industrial Estate
Chainbridge Road
Blaydon-on-Tyne
Northumberland NE21 5AB
Reinforcement
Tel: 0191 414 9600
Fax: 0191 414 9650
Website: www.rom.co.uk

Saint-Gobain Pipelines PLC
Lows Lane
Stanton-By-Dale
Derbyshire DE7 4QU
Cast iron pipes and fittings, polymer concrete channels, manhole covers and street furniture
Tel: 0115 9305000
Fax: 0115 9329513
Website. www.saint-gobain-pipelines.co.uk

Selwood Group Ltd
Bournemouth Road
Chandlers Ford
Eastleigh
Hants S053 3ZL
Plant sales and hire
Tel: 023 80266311
Fax: 023 80250423
Website: www.selwoodgroup.co.uk
e-mail: sales@selwoodgroup.co.uk

Stent
Pavilion C2
Ashwood Park
Ashwood Way
Basingstoke RG23 8BG
Piling and foundations
Tel: 01256 763161
Fax: 01256 768614
Website: www.stent.co.uk

Stocksigns Ltd
Ormside Way
Redhill
Surrey RH1 2LG
Road signs and posts
Tel: 01737 764764
Fax: 01737 763763
Website: www.stocksigns.co.uk

Tarmac Precast Concrete Ltd
Tallington
Barhom Road
Stamford
Lincs PE9 4RL
Prestressed concrete beams
Tel: 01778 381000
Fax: 01778 348041
Website: www.tarmacprecast.com
e-mail: tall@tarmac.co.uk

Tarmac Northern Ltd
P O Box 5
Fell Bank
Chester-le-Street
Birtley
Co Durham DH3 2ST
Ready mix concrete
Tel: 0191 492 4000
Fax: 0191 410 8489
Website: www.tarmac.co.uk

Tarmac Topblock Ltd
Airfield Industrial Estate
Ford
Arundel
West Sussex BN18 0HY
Blockwork
Tel: 01903 723333
Fax: 01903 711043
Website: www.topblock.co.uk

Terram Ltd
Mamhilad
Pontypool
Gwent NP4 0YR
Geotextiles
Tel: 01495 757722
Fax: 01495 762383
Website: www.terram.com
e-mail: info@terram.co.uk

Terry's Timber Ltd
35 Regent Road
Liverpool L5 9SR
Timber
Tel: 0151 207 4444
Fax: 0151 298 1443
Website: www.terrystimber.co.uk

Travis Perkins Trading Co. Ltd
Baltic Wharf
Boyn Valley Road
Maidenhead
Berkshire SL6 4EE
Builders merchants
Tel: 01628 770577
Fax: 01628 625919
Website: www.travisperkins.co.uk

Varley and Gulliver Ltd
Alfred Street
Sparkbrook
Birmingham B12 8JR
Bridge parapets
Tel: 0121 773 2441
Fax: 0121 766 6875
Website: www.v-and-g.co.uk
e-mail: sales@v-and-g.co.uk

Vibro Projects Ltd
Unit D3
Red Scar Business Park
Longridge Road
Preston PR2 5NQ
Ground Consolidation - Vibroflotation
Tel: 01772 703113
Fax: 01772 703114
Website: www.vibro.co.uk
e-mail: sales@vibro.co.uk

Volvo Construction Equipment Ltd
Duxford
Cambridge CB2 4QX
Plant manufacturers
Tel: 01223 836636
Fax: 01223 832357
Website: www.volvo.com
e-mail: sales@vcebg@volvo.com

Wavin Plastics Ltd
Parsonage Way
Chippenham
Wilts SN15 5PN
UPVC drain pipes and fittings
Tel: 01249 654121
Fax: 01249 443286
Website: www.wavin.co.uk

Wells Spiral Tubes Ltd
Prospect Works
Airedale Road
Keighley
West Yorkshire BD21 4LW
Steel culverts
Tel: 01535 664231
Fax: 01535 664235
Website: www.wells-spiral.co.uk
e-mail: sales@wells-spiral.co.uk

Winn & Coales (Denso) Ltd
Denso House
33-35 Chapel Road
West Norwood
London SE27 0TR
Anti-corrosion and sealing products
Tel: 0208 670 7511
Fax: 0208 761 2456
Website: www.denso.net
e-mail: mail@denso.net

Taylor & Francis
Taylor & Francis Group plc

PART 1

General

PURPOSE AND CONTENT OF THE BOOK

For many years the Editors have compiled a price book for use in the building industry with, more recently, companion volumes for use in connection with engineering services contracts and landscaping work. All of these price books take their reliability from the established practice within these sectors of the construction industry of pricing work by the application of unit rates to quantities measured from the designer's drawings. This practice is valid because most building work can be carried out under similar circumstances regardless of site location; a comparatively low proportion of contract value is subject to the risks that attend upon work below ground level; and once the building envelope is complete most trades can proceed without serious disruption from the weather.

This is not, however, the general method of pricing Civil Engineering work: the volume of work below ground, increased exposure to weather and the tremendous variety of projects, in terms of type, complexity and scale, makes the straightforward use of unit rates less reliable. So, whilst even in building work similar or identical measured items attract a fairly broad range of prices, the range is much greater in Civil Engineering Bills. This uncertainty is compounded by the lower number of bill items generated under Civil Engineering Methods of Measurement, so that the precise nature of the work is less apparent from the bill descriptions and the statistical effect of 'swings and roundabouts' has less scope to average out extremes of pricing.

To prepare a price for a Civil Engineering project, then, it is necessary to have regard to the method to be adopted in executing the work, draw up a detailed programme and then cost out the resources necessary to prosecute the chosen method. Because the first part of this process is the province of the Contractor's planner, there has been a tendency to postpone detailed estimating until the tendering stage itself, with the employer relying up to that point upon an estimate prepared on a 'broad brush' basis.

The result has been a growing pressure on the part of project sponsors for an improvement in budgetary advice, so that a decision to commit expenditure to a particular project is taken on firmer grounds. The absence of a detailed pricing method during the pre-contract phase also inhibits the accurate costing of alternative designs and regular cost checking to ensure that the design is being developed within the employer's budget.

This book therefore seeks to draw together the information appropriate to two methods of pricing: the cost of resources for use where an operational plan has been outlined, and unit rates for use where quantities can be taken from available drawings.

To take some note of the range of unit rates that might apply to an item, the rates themselves are in some cases related to working method - for example by identifying the different types of plant that would suit varying circumstances. Nonetheless, it would be folly to propose that all types of Civil Engineering work could be covered by a price book such as this. The Editors have therefore had in mind the type and scale of work commissioned by a local authority, a public corporation or a large private company.

This does embrace the great majority of work undertaken by the industry each year. Although almost all projects will have individual features that require careful attention in pricing, there will be some projects that are so specialist that they will not conform to standard pricing information at all.

But for most projects, within the range of work covered, this book should provide a firm foundation of cost information upon which a job-specific estimate can be built.

The contents of the book are therefore set out in a form that permits the user to follow the estimating process through in a structured way, as follows:

Part 1: General

The balance of this section describes in narrative form the work stages normally followed in a Contractor's office from receipt of the tender documents through to the submission of the tender.

Part 2: Preliminaries and General Items

Containing a checklist of items to be priced with Preliminaries and General Items (or 'Method Related Charges') and a worked example containing specific cost information.

PURPOSE AND CONTENT OF THE BOOK – continued

Part 3: Resources

This deals with the basic cost of resources, so that a resource-based system of estimating can be adopted where it is possible to develop an outline programme and method statement. Reference to this section will also assist the user to make adjustments to unit rates where different labour or material costs are thought to apply and to calculate analogous rates for work based on the hypothetical examples given. It is stressed that all of the costs given in this section are exclusive of the items costed with the preliminaries and of financing charges, head office overheads and profit. In addition, the materials and plant costs as shown are gross, with no deduction of discount.

Part 4 and 5: Unit Costs

These sections are structured around methods of measurement for Civil Engineering Work and gives 'trade by trade' unit rates for those circumstances where the application of unit rates to measured quantities is possible and practical. Again, it is stressed that the rates are exclusive of the items costed with preliminaries and of financing charges, head office overheads and profit. Both materials and plant costs are adjusted to allow a normal level of discount, with allowances for materials wastage and plant usage factors.

Part 6: Land Remediation

The Land Remediation section reviews the general background of ground contamination and discusses the impact of the introduction of the Landfill Directive in July 2004.

Part 7: Unit Costs (Ancillary Building Work)

This section is to be utilised in conjunction with Parts 4 and 5 to enable the user to incorporate within the estimate items more normally associated with Building Work rather than Civil Engineering and which do not fall readily under recognised methods of measurement for Civil Engineering Work. Due to the diversity of items that fall under such a definition, because of specification differences, the format for this section is structured to incorporate a range of items to allow the production of the estimate for such items prior to detailed design information being available. Additionally this section includes, in the same format, items covering simple Building Works which occur in connection with a Civil Engineering Contract but which do not form a significant proportion of the overall value and therefore do not need to be estimated in great detail using Parts 4 and 5 unit rates.

Part 8: Oncosts & Profit

Having produced an estimate for the predicted cost of the work, being the sum of the preliminaries and the measured work, the estimate must be converted to a tender by the application of any adjustment made by management (which follows the Management Appraisal described later in this part of the book) and by additions for financing charges, head office overheads and profit. These additions are discussed in this section and also included is a worked example of a tender summary.

Part 9: Costs and Tender Prices Indices

The cost and tender price indices included in this part of the book provide a basis for updating historical cost or price information, by presenting changes in the indices since 1984. Caution must be taken when applying these indices as individual price fluctuations outside the general trend may have significant effect on contract cost.

Part 10: Daywork

Including details of the CECA daywork schedule and advice on costing excluded items.

Part 11: Professional Fees

These contain reference to standard documentation relating to professional fees for Consulting Engineers and Quantity Surveyors.

Part 12: Approximate Estimates

The prices in this section have been assembled from a number of sources, including the relevant items in the unit costs section and recovered data from recent projects. They are intended to give broad price guides or to assist in comparison exercises.

Part 13: Outputs

Scheduled here are various types of operations and the outputs expected of them. Also listed are man hours for various trades found in Civil Engineering.

Part 14: Tables and Memoranda

These include conversion tables, formulae, and a series of reference tables structured around trade headings. It also includes a review of current Capital Allowances, Value Added Tax and The Aggregates Tax.

Part 15: Updates

This section is a pro-forma update which allows the user of the book to insert the adjustments to the published prices as and when they are advised through the free quarterly updating bulletin.

OUTLINE OF THE TENDERING AND ESTIMATING PROCESS

This section of the book outlines the nature and purpose of Civil Engineering estimating and provides background information for users. It comprises an outline of the estimating and tendering process with supporting notes and commentaries on particular aspects. Some worked examples on tender preparation referred to in this part are included at the end of Part 8.

It must be emphasised that the main purpose of this book is to aid the estimating process. Thus it is concerned more with the predicted cost of Civil Engineering work than with the prices in a bill of quantities. To ensure the correct interpretation of the information provided it is important to distinguish clearly between estimating and tendering; the following definitions are followed throughout.

The estimate is the prediction of the cost of a project to the Contractor. The Tender is the price submitted by the Contractor to the Employer.

The tender is based on the estimate with adjustments being made after review by management; these include allowances for risk, overheads, profit and finance charges. As discussed later in this section, prices inserted against individual items in a bill of quantities may not necessarily reflect the true cost of the work so described due to the view taken by the Contractor on the risks and financial aspects involved in executing the work.

Whilst projects are now constructed using many different forms of contract the core estimating process falls into two main divisions namely "Design & Construct" and "Construct only". The following list summarises the activities involved in the preparation of a tender for a typical construct only Civil Engineering project where the client issues full drawings, specifications and Bills describing the extent of the works to be priced. The diagram that follows illustrates the relationships between the activities. After the diagram, notes on factors affecting each stage of the process are given.

1. An overall appraisal of the project is made including any variations to the standard contract form, insurance provisions and any other unusual or onerous requirement.
2. Material requirements are abstracted from the tender documents and prices are obtained from suppliers. Details of work normally sub-contracted are abstracted, together with relevant extracts from the tender documents and prices are obtained from sub-contractors.
3. The site of the works and the surrounding area is visited and studied. Local information is obtained on factors affecting the execution of the contract.
4. A programme and detailed method of working for the execution of the contract is prepared, to include details of plant requirements, temporary works, unmeasured work, appropriate supervisory staff, etc.
5. Designs are made for temporary works and other features left to be designed by the Contractor, and quantities are taken off for costing.
6. Major quantities given in the tender documents are checked.
7. The cost estimate for the project is prepared by costing out all of the resources identified by stages 2 to 6. A more detailed report is made on the conditions of contract, financial requirements, etc. and an assessment of risk/opportunity is prepared.
8. The tender documents are priced.
9. Management reviews the estimate, evaluates the risks and makes allowances for overheads, profit and finance.
10. The tender is completed and submitted.

OUTLINE OF THE TENDERING AND ESTIMATING PROCESS – continued

1. INITIAL APPRAISAL

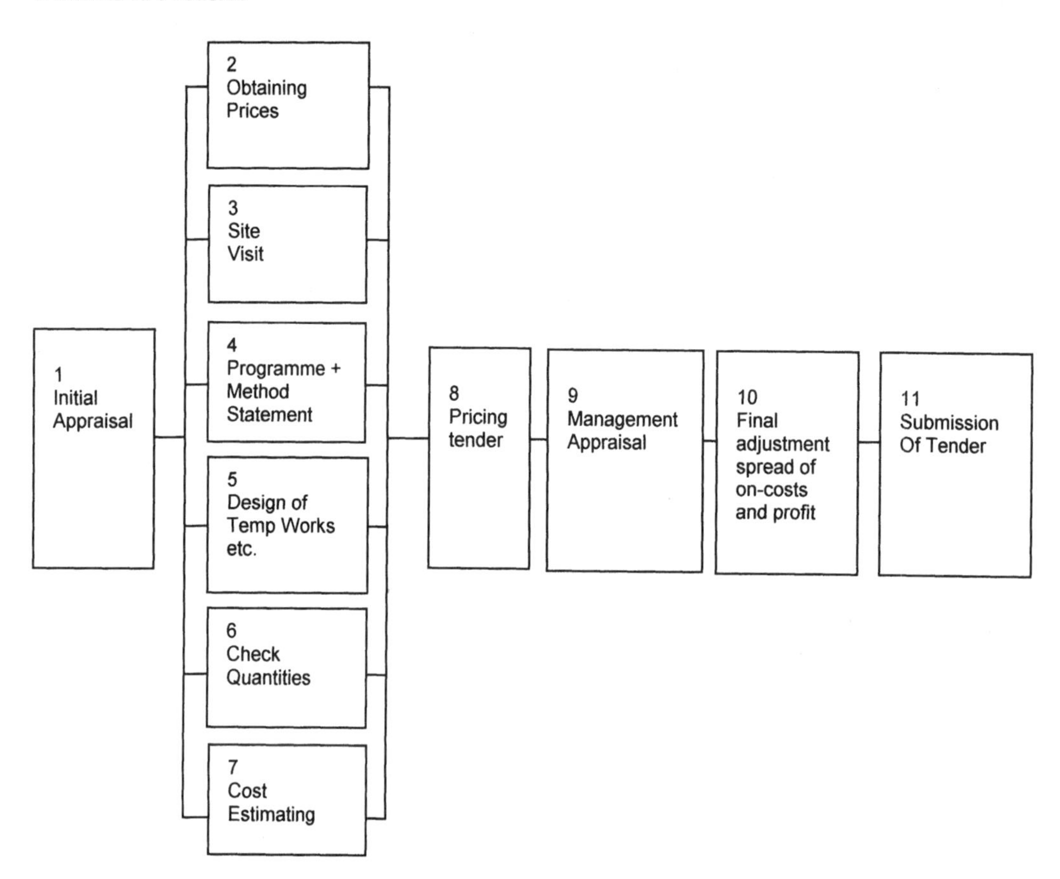

The purpose of the initial overall appraisal is to highlight any high value areas or any particular problems which may require specialist attention, it can also identify possible alternative methods of construction or temporary works.

Other points to be considered:

- the location and type of project and its suitability to the tenderer's particular expertise
- the size of project, its financing requirement, the proportion of annual turnover it would represent and the availability of resources
- the size of the tender list and the nature of the competition
- the identity of the employer and his professional consultants
- the adequacy of the tender documents
- an initial appraisal of risk and opportunity (see Section 8)

Corporate governance requires that the directors of companies are aware of all the liabilities inherent in the contract being sought. Few contracts are offered on totally unamended standard forms and specifications and it is the estimator's duty to ensure that a report is prepared to advise on the precise terms and conditions being offered and their potential implications.

It is essential for the estimator to study the contract documents issued with the enquiry or made available for inspection and to note those parts which will affect pricing or involve the Contractor in liabilities which must be evaluated and drawn to the attention of management. The following comments are indicative only:

Conditions of Contract

For Civil Engineering work these are normally, but not exclusively, based on either the I.C.E. or ECC (NEC) standard forms. However these forms are rarely offered without addition and amendment and it is imperative that the full implications are understood and directors informed. Any required bonds, guarantees and warranties must be identified, reported and included in subcontract enquiries where appropriate. Insurance requirements and excesses must be checked against company policies.

Bill of Quantities

Where a Bill is provided it serves three purposes: first and foremost it must be prepared with the objective of providing the estimator with as accurate a picture of the project as possible, so as to provide a proper basis for pricing. Second, it should enable the employer to compare tenders on an equal basis and third it will be used to evaluate the work executed for payment purposes. Individual items in the Bill do not necessarily describe in detail the work to be carried out under a particular item; reference must be made to the specification and the drawings to ascertain the full scope of the work involved.

The method of preparing the Bill may be based on the 'Civil Engineering Standard Method of Measurement' issued by the Institution of Civil Engineers or the 'Method of Measurement for Highway Works' issued by the Department for Transport, but some employing authorities have evolved their own methods and it is important for the estimator to study the Bill and its preambles to ensure that his rates and prices are comprehensive.

In all cases the quantities given in the bill are not a guarantee and the drawings usually have precedence. The estimator must understand whether he is pricing a re-measurable or fixed price contract and make due allowance.

Specification

This gives a detailed description of the workmanship, finish and materials to be used in the construction of the work. It may also give completion periods for sections and/or the whole of the work together with details of the requirements for the employer and/or the Consulting Engineer in connection with their supervision on site.

Water Authority and Highway Works in particular are based around a standard specification. However even standard specifications will have contract specific appendices and tables. The estimator must take due note of these requirements and ensure that this information is issued and taken into account by all potential subcontractors and material suppliers.

Drawings

These give details of the work to be carried out and must be read in conjunction with the specification. It is important for the estimator to study the notes and descriptions given on the drawings as these amplify the specification.

Should the estimator discover any conflict between the various documents, it is important to have such discrepancies clarified by the Employer or Engineer prior to submission of any offer.

2. OBTAINING PRICES

(a) Materials

When pricing materials, the following points must be noted:

- checks must be made to ensure that the quality of materials to be supplied meets with the requirements of the specification. If necessary, samples should be obtained and tested for compliance
- checks must be made to ensure that the rates of delivery and fabrication periods can meet the demands of the programme. It is sometimes necessary to use more than one supplier, with differing prices, to ensure a sufficient flow of materials
- tests should be carried out to ascertain allowances to be made for operations such as compaction of soils and aggregates. Records of past contracts using similar materials can give this information, providing such records are accurate and reliable

OUTLINE OF THE TENDERING AND ESTIMATING PROCESS – continued

2. OBTAINING PRICES - continued

(b) Sub-contractors

- It is common practice among Civil Engineering Contractors to sub-contract a significant proportion of their work.

Sub-contracted work can represent the bulk of the value of measured work.

When utilising sub-contractors' prices it is extremely important to ensure that the rates given cover the full extent of the work described in the main contract, and that the sub-contractors quotation allows for meeting the main Tender programme and methods of working.

Unless an exclusive relationship has been entered into prior to the tendering process it is likely that the same subcontractor will submit prices to a number of competing tenderers. It is important for the estimator ensures that the price offered represents the method intended and is not a generic sum which is subject to variation if successful.

3. SITE VISIT

Factors to check during the site visit include:

- access
- limitations of working space
- existing overhead and underground services
- nearby public transport routes
- availability of services - water, gas, electricity, telephone, etc.
- availability of labour and sub-contractors
- availability of materials - particularly aggregates and fill materials location of nearest tipping facility
- nature of the ground, including inspection of trial bores / pits if dug ground water level
- presence of other Contractors working on or adjacent to the site.

4. METHOD STATEMENT AND PROGRAMME

As previously stated whilst an estimate is being prepared it is necessary that a detailed method of working and a programme for the execution of the works is drawn up; the latter can take the form of a bar chart or, for large and more complex projects, may be prepared on more sophisticated computerised platforms. Compliance with the employer's target completion dates is, of course, essential. The method of working will depend on this programme in so far as the type and size of plant and the gang sizes to be used must be capable of achieving the output necessary to meet the programmed times. Allowance must be made for delays due to adverse weather, other hazards particular to the site and the requirements of the specification, particularly with regard to periods to be allowed for service diversions and other employer's requirements.

A method statement is prepared in conjunction with the programme, setting out the resources required, outputs to be achieved and the requirements in respect of temporary works, etc.

At the same time, separate bar charts may be produced giving:

- plant requirements
- staff and site supervision requirements
- labour requirements

Free Updates

with three easy steps...

1. Register today on www.pricebooks.co.uk/updates

2. We'll alert you by email when new updates are posted on our website

3. Then go to www.pricebooks.co.uk/updates and download.

All four Spon Price Books – *Architects' and Builders'*, *Civil Engineering and Highway Works*, *External Works and Landscape* and *Mechanical and Electrical Services* – are supported by an updating service. Three updates are loaded on our website during the year, in November, February and May. Each gives details of changes in prices of materials, wage rates and other significant items, with regional price level adjustments for Northern Ireland, Scotland and Wales and regions of England. The updates terminate with the publication of the next annual edition.

As a purchaser of a Spon Price Book you are entitled to this updating service for the 2007 edition – free of charge. Simply register via the website www.pricebooks.co.uk/updates and we will send you an email when each update becomes available.

If you haven't got internet access we can supply the updates by an alternative means. Please write to us for details: Spon Price Book Updates, Spon Press, 2 Park Square, Milton Park, Abingdon, Oxfordshire, OX14 4RN.

Find out more about spon books
Visit www.sponpress.com for more details.

These programmes and method statements will form the basis of the actual contract programme should the tender be accepted. They will also enable the Contractor to assure himself that he has available or can gain access to the necessary resources in plant, labour, materials and supervision to carry out the work should he be awarded the contract.

5. DESIGN OF TEMPORARY WORKS, ETC

Normally the period of time allowed for tendering is relatively short and therefore it is important that those aspects requiring design work are recognised as early as possible in the tender period. Design can be carried out either by the Contractor using his own engineers or by utilising the services of a Consulting Engineer. There are three aspects of design to be considered:

A. Temporary works to the Contractor's requirements to enable the works to be constructed

Design of temporary works covers the design of those parts of the work for which the Contractor accepts full responsibility in respect of stability and safety during construction. Such parts include support structures, cofferdams, temporary bridges, jetties and river diversions, special shuttering, scaffolding, haul roads and hardstandings, compounds, traffic management etc. Design must be in sufficient detail to enable materials, quantities and work content to be assessed and priced by the estimator. In designing such work, it is important that adequate attention is given to working platforms and access for labour and plant and also to ease of dismantling and re-erection for further uses without damage.

It should be noted that many specialist sub-contractors will provide a design service when submitting quotations. For example, scaffolding Contractors will design suitable support work for soffit shuttering, etc.

B. Specific items of the permanent works to meet a performance specification set out by the client or the Consulting Engineer

It is common practice for certain parts of the work to be specified by means of a performance specification. For example, concrete is specified by strength only, piles by load carrying capacity, etc. It is then left to the Contractor to use those materials, workmanship and design which he feels are most suited to the particular site and conditions.

In many cases such design will be carried out by specialist suppliers or sub-contractors.

C. Alternative designs for sections of the permanent works where the Contractor's experience leads him to consider that a more economical design could be used

It is possible for a Contractor to use his expertise and experience to design and submit alternative proposals and prices for complete sections of the permanent work without, of course, altering the basic requirements of the original design. Examples would be foundations in difficult ground conditions, bridge superstructures, use of precast in place of in situ concrete, etc. Such designs may be carried out by the Contractor's own staff or in conjunction with Consulting Engineers or specialist Contractors

Obviously this is only done when the Contractor can offer considerable savings in cost and/or a reduced construction period. It is necessary to include a price for the original design in the tender, but the decision to submit a keen tender may be underpinned by the hope of sharing such savings with the Employer.

In all cases, the Contractor's designs and calculations must be checked and approved by the Employer's Consulting Engineer, but such checks and approvals do not relieve the Contractor of his responsibilities for the safety and stability of the work.

Where any design work is undertaken it is important to ensure that adequate design insurance is maintained.

OUTLINE OF THE TENDERING AND ESTIMATING PROCESS – continued

6. QUANTITIES CHECK

Working within obvious time constraints, the estimating team will endeavour to complete a quantities check on at least the major and high price quantities of the Bill, as this could affect pricing strategy, for example, in pricing provisional items. Any major discrepancies noted should be referred to the management appraisal agenda.

7. COST ESTIMATE

At this stage the estimator draws together the information and prepares a cost estimate made up of:

- preliminaries and general items (see Part 2)
- temporary works
- labour, costed by reference to the method statement, with appropriate allowances for labour oncosts (non-productive overtime, travelling time / fares, subsistence, guaranteed time, bonus, Employer's Liability Insurance, training, statutory payments, etc.) (see page 33-37)
- material costs taken from price lists or suppliers' quotations, with appropriate allowances for waste
- plant, whether owned or hired, with appropriate allowances for transport to / from site, erection/dismantling, maintenance, fuel and operation. Heavy static plant (batching plant, tower cranes, etc.) will normally be priced with general items; the remainder will normally be allocated to the unit rates (see page 117-146)
- sublet work, as quoted by specialist sub-contractors, with appropriate allowances for all attendances required to be provided by the main Contractor

At the same time, a preliminary assessment of risk/opportunity will be made for consideration with the Management Appraisal (see Section 8 below). On major tenders a formal quantified risk assessment (QRA) will be undertaken – the estimator will be expected to evaluate the effects of the risks for inclusion in the calculations. This will include a look at:

- weather conditions - costs not recoverable by claims under the Conditions of Contract, ground becoming unsuitable for working due to the effect of weather, etc.
- flooding - liability of site to flooding and the consequent costs
- suitability of materials - particular risk can arise if prices are based on the use of borrow pits or quarries and inadequate investigation has been carried out or development is refused by the local authorities
- reliability of sub-contractors - failure of a sub-contractor to perform can result in higher costs through delays to other operations and employing an alternative sub-contractor at higher cost
- non-recoverable costs - such as excesses on insurance claims
- estimator's ability - e.g. outputs allowed. This can only be gauged from experience
- cost increase allowances for fixed price contract
- terms and conditions contained in the contract documents
- ability to meet specification requirements for the prices allowed
- availability of adequate and suitable labour

Risk is, of course, balanced by opportunity and consideration needs to be given to areas for which particular expertise possessed by the Contractor will lead to a price advantage against other tenderers.

8. PRICING THE TENDER

Once the cost estimate is complete, the estimator prices the items in the Bill. The rate to be entered against the items at this stage should be the correct rate for doing the job, whatever the quantity shown.

Where the overall operation covers a number of differing Bill items, the estimator will allocate the cost to the various items in reasonable proportions; the majority of the work is priced in this manner. The remaining items are normally priced by unit rate calculation.

Resources schedules, based on the programme and giving details of plant, labour and staff, perform an important role in enabling the estimator to check that he has included the total cost of resources for the period they are required on site. It is not unusual for an item of plant to be used intermittently for more than one operation. A reconciliation of the total time for which the cost has been included in the estimate against the total time the item is needed on site as shown on the programme, gives a period of non-productive time. The cost of this is normally included in site oncosts and preliminaries. A similar situation can arise in the cases of skilled labour, craftsmen and plant drivers.

Having priced the Bill at cost, there will remain a sum to be spread over the Bill items. The way in which this is done depends on the view taken by the Contractor of the project; for example:

- sums can be put against the listed general items in the Preliminaries Bill
- a fixed 'Adjustment Item' can be included in the Bill for the convenience of the estimator. This can be used for the adjustment made following the Management Appraisal, and for taking advantage of any late but favourable quotations received from sub-contractors or suppliers
- the balance can be spread over the rates by equal percentage on all items, or by unequal percentages to assist in financing the contract or to take advantage of possible contract variations, or expected quantity changes (see notes against 7)

The Contractor will normally assess the financial advantages to be gained from submitting his bid in this manner and possibly enabling him to submit a more competitive offer.

After completing the pricing of all aspects of the tender, the total costs are summarised and profit, risk, etc., added to arrive at the total value of the Tender. A suggested form of this summary is set out in Part 8: Oncosts and Profit.

Finally, reasonable forecasts of cash flow and finance requirements are essential for the successful result of the project. Preliminary assessments may have been made for the information of management, but contract cash flow and the amount of investment required to finance the work can now be estimated by plotting income against expenditure using the programme of work and the priced Bill of Quantities. Payment in arrears and retentions, both from the Employer and to the suppliers and sub-contractors, must be taken into account.

It is unlikely that sufficient time will be available during the tender period to produce such information accurately, but an approximate figure, for use as a guide for finance requirements, can be assessed.

A worked example is set out in Part 8: Oncosts and Profit.

9. MANAGEMENT APPRAISAL

Clearly, as far as the detail of the tender build-up is concerned, management must rely upon its established tendering procedures and upon the experience and skill of its estimators. However the comprehensive review of tenders prior to submission is an onerous duty and the estimator should look upon the process as an opportunity to demonstrate his skill. The Management Appraisal will include a review of:

- the major quantities
- the programme and method statement
- plant usage
- major suppliers and/or sub-contractors, and discounts
- the nature of the competition
- risk and opportunity
- contract conditions, including in particular the level of damages for late completion, the minimum amount of certificates, retention and bonding requirements
- cash flow and finance
- margin for head office overheads and profit
- the weighting and spreading of the cost estimate over the measured items in the Bill

OUTLINE OF THE TENDERING AND ESTIMATING PROCESS – continued

10. SUBMISSION OF TENDER

On completion of the tender, the documents are read over, comp checked and then despatched to the employer in accordance with the conditions set down in the invitation letter.

A complete copy of the tender as submitted should be retained by the Contractor. Drawings on which the offer has been based should be clearly marked 'Tender Copy', their numbers recorded and the drawings filed for future reference. These documents will then form the basis for price variations should the design be amended during the currency of the contract.

The Contractor may wish to qualify his offer to clarify the basis of his price. Normally such qualifications are included in a letter accompanying the tender. Legally the Form of Tender constitutes the offer and it is important that reference to such a letter is made on the Form of Tender to ensure that it forms part of the offer. Wording such as 'and our letter dated..., reference...' should be added prior to quoting the Tender Sum.

Before any qualifications are quoted, careful note must be taken of the 'instructions to tenderers', as qualifications, or at least qualifications submitted without a conforming tender, may be forbidden.

11. DESIGN AND CONSTRUCT VARIATIONS

Where once the design and construct tender was the province of only very major projects, this form of procurement is now much more widespread throughout the entire spectrum of Civil Engineering works. The key difference to the construct only tender is self evident: the bidder is required to produce and price designs that will satisfy the employer's stated requirements and often need to result in a scheme in line with indicative plans.

In practice, this process runs best when managed by an overall bid manager, with the estimator organising the bill preparation and pricing from the submitted designs. The bid timescales are such that an orderly progression from design through bill preparation to pricing is rarely achieved. This process places far greater demands on the estimator's flexibility and management skills, but can prove ultimately more rewarding.

PART 2

Preliminaries and General Items

This part deals with that portion of Civil Engineering costs not, or only indirectly, related to the actual quantity of work being carried out. It comprises a definition of Method Related Charges, a checklist of items to be accounted for on a typical Civil Engineering contract and a worked example illustrating how the various items on the checklist can be dealt with.

Taylor & Francis
Taylor & Francis Group plc

GENERAL REQUIREMENTS AND METHOD RELATED CHARGES

Although the more familiar terminology of Preliminaries and General Items is used in this book the principle of Method Related Charges - separating non quantity related charges from quantity related charges - is adopted. Generally the former are dealt within this part, while the latter are dealt with in Part 4: Unit Costs. In this part non quantity related charges are further subdivided into those that are time related and those that are non time related.

The concept of METHOD RELATED CHARGES can be summarised as follows :-

In commissioning Civil Engineering work the Employer buys the materials left behind, but only hires from the Contractor the men and machines which manipulate them, and the management skills to manipulate them effectively. It is logical to assess their values in the same terms as the origin of their costs. It is illogical not to do so if the Employer is to retain the right at any time to vary what is left behind and if the financial uncertainties affecting Employer and Contractors are to be minimised.

Tenderers have the option to define a group of bill items and insert charges against them to cover those expected costs which are not proportional to the quantities of Permanent Works. To distinguish these items they are called Method Related Charges. They are themselves divided into charges for recurrent or time related cost elements, such as maintaining site facilities or operating major plant, and charges for elements which are neither recurrent nor directly related to quantities, such as setting up, bringing plant to site and Temporary Works.'

Another hope expressed with the introduction of Method Related Charges was that they should accurately reflect the work described in the item and that they should not, as had become the practice with some of the vague general items frequently included in Civil Engineering Bills, be used as a home for lump sum tender adjustments quite unrelated to the item. Where cost information is given in the worked example presented at the end of this part of the book, therefore, it must be stressed that only direct and relevant costs are quoted.

Where no detailed information is available, it is suggested that when preparing a preliminary estimate an addition of between 7½% and 15% of net contract value is made to cover Contractor's Site Oncosts both time and non time related.

CHECKLIST OF ITEMS

The following checklist is representative but not exhaustive. It lists and describes the major preliminary and general items which are included, implicitly or explicitly, in a typical Civil Engineering contract and, where appropriate, gives an indication of how they might be costed. Generally contract documents give detailed requirements for the facilities and equipment to be provided for the Employer and for the Engineer's representative and Bills of Quantities produced in conformity with CESMM3 Class A provide items against which these may be priced; no such items are provided for Contractor's site oncosts or, usually, for temporary works and general purpose plant. For completeness a checklist of both types of item is given here under the following main headings :

- Contractor's site oncosts
  - time related
  - non time related
- Employer's and consultants' site requirements
  - time related
  - non time related
- Other services, charges and fees
- Temporary works (other than those included in unit costs)
- General-purpose plant (other than that included in unit costs)

CONTRACTOR'S SITE ONCOSTS - TIME RELATED

Site staff salaries
All non-productive supervisory staff on site including; agent, sub-agent, engineers, general foremen, non-productive section foremen, clerks, typists, timekeepers, checkers, quantity surveyors, cost engineers, security guards, etc. Cost includes salaries, subsistence allowance, National Health Insurance and Pension Scheme contributions, etc. Average cost approximately 3% to 5% of contract value.

Site staff expenses
Travelling, hotel and other incidental expenses incurred by staff. Average cost approximately 1% of staff salaries.

Attendant labour
Chainmen, storemen, drivers for staff vehicles, watchmen, cleaners, etc.

General yard labour
Labour employed on loading and offloading stores, general site cleaning, removal of rubbish, etc.

Plant maintenance
Fitters, electricians, and assistants engaged on general plant maintenance on site. This excludes drivers and banksmen who are provided for specifically in the Unit Costs Sections.

Site transport for staff and general use
Vehicles provided for use of staff and others including running costs, licence and insurance and maintenance if not carried out by site fitters.

Transport for labour to and from site
Buses or coaches provided for transporting employees to and from site including cost of drivers and running costs, etc., or charges by coach hire company for providing this service

Contractor's office rental
This includes:

- rental charges for provision of offices for Contractor's staff
- main office
- section offices
- timekeepers, checkers and security
- laboratory, etc.
- an allowance of approximately 8 m^2 per staff member should be made

Contractor's site huts
Rental charges for stores and other general-use site huts

Canteen and welfare huts
Rental charges for canteen and huts for other welfare facilities required under Rule XVI of the Working Rule Agreement

Rates
Chargeable by local authorities on any site, temporary buildings or quarry

General office expenditure
Provision of postage, stationery and other consumables for general office use

Telecommunications
Rental charges and charges for calls

Furniture and equipment
Rental charges for office furniture and equipment including photocopiers, calculators, typewriters, personal computers and laser printers, etc.

Surveying equipment
Rental

Canteen and welfare equipment
Rental charges for canteen and other welfare equipment

Radio communication equipment
Rental

Testing and laboratory equipment
Rental

Lighting and heating for offices and huts
Electricity, gas or other charges in connection with lighting and heating site offices and hutting

Site lighting electrical consumption
Electricity charges in connection with general external site lighting

Water consumption
Water rates and charges

Canteen operation
Labour, consumables and subsidy costs in operating site canteens

Carpenter's shop equipment
Rental of building and mechanical equipment

Fitter's shop equipment
Rental of building and equipment

Small tools
Provision of small tools and equipment for general use on site. Average cost 5% of total labour cost

Protective clothing
Provision of protective clothing for labour including boots, safety helmets, etc. Average cost 2% of total labour cost

Traffic control
Hire and operation of traffic lights

Road lighting
Hire and operation of road lighting and traffic warning lights

Cleaning vehicles
Equipment and labour cleaning vehicles before entering public roads

Cleaning roads and footpaths
Equipment and labour cleaning public roads and footpaths

Progress photographs for Contractor's records
Cost of taking and processing photographs to demonstrate progress

Rent of additional land
For Contractor's use for erection of huts, storage of soil and other materials, etc.

CONTRACTOR'S SITE ONCOSTS - NON TIME RELATED

Staff removal expenses
Costs of staff moving house to new location. Generally only applies on longer-term contracts.

Erection of offices including drainage, paths, etc.
Construction of foundations, drainage, footpaths and parking areas, erection of huts, installation of electric wiring, in situ fittings and decorating, etc.

Dismantle offices and restore site on completion
Dismantling and taking away huts and furniture, disconnecting and removing services, removing temporary foundations etc. and re-instating ground surface to condition prevailing before construction.

Erection of general site huts
Construction of foundations, drainage, footpaths and parking areas, erection of huts, installation of electric wiring, in situ fittings and decorating, etc.

Dismantle general site huts
Dismantling and taking away huts and furniture, disconnecting and removing services, removing temporary foundations etc. and re-instating ground surface to condition prevailing before construction

Erection of canteen and welfare huts
Construction of foundations, drainage, footpaths and parking areas, erection of huts, installation of electric wiring, in situ fittings and decorating, etc.

Dismantle canteen and welfare huts
Dismantling and taking away huts and furniture, disconnecting and removing services, removing temporary foundations etc. and re-instating ground surface to condition prevailing before construction

Caravan site construction and clearance
Construction of site for employees' caravans including provision for water, electricity and drainage, and subsequently clear away and restore site on completion, allow credit for any charges to be levied

Telecommunications
Charges for initial installation and removal

Furniture and equipment
Purchase costs of furniture and equipment, allow for residual sale value

Survey equipment
Purchase costs of survey equipment including pegs, profiles, paint, etc., for setting out

Canteen and welfare equipment
Purchase costs of equipment

Testing and laboratory equipment
Purchase cost of equipment

Radio communication
Installation costs

Electrical connection and installation
Initial charges for connections to mains supply

Electrical connection site plant
Connection to site mains supply and final disconnection and removal

Electrical connection site lighting
Connection to site mains supply and final disconnection and removal

Water supply
Installation on site and connection charges

Haulage plant
Cost of transport of plant and equipment to and from site

Progress photographs

Depot loading and unloading charges

Carpenter's shop
Erection of building, installation of equipment including electrical installation, etc. Dismantle and clear away on completion

Fitter's shop
Erection of building, installation of equipment including electrical installation, etc. Dismantle and clear away on completion

Stores compound
Erect and dismantle stores compound

Notice boards and signs
Supply, erect and remove Contractor's signboards, traffic control signs, etc.

Insurances
Payment of premiums for all Contractor's insurance obligations (see separate section on insurances and bond below)

Bond
Charges for provision of bond (see separate section on insurances and bond below)

Plant erection
Cost of erection of Contractor's plant on site including foundations, hardstandings, drainage, etc.

Plant dismantling
Cost of removal of Contractor's plant on site including foundations, hardstandings, drainage, etc.

Clear site on completion including removal of rubbish and reinstatement

EMPLOYER'S AND CONSULTANTS' SITE REQUIREMENTS - TIME RELATED

Office and other huts
Rental of office accommodation, sub-offices, laboratory, etc.

Office and site attendant labour
Office cleaning, chainmen, laboratory assistants, etc.

Site transport
Rental for vehicles for use of client and engineer

Telecommunications
Rental and cost of calls (if to be borne by Contractor)

Furniture and equipment
Rental of office furniture and equipment

Survey equipment
Rental of surveying equipment

Testing and laboratory equipment
Rental of testing and laboratory equipment

Radio communication equipment
Rental and maintenance

EMPLOYER'S AND CONSULTANTS' SITE REQUIREMENTS - TIME RELATED - continued

Office lighting and heating
Cost of heating and lighting all offices and huts

Office consumables
Cost of office consumables to be provided by the Contractor

EMPLOYER'S AND CONSULTANTS' SITE REQUIREMENTS - NON TIME RELATED

Erection of huts and offices
Client's and engineer's offices and other huts including foundations, pathways, parking area, electrical installation and drainage, etc.

Dismantling huts and offices
Restoration of site on completion

Telecommunications installation charges

Furniture and equipment purchase
Purchase cost for furniture and equipment

Survey equipment purchase

Testing and laboratory equipment purchase

Radio communication equipment installation

Progress photographs
Cost of professional photographer and supplying prints as required

OTHER SERVICES, CHARGES AND FEES

Design fees for alternative designs for permanent works
Design and drawing office costs and charges for preparing alternative designs and specifications and bill of quantities for alternative designs for permanent works

Design and design office charges for temporary works
Design and drawing office costs and charges for preparing designs and drawings for temporary works

Preparation of bending schedules
Drawing office charges for preparation of bending schedules

Fees to local authorities

Legal advice and fees
Fees and charges from legal adviser

TEMPORARY WORKS - OTHER THAN THOSE INCLUDED IN UNIT COSTS

Fencing

Traffic diversions

Lighting

Traffic signs

Traffic control

Footpath diversions

Stream or river diversion

Cofferdam installation

Cofferdam removal

Support works

Jetties

Bridges

De-watering

General pumping
Including construction of collecting sumps, etc.

Site access roads and maintenance

Scaffolding

GENERAL PURPOSE PLANT - OTHER THAN THAT INCLUDED IN UNIT COSTS

Lorries and dumpers for general transport around site

Tractors and trailers

Craneage for general use

Compressed air plant

Pumps

Bowsers for fuelling plant

Bowsers for water supply

Non productive time for plant on site
Obtained by comparing plant requirements as on programme with the plant time included in the build up of bill rates

Note: For all items of plant listed above the cost of drivers and other attendants must be allowed for together with consumables and other operating costs

INSURANCES AND BOND

Contractors are legally required to insure against liability which may be incurred when employees are injured at work and when individuals are injured by owners' vehicles. There is also a statutory requirement for certain types of machinery to be inspected at regular intervals. In addition to legal requirements, companies insure against possible loss due to fire, explosion, fraud, liability incurred as a result of damage to the property of others and through serious injury to individuals.

Certain risks are excluded from insurers' policies; these include war, revolution, etc, contamination by radioactivity and risks which arise from bad management.

Generally insurance companies take into account the claims record of a Contractor when assessing premiums payable on a particular contract or policy. Premiums are related to the risks involved and, on large Civil Engineering contracts, the insurers will require full details of the work, the methods of construction, plant used and risks involved due to flood, ground conditions, etc. Insurance companies or brokers should be consulted before submitting a tender for major Civil Engineering work.

INSURANCES AND BOND – continued

The following gives an outline of the items to be allowed for in a tender.

Employer's liability insurance
This provides indemnity to the Employer against legal liability for death of or injury to employees sustained in the course of their employment. The cost is normally allowed for in the build-up of the 'all-in' labour rate as a percentage addition to the gross cost. This will vary, depending on the Contractor's record. An allowance of 2% has been made in this book.

Vehicle insurance
This can cover individual vehicles or fleets. The cost is normally covered in the rate charged to the contract for the use of the vehicles.

All risks insurance
This provides for loss of or damage to permanent and/or temporary works being executed on the contract. It also covers plant, materials, etc. The cost is allowed for in the tender as a percentage of the total contract value. This will vary depending on the Contractor's record and type of contract undertaken and also on the value of excesses included in the policy.

Public liability insurance
This provides indemnity against legal liability which arises out of business activities resulting in bodily injury to any person (other than employees), loss of or damage to property (not owned or under the control of the company), obstruction, trespass or the like.

Such insurance can be extended to include labour-only sub-contractors and self-employed persons if required. The cost is generally included with head office overheads.

Professional indemnity insurance
This provides against liability arising out of claims made against the conduct and execution of the business. This covers such items as design liability, etc. The cost may be with head office overheads where such insurance is considered desirable.

Loss of money insurance
This covers loss of money and other negotiable items and loss or damage to safes and strong rooms as a result of theft. It is necessary to cover cash in transit for wages, etc. The cost is included with head office overheads.

Fidelity guarantee insurance
This covers loss by reason of any act of fraud or dishonesty committed by employees. The cost is included with head office overheads.

Other insurances
Other insurances which may be carried by a Contractor include fire insurance on his permanent premises and contents, consequential loss insurance in relation to his permanent premises, personal accident insurance on a 24 hour per day basis for employees.

Contract bond
Where the contract calls for a bond to be provided, this is normally given by either banks or insurance companies. The total value of these guarantees available to any company is limited, depending on the goodwill and assets of the company. It will also affect the borrowing facilities available to the company and therefore, to some extent, can restrict his trading. An allowance should be made at a rate of 1½% per annum on the amount of the bond for the construction period plus a rate of ½% per annum for the maintenance period.

Worked Example - Programme of Activities, Staffing and Plant

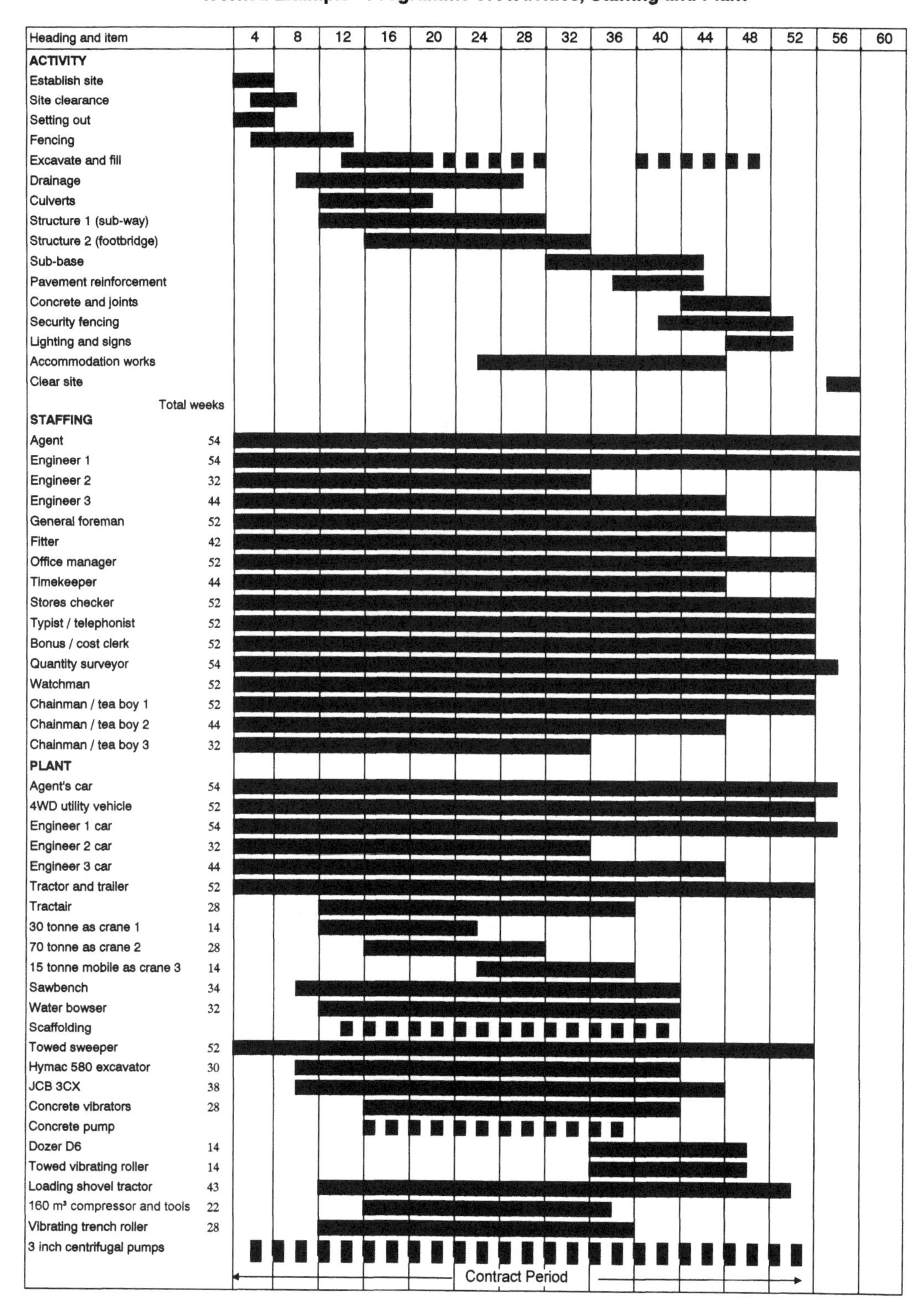
Heading and item
4
8
12
16
20
24
28
32
36
40
44
48
52
56
60
ACTIVITY
Establish site
Site clearance
Setting out
Fencing
Excavate and fill
Drainage
Culverts
Structure 1 (sub-way)
Structure 2 (footbridge)
Sub-base
Pavement reinforcement
Concrete and joints
Security fencing
Lighting and signs
Accommodation works
Clear site
Total weeks
STAFFING
Agent 54
Engineer 1 54
Engineer 2 32
Engineer 3 44
General foreman 52
Fitter 42
Office manager 52
Timekeeper 44
Stores checker 52
Typist / telephonist 52
Bonus / cost clerk 52
Quantity surveyor 54
Watchman 52
Chainman / tea boy 1 52
Chainman / tea boy 2 44
Chainman / tea boy 3 32
PLANT
Agent's car 54
4WD utility vehicle 52
Engineer 1 car 54
Engineer 2 car 32
Engineer 3 car 44
Tractor and trailer 52
Tractair 28
30 tonne as crane 1 14
70 tonne as crane 2 28
15 tonne mobile as crane 3 14
Sawbench 34
Water bowser 32
Scaffolding
Towed sweeper 52
Hymac 580 excavator 30
JCB 3CX 38
Concrete vibrators 28
Concrete pump
Dozer D6 14
Towed vibrating roller 14
Loading shovel tractor 43
160 m³ compressor and tools 22
Vibrating trench roller 28
3 inch centrifugal pumps
Contract Period

WORKED EXAMPLE

The example is of a contract for the construction of an Airport extension. The contract includes concrete surfaced aprons/runways, surface water drainage, construction of two concrete structures, minor accommodation works and two culverts. The Conditions of contract are ICE Standard conditions; the contract period is 12 months; and the approximate value is £8.0 million.

It is assumed that the main Contractor will sub-let bulk earthworks and landscape, fencing, concrete surfacing, signs and lighting as well as waterproofing to structures. It is also taken that all materials are obtained off site (including concrete).

The worked example demonstrates a method of assessing preliminary costs and is based on the programme below together with an assessment of general purpose plant (a plant reconciliation is also given).

CONTRACTOR'S SITE ONCOSTS - TIME RELATED

				£	£
Site staff salaries (see programme)					
Agent	54	wks at	£1200	64,800	
Senior Engineer	54	wks at	£960	51,840	
Engineers	76	wks at	£720	54,720	
General foreman	52	wks at	£880	45,760	
Office manager/cost clerk	52	wks at	£650	33,800	
Timekeeper/Storeman/checker	44	wks at	£450	19,800	
Typist/telephonist	52	wks at	£300	15,600	
Security guard	52	wks at	£450	23,400	
Quantity Surveyor	52	wks at	£900	46,800	
Fitter	42	wks at	£700	29,400	385,920
Site staff expenses (1% staff salaries)					3,859
Attendant labour					
Chainman	128	wks at	£520	66,560	
Driver	52	wks at	£560	29,120	
Office cleaner (part-time)	52	wks at	£150	7,800	103,480
General yard labour					
(Part-time involvement in loading and offloading, clearing site rubbish etc.)					
1 ganger	10	wks at	£560	5,600	
4 labourers	40	wks at	£520	20,800	26,400
Plant maintenance (Contractor's own plant)					
(Fitter included in Site Staff Salaries above)					
Fitter's mate	32	wks at	£520	16,640	16,640
Carried forward				£	536,299

Brought forward				£	536,299
Site transport for staff and general use					
QS/Agent's cars	54	wks at	£120	6,480	
Engineers' cars (contribution)	130	wks at	£75	9,750	
Land Rover or similar SWB	54	wks at	£325	17,550	33,780
Site transport – Labour	48	wks at	£550	26,400	26,400
Transport for labour to and from site					NIL
Contractor's office rental					
Mobile offices (10 staff x 8 m²) = 80 m²	52	wks at	£170	8,840	
Section offices (2 nr at 10 m²) = 20 m²	52	wks at	£42	2,184	11,024
Contractor's site huts					
Stores hut, 22 m²					
	52	wks at	£47	2,444	2,444
Canteen and welfare huts					
Canteen 70 m² (assume 70 men)	52	wks at	£160	8,320	
Washroom 30 m²	52	wks at	£60	3,120	
Staff toilets	52	wks at	£150	7,800	
Site toilets	52	wks at	£200	10,400	29,640
Rates					NIL
General office expenditure etc.					
Postage, stationery and other consumables	52	wks at	£110	5,720	
Telephone / fax calls / e-mail and rental	52	wks at	£120	6,240	
Furniture and equipment rental	52	wks at	£55	2,860	
Personal computers, laser printers, scanners	52	wks at	£120	6,240	
Surveying equipment rental	52	wks at	£50	2,600	
Canteen and welfare equipment rental	52	wks at	£65	3,380	
Photocopier rental	52	wks at	£100	5,200	
Testing equipment rental	52	wks at	£40	2,080	
Lighting and heating offices and huts (200m²)	52	wks at	£90	4,680	39,000
Water consumption					
2,250,000 litres at £2.00 per 5,000 litres	450	units at	£2	900	900
Small tools					
1% on labour costs of say £1,200,000					12,000
Carried forward				£	691,487

WORKED EXAMPLE - continued

CONTRACTOR'S SITE ONCOSTS - TIME RELATED - continued

Brought forward				£	691,487
Protective clothing					
½% on labour costs of say £1,200,000					6,000
Cleaning vehicles					
Cleaning roads					
Towed sweeper (tractors elsewhere)	52	wks at	£150	7,800	
Brushes			say,	1,000	
Labour (skill rate 4)	52	wks at	£560	29,120	37,920
Progress photographs			say,		1,000
Total Contractor's site oncosts - Time related				£	736,407

CONTRACTOR'S SITE ONCOSTS - NON TIME RELATED

Erect and dismantle offices				£	£
Mobile	108	m²	£8.50	918	
Site works			say,	800	
Toilets			say,	250	
Wiring, water, etc.			say,	500	2,468
Erect and dismantle other buildings					
Stores and welfare	130	m²	£15.00	1,950	
Site works			say,	900	
Toilets			say,	650	3,500
Telephone installation					500
Survey equipment and setting out					
Purchase cost including pegs, profiles, paint ranging rods, etc					1,000
Canteen and welfare equipment					
Purchase cost less residual value					1,500
Electrical installation					2,500
Water supply					
Connection charges				1,000	
Site installation				1,000	2,000
Carried forward				£	13,468

Brought forward			£	13,468
Transport of plant and equipment				3,500
Stores compound and huts				500
Sign boards and traffic signs				750
Insurances (dependent on Contractor's policy and record)				
Contractor's all risks 2% on £6,676,000			133,520	
Allow for excesses			15,000	148,520
General site clearance				3,500
Total Contractor's site oncosts - Non time			£	170,238

EMPLOYER'S AND CONSULTANT'S SITE REQUIREMENTS - TIME RELATED
(details of requirements will be defined in the contract documents)

Offices (50 m²)	52	wks at	£75		3,900
Site attendant labour (man weeks)	150	wks at	£520		78,000
Site transport (2 Land Rovers or similar)	52	wks at	£650		33,800
Telephones and calls	52	wks at	£100		5,200
Furniture and equipment	52	wks at	£40		2,080
Survey equipment	52	wks at	£50		2,600
Office heating and lighting (50 m²)	52	wks at	£30		1,560
Office consumables (provided by Contractor)	52	wks at	£40		2,080
Total employer's and consultants' requirements - time related				£	129,220

EMPLOYER'S AND CONSULTANT'S SITE REQUIREMENTS - NON TIME RELATED
(details of requirements will be defined in the contract documents)

Erection and dismantling of huts and offices	50	m² at	£8.50		425
Site works, toilets, etc.					1,000
Telephone installation					300
Electrical installation					1,000
Furniture and equipment Purchase cost less residual value					750
Progress photographs	100	sets at	£35		3,500
Total employer's and consultant's requirements - non time related				£	6,975

OTHER SERVICES, CHARGES AND FEES

Not applicable to this cost model.

TEMPORARY WORKS – OTHER THAN THOSE INCLUDED IN UNIT COSTS

Temporary Fencing					
1000m chestnut fencing					
Materials	5,000	m at	£5.85	29,250	
Labour	500	hrs at	£11.10	5,550	34,800
Traffic diversions					
Structure No. 1				8,000	
Structure No. 2				6,500	14,500
Footpath diversion					5,000
Stream diversion					15,500
Site access roads	400	m at	£120.00		48,000
Total temporary works				£	117,800

GENERAL PURPOSE PLANT – OTHER THAN THAT INCLUDED IN UNIT COSTS

Description				**Labour**	**Plant**	**Fuel etc.**	
Wheeled tractor							
hire charge	52	wks at	£550		28,600		28,600
driver (skill rate 4)	52	wks at	£520	27,040			27,040
consumables	52	wks at	£60			3,120	3,120
Trailer							
hire charge	52	wks at	£20		1,040		1,040
10 t Crawler Crane							
hire charge	40	wks at	£1,000		40,000		40,000
driver (skill rate 3)	40	wks at	£620	24,800			24,800
consumables	40	wks at	£30			1,200	1,200
Sawbench (diesel)							
hire charge	30	wks at	£45		1,350		1,350
consumables	30	wks at	£27			810	810
Carried forward						£	127,960

Brought forward						£	127,960
14.5 tonne hydraulic crawler backacter							
hire charge	6	wks at	£820		4,920		4,920
Driver (skill rate 3)	6	wks at	£620	3,720			3,720
Banksman (skill rate 4)	6	wks at	£630	3,780			3,780
consumables	6	wks at	£75			450	450
Concrete vibrators (two)							
hire charge in total	24	wks at	£100		2,400		2,400
D6 Dozer or similar							
hire charge	4	wks at	£1,400		5,600		5,600
driver (skill rate 2)	4	wks at	£698	2,792			2,792
consumables	4	wks at	£160			640	640
Towed roller BW6 or similar							
hire charge	6	wks at	£310		1,860		1,860
consumables	6	wks at	£56			336	336
Loading shovel Cat 939 or similar							
hire charge	16	wks at	£930		14,880		14,880
driver (skill rate 2)	16	wks at	£698	11,168			11,168
consumables	16	wks at	£78			1,248	1,248
Compressor 22.1m³/min (silenced)							
hire charge	12	wks at	£455		5,460		5,460
consumables	12	wks at	£333			3,996	3,996
Plate Compactor (180 kg)							
hire charge	12	wks at	£46		552		552
consumables	12	wks at	£8			96	96
75 mm 750 l/min pumps							
hire charge	25	wks at	£72		1,800		1,800
consumables	25	wks at	£8			200	200
Total costs			£	73,300	108,462	12,096	193,858

WORKED EXAMPLE

SUMMARY OF PRELIMINARIES AND GENERAL ITEMS

Contractor's site oncosts - Time related		736,407
Contractor's site oncosts - Non time related		170,238
Employer's and consultants' requirements on site - Time related		129,220
Employer's and consultants' requirements on site - Non Time related		6,975
Other services, charges and fees		
Temporary works not included in unit costs		117,800
General purpose plant and plant not included in unit costs		193,858
Total of Preliminaries and General Items	£	1,354,498

PART 3

Resources

This part comprises sections on labour, materials and plant for civil engineering work. These resources form the basis of the unit costs in Parts 4, 5, 7 and 8 and are given here so that users of the book may:

Calculate rates for work similar to, but differing in detail from, the unit costs given in Parts 4, 5, 7 and 8

Compare the costs given here with those used in their own organisation

Calculate the effects of changes in wage rates, material prices, etc.

Adjustments should be made to the rates shown to allow for time, location, local conditions, site constraints and any other factors likely to affect the cost of the specific scheme.

BASIS OF THIS SECTION

The following are brief details of the Construction Industry Joint Council agreement on pay and conditions for the Building and Civil Engineering Industry, which is effective from Monday 26th June 2006.

Copies of the Working Rule Agreement may be obtained from

Construction Industry Joint Council
55 Tufton Street
London SW1P 3QL

Rates of Pay (Rule WR.1)

The basic and additional hourly rates of pay are:-

Labourer/General Operative	£7.01
Skill Rate 4	£7.55
Skill Rate 3	£8.00
Skill Rate 2	£8.55
Skill Rate 1	£8.88
Craft Operative	£9.32

Additional Payment for Skilled Work (Rule WR.1.2.2)

Skilled Operative Additional Rate:-

These rates are to be deleted with effect from 26th June 2006.

Additional Payment (Rule WR.1.4)

The WRA recognises an entitlement to additional payments to operatives employing intermittent skill, responsibility or working in adverse conditions, for each hour so engaged as defined in Schedule 2 as follows:

	Extra £/hr		
A	£0.16	Tunnels	Operatives (other than Tunnel Machine Operators, Tunnel Miners and Tunnel Miners' Mates) wholly or mainly engaged in work of actual construction including the removal and dumping of mined materials but excluding operatives whose employment in the tunnel is occasional and temporary
B	£0.25	Sewer Work	Operatives working within a totally enclosed active surface water sewer of any nature or condition
C	£0.33	Sewer Work	Operatives working outside existing sewers excavating or removing foul materials emanating from existing sewers
C	£0.33	Working at height	Operatives (including drivers of tower cranes, but excluding the drivers of power driven derricks on high stages and linesmen-erectors and their mates and scaffolders) employed on "detached work".
D	£0.38	Sewer Work	within a totally enclosed active foul sewer of any nature or condition
E	£0.58	Stone cleaning	Operatives other than craft operatives dry-cleaning stonework by mechanical process for the removal of protective material and/or discolouration

BASIS OF THIS SECTION - continued

Bonus (Rule WR.2)

The Working Rule leaves it open to employers and employees to agree bonus scheme. Based on measured output and productivity for any operation or operations on a job.

Working Hours (Rule WR.3)

Normal working hours are unchanged at 39 hour week - Monday to Thursday at 8 hours per day and Friday at 7 hours. The working hours for operatives working shifts are 8 hours per weekday, 40 hours per week.

Rest/Meal Breaks (Rule WR.3.1)

Times fixed by the employer, not to exceed 1 hour per day in aggregate, including a meal break of not less than half an hour.

Overtime Rates (Rule WR.4)

During the period of Monday to Friday, the first 4 hours after normal working day is paid at time and a half, after 4 hours double time shall be paid. Saturday time and a half until completion of the first four hours. Remainder of Saturday and all Sunday at double time.

Daily Fare and Travelling Allowances (Rule WR.5)

This applies only to distances from home to the site of between 15 and 75 km, giving rates of 89p to £6.82 for Travelling Allowance (taxed) and £3.35 to £11.85 for Fare Allowance (not taxed). For the one way distance of 15km used in the following example. we have assumed a rate of £3.35 with effect from 26th June 2006.

Rotary Shift Working (Rule WR.6)

This relates to the situation where more than one shift of minimum 8 hours is worked on a job in a 24 hour period, and the operative rotates between the shifts either in the same or different pay weeks.

The basic rate shall be the operatives normal hourly rate plus 14%. Overtime beyond the 8 hour shift shall be at time and a half for the first 4 hours at normal rate plus 14%, thereafter double normal rate.

Night Work (Rule WR.7)

Providing night work is carried out by a separate gang from those working during daytime, an addition of 25% shall be paid on top of the normal hourly rate.

Overtime payments: during the period of Monday to Friday, the first 4 hours after normal working day is paid at time and a half plus the 25% addition of normal hourly rate, after 4 hours double time shall be paid. All hours worked on Saturday and all Sunday at double time.

Tide Work (Rule WR.9)

Where the operative is also employed on other work during the day, additional time beyond the normal working day shall be paid in accordance with the rules for overtime payments.

Where the operative is solely involved in work governed by tidal conditions, he shall be paid a minimum of 6 hours at normal rates for each tide. Payment for hours worked in excess of 8 over two tides shall be calculated proportionately.

$$\frac{\text{(Total Hours Worked - 8 hours) x Total Hours Worked}}{\text{8 hours}}$$

Work done after 4pm Saturday and during Sunday shall be at double time. Operatives are guaranteed 8 hours at ordinary rate for any time worked between 4pm and midnight on Saturdays and 16 hours for two tides worked on a Sunday.

Tunnel Work (Rule WR.10)

The first part of a shift equivalent to the length of the normal working day shall be paid at the appropriate normal rate, the first four hours thereafter at time and a half and thereafter at double time. In the case of shifts on a Saturday, the first 4 hours are at time and a half, thereafter at double time. All shifts on a Sunday at double time. Saturday shifts extending into Sunday are at double time. Sunday shifts extending into Monday, time after midnight: first 4 hours at time and a half, thereafter at double time.

Subsistence Allowance (Rule WR.15)

The 2005 allowance will increase to £27.11 per night with effect from 26th June 2006.

Annual Holidays Allowance (Rules WR.18 and 21)

Annual and Public Holiday Pay is included in accordance with the Building & Civil Engineering Benefit Schemes (B&CE Template Scheme. Allowances are calculated on total weekly earnings inclusive of overtime and bonus payments and the labour cost calculation assumes 21 days (4.2 weeks) annual and 8 days (1.6weeks) public holidays.

Easy Build Pension Contributions (Rule WR 21.3

The minimum employer contribution is £3.00 per week. Where the operative contributes between £3.01 and £10.00 per week the employer will increase the minimum contribution to match that of the operative up to a maximum of £10.00 per week. The calculations below assume 1 in 10 employees make contributions of £10.00 per week.

LABOUR COSTS

CALCULATION OF LABOUR COSTS

This section sets out a method used in calculating all-in labour rates used within this book. The calculations are based on the wage rates, plus rates and other conditions of the Working Rule Agreement (WRA); important points are discussed below. The calculations can be used as a model to enable the user to make adjustments to suit specific job conditions in respect of plus rate, bonus, subsistence and overhead allowances together with working hours etc. to produce alternative all-in rates which can be substituted for those printed.

All-in labour costs are calculated on the page overleaf for six categories of labour reflecting the different classifications set out in the WRA.

AVERAGE WORKING WEEK has been calculated to an equal balance between winter and summer (46.20 working weeks per year).

SUBSISTENCE PAYMENTS are included for key men only: for a stated percentage of the workforce.

TRAVELLING ALLOWANCES are based on rate payable for a journey of 15 Km at £3.20 per day. These allowances are adjusted to cater for operatives receiving Subsistence Payments.

BONUS PAYMENTS reflect current average payments in the industry.

WORKING HOURS & NON-PRODUCTIVE OVERTIME are calculated thus:

SUMMER (hours worked 8.00 am - 6.00pm, half hour for lunch)

	Sun	Mon	Tue	Wed	Thu	Fri	Sat	Total Paid Hours	Deductions : Lost & Wet time	Paid Breaks	Effective Total Hours
Total hrs	-	9.50	9.50	9.50	9.50	9.50	4.00	51.50	0.50	1.00	**50.00**
Overtime	-	0.75	0.75	0.75	0.75	1.25	2.00	6.25			
TOTAL HOURS						PAID :		**57.75**	EFFECTIVE :		**50.00**

WINTER (hours worked 7.30 am - 4.30pm, half hour for lunch)

	Sun	Mon	Tue	Wed	Thu	Fri	Sat	Total Paid Hours	Deductions : Lost & Wet time	Paid Breaks	Effective Total Hours
Total hrs	-	8.50	8.50	8.50	8.50	7.50	4.00	45.50	1.50	-	**44.00**
Overtime	-	0.25	0.25	0.25	0.25	0.25	2.00	3.25	-	-	-
TOTAL HOURS						PAID :		**48.75**	EFFECTIVE :		**44.00**

CALCULATION OF LABOUR COSTS - continued

AVERAGE TOTAL HOURS PAID	=	(57.75 + 48.75 / 2)	=	53.25 hours per week
AVERAGE EFFECTIVE HOURS WORKED	=	(50.00 + 44.00 / 2)	=	47.00 hours per week

			Categories of Labour					
Payment details (based on 2006 settlement)			general	skill rate 4	skill rate 3	skill rate 2	skill rate 1	craft rate
			£	£	£	£	£	£
Basic rate per hour			7.010	7.550	8.000	8.550	8.880	9.320
			£	£	£	£	£	£
53.25 Hours Paid @ Total Rate			373.28	402.04	426.00	455.29	472.86	496.29
Weekly Bonus Allowance	*		27.56	27.56	55.13	55.40	82.96	110.53
TOTAL WEEKLY EARNINGS		£	400.84	429.60	481.13	510.69	555.82	606.82
Travelling Allowances								
(say, 15km per day, i.e. £3.35 x 6 days)								
general, skill rate 4, skill rate 3 =	100%		20.10	20.10	20.10	-	-	-
skill rate 2, skill rate 1 =	80%		-	-	-	16.08	16.08	-
craft rate =	90%		-	-	-	-	-	18.09
Subsistence allowance								
(Average 6.66 nights x £27.11, + 7.00% to cover periodic travel)								
general, skill rate 4, skill rate 3 =	0%		-	-	-	-	-	-
skill rate 2, skill rate 1 =	20%		-	-	-	38.64	38.64	-
craft rate =	10%		-	-	-	-	-	19.32
TOTAL WAGES		£	420.94	449.70	501.23	565.41	610.54	644.23
National Insurance contributions (12.8% above ET except Fares)			38.89	42.57	49.17	55.01	63.67	67.73
Annual Holiday Allowance 4.2 weeks @ Total Weekly Earnings over 46.2 weeks			36.44	39.05	43.74	46.43	50.53	55.17
Public Holidays with Pay 1.6 weeks @ Total Weekly Earnings over 46.2 weeks			13.88	14.88	16.66	17.69	19.25	21.02
Easy Build pension contribution			4.00	4.00	4.00	4.00	4.00	4.00
CITB Levy (0.50% Wage Bill)			2.10	2.25	2.51	2.83	3.05	3.22
			516.25	552.45	617.30	691.37	751.05	795.36
Allowance for Employer's liability & third party insurances, safety officer's time, QA policy / inspection and all other costs and overheads :			12.91	13.81	15.43	17.28	18.78	19.88
TOTAL WEEKLY COST (47 HOURS)		£	529.16	566.26	632.73	708.65	769.83	815.24
COST PER HOUR		£	11.26	12.05	13.46	15.08	16.38	17.35
Plant Operators rates:								
Addition to Cost per Hour for Rule WR.11 - plant servicing time (6 hrs)		£	1.04	1.12	1.19	1.27	1.32	1.38
COST PER HOUR		£	12.30	13.17	14.65	16.35	17.70	18.73

Plus rates for Skill (WR 1.2.2) :
Discontinued from 26th June 2006

Additional Payments (see WR 1.4 above) :

		A	B	C	D	E
Extra rate per hour	£	0.16	0.25	0.33	0.38	0.58

* The bonus levels have been assessed to reflect the general position as at May/June 2006 regarding bonus payments limited to key personnel, as well as being site induced.

LABOUR CATEGORIES

Schedule 1 to the WRA lists "specified work establishing entitlement to the Skilled Operative Pay Rate 4, 3, 2, 1 or the "Craft Rate" as follows :-

General Operative

Unskilled general labour

Skilled Operative Rate 4

supervision	Gangers + trade chargehands
plant	Contractors plant mechanics mate; Greaser
transport	Dumper <7t; Agric. Tractor (towing use); Road going motor vehicle <10t; Loco driver
scaffolding	Trainee scaffolder
drilling	Attendee on drilling
explosives	Attendee on shot firer
piling	General skilled piling operative
tunnels	Tunnel Miner's mates; Operatives driving headings over 2 m in length from the entrance (in with drain, cable + main laying)
excavate	Banksmen for crane/hoist/derrick; Attendee at loading/tipping; Drag shovel; Trenching machine (multi bucket) <30hp; Power roller <4t; Timberman's attendee
coal	Opencast coal washeries and screening plants
tools	Compressor/generator operator; Power driven tools (breakers, tamping machines etc.); Power-driven pumps; air compressors 10 KW+; Power driven winches
concrete	Concrete leveller/vibrator operator/screeder + surface finisher, concrete placer; Mixer < 21/14 or 400 litres; Pumps/booms operator
linesmen	Linesmen-erector's mate
timber	Carpentry 1st year trainee
pipes	Pipe layers preparing beds and laying pipes <300mm diameter; Pipe jointers, stoneware or concrete pipes; pipe jointers flexible or lead joints <300mm diameter;
paving	Rolled asphalt, tar and/or bitumen surfacing: Mixing Platform Chargehand, Chipper or Tamperman; Paviors rammerman, kerb + paving jointer
dry lining	Trainee dry liners
cranes	Forklift truck <3t; Crane < 5t; Tower crane <2t; Mobile crane/hoists/fork-lifts <5t; Power driven hoist/crane

Skilled Operative Rate 3

transport	Road going motor vehicle 10t+
drilling	Drilling operator
explosives	Explosives/shot firer
piling	Piling ganger / chargehand; Pile frame winch driver
excavation	Tractors (wheeled/tracked with/without equipment) <100hp; Excavator <0.6m^3 bucket; Trenching machine (multi bucket) 30-70hp; Dumper 7-16t; Power roller 4t+; Timberman
tunnels	Tunnel Miner (working at face/operating drifter type machine
concrete	Mixer 400-1500litres; Mobile concrete pump with/without concrete placing boom; Hydraulic jacks + other tensioning devices in post-tensioning and/or pre-stressing

LABOUR CATEGORIES - continued

Skilled Operative Rate 3 - continued

formwork	Formwork carpenter 2nd year trainee
masonry	Face pitching or dry walling
linesmen	Linesmen-erectors 2nd grade
steelwork	Steelwork fixing simple work; Plate layer
pipes	Pipe jointers flexible or lead joints (300-535mm diameter)
paving	Rolled asphalt, tar and/or bitumen surfacing: Raker, Power roller 4 t+, mechanical spreader operator/leveller
dry lining	Certified dry liners
cranes	Cranes with grabs fitted; Crane 3-10t; Mobile crane/hoist/fork-lift 4-10t; Overhead/gantry crane <10t; Power-driven derrick <20t;Tower crane 2-10t; Forklift 3t+

Skilled Operative Rate 2

plant	Maintenance mechanic; Tyre fitter on heavy earthmover
transport	Dumper 16-60t
scaffolding	Scaffolder < 2 years scaffolding experience and < 1 year as Basic Scaffolder
piling	Rotary or specialist mobile piling rig driver
excavation	Tractors (wheeled/tracked with/without equipment) 100-400hp; Excavator 0.6-3.85m^3 bucket; Trenching machine (multi bucket) 70hp +; Motorised scraper; Motor grader
tunnels	Face tunnelling machine
concrete	Mixer 1500litres+; Mixer, mobile self-loading and batching <2500 litres
linesmen	Linesmen-erectors 1st grade;
welding	Gas or electric arc welder up to normal standards
pipes	Pipe jointers flexible or lead joints over 535mm diameter
cranes	Mobile cranes/hoists/fork-lifts 5-10t; Power-driven derrick 20t+, with grab <20t; Tower crane 10-20t

Skilled Operative Rate 1

plant	Contractors plant mechanic
transport	LGV driver; Lorry driver Class C+E licence; Dumper 60-125t
excavation	Excavator 3.85-7.65m^3; Tractors (wheeled/tracked with/without equipment) 400-650hp
steelwork	Steelwork assembly, erection and fixing of steel framed construction
welding	Electric arc welder up to highest standards for structural fabrication + simple pressure vessels (air receivers) including CO2 processes)
drilling	Drilling rig operator;
cranes	Power-driven derrick with grab 20t +; Tower crane 20t+

Craft Operative

transport	Dumper 125t+
scaffolding	Scaffolder a least 2 years scaffolding experience and at least 1 year as Basic and Advanced Scaffolder
excavation	Excavator 7.65t + (see WR 1.2.2 above); Tractors (wheeled/tracked with/without equipment) 650hp+
concrete	Reinforcement bender + fixer
formwork	Formwork carpenter
welding	Electric arc welder capable of all welding processes on all weldable materials including working on own initiative from drawings
cranes	Crane 10t+; Mobile cranes/hoists/fork-lifts 10t +

BASIC MATERIAL PRICES

This section comprises a price list for materials which assumes that the materials would be in quantities as required for a medium sized civil engineering project of, say, £8 - 10 million and the location of the works would be neither city centre nor excessively remote.

The material prices have been obtained from manufacturers and suppliers and are generally those prevailing at the time of preparing this edition (May/June 2006). In view of the current firming market, it is important to note that steel prices are based at May 2006. Prices are given for the units in which the materials are sold, which may not necessarily be the units used in the Unit Costs sections.

In effect these prices reflect the normal 'list price'; there are NO adjustments for the following :

- waste or loss or any offloading or distribution charges, unless specifically noted,
- trade discounts pertinent to the market, type of work involved and the relationship between contractor and supplier.

In addition, all prices quoted throughout this book are exclusive of Value Added Tax.

This section comprises :

BAR AND FABRIC REINFORCEMENT

Grade 500C deformed reinforcing bars				**Stainless steel reinforcing bars**					
High Tensile Steel to BS 4449						Stainless steel to EN 1.4301		Stainless steel to EN 1.4462	
Dia	Unit	Straight £	Bent £	Dia	Unit	Straight £	Bent £	Straight £	Bent £
T 8	t	384	424	8	t	2,500	2,600	2,600	2,700
T10	t	378	418	10	t	2,580	2,680	2,680	2,780
T12	t	375	415	12	t	2,500	2,600	2,600	2,700
T16	t	370	410	16	t	2,450	2,550	2,550	2,650
T20	t	370	410	20	t	2,360	2,460	2,460	2,560
T25	t	370	410	25	t	2,300	2,400	2,400	2,500
T32	t	375	415	32	t	2,300	2,400	2,400	2,500
T40	t	378	418						

Welded fabric to BS 4483 in sheets 4.80 x 2.40 metres	**Unit**	**£**
BS Ref. A393 6.16 kg per m²	m²	3.13
BS Ref. A252 3.95 kg per m²	m²	2.00
BS Ref. A193 3.02 kg per m²	m²	1.54
BS Ref. A142 2.22 kg per m²	m²	1.13
BS Ref. A98 1.54 kg per m²	m²	1.09
BS Ref. B113 10.90 kg per m²	m²	5.54
BS Ref. B785 8.14 kg per m²	m²	4.14
BS Ref. B503 5.93 kg per m²	m²	3.02
BS Ref. B385 4.53 kg per m²	m²	2.30
BS Ref. B283 3.73 kg per m²	m²	1.91
BS Ref. B196 3.05 kg per m²	m²	1.55
BS Ref. C785 6.72 kg per m²	m²	3.40
BS Ref. C636 5.55 kg per m²	m²	2.83
BS Ref. C503 4.34 kg per m²	m²	2.21
BS Ref. C385 3.41 kg per m²	m²	1.74
BS Ref. C283 2.61 kg per m²	m²	1.33
BS Ref. D49 0.77 kg per m²	m²	1.19

The prices shown above include delivery, but a charge of £30.00 is applicable for each delivery of under 8 tonnes

	Unit	**£**
Black annealed tying wire		
16 swg tying wire (coil)	25 kg	33.63
16 swg tying wire (coil)	2 kg	5.70
Stainless steel tying wire		
16 swg tying wire (coil)	25 kg	121.30

For reinforcement bar couplers refer to ROM Lenton reinforcement couplers detailed on page 78

BRICKWORK AND BLOCKWORK

	Unit	£
Clay bricks to BS 3921*		
Commons		
flettons	1000	224 - 270
non-flettons	1000	168 - 225
Facings		
flettons, pressed, sand-faced	1000	245 - 380
machine moulded	1000	280 - 450
extruded wire-cut	1000	224 - 500
pressed, repressed	1000	370 - 500
hand-made	1000	450 - 895
Engineering, Class A		
perforated	1000	285 - 335
solid	1000	395 - 450
solid facing	1000	460 - 560
Engineering, Class B		
perforated	1000	200 - 260
solid	1000	335 - 400
solid facing	1000	440 - 525
Concrete bricks to BS 6073*		
Commons	1000	224 - 280
Facings	1000	335 - 395
Engineering	1000	150 - 180
Calcium silicate bricks to BS 187*		
Commons	1000	220 - 280
Facings	1000	285 - 330
Glazed bricks to BS 391*		
Stretcher face only	1000	2600 - 4000
Stretcher and header faces	1000	3200 - 4800
Refractory bricks*	1000	3600 - 6100
Note:		
Prices are ex-works for full loads		
* Co-ordinating size 225 x 112.5 x 75 mm		

BRICKWORK AND BLOCKWORK - continued

	Unit	£
Concrete blocks to BS 6073		
Aerated concrete blocks		
(Strength 2.8 N/m²)		
100 mm	m²	8.00 - 9.00
140 mm	m²	10.00 - 11.50
190 mm	m²	13.50 - 15.00
(Strength 4.0 N/m²)		
100 mm	m²	6.00 - 7.50
140 mm	m²	8.50 - 10.50
190 mm	m²	13.00 - 14.50
(Strength 7.0 N/m²)		
100 mm	m²	9.00 - 10.00
140 mm	m²	12.00 - 13.50
190 mm	m²	17.50 - 18.50
Lightweight aggregate medium density blocks		
(Strength 3.5 N/m²)		
100 mm	m²	6.00- 8.00
140 mm	m²	8.00 - 10.50
190 mm	m²	10.50-15.00
Dense aggregate blocks		
(Strength 7.01 N/m²)		
100 mm	m²	5.50-7.00
140 mm	m²	7.00-9.50
190 mm	m²	11.50 - 16.00
Coloured dense concrete masonry blocks		
Hollow		
100 mm	m²	15.50-20.00
140 mm	m²	19.00-25.00
190 mm	m²	25.00-30.00
Solid		
100 mm	m²	18.00 - 24.00
140 mm	m²	25.00 - 35.00
190 mm	m²	36.00 - 46.00
Note:		
Prices are for full loads delivered to site		

CAST IRON PIPES AND FITTINGS

	Unit	£	£	£	£	£	£
Diameter in mm		50	70	100	125	150	200
Lengths in m		3	3	3	3	3	3
Cast and ductile iron above ground pipes and fittings to BS EN 877, Ensign joints							
Plain ended cast iron pipes	m	11.12	12.87	15.31	24.57	30.34	50.70
Cast iron couplings complete with stainless steel nuts and bolts and synthetic rubber gaskets	nr	5.00	5.50	7.17	8.90	14.35	32.10
Pipe brackets - ductile iron	nr	4.68	4.68	5.13	-	9.74	37.11
Bends, short radius	nr	8.66	9.74	11.53	20.44	20.71	61.64
Access bends, short radius, 88 degrees	nr	-	18.87	27.58	-	42.85	-
Bends, long radius	nr	-	-	27.30	-	84.00	-
Access bends, long radius, 88 degrees	nr	-	-	35.67	-	86.66	-
Branches, single (equal)	nr	13.88	14.65	18.67	40.29	44.67	120.63
Branches, double	nr	-	-	26.85	-	104.54	144.90
Taper pipes	nr	-	13.31	15.65	15.73	30.04	48.78
Trap, 'P', plain	nr	-	-	21.41	-	-	-
Branch trap	nr	-	-	77.57	-	-	-
Diameter in mm				100	150	225	
Lengths in m				3	3	3	
Cast and ductile iron below ground pipes and fittings to BS EN 877, Timesaver joints							
Plain ended cast iron pipes	m			27.11	50.28	150.21	
Cast iron couplings complete with stainless steel nuts and bolts and synthetic rubber gaskets	nr			17.38	21.04	48.20	
Bends, long radius, 87.5 degrees	nr			32.49	74.75	226.81	
Access bends, back door 87.5 degrees	nr			69.96	158.52	-	
Heel rest bend	nr			37.26	91.39	-	
Branches, 45 or 87.5 degrees, equal	nr			43.11	106.37	326.59	
Inspection chamber	nr			168.31	213.65	-	
Double branches, 87.5 degrees, equal	nr			86.13	116.47	-	
Pipe access, rect access	nr			80.27	146.56	356.02	
Pipes taper	nr			42.35	108.58	108.58	
(Note: as fittings are spigot ended Timesaver Couplings are required at each joint)							
Diameter in mm		700	800	900	1000	1200	
Ductile spun iron pipe to BS EN 545 with Stantyte joints for potable water applications							
Spigot and socket pipes	m	245.54	323.48	POA	POA	POA	
Stantyte gasket	nr	100.78	111.07	POA	POA	POA	
Stantyte tee, equal	nr	6603.40	10633.69	POA	POA	POA	
Stantyte bend, 90 degrees	nr	5380.42	9331.43	POA	POA	POA	
Stantyte bend, 45 degrees	nr	4400.89	5155.92	POA	POA	POA	

CAST IRON PIPES AND FITTINGS - continued

	Unit	£	£	£	£	£	£
Diameter in mm		100	150	250	350	400	600
Ductile spun iron pipe to BS EN 545 with Tyton joints							
Spigot and socket pipes	m	27.67	35.08	62.05	96.01	113.57	206.31
Tyton gasket	nr	7.67	8.25	15.61	37.25	44.35	67.68
Tyton tee, equal	nr	83.92	174.39	488.08	988.47	1032.80	3572.89
Flange on Tyton tee, equal	nr	85.76	167.03	485.37	1022.60	1528.85	4210.62
Tyton bend, 90 degrees	nr	56.52	130.14	383.90	538.82	790.91	2480.80
Tyton bend, 45 degrees	nr	53.16	83.21	247.42	419.64	549.36	1507.91
Tyton duckfoot bend, 90 degrees	nr	153.26	352.01	664.99	-	-	-
Tyton taper	nr	61.51	114.87	217.11	482.06	621.71	1681.95
Diameter in mm		100	150	250	350	450	
Couplings on BS EN 545 ductile spun iron pipe							
Viking Johnson Maxifit couplings	nr	51.31	68.39	120.74	323.59	368.39	
uLink couplings	nr	67.32	99.41	211.49	-	-	
		600	800	1000	1200		
GS LINK couplings	nr	346.45	503.06	632.74	737.47		
Size in mm			100 x 75		125 x 75		
Cast iron rainwater gutters and fittings; Classical							
1830 mm length	m		35.43		45.79		
union clips	nr		9.74		10.32		
fascia brackets	nr		8.37		10.32		
angles, 90° and 135°	nr		24.30		30.84		
running outlets	nr		9.93		11.83		
stop end outlets	nr		13.53		15.82		
stop ends	nr		8.53		10.51		
Diameter in mm			65		75		
Cast iron rainwater pipes and fittings; Classical							
single socket, 1830 mm length	m		38.58		38.58		
bends, 92.5° and 112.5°	nr		12.54		14.93		
offsets, 75 mm projection	nr		18.83		18.83		
offsets, 230 mm projection	nr		21.92		21.92		
offsets, 455 mm projection	nr		59.98		59.98		
shoe	nr		17.42		17.42		

CLAYWARE PIPES AND FITTINGS

	Unit	£	£	£	£	£	£	£
Diameter in mm		150	225	300	375	400	450	500
Standard pipe length in m		1.25	1.6	1.6	1.5	1.5	1.5	2.0
Vitrified clay spigot & socket pipes and fittings to BS EN 295 (Denseal)								
Straight pipes	m	19.17	37.03	57.89	140.25	142.85	185.55	253.60
Short length pipes (0.6 m)	nr	25.47	49.92	70.25	180.55	183.35	238.64	244.90
Bends								
11.25, 22.5, 45, 90 degrees	nr	36.32	74.08	146.19	350.05	395.94	521.39	647.31
Rest bends	nr	43.32	103.20	218.96	-	-	-	-
Saddles								
oblique/square	nr	36.17	86.21	-	-	-	-	-
Junctions								
(Main diameter stated)								
oblique 45 degrees	nr	47.44	111.42	229.69	-	-	-	-
curved square 90 degrees	nr	-	-	229.69	491.66	522.43	624.99	886.24
tumbling bay/drop back	nr	47.44	111.42	275.63	642.69	729.82	877.81	-
Tapers								
reducers/increasers	nr	85.70	175.22	272.56	-	-	-	-
Stoppers (with polyester spigot)	nr	12.36	26.68	58.81	103.08	-	-	-
Diameter in mm		100	150	225	300			
Standard pipe length in m		1.6	1.75	1.75	1.75			
Plain ended vitrified clay sewer pipes and fittings with sleeve joints to BS EN 295, including 1 nr coupling per length pipe (Densleeve)								
Straight pipes								
with coupling	m	6.25	15.16	35.37	54.06			
without coupling	m	4.62	11.10	27.93	39.32			
Bends								
11.25, 22.5, 45, 90 degrees	nr	5.34	14.89	50.98	96.47			
Rest bends	nr	11.47	20.03	57.02	158.77			
Saddles								
oblique/square	nr	11.64	21.08	66.77	-			
couplings (nitrile)	nr	4.62	11.77	20.30	48.93			

CLAYWARE PIPES AND FITTINGS - continued

	Unit	£	£	£	£	£	£	£
Taper pipes								
100 - 150 mm	nr	13.73						
150 - 225 mm	nr	55.60						
225 - 300 mm	nr	132.99						
Junctions								
oblique 45 degrees/								
curved square 90 degrees								
Main x Arm diameter								
100 x 100 mm	nr	11.36						
150 x 100 mm	nr	20.50						
150 x 150 mm	nr	22.47						
225 x 100 mm	nr	79.75						
225 x 150 mm	nr	79.75						
225 x 225 mm	nr	79.75						
300 x 100 mm	nr	165.07						
300 x 150 mm	nr	165.07						
300 x 225 mm	nr	165.07						
300 x 300 mm	nr	165.07						
Low back traps								
100 x 100 mm diameter 'P'								
trap (plain)	nr	10.85						
150 x 150 mm diameter 'P'								
trap (plain)	nr	24.82						
Diameter in mm		100	150	225	300	375	400	450
Vitrified clay socketted channel pipes and fittings to BS 65 (Naylor)								
Channel pipes (1.0m long)	nr	-	10.69	24.10	45.83	79.77	80.84	99.59
Channel bends								
11.25, 22.5, 45, 90	nr	6.29	10.42	34.98	71.36	230.20	236.85	320.38
Branch Channel bends								
three quarter section	nr	14.46	21.97	92.56	-	-	-	-
Channel Junctions								
oblique/square								
single	nr	12.75	20.94	59.92	142.59	-	-	-
double	nr	19.08	31.84	96.57	-	-	-	-
breeches	nr	19.08	31.84	96.57	-	-	-	-

	Unit	£	£	£	£	£	£	£
Channel Tapers								
increaser/reducer								
100 - 150 mm	nr	30.94						
150 - 225 mm	nr	68.38						
225 - 300 mm	nr	135.35						
300 - 375 mm	nr	209.65						
375 - 450 mm	nr	282.54						
Diameter in mm		100	150					
Standard pipe length in m		1.6	1.75					
Unglazed dense-vitrified clay plain ended cable conduit to BS65, including 1 nr coupling per length (Denduct)								
Straight conduits	m	7.13	15.32					
Bends/Bellmouths	nr	8.17	16.83					
Diameter in mm		75	100	150	225			
Standard pipe length in m		0.3	0.3	0.3	0.3			
Land drain to BS 1196 (Hepworth)								
Straight pipes	nr	1.15	1.98	4.05	10.74			
Junctions	nr	13.92	17.39	21.41				

	Unit	£	£	£	£	£	£
Diameter in mm			150	225	300	400	450
Standard pipe length in m			1.75	2.0	2.0	2.0	2.0
Vitrified clay drainage with sealing rings (Superseal, Hepseal)			s/seal	s/seal	s/seal	hepseal	hepseal
Straight pipes	m		18.69	37.00	56.75	110.58	143.63
Bends							
90 degrees	nr		34.32	76.76	145.78	415.52	547.16
45 degrees	nr		34.32	76.76	145.78	296.78	390.81
30 degrees	nr		34.32	76.76	145.78	-	-
15 degrees	nr		34.32	76.76	145.78	-	-
22.5 degrees	nr		-	-	-	207.79	273.59

CLAYWARE PIPES AND FITTINGS - continued

	Unit	£	£	£	£	£	£
Diameter in mm			150	225	300	400	450
Standard pipe length in m			1.75	2.0	2.0	2.5	2.5
Vitrified clay drainage with sealing rings (Superseal, Hepseal)			s/seal	s/seal	s/seal	hepseal	hepseal
Rest bends	nr		19.27	93.77	207.73	-	-
Stopper	nr		10.83	17.42	36.38	-	-
Double collar	nr		31.48	67.80	110.18	-	-
Saddles							
oblique	nr		21.96	79.92	139.09	-	-
square	nr		21.96	79.92	139.09	-	-
Tapers							
reducers	nr		18.59	51.52	139.09	-	-
enlargers	nr		55.38	149.52	-	-	-
Adaptors to SuperSleve	nr		16.83	-	-	-	-
Adaptors to Hepsleve	nr		16.83	32.33	63.54	-	-
		Oblique S/Sleve H/Sleve Plus	Oblique S/Seal H/Seal	Square S/Sleve H/Sleve Plus	Square S/Seal H/Seal	Tumbling Bay Square	Tumbling Bay Square
Junctions							
100 x 100 mm	nr	23.44	-	23.44	-	-	-
150 x 100 mm	nr	20.07	39.73	20.07	39.73	-	-
150 x 150 mm	nr	22.03	44.86	22.03	44.86	44.86	44.86
225 x 100 mm	nr	102.05	107.15	102.05	107.15	-	-
225 x 150 mm	nr	102.05	107.15	102.05	107.15	-	-
225 x 225 mm	nr	129.86	136.34	129.86	136.34	136.34	136.34
300 x 100 mm	nr	-	224.45	224.45	224.45	-	-
300 x 150 mm	nr	-	224.45	224.45	224.45	-	-
300 x 225 mm	nr	-	-	-	-	-	-
300 x 300 mm	nr	-	-	285.48	-	285.48	285.48
400 x 100 mm	nr	-	418.53	-	-	-	-
400 x 150 mm	nr	-	418.53	-	-	-	-
400 x 400 mm	nr	-	-	-	-	588.01	588.01
450 x 100 mm	nr	-	500.68	-	-	-	-
450 x 150 mm	nr	-	500.68	-	500.68	-	-
500 x 150 mm	nr	-	629.18	-	-	-	-

	Unit	£	£	£	£	£	£
Diameter in mm		100	150	225			
Standard pipe length in m		1.6	1.75	1.75			
Plain end vitrified clay pipes with sleeved joints BS EN 295 (SuperSleve/Hepsleve)							
Straight pipes	m	5.40	10.91	35.26			
Couplings							
standard seal ring	nr	3.98	7.23	13.88			
EPDM/Nitrile seal ring	nr	6.05	11.42	19.39			
Bends 15, 30, 45, 90 degree	nr	7.28	15.00	73.10			
Rest Bends	nr	15.51	19.27	78.21			
Saddles							
oblique	nr	15.44	21.96	79.92			
square	nr	15.44	21.96	79.92			
Adaptors Hepseal/SuperSleve	nr	6.28	16.83	13.94			
Taper pipe							
100 to 150 mm Supersleve	nr	22.19					
150 to 225 mm Supersleve	nr	55.38					
225 to 300 mm Hepsleve	nr	149.52					
Junctions, oblique/square							
100 x 100 mm	nr	15.73					
150 x 100 mm	nr	20.07					
150 x 150 mm	nr	22.03					
225 x 100 mm	nr	102.05					
225 x 150 mm	nr	102.05					
225 x 225 mm	nr	129.86					
300 x 100 mm	nr	213.76					
300 x 150 mm	nr	213.76					
300 x 225 mm	nr	256.24					
300 x 300 mm	nr	271.88					

CLAYWARE PIPES AND FITTINGS - continued

	Unit	£	£	£	£	£	£
Diameter in mm		100	150	225	300	400	450
Standard pipe length in m		1.60	1.75	1.75	2.00	2.50	2.50
Perforated vitrified clay pipes (Hepline)							
Straight pipes	m	9.05	16.45	30.24	57.61	137.99	174.87
Vitrified clay road gullies, round with rodding eye and stopper							
600 mm deep x 300 mm diameter, outlet diameter 100 mm	nr	95.41					
600 mm deep x 300 mm diameter, outlet diameter 150 mm	nr	97.70					
750 mm deep x 400 mm diameter, outlet diameter 150 mm	nr	113.31					
900 mm deep x 450 mm diameter, outlet diameter 150 mm	nr	153.31					
Vitrified clay yard gullies, round with domestic duty grating and frame (upto 1 tonne)							
225 mm diameter x 585 mm deep, 100 mm outlet diameter	nr	152.16					
225 mm diameter x 585 mm deep, 150 mm outlet diameter	nr	155.97					
Vitrified clay yard gullies, round with medium duty grating and frame (upto 5 tonnes)							
225 mm diameter x 585 mm deep, 100 mm outlet diameter	nr	196.91					
225 mm diameter x 585 mm deep, 150 mm outlet diameter	nr	200.74					
Universal grease traps with metal tray and lid							
600 x 450 x 550 mm deep	nr	855.89					
Spare filter basket and spatula	nr	136.94					

CONCRETE AND CEMENT

Ready mixed concrete supplied in full loads delivered to site within 5 miles (8 km) radius of concrete mixing plant. The figures below include for the incidence of the Aggregate Tax, levied from 1 April 2002.

Designed mixes		Aggregate size		
		10 mm	20 mm	40 mm
	Unit	£	£	£
Grade C7.5	m³	72.62	68.39	71.56
Grade C10	m³	73.64	69.41	72.48
Grade C15	m³	75.20	71.10	74.25
Grade C20	m³	76.48	72.30	75.45
Grade C25	m³	76.96	73.52	76.20
Grade C30	m³	78.22	74.75	77.35
Grade C40	m³	81.91	78.30	-
Grade C50	m³	85.10	82.51	-
Grade C60	m³	86.10	87.15	-

Prescribed mixes

Add to the above designated mix prices approximately £3.00/m³

Standard mixes	Unit	£
ST1	m³	71.89
ST2	m³	73.28
ST3	m³	74.85
ST4	m³	76.31
ST5	m³	78.16
Readymix		
Roadfill 6-1 HAM m³		66.10
Voidfill 20-1 HAM m³		56.86

	Unit	£	
Add to the above prices for			
rapid-hardening cement to BS12	m³	2.00-3.00	
sulphate resisting cement to BS4027	m³	12.50	%
polypropylene fibre additive	m³	8.94	
distance per mile in excess of 5 miles (8 km)	m³	1.16	
air entrained concrete	m³	4.66	
water repellent additive	m³	4.70	
part loads per m³ below full load	m³	21.76	
waiting time (in excess of 6 mins. / m³ unloading norm)	hr	58.07	
returned concrete	m³	110.55	

	bulk £ / t	bagged £/25 kg
Cements		
ordinary portland to BS12	115.75	2.95
high alumina	327.76	9.22
sulphate resisting	116.95	3.35
rapid hardening	110.68	3.15
white portland	181.76	5.04
masonry	112.87	3.06

	Unit	£
Cement admixtures		
Febtone colourant, brown / black	kg	8.77
Febproof waterproofer	5 l	13.32
Febspeed frostproofer	5 l	6.68
Febbond PVA bonding agent	5 l	18.77
Febmix plus plasticiser	5 l	6.35

CONCRETE MANHOLES

	Unit	£
Shaft and chamber rings		
900 mm diameter - unreinforced	m	38.50
1050 mm diameter - unreinforced	m	40.00
1200 mm diameter - unreinforced	m	50.00
1350 mm diameter - reinforced	m	70.50
1500 mm diameter - reinforced	m	90.50
1800 mm diameter - reinforced	m	107.00
2100 mm diameter - reinforced	m	200.00
2400 mm diameter - reinforced	m	255.00
2700 mm diameter - reinforced	m	325.00
3000 mm diameter - reinforced	m	462.00
Short length surcharge on 250 mm depth - 100%		
Short length surcharge on 500 mm depth - 50%		
Double steps (built-in)	nr	5.00
Soakaway perforations (75 mm dia)	nr	3.95
Reducing slabs (900 mm diameter access)		
1350 mm diameter	nr	95.00
1500 mm diameter	nr	110.00
1800 mm diameter - 1200 mm diameter access, 250 mm deep	nr	136.00
2100 mm diameter - 1200 mm diameter access, 250 mm deep	nr	302.00
2400 mm diameter - 1200 mm diameter access, 250 mm deep	nr	500.00
2700 mm diameter - 1200 mm diameter access, 250 mm deep	nr	660.00
3000 mm diameter - 1200 mm diameter access, 250 mm deep	nr	775.00
Landing slabs (900 mm diameter access)		
1500 mm diameter	nr	114.00
1800 mm diameter	nr	136.00
2100 mm diameter	nr	302.00
2400 mm diameter	nr	500.00
2700 mm diameter	nr	660.00
3000 mm diameter	nr	775.00
Cover slabs (heavy duty)		
900 mm diameter, 600 mm square access	nr	36.00
1050 mm diameter, 600 mm square access	nr	41.00
1200 mm diameter, 600 or 750 mm square access	nr	51.00
1350 mm diameter, 600 or 750 mm square access	nr	73.50
1500 mm diameter, 600 or 750 mm square access	nr	95.75
1800 mm diameter, 600 or 750 mm square access	nr	113.50

CONCRETE PIPES AND FITTINGS

	Unit	£	£	£	£	£	£
Concrete pipes with flexible joints to EN 1916 : 2002, Straight pipes (pipe length 2.5m)							
Diameter in mm		300	375	450	525	600	
Class 120	m	11.50	15.25	18.00	21.50	27.00	
Bends (22.5 or 45 degrees)							
Class 120	nr	92.00	122.00	144.00	172.00	216.00	
Concrete pipes with flexible joints to EN 1916 : 2002, Straight pipes (pipe length 2.5m)							
Diameter in mm		750	900	1200	1500	1800	2100
Class 120	m	46.00	68.00	120.00	260.00	320.00	440.00
Bends (22.5 degrees)							
Class120	nr	368.00	544.00	960.00	2100.00	2500.00	3500.00
Bends (45 degrees)							
Class 120	nr	450.00	680.00	1355.00	2835.00	3450.00	4650.00
Junctions (extra over)							
150mm	m	45.00					
225mm	m	70.00					
300mm	m	90.00					
300 x 150mm	nr	88.00					
375mm	m	110.00					
375 x 150mm	nr	75.00					
450mm	m	145.00					
450 x 150mm	nr	85.00					
525mm	m	180.00					
525 x 150mm	nr	90.00					
600mm	m	225.00					
600 x 150mm	nr	100.00					
750mm	m	395.00					
825mm	m	425.00					
900mm	m	480.00					
Concrete road gullies to BS 5911-6 : 2004							
375 x 150 x 750mm deep	nr	25.30					
450 x 150 x 750mm deep	nr	27.00					
450 x 150 x 900mm deep	nr	27.20					
450 x 150 x 1050mm deep	nr	29.00					

CONSUMABLE STORES

	Unit	£
Abrasive discs		
for disc cutters (9 inch)	nr	4.52
for angle grinders (4 inch)	nr	2.96
for floorsaws (9 inch)	nr	4.31
Bushing heads		
for breakers	nr	13.64
for scabblers	nr	50.24
Hilti gun		
Cartridges for Hilti gun (DX 450) (100 Box)	nr	11.82
Hilti nails NK 32 (100 Box)	nr	18.55
Hilti nails NK 54 (100 Box)	nr	22.15
Stone/Concrete cutting disc for Stihl saw (12 inch)	nr	3.59
Diamond blades		
asphalt, 12 inch Tyrolit	nr	66.10
concrete, 12 inch Tyrolit	nr	240.00
general, 12 inch Tyrolit	nr	44.44
Gas refills		
Propane 47 kg (fuel only)	nr	43.12
Propane 23 kg (fuel only)	nr	22.56
Butane 13kg (fuel only)	nr	19.00
Oxygen 9.66 m^3 cylinder (gas only)	nr	12.10
Acetylene 8.69 m^3 cylinder (gas only)	nr	59.22
Compressed air 8.94 m^3 cylinder (gas only)	nr	16.95
Lubrication/Oils for site vehicle use		
crankcase (25 litres)	l	1.53
transmission (25 litres)	l	1.62
hydraulic (25 litres)	l	1.40
grease (12.5 kg)	kg	2.35
two stroke (25 litres)	l	1.55
airline oil (25 litres)	l	1.70
cutting oil (25 litres)	l	1.93
Ropes and hawsers		
Rope (polypropelene)		
6 mm	220 m	11.57
8 mm	220 m	21.14
10 mm	220 m	31.02
12 mm	220 m	46.15
Drawcord Rope		
6 mm	500 m	32,38
Chains		
4 x 32 mm link, galvanised	m	1.69
6 x 42 mm link, galvanised	m	3.21

	Unit	£
Chain brothers	3 m set	21.30
Hessian		
198g 1.37 x 45 m roll	nr	42.50
283 g 1.78 x 100 m roll (Heavy Duty)	nr	138.85
Dewatering disposables		
semi rigid riser pipe	6m	4.11
flexible riser pipe	m	0.68
65 mm x 0.5 m cocos	each	2.74
65 mm x 1.0 m cocos	each	2.74
jointing tape	roll	2.74
Pneumatic breakers, steels (hex shank)		
MP1 (moil point)	nr	11.34
MP4	nr	9.45
NC1 (narrow chisel)	nr	11.87
NC4	nr	9.88
CS1 (clay spade)	nr	38.85
CS4	nr	38.85
AC1 (asphalt cutter)	nr	25.80
AC3	nr	31.42
DC1 (digging chisel)	nr	38.57
DC3	nr	39.11
P1 (Plugs)	nr	34.24
F1 (Feathers)	nr	31.94
ET1/ET2 (combined 7 inchs tampers)	nr	81.34
SM1 (speediburst demolition/rock steel)	nr	24.34
CP9 (point)	nr	5.91
CP9 (chisel)	nr	5.91
FL22 square	nr	11.33
FL22 round	nr	12.31
Kango 900 Hammer moil point 380mm diameter	nr	5.98
Kango 900 Hammer narrow point 380mm diameter	nr	5.98
Kango 900 Hammer 50mm wide chisel	nr	12.53
Kango 900 Hammer 75mm wide chisel	nr	15.66
Bosch 11304 Hammer Tarmac Cutter	nr	21.36
Bosch 11304 Hammer Asphalt Cutter	nr	28.48

CONTRACTORS SITE EQUIPMENT (PURCHASED)

	Unit	£
Traffic cones and cylinders		
thermoplastic cone 18 inches with reflective sleeve	nr	4.28
thermoplastic cone 30 inches with reflective sleeve	nr	6.75
medium density polyethylene cone 30 inches with reflective sleeve	nr	6.80
high visibility motorway cone 1 m	nr	8.67
verge post complete with metal fixing stack	nr	11.37
KINGPIN traffic cylinder 30 inches with twist and lock base	nr	13.70
no waiting' cones (standard Mk 111)	nr	9.64
no waiting' cones (police Mk 111)	nr	10.79
Pendant barrier markers on 26 m cord	nr	7.45
Crowd Barrier		
fencing 1m x 50m roll, non corrodible, reusable	nr	38.50
thermoplastic fencing foot, 720 x 230 x 160 mm	nr	12.10
galvanised mesh aperture fencing panel 3.2 x 2.0 m	nr	39.75
Temporary road signs and frames		
triangular 600 mm	nr	43.75
rectangular 1050 x 750 mm	nr	69.48
circular 1050 mm	nr	69.48
Ecolite road lamp	nr	4.02
Battery operated "Unilamp", flashing	nr	6.78
Battery operated "Unilite", photocell flashing	nr	6.84
JSP Roadwall traffic barrier 650 x 450 mm, 1.64m long, red/white	nr	79.57
Sambloc traffic separator 600 x 430 mm, 1.0 m long, red / white	nr	80.83
Hydrant stand pipe	nr	53.79
Hydrant key	nr	14.08
Drain stoppers		
100 mm steel with plastic cap	nr	5.69
150 mm steel with plastic cap	nr	8.31
225 mm steel with plastic cap	nr	18.85
300 mm steel with plastic cap	nr	32.45
500 mm aluminium with metal cap	nr	150.50
525 mm aluminium with metal cap	nr	158.07
675 mm aluminium with metal cap	nr	343.10
Air bag stopper		
100 mm	nr	12.26
150 mm	nr	13.04
225 mm	nr	19.50
300 mm	nr	24.69
Drain testing manometer kit (air gauge)	nr	36.85

	Unit	£
Bolt croppers centre cut		
24 inches (3/4 inch jaw)	nr	122.50
high tensile cut 42 inches (1 inch jaw)	nr	273.50
Soil pipe cutter (spun pipes or cast) 2 to 12 inches	nr	479.50
Debris netting (scaffolding net, 50 x 3 m roll, standard)	nr	94.89
Scaffold sheeting (weather sheet, 45 x 2 m roll)	nr	235.62
Water hose		
Rubber 3/4 inch bore x 40 m	nr	54.88
PVC 1/2 inch bore x 75 m	nr	38.98
Compressor hose (with quick release couplings)		
15m x 19mm	nr	29.55
Portable electricity supply transformer, 3Kva, two outlet sockets	nr	122.54
Temporary Lighting		
Fixed Stand with single tungsten 500 W halogen fitting	nr	32.87
Fixed Stand with twin tungsten 500 W halogen fittings	nr	53.53
Ladders		
Builders pole ladder, 15 rungs, 4.00 m long	nr	92.93
Builders pole ladder, 27 rungs, 7.00 m long	nr	153.81
Youngman 2 part push up ladder, 5.11 m extended height	nr	190.44
Youngman 2 part push up ladder, 7.43 m extended height	nr	239.06
Youngman 3 part push up ladder, 7.43 m extended height	nr	321.10
Youngman 3 way combination ladder, 7.38 m extended height	nr	289.31
Ground Stabilisation		
Tenax Turf Reinforcement Mesh 22 x 27 mm mesh, 2 x 30 m roll, std. grade	nr	140.00
Tenax Grass Protection Mesh 15 x 15 mm mesh, 2 x 20 m roll (Green)	nr	325.00
Tenax Ground Stabilisation Mesh 75 x 50 mm mesh, 2 x 50 m roll (Grey)	nr	167.00

CULVERTS

	Unit	£	£	£	£
Well-void extra plus steel culverts					
Diameter in mm		500	1000	1500	2000
1.20 mm thick	m	32.25	63.43	-	-
1.50 mm thick	m	46.23	88.15	139.75	-
2.00 mm thick	m	59.13	113.95	182.75	236.50
Asset International corrugated steel, galvanised bitumen coated culverts, supplied in 3, 4, and 6 m lengths					
Diameter in mm		1000	1600	2000	2200
1.6 mm thick	m	74.40	113.61	485.40	598.61
coupling	nr	19.80	34.64	-	-
2.0 mm thick	m	90.64	139.34	304.38	343.61
coupling	nr	20.69	38.43	43.74	46.37

	Unit	£	£	£	£
Precast concrete rectangular culverts					
1000 x 500 mm x 2.0 m long	nr	394.79			
1500 x 1000 mm x 2.0 m long	nr	727.68			
1500 x 1500 mm x 2.0 m long	nr	840.97			
2000 x 1000 mm x 2.0 m long	nr	1013.84			
2000 x 1500 mm x 2.0 m long	nr	1144.66			
2500 x 1000 mm x 2.0 m long	nr	1541.79			
2500 x 1750 mm x 1.5 m long	nr	1352.57			
3000 x 2000 mm x 1.0 m long	nr	1319.86			
3000 x 2750 mm x 1.0 m long	nr	1471.71			
4000 x 2500 mm x 1.0 m long	nr	1746.19			

EARTH RETENTION INCLUDING GABIONS, CRIB WALLING, ETC.

Woven wire mesh gabions in mesh size 75 x 75mm , wire diameter 2.7 mm galvanised to BS 443:1982

PVC coated galvanised wire; overall diameter 3.2 mm			Galvanised only	
Units size	£/each		Unit Size	£/each
1.5 x 1 x 1 m	35.73		1.5 x 1 x 1 m	27.22
2 x 1 x 1 m	48.57		2 x 1 x 1 m	37.22
3 x 1 x 1 m	68.85		3 x 1 x 1 m	52.45
2 x 1 x 0.5 m	34.35		2 x 1 x 0.5 m	26.57

Reno woven wire mattresses, wire galvanised to BS 443:1982

PVC coated galvanised wire 2.2 mm diameter, overall diameter 2.7 mm, mesh size 75 x 75 mm.			PVC coated galvanised wire 2.2 mm diameter, overall diameter 2.7 mm, mesh size 75 x 75 mm.	
Units size	£/each		Unit Size	£/each
6 x 2 x 0.17 m	94.45		6 x 2 x 0.30 m	109.81
6 x 2 x 0.23 m	103.88			
Galvanised wire 3.0 mm diameter				
6 x 2 x 0.15 m	98.28			
6 x 2 x 0.23 m	105.17			
6 x 2 x 0.30 m	112.05			

Terramesh (woven wire mesh units with 1m x 1m gabion type face in mesh size 80 x 100 wire core diameter 2.70mm, PVC coated o.d. 3.70m to BS443 & 1052)		
4 x 2 x 1m	64.75	
6 x 2 x 1m	79.19	
4 x 2 x 0.5m	47.98	
6 x 2 x 0.5m	60.53	
Green Terramesh (as for standard Terramesh but with sloping face lined with Biomac or Macmat and an inner lining of galvinised mesh and two reinforcing bars to give correct angle).		
4 x 2 x 0.6m	76.28	
6 x 2 x 0.6m	90.71	

EARTH RETENTION INCLUDING GABIONS, CRIB WALLING, ETC. - continued

		Unit	£	
Pre-cast concrete crib units				
Andacrib super maxi system		m²	84.15	
Andacrib maxi system		m²	67.32	
Timber crib walling system				
Permacrib Goliath for retaining walls upto 3.5 m high		m²	72.94	
Permacrib Hercules for retaining walls upto 5.8 m high		m²	83.42	
Precast concrete wall units				
1000 mm high x 1000 mm long		nr	70.92	
1250 mm high x 1000 mm long		nr	99.03	
1500 mm high x 1000 mm long		nr	110.29	
1750 mm high x 1000 mm long		nr	118.80	
2400 mm high x 1000 mm long		nr	173.92	
2690 mm high x 1000 mm long		nr	218.20	
3000 mm high x 1000 mm long		nr	230.03	
3750 mm high x 1000 mm long		nr	343.74	
	Unit	Toe £	Heel £	
Filler Units				
for 1000 mm high wall unit	nr	39.75	29.19	
for 1500 mm high wall unit	nr	78.31	35.89	
for 1750 mm high wall unit	nr	100.19	46.68	
for 2400 mm high wall unit	nr	170.53	70.68	
for 3000 mm high wall unit	nr	239.80	98.66	
for 3600 mm high wall unit	nr	263.14	131.53	
Mild steel straps and bolts for filler units				
1000 to 3000 mm high units	nr	121.79	118.33	
3600 mm high units	nr	181.69	177.44	

Extra over for Sulphate Resisting Cement = 10 %

FENCING (SITE BOUNDARY FENCING)

	Unit	£
Chainlink fencing (25 m rolls), including line wire		
galvanised mild steel, wire 3.55 mm thick, 50 mm mesh, 1800 mm high	nr	195.59
plastic coated bright steel, wire 2.44/3.15 mm thick, 50 mm mesh, 1800 mm high	nr	109.05
plastic coated galvanised mild steel, wire 3.00/4.00 mm thick, 50 mm mesh, 1800 mm high	nr	177.72
plastic coated galvanised mild steel, wire 3.00/4.00 mm thick, 50 mm mesh, 2400 mm high	nr	252.65
Concrete Posts		
1800 mm without leanover arms		
standard	nr	9.89
strainer	nr	39.00
end	nr	29.45
corner	nr	43.60
1800 mm with leanover arms		
standard	nr	11.87
strainer	nr	45.56
end	nr	34.64
corner	nr	53.84
2400 mm without leanover arms		
standard	nr	16.39
strainer	nr	63.67
end	nr	49.14
corner	nr	74.46
2400 mm with leanover arms		
standard	nr	18.78
strainer	nr	71.43
end	nr	55.60
corner	nr	80.94
Wire		
barbed wire (200 m roll)	nr	37.70
barbed tape		
single strand (200 m roll)	nr	56.44
concertina (14 m roll)	nr	41.86
flat/wrap (14 m roll)	nr	55.75
Chestnut pale fencing (9.1m roll)		
900 mm wide	nr	50.02
1200 mm wide	nr	56.00
Chestnut posts		
posts for 900 mm high fencing	nr	2.50
posts for 1200 mm high fencing	nr	2.82

FENCING (SITE BOUNDARY FENCING)- continued

	Unit	£
Treated softwood fencing		
100 x 100 mm posts	m	3.41
75 x 75 mm posts	m	1.92
75 x 25 mm rails	m	0.60
100 x 25 mm rails	m	0.81
125 x 25 mm rails	m	1.01
50 x 38 mm rails	m	0.60
75 x 38 mm rails	m	0.90
150 x 38 mm feather edge rails	m	1.89

GEOTEXTILES

	Unit	£
Tensar		
SS20 (50 x 4 m roll) for weak soils	nr	317.37
SS30 (50 x 4 m roll) for very weak soil	nr	489.89
SSLA20 (50 x 4 m roll) for trafficked areas	nr	426.27
SS40 (30 x 4 m roll) for over mine workings	nr	427.29
AR-G (50 x 3.8 m roll) for reinforcement of asphalt pavement	nr	478.32
GM4 (40 x 4.5 m roll) for rock fall protection	nr	451.79
Mat (30 x 3.0 m roll) for erosion protection	nr	242.90
Lotrak permeable membrane (100 x 4.5 m rolls)		
1800	nr	126.50
2800	nr	225.78
50R (100 x 5 m rolls)	nr	482.90
HF550 monofilament	nr	POA
Netlon (30 x 2 m rolls)		
CE111 (30 x 2 m roll)	nr	104.40
CE121 30 x 2 m roll)	nr	173.40
CE131 30 x 2 m roll)	nr	178.20
Terram		
500 (200 x 4.5 m roll)	nr	439.92
700 (150 x 4.5 m roll)	nr	388.19
1000 (100 x 4.5 m roll) permeable membranes	nr	205.07
1500 (100 X 4.5 m roll)	nr	429.44
2000 (100 x 4.5 m roll)	nr	593.80
Filtram laminated filter drainage membrane (25 x 1.9 m roll)		
Filtram 1B1	nr	350.01
Filtram 1BZ	nr	326.28
Fabric for fin drains		
fabric, core and fixing staples (50 m² roll)	nr	157.85
Bontec		
NW6 (5.25 x 100 m roll)	nr	122.85
NW7 (1.8 x 56 m roll)	nr	42.34
NW8 (4.5 x 100 m roll)	nr	139.39
NW9 (4.5 x 100 m roll)	nr	149.63
NW14 (5.25 x 100 m roll)	nr	275.63
Alderprufe		
1500 tanking membrane (15.5 m² roll)	nr	84.71
2000 tanking membrane (15 m² roll)	nr	105.89
GRA methane barrier (105 m² roll)	nr	395.68
MR50 methane barrier (16.5 m² roll)	nr	107.00
Gastite tape (15 m roll, 100 mm wide)	nr	11.15
Backerboard (1.9 m² sheet)	nr	10.04
Primer	5 l	12.27
Primer	25 l	46.81

GEOTEXTILES - continued

	Unit	£
Visqueen polyethylene damp proof membrane to PIFA Standard 6/83A		
250 micron 1000 gauge, black (25 x 4 m roll)	nr	60.06
300 micron 1200 gauge, black (25 x 4 m roll)	nr	72.77
500 micron 1200 gauge, blue (25 x 4 m roll)	nr	77.39
Visqueen polyethylene temporary protective cover		
light duty, clear (25 x 4 m roll)	nr	13.78
medium duty, clear (50 x 4 m roll)	nr	17.27
heavy duty, clear (25 x 4 m roll)	nr	17.27

GULLEY GRATINGS AND FRAMES TO BS EN 124

	Unit	£
Group 4; ductile iron, heavy duty pattern		
370 x 430 x 100 mm depth; hinged, non-rocking; D400	nr	72.24
434 x 434 x 100 mm depth; non-rocking; D400	nr	76.73
600 x 600 x 100 mm depth; D400	nr	196..48
600 diameter x 100 mm depth; D400	nr	196.48
Group 3; ductile iron, non-rocking		
434 x 434 x 75 mm depth; C250	nr	70.64
875 x 385 x 75 mm depth; double; C250	nr	278.35
Group 3; ductile iron, hinged		
325 x 312 x 75 mm depth; C250	nr	72.24
325 x 437 x 75 mm depth; C250	nr	96.64
400 x 432 x 75 mm depth; C250	nr	104.99
370 x 305 x 100 mm depth; C250	nr	103.70
510 x 360 x 100 mm depth; C250	nr	103.70
370 x 430 x 100 mm depth; C250	nr	103.70
Group 3; ductile iron, hinged,anti-theft		
325 x 312 x 85 mm depth; C250	nr	88.63
325 x 437 x 75 mm depth; C250	nr	96.64
400 x 532 x 75 mm depth; C250	nr	92.26
370 x 305 x 100 mm depth; C250	nr	92.26
510 x 360 x 100 mm depth; C250	nr	92.26
370 x 430 x 100 mm depth; C250	nr	92.26
Group 3; ductile iron, pedestrian grating		
325 x 312 x 75 mm depth; C250	nr	72.24
Group 3; ductile iron, pavement gratings		
250 x 250 x 39 mm depth; C250	nr	40.47
300 x 300 x 39 mm depth; C250	nr	52.01
400 x 400 x 39 mm depth; C250	nr	60.37
700 x 700 x 39 mm depth; C250	nr	183.65
Group 3; ductile iron, pavement gratings, concave		
300 x 300 x 58 mm depth; C250	nr	55.87
400 x 400 x 63 mm depth; C250	nr	73.20
500 x 500 x 68 mm depth; C250	nr	111.08
600 x 600 x 73 mm depth; C250	nr	170.81
700 x 700 x 78 mm depth; C250	nr	194.97
Group 3; ductile iron, kerb type pattern - to suit standard kerb profile		
385 x 502 mm; 37 deg kerb profile, C250.	nr	132.92

INSTRUMENTATION (SOIL INSTRUMENTATION)

	Unit	£
Tacseal pipe 450 mm diameter		
5 m long	nr	172.98
2.15 m long	nr	116.59
0.45 m long	nr	65.43
0.35 mm long	nr	57.10
joints to suit	nr	37.48
150 mm PVC-U class C (as protective sleeve)	m	7.14
32 mm mild steel rods (as settlement rod)	m	3.27
150 mm diameter steel tube (as protective sleeve)	m	17.61
Inclinometer tube	m	14.64
Piezometer tube (nylon)	m	17.02

READOUT EQUIPMENT

Prices vary depending on number of instruments, locations, distances from instrument. The following is an indicative price for modern portable equipment used for reading up to 30 separate instruments. Item complete and installed.

	Unit	£
Readout equipment for inclinometer	Item	20,000.00
Readout equipment for piezometer	Item	5,000.00

INTERLOCKING STEEL SHEET PILING AND UNIVERSAL BEARING PILES

The following prices are ex works in lots of 20 tonnes and over of one section, one size, and one specification ordered at one time, for delivery from one works to one destination. Carriage from Basing Point to destination is not included - refer to **page 110** for indicative costs.

INTERLOCKING STEEL SHEET PILING

Section	Unit	EN10248-1995 Grade S270GP £	EN10248-1995 Grade S355GP £
ProfilARBED AZ 12	t	600.00	630.00
AZ 17	t	600.00	630.00
AZ 26, AZ 36	t	600.00	630.00
AZ 38 - 700	t	610.00	640.00
AZ 48	t	610.00	640.00
AS 500 - 12.0	t	610.00	640.00
PU 6, PU 8	t	580.00	610.00
PU 12	t	580.00	610.00
PU 18 - 1.0	t	580.00	610.00
PU 22	t	610.00	640.00
PU 32	t	610.00	640.00

Steel piling sections supplied in lengths of 6 to 18 metres
Material ex basing point

Add to the above prices for:		Unit	£
Quantity	under 5 tonnes	t	45.00
	under 10 tonnes to 5 tonnes	t	30.00
	under 20 tonnes to 10 tonnes	t	15.00
Quality	copper content between 0.20% and 0.35%	t	10.00
	copper content over 0.35% not exceeding 0.50%	t	15.00
Size	the following are £ extra per tonne for various alternative lengths		

Z Sections

Section	AZ 12 £/t	AZ 17 £/t	AZ 26 £/t	AZ 36 £/t	AZ 38 £/t	AZ 48 £/t	AS 500 £/t
Length							
2 - 4 m	40.00	40.00	40.00	40.00	30.00	30.00	-
4 - 6 m	20.00	20.00	20.00	20.00	20.00	20.00	20.00
18 - 24 m	10.00	10.00	10.00	10.00	15.00	15.00	15.00

INTERLOCKING STEEL SHEET PILING AND UNIVERSAL BEARING PILES - continued

INTERLOCKING STEEL SHEET PILING - continued

U Sections

Section	PU 6 £/t	PU 8 £/t	PU 12 £/t	PU 18 £/t	PU 22 £/t	PU 32 £/t
Length						
2 - 4 m	45.00	40.00	30.00	30.00	30.00	30.00
4 - 6 m	20.00	20.00	15.00	15.00	15.00	15.00
18 - 24 m	10.00	10.00	10.00	10.00	15.00	15.00

UNIVERSAL BEARING PILES

Size mm	Weight Kg/m	Unit	EN10025,1993 Grade S275JR £	EN10025,1993 Grade S355JO £
356 x 368	174	t	843.00	873.00
356 x 368	152	t	843.00	873.00
356 x 368	133	t	843.00	873.00
356 x 368	109	t	843.00	873.00
305 x 305	223	t	823.00	853.00
305 x 305	186	t	823.00	853.00
305 x 305	149	t	823.00	853.00
305 x 305	126	t	823.00	853.00
305 x 305	110	t	823.00	853.00
305 x 305	95	t	823.00	853.00
305 x 305	88	t	823.00	853.00
305 x 305	79	t	823.00	853.00
254 x 254	85	t	803.00	833.00
254 x 254	71	t	803.00	833.00
254 x 254	63	t	803.00	833.00
203 x 203	54	t	793.00	823.00
203 x 203	45	t	793.00	823.00

Steel piling sections supplied in lengths of 5 to 15 metres

Material ex basing point - Middlesborough Railway Station (except 203 x 203 mm)
- Scunthorpe Railway Station (except 356 x 368 mm)

Add to the above prices for:		Unit	£
Quantity	under 5 tonnes	t	25.00
	under 10 tonnes to 5 tonnes	t	10.00
	over 10 tonnes	t	basis
Quality	copper content between 0.20% and 0.35%	t	10.00
	copper content over 0.35% not exceeding 0.50%	t	15.00
Size	length 3 m to under 9 m	t	15.00
	length over 18.5 m to 24 m	t	5.00
	length over 24 m	t	POA
Holes	lifting holes	nr	10.00

JOINT FILLERS AND WATERSTOPS

	Unit	£
Flexible epoxy resin joint sealant (Expoflex 800)		
2 litre tin	nr	40.73
Cold applied waterproofing membrane (Mulseal DP)		
5 litre bucket	nr	14.54
25 litre drum	nr	47.76
200 litre drum	nr	305.04
Hot bonded reinforced bitumen sheet waterproofing membrane (Famguard GS100)		
1 m x 8 m roll	nr	64.95
Non absorbent joint filler (Hydrocell XL)		
sheet 10 x 1000 x 2000 mm	nr	33.59
sheet 15 x 1000 x 2000 mm	nr	39.20
sheet 20 x 1000 x 2000 mm	nr	44.82
sheet 25 x 1000 x 2000 mm	nr	56.05
Polysulphide sealant (Thioflex 600)		
grey gun grade	l	13.68
black gun grade	l	13.68
mahogany gun grade	l	13.68
stone gun grade	l	13.68
off white gun grade	l	13.68
brick red gun grade	l	13.68
grey pouring grade	l	14.27
Hot applied sealant		
hand applied, Plastijoint (5 litre)	nr	37.50
poured, low extension grade, Pliastic N2 (15 kg sack)	nr	39.71
poured, hard grade, Pliastic 77 (15 kg sack)	nr	40.49
Epoxy resin mortars (Expocrete)		
UA (13.5 litre pack)	nr	267.38
High duty elastomeric pavement sealant (Colpor 200PF)	l	10.39
Non extruding expansion joint filler (Fibreboard)		
sheet 10 x 2440 x1200 mm	nr	17.24
sheet 12.5 x 2440 x 1200 mm	nr	18.80
sheet 20 x 2440 x 1200 mm	nr	25.80
sheet 25 x 2440 x 1200 mm	nr	29.59
Flexible expansion joint strip membrane (Expoband H45)		
100 mm x 25 m roll	nr	202.41
200 mm x 25 m roll	nr	307.44

JOINT FILLERS AND WATERSTOPS - continued

	Unit	£
Bituthene rolls		
2000 grade (20 m² - over 60 Rolls)	m²	4.15
3000 grade (20 m² - over 60 Rolls)	m²	4.96
4000 grade (20 m² - over 60 Rolls)	m²	6.52
8000 grade (20 m² - over 60 Rolls)	m²	7.00
B1 Primer (25 litre drum)	nr	71.97
B2 Primer (25 litre drum)	nr	77.04
MR dpc (30 m x 600 mm wide)	m	12.44
Biodegration resistant sealant (Nitoseal 12)		
sealant (2 litre tin) gun grade	nr	40.26
accelerator (50ml tin) gun grade	nr	11.24
Cold applied bituminous sheet waterproofing membrane (Proofex)		
3000, non reinforced (20m² roll)	nr	117.09
3000MR, reinforced (20m² roll)	nr	175.10
primer (5 litre can)	nr	20.73
Servi-dek STD (22.5 litre unit)	nr	88.10
Serviseal Pilecap(7.5 m coil - over 40 coils)	m	10.63
angles (factory labour)	nr	55.79
PVC waterstops		
Supercast 150 centre bulb (15m coil)	nr	60.83
Supercast 200 centre bulb (15 m coil)	nr	84.43
Supercast 250 centre bulb (12 m coil)	nr	86.26
Supercast rearguard 150 (12 m coil)	nr	53.56
Supercast rearguard 200 (12 m coil)	nr	84.43
Supercast rearguard 250 (12 m coil)	nr	114.39
Hydrophilic waterstops		
Supercast SW10 (15 m roll)	nr	80.23
Supercast SW20 (5 m roll)	nr	81.40
ADCOR 500S 25 mm x 20 mm strip (5 m coil)	nr	212.10
Servistrip AH 205 20 mm x 5 mm strip (10 m coil)	nr	71.80

PRECAST CONCRETE KERBS, EDGINGS AND PAVING SLABS

	Unit	£	
Hydraulically pressed kerbs (914 mm lengths)			
150 x 305 mm bullnosed	nr	6.20	
150 x 305 mm battered	nr	6.20	
150 x 305 mm half battered	nr	6.20	
125 x 255 mm bullnosed	nr	3.50	
125 x 255 mm battered	nr	3.50	
125 x 255 mm half battered	nr	3.50	
125 x 255 mm square	nr	3.50	
125 x 150 mm bullnosed	nr	2.70	
125 x 150 mm battered	nr	2.70	
125 x 150 mm half battered	nr	2.70	
125 x 150 mm square	nr	2.70	
150 x 100 mm square	nr	2.60	
Extra over for dished channel			
125 x 255 mm radius and dropper	nr	5.30	
125 x 150 mm radius	nr	4.40	
255 x 125 mm radius and dropper	nr	5.30	
150 x 125 mm radius	nr	4.40	
Quadrants	nr	7.90	
Angles	nr	7.90	
Transitions			
125 x 255 mm	nr	7.90	
Offlets (weir kerbs) laid flat radius and other transitions	nr	11.80	

	Unit	Natural £	Coloured £
Block kerbs			
main kerb 200 mm lengths	m	7.48	7.48
external angle	nr	4.25	4.25
internal angle	nr	4.25	4.25
radius unit	nr	4.25	4.25
crossover unit	nr	4.79	4.79

	Unit	£	
Crane off load			
crane off load	load	30.00	

PRECAST CONCRETE KERBS, EDGINGS AND PAVING SLABS - continued

	Full Load Unit	Up to 50 km £	
Beany block type kerbs			
base block, 500 mm long	nr	16.83	
base block, outfall	nr	28.41	
base block, outfall/junction	nr	24.71	
base block, junction	nr	24.71	
base block, bend	nr	19.38	
base block, splay cut one end for radius work	nr	23.67	
top block, 500 mm long	nr	16.30	
top block, splay cut one end for radius work	nr	24.94	
cast iron cover and frame	nr	164.47	
cable duct blocks	nr	74.25	
Standard 12.5 mm galvanised steel cover plates			
plates, 500 mm long	nr	32.27	
cover plate bend, 45 degrees	nr	47.86	
cover plate bend, 90 degrees	nr	42.67	
cover plate bend, special	nr	67.82	
offset cover plate	nr	46.41	
cover plate, splay cut one end for standard radius work	nr	46.41	
Safeticurb type kerbs	Unit	£	
Standard capacity slot units (914 mm long)			
250 x 250 mm, 125 mm bore unit	nr	28.11	
250 x 250 mm, 125 mm bore unit with cast iron insert	nr	50.05	
Standard capacity grid units (914 mm long)			
250 x 250 mm, 125 mm bore unit with pressed steel grid	nr	43.97	
250 x 250 mm, 125 mm bore unit with cast iron grid	nr	83.82	
Cast iron silt box for standard capacity units	nr	196.35	
Large capacity slot units (914 mm long)			
305 x 305 mm, 150 mm bore unit	nr	60.45	
Large capacity grid units (914 mm long)			
305 x 305 mm, 150 mm bore unit with pressed steel grid	nr	71.35	
305 x 305 mm, 150 mm bore unit with cast iron grid	nr	107.80	
Cast iron silt box for large capacity units	nr	349.44	
Charcon clearway (400 mm long)			
324 x 257 mm, slot unit	m	75.08	
324 x 257 mm, grid unit	m	121.77	
silt box top	m	309.75	
Standard capacity kerb units (914 mm long)			
250 x 320 mm and 250 x 350 mm units, 125 mm bore	nr	41.78	
inspection units	nr	117.35	
transition units	nr	49.28	
type K manhole covers	nr	422.42	

	Unit	£		
Hydraulically pressed edgings (3 feet lengths)				
50 x 150 mm square	nr	1.54		
50 x 150 mm bullnosed	nr	1.54		
50 x 205 mm square	nr	2.15		
50 x 205 mm bullnosed	nr	2.15		
50 x 255 mm square	nr	2.42		
50 x 255 mm bullnosed	nr	2.42		
Vibrated path edgings				
50 x 150 mm round top (straight)	nr	2.15		
50 x 150 mm radius	nr	2.15		
50 x 200 mm round top (straight)	nr	2.15		
50 x 200 mm radius	nr	2.15		
50 x 255 mm round top (straight)	nr	2.15		
50 x 255 mm radius	nr	2.15		
	Unit	Natural £	Coloured £	
Paving Flags BS 7263				
450 x 600 x 50 mm	nr	2.48	3.74	
600 x 600 x 50 mm	nr	2.66	4.17	
750 x 600 x 50 mm	nr	3.15	4.81	
900 x 600 x 50 mm	nr	3.44	5.13	
600 x 300 x 50 mm	nr	1.84	-	
450 x 600 x 63 mm	nr	2.92	-	
600 x 600 x 63 mm	nr	3.18	-	
750 x 600 x 63 mm	nr	3.56	-	
900 x 600 x 63 mm	nr	3.91	-	
900 x 300 x 50 mm	nr	2.26	-	
	Unit	Natural £	Coloured £	Brindle £
Block paving BS 6717				
rectangular block paving (Charcon Europa)				
200 x 100 x 60 mm	m^2	9.85	10.51	10.51
200 x 100 x 80 mm	m^2	11.88	12.32	12.32
Deterrent paving (Charcon Elite)				
format 1				
200 x 144 x 80 mm	m^2	42.63	-	-
format 2				
298 x 132 x 80 mm	m^2	53.86	-	-
format 3				
298 x 132 x 80 mm	m^2	45.10	-	-

PRECAST CONCRETE KERBS, EDGINGS AND PAVING SLABS - continued

	Unit	Natural £	Coloured £	
Block Starter Units				
herringbone 65 mm thick	pack of 80	78.57	95.72	
herringbone 80 mm thick	pack of 80	99.03	102.88	
interlocking 65 mm thick	pack of 98	52.52	61.24	
interlocking 80 mm thick	pack of 98	57.75	-	
Landscape paving				
textured (exposed aggregate finish)				
400 x 400 x 65 mm	nr	3.63	4.18	
450 x 450 x 70 mm	nr	3.74	4.24	
600 x 600 x 50 mm	nr	5.39	6.27	
ground (smooth finish)				
400 x 400 x 65 mm	nr	3.74	4.07	
600 x 600 x 50 mm	nr	7.59	7.92	
non slip (pimpled finish)				
400 x 400 x 50 mm	nr	-	3.03	

	Unit	£		
Grassgrid				
366 x 274 x 100 mm	m²	18.15		
Stone paving				
granite setts new 100 x 100 mm (approx)	t	154.26		
granite setts reclaimed random sized	t	88.50		
granite setts reclaimed 100 x 100 mm sorted	t	104.68		
random size cobbles reclaimed	t	68.07		

LANDSCAPING/SOILING

For a more detailed appraisal of landscaping prices, please refer to Spons Landscape and External Works Price Book.

CULTIVATED TURF			
	Unit	Over 1200 £	
Rolawn			
RB Medallion	m^2	3.71	
Inturf			
Inturf SS2 Standard Turf	m^2	2.86	

IMPORTED TOPSOIL

The price for topsoil is so variable depending on site location, quantity available, season, etc. The following prices are a guide for use in the unit cost build ups and reflect reasonable quality soils from a single sources delivered to site in 20 tonne tipper trucks, within a 20 km radius of source.

	Unit	Under 1000 £	1000 to 5000 £	Over 5000 £
Subsoil	m^3	6.19	4.95	4.95
Reasonable quality topsoil for seeding and general, sward establishment (1.6 t/m^3)	m^3	14.85	13.37	12.69
Better quality topsoil for species planting and for use in raised beds, planters, etc. (1.6 t/m^3)	m^3	31.24	24.99	22.50

GROWING MEDIUMS/SOIL IMPROVERS

		Unit	£	
Mulches (delivered in 70 m^3 loads) :				
Graded bark flakes		m^3	34.27	
Bark nuggets		m^3	31.24	
Ornamental bark mulch		m^3	35.37	
Amenity bark mulch		m^3	18.65	
Forest biomulch		m^3	16.72	
Decorative biomulch		m^3	19.47	
Rustic biomulch		m^3	23.60	
Mulchip		m^3	21.01	
Pulverised bark		m^3	19.39	
Woodland mulch		m^3	14.05	
Woodfibre mulch		m^3	17.78	

LANDSCAPING/SOILING - continued

GRASS SEEDS/FERTILIZERS ETC.	Unit	£
Soil Ameliorants:		
Landscape amenity	m^3	18.65
Spent Mushroom Compost	m^3	8.36
Super humus	m^3	14.80
Topgrow	m^3	17.22
Grass seed mixtures, British Seed Houses		
(Reclamation) BSH ref. A15, sowing rate 15-25 g/m^2	25 kg	96.25
(Country Park) BSH ref. A16, sowing rate 8-20 g/m^2	25 kg	99.55
(Road Verges) BSH ref. A18, sowing rate 6 g/m^2	25 kg	81.40
(Landscape) BSH ref. A3, sowing rate 25-50 g/m^2	25 kg	75.90
(Parkland) BSH ref. A4, sowing rate 17-35 g/m^2	25 kg	88.55
Fertilisers		
Fisons PS5 pre-seeding fertiliser	25 kg	10.25
Fisons "Ficote" fertiliser (per 10 kg)	10 kg	32.35
BSH amenity granular and slow release fertilisers		
BSH 1 pre-seeding granular (reasonable) 6-9-6	25 kg	14.08
BSH 2 pre-seeding granular (impoverished) 10-15-10	25 kg	15.40
BSH 4 pre-seeding mini granular (spring/summer) 11-5-5	25 kg	15.95
BSH 5 pre-seeding granular (spring/summer) 9-7-7	25 kg	14.08
BSH 7 pre-seeding mini granular (autumn/winter) 3-10-5.2	25 kg	15.95
BSH 8 outfield granular (autumn/winter fertiliser) 3-12-12	25 kg	14.08
Growmore granular fertiliser (per 25 kg)	25 kg	14.40
Compost		
tree planting and mulching compost	m^3	19.97
seeding	m^3	15.66
Selective weedkiller	25 l	177.03
Hydroseeding - specialist process; refer to unit cost landscaping section for typical specification and situation		

MANHOLE COVERS AND FRAMES TO BS EN 124

Manhole covers and frames (all dimensions are clear opening sizes)		
	Unit	£
Group 5, ductile iron		
600 x 600 x 150 mm depth; Super Heavy Duty; E600	nr	365.68
1210 x 685 x 150 mm depth; Super Heavy Duty; E600	nr	609.36
1825 x 685 x 150 mm depth; Super Heavy Duty; E600	nr	1297.06
Group 4, ductile iron, double triangular		
600 x 600 x 100 mm depth; D400	nr	192.63
600 x 600 x 100 mm depth; ventilated; D400	nr	229.71
675 x 675 x 100 mm depth; D400	nr	267.12
Group 4, ductile iron, double triangular, hinged, lockable		
600 x 600 x 100 mm depth; D400	nr	179.79
Group 4, ductile iron, double triangular		
600 mm diameter x 100 mm depth; D400	nr	146.72
600 x 600 x 100 mm depth; D400	nr	170.81
675 mm diameter x 100 mm depth; D400	nr	231.74
675 x 675 x 100 mm depth; D400	nr	240.47
900 x 600 x 100 mm depth; D400	nr	367.61
900 x 900 x 125 mm depth; D400	nr	602.63
1250 x 675 x 100 mm depth; D400	nr	390.40
Group 4, ductile iron, single triangular, hinged		
600 mm x 120 mm depth; D400	nr	279.69
Group 4, ductile iron, circular cover, hinged		
600 mm diameter x 100 mm depth; square frame; D400	nr	141.27
Group 2, ductile iron, single seal		
600 x 450 x 75 mm depth; B125	nr	104.36
600 x 600 x 75 mm depth; B125	nr	129.71
Group 2, ductile iron, single seal, ventilated		
600 x 450 x 75 mm depth: B125	nr	145.76
600 x 600 x 75 mm depth: B125	nr	203.55
Group 2, ductile iron, double seal, single piece cover		
600 x 600 mm; B125	nr	274.82
Group 2, ductile iron, double seal, single piece cover, recessed		
600 x 450 mm; B125	nr	209.97
600 x 600 mm; B125	nr	253.00
MANHOLE STEP IRONS		
Step irons, galvanised malleable iron		
115mm tail	nr	7.07
230mm tail	nr	9.00

MISCELLANEOUS AND MINOR ITEMS

		Unit	£
Concrete spacer blocks for reinforcement			
bars; 30 mm cover		500	29.87
bars; 50 mm cover		200	17.56
bars; 75 mm cover		50	9.66
mesh; 50 mm cover		100	21.12
Wire spacers			
60 mm height		m	1.40
90 mm height		m	1.43
120 mm height		m	1.49
150 mm height		m	1.61
180 mm height		m	1.91
Tying wire; 16 gauge black annealed		20 kg coil	27.80

	Coupler A £ / each	Coupler B £ / each	Thread bar (at factory) £ / end
ROM Lenton reinforcement couplers (Coupler A for use where one bar can be rotated, Coupler B where neither can be)			
12 mm diameter bar	3.21	19.28	2.81
16 mm diameter bar	4.14	22.46	3.00
20 mm diameter bar	7.41	24.56	3.51
25 mm diameter bar	11.56	28.48	3.98
32 mm diameter bar	15.78	38.21	5.55
40 mm diameter bar	22.83	58.02	7.11

		Unit	£
Dowel bars round			
16 x 800 mm		nr	1.34
20 x 800 mm		nr	2.10
25 x 800 mm		nr	3.25
32 x 800 mm		nr	5.83
Expansion dowel caps			
12 x 100 mm		100	15.73
16 x 100 mm		100	22.98
20 x 100 mm		100	23.61
25 x 100 mm		100	33.04
32 x 100 mm		100	42.48
Debonding sleeves for dowel bars			
12 x 450 mm		100	26.57
16 x 200 mm		100	28.54
20 x 300 mm		100	29.28
25 x 300 mm		100	38.37
32 x 375 mm		100	50.92

	Unit	£
Debonding compound	5 l	19.28
Crack inducers (5 m lengths)		
10 mm wide x 50 mm deep top type (two piece)	5m	9.40
10 mm wide x 75 mm deep top type (two piece)	5m	15.76
10 mm wide x 40 mm deep bottom Y type	5m	7.11
10 mm wide x 75 mm deep bottom Y type	5m	13.36
Silver sand (bagged)	25kg	3.90
Formwork release agents		
General purpose mould oil	25 l	27.03
Chemical release agent	25 l	26.37
ROMLEASE chemical release agent (35-50 m^2/l)	25 l	29.81
Retarders		
ROMTARD CF retarder, subsequently brush to expose aggregate/form key ($5m^2/l$)	25 l	75.71
ROMTARD MA retarder, apply to face of formwork (15-16 m^2/l)	25 l	150.72
Surface Hardeners and Sealers		
ROMTUF concrete surface hardener and dustproofer (2 coats, $6m^2/l$/coat)	15 l	39.67
ROMCURE Sealer RS resin based concrete floor sealer (2 coats, 5.5 m^2/l/coat)	25 l	95.55
ROMCURE Standard, resin based curing membrane (5.5 m^2/l)	25 l	51.55
Mortars and grouts		
Epoxy grout high density	kg	4.38
Epoxy mortar general purpose	kg	4.86
5 star grout	kg	4.34
ROMGROUT grout admixture for infilling bolt boxes (1 packet / 50 kg OPC)	20 packets	48.58
Air entraining agents		
Rockbond Admix RB A1901, non flow/air entraining	kg	5.04
Rockbond Admix RB A2001, flowing/air entraining	kg	4.73
Resin bonded anchors :		
Resin capsules		
R-HAC 8mm (10 per box)	100	138.60
R-HAC 10mm (10 per box)	100	146.90
R-HAC 12mm (10 per box)	100	177.40
R-HAC 16mm (10 per box)	100	220.70
R-HAC 20mm (6 per box)	100	254.16
R-HAC 24mm (6 per box)	100	316.33
R-HAC 30mm (6 per box)	100	605.00

MISCELLANEOUS AND MINOR ITEMS - continued

	Unit	£
Zinc plated steel threaded rod with nut and washer		
8 x 110 mm (10 per box)	100	66.00
10 x 130 mm (10 per box)	100	102.96
12 x 165 mm (10 per box)	100	161.04
16 x 190 mm (10 per box)	100	297.00
20 x 250 mm (6 per box)	100	576.84
24 x 300 mm (6 per box)	100	1069.20
30 x 380 mm (6 per box)	100	3007.09
Stainless steel threaded rod with nut and washer		
8 x 110 mm (10 per box)	100	167.04
10 x 130 mm (10 per box)	100	254.88
12 x 165 mm (10 per box)	100	408.96
16 x 190 mm (10 per box)	100	860.04
20 x 250 mm (6 per box)	100	1569.60
24 x 300 mm (6 per box)	100	2962.01
Rawl SafetyPlus Anchors		
Loose Bolt Anchors		
M8 15L (50 per box)	100	137.03
M8 40L (50 per box)	100	163.11
M10 40L (50 per box)	100	211.39
M12 25L (25 per box)	100	276.65
M16 25L (10 per box)	100	550.62
M16 50L (10 per box)	100	637.98
M20 30L (10 per box)	100	1072.68
Bolt Projecting type Anchors		
M8 15P (50 per box)	100	159.18
M10 20P (50 per box)	100	213.99
M12 25P (25 per box)	100	307.94
M12 50P (25 per box)	100	349.73
M16 25Pm (10 per box)	100	626.33
M16 50P (10 per box)	100	709.80
M20 30P (10 per box)	100	1174.43
Hyrib permanent shuttering		
ref. 2411	m^2	20.79
ref. 2611	m^2	16.29
ref. 2811	m^2	15.21

	Unit	£
Exmet galvanised brickwork reinforcement (20 m rolls)		
65 mm wide	m	0.62
115 mm wide	m	1.07
175 mm wide	m	1.67
225 mm wide	m	2.15
305 mm wide	m	2.71
Expamet stainless steel bed joint brickwork reinforcement (3050 mm rolls)		
60 mm wide	m	2.29
100 mm wide	m	2.44
150 mm wide	m	2.92
175 mm wide	m	3.90
200 mm wide	m	4.59

PAINT/STAINS/PROTECTIVE COATINGS

	Unit	£
Crown trade gloss		
magnolia/ brilliant white	5 l	22.92
white	5 l	21.86
colours	5 l	29.26
Crown trade undercoat		
white	5 l	21.86
colours	5 l	29.26
Crown trade eggshell		
magnolia	5 l	34.77
colours	5 l	42.36
Crown trade matt emulsion		
magnolia and brilliant white	5 l	16.85
white	5 l	16.07
colours	5 l	21.99
Crown trade silk emulsion		
magnolia and brilliant white	5 l	20.36
white	5 l	19.38
colours	5 l	23.09
Crown trade masonry paint, colours	5 l	24.07
Crown trade Timberguard	5 l	30.02
Creosote; dark / light	25 l	15.32
Cuprinol preservers		
wood preservers	25 l	84.13
Exterior Wood; garden shed + fence preservers	25 l	47.84
Cuprinol woodstains		
Select Woodstain; matt finish - semi-transparent	2.5 l	28.19
Select Woodstain; matt finish - coloured	5 l	40.58
Premier Woodstain; satin finish - semi-transparent	1 l	9.63
Premier Woodstain; satin finish - coloured	5 l	44.71
5 Year QD woodstain	2.5 l	20.75
5 Star wood treatment	25 l	107.14
Crown trade varnishes		
Gloss varnish	5 l	50.88
Trade Yacht varnish; clear gloss	5 l	49.56

	Unit	£
Crown Primers		
alkali resisting plaster primer	5 l	30.43
oil based non-toxic wood primer	5 l	29.97
acrylic wood primer	5 l	30.83
aluminium wood primer	5 l	38.54
zinc phosphate metal primer	5 l	28.76
red oxide metal primer	5 l	46.39
acrylic primer sealer undercoat	5 l	30.33
stabilising primer, pigmented	5 l	29.06
Crown anti-condensation paint	5 l	52.56
Crown road marking paint, white/yellow	5 l	56.75
Sadolin		
high performance clear varnish, internal woodwork	5 l	32.84
Classic, translucent matt finish, external woodwork	5 l	39.92
Extra, translucent, semi-gloss for all external joinery	5 l	47.36
Superdec, opaque, semi gloss, waterborne for all external joinery	5 l	46.41
Supercoat wood stain	5 l	54.44
Dulux Weathershield masonry paint		
Weathershield masonry, coloured	5 l	30.46
All Seasons masonry, coloured	5 l	34.20
Fine Textured masonry, coloured	5 l	30.46
masonry stabilising solution	5 l	26.69
masonry fungicidal solution	5 l	25.82
Sandtex masonry paint		
Matt Exterior / Hi Cover Smooth, coloured	5 l	22.45
fungicidal	5 l	15.82
stabilising solution, clear	5 l	22.45
Solignum Architectural wood stain	5 l	56.50
Aluminium paint	5 l	62.16
Heat resisting aluminium paint	5 l	63.23
Anti condensation paint	5 l	56.75
Red oxide paint	5 l	48.71
General purpose bituminous paint	5 l	26.22
Concrete and floor paint	5 l	38.90

PAINT/STAINS/PROTECTIVE COATINGS - continued

PROTECTIVE COATINGS

	Unit	£	Price per theoretical m² £
For fabricated steel, lighting columns, parapets and guard railings, etc.			
Prefabrication primers			
Metagard G280 primer	l	6.53	0.60
Metagard K232 zinc rich epoxy blast primer	l	16.38	0.78
Metagard L574 blast primer	l	7.98	0.69
Alkyd primers			
Leighs L489 zinc phosphate primer	l	7.90	1.03
Leighs M583 quick drying alkyd primer	l	8.73	1.09
Leighs M600 quick drying zinc phosphate primer	l	6.78	1.19
Alkyd intermediates & top coats			
Leighs C530 QD high gloss finish	l	11.40	0.67
Metagrip L654 MIO paint	l	8.38	1.05
Leighs M671 undercoat	l	7.35	0.49
Protective finishes - alkyd			
Leighs M155 matt protection finish	l	11.63	1.62
Leighs M255 gloss protective finish	l	14.24	2.64
Protective finishes - epoxy			
Epigrip M455 sheen protective finish	l	16.54	2.07
Epigrip M555 HB sheen protective finish	l	19.39	4.04
Fire protection			
Firetex FX1000/FX2000 intumescent coating	l	27.53	Varies
Firetex FX3000/FX4000 intumescent coating	l	27.42	Varies
Concrete protection			
RIW liquid asphatic composition (two coats) (25 litre drum)	l	4.36	3.81
RIW Heviseal (two coats) (20 litre drum)	l	4.97	6.30
RIW hydrocoat (25 litre drum)	l	2.00	0.57
RIW bitumen protection board, 1.625 x 1.285 m x 3 mm thick sheet	nr	12.35	-
RIW sheet seal, grade 9000/300, 20 x 0.3 m roll	nr	78.12	-

POLYMER CHANNELS AND FITTINGS

	Unit	£
ACO S100 interlocking,stepped channel with locked ductile iron grating; Class F900		
length 1000 mm	nr	85.75
length 500 mm	nr	59.60
ACO N100KS interlocking, pre-sloped channel with stainless steel frame; Class C250		
length 1000 mm	nr	41.79
length 500 mm	nr	31.28
ACO ParkDrain one piece polymer concrete channel system with integral grating; Class C250		
length 500 mm	nr	30.00
ACO KerbDrain 305 one piece polymer concrete combined kerb drainage system Class D400		
length 500 mm	nr	28.00
access unit	nr	88.96
drop kerb unit	nr	49.92
mitre unit	nr	28.27
endcaps; closing	nr	11.35

	Unit	S100 & S100K (heavy duty Class F)	N100K (normal duty Class A-C)
ACO end caps		£	£
end cap	nr	9.51	16.64
outlet end cap	nr	16.42	22.21
inlet end cap	nr	16.42	22.21

	Unit	£
ACO universal gullies complete with grating and galvanised bucket		
gulley	nr	273.75
ACO sump units with galvanised steel bucket		
for S100 channel with ductile iron grating	nr	162.89
for N100KS channel with stainless steel frame	nr	106.60
for K100S channel with galvanised steel frame, roddable	nr	53.55
ACO Quicklock composite grating; Class C250		
length 500 mm	nr	13.30
ACO KerbDrain 305		
length 500mm	nr	28.00
access unit	nr	88.96
drop kerb unit; length 915mm	nr	49.92
mitre unit	nr	28.27
endcaps	nr	11.35

POLYMER CHANNELS AND FITTINGS - continued

	Unit	£
ACO RoadDrain interlocking channel system		
RoadDrain 100; length 500mm	nr	25.21
RoadDrain 200; length 500mm	nr	43.58
Inter-connecting channel with inbuilt fall		
1000 mm long	nr	24.77
1000 mm long, galvanised steel edging	nr	35.68
Drain end caps	nr	4.31
Drain end caps wiith 100 mm PVC-U outlet	nr	8.09
PVC accessories		ACO
100 mm drain union	nr	2.61
150 mm drain union	nr	10.96
200 mm drain union	nr	21.86
150 mm oval to round union	nr	11.52
100 mm foul air trap	nr	17.54
150 mm foul air trap	nr	42.24
Channel gratings		
Class C ductile iron, slotted, 500 mm long	nr	21.14
Class E ductile iron, slotted, 500 mm long	nr	21.14
Class A galvanised steel, slotted, 500 mm long	nr	9.82
Class A galvanised steel, slotted, 1000 mm long	nr	13.75
Class C galvanised steel, mesh, 500 mm long	nr	14.19
Class C galvanised steel, mesh, 1000 mm long	nr	20.64
Class A stainless steel, slotted, 500 mm long	nr	36.45
Class A stainless steel, slotted, 1000 mm long	nr	46.40
Class A stainless steel, mesh, 500 mm long	nr	82.76
Class A stainless steel, mesh, 1000 mm long	nr	92.55
Ductile iron or steel locking bar and mild steel bolt	nr	5.63
Steel locking bar to suit mesh gratings	nr	5.63
Silt boxes		
standard	nr	83.12
roddable with in-line outlet	nr	114.35
roddable with side outlet	nr	114.42
roddable with in-line outlet (galvanised)	nr	117.23
roddable with side inlet (galvanised)	nr	117.57
standard galvanised	nr	99.94
Gullies with ductile iron grating, galvanised steel bucket and locking mechanism		
shallow	nr	396.69
deep	nr	448.51
deep, roddable	nr	474.76

QUARRY PRODUCTS, AGGREGATES, ETC.

Cost per tonne delivered to site within a 20 mile radius of quarry or production centres, cost based on most economically available materials, minimum order 5,000 tonnes (average prices - see footnote).

	Scotland, North, and Wales	Midlands and South West	South General
Graded granular material			
natural gravels (DTp class 1 A, B, or C) general fill	9.47	10.26	10.33
crushed gravel or rock ditto	11.01	13.12	13.78
Reclaimed material (DTp class 2E)			
blast furnace slag; general fill	6.44	7.55	9.56
quarry waste	8.41	9.62	11.56
quarry waste from processing of Aggregate Tax excluded minerals	6.47	7.68	9.62
Well graded granular material			
natural gravel or sand (DTp class 6A) fill below water	11.06	12.40	11.53
crushed gravel or rock; ditto (free draining)	10.66	13.08	15.69
Selected granular material			
natural gravel (DTp class 6F) capping layers	9.71	11.11	11.83
crushed rock; ditto	10.87	12.43	16.99
Selected uniformly graded material			
natural gravel or crushed rock (DTp class 6I,J) fill to reinforced earth-blinding material	9.72	10.12	10.76
pea gravel fill on bridges and other structures	19.63	20.62	20.62
Selected graded material			
natural gravel (DTp class 6N, P) fill to structure	9.64	11.95	12.51
natural rock; ditto	11.64	15.31	14.27
Selected graded material			
natural gravel (DTp class 6C) rock fill	11.18	11.94	14.25
crushed gravel; ditto	11.47	12.23	15.03
Wet cohesive materials			
natural boulder clay (DTp class 2A) general fill	9.93	10.95	10.95
oxford clays; ditto	12.03	12.77	12.77
Dry cohesive materials			
natural boulder clay (DTp class 2B) general fill	9.93	10.95	10.95
oxford clays; ditto	12.03	12.77	12.77
Chalk fill			
chalk & associated materials (DTp class 3) general fill	-	-	9.26
Rock fill			
crushed rock, core fill, rock punching	7.67	8.37	14.82

QUARRY PRODUCTS, AGGREGATES, ETC. - continued

	Scotland, North, and Wales	Midlands and South West	South General
Armour stone single size (minimum costs) sea defence shoreprotection			
0.5 tonne	16.86	19.60	19.60
1 tonne	24.66	31.93	31.93
3 tonne	39.06	40.15	40.15
Granular material single sized aggregate			
natural gravel (DTp Clause NG503) pipe bedding/haunching	11.19	11.98	13.76
crushed rock; ditto	11.36	11.40	15.90
Filter material			
natural or crushed other than unburned slag or chalk			
(DTp Clause NG505) backfill to drain trenches and filter drains			
type A	10.02	12.00	19.65
type B	9.09	11.70	17.33
type C	10.31	12.56	20.63
Graded granular material			
crushed rock	11.11	13.33	13.25
natural sand or gravels (DTp Clauses NG803/804) sub base			
type 1/2	13.13	15.67	15.55
crushed concrete (reclaimed)	6.30	8.49	9.26
Coated roadstone products			
DBM roadbases (DTp class 903) carriageway construction	46.26	37.40	41.78
HRM base and wearing; carriageway construction	55.51	44.61	46.67
Bitumen Macadam			
40 mm aggregate	37.29	40.89	42.41
20 mm aggregate	38.31	41.90	43.55
10 mm aggregate	40.57	44.82	46.67
Concrete aggregates			
all in aggregate			
10 mm for concrete production	11.93	13.33	14.22
20 mm for concrete production	11.57	12.44	15.11
40 mm for concrete production	11.93	13.33	15.46
fine screen sharp sand for concrete production	10.67	12.08	13.33

PLEASE NOTE:

The above prices are based upon information obtained from a number of companies or individual quarries. An average of the above prices is used in Part 4 and 5 (Unit Cost compilation), and provides a reasonable guide as to comparative costs of common or DfT Specification materials.

It must be remembered that transport costs form a high proportion of delivered stone prices: if a 20 tonne tipper costs £40.00 per hour to operate then it follows that for each hour journey time £2.00 will be added to each tonne delivered.

The above costs are based on approximately £3.00 per tonne transport cost. Minimum 16 tonne loads.

Allow 15-20p / tonne per mile adjustment to 20 mile inclusion if required.

Market conditions and commercial considerations could also affect these prices to a surprising degree - up to double for the more selective materials. We have included such variations only where we feel that there is a sound reason for them and so have attempted to produce a reasonable guide to market conditions.

If specific requirements are known then local quarries are usually helpful with information and guide prices.

Refer to section dealing with the Aggregate Tax on page 695. The above costs exclude the tax.

SCAFFOLDING AND PROPS

	Unit	£
Scaffold tube		
black tube	ft	0.87
galvanised tube	ft	0.76
alloy tube	ft	1.90
Scaffold fittings		
double	nr	2.25
single	nr	1.50
swivel	nr	1.82
sleeve	nr	1.74
joint pin	nr	1.50
baseplate	nr	0.48
Scaffold boards		
grade A selected (3.97 m long)	nr	8.42
grade A selected (3.00 m long)	nr	6.68
Scaffold towers, alloy,coming with ladder access		
2.5 x 1.35 x 6.05 m	nr	2901.78
2.5 x 1.35 x 9.30 m	nr	3650.22
Adjustable struts		
size 0	nr	15.59
size 1	nr	18.43
size 2	nr	21.34
size 3	nr	23.28
Adjustable props		
size 0	nr	19.68
size 1	nr	23.98
size 2	nr	24.74
size 3	nr	26.64
size 4	nr	31.96
Lightweight stagings		
3.60 m	nr	137.21
4.20 m	nr	149.69
4.80 m	nr	192.65
5.40 m	nr	223.15
6.00 m	nr	238.39
6.60 m	nr	285.52
7.20 m	nr	295.22

	Unit	£
Reinforced timber pole ladders		
4 m	nr	90.22
5 m	nr	110.56
6 m	nr	130.00
7 m	nr	149.33
8 m	nr	174.56
9 m	nr	200.00
10 m	nr	220.50
Health & Safety Signs	nr	5 - 50
Trench sheeting		
standard overlapping	m	9.29
L8 interlocking	m	15.60
Fence panels		
anti-climb, 2m high x 3.5m with block and coupler	nr	62.90

SEPTIC TANKS, CESSPOOLS AND INTERCEPTORS

	Unit	£
Klargester 'Alpha' GRP septic tanks excluding cover and frame		
2800 litre capacity; 1000 mm invert	nr	550.00
2800 litre capacity; 1500 mm invert	nr	596.00
3800 litre capacity; 1000 mm invert	nr	750.00
3800 litre capacity; 1500 mm invert	nr	796.00
4600 litre capacity; 1000 mm invert	nr	850.00
4600 litre capacity; 1500 mm invert	nr	896.00
pedestrian cover and frame	nr	60.00
Klargester GRP cesspools including covers and frames		
18200 litre capacity; 1000mm invert	nr	2011.00
Klargester 3 stage GRP petrol interceptors		
2000 litre capacity	nr	875.70
2500 litre capacity	nr	1101.45
4000 litre capacity	nr	1271.55
Klargester Biodisc; self contained sewage treatment plant		
population equivalent of 6, 450 mm invert	nr	2637.00
population equivalent of 6, 750 mm invert	nr	2715.00
population equivalent of 6, 1250 mm invert	nr	2767.00
population equivalent of 12, 450 mmm invert	nr	3434.00
population equivalent of 12, 750 mm invert	nr	3518.00
population equivalent of 12, 1250 mm invert	nr	3576.00
population equivalent of 18, 600 mm invert	nr	4390.00
population equivalent of 18, 1100 mm invert	nr	4774.00
Klargester Single Effluent Pump Stations, GRP sump and cover		
1000 mm diameter, 1.0 m invert, 2340 mm total depth	nr	645.00
Klargester Bypass Interceptors, suitable for areas of low risk such as carparks		
NSB3 160 mm pvc bypass separator; 1,670 m² drainage area	nr	647.00
NSB4 160 mm pvc bypass separator; 2,500 m² drainage area	nr	730.00
NSB6 300 mm grp bypass separator; 3,335 m² drainage area	nr	777.00
NSB8 300 mm grp bypass separator; 4,445 m² drainage area	nr	953.00

SHUTTERING, TIMBER AND NAILS

	Unit	£
Finnish Birch faced plywood exterior WBP size 1220 x 2440 mm, full pallets		
6.5 mm thick	m²	5.52
12 mm thick	m²	9.13
18 mm thick	m²	13.68
24 mm thick	m²	18.24
Far eastern redwood faced plywood exterior WBP size 1220 x 2440 mm, full pallets		
4.0 mm thick	m²	2.14
6.0 mm thick	m²	2.73
12.0 mm thick	m²	5.23
18.0 mm thick	m²	7.58
25.0 mm thick	m²	10.54
Coated shuttering plywood. Phenolic film faced, edges sealed ext WBP, Finnish Birch faced (Combi)		
size 1220 x 2440 x 12.0 mm	m²	13.20
size 1220 x 2440 x 18.0 mm	m²	18.12
Coated shuttering plywood. Phenolic film faced, edges sealed ext WBP, Eastern European Birch throughout		
size 1220 x 2440 x 12.0 mm	m²	8.10
size 1220 x 2440 x 18.0 mm	m²	12.03
Resin impregnated overlay edges sealed Pourform/Ultraform/ B Matte/ Channelform		
size 1220 x 2440 x 17.5 mm	m²	9.26

Thickness	£/m 25 mm	£/m 50 mm	£/m 75 mm	
Sawn softwood random lengths (used); good quality timber for formwork, falseworks, temporary works and trench supports, delivered to site				
50 mm wide	0.32	0.65	0.97	
75 mm wide	0.49	0.97	1.46	
100 mm wide	0.65	1.30	1.94	
150 mm wide	0.97	1.94	2.92	

SHUTTERING, TIMBER AND NAILS - continued

Thickness	£/m 12.5 mm		£/m 25 mm	£/m 38 mm	£/m 50 mm	£/m 75 mm	
Wrought (planed) timber; European softwood (new)							
25 mm wide	0.15		0.28	0.51	0.53	0.77	
50 mm wide	0.28		0.53	1.00	1.03	1.49	
75 mm wide	0.41		0.77	1.39	1.49	2.53	
100 mm wide	0.53		1.03	1.84	1.99	3.37	
125 mm wide	0.65		1.27	2.31	2.54	4.04	
150 mm wide	0.77		1.37	2.77	3.25	4.77	
175 mm wide	0.94		1.81	3.44	3.77	5.66	
200 mm wide	1.23		2.27	4.37	4.30	6.63	
225 mm wide	1.37		2.66	5.11	4.89	7.45	
250 mm wide	1.74		3.35	5.71	6.40	9.82	
275 mm wide	2.08		4.22	6.35	8.01	12.24	

	Unit	New £		Used £
Structural softwood graded SC3 / 4				
100 x 225 mm	m^3	307.29		204.86
100 x 200 mm	m^3	288.66		169.94
75 x 175 mm	m^3	271.21		183.91
75 x 125 mm	m^3	288.66		181.58
50 x 225 mm	m^3	252.59		160.63
50 x 175 mm	m^3	235.12		160.63

	Unit	Greenheart £	Opepe £	Ekki £
Construction hardwood (for piers, jetties, groynes, dolphins, etc.) Note: Medium price shown, actual can vary considerably dependent on quantity and specification; larger sizes may prove difficult if not impossible to obtain				
100 x 75 mm	m		6.36	6.36
150 x 75 mm	m	7.38	9.58	9.59
200 x 100 mm	m	14.76	17.03	17.03
200 x 200 mm	m	25.55	34.07	34.07
225 x 100 mm	m	14.76	19.17	19.17
300 x 100 mm	m	19.30	25.55	25.55
300 x 200 mm	m	38.61	51.10	51.10
300 x 300 mm	m	57.91	76.65	76.65
400 x 400 mm	m	106.74	147.62	153.30
600 x 300 mm	m		153.30	170.10

	Unit	£	£
Prime Parana pine			
50 x 300 mm	m³	510.23	
38 x 300 mm	m³	510.23	
38 x 300 mm (planed)	m³	617.01	
32 x 300 mm	m³	545.83	
32 x 300 mm (planed)	m³	652.61	
25 x 300 mm	m³	486.49	
25 x 300 mm (planed)	m³	587.36	
Used timber			
beams, baulks, pit props (used for kerbing timbers, temporary barriers, plant supports, etc)	m³	130 - 150	
Nails; steel		bright	sherardised
annular ring shank			
100 mm long	kg	1.79	3.76
50 mm long	kg	1.84	4.13
round plain head			
150 mm long	kg	1.46	3.15
100 mm long	kg	1.51	4.43
oval lost head			
65 mm long	kg	2.30	5.83
75 mm long	kg	2.30	5.83
oval brad head			
150 mm long	kg	2.01	4.71
100 mm long	kg	1.96	4.74
clout, plain head nails			
75 mm long	kg	3.50	
panel pins			
15 mm long	kg	6.51	14.29
50 mm long	kg	3.22	7.09

	Unit	£
Coach screws; stainless steel; hexagon head		
5 x 75 mm	100	123.87
7 x 105 mm	100	298.77
7 x 140 mm	100	475.90
10 x 95 mm	100	653.66
10 x 165 mm	100	1233.92

SHUTTERING, TIMBER AND NAILS - continued

	Unit	£
Metric mild steel bolts and nuts		
M6 x 50 mm	100	5.09
M6 x 100 mm	100	9.06
M8 x 50 mm	100	8.06
M8 x 100 mm	100	13.76
M10 x 50 mm	100	12.45
M10 x 100 mm	100	21.22
M12 x 50 mm	100	17.51
M12 x 100 mm	100	27.80
M12 x 200 mm	100	73.23
M12 x 300 mm	100	130.59
M16 x 50 mm	100	31.46
M16 x 100 mm	100	46.48
M16 x 200 mm	100	106.33
M16 x 300 mm	100	181.04
M20 x 50 mm	100	54.53
M20 x 100 mm	100	76.40
M20 x 200 mm	100	159.17
M20 x 300 mm	100	256.50
M24 x 50 mm	100	120.29
M24 x 100 mm	100	143.89
M24 x 200 mm	100	234.30
M24 x 300 mm	100	369.83
M30 x 100 mm	100	388.31
M30 x 150 mm	100	445.21
M30 x 200 mm	100	502.13
Washers for metric mild steel bolts and nuts		
12.5 o/d x 1.6 (M6 bolt)	100	1.21
17 o/d x 1.6 (M8 bolt)	100	1.46
21 o/d x 2.0 (M10 bolt)	100	2.07
24 o/d x 2.5 (M12 bolt)	100	2.67
30 o/d x 3.0 (M16 bolt)	100	4.25
37 o/d x 3.0 (M20 bolt)	each	1.16

	Unit	£
Metric mild steel setscrews		
M6 x 20 mm	100	2.69
M6 x 60 mm	100	6.82
M8 x 20 mm	100	3.67
M8 x 60 mm	100	10.76
M10 x 20 mm	100	6.70
M10 x 60 mm	100	14.04
M12 x 40 mm	100	11.68
M12 x 80 mm	100	50.44
M16 x 40 mm	100	23.09
M16 x 80 mm	100	70.26
M20 x 40 mm	100	45.74
M20 x 80 mm	100	118.16
M24 x 50 mm	100	127.05
M24 x 80 mm	100	186.76
Multiholed Straps (galvanised)		
30 x 2.5 mm vertical restraint strap		
800 mm long	nr	2.00
1000 mm long	nr	2.59
1200 mm long	nr	3.15
1600 mm long	nr	4.05
30 x 2.5 mm lateral restraint strap		
800 mm long	nr	3.70
1000 mm long	nr	4.75
1200 mm long	nr	5.67
1600 mm long	nr	7.52
Joist Fittings (galvanised)		
Splice plate		
400 x 61 x 18 mm	each	2.71
560 x 80 x 18 mm	each	4.34
560 x 98 x 18 mm	each	5.25
Junction clip		
generally	250	68.62
Toothplate single sided timber connector (galvanised)		
50mm diameter	200	81.48
63mm diameter	150	89.35
75mm diameter	100	87.76

SHUTTERING, TIMBER AND NAILS - continued

	Unit	£
Joist Fittings (galvanised) - continued		
Toothplate double sided timber connector (galvanised)		
50mm diameter (holed for M12 bolts)	200	91.08
63mm diameter (holed for M12 bolts)	150	98.93
75mm diameter (holed for M12 bolts)	100	91.51
Split ring connector (galvanised)		
63mm diameter x 5mm thick	250	477.11
101mm diameter x 5mm thick	100	509.44
Shear plate connector (galvanised)		
67mm diameter x 4mm thick	150	458.66
101mm diameter x 6mm thick	50	662.49
Bonded polystyrene void formers circular in section		
225 mm diameter	m	5.57
300 mm diameter	m	6.54
450 mm diameter	m	14.65
600 mm diameter	m	17.81
750 mm diameter	m	30.17
Expanded polystyrene in blocks	m^3	48.74

STEEL PIPES AND FITTINGS

	Unit	£	£	£	£	£	£
Diameter in mm		100	125	150	200	250	300
Steel pipe to BS 3601 or equivalent	m	24.25	32.54	37.39	39.42	49.56	56.79
Fittings to BS 1640, standard strength							
90 degrees bend, long radius	nr	21.52	47.89	52.13	103.94	188.71	278.15
45 degrees bend, long radius	nr	16.08	33.94	38.66	81.65	148.25	199.63
single junction, equal	nr	54.19	95.78	98.06	192.74	367.78	573.77
single junction, 100 mm branch	nr	-	-	112.97	229.87	-	-
single junction, 150 mm branch	nr	-	-	-	229.87	-	-
single junction, 200mm branch	nr	-	-	-	-	-	-
concentric reducers, reducing to 100 mm	nr	-	-	29.45	68.54	-	-
concentric reducers, reducing to 150 mm	nr	-	-	-	45.29	83.06	-
concentric reducers, reducing to 200 mm	nr	-	-	-	-	74.42	130.68
concentric reducers, reducing to 250 mm	nr	-	-	-	-	-	131.77

	Unit	£	£	£	£	£	£
Diameter in mm		75	110	160	200		
Stainless steel pipe, grade AISI 316	m	26.75	37.90	59.14	87.37		
Fittings							
90 degree bend	m	33.67	43.00	108.88	244.50		
45 degree bend	m	24.61	32.75	92.18	174.62		
90 degree branch	m	36.47	46.38	111.79	164.68		

	Unit	£	£	£	£	£	£
Diameter in mm		100	150	200	250	300	
Fusion bonded epoxy coated steel pipe	m	93.88	161.25	181.66	230.62	295.95	
Fittings							
90 degrees bend; flanged	nr	208.18	253.08	381.66	602.10	781.69	
45 degrees bend; flanged	nr	128.57	228.59	876.92	516.38	651.07	
Single junctions; flanged							
branch size 100 mm	nr	-	348.14	406.25	536.49	656.99	
branch size 150 mm	nr	-	85.36	485.75	626.58	687.81	
branch size 200 mm	nr	-	-	536.77	675.56	802.10	
branch size 250 mm	nr	-	-	-	753.12	881.71	
branch size 300 mm	nr	-	-	-	-	949.05	

STRUCTURAL STEELWORK

Note: The following basic prices are for basic quantities of BS EN10025 1993 grade 275 JR steel (over 10 tonnes of one quality, one serial size of section and one thickness in lengths between 9 m and 18.5 m, for delivery to one destination). In view of firming prices in the steel market, we would note that these prices are based at April 2006 and may be subject to surcharge. Transport charges are additional.

See page 102 for other extra charges

Based on delivery Middlesborough/Scunthorpe Railway Stations - refer to supplier for section availability at each location.

See page 110 for delivery charges.

	£/tonne		£/tonne
Universal beams (kg/m)			
1016 x 305 mm (222,249,272,314,349, 393,438,487)	813.00	406 x 140 mm (39,46)	763.00
		356 x 171 mm (45,51,57,67)	763.00
914 x 419 mm (343,388)	803.00	356 x 127 mm (33,39)	763.00
914 x 305 mm (201,224,253,289)	798.00	305 x 165 mm (40,46,54)	758.00
838 x 292 mm (176,194,226)	793.00	305 x 127 mm (37,42,48)	758.00
762 x 267 mm (134,147,173,197)	793.00	305 x 102 mm (25,28,33)	758.00
686 x 254 mm (125,140,152,170)	793.00	254 x 146 mm (31,37,43)	773.00
610 x 305 mm (149,179,238)	783.00	254 x 102 mm (22,25,28)	763.00
610 x 229 mm (101,113,125,140)	783.00	203 x 133 mm (25,30)	773.00
533 x 210 mm (82,92,101,109,122)	768.00	203 x 102 mm (23)	773.00
457 x 191 mm (67,74,82,89,98)	758.00	178 x 102 mm (19)	773.00
457 x 152 mm (52,60,67,74,82)	758.00	152 x 89 mm (16)	783.00
406 x 178 mm (54,60,67,74)	763.00	127 x 76 mm (13)	783.00
Universal columns (kg/m)			
356 x 406 mm (235,287,340,393,467, 551,634)	803.00	254 x 254 mm (73,89,107,132,167)	763.00
356 x 368 mm (129,153,177,202)	803.00	203 x 203 mm (46,52,60,71,86)	753.00
305 x 305 mm (97, 118, 137, 158, 198, 240 283)	783.00	152 x 152 mm (23,30,37)	773.00
Joists (kg/m)			
254 x 203 mm (82.0)	POA	114 x 114 mm (26.9)	POA
203 x 152 mm (60.2)	763.00	102 x 102 mm (23.0)	POA
203 x 152 mm (52.3)	738.00	102 x 44 mm (7.5)	POA
152 x 127 mm (37.3)	738.00	89 x 89 mm (19.5)	POA
127 x 114 mm (26.9,29.3)	POA	76 x 76 mm (12.8)	POA
Channels (kg/m)			
430 x 100 mm (64.4)	813.00	200 x 90 mm (29.7)	783.00
380 x 100 mm (54.0)	813.00	200 x 75 mm (23.4)	748.00
300 x 100 mm (45.5)	783.00	180 x 90 mm (26.1)	783.00
300 x 90 mm (41.4)	783.00	180 x 75 mm (20.3)	748.00
260 x 90 mm (34.8)	783.00	150 x 90 mm (23.9)	783.00
260 x 75 mm (27.6)	783.00	150 x 75 mm (17.9)	748.00
230 x 90 mm (32.2)	783.00	125 x 65 mm (14.8)	748.00
230 x 75 mm (25.7)	783.00	100 x 50 mm (10.2)	748.00

	£/tonne		£/tonne		£/tonne
Equal angles					
200 x 200 x 16 mm	738.00	150 x 150 x 18 mm	738.00	100 x 100 x 12 mm	723.00
200 x 200 x 18 mm	738.00	120 x 120 x 8 mm	738.00	100 x 100 x 15 mm	723.00
200 x 200 x 20 mm	738.00	120 x 120 x 10 mm	738.00	90 x 90 x 6 mm	723.00
200 x 200 x 24 mm	738.00	120 x 120 x 12 mm	738.00	90 x 90 x 8 mm	723.00
150 x 150 x 10 mm	738.00	120 x 120 x 15 mm	738.00	90 x 90 x 10 mm	723.00
150 x 150 x 12 mm	738.00	100 x 100 x 8 mm	723.00	90 x 90 x 12 mm	723.00
150 x 150 x 15 mm	738.00	100 x 100 x 10 mm	723.00		
Unequal angles					
200 x 150 x 12 mm	753.00	150 x 90 x 12 mm	733.00	125 x 75 x 12 mm	723.00
200 x 150 x 15 mm	753.00	150 x 90 x 15 mm	733.00	100 x 75 x 8 mm	723.00
200 x 150 x 18 mm	753.00	150 x 75 x 10 mm	723.00	100 x 75 x 10 mm	723.00
200 x 100 x 10 mm	748.00	150 x 75 x 12 mm	723.00	100 x 75 x 12 mm	723.00
200 x 100 x 12 mm	748.00	150 x 75 x 15 mm	723.00	100 x 65 x 7 mm	723.00
200 x 100 x 15 mm	748.00	125 x 75 x 8 mm	723.00	100 x 65 x 8 mm	723.00
150 x 90 x 10 mm	733.00	125 x 75 x 10 mm	723.00	100 x 65 x 10 mm	723.00

STRUCTURAL STEELWORK - continued

Add to the aforementioned prices for:

		Unit	£
Universal beams, columns, joists, channels and angles non-standard sizes		t	25.00
Quantity	under 10 tonnes to 5 tonnes	t	10.00
	under 5 tonnes to 2 tonnes	t	25.00
	under 2 tonnes	t	50.00
	order under 1 tonnes are not normally accepted		
Size	lengths 3,000 mm to under 9,000 mm in 100 mm increments	t	15.00
	lengths over 18,500 mm to 24,000 mm in 100 mm increments	t	5.00
	lengths over 24,000 mm are subject to referral	t	POA
Tees cut from universal beams and columns and joists			
	weight per metre of rolled section before splitting		
	up to 25 kg per metre	t	150.00
	25 - 40 kg per metre	t	120.00
	40 - 73 kg per metre	t	100.00
	73 - 125 kg per metre	t	90.00
	125 kg + per metre	t	85.00
	impact testing within the specification	t	10.00
Shotblasting and priming			
	Epoxy Zinc Phosphate primer to universal beams and columns	m²	2.60
	Epoxy Zinc Phosphate primer to channels and angles	m²	2.70
	Zinc rich epoxy primer to universal beams and columns	m²	3.60
	Zinc rich epoxy primer to channels and angles	m²	3.50
Surface quality			
	Specification of class D in respect of EN 10163-3: 1991	t	40.00

Note: The following prices include end user discounts and are for quantities of 10 tonnes and over in one size, thickness, length,steel grade and surface finish and include delivery to mainland of Great Britain to one destination. Additional costs for variations to these factors vary between sections and should be ascertained from the supplier.

The following lists are not fully comprehensive and for other sections manufacturer's price lists should be consulted.

Hot formed structural hollow section	Approx metres per tonne (m)	S355J2H Grade 50D £/100m
Circular (kg/m)		
26.9 x 3.2 mm (1.87)	535	143.40
33.7 x 3.2 mm (2.41)	415	142.98
33.7 x 4.0 mm (2.93)	342	240.46
42.4 x 3.2 mm (3.09)	324	277.17
42.4 x 4.0 mm (3.79)	264	309.39
48.3 x 3.2 mm (3.56)	281	280.57
48.3 x 4.0 mm (4.37)	229	347.26
48.3 x 5.0 mm (5.34)	188	424.34
60.3 x 3.2 mm (4.51)	222	355.41
60.3 x 4.0 mm (5.55)	181	455.48
60.3 x 5.0 mm (6.82)	147	559.69
76.1 x 3.2 mm (5.75)	174	453.17
76.1 x 4.0 mm (7.11)	141	583.50
76.1 x 5.0 mm (8.77)	115	719.72
88.9 x 3.2 mm (6.76)	148	532.78
88.9 x 4.0 mm (8.38)	120	660.46
88.9 x 5.0 mm (10.3)	97.1	845.30
114.3 x 3.6 mm (9.83)	102	806.73
114.3 x 5.0 mm (13.5)	74.1	1107.91
114.3 x 6.3 mm (16.8)	59.6	1377.73
139.7 x 5.0 mm (16.6)	60.3	1355.47
139.7 x 6.3 mm (20.7)	48.4	1690.26
139.7 x 8.0 mm (26.0)	38.5	2123.03
139.7 x 10.0 mm (32.0)	31.3	2687.15
168.3 x 5.0 mm (20.1)	49.8	1643.27
168.3 x 6.3 mm (25.2)	39.7	2057.71
168.3 x 8.0 mm (31.6)	31.7	2589.55
168.3 x 10.0 mm (39.0)	25.7	3286.21
193.7 x 5.0 mm (23.3)	42.9	1902.55
193.7 x 6.3 mm (29.1)	34.4	2376.16
193.7 x 8.0 mm (36.6)	27.3	2988.57
193.7 x 10.0 mm (45.3)	22.1	3804.00

STRUCTURAL STEELWORK - continued

Hot formed structural hollow section	Approx metres per tonne (m)	S355J2H Grade 50D £/100m
Circular (kg/m)		
219.1 x 5.0 mm (26.4)	37.9	2280.39
219.1 x 6.3 mm (33.1)	30.2	2859.14
219.1 x 8.0 mm (41.6)	24.1	3593.35
219.1 x 10.0 mm (51.6)	19.4	4458.03
219.1 x 12.5 mm (63.7)	15.7	5881.78
244.5 x 8.0 mm (46.7)	21.5	4033.88
244.5 x 10.0 mm (57.8)	17.4	4992.68
244.5 x 12.5 mm (71.5)	14.0	6437.13
244.5 x 16.0 mm (90.2)	11.1	8285.34
273.0 x 6.3 mm (41.4)	24.2	3576.07
273.0 x 8.0 mm (52.3)	19.1	4517.60
273.0 x 10.0 mm (64.9)	15.4	5605.97
273.0 x 12.5 mm (80.3)	12.5	7229.38
273.0 x 16.0 mm (101)	9.91	9277.38
323.9 x 6.3 mm (49.3)	20.3	4258.46
323.9 x 8.0 mm (62.3)	16.1	5381.38
323.9 x 10.0 mm (77.4)	12.9	6685.70
323.9 x 12.5 mm (96.0)	10.4	8642.85
323.9 x 16.0 mm (121)	8.27	11114.48
355.6 x 16.0 mm (134)	7.47	12308.59
406.4 x 10.0 mm (97.8)	10.2	8447.82
406.4 x 12.5 mm (121)	8.27	10893.59
406.4 x 16.0 mm (154)	6.50	14145.70
457.0 x 10.0 mm (110)	9.09	9521.64
457.0 x 12.5 mm (137)	7.30	11833.86
457.0 x 16.0 mm (174)	5.75	15982.80
508.0 x 10.0 mm (123)	8.13	10624.56
508.0 x 12.5 mm (153)	6.54	13774.53
508.0 x 16.0 mm (194)	5.16	17819.55

Hot formed structural hollow section	Approx metres per tonne (m)	S355J2H Grade 50D £/100m
Square (kg/m)		
40 x 40 x 3.0 mm (3.41)	293.3	262.70
40 x 40 x 3.2 mm (3.61)	277.0	278.11
40 x 40 x 4.0 mm (4.39)	227.8	338.21
40 x 40 x 5.0 mm (5.28)	189.4	406.77
50 x 50 x 3.2 mm (4.62)	216.5	355.93
50 x 50 x 4.0 mm (5.64)	177.3	434.51
50 x 50 x 5.0 mm (6.85)	146.0	527.75
50 x 50 x 6.3 mm (8.31)	120.3	640.20
60 x 60 x 3.0 mm (5.29)	189.0	407.54
60 x 60 x 3.2 mm (5.62)	177.9	432.96
60 x 60 x 4.0 mm (6.90)	145.0	552.75
60 x 60 x 5.0 mm (8.42)	118.8	694.52
60 x 60 x 6.3 mm (10.3)	97.1	825.12
60 x 60 x 8.0 mm (12.5)	80.0	1001.36
70 x 70 x 3.6 mm (7.40)	135.1	594.64
70 x 70 x 5.0 mm (9.99)	100.1	800.29
70 x 70 x 6.3 mm (12.3)	81.3	985.34
70 x 70 x 8.0 mm (15.0)	66.7	1201.63
80 x 80 x 3.6 mm (8.53)	117.2	660.70
80 x 80 x 5.0 mm (11.6)	86.2	929.26
80 x 80 x 6.3 mm (14.2)	70.4	1137.54
80 x 80 x 8.0 mm (17.5)	57.1	1401.90
90 x 90 x 5.0 mm (13.1)	76.3	1049.43
90 x 90 x 6.3 mm (16.2)	61.7	1297.76
90 x 90 x 8.0 mm (20.1)	49.8	1610.19
100 x 100 x 4.0 mm (11.9)	84.0	904.61
100 x 100 x 5.0 mm (14.7)	68.0	1177.60
100 x 100 x 6.3 mm (18.2)	54.9	1457.97
100 x 100 x 8.0 mm (22.6)	44.2	1875.17
100 x 100 x 10.0 mm (27.4)	36.5	2153.53
120 x 120 x 5.0 mm (17.8)	56.2	1356.25
120 x 120 x 6.3 mm (22.2)	45.0	1891.50
120 x 120 x 8.0 mm (27.6)	36.2	2102.95
120 x 120 x 10.0 mm (33.7)	29.7	2848.70
120 x 120 x 12.5 mm (40.9)	24.4	3010.48
140 x 140 x 5.0 mm (21.0)	47.6	1600.07
140 x 140 x 6.3 mm (26.1)	38.3	1988.53
140 x 140 x 8.0 mm (32.6)	30.7	2493.17
140 x 140 x 10.0 mm (40.0)	25.0	3143.84
140 x 140 x 12.5 mm (48.7)	20.5	4917.87

STRUCTURAL STEELWORK - continued

Hot formed structural hollow section	Approx metres per tonne (m)	S355J2H Grade 50D £/100m
Square (kg/m)		
150 x 150 x 5.0 mm (22.6)	44.2	1721.98
150 x 150 x 6.3 mm (28.1)	35.6	2141.04
150 x 150 x 8.0 mm (35.1)	28.5	2674.41
150 x 150 x 10.0 mm (43.1)	23.2	3387.49
150 x 150 x 12.5 mm (52.7)	19.0	4982.71
160 x 160 x 5.0 mm (24.1)	41.5	1901.45
160 x 160 x 6.3 mm (30.1)	33.2	2374.84
160 x 160 x 8.0 mm (37.6)	26.6	2966.57
160 x 160 x 10.0 mm (46.3)	21.6	3653.99
160 x 160 x 12.5 mm (56.6)	17.7	4644.13
180 x 180 x 6.3 mm (34.0)	29.4	2682.53
180 x 180 x 8.0 mm (42.7)	23.4	3368.96
180 x 180 x 10.0 mm (52.5)	19.0	4142.15
180 x 180 x 12.5 mm (64.4)	15.5	5286.63
180 x 180 x 16.0 mm (80.2)	12.5	7066.76
200 x 200 x 5.0 mm (30.4)	32.9	2398.50
200 x 200 x 6.3 mm (38.0)	26.3	2998.13
200 x 200 x 8.0 mm (47.7)	21.0	3763.45
200 x 200 x 10.0 mm (58.8)	17.0	11870.71
200 x 200 x 12.5 mm (72.3)	13.8	5932.35
200 x 200 x 16.0 mm (90.3)	11.1	7956.72
250 x 250 x 6.3 mm (47.9)	20.9	3779.22
250 x 250 x 8.0 mm (60.3)	16.6	4757.56
250 x 250 x 10.0 mm (74.5)	13.4	5877.91
250 x 250 x 12.5 mm (91.9)	10.9	7540.56
250 x 250 x 16.0 mm (115)	8.70	10133.13
300 x 300 x 6.3 mm (57.8)	17.3	4560.31
300 x 300 x 8.0 mm (72.8)	13.7	5743.78
300 x 300 x 10.0 mm (90.2)	11.1	7116.61
300 x 300 x 12.5 mm (112)	8.93	9189.79
300 x 300 x 16.0 mm (141)	7.09	12424.10
350 x 350 x 8.0 mm (85.4)	11.7	9891.88
350 x 350 x 10.0 mm (106)	9.43	8363.20
350 x 350 x 12.5 mm (131)	7.63	10748.78
350 x 350 x 16.0 mm (166)	6.02	14626.96
400 x 400 x 10.0 mm (122)	8.20	9625.57
400 x 400 x 12.5 mm (151)	6.62	12389.83
400 x 400 x 16.0 mm (191)	5.24	16829.81

Hot formed structural hollow section	Approx metres per tonne (m)	S355J2H Grade 50D £/100m
Rectangular (kg/m)		
50 x 30 x 3.2 mm (3.61)	277.0	278.11
60 x 40 x 3.0 mm (4.35)	229.9	335.12
60 x 40 x 4.0 mm (5.64)	177.3	434.51
60 x 40 x 5.0 mm (6.85)	146.0	527.72
80 x 40 x 3.2 mm (5.62)	177.9	434.50
80 x 40 x 4.0 mm (6.90)	144.9	552.75
80 x 40 x 5.0 mm (8.42)	118.8	674.52
80 x 40 x 6.3 mm (10.3)	97.1	825.12
80 x 40 x 8.0 mm (12.5)	80.0	1001.36
90 x 50 x 3.6 mm (7.40)	135.1	574.64
90 x 50 x 5.0 mm (9.99)	100.1	800.29
90 x 50 x 6.3 mm (12.3)	81.3	985.34
100 x 50 x 3.0 mm (6.71)	149.0	501.15
100 x 50 x 3.2 mm (7.13)	140.3	532.52
100 x 50 x 4.0 mm (8.78)	113.9	681.80
100 x 50 x 5.0 mm (10.8)	92.6	865.17
100 x 50 x 6.3 mm (13.3)	75.2	1065.45
100 x 50 x 8.0 mm (16.3)	61.3	1305.77
100 x 60 x 3.6 mm (8.53)	117.2	662.38
100 x 60 x 5.0 mm (11.6)	86.2	929.25
100 x 60 x 6.3 mm (14.2)	70.4	1137.54
100 x 60 x 8.0 mm (20.1)	57.1	1401.90
120 x 60 x 3.6 mm (9.70)	103.1	753.24
120 x 60 x 5.0 mm (13.1)	76.3	1049.43
120 x 60 x 6.3 mm (16.2)	61.7	1297.76
120 x 60 x 8.0 mm (17.5)	49.8	1610.21
120 x 80 x 5.0 mm (14.7)	68.0	1177.60
120 x 80 x 6.3 mm (18.2)	54.9	1457.97
120 x 80 x 8.0 mm (22.6)	44.2	1875.67
120 x 80 x 10.0 mm (27.4)	36.5	2153.53
150 x 100 x 5.0 mm (18.6)	53.8	1417.21
150 x 100 x 6.3 mm (23.1)	43.3	1760.08
150 x 100 x 8.0 mm (28.9)	34.6	2202.01
150 x 100 x 10.0 mm (35.3)	28.3	2774.44
150 x 100 x 12.5 mm (42.8)	23.4	3150.33
160 x 80 x 4.0 mm (14.4)	69.4	1097.19
160 x 80 x 5.0 mm (17.8)	56.2	1356.25
160 x 80 x 6.3 mm (22.2)	45.0	1691.50
160 x 80 x 8.0 mm (27.6)	36.2	2102.95

STRUCTURAL STEELWORK - continued

Hot formed structural hollow section	Approx metres per tonne (m)	S355J2H Grade 50D £/100m
Rectangular (kg/m)		
200 x 100 x 5.0 mm (22.6)	44.2	2232.88
200 x 100 x 6.3 mm (28.1)	35.6	2776.28
200 x 100 x 8.0 mm (35.1)	28.5	3467.88
200 x 100 x 10.0 mm (43.1)	23.2	4398.62
200 x 100 x 12.5 mm (52.7)	19.0	7111.34
250 x 150 x 5.0 mm (30.4)	32.9	2398.50
250 x 150 x 6.3 mm (38.0)	26.3	2998.13
250 x 150 x 8.0 mm (47.7)	21.0	3763.45
250 x 150 x 10.0 mm (58.8)	17.0	4639.22
250 x 150 x 12.5 mm (72.3)	13.8	5932.35
250 x 150 x 16.0 mm (90.3)	11.1	7956.72
300 x 200 x 6.3 mm (47.9)	20.9	3779.22
300 x 200 x 8.0 mm (60.3)	16.6	4957.56
300 x 200 x 10.0 mm (74.5)	13.4	5877.91
300 x 200 x 12.5 mm (91.9)	10.9	7540.56
300 x 200 x 16.0 mm (115)	8.70	10133.13
400 x 200 x 8.0 mm (72.8)	13.7	5943.78
400 x 200 x 10.0 mm (90.2)	11.1	7116.63
400 x 200 x 12.5 mm (112)	8.93	9189.79
400 x 200 x 16.0 mm (141)	7.09	12424.10
450 x 250 x 8.0 mm (85.4)	11.7	6937.90
450 x 250 x 10.0 mm (106)	9.43	8363.20
450 x 250 x12.5 mm (131)	7.63	10748.78
450 x 250 x16.0 mm (166)	6.02	14626.96
500 x 300 x 10.0 mm (122)	8.20	9625.57
500 x 300 x 12.5 mm (151)	6.62	12389.81
500 x 300 x 16.0 mm (191)	5.24	16829.81
Ovals (kg/m)		
150 x 75 x 5.0 mm (13.3)	75.19	
200 x 100 x 6.3 mm (22.3)	44.84	
250 x 125 x 8.0 mm (34.4)	29.07	
300 x 150 x 10 mm (53.0)	18.87	
400 x 200 x 12.5 mm (88.6)	11.29	
400 x 200 x 16.0 mm (112)	8.93	
500 x 250 x 12.5 mm (112)	8.93	
500 x 250 x 16 mm (142)	7.04	

Add to the aforementioned prices for:	Percentage extra %
Quantity	
- Work despatches	
Orders for the following hollow sections of one size, thickness, length, steel grade and surface finish.	
(a) circular hollow sections over 200 mm diameter	
(b) square hollow sections over 150 x 150 mm	
(c) rectangular hollow sections over 600 mm girth	
4 tonnes to under 10 tonnes	15.0
2 tonnes to under 4 tonnes	20.0
1 tonne to under 2 tonnes	25.0
orders under 1 tonne are not supplied	
- Warehouse despatches	
Orders for the following hollow sections in one steel grade, for delivery to one destination in one assignment.	
(a) circular hollow sections less than 200 mm diameter	
(b) square hollow sections up to and including 150 x 150 mm	
(c) rectangular hollow sections up to and including 600 mm girth	
10 tonnes and over	7.5
4 tonnes to under 10 tonnes	10.0
2 tonnes to under 4 tonnes	12.5
1 tonne to under 2 tonnes	17.5
500 kg to under 1 tonne	22.5
250 kg to under 500 kg	35.0
100 kg to under 250 kg	50.0
under 100 kg	100.0
(a) circular hollow sections over 200 mm diameter	
(b) square hollow sections over 150 x 150 mm	
(c) rectangular hollow sections over 600 mm girth	
10 tonnes and over	12.5
4 tonnes to under 10 tonnes	15.00
2 tonnes to under 4 tonnes	20.00
1 tonne to under 2 tonnes	25.00
500 kg to under 1 tonne	45.00
250 kg to under 500 kg	75.00
Finish	
Self colour is supplied unless otherwise specified.	
Transit primer painted (All sections except for circular hollow sections over 200 mm dia)	5.0
Test Certificates	
Test certificates will be charged at a rate of £25 per certificate	

STRUCTURAL STEELWORK - continued

TRANSPORT CHARGES BASED ON RADIAL DISTANCES AND QUANTITIES

RADIAL DISTANCES MILES FROM BASING POINT	0 - under 5 tonnes £/tonne	5 - under 10 tonnes £/tonne	10 - under 20 tonnes £/tonne	20 tonnes and over £/tonne
Schedule IX - Iron and Steel carriage Tariff (Structural sections and steel piling)				
up to 10	8.97	5.92	4.48	4.08
over 10 up to 15	9.83	6.44	5.00	4.48
over 15 up to 20	10.46	7.02	5.58	4.94
over 20 up to 25	11.10	7.53	5.98	5.52
over 25 up to 30	11.62	8.10	6.56	5.92
over 30 up to 35	12.19	8.57	6.96	6.44
over 35 up to 40	12.82	9.14	7.41	6.84
over 40 up to 45	13.28	9.66	7.93	7.30
over 45 up to 50	13.74	10.23	8.40	7.76
over 50 up to 60	15.76	10.93	9.09	8.39
over 60 up to 70	16.44	11.90	10.00	9.32
over 70 up to 80	17.54	12.82	10.81	10.23
over 80 up to 90	18.80	13.74	11.67	10.98
over 90 up to 100	19.89	14.61	12.54	11.90
over 100 up to 110	22.54	15.46	13.39	12.70
over 110 up to 120	23.35	16.33	14.20	13.57
over 120 up to 130	24.09	17.14	15.07	14.37
over 130 up to 140	25.01	18.00	15.81	15.18
over 140 up to 150	25.24	18.80	16.67	15.99
over 150 up to 160	27.77	19.67	17.48	16.79
over 160 up to 170	28.75	20.41	18.29	17.60
over 170 up to 180	29.50	21.22	19.03	18.40
over 180 up to 190	30.53	22.02	19.89	19.09
over 190 up to 200	31.51	22.77	20.64	19.55
over 200 up to 210	34.67	23.52	21.39	20.36
over 210 up to 220	35.71	24.27	22.20	21.05
over 220 up to 230	36.57	25.02	22.77	21.85
over 230 up to 240	37.95	25.76	23.40	22.48
over 240 up to 250	38.76	26.51	24.04	22.94
over 250 up to 275	43.01	27.72	25.07	24.15
over 275 up to 300	45.48	29.56	26.56	25.59
over 300 up to 325	47.78	31.28	28.11	27.26
over 325 up to 350	50.26	33.06	29.67	28.75
over 350 up to 375	52.73	34.73	30.93	30.42
over 375 up to 400	55.20	36.39	32.25	31.80

Note: - The minimum charge under this schedule will be as for 2 tonnes

Lengths over 14 metres and up to 18 metres will be subject to a surcharge of 20%

Lengths over 18 metres and up to 24 metres will be subject to a surcharge of 40%

Lengths over 24 metres will be subject to a surcharge of 70%

Collection by customer from works will incur an extra cost of £3.00/tonne in addition to any additional transport charges between the basing point and the producing works.

PVC-U DRAIN PIPES AND FITTINGS

	Unit	£	£	£
Diameter in mm		82	110	160
Plain ended pipes (Osmadrain)				
3 m lengths	m	12.29	8.98	20.28
6 m lengths	m	12.29	8.98	20.28
Couplers				
for jointing pipes	nr	9.96	8.28	16.68
for new branch entry connections	nr	13.28	13.73	30.86
Reducers, single socket	nr	-	16.09	27.38
Junctions, equal, single socket				
87.5 degrees	nr	27.90	28.33	85.51
45 degrees	nr	-	29.57	84.52
Short radius bends, single socket				
up to 30 degrees	nr	-	19.32	41.96
45 to 90 degrees	nr	18.03	20.07	47.35
Suspended bracketing system adjustable pipe bracket assembly (pack containing threaded bracket, bracket plateand pipe/socket bracket)	nr	-	25.16	35.87
Adjustable socket bracket and brace assembly (pack containing threaded rod, threaded bracket, bracket plate, two adjustable braces and pipe/socket bracket)	nr	-	51.78	60.80
Marley underground drainage system				
Straight pipes				
ring seal socket 6 m	m	-	7.26	19.34
ring seal socket 3 m	m	11.29	7.85	21.81
double spigot 6 m	m	-	6.69	16.86
double spigot 3 m	m	-	6.69	-
Couplings				
double ring seal straight, polypropylene	nr	7.44	6.18	20.56
loose pipe socket	nr	7.44	6.18	16.51
triple socket	nr	-	19.95	-
Bends, socket/spigot				
short radius (87.5 degrees)	nr	15.02	15.72	39.05
short radius (45 degrees)	nr	15.02	17.00	35.31
adjustable (11 to 87.5 degrees)	nr	25.35	-	-
adjustable (21 to 90 degrees)	nr	-	29.85	-
adjustable (15 to 90 degrees)	nr	-	-	57.03
Bends, socket/socket				
short radius (87.5 degrees)	nr	-	15.81	44.93
short radius (15, 30, 45 degrees)	nr	-	13.65	40.56
long radius (87.5 degrees)	nr	-	27.39	-

PVC-U DRAIN PIPES AND FITTINGS - continued

	Unit	£	£	£
Diameter in mm		82	110	160
Branches				
socket/spigot 45 degrees equal	nr	23.77	23.43	45.70
socket/spigot 87.5 degrees equal	nr	-	23.43	45.65
socket/spigot 45 degrees, 160 x 110 mm	nr	-	-	39.64
socket/socket 87.5 degrees equal	nr	-	25.36	64.00
socket/socket 45 degrees equal	nr	-	25.36	64.00
Access components				
socket/spigot pipe	nr	-	59.13	-
socket/spigot bend 87.5 degrees rear access	nr	-	31.76	-
socket/spigot branches				
45 degrees branch	nr	-	86.94	-
45 degrees double branch	nr	-	82.57	-
rodding point 45 degrees socketed	nr	-	51.33	133.39
cap and pressure plug	nr	-	16.19	41.29
cap	nr	13.49	-	-
pressure plug	nr	-	12.39	21.41
socket plug, 1 boss upstand	nr	-	8.15	-
Open channels				
straight double spigot (1500 mm long)	nr	-	49.19	-
long radius channel bend 87.5 degrees	nr	-	70.55	-
slipper bend	nr	-	22.69	-
Bottle Gully				
bottle gully	nr	-	59.15	-
sealed acces lid	nr	-	9.58	-
Reducers				
level invert				
110 mm spigot to 82 mm socket	nr	-	15.02	-
160 mm spigot to 110 mm socket	nr	-	-	16.85
eccentric				
82 mm socket to boss upstand	nr	12.62	-	-
82 mm socket to 68 mm socket	nr	5.41	-	-
110 mm socket to boss upstand	nr	-	9.15	-
110 mm socket to 68 mm socket	nr	-	7.56	-
concentric				
110 mm socket to boss upstand	nr	-	8.15	-
Adapters				
PVC-U spigot to salt glazed/pitch fibre socket	nr	-	10.65	-
PVC-U spigot to salt glazed/cast iron to pvc-u socket	nr	-	20.69	-
PVC-U spigot to thin wall clay spigot	nr	-	24.19	-
PVC-U spigot to thick wall clay spigot	nr	-	24.19	-
				-

	Unit	£	£	£
Diameter in mm		82	110	160
Access point covers				
450 mm lid and frame (A15 loading)	nr	-	71.50	-
Inspection chambers				
450 mm chamber base 230 mm high	nr	-	130.74	163.07
450 mm chamber riser 400 mm high	nr	-	49.46	-
450 mm cast iron cover and frame	nr	-	83.67	-
450 mm ductile iron lid and cast iron frame	nr	-	119.86	-
250 mm equal double branch chamber base	nr	-	88.65	-
250 mm double branch chamber base	nr	-	71.81	-
250 mm bottom outlet chamber body 4 x 110 mm upstands	nr	-	90.29	-
250 mm pressure plug	nr	-	44.11	-
250 mm chamber riser 375 mm long	nr	-	27.22	-
lifting handle	nr	-	14.59	-
square lid and frame	nr	-	71.77	-
PVC-U lid and frame	nr	-	35.32	-
Gully components				
compact gully, 45 degrees outlet	nr	-	68.05	-
gully trap base, 45 degrees outlet	nr	-	17.52	-
P' trap gully 81.5 degrees outlet	nr	-	43.96	-
hoppers (rectangular)	nr	-	17.52	-
hoppers (square)	nr	-	13.47	-
inlet raising pieces				
2 x 82 mm upstands	nr	-	8.69	-
4 boss upstands	nr	-	15.00	-
grating assembly	nr	-	9.40	-
Diameter in mm		150	225	300
Marley Quantum underground drainage system				
Straight pipes				
plain pipe 6 m	m	6.44	17.13	26.06
pipe 6 m with coupling and seals	m	6.43	17.13	26.06
Couplings				
straight or slip	nr	8.94	21.83	42.40
Bends				
double socket 87.5 degrees	nr	17.99	86.48	155.05
double socket, 15, 30 and 45 degrees	nr	15.63	62.05	115.39
Branches				
all socket, equal	nr	31.26	145.73	337.04
all socket, 150 x 110 mm	nr	27.46	-	
all socket, 225 x 110 mm	nr	-	89.00	
all socket, 225 x 150 mm	nr	-	74.07	
all socket, 300 x 110 mm	nr	-	-	140.58
all socket, 300 x 150 mm	nr	-	-	120.47
all socket, 300 x 225 mm	nr			337.04

PVC-U DRAIN PIPES AND FITTINGS - continued

	Unit	£	£	£
Diameter in mm		150	225	300
Reducers				
level invert 225 mm to 150 mm	nr	-	45.24	-
level invert 300 mm to 225 mm	nr	-	-	98.22
Plugs				
socket plug	nr	12.13	-	-
end cap	nr	5.90	34.74	82.89
Adapters				
pipe to clayware	nr	24.96	-	-
pipe spigot to pvc socket	nr	15.51	-	-
flexible adaptors	nr	50.58	62.25	154.87

PVC-U RAINWATER GUTTERS AND PIPES	Unit	£		
Gutters, Marley Deep Flow				
110 x 75 mm gutters (3 m length)	m	5.02		
110 x 75 mm gutters (4 m length)	m	5.02		
union bracket	nr	5.97		
fascia brackets	nr	1.58		
angles, 90 degrees	nr	7.75		
angles, 45 degrees	nr	11.64		
running outlets	nr	7.09		
stopend outlets	nr	6.23		
stopends	nr	3.82		
Gutters, Marley Industrial				
150 mm half round gutters (4 m length)	m	7.36		
union clips	nr	7.01		
fascia brackets	nr	2.41		
angles, 90 degrees	nr	12.43		
running outlets	nr	13.61		
stopend, internal	nr	2.68		
stopend, external	nr	4.91		

	Unit	£	£	
Diameter in mm		68	110	
Downpipes, Marley circular				
ring seal socket 2.5 m	m	5.42	-	
ring seal socket 3 m	m	5.00	10.73	
ring seal socket 5.5 m	m	4.96	-	
sockets, loose pipe	nr	4.04	8.35	
sockets, solvent weld pipe socket with ring seal	nr	6.61	-	
bends	nr	7.30	18.80	
offset bends	nr	4.04	17.31	
branches	nr	14.53	25.31	
shoes	nr	6.31	14.83	
hopper heads	nr	23.37	83.51	
one piece pipe clip	nr	2.06	-	
socket/pipe clip	nr	1.67	7.12	
backplate	nr	1.05	-	

PVC-U SUBSOIL DRAIN PIPES AND FITTINGS

		Unit	£	£
Diameter in mm			80	100
Wavin Land subsoil drainage system				
S/S coils, 25m coils		m	1.33	2.34
end caps		nr	1.92	2.18
couplers		nr	1.78	1.98
reducers 100 to 80mm		nr	-	2.13
equal junctions, 67.5 degree		nr	4.34	4.85
unequal junctions, 67.5 degree		nr	-	4.62

	Unit	£	£	£
Diameter in mm		150	225	300
Ultra-Rib system				
S/S pipe (3m length)	m	9.26	21.24	31.86
S/S pipe (6m length)	m	9.26	21.45	-
D/S pipe coupler	nr	15.72	32.88	66.30
D/S short radius bend, 45 degree	nr	21.13	85.04	154.43
D/S equal junction, 45 degree	nr	51.10	169.58	361.70
Ultra-Rib Inspection Chambers, 450mm diameter				
D/S base, 150 x 150mm	nr	129.57	-	-
shaft, 230mm effective length	nr	35.83	-	-
DI cover and frame, single seal, Class BS EN 124-B125 Medium Duty	nr	112.86	-	-
Steel cover and polypropylene frame, to 25kN loading	nr	67.50	-	-
Ultra-Rib manhole basis, 750mm diameter				
P/E unequal, 150 x 110mm	nr	229.28	-	-
P/E unequal, 225 x 150mm	nr	-	304.19	-
P/E equal, 150 x 150mm	nr	247.92	-	-
P/E equal, 225 x 225mm	nr	-	245.46	-
P/E channel access pipe	nr	50.77	116.39	195.50
Ultra-Rib sealed rodding eye				
oval cover, upto 35 kN loading	nr	119.63	-	-

INTRODUCTION

The information, rates and prices included in this section are calculated examples of actual owning and operating costs of a range of construction plant and equipment. To an extent they will serve as a guide to prevailing commercial plant hire rates, but be aware that many factors will influence actual hire rates. For example, rates could be lower because of long term hire, use of older or second hand machines, low market demands/loss leaders. Easy tasks or sites near to the depot. Rates could be higher due to short term hire, high production demands, low utilisation factors, specialist operations, restricted working and restricted accessibility problems, high profit or overhead demands and finance fluctuations. The use factors will mean a hired machine may not be used productively for each hour on site, it is therefore up to the estimator to allow for reasonable outputs in his unit rate calculations.

These an other considerations must be borne in mind if using these costs as a comparison to hire rates especially operated plant.

All rates quoted EXCLUDE the following:

- Cost of drivers / operators
- VAT

CONTRACTOR OWNED PLANT

Plant owned by a Contractor generally falls under two headings:

- Small plant and tools which are the subject of a direct charge to contracts and for estimating purposes are normally allowed for as a percentage of the labour cost in site on-costs (see Preliminaries and General Items), although many items are shown in this section for information.
- Power driven plant and major non-mechanical plant such as steel trestling, scaffolding, gantries etc. Such plant is normally charged to the contract on a rental basis except in the case of purpose made or special plant bought specifically for a particular operation; this latter is normally charged in full to contracts and allowance made for disposal on completion (often at scrap value).

A very wide range of plant is readily available from plant hire companies; it is not usually economical for Contractors to own plant unless they can ensure at least 75 to 80% utilisation factor based on the Contractor's normal working hours. Where a Contractor does own plant, however, it is essential that he maintains reasonably accurate records of the working hours and detailed costs of maintenance and repairs in order that he may estimate the charges to be made for each item of plant. For this reason, where a Contractor owns a large quantity of plant, it is normal for a separate plant hire department or company to be formed.

This department or company should be financially self supporting and may hire plant to other Contractors when it is not needed. Maintenance of contractor owned plant can be carried out on site; it is not necessary to maintain a centrally based repair workshop. It is, however, desirable to have some storage facilities available for plant when not in use. The cost of owning plant and hence the rental charges to be made to contracts must take into account:

- Capital cost
- Depreciation charges
- Maintenance and repairs
- Cost of finance
- Insurances and licences
- Administration, head office, depot and other overhead charges

CONTRACTOR OWNED PLANT - continued

Calculated examples of hourly owning costs of a range of plant and equipment:

PLANT ITEMS	A UTILIZATION FACTOR %	B P.A. HOURS	C PLANT Years	D LIFE Hours	E PURCHASE PRICE £	F RESALE %	G RESALE PRICE £	H LOSS IN VALUE £	I DEPREC. £	J MAINTEN. £	K FINANCE @5% £	L INSUR. £	M ADMIN. £	N TOTAL COST/HR £
ACCESS PLATFORMS														
JLG 450AJ, diesel	65	1,365	10	13,650	35,000	20.00	7,000	28,000	2.05	2.56	0.51	0.51	1.13	6.76
JLG 600AJ, diesel	65	1,365	10	13,650	55,000	20.00	11,000	44,000	3.22	4.03	0.81	0.81	1.77	10.64
ASPHALT PAVERS														
BK 151	65	1,365	7	9,555	75,000	27.50	20,625	54,375	5.69	5.49	1.00	1.10	2.66	15.94
BK 181	65	1,365	7	9,555	115,000	27.50	31,625	83,375	8.73	8.42	1.53	1.68	4.07	24.43
BK 191	65	1,365	7	9,555	120,000	27.50	33,000	87,000	9.11	8.79	1.59	1.76	4.25	25.50
COMPRESSORS (PORTABLE)														
17 m³/min, electric motor	85	1,785	9	16,065	50,000	20.00	10,000	40,000	2.49	2.80	0.56	0.56	1.28	7.69
7.3 m³/min, electric motor	85	1,785	9	16,065	25,000	20.00	5,000	20,000	1.24	1.40	0.28	0.28	0.64	3.84
CRANES														
Kato 16 t mobile	60	1,260	12	15,120	180,000	25.00	45,000	135,000	8.93	14.29	2.68	2.86	5.75	34.51
Grove 35 t mobile	60	1,260	15	18,900	200,000	27.50	55,000	145,000	7.67	15.87	2.88	3.17	5.92	35.51
GENERATORS														
10 kVA, water cooled	85	1,785	5	8,925	6,000	15.00	900	5,100	0.57	0.34	0.07	0.07	0.21	1.26
10 kVA, air cooled	85	1,785	5	8,925	5,000	15.00	750	4,250	0.48	0.28	0.06	0.06	0.18	1.06
EXCAVATORS														
Hitachi ZX130 B/A track	80	1,680	9	15,120	80,000	27.50	22,000	58,000	3.84	4.76	0.86	0.95	2.08	12.49
Hitachi ZX800 B/A track	75	1,575	9	14,175	375,000	27.50	103,125	271,875	19.18	23.81	4.32	4.76	10.41	62.48
JCB 3CX	85	1,785	6	10,710	48,000	27.50	13,200	34,800	3.25	2.69	0.49	0.54	1.39	8.36
ROLLERS														
Bomag BW65H	85	1,785	6	10,710	15,000	27.50	4,125	10,875	1.02	0.84	0.15	0.17	0.44	2.62
Bomag BW80ADH (narrow, heavy duty)	85	1,785	6	10,710	28,000	27.50	7,700	20,300	1.90	1.57	0.28	0.31	0.81	4.87
Bomag BW75H	85	1,785	6	10,710	28,000	27.50	7,700	20,300	1.90	1.57	0.28	0.31	0.81	4.87
CRAWLER LOADERS														
CAT 963 C	80	1,680	9	15,120	140,000	27.50	38,500	101,500	6.71	8.33	1.51	1.67	3.64	21.86
TIPPERS														
Scania P94 CB 4	90	1,890	6	11,340	45,000	22.00	9,900	35,100	3.10	2.38	0.46	0.48	1.28	7.70
Scania P114 CB 8	90	1,890	6	11,340	70,000	22.00	15,400	54,600	4.81	3.70	0.72	0.74	1.99	11.96
DUMP TRUCKS														
Volvo A35 D	75	1,575	10	15,750	290,000	30.00	87,000	203,000	12.89	18.41	3.22	3.68	7.64	45.84

NOTES: The price of foreign manufactured plant will vary according to the strength of the £ Sterling internationally.
No allowance is included for Road Fund Licence on road-going vehicles.

Notes – example plant ownership hourly costs

- ☐ The above costs would be updated annually or bi-annually, the rental rate being revised to ensure complete recovery of the costs associated with the item of plant and return of capital to enable the machine to be replaced at the end of its life with the Contractor. The purchase price must also be adjusted to ensure recovery of the replacement cost and not the original cost of purchase.
- ☐ Driver's wages and costs should be charged direct to site wages.
- ☐ Column A: The utilisation factors used in conjunction with the period of ownership is the percentage of time (per annum) that an item of plant can be expected to be used productively on a site or job and therefore is a very important influence of hourly costs.

  (Note: Utilisation factors are not the same as site utilisation rates - see note overleaf regarding Fuel Consumption)

Economical owning periods/use factors:-	Life (years)	Usage (%)
Hydraulic Excavators, large	9	75
Hydraulic Excavators, medium	7	76
Hydraulic Excavators, mini	6	77
Dozers/Scrapers	10	78
Loaders/Shovels	9	79
Mobile Cranes	12	80
Crawler Cranes	15	81
Dump Trucks	8	82
Dumpers	6	83
Rollers/Compaction	6	84
Compressors/Generators	7 - 9	85
Diesel engine road vehicles	6 - 7	86
Petrol engine road vehicles	5 - 6	87

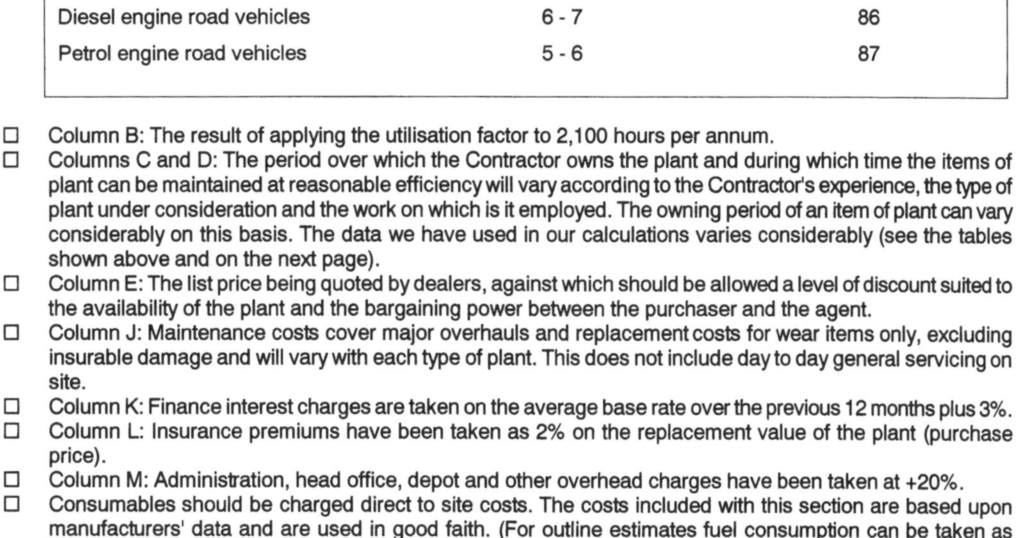

- ☐ Column B: The result of applying the utilisation factor to 2,100 hours per annum.
- ☐ Columns C and D: The period over which the Contractor owns the plant and during which time the items of plant can be maintained at reasonable efficiency will vary according to the Contractor's experience, the type of plant under consideration and the work on which is it employed. The owning period of an item of plant can vary considerably on this basis. The data we have used in our calculations varies considerably (see the tables shown above and on the next page).
- ☐ Column E: The list price being quoted by dealers, against which should be allowed a level of discount suited to the availability of the plant and the bargaining power between the purchaser and the agent.
- ☐ Column J: Maintenance costs cover major overhauls and replacement costs for wear items only, excluding insurable damage and will vary with each type of plant. This does not include day to day general servicing on site.
- ☐ Column K: Finance interest charges are taken on the average base rate over the previous 12 months plus 3%.
- ☐ Column L: Insurance premiums have been taken as 2% on the replacement value of the plant (purchase price).
- ☐ Column M: Administration, head office, depot and other overhead charges have been taken at +20%.
- ☐ Consumables should be charged direct to site costs. The costs included with this section are based upon manufacturers' data and are used in good faith. (For outline estimates fuel consumption can be taken as approximately 0.15 - 0.20 litres per kW per working hour; lube oils, filters, grease etc., can be taken as 2% to 10% of the fuel cost, depending upon the working conditions of the machine and its attachments). See below for details.

CONTRACTOR OWNED PLANT – continued

Notes – example plant ownership hourly costs – continued

- ☐ Please note that the costs in this section have been completely reviewed for recent editions, based on the latest available advice and data from manufacturers and Plant Hire firms, the basis of calculation, including retention period of the plant, has been changed in a number of cases, providing a realistic overall owning and operating cost of the machine.
- ☐ These examples are for machine costs only and do not include for operator costs or profit element or for transport costs to and from site.
- ☐ Fuel consumption is based on the following site utilisation rates. (Site utilisation being the percentage of time that the machine is operating at its average fuel consumption during a working day).

Typical utilization factors:	Utilisation Factor	Typical utilization factors:	Utilisation Factor
Excavators	0.75	Compressors, mixers, generators	
Dozers/Scrapers	0.80	Tractors	0.75
Loaders/Shovels/Graders	0.75	Hoists, etc.	0.65
Mobile Cranes	0.50	Drills and Saws	0.90
Crawler Cranes	0.25	Rollers/Compaction Plant	0.75
Dump Trucks (not Tippers)	0.75	Piling/Asphalt Equipment	0.65
Dumpers	0.80	All other items	1.00

- ☐ Approximate fuel consumption in litres per hour and percentage addition for consumables (examples only) based on the above site utilisation rates. assuming that reasonably new and well maintained plant would be used.

		Fuel l/hr	Consumables % of fuel			Fuel l/hr	Consumables % of fuel
Tractor	116 hp	9.0	2.0%	Rollers	BW 90AD	2.0	7.5%
Paver	Bitelli BB650	16.0	5.0%		BW 120AD	4.4	7.5%
Compactor plate	188kg	0.9	3.0%		BW6 (towed)	7.6	7.5%
Compressor	3.5 m³/min	5.9	3.0%	Dozers	CAT D6E	21.0	6.0%
	10 m³/min	18.0	3.0%		CAT D8N	38.0	3.0%
	Tractor mounted	9.0	5.0%	Loaders	CAT 953B	18.0	4.0%
Mixer	5/3.5	1.2	2.0%	Scrapers	CAT 621F	41.5	3.0%
Mobile crane	15t	5.1	10.0%		CAT 657E	130.0	6.0%
Mobile crane	40t	6.3	10.0%	Skidsteer	Bobcat 553	3.4	5.0%
Generator	4 kVA	1.6	3.0%	Graders	CAT 140H	23.0	5.0%
	25 kVA	4.9	5.0%	Dumpers	2 tonne	3.0	3.0%
Backacter	11.5 t	9.5	7.5%		Volvo A25	17.0	3.0%
	19 t	14.0	7.5%		36 tonne	25.0	5.0%
	30 t	24.0	7.5%	Pumps	75 mm 65 m³/hr	1.1	2.0%
Backhoes	JCB 3CX	7.5	6.0%		150 mm 360 m³/hr	4.2	3.0%

- ☐ Note that the above consumption figures represent 'medium' plant operation, i.e. the plant would not be operating at full throttle and working conditions such as the soil and grades are average. A study of manufacturers' figures has indicated that 'high' or 'low' usage can affect consumption on average by ± 25%; 'high' usage would involve full loads, continuous working at full throttle with difficult ground and adverse grades, whereas 'low' usage would involve more intermittent working with perhaps considerable idling periods, more easily worked ground and easier grades.

PLANT COSTS

The costs included in this section are based upon the methodology of the worked examples, these have then been compared to average costs, allowances and hire rates in force at the time of writing (Summer 2006). These costs then form the basis for plant costs included in parts 4 and 5 (except for specialist advice). Weekly costs are based upon a 40 hour week, daily costs are based on an 8 hour day, all costs are exclusive of labour for operation. Also given here is a reference to the November 2002 CECA Daywork Schedule (Plant Section).

Rates given for consumables are in line with the notes on the preceding page without loss or wastage and are based on the following (All plant except those noted are priced on Gas Oil usage):

Petrol ULS	99.00	pence/litre
Petrol Unleaded	99.00	pence/litre
Fuel Oil (taxed for use in licensed vehicles) DERV	100.00	pence/litre
Fuel Oil (lower tax, for use on site) Gas Oil	39.00	pence/litre
Lubrication Oil (15/40)	71.15	pence/litre
Mains electric power	13.23	pence/kWh
Oxygen	£18.98	/10 m³
	0.19	pence/litre
Acetylene	£55.62	/8.4 m³
	0.66	pence/litre
Propane	122.56	pence/kg

Also included here are allowances for Transmission Oils, Hydraulic Oil, Filters, Grease, etc.
N/A indicates that information not available.

Imperial units of measurements are used frequently in this section reflecting their continuing usage in this sector of the industry.

	Hire Rate £	Hire Period	Cost per working hour					DSR
			Plant £	Usage %	Fuel l/hr	Oil etc. % fuel	Total £	
ACCESS PLATFORMS								
Access platform, scissor type, towed; height of platform 5 m	4.47	hour	4.47	-	-	-	4.47	1(8)
Access platform, scissor type, electric; height of platform 5 m	5.58	hour	5.58	-	-	-	5.58	1(1)
Access platform, scissor type, rough terrain, petrol driven; height of platform 8 m	8.56	hour	8.56	50	1.5	2.0	9.32	1(6)
Access platform, telescopic, towed; height of platform 12 m	6.73	hour	6.73	-	-	-	6.73	1(13)
Access platform telescopic, petrol; height of platform 14 m	8.63	hour	8.63	50	2.0	3.0	9.65	1(11)

PLANT COSTS - continued

	Hire Rate £	Hire Period	Cost per working hour Plant £	Usage %	Fuel l/hr	Oil etc. % fuel	Total £	DSR
ACCESS PLATFORMS - continued								
Access platform, telescopic, vehicle mounted (DERV)								
height of platform 12 m	18.42	hour	18.42	50	2.0	3.0	19.45	1(16)
height of platform 15 m	19.27	hour	19.27	50	3.0	3.0	20.82	1(*)
Access platform underbridge type (Simon UB40)	29.97	hour	29.97	-	-	-	29.97	1(13)
AGRICULTURAL TYPE TRACTORS								
Tractor 4.82 t, 4WD (74 kW)	11.63	hour	11.63	75	9.0	2.0	14.32	32(63)
Tractor (68 HP) c/w 1T Hiab lift	13.44	hour	13.44	75	9.0	2.0	16.13	32(62)
Tractor (68 HP) with front end loading bucket (0.33 m³)	12.76	hour	12.76	75	9.0	2.0	15.45	37(51)
Tractor (68 HP) with Hydro seeding equipment	13.59	hour	13.59	75	9.0	2.0	16.28	37(*)
Tractor (68 HP) includes 2 tool compressor and front end loading bucket	13.97	hour	13.97	75	10.5	3.5	17.15	-
Tractor (68HP) with Fencing Auger	12.69	hour	12.69	75	9.0	3.5	15.41	-
ASPHALT/ROAD CONSTRUCTION								
Asphalt pavers								
maximum paving width 3.60 m, 24kW engine	22.68	hour	22.68	80	6.2	4.0	24.69	2(1)
extending up to 4 m, 35 kW engine	40.78	hour	40.78	80	8.2	5.0	43.47	2(1)
maximum paving width 9 m, 80kW engine	73.95	hour	73.95	80	17.0	3.0	79.41	2(3)
Associated equipment								
tar sprayer (100 litre)	1.54	hour	1.54	80	9.0	3.0	4.43	29(6)
self propelled chip spreader	6.00	hour	6.00	80	9.0	3.0	8.89	2(9)
heating iron	1.10	hour	1.10	-	-	-	1.10	-
insulated (20 tonne) tipper (DERV)	19.58	hour	19.58	80	14.0	7.0	31.56	14(18)
3 point roller	7.31	hour	7.31	80	6.0	7.5	9.32	23(4)
Surface planers, cold plane								
planing width 0.5 m, 72 kW	11.25	hour	11.25	80	6.0	5.0	13.22	2(12)
planing width 2.1 m, 390 kW	46.87	hour	46.87	80	12.0	5.0	50.80	2(14)

	Hire Rate £	Hire Period	Cost per working hour					DSR
			Plant £	Usage %	Fuel l/hr	Oil etc. % fuel	Total £	
Surface planers, heat plane								
planing width up to 1.0 m								
re-mixer	22.50	hour	22.50	80	8.0	3.0	25.07	2(11)
planing width up to 4.5 m								
remixer	69.35	hour	69.35	80	13.0	5.0	73.61	2(-)
Concrete pavers								
maximum paving width 6.00 m,								
depth 300 mm, 123 kW	65.60	hour	65.60	80	17.0	3.0	71.06	-
maximum paving width 12.8 m								
depth 500 mm 340 kW	95.59	hour	95.59	80	72.0	3.0	118.73	-
Concrete Slipform paver/trimmer								
maximum paving width 5 m	78.72	hour	78.72	80	24.0	3.0	86.43	-
Slipformed Concrete joint and bar								
inserter	9.37	hour	9.37	80	4.0	3.0	10.66	-
Concrete Slipform finisher	14.15	hour	14.15	80	9.0	3.0	17.04	-
COMPACTION								
Plate compactors								
Vibrating compaction plate 64 kg								
unit weight 360 mm wide								
(petrol)	43.27	week	1.08	85	0.8	3.0	1.77	22(6)
Vibrating compaction plate 140 kg								
unit weight 400 mm wide								
(petrol)	48.50	week	1.21	85	1.2	3.0	2.25	22(8)
Vibrating compaction plate 180 kg								
unit weight 600 mm wide	47.44	week	1.19	85	0.9	3.0	1.50	22(10)
Vibrating compaction plate 345 kg								
unit weight 600 mm wide	51.25	week	1.28	85	1.2	3.0	1.69	22(12)
Vibrating compaction trench plate								
300 mm wide	38.88	week	0.97	85	0.8	3.0	1.24	22(*)
Tamper BT58 62 kg (petrol)	43.27	week	1.08	85	0.8	3.0	1.77	22(1)
CLEANERS/SWEEPERS								
Front Tractor or Truck mounted								
2 m wide (excluding vehicle)	66.81	week	1.67	-	-	-	1.67	14(36)
Towed sweeper diesel engined	99.83	week	2.50	75	4.0	4.0	5.62	-
Glutton type mobile gully sucker	12.92	hour	12.92	75	12.0	3.0	22.19	-

PLANT COSTS - continued

	Hire Rate £	Hire Period	Cost per working hour Plant £	Usage %	Fuel l/hr	Oil etc. % fuel	Total £	DSR
MOBILE COMPRESSORS (SILENCED OR SUPER SILENCED)								
Compressors, single tool								
1.80 m³/min, 13 kW	107.99	week	2.70	85	3.2	4.0	3.80	4(1)
Compressors, two tool								
2.80 m³/min, 23 kW, two tool	122.68	week	3.07	85	4.5	3.0	4.61	4(2)
3.70 m³/min, 28 kW, two tool	144.98	week	3.62	85	5.9	3.0	5.63	4(3)
4.80 m³/min, 34 kW, two tool	167.76	week	4.19	85	6.8	3.0	6.51	4(4)
7.30 m³/min, 53 kW, two tool	218.32	week	5.46	85	10.2	3.0	8.94	4(6)
Compressors, four tool								
11.30 m³/min, 95 kW, four tool	354.25	week	8.86	85	19.0	3.0	15.35	4(8)
15.50 m³/min, 160 kW, four tool	393.62	week	9.84	85	32.0	3.0	20.77	4(9)
16.00 m³/min, 160 kW, four tool	446.10	week	11.15	85	32.0	3.0	22.08	4(9)
18.40 m³/min, 235 kW, four tool	384.24	week	9.61	85	47.0	3.0	25.66	4(10)
22.10 m³/min, 216 kW, four tool	468.59	week	11.71	85	43.0	3.0	26.39	4(11)
25.30 m³/min, 216 kW	562.30	week	14.06	85	43.0	3.0	28.74	4(11)
Reciprocating compressors								
up to 4 cfm for small air tools 240v	40.04	week	1.00	85	3.0	3.0	2.02	4(16)
up to 9 cfm for small air tools 240v	71.70	week	1.79	85	3.0	3.0	2.81	4(16)
up to 13 cfm for small air tools 240v	79.04	week	1.98	85	4.0	3.0	3.35	4(16)
Petrol driven hydraulic power unit c/w breaker (63 kg)	68.32	week	1.71	85	1.5	3.0	2.22	-
Metreater self propelled hydraulic asphalt cutter	106.28	week	2.66	75	9.0	3.0	5.37	-
Vehicle compressor CompAir 4m³/min on Ford Transit chassis	9.46	hour	9.46	85	12.0	3.0	13.56	4(15)
Bristol MF-4 on International Tractor	9.87	hour	9.87	85	12.0	3.0	13.97	4(*)
Bristol MF-4 on International Tractor with loading bucket	11.08	hour	11.08	85	12.0	3.0	15.18	4(*)

	Hire Rate £	Hire Period	Cost per working hour Plant £	Usage %	Fuel l/hr	Oil etc. % fuel	Total £	DSR
COMPRESSOR TOOLS (cw 50 ft/15 m HOSE)								
chipping hammer, 0.7 m³/min	32.81	week	0.82	-	-	-	0.82	30(4)
brick hammer/demolition pick,1.4 m³/ min	12.18	week	0.30	-	-	-	0.30	-
clay spade,1.6 m³/min	15.93	week	0.40	-	-	-	0.40	30(3)
road breaker, 2.4 m³/min	30.28	week	0.76	-	-	-	0.76	30(1)
hand hammer drill, 0.9 m³/min	34.73	week	0.87	-	-	-	0.87	30(*)
light rock drill 28, 1.2 m³/min	41.59	week	1.04	-	-	-	1.04	30(5)
medium rock drill 30, 1.5 m³/min	57.29	week	1.43	-	-	-	1.43	30(6)
heavy rock drill 33, 2.4 m³/min	76.27	week	1.91	-	-	-	1.91	30(7)
scabbler 1 head hand, 0.3 m³/min	69.66	week	1.74	-	-	-	1.74	-
scabbler 3 head hand, 0.7 m³/min	83.58	week	2.09	-	-	-	2.09	-
scabbler 5 head trolley, 3.1 m³/min	184.35	week	4.61	-	-	-	4.61	-
scabbler 7 head trolley, 4.5 m³/min	216.33	week	5.41	-	-	-	5.41	-
needle gun/descaler, 0.3 m³/min	44.40	week	1.11	-	-	-	1.11	30(4)
concrete crack cutter, 0.7 m³/min	98.55	week	2.46	-	-	-	2.46	-
rammer, 1.5 m³/min	16.22	week	0.41	-	-	-	0.41	22(1)
air lance, variable valve, 0.3 m³/min	15.65	week	0.39	-	-	-	0.39	-
impact wrench 12 mm drive, 0.4m³/min	36.58	week	0.91	-	-	-	0.91	-
impact wrench 25 mm drive, 0.4m³/min	61.59	week	1.54	-	-	-	1.54	-
trench sheet driver, 2.4 m³/min	79.79	week	1.99	-	-	-	1.99	-
steel post driver (crash barrier), 2.4 m³/min	79.94	week	2.00	-	-	-	2.00	-
angle grinder 9", 0.8 m³/min	41.95	week	1.05	-	-	-	1.05	-
disc cutter 12", 1.4 m³/min	52.31	week	1.31	-	-	-	1.31	-
poker vibrator P35 (35 mm), 2.4 m³/min	38.50	week	0.96	-	-	-	0.96	6(24)
poker vibrator P54 (54 mm), 2.4 m³/min	48.65	week	1.22	-	-	-	1.22	6(24)
poker vibrator P70 (70 mm), 2.4 m³/min	56.64	week	1.42	-	-	-	1.42	6(24)
rotary drill, 0.3 m³/min	38.50	week	0.96	-	-	-	0.96	-
extra 50 ft/15 m hose	9.84	week	0.25	-	-	-	0.25	-

PLANT COSTS - continued

	Hire Rate £	Hire Period	Cost per working hour					DSR
			Plant £	Usage %	Fuel l/hr	Oil etc. % fuel	Total £	
CONCRETE EQUIPMENT								
Poker vibrators								
air pokers (see compressor tables)								6(24)
50 mm, petrol	45.10	week	1.13	75	2.0	2.0	2.64	6(21)
50 mm, electric	45.10	week	1.13	-	-	-	1.13	6(23)
75 mm, diesel	45.10	week	1.13	75	1.5	2.0	1.58	6(22)
75 mm, electric	45.94	week	1.15	-	-	-	1.15	6(23)
Tampers and screeders								
vibrating tamper and handles (petrol)	64.16	week	1.60	75	2.0	2.0	3.11	6(32)
razor-back screeder (per metre)	40.09	week	1.00	75	1.5	2.0	1.45	-
double beam screeder (7.6 m wide)	89.03	week	2.23	75	1.5	2.0	2.68	6(34)
Rotary power float (petrol),								
687 mm diameter	45.94	week	1.15	75	2.0	2.0	2.66	6(36)
865 mm diameter	55.44	week	1.39	75	2.0	2.0	2.90	6(36)
Reinforcement								
power bar cropper (electric)	72.95	week	1.82	-	-	-	1.82	3(6)
Reinforcement								
power bar bender (electric)	90.25	week	2.26	-	-	-	2.26	3(3)
Compaction								
stud roller	15.48	week	0.39	75	1.5	2.0	0.84	-
MIXERS								
Concrete mixer								
3/2 tip up, 240 v	33.68	week	0.84	-	-	-	0.84	5(1)
4/3, petrol	37.38	week	0.93	75	1.5	2.0	2.07	5(1)
5/3½, diesel	56.63	week	1.42	75	1.2	2.0	1.78	5(2)
8½/6, diesel, skip fed	71.70	week	1.79	75	1.3	2.0	2.18	5(6)
Paddle screed mixer 5/3½ diesel	41.18	week	1.03	75	1.2	2.0	1.39	5(13)
Truck mixer 6 m³	29.12	hour	29.12	75	14.0	3.0	33.34	5(40)
			mixer :	75	5.5	2.0	1.64	
CONCRETE PUMPS								
Air operated 'SEM' pump	6.81	hour	6.81	-	-	-	6.81	6(1)
Truck mounted 45 cm/hr (air pump)	30.53	hour	30.53	75	8.0	3.0	32.94	6(6)
Schwing slurry pump	10.94	hour	10.94	75	1.5	3.0	11.39	6(1)

	Hire Rate £	Hire Period	Cost per working hour					DSR
			Plant £	Usage %	Fuel l/hr	Oil etc. % fuel	Total £	
CONCRETE SKIPS								
Rollover type								
0.5 yd³ (0.38 m³)	24.26	week	0.61	-	-	-	0.61	8(22)
Dual flow type								
0.5 yd³ (0.38 m³)	26.86	week	0.67	-	-	-	0.67	8(22)
0.75 yd³ (0.57 m³)	31.73	week	0.79	-	-	-	0.79	8(23)
1.00 yd³ (0.76 m³)	36.09	week	0.90	-	-	-	0.90	8(24)
CRANES								
7 t mobile (site use)	16.54	hour	16.54	60	3.4	10.0	17.42	7(2)
Truck mounted mobile (wheeled) cranes								
15 t SWL	27.55	hour	27.55	60	5.1	10.0	28.86	7(20)
20 t SWL	33.10	hour	33.10	60	5.5	10.0	34.52	7(20)
25 t SWL	41.82	hour	41.82	60	5.8	10.0	43.31	7(21)
40 t SWL	59.76	hour	59.76	60	6.3	10.0	61.38	7(23)
50 t SWL	79.88	hour	79.88	60	8.5	10.0	82.07	7(24)
Truck mounted mobile (wheeled) cranes								
70 t SWL	112.64	hour	112.64	60	10.0	10.0	115.21	7(25)
90 t SWL	119.25	hour	119.25	60	12.0	10.0	122.34	7(25)
Crawler cranes								
10 t	23.25	hour	23.25	60	3.0	10.0	24.02	7(12)
20 t	25.24	hour	25.24	60	3.5	10.0	26.14	7(12)
30 t	27.67	hour	27.67	60	3.5	10.0	28.57	7(13)
DIESEL GENERATORS								
500 w Petrol 110 or 240 V	46.02	week	1.15	85	0.8	3.0	1.84	11(1)
1.5 kVA Petrol 110 or 240 V	57.48	week	1.44	85	1.1	3.0	2.39	11(1)
2.5 kVA Petrol 110 or 240 V	68.44	week	1.71	85	1.7	3.0	3.18	11(2)
4 kVA diesel dual voltage	81.73	week	2.04	85	1.6	3.0	2.59	11(2)
7 kVA diesel dual voltage	108.35	week	2.71	85	1.8	3.0	3.32	11(3)
7.5 kVA diesel dual voltage	120.55	week	3.01	85	1.8	3.0	3.62	11(3)
7.5 kVA diesel (super silenced)	123.82	week	3.10	85	1.8	3.0	3.71	11(3)
10 kVA diesel dual voltage	151.67	week	3.79	85	2.1	4.0	4.51	11(3)
15 kVA diesel dual voltage	189.61	week	4.74	85	3.0	4.0	5.77	11(4)
25 kVA diesel 3 phase 440 V	238.34	week	5.96	85	4.9	5.0	7.67	11(5)
55 kVA diesel 3 phase 440 V	287.11	week	7.18	85	10.9	5.0	10.97	11(7)
75 kVA diesel 3 phase 440 V	335.19	week	8.38	85	17.0	5.0	14.30	11(8)

PLANT COSTS - continued

	Hire Rate £	Hire Period	Cost per working hour Plant £	Usage %	Fuel l/hr	Oil etc. % fuel	Total £	DSR
DIESEL GENERATORS - continued								
125 kVA diesel 3 phase 440 V	479.94	week	12.00	85	21.8	5.0	19.59	11(11)
200 kVA diesel 3 phase 440 V	652.01	week	16.30	85	40.0	5.0	30.22	11(12)
250 kVA diesel 3 phase 440 V	840.73	week	21.02	85	48.0	5.0	37.73	11(13)
EXCAVATORS								
(Bucket capacity refers to SAE heaped)								
Rope operated, crawler mounted dragline equipment								
R-B Lincoln 22RB HD, 27.6 t, 1.15 m³, 56 kW	34.17	hour	34.17	75	9.0	7.5	37.00	10(4)
R-B Lincoln 51-60, 65.3 t, 2.0 m³,172 kW	79.17	hour	79.17	75	20.0	7.5	85.46	10(7)
R-B Lincoln 61RB HD, 93.2 t, 3.0 m³,194 kW	113.09	hour	113.09	75	32.0	7.5	123.15	10(*)
Mini Excavators (tracked)								
Kubota KX36, 1435 kg (0.04 m³)	341.98	week	8.55	85	4.0	5.0	9.94	10(15)
Kubota KX41, 1565 kg (0.045 m³)	341.98	week	8.55	85	4.0	5.0	9.94	10(15)
Kubota KX71, 2835 kg (0.075 m³)	451.43	week	11.29	85	5.0	5.0	13.03	10(16)
Kubota U45, 4500 kg (0.12 m³)	451.43	week	11.29	85	5.0	5.0	13.03	10(16)
Hitachi ZX16, 1660 kg (0.05 m³)	335.03	week	8.38	85	4.0	5.0	9.77	10(15)
Mini Excavators (wheeled)								
Komatsu PW30, 3000 kg	391.70	week	9.79	85	5.0	5.0	11.53	10(15)
Hydraulic crawler mounted backacter approximate data:								
weight up to SAE bucket kW								
3.5 tonne 0.10 m³ 25	12.32	hour	12.32	80	4.0	7.5	13.66	10(17)
6.5 tonne 0.25 m³ 40	13.06	hour	13.06	80	5.0	7.5	14.74	10(19)
8.5 tonne 0.30 m³ 50	13.74	hour	13.74	80	8.0	7.5	16.42	10(19)
11.5 tonne 0.40 m³ 63	15.09	hour	15.09	80	9.5	7.5	18.28	10(20)
14.5 tonne 0.45 m³ 70	17.41	hour	17.41	80	10.5	7.5	20.93	10(21)
16 tonne 0.55 m³ 80	19.78	hour	19.78	80	12.0	7.5	23.80	10(21)
19 tonne 0.70 m³ 90	24.17	hour	24.17	80	14.0	7.5	28.87	10(22)
21 tonne 1.00 m³ 100	26.54	hour	26.54	80	16.0	7.5	31.91	10(22)
23.5 tonne 1.00 m³ 115	30.63	hour	30.63	80	19.0	7.5	37.00	10(23)

	Hire Rate £	Hire Period	Cost per working hour Plant £	Usage %	Fuel l/hr	Oil etc. % fuel	Total £	DSR
Hydraulic crawler mounted backacter; approximate data:								
weight up to SAE bucket kW								
28.5 tonne 1.20 m³ 130	35.40	hour	35.40	75	24.0	7.5	42.95	10(24)
35.5 tonne 1.20 m³ 160	66.98	hour	66.98	75	31.0	7.5	76.73	10(25)
40 tonne 1.60 m³ 180	67.92	hour	67.92	75	36.0	7.5	79.24	10(26)
50 tonne 2.00 m³ 205	73.67	hour	73.67	75	45.0	7.5	87.82	10(26)
60 tonne 2.50 m³ 235	88.67	hour	88.67	75	52.0	7.5	105.02	10(27)
20 tonne (0.6 m³) Priestman VC15 15 m reach, cable operated	32.95	hour	32.95	80	16.0	7.5	38.32	10(22)
30 tonne (0.8 m³) Priestman VC20 20 m reach, cable operated	57.41	hour	57.41	75	24.0	7.5	64.96	10(24)
Hydraulic wheeled backacter								
weight up to SAE bucket								
3.5 tonne 0.10 m³	11.55	hour	11.55	80	4.0	7.5	12.89	10(17)
6.5 tonne 0.27 m³	13.58	hour	13.58	80	5.0	7.5	15.26	10(19)
8 tonne 0.34 m³	15.16	hour	15.16	80	8.0	7.5	17.84	10(19)
11 tonne 0.43 m³	13.91	hour	13.91	80	9.5	7.5	17.10	10(19)
15 tonne 0.47 m³	19.76	hour	19.76	80	10.5	7.5	23.28	10(21)
19 tonne 1.00 m³	23.90	hour	23.90	80	14.0	7.5	28.60	10(22)
Backhoe loader, wheeled								
JCB 3CX, 0.9 m³	14.47	hour	14.47	75	7.5	6.0	16.80	10(36)
Case 580M, 0.96 m³	14.80	hour	14.80	75	7.5	6.0	17.13	10(37)
Cat 416, 0.76 m³	12.07	hour	12.07	75	7.5	6.0	14.40	10(35)
Cat 428, 1.00 m³	14.31	hour	14.31	75	7.5	6.0	16.64	10(36)
Hymac 180C, 0.95 m³	15.16	hour	15.16	75	7.5	6.0	17.49	10(36)
Percussion breaker attachments								
BRH91 (142 kg) with steels	140.69	week	3.52	-	-	-	3.52	10(30)
BRH125 (310 kg) with steels	198.52	week	4.96	-	-	-	4.96	10(30)
HB501 (1000 kg) with steels	297.01	week	7.43	-	-	-	7.43	10(31)
Krupp HM 1500 (2100 kg) with steels	397.05	week	9.93	-	-	-	9.93	10(34)

PLANT COSTS - continued

	Hire Rate £	Hire Period	Cost per working hour Plant £	Usage %	Fuel l/hr	Oil etc. % fuel	Total £	DSR
EXCAVATORS - continued								
Other hydraulic attachments								
Rock backacter bucket (ripper teeth)	109.43	week	2.74	-	-	-	2.74	-
Ditch bucket (c/w side cutters)	73.47	week	1.84	-	-	-	1.84	-
Trapezoidal bucket	81.29	week	2.03	-	-	-	2.03	-
Clamshell Grab (1.0 m³)	146.94	week	3.67	-	-	-	3.67	-
Rock handling Grab (up to 3 t)	137.55	week	3.44	-	-	-	3.44	-
Scrap handling Grab (up to 1 t)	137.55	week	3.44	-	-	-	3.44	-
Scrap shears	182.89	week	4.57	-	-	-	4.57	-
"Grab John" bucket	81.29	week	2.03	-	-	-	2.03	-
Demolition ball	26.58	week	0.66	-	-	-	0.66	-
Large scoop bucket 1.40 m³	65.65	week	1.64	-	-	-	1.64	-
Single shank ripper	34.40	week	0.86	-	-	-	0.86	-
Flat track shoes (highway use)	18.76	week	0.47	-	-	-	0.47	-
HOISTS, LIFTING AND HANDLING								
Telescopic handler 4WD lifting up to 2.50 tonnes	10.07	hour	10.07	25	75.0	4.0	17.68	15(12)
Electric forklift 2.5 tonnes	4.45	hour	4.45	-	-	-	4.45	
Rough terrain forklift (Manitou M50) 5.0 tonnes	13.32	hour	13.32	75	7.5	4.0	15.60	15(20)
Lorry (8T) with 1T Hiab	19.05	hour	19.05	75	7.5	4.0	21.33	14(10)
1200 kg Tower platform hoist	217.92	week	5.45	75	3.0	3.0	6.35	-
Mast section per 1.5 m (max x 91 m)	8.11	week	0.20	-	-	-	0.20	-
Tirfor winch TU 32 (3T SWL)	38.69	week	0.97	-	-	-	0.97	36(14)
Tirfor winch TU 16	27.86	week	0.70	-	-	-	0.70	36(12)
Chain hoist up to 2000 kg capacity	26.16	week	0.65	-	-	-	0.65	-
Chain hoist up to 3000 kg capacity	27.86	week	0.70	-	-	-	0.70	-
Electric cable hoist 200 kg SWL with gantry	60.00	week	1.50	-	-	-	1.50	36(6)
Engine crane up to 10 CWT (500 kg SWL)	27.16	week	0.68	75	2.0	3.0	1.28	36(1)

	Hire Rate £	Hire Period	Cost per working hour					DSR
			Plant £	Usage %	Fuel l/hr	Oil etc. % fuel	Total £	
PILING PLANT								
(weight = piston weight)								
Piling hammer, double acting (air)								
weight up to 2,500 kg	795.52	week	19.89	-	-	-	19.89	17(5)
weight up to 3,500 kg	1,006.95	week	25.17	-	-	-	25.17	17(6)
weight up to 4,500 kg	1,125.18	week	28.13	-	-	-	28.13	17(7)
Piling hammer, double acting hydraulic								
weight up to 5,000 kg	386.93	week	9.67	-	-	-	9.67	-
weight up to 10,000 kg	417.88	week	10.45	-	-	-	10.45	-
weight up to 50,000 kg	580.39	week	14.51	-	-	-	14.51	-
weight up to 100,000 kg	1,207.21	week	30.18	-	-	-	30.18	-
Piling hammer, single acting diesel								
weight up to 1,500 kg	626.55	week	15.66	75	6.0	5.0	17.50	17(30)
weight up to 2,500 kg	1,007.99	week	25.20	75	11.0	5.0	28.58	17(32)
weight up to 3,500 kg	1,523.86	week	38.10	75	14.0	5.0	42.40	17(34)
weight up to 5,000 kg	1,656.55	week	41.41	75	16.0	5.0	46.32	-
weight up to 5,700 kg	1,760.04	week	44.00	75	18.0	5.0	49.53	17(35)
weight up to 6,200 kg	2,394.02	week	59.85	75	26.0	5.0	67.84	17(36)
weight up to 7,500 kg	2,785.87	week	69.65	75	30.0	5.0	78.86	17(37)
weight up to 8,800 kg	3,057.49	week	76.44	75	35.0	5.0	87.19	17(38)
Hanging leaders for use with diesel hammers								
length 8.5 m	255.33	week	6.38	-	-	-	6.38	17(*)
length 15.5 m	437.04	week	10.93	-	-	-	10.93	17(53)
length 22.5 m	501.74	week	12.54	-	-	-	12.54	17(54)
Flexible hose for compressed air (per 30 ft length)								
38 mm diameter	23.21	week	0.58	-	-	-	0.58	17(24)
50 mm diameter	29.40	week	0.74	-	-	-	0.74	17(25)
Piling extractors (including compressor)								
BSP HD10 unit weight 3,000 kg	1,284.55	week	32.11	75	20.0	4.0	38.19	17(20)
BSP HD15 unit rate 4,580 kg	1,498.84	week	37.47	75	30.0	4.0	46.60	17(21)
Vibratory hammer/extractor, hydraulic								
cent. force - 16 tonne, pulling force -12.5 tonne	1,515.68	week	37.89	75	30.0	7.5	47.32	17(39)
cent. force - 59 tonne, pulling force -36 tonne	3,416.96	week	85.42	75	30.0	7.5	94.85	17(41)
cent. force - 200 tonne, pulling force -80 tonne	10,948.29	week	273.71	75	30.0	7.5	283.14	17(43)

PLANT COSTS - continued

	Hire Rate £	Hire Period	Cost per working hour Plant £	Usage %	Fuel l/hr	Oil etc. % fuel	Total £	DSR
ROLLERS								
Pedestrian rollers								
475 kg trench compactor	2.25	hour	2.25	80	0.9	7.5	2.55	23(11)
600 kg double drum vibratory	3.27	hour	3.27	80	1.0	7.5	3.60	23(11)
928 kg double drum vibratory	3.64	hour	3.64	80	1.6	7.5	4.17	23(12)
1300 kg double drum vibratory	4.01	hour	4.01	80	2.3	7.5	4.78	23(13)
159 kg single drum vibratory	2.16	hour	2.16	80	0.8	7.5	2.43	23(9)
466 kg single drum vibratory	2.25	hour	2.25	80	0.9	7.5	2.56	23(9)
Self propelled rollers								
2½ tonne double vibratory	6.88	hour	6.88	80	3.6	7.5	8.10	23(21)
5 tonne double vibratory	9.29	hour	9.29	80	3.8	7.5	10.58	23(22)
Self propelled rollers								
9 tonne single drum vibratory	14.12	hour	14.12	80	10.0	7.5	17.47	23(23)
Bomag BW55E single drum vibratory	14.95	hour	14.95	80	11.0	7.5	18.64	23(25)
Bomag BW71E single drum vibratory	15.98	hour	15.98	80	16.3	7.5	21.44	23(26)
Bomag BW217D single drum vibratory	23.04	hour	23.04	80	23.8	7.5	31.01	23(*)
Bomag BC601RB Refuse compactor	19.48	hour	19.48	80	42.0	7.5	33.57	23(*)
3 wheel dead weight								
10 tonne	9.84	hour	9.84	80	16.0	7.5	15.21	23(3)
10 tonne	9.09	hour	9.09	80	15.0	7.5	14.12	23(3)
12 tonne	9.84	hour	9.84	80	17.0	7.5	15.54	23(4)
Rubber tyred rollers								
10 tonne	13.00	hour	13.00	80	17.0	7.5	18.70	23(7)
6 tonne	12.57	hour	12.57	80	12.0	7.5	16.59	23(6)
Towed vibratory rollers								
Bomag BW6 (6 tonne)	7.21	hour	7.21	80	7.5	7.5	9.73	23(17)
Bomag BW6S Sheepsfoot	7.59	hour	7.59	80	4.0	7.5	8.93	23(30)
Dual Purpose Rollers								
Benford MBR71B with breaker	3.33	hour	3.33	80	4.0	7.5	4.67	-
Bomag BW138 with breaker	3.78	hour	3.78	80	4.0	7.5	5.12	-

	Hire Rate £	Hire Period	Cost per working week Plant £	Usage %	Fuel l/hr	Oil etc. % fuel	Total £	DSR
SCAFFOLDING AND ACCESSORIES								
Tube alloy per metre	0.73	4 week	0.18	-	-	-	0.18	25(2)
Tube galvanised per metre	0.29	4 week	0.07	-	-	-	0.07	25(1)
Boards - per 4 metre length	0.24	4 week	0.06	-	-	-	0.06	25(19)
Base plates adjustable each	0.44	4 week	0.11	-	-	-	0.11	25(8)
Base plates fixed each	0.23	4 week	0.06	-	-	-	0.06	25(7)
Clips each (all types)	0.23	4 week	0.06	-	-	-	0.06	25(7)
Couplers each (all types)	0.74	4 week	0.18	-	-	-	0.18	25(7)
Spigots each ledgers and braces/m	0.46	4 week	0.12	-	-	-	0.12	25(7)
Putlog 1.5 m blade end each steel	0.65	4 week	0.16	-	-	-	0.16	25(3)
Putlog 1.8 m blade end each steel	0.65	4 week	0.16	-	-	-	0.16	25(5)
Reveal screws each	0.15	4 week	0.04	-	-	-	0.04	25(7)
Castor wheel rubber tyred w/brake	5.54	4 week	1.39	-	-	-	1.39	25(16)
ACCESS STAGING AND TOWERS								
Alloy towers base size 1.8 x 0.80 height to platform 2.6 m	42.98	week	42.98	-	-	-	42.98	-
Extra sections per 2 m rise (max 3)	25.55	week	25.55	-	-	-	25.55	-
Alloy towers base size 1.8 x 1.80 height to platform 2.6 m	48.67	week	48.67	-	-	-	48.67	-
Extra sections per 2 m rise (max 7)	29.85	week	29.85	-	-	-	29.85	-
Alloy stairwell tower per 2 m rise	35.96	week	35.96	-	-	-	35.96	-
Steel towers base size 1.80 x 1.80 height to platform 1.80 m	22.06	week	22.06	-	-	-	22.06	-
Extra sections per 0.60 m rise	5.87	week	5.87	-	-	-	5.87	-
Steel towers base size 3.80 x 3.80 height to platform 1.80 m	38.70	week	38.70	-	-	-	38.70	-
Extra sections per 1.20 m rise	13.17	week	13.17	-	-	-	13.17	-
Castor wheels	5.54	4 week	1.39	-	-	-	1.39	25(14)
Toe boards per 4 metre length	0.69	4 week	0.17	-	-	-	0.17	25(19)
Toe boards hinged per 4 metre length	0.34	4 week	0.09	-	-	-	0.09	-
Handrail (included with towers)	0.23	4 week	0.06	-	-	-	0.06	-
Stair tread complete	0.37	4 week	0.09	-	-	-	0.09	-
Brick guard	3.22	4 week	0.81	-	-	-	0.81	-
Bridging Unit per 1.00 m	4.09	week	4.09	-	-	-	4.09	-

PLANT COSTS - continued

	Hire Rate £	Hire Period	Cost per working hour Plant £	Usage %	Fuel l/hr	Oil etc. % fuel	Total £	DSR
TRACTORS								
Tractor dozers with single equipment								
50 kW	13.45	hour	13.45	80	11.0	7.5	17.14	32(1)
74 kW	18.54	hour	18.54	80	14.5	7.5	23.40	32(2)
104 kW	32.28	hour	32.28	80	21.0	6.0	39.23	32(4)
212 kW	76.49	hour	76.49	80	38.0	6.0	89.06	32(7)
276 kW	88.59	hour	88.59	80	53.0	7.5	106.37	32(9)
67 kW	18.79	hour	18.79	80	14.5	7.5	23.65	32(1)
123 kW	36.52	hour	36.52	80	25.0	7.5	44.91	32(5)
Dozer attachments								
single shank ripper	3.41	hour	3.41	-	-	-	3.41	-
triple shank ripper	6.19	hour	6.19	-	-	-	6.19	-
'U' dozer blade	1.86	hour	1.86	-	-	-	1.86	-
angle dozer blade	1.93	hour	1.93	-	-	-	1.93	-
skeleton blade	1.93	hour	1.93	-	-	-	1.93	-
Tractor loaders								
0.8 m^3	16.56	hour	16.56	75	11.0	7.5	20.02	32(11)
1.5 m^3	25.71	hour	25.71	75	19.0	7.5	31.68	32(15)
2.0 m^3	43.79	hour	43.79	75	25.0	7.5	51.65	32(16)
3.0 m^3	56.07	hour	56.07	75	34.0	7.5	66.76	32(18)
3.0 m^3	56.35	hour	56.35	75	34.0	7.5	67.04	32(18)
Wheeled loaders								
0.55 m^3 33 kW	10.73	hour	10.73	75	8.0	7.5	13.25	32(28)
0.75 m^3 41 kW	12.69	hour	12.69	75	9.5	7.5	15.68	32(28)
1.60 m^3 83 kW	25.88	hour	25.88	75	11.0	7.5	29.34	32(32)
2.70 m^3 115 kW	30.70	hour	30.70	75	19.0	7.5	36.67	32(36)
3.10 m^3 135 kW	45.59	hour	45.59	75	23.0	7.5	52.82	32(37)
3.85 m^3 194 kW	60.69	hour	60.69	80	30.0	7.5	70.75	32(39)
6.00 m^3 310 kW	128.49	hour	128.49	80	53.0	7.5	146.27	32(41)
Wheeled loaders								
10.50 m^3 588 kW	87.57	hour	87.57	80	90.0	7.5	117.76	32(*)
Long reach loader; 1.00 m^3, 54 kW	23.89	hour	23.89	75	11.0	7.5	27.35	-
Wheel dozer, blade capacity 7.5 m^3 338 kW	48.43	hour	48.43	80	68.0	7.5	71.24	32(9)
Loader Attachments								
4 in 1 bucket	2.76	hour	2.76	-	-	-	2.76	-
side dumping bucket	2.44	hour	2.44	-	-	-	2.44	-
rock bucket	2.11	hour	2.11	-	-	-	2.11	-
skeleton bucket	2.19	hour	2.19	-	-	-	2.19	-

	Hire Rate £	Hire Period	Cost per working hour					DSR
			Plant £	Usage %	Fuel l/hr	Oil etc. % fuel	Total £	
single shank ripper (D6)	3.56	hour	3.56	-	-	-	3.56	-
multi shank ripper (D8)	6.48	hour	6.48	-	-	-	6.48	-
rear mounted backacter	9.25	hour	9.25	-	-	-	9.25	-
Hydraulic face shovels								
32 tonne, 1.50 m³	54.40	hour	54.40	80	34.0	7.5	65.80	10(25)
42 tonne, 2.60 m³	73.99	hour	73.99	80	44.0	7.5	88.75	10(26)
62 tonne, 1.30 - 3.40 m³	87.36	hour	87.36	80	64.0	7.5	108.83	10(27)
Skid Steer Loaders								
Bobcat 463 or similar	6.57	hour	6.57	75	3.5	5.0	7.64	32(48)
Bobcat 553 or similar	7.47	hour	7.47	75	4.5	5.0	8.85	32(49)
Case 85 XT or similar	8.59	hour	8.59	75	8.3	5.0	11.14	32(49)
Motor Scraper - single engine								
15.30 m³	72.20	hour	72.20	80	44.0	3.0	86.34	32(69)
33.60 m³	164.49	hour	164.49	80	20.0	3.0	170.92	32(*)
Motor Scraper - twin engine								
16.00 m³	85.21	hour	85.21	80	70.0	3.0	107.71	32(71)
33.60 m³	187.71	hour	187.71	80	125.5	6.0	229.22	32(73)
Motor scraper - elevating								
8.40 m³	52.05	hour	52.05	80	25.0	3.0	60.08	32(74)
13.00 m³	61.75	hour	61.75	80	36.0	3.0	73.32	32(75)
16.80 m³	70.15	hour	70.15	80	46.0	3.0	84.93	32(76)
Tractor (tracked) pipe layer complete with counterweight and boom								
up to 145 kW lifting max 28 t (Komatsu/CAT or similar)	41.46	hour	41.46	80	17.0	4.0	46.98	-
Tractor (tracked) pipe layer complete with counterweight and boom								
up to 149 kW lifting max 41 t (CAT or similar)	48.75	hour	48.75	80	17.0	4.0	54.27	-
up to 224 kW, lifting max 70 t (CAT or similar)	59.56	hour	59.56	80	26.0	4.0	68.00	-
Tractor (tracked) no equipment, for Agro/towing use								
75 kW	12.97	hour	12.97	75	10.0	5.0	16.04	-
90 kW	16.54	hour	16.54	75	12.0	5.0	20.23	-
125 kW	21.90	hour	21.90	75	18.0	5.0	27.43	-

PLANT COSTS - continued

	Hire Rate £	Hire Period	Cost per working hour Plant £	Usage %	Fuel l/hr	Oil etc. % fuel	Total £	DSR
TRACTORS - continued								
Motor Grader (6 wheel)								
up to 93kW blade 3.66m (Cat or similar)	19.29	hour	19.29	80	17.0	7.5	24.99	32(55)
up to 113kW blade 3.66m (Cat or similar)	40.47	hour	40.47	80	23.0	7.5	48.18	32(56)
up to 158kW blade 3.96m	31.42	hour	31.42	80	27.0	7.5	40.48	32(57)
up to 205kW blade 4.88m (Cat or similar)	41.68	hour	41.68	80	32.0	7.5	52.41	33(*)
Trench excavator (Trencher)								
up to 100kW max width 0.60m maxdepth 2.40m (Vermeer or simila	54.47	hour	54.47	75	22.0	3.0	61.10	33(5)
up to 230kW max width 1.07m maxdepth 3.66m (Vermeer or simila	83.14	hour	83.14	75	40.0	3.0	95.19	33(7)
towed type from Agro Tractor maxwidth 0.13m max depth 0.55m	8.82	hour	8.82	75	3.0	3.0	9.72	33(*)
TRANSPORT (TIPPERS AND DUMPERS)								
Tipping lorries								
4 x 2 wheel tippers								
gross weight up to 8.5 tonnes payloadup to 5.5 tonnes - side tipping	11.43	hour	11.43	80	30.0	4.0	21.16	14(15/ 19
gross weight up to 11 tonnes payloadup to 8 tonnes	13.95	hour	13.95	80	25.0	4.0	22.06	14(15)
gross weight up to 14 tonnes payloadup to 10 tonnes	14.51	hour	14.51	80	30.0	4.0	24.24	14(16)
gross weight up to 17 tonnes payloadup to 12 tonnes	15.41	hour	15.41	80	33.0	4.0	26.12	14(16)
6 x 4 wheel tippers								
gross weight up to 24 tonnes payload up to 18 tonnes	24.82	hour	24.82	80	35.0	4.0	36.18	14(17)
gross weight up to 26 tonnes payload up to 20 tonnes	27.57	hour	27.57	80	40.0	4.0	40.55	14(18)
8 x 4 wheel tippers								
gross weight up to 27 tonnes payload up to 20 tonnes	30.22	hour	30.22	80	40.0	4.0	43.20	14(18)
gross weight up to 30 tonnes payload up to 22 tonnes	32.30	hour	32.30	80	42.0	4.0	45.93	14(18)
gross weight up to 33 tonnes payload up to 24 tonnes	35.06	hour	35.06	80	44.0	4.0	49.34	14(*)

	Hire Rate £	Hire Period	Cost per working hour Plant £	Usage %	Fuel l/hr	Oil etc. % fuel	Total £	DSR
Articulated tipper trailers (Highway type excluding tractor)								
twin axle (Bogie) payload up to 20 tonnes (40' length)	19.19	hour	19.19	80	40.0	4.0	32.17	-
triple axle (Bogie) payload up to 25 tonnes (50' length)	27.57	hour	27.57	80	44.0	4.0	41.85	-
Dump truck (on site use only)								
10.5 m³ heaped capacity, 15.5 tonne payload	14.11	hour	14.11	80	15.0	5.0	19.02	9(13)
14 m³ heaped capacity, 18 tonne payload	20.26	hour	20.26	80	31.0	5.0	30.42	9(15)
Dump truck (on site use only)								
19 m³ heaped capacity, 31 tonne payload	36.82	hour	36.82	80	37.5	5.0	49.11	9(18)
24 m³ heaped capacity, 36 tonne payload	55.80	hour	55.80	80	37.9	5.0	68.22	9(18)
57 m³ heaped capacity, 88 tonne payload	73.25	hour	73.25	80	112.0	5.0	109.94	9(*)
78 m³ heaped capacity, 136 tonne payload	86.02	hour	86.02	80	122.0	5.0	125.99	9(*)
Dump truck (articulated type ADT's)								
Volvo 6 x 6, 18.5 tonne or similar, payload11 m³ heaped capacity	21.03	hour	21.03	80	21.4	3.0	27.91	9(23)
Volvo 6 x 6, 22.5 tonne or similar, payload12 m³ heaped capacity	33.30	hour	33.30	80	29.2	3.0	42.68	9(24)
Dump truck (articulated type ADT's)								
23 tonne payload, 13 m³ heaped capacity	33.63	hour	33.63	80	26.9	3.0	42.27	9(24)
25 tonne payload, 13.6 m³ heaped capacity	33.63	hour	33.63	80	29.5	3.0	43.11	9(25)
25 tonne payload, 14.2 m³ heaped capacity	33.63	hour	33.63	80	25.0	3.0	41.66	9(25)
32 tonne payload, 19 m³ heaped capacity	46.76	hour	46.76	80	36.5	3.0	58.49	9(26)

PLANT COSTS - continued

	Hire Rate £	Hire Period	Cost per working hour					DSR
			Plant £	Usage %	Fuel l/hr	Oil etc. % fuel	Total £	
TRANSPORT (TIPPERS AND DUMPERS) - continued								
Small dumpers								
750kg payload, heaped capacity 0.55 m³	2.12	hour	2.12	80	1.5	3.0	2.60	9(3)
1000kg payload, heaped capacity 0.75 m³	2.29	hour	2.29	80	2.0	3.0	2.93	9(3)
2000kg payload, heaped capacity 1.55 m³ (Thwaites Alldrive 4000, 4 x 4)	3.31	hour	3.31	80	3.0	3.0	4.27	9(5)
2500kg payload, heaped capacity 1.80 m³ (Thwaites Alldrive 6000, 4 x 4)	4.25	hour	4.25	80	4.0	3.0	5.54	9(6)
3000kg payload, heaped capacity 2.00 m³ (Thwaites Alldrive 7000, 4 x 4)	5.10	hour	5.10	80	5.0	3.0	6.71	9(7)
4000kg payload, heaped capacity 2.50 m³ (Thwaites Alldrive 4t, 4 x 4)	6.12	hour	6.12	80	7.0	3.0	8.37	9(8)
Power barrow	3.31	hour	3.31	80	0.8	2.0	3.56	9(49)
Transport (trucks and vans)								
Truck, 5t chassis	6.48	hour	6.48	80	7.0	3.0	8.73	
Truck, 8t chassis	7.95	hour	7.95	80	8.0	3.0	10.52	
Truck, 10t chassis	10.04	hour	10.04	80	10.0	3.0	13.25	
Truck, 16t chassis	11.83	hour	11.83	80	15.0	3.0	16.65	
Skip Loader, 17t chassis	17.68	hour	17.68	80	15.0	3.0	22.50	
Small van (petrol)	5.92	hour	5.92	80	8.0	3.0	8.49	14(20)
Pick up (DERV) (1.10 tonnes)	6.51	hour	6.51	80	8.0	3.0	9.08	14(21)
Personnel carrier (12 seat transit)	6.59	hour	6.59	80	10.0	3.0	9.80	14(30)
SWB utility (DERV)	7.95	hour	7.95	80	7.0	3.0	10.20	14(26)
LWB utility (DERV)	10.11	hour	10.11	80	7.0	3.0	12.36	14(27)
Trailers								
Flat trailer (2 axles)	40.53	4 week	0.25	-	-	-	0.25	-
Drop side trailer (2 axles)	48.64	4 week	0.30	-	-	-	0.30	-
Plant Trailer up to 10 ton	77.19	4 week	0.48	-	-	-	0.48	-
Plant Trailer up to 20 ton	110.27	4 week	0.69	-	-	-	0.69	-

	Hire Rate £	Hire Period	Cost per working week Plant £	Usage %	Fuel l/hr	Oil etc. % fuel	Total £	DSR
TRENCH SHEETS ETC. (MINIMUM 50)								
Trench struts (each)								
Nr 0; range 0.30 m - 0.40 m	0.69	week	0.69	-	-	-	0.69	-
Nr 1; range 0.48 m - 0.70 m	0.76	week	0.76	-	-	-	0.76	26(3)
Nr 2; range 0.71 m - 1.12 m	0.80	week	0.80	-	-	-	0.80	26(4)
Nr 3; range 1.02 m - 1.73 m	0.83	week	0.83	-	-	-	0.83	26(5)
Standard props (each)								
Nr 0 3ft 6in - 6ft 0in	0.79	week	0.79	-	-	-	0.79	26(13)
Nr 1 5ft 9in - 10ft 3in	0.82	week	0.82	-	-	-	0.82	26(14)
Nr 2 6ft 6in - 11ft 0in	0.84	week	0.84	-	-	-	0.84	26(15)
Nr 3 8ft 6in - 13ft 0in	0.87	week	0.87	-	-	-	0.87	-
Nr 4 10ft 6in - 16ft 0in	0.96	week	0.96	-	-	-	0.96	-
Nr 5 16ft 0in - 21ft 0in	1.06	week	1.06	-	-	-	1.06	-
Beam heads (each)								
Nr 1	0.48	week	0.48	-	-	-	0.48	-
Nr 2	0.56	week	0.56	-	-	-	0.56	-
Nr 3	0.72	week	0.72	-	-	-	0.72	-
Nr 4	0.82	week	0.82	-	-	-	0.82	-
Split heads (each)								
Nr 1 1ft 9in - 2ft 9in	0.64	week	0.64	-	-	-	0.64	-
Nr 2 3ft 9in - 4ft 6in	0.72	week	0.72	-	-	-	0.72	-
Nr 3 4ft 6in - 6ft 3in	0.90	week	0.90	-	-	-	0.90	-
Trench Sheets								
6'0"	1.14	week	1.14	-	-	-	1.14	-
8'0"	1.31	week	1.31	-	-	-	1.31	-
10'0"	1.62	week	1.62	-	-	-	1.62	-
12'0"	2.02	week	2.02	-	-	-	2.02	-
14'0"	2.26	week	2.26	-	-	-	2.26	-
16'0"	2.60	week	2.60	-	-	-	2.60	-
Trench boxes (approximate cost per m²) - refer to Approximate Estimates								
PUMPING AND DEWATERING								
Self priming centrifugal diesel pumps(inc. 8 m suction and delivery pipes)								
50 mm 550 l/min	52.02	week	1.30	80	0.8	2.0	1.55	20(10)
75 mm Flygt submersible 1050 l/min (4 kW)	52.29	week	1.31	80	-	-	1.31	20(24)
75 mm 750 l/min	68.74	week	1.72	80	1.0	2.0	2.04	20(11)

PLANT COSTS - continued

	Hire Rate £	Hire Period	Cost per working hour Plant £	Usage %	Fuel l/hr	Oil etc. % fuel	Total £	DSR
PUMPING AND DEWATERING - continued								
100 mm 1500 l/min	94.41	week	2.36	80	2.8	2.0	3.25	20(12)
150 mm Sykes UV5 head pump / high capacity	163.90	week	4.10	80	6.8	2.0	6.26	20(13)
76 mm Sykes VCD3 jetting pump	195.01	week	4.88	80	2.0	2.0	5.52	-
100 mm Sykes WP100/60 wellpointing pump	179.84	week	4.50	80	4.5	2.0	5.93	-
150 mm Sykes WP150/60 wellpointing pump	222.11	week	5.55	80	6.8	2.0	7.71	-
Extra hose (per metre)								
50 mm Suction	1.40	week	0.04	-	-	-	0.04	21(2)
50 mm Delivery	1.02	week	0.03	-	-	-	0.03	21(8)
75 mm Suction	1.55	week	0.04	-	-	-	0.04	21(3)
75 mm Delivery	0.94	week	0.02	-	-	-	0.02	21(9)
100 mm Suction	1.93	week	0.05	-	-	-	0.05	21(4)
100 mm Delivery	1.47	week	0.04	-	-	-	0.04	21(10)
150 mm Suction	2.94	week	0.07	-	-	-	0.07	21(5)
150 mm Delivery	2.86	week	0.07	-	-	-	0.07	21(11)
150 mm Duraline (lay flat fine hose)	-	week	(included with		-	-	-	-
150 mm Jetting tube	-	week	Jetting Pump)		-	-	-	-
150 mm PVC header pipe and attachments at 1.50m centres including jet wells for wellpointing to 6.5 m deep (per well- point)	3.72	week	0.09	-	-	-	0.09	-
MISCELLANEOUS								
Road form 10m long (each)								
100mm high	3.91	week	0.10	-	-	-	0.10	27(2)
125mm high	4.16	week	0.10	-	-	-	0.10	-
150mm high	4.19	week	0.10	-	-	-	0.10	27(3)
200mm high	4.47	week	0.11	-	-	-	0.11	27(4)
225mm high	4.73	week	0.12	-	-	-	0.12	27(7)
Flexible road form 3m long (each)								
150mm high	1.80	week	0.05	-	-	-	0.05	-
200mm high	1.88	week	0.05	-	-	-	0.05	-
225mm high	1.96	week	0.05	-	-	-	0.05	-

	Hire Rate £	Hire Period	Cost per working hour Plant £	Usage %	Fuel l/hr	Oil etc. % fuel	Total £	DSR
Road signs on stands (each)								
600mm diameter	8.17	week	0.20	-	-	-	0.20	37(43)
780mm diameter	8.17	week	0.20	-	-	-	0.20	37(44)
900mm diameter	10.22	week	0.26	-	-	-	0.26	37(45)
1200mm diameter	11.92	week	0.30	-	-	-	0.30	-
Road safety								
flashing hazard lamps (Batteries inc.)	1.78	week	0.04	-	-	-	0.04	37(22)
traffic lamps static	1.37	week	0.03	-	-	-	0.03	37(10)
standard cone	1.37	week	0.03	-	-	-	0.03	37(39)
road pins	0.68	week	0.02	-	-	-	0.02	-
PVC-U barrier	4.78	week	0.12	-	-	-	0.12	37(6)
cone converter	1.70	week	0.04	-	-	-	0.04	-
railway sleepers - baulks - each	2.57	week	0.06	-	-	-	0.06	-
traffic light systems								
two way radar, 110V	124.58	week	3.11	-	-	-	3.11	-
two way, mains, timer operated	106.79	week	2.67	-	-	-	2.67	37(9)
Transformers/cables								
2kVA	10.69	week	0.27	-	-	-	0.27	11(16)
4kVA	19.58	week	0.49	-	-	-	0.49	11(17)
6kVA	31.16	week	0.78	-	-	-	0.78	11(18)
Extension cable 240V/110V								
15 m plug and socket	5.45	week	0.14	-	-	-	0.14	-
30 m plug and socket	6.82	week	0.17	-	-	-	0.17	-
15 m 4 mm	9.40	week	0.23	-	-	-	0.23	-
Transformers/cables								
240V junction box	9.82	week	0.25	-	-	-	0.25	-
110V junction box	9.82	week	0.25	-	-	-	0.25	-
Welding & cutting sets								
ARC Mains 140 amps portable	23.83	week	0.60	-	-	-	0.60	35(5)
ARC Petrol 200 amps mobile	78.97	week	1.97	80	1.8	-	3.40	-
ARC Diesel 300 amps trailer mounted	102.15	week	2.55	80	2.3	-	3.27	35(3)
Oxy acetylene set portable (cutting)	51.07	week	1.28	75	650/325	-	3.82	35(1)
Oxy acetylene set portable (welding)	57.71	week	1.44	75	500/325	-	3.98	35(2)
MIG welder 120 amp	31.51	week	0.79	-	-	-	0.79	-

PLANT COSTS - continued

	Hire Rate £	Hire Period	Cost per working hour					DSR
			Plant £	Usage %	Fuel l/hr	Oil etc. % fuel	Total £	
MISCELLANEOUS - continued								
Lighting							0.66	
floodlight 6' stand 1000 watt	32.41	week	0.81	-	-	-	0.81	-
17' floodlight tower (cw 2kVA generator)	62.35	week	1.56	-	-	-	1.56	-
40' floodlight tower (cw 6kVA generator)	136.20	week	3.41	-	-	-	3.41	-
Festoon lighting set (34m)	22.13	week	0.55	-	-	-	0.55	-
Tripod floodlight 1.8m stand 500 watt	21.79	week	0.54	-	-	-	0.54	-
Drain tools								
drain rods (10m set) including equipment	11.48	week	0.29	-	-	-	0.29	37(24)
drain plug (rubber diaphragm) 100	4.09	week	0.10	-	-	-	0.10	37(26)
drain plug (rubber diaphragm) 150	5.11	week	0.13	-	-	-	0.13	37(27)
drain plug (rubber diaphragm) 300	5.45	week	0.14	-	-	-	0.14	37(28)
mains powered drain cleaner	61.29	week	1.53	-	-	-	1.53	-
drain tester - smoke	8.17	week	0.20	-	-	-	0.20	-
drain tester - U gauge	9.19	week	0.23	-	-	-	0.23	-
drain tester - pressure pump	27.24	week	0.68	-	-	-	0.68	-
drain tester - mandrel	5.11	week	0.13	-	-	-	0.13	-
stilson wrench 24"	7.08	week	0.18	-	-	-	0.18	-
stilson wrench 36"	10.89	week	0.27	-	-	-	0.27	-
clay pipe cutter to 9"	20.42	week	0.51	-	-	-	0.51	-
steel pipe cutter to 6"	15.49	week	0.39	-	-	-	0.39	-
drain bag 4" or 6"	6.82	week	0.17	-	-	-	0.17	-
Tar boilers								
lorry mounted <400 litres	15.68	hour	15.68	-	-	-	15.68	29(1)
power operated <1200 litres	42.57	week	1.06	-	-	-	1.06	29(2)
Shot blasting equipment								
150 lb	112.37	week	2.81	-	-	-	2.81	-
350 lb	155.34	week	3.88	-	-	-	3.88	-
Grit blaster ICE160, 72kg	125.30	week	3.13	-	-	-	3.13	-
Stone splitter 25" x 4"	51.58	week	1.29	-	-	-	1.29	-
Block splitter	27.24	week	0.68	-	-	-	0.68	-
Slab splitter	34.05	week	0.85	-	-	-	0.85	-
Slab lifter	7.16	week	0.18	-	-	-	0.18	-

	Hire Rate £	Hire Period	Cost per working hour Plant £	Usage %	Fuel l/hr	Oil etc. % fuel	Total £	DSR
Cartridge guns								
Hilti DX650	29.19	week	0.73	-	-	-	0.73	-
Hilti DX450	25.94	week	0.65	-	-	-	0.65	-
Heating/Drying					Propane			
radiant heater (12,500 BTU)	13.26	week	0.33	80	0.3	-	0.58	37(11)
forced air heater (60,000 BTU)	42.16	week	1.05	80	1.2	-	2.23	37(12)
forced air heater (150,000 BTU)	58.37	week	1.46	80	3.0	-	4.40	37(14)
forced air heater (325,000 BTU)	84.28	week	2.11	80	6.5	-	8.48	37(16)
dehumidifier (40 litres/day)	51.89	week	1.30	-	-	-	1.30	37(34)
fume extractor (6" diameter)	56.75	week	1.42	-	-	-	1.42	-
Drills/Saws/Tools								
diamond drill, 6" max. diameter	84.96	week	2.12	-	-	-	2.12	-
masonry drill & core bits, 10mm	27.24	week	0.68	-	-	-	0.68	-
masonry drill & core bits, 32mm	52.21	week	1.31	-	-	-	1.31	31(1)
angle head drill	16.22	week	0.41	-	-	-	0.41	31(4)
Roto broach, milling to 34mm	58.63	week	1.47	-	-	-	1.47	-
rotary hammer drill, light weight	27.24	week	0.68	-	-	-	0.68	31(15)
rotary hammer drill, heavy weight	24.00	week	0.60	-	-	-	0.60	31(16)
demolition hammer, heavy duty	25.94	week	0.65	-	-	-	0.65	-
grinder, 225mm, electric	19.45	week	0.49	-	-	-	0.49	-
grinder, 114mm, electric	14.27	week	0.36	-	-	-	0.36	31(11)
saw bench, 410mm diameter, petrol	45.39	week	1.13	75	1.0	2.0	1.89	24(10)
saw bench, 610mm diameter, diesel	48.64	week	1.22	75	1.0	2.0	1.52	24(13)
chain saw, up to 410mm	38.92	week	0.97	75	0.4	2.0	1.09	24(2)
225mm electric grinder (angle)	24.65	week	0.62	-	-	-	0.62	31(10)
450mm conc saw 25hp diesel	113.50	week	2.84	75	0.3	2.0	2.93	6(41)
600mm conc saw 40hp diesel	134.74	week	3.37	75	0.3	2.0	3.46	6(42)
Stihl saw disc 305mm petrol	35.67	week	0.89	75	0.2	2.0	1.04	37(18)
Drills/Saws/Tools								
Stihl saw disc 12" air	33.24	week	0.83	-	-	-	0.83	37(18)
8" floor sander (electric)	44.76	week	1.12	-	-	-	1.12	31(13)
Water/Fuel Supply								
towed (trailer) water/diesel bowser, 275 gallon	24.33	week	0.61	-	-	-	0.61	34(8)
bowser, 500 gallon	44.60	week	1.12	-	-	-	1.12	34(9)
lorry-mounted tanker, 1000 gallon (DERV) 8.5t chassis	12.89	hour	12.89	75	9.0	2.0	19.78	34(10)

PLANT COSTS - continued

	Hire Rate £	Hire Period	Cost per working hour Plant £	Usage %	Fuel l/hr	Oil etc. % fuel	Total £	DSR
MISCELLANEOUS - continued								
lorry-mounted tanker,1500 gallon (DERV) 12t chassis	15.24	hour	15.24	75	13.0	2.0	25.19	34(11)
storage tanks, up to 250 gallon	37.30	4 week	0.23	-	-	-	0.23	34(1)
storage tanks, up to 500 gallon	47.02	4 week	0.29	-	-	-	0.29	34(2)
up to 1000 gallon	58.37	4 week	0.36	-	-	-	0.36	34(3)
up to 2000 gallon	64.86	4 week	0.41	-	-	-	0.41	34(4)
Landscaping Items and Tools								
post hole auger 6"/9" hand	9.88	week	0.25	-	-	-	0.25	-
powered auger petrol	64.86	week	1.62	75	2.0	2.0	3.13	-
post driver for 4" post	8.11	week	0.20	-	-	-	0.20	-
post driver for 6" posts	8.93	week	0.22	-	-	-	0.22	-
linemarker 2", 3" or 4" wide	18.15	week	0.45	-	-	-	0.45	
bolt croppers 24" or 36"	11.36	week	0.28	-	-	-	0.28	
tipping and loading skip 6 tonne	38.92	week	0.97	-	-	-	0.97	8(13)
cultivator self propelled	142.69	week	3.57	75	1.0	2.0	3.87	-
Boilers and Sprays								
bitumen boiler 10 gallon (45 litres)	25.94	week	0.65	75	1.0	2.0	1.41	-
bitumen boiler 15 gallon (68 litres)	46.18	week	1.15	75	1.0	2.0	1.91	-
cold tar spray	46.18	week	1.15	-	-	-	1.15	-
Flame gun propane (road burner)	14.27	week	0.36	75	N/a	N/a	0.36	2(15)
Air tool bits-landscaping (see also compressor tools)								
shove holer	9.33	week	0.23	-	-	-	0.23	-
points and chisels	1.54	week	0.04	-	-	-	0.04	-
clayspade	2.11	week	0.05	-	-	-	0.05	30(3)
tarmac cutter	3.64	week	0.09	-	-	-	0.09	-
comb holder	1.62	week	0.04	-	-	-	0.04	-
self-drill anchor holder	2.11	week	0.05	-	-	-	0.05	-
rammer foot	3.09	week	0.08	-	-	-	0.08	-
Floor sander (belt)	27.56	week	0.69	-	-	-	0.69	31(13)
Pinking roller	16.22	week	0.41	-	-	-	0.41	-
Floor grinder/scarifier, electric	61.61	week	1.54	-	-	-	1.54	-
Floor saw, 450mm petrol	81.06	week	2.03	75	1.0	2.0	2.79	6(38)
Floor plane, petrol	72.97	week	1.82	75	1.0	2.0	2.58	-
Floor scaler, petrol	89.17	week	2.23	75	1.0	2.0	2.99	-
Rubbish chute (sections)	5.18	week	0.13	-	-	-	0.13	-
Diesel elevators								
up to 15 m	64.86	week	1.62	75	3.0	2.0	2.52	-
up to 30 m	89.17	week	2.23	75	5.0	2.0	3.72	-

	Hire Rate £	Hire Period	Cost per working week Plant £	Usage %	Fuel l/hr	Oil etc. % fuel	Total £	DSR
FORMWORK EQUIPMENT								
Adjustable props								
Nr 0 1.07 - 1.82	3.01	week	3.01	-	-	-	3.01	
Nr 1 1.75 - 3.12	3.01	week	3.01	-	-	-	3.01	
Nr 2 1.98 - 3.35	3.50	week	3.50	-	-	-	3.50	
Nr 3 2.59 - 3.95	3.50	week	3.50	-	-	-	3.50	
Nr 4 3.20 - 4.88	3.50	week	3.50	-	-	-	3.50	
MISCELLANEOUS								
Decking per SM (Kwikstage) (excluding props)	8.37	week	8.37	-	-	-	8.37	
Column clamps (set)								
Nr 1 0.3 - 0.53	1.68	week	1.68	-	-	-	1.68	27(17)
Nr 2 0.46 - 0.84	2.08	week	2.08	-	-	-	2.08	27(18)
Nr 3 0.66 - 1.222	2.86	week	2.86	-	-	-	2.86	27(19)
Beam Clamps								
300mm arms	1.41	week	1.41	-	-	-	1.41	27(20)
450mm arms	1.54	week	1.54	-	-	-	1.54	27(21)
600mm arms	1.75	week	1.75	-	-	-	1.75	27(22)
Manhole shutters - 1800mm high								
675 mm internal diameter	34.79	week	34.79	-	-	-	34.79	
900 mm internal diameter	41.75	week	41.75	-	-	-	41.75	
1050 mm internal diameter	44.99	week	44.99	-	-	-	44.99	
1200 mm internal diameter	48.70	week	48.70	-	-	-	48.70	
1500 mm internal diameter	55.66	week	55.66	-	-	-	55.66	
1800 mm internal diameter	58.91	week	58.91	-	-	-	58.91	
2100 mm internal diameter	69.57	week	69.57	-	-	-	69.57	
2400 mm internal diameter	76.52	week	76.52	-	-	-	76.52	
2700 mm internal diameter	79.78	week	79.78	-	-	-	79.78	
Portable Accommodation								
(including electrical fittings but excluding additional security items, shutters, grilles etc. per unit delivered and set up on site)								
Offices								
Jack Leg hutment 12'2" x 8'8"	26.02	week	26.02	-	-	-	26.02	16(1)
Jack Leg hutment 16'2" x 8'8"	32.03	week	32.03	-	-	-	32.03	16(1)
Jack Leg hutment 24'2" x 10'2"	40.03	week	40.03	-	-	-	40.03	16(1)
Jack Leg hutment 32'2" x 10'2"	54.05	week	54.05	-	-	-	54.05	16(1)

PLANT COSTS - continued

	Hire Rate £	Hire Period	Cost per working week Plant £	Usage %	Fuel l/hr	Oil etc. % fuel	Total £	DSR
MISCELLANEOUS - continued								
Portable Accommodation - continued								
Jack Leg hutment 48'2" x 12'2"	70.07	week	70.07	-	-	-	70.07	16(1)
Wheeled cabin 12'1" x 7'6"	19.45	week	19.45	-	-	-	19.45	16(6)
Wheeled cabin 22'1" x 7'6"	38.92	week	38.92	-	-	-	38.92	16(7)
Stores								
Jack Leg hutment 12'2" x 8'8"	16.02	week	16.02	-	-	-	16.02	16(1)
Jack Leg hutment 16'2" x 8'8"	18.02	week	18.02	-	-	-	18.02	16(1)
Jack Leg hutment 24'2" x 10'2"	22.02	week	22.02	-	-	-	22.02	16(1)
Jack Leg hutment 32'2" x 10'2"	34.03	week	34.03	-	-	-	34.03	16(1)
Sundries								
Toilet, chemical	25.16	week	25.16	-	-	-	25.16	16(5)
Toilet unit, water flushing	32.03	week	32.03	-	-	-	32.03	16(5)
Toilet unit, VIP, water flushing	52.05	week	52.05	-	-	-	52.05	16(5)
Pollution Decontamination unit 22'1" X 7'6"	72.07	week	72.07	-	-	-	72.07	16(5)
Canteen Unit 32'4" x 8'8"	60.05	week	60.05	-	-	-	60.05	16(2)

PART 4

Unit Costs (Civil Engineering Works)

INTRODUCTORY NOTES

The Unit Costs in this part represent the net cost to the Contractor of executing the work on site; they are not the prices which would be entered in a tender Bill of Quantities.

It must be emphasised that the unit rates are averages calculated on unit outputs for typical site conditions. Costs can vary considerably from contract to contract depending on individual Contractors, site conditions, methods of working and various other factors. Reference should be made to Part 1 for a general discussion on Civil Engineering Estimating.

Guidance prices are included for work normally executed by specialists, with a brief description where necessary of the assumptions upon which the costs have been based. Should the actual circumstances differ, it would be prudent to obtain check prices from the specialists concerned, on the basis of actual / likely quantity of the work, nature of site conditions, geographical location, time constraints, etc.

The method of measurement adopted in this section is the **CESMM3**, *subject to variances where this has been felt to produce more helpful price guidance.*

We have structured this Unit Cost section to cover as many aspects of Civil works as possible.

The Gang hours column shows the output per measured unit in actual time, not the total labour hours; thus for an item involving a gang of 5 men each for 0.3 hours, the total labour hours would naturally be 1.5, whereas the Gang hours shown would be 0.3.

This section is structured to provide the User with adequate background information on how the rates have been calculated, so as to allow them to be readily adjusted to suit other conditions to the example presented:

- *alternative gang structures as well as the effect of varying bonus levels, travelling costs etc.*
- *Other types of plant or else different running costs from the medium usage presumed*
- *Other types of materials or else different discount / waste allowances from the levels presumed*

Reference to Part 3 giving basic costs of labour, materials and plant together with Parts 13 and 14 will assist the reader in making adjustments to the unit costs.

GUIDANCE NOTES

Generally

Adjustments should be made to the rates shown for time, location, local conditions, site constraints and any other factors likely to affect the costs of a specific scheme.

Method of Measurement

Although this part of the book is primarily based on CESMM3, the specific rules have been varied from in cases where it has been felt that an alternative presentation would be of value to the book's main purpose of providing guidance on prices. This is especially so with a number of specialist contractors but also in the cases of work where a more detailed presentation will enable the user to allow for ancillary items.

Materials cost

Materials costs within the rates have been calculated using the 'list prices' contained in Part 3: Resources (pages 40 to 116), with an index appearing on page 39), adjusted to allow for delivery charges (if any) and a 'reasonable' level of discount obtainable by the contractor, this will vary very much depending on the contractor's standing, the potential size of the order and the supplier's eagerness and will vary also between raw traded goods such as timber which will attract a low discount of perhaps 3%, if at all, and manufactured goods where the room for bargaining is much greater and can reach levels of 30% to 40%. High demand for a product at the time of pricing can dramatically reduce the potential discount, as can the world economy in the case of imported goods such as timber and copper. Allowance has also been made for wastage on site (generally 2½% to 5%) dependant upon the risk of damage, the actual level should take account of the nature of the material and its method of storage and distribution about the site.

Labour cost

The composition of the labour and type of plant is generally stated at the beginning of each section, more detailed information on the calculation of the labour rates is given in Part 3: Resources, pages 33 to 37. In addition on pages 37 and 38 is a summary of labour grades and responsibilities extracted from the Working Rule Agreement. Within Parts 4 and 5, each section is prefaces by a detailed build-up of the labour gang assumed for each type of work. This should allow the user to see the cost impact of a different gang as well as different levels of bonus payments, allowances for skills and conditions, travelling allowances etc. The user should be aware that the output constants are based on the gangs shown and would also need to be changed.

Plant cost

A rate build-up of suitable plant is generally stated at the beginning of each section, with more detailed information on alternative machines and their average fuel costs being given in Part 3: Resources, pages 121 to 146. Within Parts 4 and 5, each section is prefaces by a detailed build-up of plant assumed for each type of work. This should allow the user to see the cost impact of using alternative plant as well as different levels of usage (see notes on pages 117 and 118). The user should be aware that the output constants are based on the plant shown and would also need to be changed.

Outputs

The user is directed to Part 13: Outputs (pages 591 to 600) which contains a selection of output constants, in particular a chart of haulage times for various capacities of Tippers on page 593.

CLASS A: GENERAL ITEMS

Item	Gang hours	Labour £	Plant £	Material £	Unit	**Total rate £**
NOTES						
Refer also to the example calculation of Preliminaries in Part 2 and also to Part 7 Oncosts and Profit.						
CONTRACTUAL REQUIREMENTS						
Performance bond						
The cost of the bond will relate to the nature and degree of difficulty and risk inherent in the type of works intended, the perceived ability and determination of the Contractor to complete them, his financial status and whether he has a proven track record in this field with the provider.						
Refer to the discussion of the matter on page 22.						
Insurance of the Works						
Refer to the discussion of the various insurances on pages 21 and 22 as well as the example on page 27.						
Third party insurance						
Refer to the discussion of the various insurances on pages 21 and 22 as well as the example on page 27.						
SPECIFIED REQUIREMENTS						
General						
This section entails the listing of services and facilities over and above the 'Permanent Works' which the Contractor would be instructed to provide in the Contract Documents.						
Accommodation for the Engineer's Staff						
Refer to resources - Plant page 145 and 146 for a list of accommodation types.						
Services for the Engineer's staff						
Transport vehicles						
4 WD utility short wheelbase	-	-	399.84	-	week	**399.84**
4 WD long wheelbase	-	-	486.36	-	week	**486.36**
(for other vehicles, refer to Resources - Plant page 138)						
Telephones						
(allow for connection charges and usage of telephones required for use by the Engineer's Staff)						
Equipment for use by the Engineer's staff						
Allow for equipment specifically required; entailing Office Equipment, Laboratory Equipment and Surveying Equipment.						

CLASS A: GENERAL ITEMS

Item	Gang hours	Labour £	Plant £	Material £	Unit	Total rate £
SPECIFIED REQUIREMENTS – cont'd						
Attendance upon the Engineer's staff						
Driver	40.00	466.62	-	-	week	**466.62**
Chainmen	40.00	466.62	-	-	week	**466.62**
Laboratory assistants	40.00	315.00	-	-	week	**315.00**
Testing of materials						
Testing of the Works						
Temporary Works						
Temporary Works relate to other work which the contractor may need to carry out due to his construction method. These are highly specific to the particular project and envisaged construction method. Examples are: • access roads and hardstandings/ bases for plant and accommodation as well as to the assembly/working area generally • constructing ramps for access to low excavations • steel sheet, diaphragm wall or secant pile cofferdam walling to large excavations subject to strong forces / water penetration • bridges • temporary support works and decking The need for these items should be carefully considered and reference made to the other sections of the book for guidance on what costs should be set against the design assumptions made. Extensive works could well call for the involvement of a contractor's temporary works engineer for realistic advice.						
METHOD-RELATED CHARGES						
Accommodation and buildings						
Offices; establishment and removal; Fixed Charge						
80 m2 mobile unit (10 staff x 8 m2)	-	-	296.30	-	sum	**296.30**
10 m2 section units; two	-	-	148.15	-	sum	**148.15**
Offices; maintaining; Time-Related Charge						
80 m2 mobile unit (10 staff x 8 m2)	-	-	48.90	-	week	**48.90**
10 m2 section units; two	-	-	15.66	-	week	**15.66**
Stores; establishment and removal; Fixed Charge						
22 m2 section unit	-	-	124.44	-	sum	**124.44**
Stores; maintaining; Time-Related Charge						
22 m2 section unit	-	-	24.49	-	sum	**24.49**
Canteens and messrooms; establishment and removal; Fixed Charge						
70 m2 mobile unit (70 men)	-	-	296.30	-	sum	**296.30**
Canteens and messrooms; maintaining; Time-Related Charge						
70 m2 mobile unit (70 men)	-	-	110.34	-	week	**110.34**

CLASS A: GENERAL ITEMS

Item	Gang hours	Labour £	Plant £	Material £	Unit	**Total rate £**
Plant						
General purpose plant not included in Unit Costs:-						
Transport						
wheeled tractor	40.00	450.40	465.20	-	week	**915.60**
trailer	-	-	12.26	-	week	**12.26**
Tractair						
hire charge	-	-	558.80	-	week	**558.80**
driver (skill rate 4)	40.00	526.80	-	-	week	**526.80**
fuel and consumables	-	-	122.85	-	week	**122.85**
Cranes						
22RB Crane	40.00	586.00	1060.44	-	week	**1646.44**
Miscellaneous						
sawbench	-	-	62.82	-	week	**62.82**
concrete vibrator	-	-	90.10	-	week	**90.10**
75 mm 750 l/min pump	-	-	86.34	-	week	**86.34**
compressor	-	-	1066.10	-	week	**1066.10**
plate compactor; 180 kg	-	-	57.97	-	week	**57.97**
towed roller; BW6	-	-	388.74	-	week	**388.74**
Excavators etc.						
hydraulic backacter; 14.5 tonne; driver + banksman	80.00	1068.00	819.21	-	week	**1887.21**
bulldozer; D6; driver	40.00	654.00	1536.78	-	week	**2190.78**
loading shovel; CAT 939; driver	40.00	586.00	791.46	-	week	**1377.46**
Temporary Works						
Supervision and labour						
Supervision for the duration of construction;						
Time-Related Charge						
Agent	40.00	1260.00	-	-	week	**1260.00**
Senior Engineer	40.00	1008.00	-	-	week	**1008.00**
Engineers	40.00	756.00	-	-	week	**756.00**
General Foreman	40.00	924.00	-	-	week	**924.00**
Administration for the duration of construction;						
Time-Related Charge						
Office manager / cost clerk	40.00	682.50	-	-	sum	**682.50**
Timekeeper / Storeman / Checker	40.00	472.50	-	-	sum	**472.50**
Typist / telephonist	40.00	315.00	-	-	sum	**315.00**
Security guard	40.00	472.50	-	-	sum	**472.50**
Quantity Surveyor	40.00	945.00	-	-	sum	**945.00**
Labour teams for the duration of construction;						
Time-Related Charge						
General yard labour (part time); loading and offloading, clearing site rubbish etc.; ganger	40.00	556.08	-	-	sum	**556.08**
General yard labour (part time); loading and offloading, clearing site rubbish etc.; four unskilled operatives	160.00	1866.48	-	-	sum	**1866.48**
Maintenance of Contractor's own plant; Fitter	40.00	735.00	-	-	sum	**735.00**
Maintenance of Contractor's own plant; Fitter's Mate	40.00	624.12	-	-	sum	**624.12**

CLASS B: GROUND INVESTIGATION

Item	Gang hours	Labour £	Plant £	Material £	Unit	Total rate £
RESOURCES - LABOUR						
Trial hole gang						
1 ganger or chargehand (skill rate 4)		12.05				
1 skilled operative (skill rate 4)		12.05				
2 unskilled operatives (general)		22.52				
1 plant operator (skill rate 3)		14.65				
Total Gang Rate / Hour	£	**61.27**				
RESOURCES - PLANT						
Trial holes						
8 tonne wheeled backacter			18.21			
3 tonne dumper			6.80			
3.7 m3/min compressor, 2 tool			6.81			
two 2.4m3/min road breakers			1.51			
extra 50 ft / 15m hose			0.24			
plate compactors; vibrating compaction; plate 180kg/600mm			1.45			
Total Rate / Hour		£	**35.02**			
TRIAL PITS AND TRENCHES						
The following costs assume the use of mechanical plant and excavating and backfilling on the same day						
Trial holes measured by number						
Excavating trial hole; plan size 1.0 x 2.0 m; supports, backfilling						
ne 1.0 m deep	0.24	14.70	8.41	-	nr	**23.11**
1.0 - 2.0 m deep	0.47	28.80	16.47	-	nr	**45.27**
over 2.0 m deep	0.53	32.47	18.58	-	nr	**51.05**
Excavating trial hole in rock or similar; plan size 1.0 x 2.0 m; supports, backfilling						
ne 1.0 m deep	0.28	17.16	9.81	-	nr	**26.96**
1.0 - 2.0 m deep	0.51	31.25	17.87	-	nr	**49.12**
over 2.0 m deep	0.58	35.54	20.31	-	nr	**55.85**
Trial holes measured by depth						
Excavating trial hole; plan size 1.0 x 2.0 m; supports, backfilling						
ne 1.0 m deep	0.24	14.70	8.41	-	m	**23.11**
1.0 - 2.0 m deep	0.53	32.47	18.58	-	m	**51.05**
2.0 - 3.0 m deep	0.59	36.15	20.68	-	m	**56.83**
3.0 - 5.0 m deep	0.65	39.83	22.78	-	m	**62.60**
Excavating trial hole in rock or similar; plan size 1.0 x 2.0 m; supports, backfilling						
ne 1.0 m deep	2.95	180.75	103.33	-	m	**284.08**
1.0 - 2.0 m deep	3.65	223.64	127.85	-	m	**351.48**
2.0 - 3.0 m deep	3.95	242.02	138.35	-	m	**380.37**
3.0 - 5.0 m deep	4.50	275.71	157.60	-	m	**433.32**

CLASS B: GROUND INVESTIGATION

Item	Gang hours	Labour £	Plant £	Material £	Unit	**Total rate £**
Sundries in trial holes						
Removal of obstructions from trial holes irrespective of depth	1.00	61.27	35.02	-	hr	**96.29**
Pumping; maximum depth 4.0 m minimum 750 litres per hour	0.06	1.18	8.09	-	hr	**9.27**
LIGHT CABLE PERCUSSION BOREHOLES						
The following costs are based on using Specialist Contractors and are for guidance only.						
Establishment of standard plant and equipment and removal on completion	-	-	-	-	sum	**517.50**
Number; 150 mm nominal diameter at base	-	-	-	-	nr	**43.98**
Depth; 150 mm nominal diameter of base						
in holes of maximum depth not exceeding 5 m	-	-	-	-	m	**18.85**
in holes of maximum depth 5 - 10 m	-	-	-	-	m	**21.36**
in holes of maximum depth 10 - 20 m	-	-	-	-	m	**27.65**
in holes of maximum depth 20 - 30 m	-	-	-	-	m	**32.67**
Depth backfilled; selected excavated material	-	-	-	-	m	**1.58**
Depth backfilled; imported pulverised fuel ash	-	-	-	-	m	**5.03**
Depth backfilled; imported gravel	-	-	-	-	m	**7.53**
Depth backfilled; bentonite grout	-	-	-	-	m	**12.56**
Chiselling to prove rock or to penetrate obstructions	-	-	-	-	hr	**50.26**
Standing time of rig and crew	-	-	-	-	hr	**43.98**
ROTARY DRILLED BOREHOLES						
The following costs are based on using Specialist Contractors and are for guidance only.						
Establishment of standard plant and equipment and removal on completion	-	-	-	-	sum	**1769.76**
Setting up at each borehole position	-	-	-	-	nr	**153.63**
Depth without core recovery; nominal minimum core diameter 100 mm						
ne 5.0 m deep	-	-	-	-	m	**30.23**
5 - 10 m deep	-	-	-	-	m	**30.23**
10 - 20 m deep	-	-	-	-	m	**30.23**
20 - 30 m deep	-	-	-	-	m	**30.23**
Depth with core recovery; nominal minimum core diameter 75 mm						
ne 5.0 m deep	-	-	-	-	m	**90.66**
5 - 10 m deep	-	-	-	-	m	**96.70**
10 - 20 m deep	-	-	-	-	m	**102.74**
20 - 30 m deep	-	-	-	-	m	**111.05**
Depth cased; semi-rigid plastic core barrel liner	-	-	-	-	m	**8.80**
Depth backfilled, selected excavated material	-	-	-	-	m	**1.58**

CLASS B: GROUND INVESTIGATION

Item	Gang hours	Labour £	Plant £	Material £	Unit	Total rate £
SAMPLES						
From the surface or from trial pits and trenches						
undisturbed soft material; minimum 200 mm cube	-	-	-	-	nr	**11.69**
disturbed soft material; minimum 5 kg	-	-	-	-	nr	**2.60**
rock; minimum 5 kg	-	-	-	-	nr	**19.49**
groundwater; miniumum 1 l	-	-	-	-	nr	**4.54**
From boreholes						
open tube; 100 mm diameter X 450 mm long	-	-	-	-	nr	**11.69**
disturbed; minimum 5 kg	-	-	-	-	nr	**2.60**
groundwater; minimum 1 l	-	-	-	-	nr	**4.55**
stationary piston	-	-	-	-	nr	**4.55**
Swedish foil	-	-	-	-	nr	**19.49**
Delft	-	-	-	-	nr	**19.49**
Bishop sand	-	-	-	-	nr	**19.49**
SITE TESTS AND OBSERVATIONS						
Groundwater level						
Standard penetration						
in light cable percussion boreholes	-	-	-	-	nr	-
Vane in borehole						
Plate bearing						
in pits and trenches; loading table	-	-	-	-	nr	-
in pits and trenches; hydraulic jack and kentledge	-	-	-	-	nr	-
California bearing ratio	-	-	-	-	nr	-
Mackintosh probe						
Hand auger borehole						
mm minimum diameter; 6 m maximum depth	-	-	-	-	nr	-
INSTRUMENTAL OBSERVATIONS						
General						
Pressure head						
standpipe; 75 mm diameter HDPE pipe	-	-	-	-	m	**26.58**
piezometer	-	-	-	-	m	**39.87**
install protective cover	-	-	-	-	nr	**99.66**
readings	-	-	-	-	nr	**19.93**
LABORATORY TESTS						
General						
Classification						
moisture content	-	-	-	-	nr	**3.78**
specific gravity	-	-	-	-	nr	**6.30**
particle size analysis by sieve	-	-	-	-	nr	**25.22**
particle size analysis by pipette or hydrometer	-	-	-	-	nr	**35.32**
Chemical content						
organic matter	-	-	-	-	nr	**27.74**
sulphate	-	-	-	-	nr	**18.92**
pH value	-	-	-	-	nr	**6.09**

CLASS B: GROUND INVESTIGATION

Item	Gang hours	Labour £	Plant £	Material £	Unit	**Total rate £**
contaminants; Interdepartmental Committee for the Redevelopment of Contaminated Land Maxi Comprehensive, Guidance Note 59/83	-	-	-	-	nr	**220.62**
contaminants; Interdepartmental Committee for the Redevelopment of Contaminated Land Midi, Abbreviated, Guidance Note 59/83	-	-	-	-	nr	**157.66**
contaminants; Interdepartmental Committee for the Redevelopment of Contaminated Land Mini, Screening, Guidance Note 59/83	-	-	-	-	nr	**126.13**
contaminants; nitrogen herbicides	-	-	-	-	nr	**113.52**
contaminants; organophosphorus pesticides	-	-	-	-	nr	**94.60**
contaminants; organochlorine pesticides	-	-	-	-	nr	**94.60**
Compaction						
standard	-	-	-	-	nr	**113.51**
heavy	-	-	-	-	nr	**113.51**
vibratory	-	-	-	-	nr	**126.13**
Permeability						
falling head	-	-	-	-	nr	**50.45**
Soil strength						
quick undrained triaxial; set of three 38 mm diameter specimens	-	-	-	-	nr	**25.22**
shear box; peak only; size of shearbox 100 x 100 mm	-	-	-	-	nr	**31.54**
California bearing ratio; typical	-	-	-	-	nr	**37.84**
Rock strength						
point load test; minimum 5 kg sample	-	-	-	-	nr	**63.06**
PROFESSIONAL SERVICES						
General						
Technician	-	-	-	-	h	**23.16**
Technician engineer	-	-	-	-	h	**33.06**
Engineer or geologist						
graduate	-	-	-	-	h	**33.06**
Chartered	-	-	-	-	h	**46.28**
principal or consultant	-	-	-	-	h	**59.51**
Visits to the Site						
technician	-	-	-	-	nr	**23.15**
technician engineer / graduate engineer or geologist	-	-	-	-	nr	**33.06**
chartered engineer	-	-	-	-	nr	**46.28**
principal or consultant	-	-	-	-	nr	**59.51**
Overnight stays in connection with visits to the site						
technician	-	-	-	-	nr	**33.06**
technician engineer / graduate engineer or geologist	-	-	-	-	nr	**33.06**
Chartered engineer / senior geologist	-	-	-	-	nr	**46.28**
principal or consultant	-	-	-	-	nr	**59.51**

CLASS C: GEOTECHNICAL AND OTHER SPECIALIST PROCESSES

Item	Gang hours	Labour £	Plant £	Material £	Unit	Total rate £
NOTE						
The processes referred to in this Section are generally carried out by Specialist Contractors and the costs are therefore an indication of the probable costs based on average site conditions.						
DRILLING FOR GROUT HOLES						
The following unit costs are based on drilling 100 grout holes on a clear site with reasonable access						
Establishment of standard drilling plant and equipment and removal on completion	-	-	-	-	sum	**8300.00**
Standing time	-	-	-	-	hour	**155.00**
Drilling through material other than rock or artificial hard material						
vertically downwards						
depth ne 5 m	-	-	-	-	m	**17.48**
depth 5 - 10 m	-	-	-	-	m	**20.29**
depth 10 - 20 m	-	-	-	-	m	**23.78**
depth 20 - 30 m	-	-	-	-	m	**29.37**
downwards at an angle 0-45 degrees to the vertical						
depth ne 5 m	-	-	-	-	m	**17.48**
depth 5 - 10 m	-	-	-	-	m	**20.29**
depth 10 - 20 m	-	-	-	-	m	**23.78**
depth 20 - 30 m	-	-	-	-	m	**29.37**
horizontally or downwards at an angle less than 45 degrees to the horizontal						
depth ne 5 m	-	-	-	-	m	**17.48**
depth 5 - 10 m	-	-	-	-	m	**20.29**
depth 10 - 20 m	-	-	-	-	m	**23.78**
depth 20 - 30 m	-	-	-	-	m	**29.37**
upwards at an angle 0-45 degrees to the horizontal						
depth ne 5 m	-	-	-	-	m	**29.37**
depth 5 - 10 m	-	-	-	-	m	**33.58**
depth 10 - 20 m	-	-	-	-	m	**36.38**
depth 20 - 30 m	-	-	-	-	m	**39.88**
upwards at an angle less than 45 degrees to the vertical						
depth ne 5 m	-	-	-	-	m	**29.37**
depth 5 - 10 m	-	-	-	-	m	**33.58**
depth 10 - 20 m	-	-	-	-	m	**36.38**
depth 20 - 30 m	-	-	-	-	m	**39.88**
Drilling through rock or artificial hard material						
Vertically downwards						
depth ne 5 m	-	-	-	-	m	**19.43**
depth 5 - 10 m	-	-	-	-	m	**23.84**
depth 10 - 20 m	-	-	-	-	m	**29.17**
depth 20 - 30 m	-	-	-	-	m	**36.19**

CLASS C: GEOTECHNICAL AND OTHER SPECIALIST PROCESSES

Item	Gang hours	Labour £	Plant £	Material £	Unit	**Total rate £**
Downwards at an angle 0-45 degrees to the vertical						
depth ne 5 m	-	-	-	-	m	**19.43**
depth 5 - 10 m	-	-	-	-	m	**23.84**
depth 10 - 20 m	-	-	-	-	m	**29.17**
depth 20 - 30 m	-	-	-	-	m	**36.19**
Horizontally or downwards at an angle less than 45 degrees to the horizontal						
depth ne 5 m	-	-	-	-	m	**19.43**
depth 5 - 10 m	-	-	-	-	m	**23.84**
depth 10 - 20 m	-	-	-	-	m	**29.17**
depth 20 - 30 m	-	-	-	-	m	**36.19**
Upwards at an angle 0-45 degrees to the horizontal						
depth ne 5 m	-	-	-	-	m	**35.33**
depth 5 - 10 m	-	-	-	-	m	**39.73**
depth 10 - 20 m	-	-	-	-	m	**45.03**
depth 20 - 30 m	-	-	-	-	m	**50.33**
Upwards at an angle less than 45 degrees to the horizontal						
depth ne 5 m	-	-	-	-	m	**35.33**
depth 5 - 10 m	-	-	-	-	m	**39.73**
depth 10 - 20 m	-	-	-	-	m	**45.03**
depth 20 - 30 m	-	-	-	-	m	**50.33**
GROUT HOLES						
The following unit costs are based on drilling 100 grout holes on a clear site with reasonable access						
Grout holes						
number of holes	-	-	-	-	nr	**96.64**
multiple water pressure tests	-	-	-	-	nr	**6.92**
GROUT MATERIALS AND INJECTION						
The following unit costs are based on drilling 100 grout holes on a clear site with reasonable access						
Materials						
ordinary portland cement	-	-	-	-	tonne	**115.50**
sulphate resistant cement	-	-	-	-	tonne	**117.60**
cement grout	-	-	-	-	tonne	**142.95**
pulverised fuel ash	-	-	-	-	tonne	**16.75**
sand	-	-	-	-	tonne	**16.15**
pea gravel	-	-	-	-	tonne	**16.15**
bentonite (2:1)	-	-	-	-	tonne	**149.00**
Injection						
Establishment of standard injection plant and removal on completion	-	-	-	-	sum	**8022.00**
Standing time	-	-	-	-	hr	**126.47**
number of injections	-	-	-	-	nr	**59.36**
neat cement grout	-	-	-	-	tonne	**103.16**
cement / P.F.A. grout	-	-	-	-	tonne	**59.93**

CLASS C: GEOTECHNICAL AND OTHER SPECIALIST PROCESSES

Item	Gang hours	Labour £	Plant £	Material £	Unit	**Total rate £**
DIAPHRAGM WALLS						
Notes						
Diaphragm walls are the construction of vertical walls, cast in place in a trench excavation. They can be formed in reinforced concrete to provide structural elements for temporary or permanent retaining walls. Wall thicknesses of 500 mm to 1.50 m and up to 40 m deep may be constructed. Special equipment such as the Hydrofraise can construct walls up to 100 m deep. Restricted urban sites will significantly increase the costs.						
The following costs are based on constructing a diaphragm wall with an excavated volume of 4000 m² using standard equipment Typical progress would be up to 500 m per week.						
Establishment of standard plant and equipment including bentonite storage tanks and removal on completion	-	-	-	-	sum	**110000.00**
Standing time	-	-	-	-	hr	**875.00**
Excavation, disposal of soil and placing of concrete	-	-	-	-	m³	**400.00**
Provide and place reinforcement cages	-	-	-	-	tonne	**650.00**
Excavate/chisel in hard material/rock	-	-	-	-	hr	**950.00**
Waterproofed joints	-	-	-	-	m	**5.00**
Guide walls						
guide walls (twin)	-	-	-	-	m	**305.00**

CLASS C: GEOTECHNICAL AND OTHER SPECIALIST PROCESSES

Item	Gang hours	Labour £	Plant £	Material £	Unit	**Total rate £**
GROUND ANCHORAGES						
Notes Ground anchorages consist of the installation of a cable or solid bar tendon fixed in the ground by grouting and tensioned to exceed the working load to be carried. Ground anchors may be of a permanent or temporary nature and can be used in conjunction with diaphragm walls or sheet piling to eliminate the use of strutting etc..						
The following costs are based on the installation of 50 nr ground anchors.						
Establishment of standard plant and equipment and removal on completion	-	-	-	-	sum	**10000.00**
Standing time	-	-	-	-	hr	**150.00**
Ground anchorages; temporary or permanent						
15.0 m maximum depth; in rock, alluvial or clay; 0 - 50 t load	-	-	-	-	nr	**70.80**
15.0 m maximum depth; in rock or alluvial; 50 – 90 t load	-	-	-	-	nr	**84.68**
15.0 m maximum depth; in rock only; 90 - 150 t load	-	-	-	-	nr	**99.97**
Temporary tendons						
in rock, alluvial or clay; 0 - 50 t load	-	-	-	-	nr	**54.46**
in rock or alluvial; 50 - 90 t load	-	-	-	-	nr	**83.08**
in rock only; 90 - 150 t load	-	-	-	-	nr	**111.67**
Permanent tendons						
in rock, alluvial or clay; 0 - 50 t load	-	-	-	-	nr	**81.70**
in rock or alluvial; 50 - 90 t load	-	-	-	-	nr	**111.68**
in rock only; 90 - 150 t load	-	-	-	-	nr	**138.93**

CLASS C: GEOTECHNICAL AND OTHER SPECIALIST PROCESSES

Item	Gang hours	Labour £	Plant £	Material £	Unit	Total rate £
SAND, BAND AND WICK DRAINS						
Notes Vertical drains are a technique by which the rate of consolidation of fine grained soils can be considerably increased by the installation of vertical drainage paths commonly in the form of columns formed by a high-quality plastic material encased in a filter sleeve. Columns of sand are rarely used in this country these days. Band drains are generally 100 mm wide and 3 - 5 mm thick. water is extracted through the drain from the soft soils when the surface is surcharged. The rate of consolidation is dependent on the drain spacing and the height of surcharge. Drains are usually quickly installed up to depths of 25 m by special lances either pulled or vibrated into the ground. typical drain spacing would be one per 1 - 2 m with the rate of installation varying between 1,500 to 6,000 m per day depending on ground conditions and depths.						
The following costs are based on the installation of 2,000 nr vertical band drains to a depth of 12 m						
Establishment of standard plant and equipment and removal on completion	-	-	-	-	sum	**6000.00**
Standing time	-	-	-	-	hr	**150.00**
Set up installation equipment at each drain position	-	-	-	-	nr	**2.58**
Install drains maximum depth 10 - 15 m	-	-	-	-	m	**0.67**
Additional costs in pre-drilling through hard upper strata at each drain position :						
establishment of standard drilling plant and equipment and removal on completion	-	-	-	-	sum	**3500.00**
set up at each drain position	-	-	-	-	nr	**2.63**
drilling for vertical band drains up to a maximum depth of 3 m	-	-	-	-	m	**2.53**

CLASS C: GEOTECHNICAL AND OTHER SPECIALIST PROCESSES

Item	Gang hours	Labour £	Plant £	Material £	Unit	**Total rate £**
GROUND CONSOLIDATION – VIBRO-REPLACEMENT						
Notes Vibroreplacement is a method of considerably increasing the ground bearing pressure and consists of a specifically designed powerful poker vibrator penetrating vertically into the ground hyderaulically. Air and water jets may be used to assist penetration. In cohesive soils a hole is formed into which granular backfill is placed and compacted by the poker, forming a dense stone column. In natural sands and gravels the existing loose deposits may be compacted without the addition of extra material other than making up levels after settlement resulting from compaction. There are many considerations regarding the soil types to be treaten, whether cohesive or non-cohesive, made-up or natural ground, which influence the choice of wet or dry processes, pure densification or stone column techniques with added granular backfill. It is therefore possible only to give indicative costs; a Specialist Contractor should be consulted for more accurate costs for a particular site. Testing of conditions after consolidation can be static or dynamic penetration tests, plate bearing tests or zone bearing tests. A frequently adopted specification calls for plate bearing tests at 1 per 1000 stone columns. Allowable bearing pressures of up to 400 kN/m^2 by the installation of stone columns in made or natural ground.						
The following costs are typical rates for this sort of work						
Establishment of standard plant and equipment and removal on completion	-	-	-	-	sum	**4200.00**
Standing time	-	-	-	-	hr	**262.50**
Construct stone columns to a depth ne 4 m						
dry formed	-	-	-	-	m	**18.36**
water jet formed	-	-	-	-	m	**26.81**
Plate bearing test						
ne 11 t or 2 hour duration	-	-	-	-	nr	**437.45**
Zone loading test to specification	-	-	-	-	nr	**7950.00**

CLASS C: GEOTECHNICAL AND OTHER SPECIALIST PROCESSES

Item	Gang hours	Labour £	Plant £	Material £	Unit	**Total rate £**
GROUND CONSOLIDATION – DYNAMIC COMPACTION						
Notes Ground consolidation by dynamic compaction is a technique which involves the dropping of a steel or concrete pounder several times in each location on a grid pattern that covers the whole site. For ground compaction up to 10 m, a 15 t pounder from a free fall of 20 m would be typical. Several passes over the site are normally required to achieve full compaction. The process is recommended for naturally cohesive soils and is usually uneconomic for areas of less than 4,000 m² for sites with granular or mixed granular cohesive soils and 6,000 m for a site with weak cohesive soils. The main considerations to be taken into account when using this method of consolidation are: • sufficient area to be viable • proximity and condition of adjacent property and services • need for blanket layer of granular material for a working surface and as backfill to offset induced settlement • water table level The final bearing capacity and settlement criteria that can be achieved depends on the nature of the material being compacted. Allowable bearing capacity may be increased by up to twice the pre-treated value for the same settlement. Control testing can be by crater volume measurements, site levelling between passes, penetration tests or plate loading tests.						
The following range of costs are average based on treating an area of about 10,000 m² for a 5 - 6 m compaction depth. Typical progress would be 1,500 - 2,000 m² per week						
Establishment of standard plant and equipment and removal on completion.	-	-	-	-	sum	**35000.00**
Ground treatment	-	-	-	-	m²	**8.12**
Laying free-draining granular blanket layer as both working surface and backfill material (300 mm thickness required of filter material)	-	-	-	-	m²	**12.19**
Control testing including levelling, piezometers and penetrameter testing	-	-	-	-	m²	**3.33**
Kentledge load test	-	-	-	-	nr	**10652.09**

CLASS C: GEOTECHNICAL AND OTHER SPECIALIST PROCESSES

Item	Gang hours	Labour £	Plant £	Material £	Unit	Total rate £
CONSOLIDATION OF ABANDONED MINE WORKINGS						
The following costs are based on using Specialist Contractors and are for guidance only						
Transport plant, labour and all equipment to and from site (max. 100 miles)	-	-	-	-	sum	**226.38**
Drilling bore holes						
move to each seperate bore position; erect equipment; dismantle prior to next move	-	-	-	-	nr	**33.87**
drill 50 mm diameter bore holes	-	-	-	-	m	**10.04**
drill 100 mm diameter bore holes for pea gravel injection	-	-	-	-	m	**16.65**
extra for casing, when required	-	-	-	-	m	**13.71**
standing time for drilling rig and crew	-	-	-	-	hr	**88.19**
Grouting drilled bore holes						
connecting grout lines	-	-	-	-	m	**17.51**
injection of grout	-	-	-	-	tonne	**68.59**
add for pea gravel injection	-	-	-	-	tonne	**82.31**
standing time for grouting rig and crew	-	-	-	-	hr	**84.50**
Provide materials for grouting						
ordinary portland cement	-	-	-	-	tonne	**115.50**
sulphate resistant cement	-	-	-	-	tonne	**117.60**
pulverised fuel ash (PFA)	-	-	-	-	tonne	**16.75**
sand	-	-	-	-	tonne	**16.15**
pea gravel	-	-	-	-	tonne	**16.16**
bentonite (2:1)	-	-	-	-	tonne	**149.00**
cement grout	-	-	-	-	tonne	**142.95**
Capping to old shafts or similar; reinforced concrete grade C20P, 20 mm aggregate; thickness						
ne 150 mm	-	-	-	-	m^3	**168.00**
150-300 mm	-	-	-	-	m^3	**162.75**
300-500 mm	-	-	-	-	m^3	**157.50**
over 500 mm	-	-	-	-	m^3	**126.00**
Mild steel bars BS4449; supplied in bent and cut lengths						
6 mm nominal size	-	-	-	-	tonne	**871.20**
8 mm nominal size	-	-	-	-	tonne	**733.92**
10 mm nominal size	-	-	-	-	tonne	**702.24**
12 mm nominal size	-	-	-	-	tonne	**686.40**
16 mm nominal size	-	-	-	-	tonne	**654.72**
20 mm nominal size	-	-	-	-	tonne	**554.40**
High yield steel bars BS4449 or 4461; supplied in bent and cut lengths						
6 mm nominal size	-	-	-	-	tonne	**855.36**
8 mm nominal size	-	-	-	-	tonne	**720.58**
10 mm nominal size	-	-	-	-	tonne	**689.47**
12 mm nominal size	-	-	-	-	tonne	**675.40**
16 mm nominal size	-	-	-	-	tonne	**644.16**
20 mm nominal size	-	-	-	-	tonne	**543.84**

CLASS D: DEMOLITION AND SITE CLEARANCE

Item	Gang hours	Labour £	Plant £	Material £	Unit	**Total rate £**
GENERAL CLEARANCE						
The rates for site clearance include for all sundry items, small trees (i.e. under 500 mm diameter), hedges etc., but exclude items that are measured separately; examples of which are given in this section						
Clear site vegetation						
generally	-	-	-	-	ha	**895.00**
wooded areas	-	-	-	-	ha	**2755.00**
areas below tidal level	-	-	-	-	ha	**3300.00**
TREES						
The following rates are based on removing a minimum of 100 trees, generally in a group. Cutting down a single tree on a site would be many times these costs						
Remove trees						
girth 500 mm - 1 m	-	-	-	-	nr	**33.50**
girth 1 - 2 m	-	-	-	-	nr	**55.00**
girth 2 - 3 m	-	-	-	-	nr	**198.50**
girth 3 - 5 m	-	-	-	-	nr	**827.00**
girth 7 m	-	-	-	-	nr	**995.00**
STUMPS						
Clearance of stumps						
diameter 150 - 500 mm	-	-	-	-	nr	**24.75**
diameter 500 mm - 1 m	-	-	-	-	nr	**45.50**
diameter 2 m	-	-	-	-	nr	**100.20**
Clearance of stumps; backfilling holes with topsoil from site						
diameter 150 - 500 mm	-	-	-	-	nr	**28.75**
diameter 500 mm - 1 m	-	-	-	-	nr	**72.80**
diameter 2 m	-	-	-	-	nr	**230.20**
Clearance of stumps; backfilling holes with imported hardcore						
diameter 150 - 500 mm	-	-	-	-	nr	**46.80**
diameter 500 mm - 1 m	-	-	-	-	nr	**175.50**
diameter 2 m	-	-	-	-	nr	**620.40**
BUILDINGS						
The following rates are based on assuming a non urban location where structure does not take up a significant area of the site						
Demolish building to ground level and dispose off site						
brickwork with timber floor and roof	-	-	-	-	m^3	**5.35**
brickwork with concrete floor and roof	-	-	-	-	m^3	**8.80**
masonry with timber floor and roof	-	-	-	-	m^3	**6.90**
reinforced concrete frame with brick infill	-	-	-	-	m^3	**9.20**

CLASS D: DEMOLITION AND SITE CLEARANCE

Item	Gang hours	Labour £	Plant £	Material £	Unit	**Total rate £**
steel frame with brick cladding	-	-	-	-	m³	**5.00**
steel frame with sheet cladding	-	-	-	-	m³	**4.74**
timber	-	-	-	-	m³	**4.25**
Demolish buildings with asbestos linings to ground level and dispose off site						
brick with concrete floor and roof	-	-	-	-	m³	**20.47**
reinforced concrete frame with brick infill	-	-	-	-	m³	**21.30**
steel frame with brick cladding	-	-	-	-	m³	**11.65**
steel frame with sheet cladding	-	-	-	-	m³	**11.25**
OTHER STRUCTURES						
The following rates are based on assuming a non urban location where structure does not take up a significant area of the site						
Demolish walls to ground level and dispose off site						
reinforced concrete wall	-	-	-	-	m³	**121.55**
brick or masonry wall	-	-	-	-	m³	**54.70**
brick or masonry retaining wall	-	-	-	-	m³	**66.85**
PIPELINES						
Removal of redundant services						
electric cable; LV	-	-	-	-	m	**2.15**
75 mm diameter water main; low pressure	-	-	-	-	m	**2.86**
150 mm diameter gas main; low pressure	-	-	-	-	m	**4.16**
earthenware ducts; one way	-	-	-	-	m	**2.87**
earthenware ducts; two way	-	-	-	-	m	**3.58**
100 or 150 mm diameter sewer or drain	-	-	-	-	m	**4.35**
225 mm diameter sewer or drain	-	-	-	-	m	**5.05**
300 mm diameter sewer or drain	-	-	-	-	m	**8.34**
450 mm diameter sewer or drain	-	-	-	-	m	**12.93**
750 mm diameter sewer or drain	-	-	-	-	m	**23.65**
Extra for breaking up concrete surround	-	-	-	-	m	**5.73**
Grouting redundant drains or sewers						
100 mm diameter	-	-	-	-	m	**5.05**
150 mm diameter	-	-	-	-	m	**7.55**
225 mm diameter	-	-	-	-	m	**13.27**
manhole chambers	-	-	-	-	m³	**107.60**

CLASS E: EXCAVATION

Item	Gang hours	Labour £	Plant £	Material £	Unit	**Total rate £**
NOTES						
Ground conditions The following unit costs for 'excavation in material other than topsoil, rock or artificial hard material' are based on excavation in firm sand and gravel soils. For alternative types of soil, multiply the following rates by:						
Scrapers						
Stiff clay 1.5						
Chalk 2.5						
Soft rock 3.5						
Broken rock 3.7						
Tractor dozers and loaders						
Stiff clay 2.0						
Chalk 3.0						
Soft rock 2.5						
Broken rock 2.5						
Backacter (minimum bucket size 0.5 m^3)						
Stiff clay 1.7						
Chalk 2.0						
Soft rock 2.0						
Broken rock 1.7						
Basis of disposal rates All pricing and estimating for disposal is based on the volume of solid material excavated and rates for disposal should be adjusted by the following factors for bulkage. Multiply the rates by:						
Sand bulkage 1.10						
Gravel bulkage 1.20						
Compacted soil bulkage 1.30						
Compacted sub-base, suitable fill etc. bulkage 1.30						
Stiff clay bulkage 1.20						
See also Part 14: Tables and Memoranda						
Basis of rates generally To provide an overall cost comparison, rates, prices and outputs have been based on a medium sized Civil Engineering project of £ 6 - 8 million, location neither in city centre nor excessively remote, with no abnormal ground conditions that would affect the stated output and consistency of work produced. The rates are optimum rates and assume continuous output with no delays caused by other operations or works.						

CLASS E: EXCAVATION

Item	Gang hours	Labour £	Plant £	Material £	Unit	**Total rate £**
RESOURCES - LABOUR						
Excavation for cuttings gang						
1 plant operator (skill rate 1) - 33% of time		17.70				
1 plant operator (skill rate 2) - 66% of time		10.79				
1 banksman (skill rate 4)		12.05				
Total Gang Rate / Hour	£	**27.68**				
Excavation for foundations gang						
1 plant operator (skill rate 3) - 33% of time		4.83				
1 plant operator (skill rate 2) - 66% of time		10.79				
1 banksman (skill rate 4)		12.05				
Total Gang Rate / Hour	£	**27.68**				
General excavation gang						
1 plant operator (skill rate 3)		14.65				
1 plant operator (skill rate 3) - 25% of time		3.66				
1 banksman (skill rate 4)		12.05				
Total Gang Rate / Hour	£	**30.36**				
Filling gang						
1 plant operator (skill rate 4)		13.17				
2 unskilled operatives (general)		22.52				
Total Gang Rate / Hour	£	**35.69**				
Treatment of filled surfaces gang						
1 plant operator (skill rate 2)		16.35				
Total Gang Rate / Hour	£	**16.35**				
Geotextiles (light sheets) gang						
1 ganger/chargehand (skill rate 4) - 20% of time		2.56				
2 unskilled operatives (general)		22.52				
Total Gang Rate / Hour	£	**25.08**				
Geotextiles (medium sheets) gang						
1 ganger/chargehand (skill rate 4) - 20% of time		2.56				
3 unskilled operatives (general)		33.78				
Total Gang Rate / Hour	£	**36.34**				
Geotextiles (heavy sheets) gang						
1 ganger/chargehand (skill rate 4) - 20% of time		2.56				
2 unskilled operatives (general)		22.52				
1 plant operator (skill rate 3)		15.45				
Total Gang Rate / Hour	£	**39.73**				
Horticultural works gang						
1 skilled operative (skill rate 4)		12.05				
1 unskilled operative (general)		11.26				
Total Gang Rate / Hour	£	**23.31**				
RESOURCES - PLANT						
Excavation for foundations						
Hydraulic Backacter - 21 tonne (33% of time)			10.82			
Hydraulic Backacter - 14.5 tonne (33% of time)			6.83			
1000 kg hydraulic breaker (33% of time)			2.47			
Tractor loader - 0.80 m³ (33% of time)			6.59			
Total Rate / Hour		£	**26.71**			
Filling						
1.5 m³ tractor loader			31.25			
6t vibratory roller			9.72			
Pedestrian Roller, Bomag BW35			2.56			
Total Rate / Hour		£	**43.53**			

CLASS E: EXCAVATION

Item	Gang hours	Labour £	Plant £	Material £	Unit	**Total rate £**
RESOURCES - PLANT – cont'd						
Treatment of filled surfaces						
1.5 m³ tractor loader			31.25			
Pedestrian Roller, Bomag BW35			2.56			
Total Rate / Hour		£	**33.81**			
Geotextiles (heavy sheets)						
1.5 m³ tractor loader			31.25			
Total Rate / Hour		£	**31.25**			
EXCAVATION BY DREDGING						
Notes						
Dredging can be carried out by land based machines or by floating plant. The cost of the former can be assessed by reference to the excavation costs of the various types of plant given below, suitably adjusted to take account of the type of material to be excavated, depth of water and method of disposal. The cost of the latter is governed by many factors which affect the rates and leads to wide variations.						
For reliable estimates it is advisable to seek the advice of a Specialist Contractor. The prices included here are for some typical dredging situations and are shown for a cost comparison and EXCLUDE initial mobilisation charges which can range widely between £2,000 and £8,000 depending on plant, travelling distance etc.. Some clients schedule operations for when the plant is passing so as to avoid the large travelling cost.						
Of the factors affecting the cost of floating plant, the matter of working hours is by far one of the most important. The customary practice in the dredging industry is to work 24 hours per day, 7 days per week. Local constraints, particularly noise restrictions, will have a significant impact. Other major factors affecting the cost of floating plant are :						
• type of material to be dredged						
• depth of water						
• depth of cut						
• tidal range						
• disposal location						
• size and type of plant required						
• current location of plant						
• method of disposal of dredged material						

CLASS E: EXCAVATION

Item	Gang hours	Labour £	Plant £	Material £	Unit	**Total rate £**
In tidal locations, creating new channels on approaches to quays and similar locations or within dock systems						
Backhoe dredger loading material onto two hopper barges with bottom dumping facility maximum water depth 15 m, distance to disposal site less than 20 miles						
approximate daily cost of backhoe dredger and two hopper barges £9,500; average production 100 m³/hr, 1,500 m³/day; locate, load, deposit and relocate	-	-	-	-	m³	**7.69**
For general bed lowering or in maintaining shipping channels in rivers, estuaries or deltas						
Trailer suction hopper dredger excavating non-cohesive sands, grits or silts; hopper capacity 2000m³, capable of dredging to depths of 25 m with ability either to dump at disposal site or pump ashore for reclamation. Costs totally dependent on nature of material excavated and method of disposal. Approximate cost per tonne £2.50, which should be adjusted by the relative density of the material for conversion to m³						
Locate, load, deposit and relocate	-	-	-	-	tonne	**1.93**
Harbour bed control						
Maintenance dredging of this nature would in most cases be carried out by trailing suction hopper dredger as detailed above, at similar rates. The majority of the present generation of trailers have the ability to dump at sea or discharge ashore.						
Floating craft using diesel driven suction method with a 750 mm diameter flexible pipe for a maximum distance of up to 5000 m from point of suction to point of discharge using a booster (standing alone, the cutter suction craft should be able to pump up to 2000 m). Maximum height of lift 10 m.						
Average pumping capacity of silt/sand type materials containing maximum 30% volume of solids would be about 8,000 m³/day based on 24 hour working, Daily cost (hire basis) in the region of £20,000 including all floating equipmnet and discharge pipes, maintenance and all labour and plant to service but excluding mobilisation/initial set-up and demobilistaion costs (minimum £10,000).	-	-	-	-	m³	**2.98**

CLASS E: EXCAVATION

Item	Gang hours	Labour £	Plant £	Material £	Unit	**Total rate £**
EXCAVATION BY DREDGING – cont'd						
For use in lakes, canals, rivers, industrial lagoons and from silted locations in dock systems						
Floating craft using diesel driven suction method with a 200 mm diameter flexible pipe maximum distance from point of suction to point of discharge 1500 m. Maximum dredge depth 5 m.						
Average pumping output 30 m³ per hour and approximate average costs (excluding mobilisation and demobilisation costs ranging between £3,000and £6,000 apiece)	-	-	-	-	m³	**5.53**
EXCAVATION FOR CUTTINGS						
The following unit costs are based on backacter and tractor loader machines						
Excavate topsoil						
maximum depth ne 0.25 m	0.03	0.86	0.82	-	m³	**1.67**
Excavate material other than topsoil, rock or artificial hard material						
ne 0.25 m maximum depth	0.03	0.86	0.82	-	m³	**1.67**
0.25 - 0.5 m maximum depth	0.03	0.86	0.82	-	m³	**1.67**
0.5 - 1.0 m maximum depth	0.04	1.14	1.06	-	m³	**2.20**
1.0 - 2.0 m maximum depth	0.06	1.71	1.60	-	m³	**3.32**
2.0 - 5.0 m maximum depth	0.10	2.86	2.66	-	m³	**5.52**
5.0 - 10.0 m maximum depth	0.20	5.71	5.36	-	m³	**11.07**
10.0 - 15.0 m maximum depth	0.29	8.28	7.75	-	m³	**16.03**
The following unit costs are based on backacter machines fitted with hydraulic breakers and tractor loader machines						
Excavate rock (medium hard)						
ne 0.25 m maximum depth	0.31	8.86	8.29	-	m³	**17.14**
0.25 - 0.5 m maximum depth	0.43	12.28	11.49	-	m³	**23.78**
0.5 - 1.0 m maximum depth	0.58	16.57	15.49	-	m³	**32.06**
1.0 - 2.0 m maximum depth	0.80	22.85	21.41	-	m³	**44.27**
Excavate unreinforced concrete exposed at the commencing surface						
ne 0.25 m maximum depth	0.57	16.28	15.25	-	m³	**31.53**
0.25 - 0.5 m maximum depth	0.60	17.14	16.03	-	m³	**33.17**
0.5 -1.0 m maximum depth	0.67	19.14	17.91	-	m³	**37.05**
1.0 - 2.0 m maximum depth	0.70	20.00	18.69	-	m³	**38.69**
Excavate reinforced concrete exposed at the commencing surface						
ne 0.25 m maximum depth	0.90	25.71	24.05	-	m³	**49.76**
0.25 - 0.5 m maximum depth	0.90	25.71	24.05	-	m³	**49.76**
0.5 - 1.0 m maximum depth	0.95	27.14	25.38	-	m³	**52.52**
1.0 - 2.0 m maximum depth	1.08	30.85	28.86	-	m³	**59.71**
Excavate unreinforced concrete not exposed at the commencing surface						
ne 0.25 m maximum depth	0.65	18.57	17.37	-	m³	**35.94**
0.25 - 0.5 m maximum depth	0.67	19.14	17.91	-	m³	**37.05**
0.5 -1.0 m maximum depth	0.69	19.71	18.45	-	m³	**38.16**
1.0 - 2.0 m maximum depth	0.72	20.57	19.24	-	m³	**39.81**

CLASS E: EXCAVATION

Item	Gang hours	Labour £	Plant £	Material £	Unit	**Total rate £**
EXCAVATION FOR FOUNDATIONS						
The following unit costs are based on the use of backacter machines						
Excavate topsoil						
maximum depth ne 0.25 m	0.03	0.83	0.82	-	m³	**1.65**
Excavate material other than topsoil, rock or artificial hard material						
0.25 - 0.5 m deep	0.05	1.38	1.36	-	m³	**2.74**
0.5 - 1.0 m deep	0.06	1.66	1.60	-	m³	**3.26**
1.0 - 2.0 m deep	0.07	1.94	1.85	-	m³	**3.79**
2.0 - 5.0 m deep	0.12	3.32	3.21	-	m³	**6.53**
5.0 - 10.0 m deep	0.23	6.37	6.17	-	m³	**12.53**
The following unit costs are based on backacter machines fitted with hydraulic breakers						
Excavate unreinforced concrete exposed at the commencing surface						
ne 0.25 m maximum depth	0.60	16.61	16.03	-	m³	**32.64**
0.25 - 0.5 m maximum depth	0.66	18.27	17.64	-	m³	**35.90**
0.5 - 1.0 m maximum depth	0.72	19.93	19.24	-	m³	**39.17**
1.0 - 2.0 m maximum depth	0.80	22.14	21.41	-	m³	**43.55**
Excavate reinforced concrete exposed at the commencing surface						
ne 0.25 m maximum depth	0.94	26.02	25.09	-	m³	**51.10**
0.25 - 0.5 m maximum depth	0.98	27.12	26.22	-	m³	**53.34**
0.5 - 1.0 m maximum depth	1.00	27.68	26.71	-	m³	**54.39**
1.0 - 2.0 m maximum depth	1.04	28.78	27.83	-	m³	**56.61**
Excavate tarmacadam exposed at the commencing surface						
ne 0.25 m maximum depth	0.30	8.30	8.02	-	m³	**16.32**
0.25 - 0.5 m maximum depth	0.34	9.41	9.05	-	m³	**18.46**
0.5 - 1.0 m maximum depth	0.37	10.24	9.87	-	m³	**20.11**
1.0 - 2.0 m maximum depth	0.40	11.07	10.65	-	m³	**21.73**
Excavate unreinforced concrete not exposed at the commencing surface						
ne 0.25 m maximum depth	0.70	19.37	18.67	-	m³	**38.04**
0.25 - 0.5 m maximum depth	0.85	23.52	22.70	-	m³	**46.22**
0.5 -1.0 m maximum depth	0.96	26.57	25.65	-	m³	**52.22**
1.0 - 2.0 m maximum depth	1.02	28.23	27.26	-	m³	**55.49**

CLASS E: EXCAVATION

Item	Gang hours	Labour £	Plant £	Material £	Unit	**Total rate £**
GENERAL EXCAVATION						
The following unit costs are based on backacter and tractor loader machines						
Excavate topsoil						
maximum depth ne 0.25 m	0.03	0.91	0.82	-	m³	**1.73**
Excavate material other than topsoil, rock or artificial hard material						
ne 0.25 m maximum depth	0.03	0.91	0.80	-	m³	**1.71**
0.25 - 0.5 m maximum depth	0.03	0.91	0.82	-	m³	**1.73**
0.5 - 1.0 m maximum depth	0.04	1.21	1.06	-	m³	**2.27**
1.0 - 2.0 m maximum depth	0.06	1.82	1.60	-	m³	**3.43**
2.0 - 5.0 m maximum depth	0.10	3.04	2.66	-	m³	**5.70**
5.0 - 10.0 m maximum depth	0.20	6.07	5.36	-	m³	**11.43**
10.0 - 15.0 m maximum depth	0.29	8.81	7.75	-	m³	**16.55**
The following unit costs are based on backacter machines fitted with hydraulic breakers and tractor loader machines						
Excavate rock (medium hard)						
ne 0.25 m maximum depth	0.31	9.41	8.29	-	m³	**17.70**
0.25 - 0.5 m maximum depth	0.43	13.06	11.49	-	m³	**24.55**
0.5 - 1.0 m maximum depth	0.58	17.61	15.49	-	m³	**33.10**
1.0 - 2.0 m maximum depth	0.80	24.29	21.41	-	m³	**45.70**
Excavate unreinforced concrete exposed at the commencing surface						
ne 0.25 m maximum depth	0.57	17.31	15.25	-	m³	**32.55**
0.25 - 0.5 m maximum depth	0.60	18.22	16.03	-	m³	**34.25**
0.5 -1.0 m maximum depth	0.67	20.34	17.91	-	m³	**38.25**
1.0 - 2.0 m maximum depth	0.70	21.25	18.69	-	m³	**39.95**
Excavate reinforced concrete exposed at the commencing surface						
ne 0.25 m maximum depth	0.90	27.33	24.05	-	m³	**51.38**
0.25 - 0.5 m maximum depth	0.90	27.33	24.05	-	m³	**51.38**
0.5 - 1.0 m maximum depth	0.95	28.84	25.38	-	m³	**54.23**
1.0 - 2.0 m maximum depth	1.08	32.79	28.86	-	m³	**61.65**
Excavate unreinforced concrete not exposed at the commencing surface						
ne 0.25 m maximum depth	0.65	19.74	17.37	-	m³	**37.10**
0.25 - 0.5 m maximum depth	0.67	20.34	17.91	-	m³	**38.25**
0.5 -1.0 m maximum depth	0.69	20.95	18.45	-	m³	**39.40**
1.0 - 2.0 m maximum depth	0.72	21.86	19.24	-	m³	**41.10**
EXCAVATION ANCILLARIES						
The following unit costs are for various machines appropriate to the work						
Trimming topsoil; using D4H dozer						
horizontal	0.04	0.53	0.46	-	m²	**0.99**
10 - 45 degrees to horizontal	0.05	0.67	0.57	-	m²	**1.24**
Trimming material other than topsoil, rock or artificial hard material; using D4H dozer, tractor loader or motor grader average rate						
horizontal	0.04	0.53	0.68	-	m²	**1.21**
10 - 45 degrees to horizontal	0.04	0.53	0.68	-	m²	**1.21**
45 - 90 degrees to horizontal	0.06	0.80	1.01	-	m²	**1.81**

CLASS E: EXCAVATION

Item	Gang hours	Labour £	Plant £	Material £	Unit	**Total rate £**
Trimming rock; using D6E dozer						
horizontal	0.08	1.07	15.37	-	m²	**16.44**
10 - 45 degrees to horizontal	0.80	10.68	15.37	-	m²	**26.05**
45 - 90 degrees to horizontal	1.06	14.15	20.36	-	m²	**34.51**
vertical	1.06	14.15	20.36	-	m²	**34.51**
Preparation of topsoil; using D4H dozer, tractor loader or motor grader average rate						
horizontal	0.06	0.80	0.81	-	m²	**1.61**
10 - 45 degrees to horizontal	0.06	0.80	0.81	-	m²	**1.61**
Preparation of material other than rock or artificial hard material; using D6E dozer, tractor loader or motor grader average rate						
horizontal	0.06	0.80	1.17	-	m²	**1.97**
10 - 45 degrees to horizontal	0.06	0.80	1.17	-	m²	**1.97**
45 - 90 degrees to horizontal	0.06	0.80	1.17	-	m²	**1.97**
Preparation of rock; using D6E dozer						
horizontal	0.80	10.68	15.37	-	m²	**26.05**
10 - 45 degrees to horizontal	0.60	8.01	11.53	-	m²	**19.54**
45 - 90 degrees to horizontal	0.60	8.01	11.53	-	m²	**19.54**
The following unit costs for disposal are based on using a 22.5 t ADT for site work and 20 t tipper for off-site work. The distances used in the calculation are quoted to assist estimating, although this goes beyond the specific requirements of CESMM3						
Disposal of excavated topsoil						
storage on site; 100 m maximum distance	0.05	0.82	2.09	-	m³	**2.91**
removal; 5 km distance	0.11	1.80	4.61	-	m³	**6.41**
removal; 15 km distance	0.21	3.43	8.80	-	m³	**12.24**
Disposal of excavated earth other than rock or artificial hard material						
storage on site; 100 m maximum distance; using 22.5 t ADT	0.05	0.82	2.09	-	m³	**2.91**
removal; 5 km distance; using 20 t tipper	0.12	1.96	5.03	-	m³	**6.99**
removal; 15 km distance; using 20 t tipper	0.22	3.60	9.22	-	m³	**12.82**
Disposal of excavated rock or artificial hard material						
storage on site; 100 m maximum distance; using 22.5 t ADT	0.06	0.98	2.51	-	m³	**3.49**
removal; 5 km distance; using 20 t tipper	0.13	2.13	5.45	-	m³	**7.58**
removal; 15 km distance; using 20 t tipper	0.23	3.76	9.64	-	m³	**13.40**
Add to the above rates where tipping charges apply:						
non-hazardous waste	-	-	-	-	m³	**26.93**
hazardous waste	-	-	-	-	m³	**32.92**
special waste	-	-	-	-	m³	**55.38**
contaminated liquid	-	-	-	-	m³	**82.31**
contaminated sludge	-	-	-	-	m³	**134.66**
Add to the above rates where Landfill Tax applies:						
exempted material	-	-	-	-	m³	**-**
inactive or inert material	-	-	-	-	m³	**3.15**
other material	-	-	-	-	m³	**22.05**

CLASS E: EXCAVATION

Item	Gang hours	Labour £	Plant £	Material £	Unit	Total rate £
EXCAVATION ANCILLARIES – cont'd						
The following unit costs for double handling are based on using a 1.5 m³ tractor loader and 22.5 t ADT. A range of distances is listed to assist in estimating, although this goes beyond the specific requirements of CESMM3.						
Double handling of excavated topsoil; using 1.5 m³ tractor loader and 22.5 t ADT						
300 m average distance moved	0.08	1.31	2.92	-	m³	**4.23**
600 m average distance moved	0.10	1.64	3.65	-	m³	**5.29**
1000 m average distance moved	0.12	1.96	4.39	-	m³	**6.35**
Double handling of excavated earth other than rock or artificial hard material; using 1.5 m³ tractor loader and 22.5 t ADT						
300 m average distance moved	0.08	1.31	2.92	-	m³	**4.23**
600 m average distance moved	0.10	1.64	3.65	-	m³	**5.29**
1000 m average distance moved	0.12	1.96	4.39	-	m³	**6.35**
Double handling of rock or artificial hard material; using 1.5 m³ tractor loader and 22.5 t ADT						
300 m average distance moved	0.16	2.62	5.85	-	m³	**8.46**
600 m average distance moved	0.18	2.94	6.58	-	m³	**9.52**
1000 m average distance moved	0.20	3.27	7.31	-	m³	**10.58**
The following unit rates for excavation below Final Surface are based on using a 16 t backacter machine						
Excavation of material below the Final Surface and replacement of with:						
granular fill	1.20	15.55	13.97	17.80	m³	**47.31**
concrete Grade C7.5P	0.60	7.77	6.98	75.89	m³	**90.65**
Timber supports left in	0.32	5.47	-	6.90	m²	**12.37**
Metal supports left in	0.89	14.86	-	30.56	m²	**45.42**
FILLING						
Excavated topsoil; DTp specified type 5A						
Filling						
to structures	0.07	2.55	3.11	-	m³	**5.66**
embankments	0.03	0.89	1.09	-	m³	**1.98**
general	0.02	0.75	0.91	-	m³	**1.66**
150 mm thick	0.01	0.25	0.30	-	m²	**0.55**
250 mm thick	0.01	0.43	0.52	-	m²	**0.95**
400 mm thick	0.02	0.61	0.74	-	m²	**1.35**
600 mm thick	0.02	0.86	1.04	-	m²	**1.90**
Imported topsoil DTp specified type 5B;						
Filling						
embankments	0.03	0.89	1.09	19.84	m³	**21.82**
general	0.02	0.75	0.91	19.84	m³	**21.50**
150 mm thick	0.01	0.25	0.30	2.98	m²	**3.53**
250 mm thick	0.01	0.43	0.52	4.96	m²	**5.91**
400 mm thick	0.02	0.61	0.74	7.94	m²	**9.28**
600 mm thick	0.02	0.86	1.04	11.90	m²	**13.80**

CLASS E: EXCAVATION

Item	Gang hours	Labour £	Plant £	Material £	Unit	**Total rate £**
Non-selected excavated material other than topsoil or rock						
Filling						
to structures	0.04	1.28	1.57	-	m³	**2.85**
embankments	0.02	0.54	0.65	-	m³	**1.19**
general	0.01	0.39	0.48	-	m³	**0.87**
150 mm thick	-	0.14	0.17	-	m²	**0.32**
250 mm thick	0.01	0.21	0.26	-	m²	**0.48**
400 mm thick	0.01	0.32	0.39	-	m²	**0.71**
600 mm thick	0.01	0.43	0.52	-	m²	**0.95**
Selected excavated material other than topsoil or rock						
Filling						
to structures	0.04	1.46	1.78	-	m³	**3.25**
embankments	0.01	0.50	0.61	-	m³	**1.11**
general	0.01	0.43	0.52	-	m³	**0.95**
150 mm thick	-	0.14	0.17	-	m²	**0.32**
250 mm thick	0.01	0.25	0.30	-	m²	**0.55**
400 mm thick	0.01	0.36	0.44	-	m²	**0.79**
600 mm thick	0.01	0.46	0.57	-	m²	**1.03**
Imported natural material other than topsoil or rock; subsoil						
Filling						
to structures	0.04	1.28	1.57	18.47	m³	**21.33**
embankments	0.02	0.54	0.65	18.47	m³	**19.66**
general	0.02	0.54	0.65	18.47	m³	**19.66**
150 mm thick	0.01	0.32	0.39	2.77	m²	**3.48**
250 mm thick	0.01	0.46	0.57	4.62	m²	**5.65**
400 mm thick	0.02	0.54	0.65	7.39	m²	**8.58**
600 mm thick	0.02	0.61	0.74	11.08	m²	**12.43**
Imported natural material other than topsoil or rock; granular graded material						
Filling						
to structures	0.04	1.28	1.57	20.19	m³	**23.04**
embankments	0.02	0.61	0.74	20.19	m³	**21.54**
general	0.02	0.61	0.74	20.19	m³	**21.54**
150 mm thick	0.01	0.43	0.52	3.03	m²	**3.98**
250 mm thick	0.02	0.61	0.74	5.05	m²	**6.39**
400 mm thick	0.02	0.71	0.87	8.08	m²	**9.66**
600 mm thick	0.02	0.79	0.96	12.12	m²	**13.86**
Imported natural material other than topsoil or rock; granular selected material						
Filling						
to structures	0.04	1.28	1.57	18.80	m³	**21.65**
embankments	0.02	0.61	0.74	18.80	m³	**20.15**
general	0.02	0.61	0.74	18.80	m³	**20.15**
150 mm thick	0.01	0.43	0.52	2.82	m²	**3.77**
250 mm thick	0.02	0.61	0.74	4.70	m²	**6.05**
400 mm thick	0.02	0.71	0.87	7.52	m²	**9.11**
600 mm thick	0.02	0.79	0.96	11.28	m²	**13.02**

CLASS E: EXCAVATION

Item	Gang hours	Labour £	Plant £	Material £	Unit	**Total rate £**
FILLING – cont'd						
Excavated rock						
Filling						
to structures	0.04	1.43	1.74	-	m³	**3.17**
embankments	0.05	1.78	2.18	-	m³	**3.96**
general	0.05	1.78	2.18	-	m³	**3.96**
150 mm thick	0.02	0.61	0.74	-	m²	**1.35**
250 mm thick	0.03	0.93	1.13	-	m²	**2.06**
400 mm thick	0.04	1.43	1.74	-	m²	**3.17**
600 mm thick	0.06	1.96	2.39	-	m²	**4.36**
Imported rock						
Filling						
to structures	0.04	1.43	1.74	40.80	m³	**43.97**
embankments	0.02	0.86	1.04	40.80	m³	**42.70**
general	0.02	0.71	0.87	40.80	m³	**42.39**
150 mm thick	0.01	0.43	0.52	6.12	m²	**7.07**
250 mm thick	0.02	0.71	0.87	10.20	m²	**11.79**
400 mm thick	0.03	1.07	1.31	16.32	m²	**18.70**
600 mm thick	0.03	1.07	1.31	24.48	m²	**26.86**
FILLING ANCILLARIES						
Trimming of filled surfaces						
Topsoil						
horizontal	0.02	0.34	0.71	-	m²	**1.05**
inclined at an angle of 10 - 45 degrees to horizontal	0.02	0.34	0.71	-	m²	**1.05**
inclined at an angle of 45 - 90 degrees to horizontal	0.03	0.44	0.91	-	m²	**1.35**
Material other than topsoil, rock or artificial hard material						
horizontal	0.02	0.34	0.71	-	m²	**1.05**
inclined at an angle of 10 - 45 degrees to horizontal	0.02	0.34	0.71	-	m²	**1.05**
inclined at an angle of 45 - 90 degrees to horizontal	0.03	0.44	0.91	-	m²	**1.35**
Rock						
horizontal	0.51	8.34	17.24	-	m²	**25.58**
inclined at an angle of 10 - 45 degrees to horizontal	0.52	8.50	17.58	-	m²	**26.08**
inclined at an angle of 45 - 90 degrees to horizontal	0.70	11.45	23.67	-	m²	**35.11**
Preparation of filled surfaces						
Topsoil						
horizontal	0.03	0.49	1.01	-	m²	**1.50**
inclined at an angle of 10 - 45 degrees to horizontal	0.03	0.49	1.01	-	m²	**1.50**
inclined at an angle of 45 - 90 degrees to horizontal	0.04	0.60	1.25	-	m²	**1.86**

CLASS E: EXCAVATION

Item	Gang hours	Labour £	Plant £	Material £	Unit	**Total rate £**
Material other than topsoil, rock or artificial hard material						
horizontal	0.03	0.49	1.01	-	m²	**1.50**
inclined at an angle of 10 - 45 degrees to horizontal	0.03	0.49	1.01	-	m²	**1.50**
inclined at an angle of 45 - 90 degrees to horizontal	0.04	0.60	1.25	-	m²	**1.86**
Rock						
horizontal	0.33	5.40	11.16	-	m²	**16.55**
inclined at an angle of 10 - 45 degrees to horizontal	0.33	5.40	11.16	-	m²	**16.55**
inclined at an angle of 45 - 90 degrees to horizontal	0.52	8.50	17.58	-	m²	**26.08**
GEOTEXTILES						
NOTES						
The geotextile products mentioned below are not specifically confined to the individual uses stated but are examples of one of many scenarios to which they may be applied. Conversely, the scenarios are not limited to the geotextile used as an example.						
Geotextiles; stabilisation applications for reinforcement of granular sub-bases, capping layers and railway ballast placed over weak and variable soils						
For use over weak soils with moderate traffic intensities e.g. car parks, light access roads; Tensar SS20 Polypropylene Geogrid						
horizontal	0.04	1.05	-	1.54	m²	**2.60**
inclined at an angle of 10 to 45 degrees to the horizontal	0.05	1.33	-	1.54	m²	**2.87**
For use over weak soils with high traffic intensities and/or high axle loadings; Tensar SS30 Polypropylene Geogrid						
horizontal	0.05	1.64	-	2.35	m²	**3.98**
inclined at an angle of 10 to 45 degrees to the horizontal	0.06	2.04	-	2.35	m²	**4.38**
For use over very weak soils e.g. alluvium, marsh or peat or firmer soil subject to exceptionally high axle loadings;Tensar SS40 Polypropylene Geogrid						
horizontal	0.05	1.79	1.41	3.84	m²	**7.04**
inclined at an angle of 10 to 45 degrees to the horizontal	0.06	2.22	1.75	3.84	m²	**7.82**
For trafficked areas where fill comprises aggregate exceeding 100mm; Tensar SSLA20 Polypropylene Geogrid						
horizontal	0.04	1.05	-	2.11	m²	**3.16**
inclined at an angle 10 - 45 degrees to the horizontal	0.05	1.33	-	2.11	m²	**3.44**

CLASS E: EXCAVATION

Item	Gang hours	Labour £	Plant £	Material £	Unit	**Total rate £**
GEOTEXTILES – cont'd						
Geotextiles; stabilisation applications for reinforcement of granular sub-bases, capping layers and railway ballast placed over weak and variable soils – cont'd						
Stabilisation and separation of granular fill from soft sub grade to prevent intermixing: Terram 1000						
horizontal	0.05	1.85	-	0.44	m²	**2.30**
inclined at an angle of 10 to 45 degrees to the horizontal	0.06	2.33	-	0.44	m²	**2.77**
Stabilisation and separation of granular fill from soft sub grade to prevent intermixing: Terram 2000						
horizontal	0.04	1.67	1.31	1.28	m²	**4.26**
inclined at an angle of 10 to 45 degrees to the horizontal	0.05	2.11	1.66	1.28	m²	**5.04**
Geotextiles; reinforcement applications for asphalt pavements						
For roads, hardstandings and airfield pavements; Tensar AR-G composite comprising Tensar AR-1 grid bonded to a geotextile, laid within asphalt						
horizontal	0.05	1.64	-	3.58	m²	**5.22**
inclined at an angle of 10 to 45 degrees to the horizontal	0.06	2.04	-	3.58	m²	**5.62**
Geotextiles; slope reinforcement and embankment support; for use where soils can only withstand limited shear stresses, therefore steep slopes require external support						
Paragrid 30/155; 330g/m²						
horizontal	0.04	1.05	-	2.03	m²	**3.09**
inclined at an angle of 10-45 degrees to the horizontal	0.05	1.33	-	2.03	m²	**3.36**
Paragrid 100/255; 330g/m²						
horizontal	0.04	1.05	-	2.77	m²	**3.82**
inclined at an angle of 10-45 degrees to the horizontalhorizontal	0.05	1.33	-	2.77	m²	**4.10**
Paralink 200s; 1120g/m²						
horizontal	0.05	2.15	1.69	5.44	m²	**9.27**
inclined at an angle of 10-45 degrees to the horizontal	0.07	2.70	2.13	5.44	m²	**10.27**
Paralink 600s; 2040g/m²						
horizontal	0.06	2.50	1.97	11.49	m²	**15.96**
inclined at an angle of 10-45 degrees to the horizontal	0.08	3.14	2.47	11.49	m²	**17.10**
Terram grid 30/30						
horizontal	0.06	2.26	1.78	0.01	m²	**4.05**
inclined at an angle of 10-45 degrees to the horizontal	0.07	2.82	2.22	0.01	m²	**5.05**

CLASS E: EXCAVATION

Item	Gang hours	Labour £	Plant £	Material £	Unit	**Total rate £**
Geotextiles; scour and erosion protection						
For use where erosion protection is required to the surface of a slope once its geotechnical stability has been achieved, and to allow grass establishment; Tensar 'Mat' Polyethelene mesh; fixed with Tensar pegs						
horizontal	0.04	1.49	-	5.40	m²	**6.89**
inclined at an angle of 10-45 degrees to the horizontal	0.05	1.85	-	5.40	m²	**7.26**
For use where hydraulic action exists, such as coastline protection from pressures exerted by waves, currents and tides; Typar SF56						
horizontal	0.05	2.03	1.59	0.53	m²	**4.15**
inclined at an angle of 10-45 degrees to the horizontal	0.06	2.54	2.00	0.53	m²	**5.07**
For protection against puncturing to reservoir liner; Typar SF563						
horizontal	0.05	2.03	1.59	0.53	m²	**4.15**
inclined at an angle of 10-45 degrees to the horizontal	0.06	2.54	2.00	0.53	m²	**5.07**
Geotextiles; temporary parking areas						
For reinforcement of grassed areas subject to wear from excessive pedestrian and light motor vehicle traffic; Netlon CE131 high density polyethelyene geogrid						
horizontal	0.04	1.10	-	3.59	m²	**4.70**
Geotextiles; landscaping applications						
For prevention of weed growth in planted areas by incorporating a geotextile over top soil and below mulch or gravel; Typar SF20						
horizontal	0.08	2.73	-	0.31	m²	**3.04**
inclined at an angle of 10-45 degrees to the horizontal	0.09	3.42	-	0.31	m²	**3.73**
For root growth control-Prevention of lateral spread of roots and mixing of road base and humus; Typar SF20						
horizontal	0.08	2.73	-	0.31	m²	**3.04**
inclined at an angle of 10-45 degrees to the horizontal	0.09	3.42	-	0.31	m²	**3.73**
Geotextiles; drainage applications						
For clean installation of pipe support material and to prevent siliting of the drainage pipe and minimising differential settlement; Typar SF10						
horizontal	0.04	1.49	-	0.34	m²	**1.83**
inclined at an angle of 10-45 degrees to the horizontal	0.05	1.85	-	0.34	m²	**2.20**
For wrapping to prevent clogging of drainage pipes surrounded by fine soil; Typar SF10						
sheeting	0.08	2.91	-	0.68	m²	**3.58**
For wrapping to prevent clogging of drainage pipes surrounded by fine soil; Terram 1000						
sheeting	0.08	2.91	-	0.48	m²	**3.39**

CLASS E: EXCAVATION

Item	Gang hours	Labour £	Plant £	Material £	Unit	**Total rate £**
GEOTEXTILES – cont'd						
Geotextiles; drainage applications – cont'd						
For vertical structure drainage to sub-surface walls, roofs and foundations; Filtram 1B1						
sheeting	0.08	2.91	-	7.65	m²	**10.56**
For waterproofing to tunnels, buried structures, etc. where the membrane is buried, forming part of the drainage system; Filtram 1BZ						
sheeting	0.08	2.91	-	7.13	m²	**10.04**
Geotextiles; roofing insulation and protection						
Protection of waterproofing membrane from physical damage and puncturing; Typar SF56						
sheeting	0.05	2.03	1.59	0.53	m²	**4.15**
Internal reinforcement of in situ spread waterproof bitumen emulsion; Typar SF10						
sheeting	0.07	2.40	-	0.34	m²	**2.74**
LANDSCAPING						
Preparatory operations prior to landscaping						
Supply and apply granular cultivation treatments by hand						
35 grammes/m²	0.50	11.65	-	4.85	100 m²	**16.51**
50 grammes/m²	0.65	15.15	-	6.93	100 m²	**22.08**
75 grammes/m²	0.85	19.81	-	10.39	100 m²	**30.21**
100 grammes/m²	1.00	23.31	-	13.86	100 m²	**37.17**
Supply and apply granular cultivation treatments by machine in suitable economically large areas						
100 grammes/m²	0.25	3.29	0.94	13.86	100m²	**18.09**
Supply and incorperate cultivation additives into top 150 mm of topsoil by hand						
1 m³ / 10 m²	20.00	466.20	-	11.55	100m²	**477.75**
1 m³ / 13 m²	20.00	466.20	-	8.88	100m²	**475.08**
1 m³ / 20 m²	19.00	442.89	-	5.78	100m²	**448.66**
1 m³ / 40 m²	17.00	396.27	-	2.89	100m²	**399.16**
Supply and incorporate cultivation additives into top 150 mm of topsoil by machine in suitable economically large areas						
1 m³ / 10 m²	-	-	139.56	11.55	100m²	**151.11**
1 m³ / 13 m²	-	-	128.82	8.88	100m²	**137.70**
1 m³ / 20 m²	-	-	114.52	5.78	100m²	**120.29**
1 m³ / 40 m²	-	-	105.57	2.89	100m²	**108.45**
Turfing						
Turfing						
horizontal	0.12	2.80	-	1.97	m²	**4.77**
10 - 45 degress to horizontal	0.17	3.96	-	1.97	m²	**5.93**
45 - 90 degrees to horizontal; pegging down	0.19	4.43	-	1.97	m²	**6.40**

CLASS E: EXCAVATION

Item	Gang hours	Labour £	Plant £	Material £	Unit	**Total rate £**
Hydraulic mulch grass seeding						
Grass seeding						
horizontal	0.01	0.12	-	0.13	m²	**0.24**
10 - 45 degrees to horizontal	0.01	0.16	-	0.13	m²	**0.29**
45 - 90 degrees to horizontal	0.01	0.21	-	0.13	m²	**0.34**
Selected grass seeding						
Grass seeding; sowing at the rate of 0.050 kg/m² in two operations						
horizontal	0.01	0.23	-	0.18	m²	**0.41**
10 - 45 degrees to horizontal	0.02	0.35	-	0.18	m²	**0.53**
45 - 90 degrees to horizontal	0.02	0.47	-	0.18	m²	**0.64**
Plants						
Form planting hole in previously cultivated area, supply and plant specified herbaceous plants and backfill with excavated material						
5 plants/m²	0.01	0.23	-	3.54	m²	**3.77**
10 plants/m²	0.02	0.51	-	12.96	m²	**13.48**
25 plants/m²	0.05	1.17	-	30.19	m²	**31.36**
35 plants/m²	0.07	1.63	-	42.26	m²	**43.89**
50 plants/m²	0.10	2.33	-	60.38	m²	**62.71**
Supply and fix plant support netting on 50 mm diameter stakes 750 mm long driven into the ground at 1.5 m centres						
1.15m high green extruded plastic mesh, 125mm square mesh	0.06	1.40	-	1.71	m²	**3.11**
Form planting hole in previously cultivated area; supply and plant bulbs and backfill with excavated material						
small	0.01	0.23	-	0.16	each	**0.39**
medium	0.01	0.23	-	0.25	each	**0.49**
large	0.01	0.23	-	0.30	each	**0.54**
Supply and plant bulbs in grassed area using bulb planter and backfill with screened topsoil or peat and cut turf plug						
small	0.01	0.23	-	0.16	each	**0.39**
medium	0.01	0.23	-	0.25	each	**0.49**
large	0.01	0.23	-	0.30	each	**0.54**
Shrubs						
Form planting hole in previously cultivated area, supply and plant specified shrub and backfill with excavated material						
shrub 300 mm high	0.01	0.23	-	2.39	each	**2.63**
shrub 600 mm high	0.01	0.23	-	3.44	each	**3.67**
shrub 900 mm high	0.01	0.23	-	4.11	each	**4.35**
shrub 1 m high and over	0.01	0.23	-	5.49	each	**5.73**
Supply and fix shrub stake including two ties						
one stake; 1.5 m long, 75 mm diameter	0.12	2.80	-	5.08	each	**7.88**

CLASS E: EXCAVATION

Item	Gang hours	Labour £	Plant £	Material £	Unit	**Total rate £**
LANDSCAPING – cont'd						
Trees						
The cost of planting semi-mature trees will depend on the size and species, and on the access to the site for tree handling machines. Prices should be obtained for individual trees and planting.						
Break up subsoil to a depth of 200 mm						
in tree pit	0.05	1.17	-	-	each	**1.17**
Supply and plant tree in prepared pit; backfill with excavated topsoil minimum 600 mm deep						
light standard tree	0.25	5.83	-	10.08	each	**15.91**
standard tree	0.45	10.49	-	15.12	each	**25.61**
selected standard tree	0.75	17.48	-	20.16	each	**37.64**
heavy standard tree	0.85	19.81	-	35.28	each	**55.09**
extra heavy standard tree	1.50	34.97	-	63.00	each	**97.97**
Extra for filling with topsoil from spoil ne 100 m	0.15	3.50	-	-	m³	**3.50**
Extra for filling with imported topsoil	0.08	1.86	-	18.17	m³	**20.03**
Supply tree stake and drive 500 mm into firm ground and trim to approved height, including two tree ties to approved pattern						
one stake; 2.4 m long, 100 mm diameter	0.16	3.73	-	6.49	each	**10.22**
one stake; 3.0 m long, 100 mm diameter	0.20	4.66	-	8.18	each	**12.84**
two stakes; 2.4 m long, 100 mm diameter	0.24	5.59	-	12.98	each	**18.57**
two stakes; 3.0 m long, 100 mm diameter	0.30	6.99	-	16.35	each	**23.35**
Supply and fit tree support comprising three collars and wire guys; including pickets						
galvanised steel 50 x 600 mm	1.50	34.97	-	27.19	each	**62.16**
hardwood 75 x 600 mm	1.50	34.97	-	27.82	each	**62.78**
Supply and fix standard steel tree guard	0.30	6.99	-	22.72	each	**29.71**
Hedges						
Excavate trench by hand for hedge and deposit soil alongside trench						
300 wide x 300 mm deep	0.10	2.33	-	-	m	**2.33**
450 wide x 300 mm deep	0.13	3.03	-	-	m	**3.03**
Excavate trench by machine for hedge and deposit soil alongside trench						
300 wide x 300 mm deep	0.02	0.47	-	-	m	**0.47**
450 wide x 300 mm deep	0.02	0.47	-	-	m	**0.47**
Set out, nick out and excavate trench and break up subsoil to minimum depth of 300 mm						
400 mm minimum deep	0.28	6.50	-	-	m	**6.50**
Supply and plant hedging plants, backfill with excavated topsoil						
single row; plants at 200 mm centres	0.25	5.83	-	3.02	m	**8.85**
single row; plants at 300 mm centres	0.17	3.96	-	2.01	m	**5.97**
single row; plants at 400 mm centres	0.13	2.91	-	1.51	m	**4.42**
single row; plants at 500 mm centres	0.10	2.33	-	1.21	m	**3.54**
single row; plants at 600 mm centres	0.08	1.86	-	1.00	m	**2.87**
double row; plants at 200 mm centres	0.50	11.65	-	6.04	m	**17.69**
double row; plants at 300 mm centres	0.34	7.93	-	4.02	m	**11.95**
double row; plants at 400 mm centres	0.25	5.83	-	3.02	m	**8.85**
double row; plants at 500 mm centres	0.20	4.66	-	2.42	m	**7.08**
Extra for incorporating manure at the rate of 1m³ per 30 m of trench	0.12	2.80	-	0.10	m	**2.90**

CLASS E: EXCAVATION

Item	Gang hours	Labour £	Plant £	Material £	Unit	**Total rate £**
COMPARATIVE COSTS - EARTH MOVING						
Notes						
The cost of earth moving and other associated works is dependent on matching the overall quantities and production rate called for by the programme of works with the most appropriate plant and assessing the most suitable version of that plant which will: • deal with the site conditions (e.g. type of ground, type of excavation, length of haul, prevailing weather, etc.) • comply with the specification requirements (e.g. compaction, separation of materials, surface tolerances, etc.) • complete the work economically (e.g. provide surface tolerances which will avoid undue expense of imported materials)						
Labour costs are based on a plant operative skill rate 3 unless otherwise stated						
Comparative costs of excavation equipment						
The following are comparative costs using various types of excavation equipment and include loading into transport. All costs assume 50 minutes productive work in every hour and adequate disposal transport being available to obviate any delay.						
Dragline (for excavations requiring a long reach and long discharge, mainly in clearing streams and rivers)						
bucket capacity ne 1.15 m^3	0.06	0.98	2.21	-	m^3	**3.19**
bucket capacity ne 2.00 m^3	0.03	0.49	2.55	-	m^3	**3.04**
bucket capacity ne 3.00 m^3	0.02	0.33	2.45	-	m^3	**2.78**
Hydraulic backacter (for all types of excavation and loading, including trenches, breaking hard ground, etc.)						
bucket capacity ne 0.40 m^3	0.05	0.73	0.89	-	m^3	**1.63**
bucket capacity ne 1.00 m^3	0.02	0.33	0.65	-	m^3	**0.98**
bucket capacity ne 1.60 m^3	0.01	0.16	0.82	-	m^3	**0.98**
Hydraulic face shovel (predominantly for excavating cuttings and embankments over 2 m high requiring high output)						
bucket capacity ne 1.50 m^3	0.04	0.65	2.57	-	m^3	**3.23**
bucket capacity ne 2.60 m^3	0.02	0.33	1.74	-	m^3	**2.06**
bucket capacity ne 3.40 m^3	0.01	0.16	1.06	-	m^3	**1.22**
Tractor loader (for loading, carrying, placing materials, spreading and levelling, some site clearance operations and reducing levels)						
bucket capacity ne 0.80 m^3	0.02	0.33	0.40	-	m^3	**0.72**
bucket capacity ne 1.50 m^3	0.01	0.16	0.31	-	m^3	**0.48**
bucket capacity ne 3.00 m^3	0.01	0.11	0.46	-	m^3	**0.58**
Multipurpose wheeled loader / backhoe (versatile machine for small to medium excavations, trenches, loading, carrying and back filling)						
bucket capacity ne 0.76 m^3	0.08	1.31	1.19	-	m^3	**2.50**
bucket capacity ne 1.00 m^3	0.06	0.98	1.03	-	m^3	**2.01**

CLASS E: EXCAVATION

Item	Gang hours	Labour £	Plant £	Material £	Unit	Total rate £
COMPARATIVE COSTS - EARTH MOVING – cont'd						
Comparative costs of transportation equipment						
The following are comparative costs for using various types of transportation equipment to transport excavated loose material. The capacity of the transport must be suitable for the output of the loading machine. The cost will vary depending on the number of transport units required to meet the output of the loading unit and the distance to be travelled.						
Loading loose material into transport by wheeled loader						
ne 2.1 m³ capacity	0.02	0.33	0.73	-	m³	**1.05**
ne 5.4 m³ capacity	0.01	0.16	1.44	-	m³	**1.60**
ne 10.5 m³ capacity	0.01	0.11	0.80	-	m³	**0.91**
Transport material within site by dump truck (rear dump)						
ne 24 m³ heaped capacity, distance travelled ne 0.5 Km	0.03	0.49	2.01	-	m³	**2.50**
Add per 0.5 Km additional distance	0.01	0.16	0.67	-	m³	**0.83**
ne 57 m³ heaped capacity, distance travelled ne 0.5 Km	0.03	0.53	2.12	-	m³	**2.65**
Add per 0.5 Km additional distance	0.03	0.53	1.06	-	m³	**1.59**
Transport material within site by dump truck (articulated)						
ne 32 t payload, distance travelled ne 0.5 Km	0.04	0.71	2.30	-	m³	**3.01**
Add per 0.5 Km additional distance	0.02	0.35	1.15	-	m³	**1.50**
Transport material within or off site by tipping lorry						
ne 10 t payload, distance travelled ne 1 Km	0.05	0.73	1.16	-	m³	**1.90**
Add per 1 Km additional distance	0.03	0.37	0.58	-	m³	**0.95**
10 - 15 t payload, distance travelled ne 1 Km	0.04	0.59	1.00	-	m³	**1.59**
Add per 1 Km additional distance	0.02	0.29	0.50	-	m³	**0.79**
15 - 25 t payload, distance travelled ne 1 Km	0.03	0.49	1.05	-	m³	**1.54**
Add per 1 Km additional distance	0.02	0.25	0.53	-	m³	**0.77**
Comparative costs of earth moving equipment						
The following are comparative costs using various types of earth moving equipment to include excavation, transport, spreading and levelling.						
Bulldozer up to 74 KW (CAT D4H sized machine used for site strip, reducing levels or grading and spreading materials over smaller sites)						
average push one way 10 m	0.03	0.44	0.68	-	m³	**1.12**
average push one way 30 m	0.08	1.17	1.82	-	m³	**2.99**
average push one way 50 m	0.14	2.05	3.19	-	m³	**5.24**
average push one way 100 m	0.29	4.25	6.60	-	m³	**10.85**
Bulldozer up to 104 KW (CAT D6E sized machine for reducing levels, excavating to greater depths at steeper inclines, grading surfaces, small cut and fill operations; maximum push 100 m)						
average push one way 10 m	0.03	0.49	1.15	-	m³	**1.64**
average push one way 30 m	0.07	1.14	2.69	-	m³	**3.83**
average push one way 50 m	0.09	1.47	3.46	-	m³	**4.93**
average push one way 100 m	0.19	3.11	7.30	-	m³	**10.41**

CLASS E: EXCAVATION

Item	Gang hours	Labour £	Plant £	Material £	Unit	**Total rate £**
Bull or Angle Dozer up to 212 KW (CAT D8N sized machine for high output, ripping and excavating by reducing levels at steeper inclines or in harder material than with D6E, larger cut and fill operations used in conjunction with towed or S.P. scrapers. Spreading and grading over large areas; maximum push 100 m)						
average push one way 10 m	0.03	0.49	2.63	-	m^3	**3.12**
average push one way 30 m	0.06	0.98	5.26	-	m^3	**6.24**
average push one way 50 m	0.07	1.14	6.13	-	m^3	**7.28**
average push one way 100 m	0.11	1.80	9.64	-	m^3	**11.44**
Motorised scraper, 15 m^3 capacity (for excavating larger volumes over large haul lengths, excavating to reduce levels and also levelling ground, grading large sites, moving and tipping material including hard material - used in open cast sites)						
average haul one way 500 m	-	0.07	0.34	-	m^3	**0.41**
average haul one way 1,000 m	0.01	0.10	0.51	-	m^3	**0.61**
average haul one way 2,000 m	0.01	0.13	0.68	-	m^3	**0.81**
average haul one way 3,000 m	0.01	0.16	0.85	-	m^3	**1.01**
Twin engined motorised scraper, 16 m^3 capacity						
average haul one way 500 m	-	0.05	0.32	-	m^3	**0.37**
average haul one way 1,000 m	-	0.07	0.42	-	m^3	**0.49**
average haul one way 2,000 m	0.01	0.08	0.53	-	m^3	**0.61**
average haul one way 3,000 m	0.01	0.11	0.74	-	m^3	**0.85**
Twin engined motorised scraper, 34 m^3 capacity						
average haul one way 500 m	-	0.03	0.45	-	m^3	**0.48**
average haul one way 1,000 m	-	0.05	0.67	-	m^3	**0.72**
average haul one way 2,000 m	-	0.05	0.67	-	m^3	**0.72**
average haul one way 3,000 m	-	0.07	0.90	-	m^3	**0.96**
Excavation by hand						
Desirable for work around live services or in areas of highly restricted access.						
Excavate and load into skip or dumper bucket						
loose material	2.02	22.75	-	-	m^3	**22.75**
compacted soil or clay	3.15	35.47	-	-	m^3	**35.47**
mass concrete or sandstone	3.00	33.78	-	-	m^3	**33.78**
broken rock	2.95	33.22	-	-	m^3	**33.22**
existing sub-base or pipe surrounds	3.12	35.13	-	-	m^3	**35.13**
Excavate by hand using pneumatic equipment						
Excavate below ground using 1.80 m^3/min single tool compressor and pneumatic breaker and load material into skip or dumper bucket						
rock (medium drill)	2.12	23.87	12.83	-	m^3	**36.70**
brickwork or mass concrete	2.76	31.08	13.61	-	m^3	**44.69**
reinforced concrete	3.87	43.58	21.92	-	m^3	**65.50**
asphalt in carriageways	1.15	12.95	6.19	-	m^3	**19.14**

CLASS E: EXCAVATION

Item	Gang hours	Labour £	Plant £	Material £	Unit	**Total rate £**
COMPARATIVE COSTS - EARTH MOVING – cont'd						
Comparative prices for ancillary equipment						
Excavate using 6 tonne to break out (JCB 3CX and Montalbert 125 breaker)						
medium hard rock	0.43	6.30	9.61	-	m^3	**15.90**
brickwork or mass concrete	0.54	7.91	12.05	-	m^3	**19.96**
reinforced concrete	0.64	9.38	14.28	-	m^3	**23.66**
Load material into skip or dumper						
Load material into skip or dumper bucket using a 11.5 tonne crawler backacter with 0.80 m^3 rock bucket						
medium hard rock	0.07	1.03	1.45	-	m^3	**2.47**
brickwork or mass concrete	0.08	1.17	1.65	-	m^3	**2.82**
reinforced concrete	0.09	1.32	1.86	-	m^3	**3.18**
RILLING AND BLASTING IN ROCK						
The cost of blasting is controlled by the number of holes and the length of drilling required to achieve the tolerances and degree of shatter required, e.g. line drilling to trenches, depth of drilling to control horizontal overbreak.						
Drilling with rotary percussion drills						
105 - 110 mm diameter; hard rock	-	-	-	-	m	**12.69**
105 - 110 mm diameter; sandstone	-	-	-	-	m	**6.42**
125 mm diameter; hard rock	-	-	-	-	m	**10.08**
125 mm diameter; sandstone	-	-	-	-	m	**9.06**
165 mm diameter; hard rock	-	-	-	-	m	**11.55**
165 mm diameter; sandstone	-	-	-	-	m	**11.55**
Drilling and blasting in open cut for bulk excavation excluding cost of excavatio or trimming						
hard rock	-	-	-	-	m	**2.92**
sandstone	-	-	-	-	m	**3.95**
Drilling and blasting for quarry operations with face height exceeding 10 m						
hard rock	-	-	-	-	m	**2.79**
sandstone	-	-	-	-	m	**2.79**
Drilling and blasting in trenches excluding cost of excavation or trimming						
trench width ne 1.0 m	-	-	-	-	m	**21.92**
trench width 1.0 - 1.5 m	-	-	-	-	m	**19.00**
trench width over 1.5 m	-	-	-	-	m	**16.10**
Secondary blasting to boulders						
pop shooting	-	-	-	-	m	**6.87**
plaster shooting	-	-	-	-	m	**3.66**

CLASS E: EXCAVATION

Item	Gang hours	Labour £	Plant £	Material £	Unit	Total rate £
DEWATERING						
The following unit costs are for dewatering pervious ground only and are for sets of equipment comprising :						
1 nr diesel driven pump (WP 150/60 or similar) complete with allowance of £30 for fuel	-	-	-	-	day	**74.70**
50 m of 150 mm diameter header pipe	-	-	-	-	day	**18.19**
35 nr of disposable well points	-	-	-	-	buy	**278.93**
18 m of delivery pipe	-	-	-	-	day	**7.17**
1 nr diesel driven standby pump	-	-	-	-	day	**33.24**
1 nr jetting pump with hoses (for installation of wellpoints only)	-	-	-	-	-	**58.17**
attendant labour and plant (2 hrs per day) inclusive of small dumper and bowser)	-	-	-	-	-	**93.06**
Costs are based on 10 hr shifts with attendant labour and plant (specialist advice)						
Guide price for single set of equipment comprising pump, 150 mm diameter header pipe, 35 nr well points, delivery pipes and attendant labour and plant						
Bring to site equipment and remove upon completion	-	-	-	-	sum	**2152.50**
Installation costs						
hire of jetting pump with hoses; 1 day	-	-	-	-	sum	**58.17**
purchase of well points; 35 Nr	-	-	-	-	sum	**278.93**
labour and plant; 10 hours	-	-	-	-	sum	**930.62**
Operating costs						
hire of pump, header pipe, delivery pipe and standby pump complete with fuel etc. and 2 hours attendant labour and plant	-	-	-	-	day	**319.42**

CLASS F: IN SITU CONCRETE

Item	Gang hours	Labour £	Plant £	Material £	Unit	**Total rate £**
NOTES						
The unit rates in this Section are based on nett measurements and appropriate adjustments should be made for unmeasured excess (e.g. additional blinding thickness as a result of ground conditions).						
The unit rates for the provision of concrete are based on ready mixed concrete in full loads delivered to site within 5 miles (8km) of the concrete mixing plant and include an allowance for wastage prior to placing.						
This section assumes optimum outputs of an efficiently controlled pour with no delays caused by out of sequence working and no abnormal conditions that would affect continuity of work.						
RESOURCES - LABOUR						
Concrete gang						
1 ganger or chargehand (skill rate 4)		12.80				
2 skilled operatives (skill rate 4)		24.10				
4 unskilled operatives (general)		45.04				
1 plant operator (skill rate 3) - 25% of time		3.66				
Total Gang Rate / Hour	£	**85.60**				
RESOURCES - MATERIALS						
The following costs do not reflect in the rates and should be considered separately						
Delivery to site for each additional mile from the concrete plant further than 5 miles (8km)	-	-	-	1.10	m^3	**1.10**
Mix design, per trial mix	-	-	-	162.02	mix	**162.02**
Part loads, per m^3 below full load	-	-	-	24.80	m^3	**24.80**
Waiting time (in excess of 6 mins/m^3 'norm' discharge time)	-	-	-	51.70	hr	**51.70**
Making and testing concrete cube	0.69	7.75	-	-	nr	**7.75**
Pumping from ready mix truck to point of placing at the rate of 25 m^3 / hour	0.11	1.18	1.68	-	m^3	**2.86**
Pumping from ready mix truck to point of placing at the rate of 45 m^3 / hour	0.05	0.57	1.77	-	m^3	**2.34**
RESOURCES - PLANT						
Concrete						
22RB Crane (50% of time)			13.26			
1.00 m^3 concrete skip (50% of time)			0.45			
11.30 m^3/min compressor, 2 tool			16.03			
four 54 mm poker vibrators			4.84			
Total Rate / Hour		£	**34.58**			

CLASS F: IN SITU CONCRETE

Item	Gang hours	Labour £	Plant £	Material £	Unit	**Total rate £**
PROVISION OF CONCRETE						
Standard mix; cement to BS 12						
Type ST1	-	-	-	75.45	m³	-
Type ST2	-	-	-	75.89	m³	-
Type ST3	-	-	-	78.59	m³	-
Type ST4	-	-	-	80.13	m³	-
Type ST5	-	-	-	82.07	m³	-
Standard mix; sulphate resisting cement to BS 4027						
Type ST1	-	-	-	84.38	m³	-
Type ST2	-	-	-	84.82	m³	-
Type ST3	-	-	-	87.52	m³	-
Type ST4	-	-	-	89.05	m³	-
Type ST5	-	-	-	90.99	m³	-
Designed mix; cement to BS 12						
Grade C7.5						
10 mm aggregate	-	-	-	76.25	m³	-
20 mm aggregate	-	-	-	71.77	m³	-
40 mm aggregate	-	-	-	75.14	m³	-
Grade C10						
10 mm aggregate	-	-	-	77.32	m³	-
20 mm aggregate	-	-	-	72.88	m³	-
40 mm aggregate	-	-	-	76.10	m³	-
Grade C15						
10 mm aggregate	-	-	-	78.96	m³	-
20 mm aggregate	-	-	-	74.66	m³	-
40 mm aggregate	-	-	-	77.96	m³	-
Grade C20						
10 mm aggregate	-	-	-	80.30	m³	-
20 mm aggregate	-	-	-	75.92	m³	-
40 mm aggregate	-	-	-	79.22	m³	-
Grade C25						
10 mm aggregate	-	-	-	80.81	m³	-
20 mm aggregate	-	-	-	77.20	m³	-
40 mm aggregate	-	-	-	80.01	m³	-
Grade C30						
10 mm aggregate	-	-	-	82.13	m³	-
20 mm aggregate	-	-	-	78.49	m³	-
40 mm aggregate	-	-	-	81.22	m³	-
Grade C40						
10 mm aggregate	-	-	-	86.01	m³	-
20 mm aggregate	-	-	-	82.22	m³	-
Grade C50						
10 mm aggregate	-	-	-	89.36	m³	-
20 mm aggregate	-	-	-	86.64	m³	-
Grade C60						
10 mm aggregate	-	-	-	90.41	m³	-
20 mm aggregate	-	-	-	91.51	m³	-
Designed mix; sulphate resisting cement to BS 4027						
Grade C7.5						
10 mm aggregate	-	-	-	84.97	m³	-
20 mm aggregate	-	-	-	80.49	m³	-
40 mm aggregate	-	-	-	83.86	m³	-

CLASS F: IN SITU CONCRETE

Item	Gang hours	Labour £	Plant £	Material £	Unit	Total rate £
PROVISION OF CONCRETE – cont'd						
Standard mix; sulphate resisting cement to BS 4027 – cont'd						
Grade C10						
10 mm aggregate	-	-	-	86.18	m³	-
20 mm aggregate	-	-	-	81.74	m³	-
40 mm aggregate	-	-	-	84.96	m³	-
Grade C15						
10 mm aggregate	-	-	-	87.97	m³	-
20 mm aggregate	-	-	-	83.66	m³	-
40 mm aggregate	-	-	-	86.97	m³	-
Grade C20						
10 mm aggregate	-	-	-	89.40	m³	-
20 mm aggregate	-	-	-	85.01	m³	-
40 mm aggregate	-	-	-	88.32	m³	-
Grade C25						
10 mm aggregate	-	-	-	90.20	m³	-
20 mm aggregate	-	-	-	86.59	m³	-
40 mm aggregate	-	-	-	89.40	m³	-
Grade C30						
10 mm aggregate	-	-	-	91.67	m³	-
20 mm aggregate	-	-	-	88.03	m³	-
40 mm aggregate	-	-	-	90.76	m³	-
Prescribed mix; cement to BS 12						
Grade C7.5						
10 mm aggregate	-	-	-	68.45	m³	-
20 mm aggregate	-	-	-	64.23	m³	-
40 mm aggregate	-	-	-	67.39	m³	-
Grade C10						
10 mm aggregate	-	-	-	69.47	m³	-
20 mm aggregate	-	-	-	65.25	m³	-
40 mm aggregate	-	-	-	68.31	m³	-
Grade C15						
10 mm aggregate	-	-	-	71.02	m³	-
20 mm aggregate	-	-	-	66.93	m³	-
40 mm aggregate	-	-	-	70.01	m³	-
Grade C20						
10 mm aggregate	-	-	-	72.30	m³	-
20 mm aggregate	-	-	-	68.64	m³	-
40 mm aggregate	-	-	-	71.25	m³	-
Grade C25						
10 mm aggregate	-	-	-	72.78	m³	-
20 mm aggregate	-	-	-	69.35	m³	-
40 mm aggregate	-	-	-	71.92	m³	-
Grade C30						
10 mm aggregate	-	-	-	74.03	m³	-
20 mm aggregate	-	-	-	70.55	m³	-
40 mm aggregate	-	-	-	73.17	m³	-
Grade C40						
10 mm aggregate	-	-	-	77.72	m³	-
20 mm aggregate	-	-	-	74.10	m³	-
Grade C50						
10 mm aggregate	-	-	-	80.90	m³	-
20 mm aggregate	-	-	-	78.31	m³	-
Grade C60						
10 mm aggregate	-	-	-	81.83	m³	-
20 mm aggregate	-	-	-	82.94	m³	-

CLASS F: IN SITU CONCRETE

Item	Gang hours	Labour £	Plant £	Material £	Unit	**Total rate £**
Prescribed mix; sulphate resisting cement to BS 4027						
Grade C7.5						
10 mm aggregate	-	-	-	77.17	m³	-
20 mm aggregate	-	-	-	72.95	m³	-
40 mm aggregate	-	-	-	76.11	m³	-
Grade C10						
10 mm aggregate	-	-	-	78.33	m³	-
20 mm aggregate	-	-	-	74.11	m³	-
40 mm aggregate	-	-	-	77.17	m³	-
Grade C15						
10 mm aggregate	-	-	-	80.03	m³	-
20 mm aggregate	-	-	-	75.94	m³	-
40 mm aggregate	-	-	-	79.02	m³	-
Grade C20						
10 mm aggregate	-	-	-	81.39	m³	-
20 mm aggregate	-	-	-	77.73	m³	-
40 mm aggregate	-	-	-	80.35	m³	-
Grade C25						
10 mm aggregate	-	-	-	82.17	m³	-
20 mm aggregate	-	-	-	78.74	m³	-
40 mm aggregate	-	-	-	81.31	m³	-
Grade C30						
10 mm aggregate	-	-	-	83.57	m³	-
20 mm aggregate	-	-	-	80.09	m³	-
40 mm aggregate	-	-	-	82.71	m³	-
PLACING OF CONCRETE; MASS						
Blinding; thickness						
ne 150 mm	0.18	15.41	0.36	-	m³	**15.77**
150 - 300 mm	0.16	13.70	1.06	-	m³	**14.75**
300 - 500 mm	0.14	11.98	1.75	-	m³	**13.73**
exceeding 500 mm	0.12	10.27	2.44	-	m³	**12.71**
Bases, footings, pile caps and ground slabs; thickness						
ne 150 mm	0.20	17.12	6.91	-	m³	**24.04**
150 - 300 mm	0.17	14.55	5.90	-	m³	**20.45**
300 - 500 mm	0.15	12.84	5.21	-	m³	**18.05**
exceeding 500 mm	0.14	11.98	4.84	-	m³	**16.83**
ADD to the above for placing against an excavated surface	0.03	2.14	0.85	-	m³	**2.99**
Walls; thickness						
ne 150 mm	0.21	17.98	7.28	-	m³	**25.26**
150 - 300 mm	0.15	12.84	5.21	-	m³	**18.05**
300 - 500 mm	0.13	11.13	4.51	-	m³	**15.64**
exceeding 500 mm	0.12	10.27	4.15	-	m³	**14.42**
ADD to the above for placing against an excavated surface	0.03	2.57	1.06	-	m³	**3.62**
Other concrete forms						
plinth 1000 x 1000 x 600 mm	0.33	28.25	11.43	-	m³	**39.68**
plinth 1500 x 1500 x 750 mm	0.25	21.40	8.66	-	m³	**30.06**
plinth 2000 x 2000 x 600 mm	0.20	17.12	6.91	-	m³	**24.04**
surround to precast concrete manhole chambers 200 mm thick	0.29	24.82	10.04	-	m³	**34.87**

CLASS F: IN SITU CONCRETE

Item	Gang hours	Labour £	Plant £	Material £	Unit	Total rate £
PLACING OF CONCRETE; REINFORCED						
Bases, footings, pile caps and ground slabs; thickness						
ne 150 mm	0.21	17.98	7.28	-	m^3	**25.26**
150 - 300 mm	0.18	15.41	6.23	-	m^3	**21.63**
300 - 500 mm	0.16	13.70	5.53	-	m^3	**19.23**
exceeding 500 mm	0.15	12.84	5.21	-	m^3	**18.05**
Suspended slabs; thickness						
ne 150 mm	0.27	23.11	9.35	-	m^3	**32.47**
150 - 300 mm	0.21	17.98	7.28	-	m^3	**25.26**
300 - 500 mm	0.19	16.26	6.59	-	m^3	**22.85**
exceeding 500 mm	0.19	16.26	6.59	-	m^3	**22.85**
Walls; thickness						
ne 150 mm	0.29	24.82	10.04	-	m^3	**34.87**
150 - 300 mm	0.22	18.83	7.61	-	m^3	**26.44**
300 - 500 mm	0.20	17.12	6.91	-	m^3	**24.04**
exceeding 500 mm	0.20	17.12	6.91	-	m^3	**24.04**
Columns and piers; cross-sectional area						
ne 0.03 m^2	0.50	42.80	17.29	-	m^3	**60.09**
0.03 - 0.10 m^2	0.40	34.24	13.83	-	m^3	**48.08**
0.10 - 0.25 m^2	0.35	29.96	12.12	-	m^3	**42.08**
0.25 - 1.00 m^2	0.35	29.96	12.12	-	m^3	**42.08**
exceeding 1 m^2	0.28	23.97	9.68	-	m^3	**33.65**
Beams; cross-sectional area						
ne 0.03 m^2	0.50	42.80	17.29	-	m^3	**60.09**
0.03 - 0.10 m^2	0.40	34.24	13.83	-	m^3	**48.08**
0.10 - 0.25 m^2	0.35	29.96	12.12	-	m^3	**42.08**
0.25 - 1.00 m^2	0.35	29.96	12.12	-	m^3	**42.08**
exceeding 1 m^2	0.28	23.97	9.68	-	m^3	**33.65**
Casings to metal sections; cross-sectional area						
ne 0.03 m^2	0.47	40.23	16.27	-	m^3	**56.50**
0.03 - 0.10 m^2	0.47	40.23	16.27	-	m^3	**56.50**
0.10 - 0.25 m^2	0.40	34.24	13.83	-	m^3	**48.07**
0.25 - 1.00 m^2	0.40	34.24	13.83	-	m^3	**48.07**
exceeding 1 m^2	0.35	29.96	12.12	-	m^3	**42.08**
PLACING OF CONCRETE; PRESTRESSED						
Suspended slabs; thickness						
ne 150 mm	0.28	23.97	9.68	-	m^3	**33.65**
150 - 300 mm	0.22	18.83	7.61	-	m^3	**26.44**
300 - 500 mm	0.20	17.12	6.91	-	m^3	**24.04**
exceeding 500 mm	0.19	16.26	6.59	-	m^3	**22.85**
Beams; cross-sectional area						
ne 0.03 m^2	0.50	42.80	17.29	-	m^3	**60.09**
0.03 - 0.10 m^2	0.40	34.24	13.83	-	m^3	**48.08**
0.10 - 0.25 m^2	0.35	29.96	12.12	-	m^3	**42.08**
0.25 - 1.00 m^2	0.35	29.96	12.12	-	m^3	**42.08**
exceeding 1 m^2	0.28	23.97	9.68	-	m^3	**33.65**

CLASS G: CONCRETE ANCILLARIES

Item	Gang hours	Labour £	Plant £	Material £	Unit	Total rate £
RESOURCES - LABOUR						
Formwork gang - small areas						
1 foreman joiner (craftsman)		18.47				
1 joiner (craftsman)		17.35				
1 unskilled operative (general)		11.26				
1 plant operator (craftsman) - 50% of time		9.37				
Total Gang Rate / Hour	£	**56.45**				
Formwork gang - large areas						
1 foreman joiner (craftsman)		18.47				
2 joiners (craftsman)		34.70				
1 unskilled operative (general)		11.26				
1 plant operator (craftsman) - 25% of time		4.68				
Total Gang Rate / Hour	£	**69.11**				
Reinforcement gang						
1 foreman steel fixer (craftsman)		18.47				
4 steel fixers (craftsman)		69.40				
1 unskilled operative (general)		11.26				
1 plant operator (craftsman) - 25% of time		4.68				
Total Gang Rate / Hour	£	**103.81**				
Reinforcement - on-site bending/baling gang						
1 steel fixer (craftsman)		17.35				
1 unskilled operative (general)		11.26				
1 plant operator (craftsman) - 25% of time		4.68				
Total Gang Rate / Hour	£	**33.29**				
Joints gang						
1 ganger/chargehand (skill rate 4)		12.05				
1 skilled operative (skill rate 4)		12.05				
1 unskilled operative (general)		11.26				
Total Gang Rate / Hour	£	**35.36**				
Concrete accessories gang						
1 ganger/chargehand (skill rate 4)		12.05				
1 skilled operative (skill rate 4)		12.05				
1 unskilled operative (general)		11.26				
Total Gang Rate / Hour	£	**35.36**				
RESOURCES - PLANT						
Formwork - small areas						
20t crawler crane - 50% of time			14.43			
22" diameter saw bench			1.51			
allowance for small power tools			2.04			
Total Rate / Hour		£	**17.98**			
Formwork - large areas						
20t crawler crane - 25% of time			7.21			
22" diameter saw bench			1.51			
allowance for small power tools			2.04			
Total Rate / Hour		£	**10.77**			
Reinforcement						
30 t crawler crane - 25% of time			7.88			
bar cropper			2.26			
small power tools			0.66			
support acrows, tirfors, kentledge, etc.			0.80			
Total Rate / Hour		£	**11.60**			

CLASS G: CONCRETE ANCILLARIES

Item	Gang hours	Labour £	Plant £	Material £	Unit	**Total rate £**
RESOURCES - MATERIALS						
Formwork						
Formwork materials include for shutter, bracing, ties, support, kentledge and all consumables. The following unit costs do not include for formwork outside the payline and are based on an optimum of a minimum 8 uses with 10% per use towards the cost of repairs / replacement of components damaged during disassembly						
ADD to formwork material costs generally depending on the number of uses :						
Nr of uses — % Addition — % Waste						
1 — + 90 to 170 — +7						
2 — + 50 to 80 — +7						
3 — + 15 to 30 — +6						
6 — + 5 to 10 — +6						
8 — No change — +5						
10 — - 5 to 7 — +5						
Reinforcement						
Reinforcement materials include for bars, tying wire, spacers, couplers and steel supports for bottom layer reinforcement (stools, chairs and risers).						
FORMWORK; ROUGH FINISH						
Plane horizontal, width						
ne 0.1 m	0.10	5.64	1.80	1.99	m	**9.43**
0.1- 0.20 m	0.18	10.16	3.24	2.26	m	**15.66**
0.2 - 0.40 m	0.40	27.64	4.31	5.94	m²	**37.89**
0.4 - 1.22 m	0.40	27.64	4.31	5.94	m²	**37.89**
exceeding 1.22 m	0.38	26.26	4.09	5.94	m²	**36.29**
Plane sloping, width						
ne 0.1 m	0.11	6.21	1.98	1.99	m	**10.18**
0.1- 0.20 m	0.20	11.29	3.60	2.26	m	**17.15**
0.2 - 0.40 m	0.43	29.72	4.60	8.08	m²	**42.39**
0.4 - 1.22 m	0.43	29.72	4.60	8.08	m²	**42.39**
exceeding 1.22 m	0.38	26.26	4.09	8.08	m²	**38.43**
Plane battered, width						
ne 0.1 m	0.12	6.77	2.16	3.70	m	**12.64**
0.1- 0.20 m	0.20	11.29	3.60	3.97	m	**18.86**
0.2 - 0.40 m	0.43	29.72	4.70	10.33	m²	**44.75**
0.4 - 1.22 m	0.43	29.72	4.70	10.33	m²	**44.75**
exceeding 1.22 m	0.40	27.64	4.31	10.33	m²	**42.28**
Plane vertical, width						
ne 0.1 m	0.12	6.77	2.16	3.15	m	**12.08**
0.1- 0.20 m	0.19	10.73	3.42	3.42	m	**17.56**
0.2 - 0.40 m	0.70	48.38	7.54	8.70	m²	**64.62**
0.4 - 1.22 m	0.51	35.25	5.42	8.70	m²	**49.37**
exceeding 1.22 m	0.47	32.48	5.14	8.70	m²	**46.32**
Curved to one radius in one plane, 0.5 m radius, width						
ne 0.1 m	0.19	10.73	3.42	4.22	m	**18.37**
0.1- 0.20 m	0.25	14.11	4.50	4.49	m	**23.10**
0.2 - 0.40 m	0.90	62.20	9.69	9.77	m²	**81.66**
0.4 - 1.22 m	0.72	49.76	7.75	9.77	m²	**67.29**
exceeding 1.22 m	0.65	44.92	6.93	9.77	m²	**61.62**

CLASS G: CONCRETE ANCILLARIES

Item	Gang hours	Labour £	Plant £	Material £	Unit	**Total rate £**
Curved to one radius in one plane, 2 m radius, width						
ne 0.1 m	0.18	10.16	3.24	4.22	m	**17.62**
0.1- 0.20 m	0.22	12.42	3.96	4.49	m	**20.87**
0.2 - 0.40 m	0.84	58.05	9.05	9.77	m²	**76.87**
0.4 - 1.22 m	0.66	45.61	7.11	9.77	m²	**62.49**
exceeding 1.22 m	0.52	35.94	5.60	9.77	m²	**51.31**
For voids						
small void; depth ne 0.5 m	0.07	3.95	-	3.44	nr	**7.40**
small void; depth 0.5 - 1.0 m	0.12	6.77	-	6.70	nr	**13.48**
small void; depth ne 1.0 - 2.0 m	0.16	9.03	-	12.44	nr	**21.47**
large void; depth ne 0.5 m	0.14	7.90	-	10.22	nr	**18.12**
large void; depth 0.5 - 1.0 m	0.33	18.63	-	20.19	nr	**38.82**
large void; depth 1.0 - 2.0 m	0.58	32.74	-	39.84	nr	**72.58**
For concrete components of constant cross-section						
beams; 200 x 200 mm	0.48	27.10	8.63	4.29	m	**40.01**
beams; 500 x 500 mm	0.55	31.05	9.89	9.18	m	**50.12**
beams; 500 x 800 mm	0.67	37.82	12.05	10.15	m	**60.03**
columns; 200 x 200 mm	0.55	31.05	9.89	6.04	m	**46.98**
columns; 300 x 300 mm	0.55	31.05	9.89	8.45	m	**49.39**
columns; 300 x 500 mm	0.62	35.00	11.15	9.73	m	**55.88**
to walls; 1.0 m high thickness 250 mm	1.10	76.02	11.85	11.30	m	**99.16**
to walls; 1.5 m high thickness 300 mm	1.50	103.67	16.15	16.04	m	**135.85**
box culvert; 2 x 2 m internally and wall thickness 300 mm	4.50	311.00	48.46	104.89	m	**464.34**
projections (100 mm deep)	0.10	5.64	1.80	2.00	m	**9.44**
intrusions (100 mm deep)	0.10	5.64	1.80	2.00	m	**9.44**
Allowance for additional craneage and rub up where required						
ADD to items measures linear	0.04	2.26	0.72	0.03	m	**3.00**
ADD to items measures m²	0.12	6.77	2.16	0.09	m²	**9.02**
FORMWORK; FAIR FINISH						
Plane horizontal, width						
ne 0.1 m	0.10	5.64	1.80	3.07	m	**10.51**
0.1- 0.20 m	0.18	10.16	3.24	3.45	m	**16.85**
0.2 - 0.40 m	0.40	27.64	4.31	10.92	m²	**42.88**
0.4 - 1.22 m	0.40	27.64	4.31	10.92	m²	**42.88**
exceeding 1.22 m	0.38	26.26	4.09	10.92	m²	**41.28**
Plane sloping, width						
ne 0.1 m	0.11	6.21	1.98	3.07	m	**11.26**
0.1- 0.20 m	0.20	11.29	2.15	3.45	m	**16.90**
0.2 - 0.40 m	0.43	29.72	12.96	16.93	m²	**59.61**
0.4 - 1.22 m	0.43	29.72	12.96	16.93	m²	**59.61**
exceeding 1.22 m	0.38	26.26	4.09	16.93	m²	**47.28**
Plane battered, width						
ne 0.1 m	0.12	6.77	2.16	5.75	m	**14.68**
0.1- 0.20 m	0.20	11.29	3.60	6.13	m	**21.02**
0.2 - 0.40 m	0.43	29.72	4.63	21.11	m²	**55.46**
0.4 - 1.22 m	0.43	29.72	4.63	21.11	m²	**55.46**
exceeding 1.22 m	0.40	27.64	4.31	21.11	m²	**53.06**

CLASS G: CONCRETE ANCILLARIES

Item	Gang hours	Labour £	Plant £	Material £	Unit	**Total rate £**
FORMWORK; FAIR FINISH – cont'd						
Plane vertical, width						
ne 0.1 m	0.12	6.77	2.16	5.19	m	**14.13**
0.1- 0.20 m	0.20	11.29	3.60	5.58	m	**20.47**
0.2 - 0.40 m	0.72	49.76	7.75	17.55	m²	**75.07**
0.4 - 1.22 m	0.53	36.63	5.71	17.55	m²	**59.89**
exceeding 1.22 m	0.48	33.17	5.17	17.55	m²	**55.90**
Curved to one radius in one plane, 0.5 m radius, width						
ne 0.1 m	0.20	11.29	3.60	8.20	m	**23.08**
0.1- 0.20 m	0.25	14.11	4.50	8.58	m	**27.19**
0.2 - 0.40 m	0.90	62.20	9.69	20.56	m²	**92.45**
0.4 - 1.22 m	0.72	49.76	7.75	20.56	m²	**78.07**
exceeding 1.22 m	0.65	44.92	7.00	20.56	m²	**72.48**
Curved to one radius in one plane, 2.0 m radius, width						
ne 0.1 m	0.18	10.16	3.24	8.20	m	**21.60**
0.1- 0.20 m	0.22	12.42	3.96	8.58	m	**24.96**
0.2 - 0.40 m	0.84	58.05	9.05	20.56	m²	**87.66**
0.4 - 1.22 m	0.66	45.61	7.11	20.56	m²	**73.28**
exceeding 1.22 m	0.52	35.94	5.60	20.56	m²	**62.09**
For voids, using void former						
small void; depth ne 0.5 m	0.07	3.95	-	5.02	nr	**8.97**
small void; depth 0.5 - 1.0 m	0.12	6.77	-	9.68	nr	**16.45**
small void; depth ne 1.0 - 2.0 m	0.16	9.03	-	18.44	nr	**27.47**
large void; depth ne 0.5 m	0.14	7.90	-	10.81	nr	**18.72**
large void; depth 0.5 - 1.0 m	0.33	18.63	-	21.61	nr	**40.24**
large void; depth 1.0 - 2.0 m	0.58	32.74	-	38.20	nr	**70.94**
For concrete components of constant cross-section						
beams; 200 x 200 mm	0.49	27.66	8.81	8.83	m	**45.30**
beams; 500 x 500 mm	0.57	32.18	10.25	15.79	m	**58.22**
beams; 500 x 800 mm	0.69	38.95	12.41	19.91	m	**71.27**
columns; 200 x 200 mm	0.57	32.18	10.25	10.81	m	**53.23**
columns; 300 x 300 mm	0.57	32.18	10.25	15.30	m	**57.72**
columns; 300 x 500 mm	0.64	36.13	11.51	18.28	m	**65.92**
to walls; 1.0 m high thickness 250 mm	1.20	82.93	12.92	21.27	m	**117.13**
to walls; 1.5 m high thickness 300 mm	1.60	110.58	17.23	31.00	m	**158.81**
box culvert; 2 x 2 m internally and wall thickness 300 mm	4.60	317.91	49.53	213.42	m	**580.87**
projections (100 mm deep)	0.10	5.64	1.80	4.34	m	**11.79**
intrusions (100 mm deep)	0.10	5.64	1.80	4.34	m	**11.79**
Allowance for additional craneage and rub up where required						
ADD to items measures linear	0.14	7.90	2.52	0.03	m	**10.45**
ADD to items measures m²	0.12	6.77	2.16	0.12	m²	**9.05**
FORMWORK; EXTRA SMOOTH FINISH						
Plane horizontal, width						
ne 0.1 m	0.10	5.64	1.80	3.28	m	**10.73**
0.1- 0.20 m	0.18	10.16	3.24	3.85	m	**17.24**
0.2 - 0.40 m	0.40	27.64	4.31	12.94	m²	**44.89**
0.4 - 1.22 m	0.40	27.64	4.31	12.94	m²	**44.89**
exceeding 1.22 m	0.38	26.26	4.08	12.94	m²	**43.28**

CLASS G: CONCRETE ANCILLARIES

Item	Gang hours	Labour £	Plant £	Material £	Unit	**Total rate £**
Plane sloping, width						
ne 0.1 m	0.11	6.21	1.98	3.28	m	**11.47**
0.1- 0.20 m	0.20	11.29	3.60	3.85	m	**18.73**
0.2 - 0.40 m	0.43	29.72	4.65	18.94	m²	**53.31**
0.4 - 1.22 m	0.43	29.72	4.65	18.94	m²	**53.31**
exceeding 1.22 m	0.38	26.26	4.09	18.94	m²	**49.29**
Plane battered, width						
ne 0.1 m	0.12	6.77	2.16	5.96	m	**14.89**
0.1- 0.20 m	0.20	11.29	3.60	6.53	m	**21.41**
0.2 - 0.40 m	0.43	29.72	4.65	23.12	m²	**57.48**
0.4 - 1.22 m	0.43	29.72	4.65	23.12	m²	**57.48**
exceeding 1.22 m	0.40	27.64	4.31	23.12	m²	**55.07**
Plane vertical, width						
ne 0.1 m	0.12	8.29	2.16	5.41	m	**15.86**
0.1- 0.20 m	0.20	11.29	3.60	5.97	m	**20.86**
0.2 - 0.40 m	0.74	51.14	7.97	19.56	m²	**78.67**
0.4 - 1.22 m	0.53	36.63	5.72	19.56	m²	**61.92**
exceeding 1.22 m	0.48	33.17	5.17	19.56	m²	**57.91**
Curved to one radius in one plane, 0.5 m radius, width						
ne 0.1 m	0.20	11.29	3.60	8.41	m	**23.30**
0.1- 0.20 m	0.25	14.11	4.49	5.97	m	**24.58**
0.2 - 0.40 m	0.90	62.20	9.69	22.57	m²	**94.46**
0.4 - 1.22 m	0.72	49.76	7.75	22.57	m²	**80.08**
exceeding 1.22 m	0.65	44.92	7.02	22.57	m²	**74.51**
Curved to one radius in one plane, 1.0 m radius, width						
ne 0.1 m	0.18	10.16	3.24	8.41	m	**21.81**
0.1- 0.20 m	0.22	12.42	3.96	5.97	m	**22.35**
0.2 - 0.40 m	0.84	58.05	14.82	22.57	m²	**95.44**
0.4 - 1.22 m	0.66	45.61	7.11	22.57	m²	**75.29**
exceeding 1.22 m	0.52	35.94	5.60	22.57	m²	**64.10**
For voids, using void former						
small void; depth ne 0.5 m	0.07	3.95	-	8.73	nr	**12.69**
small void; depth 0.5 - 1.0 m	0.12	6.77	-	17.15	nr	**23.92**
small void; depth ne 1.0 - 2.0 m	0.16	9.03	-	32.24	nr	**41.28**
large void; depth ne 0.5 m	0.14	7.90	-	19.22	nr	**27.12**
large void; depth 0.5 - 1.0 m	0.33	18.63	-	37.52	nr	**56.15**
large void; depth 1.0 - 2.0 m	0.58	32.74	-	65.31	nr	**98.05**
For concrete components of constant cross-section						
beams; 200 x 200 mm	0.50	28.23	8.99	10.03	m	**47.25**
beams; 500 x 500 mm	0.57	32.18	6.13	18.80	m	**57.11**
beams; 500 x 800 mm	0.69	38.95	7.45	23.52	m	**69.92**
columns; 200 x 200 mm	0.57	32.18	6.13	12.41	m	**50.72**
columns; 300 x 300 mm	0.57	32.18	6.13	17.71	m	**56.01**
columns; 300 x 500 mm	0.64	36.13	6.89	21.50	m	**64.52**
to walls; 1.0 m high thickness 250 mm	1.20	82.93	12.92	25.29	m	**121.15**
to walls; 1.5 m high thickness 300 mm	1.60	110.58	17.23	37.03	m	**164.84**
box culvert; 2 x 2 m internally and wall thickness 300 mm	4.60	317.91	49.53	234.77	m	**602.21**
projections (100 mm deep)	0.10	5.64	1.80	4.74	m	**12.18**
intrusions (100 mm deep)	0.10	5.64	1.80	4.74	m	**12.18**
Allowance for additional craneage and rub up where required						
ADD to items measures linear	0.14	7.90	2.52	0.03	m	**10.45**
ADD to items measures m²	0.12	6.77	2.16	0.15	m²	**9.08**

CLASS G: CONCRETE ANCILLARIES

Item	Gang hours	Labour £	Plant £	Material £	Unit	Total rate £
REINFORCEMENT						
Plain round mild steel bars to BS 4449						
Bars; supplied in straight lengths						
6 mm nominal size	8.00	830.48	92.82	428.24	tonne	**1351.54**
8 mm nominal size	6.74	699.68	78.20	418.14	tonne	**1196.02**
10 mm nominal size	6.74	699.68	78.20	409.05	tonne	**1186.93**
12 mm nominal size	6.74	699.68	78.20	407.03	tonne	**1184.91**
16 mm nominal size	6.15	638.43	71.36	404.00	tonne	**1113.79**
20 mm nominal size	4.44	460.92	51.52	404.00	tonne	**916.43**
25 mm nominal size	4.44	460.92	51.52	404.00	tonne	**916.43**
32 mm nominal size	4.44	460.92	51.52	407.03	tonne	**919.46**
40 mm nominal size	4.44	460.92	51.52	420.16	tonne	**932.59**
Bars; supplied in bent and cut lengths						
6 mm nominal size	8.00	830.48	92.82	467.63	tonne	**1390.93**
8 mm nominal size	6.74	699.68	78.20	456.52	tonne	**1234.40**
10 mm nominal size	6.74	699.68	78.20	446.42	tonne	**1224.30**
12 mm nominal size	6.74	699.68	78.20	444.40	tonne	**1222.28**
16 mm nominal size	6.15	638.43	71.36	441.37	tonne	**1151.16**
20 mm nominal size	4.44	460.92	51.52	441.37	tonne	**953.80**
25 mm nominal size	4.44	460.92	51.52	441.37	tonne	**953.80**
32 mm nominal size	4.44	460.92	51.52	434.30	tonne	**946.73**
40 mm nominal size	4.44	460.92	51.52	459.55	tonne	**971.98**
Deformed high yield steel bars to BS 4449						
Bars; supplied in straight lengths						
6 mm nominal size	8.00	830.48	92.82	397.94	tonne	**1321.24**
8 mm nominal size	6.74	699.68	78.20	387.84	tonne	**1165.72**
10 mm nominal size	6.74	699.68	78.20	381.78	tonne	**1159.66**
12 mm nominal size	6.74	699.68	78.20	378.75	tonne	**1156.63**
16 mm nominal size	6.15	638.43	71.36	373.70	tonne	**1083.49**
20 mm nominal size	4.44	460.92	51.52	373.70	tonne	**886.13**
25 mm nominal size	4.44	460.92	51.52	373.70	tonne	**886.13**
32 mm nominal size	4.44	460.92	51.52	378.75	tonne	**891.18**
40 mm nominal size	4.44	460.92	51.52	381.78	tonne	**894.21**
Bars; supplied in bent and cut lengths						
6 mm nominal size	8.00	830.48	92.82	438.34	tonne	**1361.64**
8 mm nominal size	6.74	699.68	78.20	428.24	tonne	**1206.12**
10 mm nominal size	6.74	699.68	78.20	417.13	tonne	**1195.01**
12 mm nominal size	6.74	699.68	78.20	419.15	tonne	**1197.03**
16 mm nominal size	6.15	638.43	71.36	414.10	tonne	**1123.89**
20 mm nominal size	4.44	460.92	51.52	414.10	tonne	**926.53**
25 mm nominal size	4.44	460.92	51.52	414.10	tonne	**926.53**
32 mm nominal size	4.44	460.92	51.52	414.10	tonne	**926.53**
40 mm nominal size	4.44	460.92	51.52	422.18	tonne	**934.61**
Stainless steel bars						
Bars; supplied in straight lengths						
8 mm nominal size	6.74	699.68	78.20	2500.00	tonne	**3277.88**
10 mm nominal size	6.74	699.68	78.20	2580.00	tonne	**3357.88**
12 mm nominal size	6.74	699.68	78.20	2500.00	tonne	**3277.88**
16 mm nominal size	6.15	638.43	71.36	2450.00	tonne	**3159.79**
20 mm nominal size	4.44	460.92	51.52	2360.00	tonne	**2872.43**
25 mm nominal size	4.44	460.92	51.52	2300.00	tonne	**2812.43**
32 mm nominal size	4.44	460.92	51.52	2300.00	tonne	**2812.43**

CLASS G: CONCRETE ANCILLARIES

Item	Gang hours	Labour £	Plant £	Material £	Unit	**Total rate £**
Bars; supplied in bent and cut lengths						
8 mm nominal size	6.74	699.68	78.20	2600.00	tonne	**3377.88**
10 mm nominal size	6.74	699.68	78.20	2680.00	tonne	**3457.88**
12 mm nominal size	6.74	699.68	78.20	2600.00	tonne	**3377.88**
16 mm nominal size	6.15	638.43	71.36	2550.00	tonne	**3259.79**
20 mm nominal size	4.44	460.92	51.52	2460.00	tonne	**2972.43**
25 mm nominal size	4.44	460.92	51.52	2400.00	tonne	**2912.43**
32 mm nominal size	4.44	460.92	51.52	2400.00	tonne	**2912.43**
Additional allowances to bar reinforcement						
Add to the above bars						
12-15 m long; mild steel to BS 4449	-	-	-	20.50	tonne	**20.50**
12-15 m long; high yield steel to BS 4449	-	-	-	15.25	tonne	**15.25**
Over 15 m long, per 500 mm increment; mild steel to BS 4449	-	-	-	5.00	tonne	**5.00**
Over 15 m long, per 500 mm increment; high yield steel to BS 4449	-	-	-	3.40	tonne	**3.40**
Add for cutting, bending, tagging and baling reinforcement on site						
6 mm nominal size	4.87	162.12	56.51	1.20	tonne	**219.83**
8 mm nominal size	4.58	152.47	53.14	1.20	tonne	**206.81**
10 mm nominal size	3.42	113.85	39.68	1.20	tonne	**154.73**
12 mm nominal size	2.55	84.89	29.59	1.20	tonne	**115.68**
16 mm nominal size	2.03	67.58	23.56	1.20	tonne	**92.34**
20 mm nominal size	1.68	55.93	19.49	1.20	tonne	**76.62**
25 mm nominal size	1.68	55.93	19.49	1.20	tonne	**76.62**
32 mm nominal size	1.39	46.27	16.13	1.20	tonne	**63.61**
40 mm nominal size	1.39	46.27	16.13	1.20	tonne	**63.61**
Special joints						
Lenton type A couplers; threaded ends on reinforcing bars						
12 mm	0.09	9.34	-	8.96	nr	**18.30**
16 mm	0.09	9.34	-	10.31	nr	**19.65**
20 mm	0.09	9.34	-	14.68	nr	**24.02**
25 mm	0.09	9.34	-	19.89	nr	**29.24**
32 mm	0.09	9.34	-	27.39	nr	**36.73**
40 mm	0.09	9.34	-	37.78	nr	**47.13**
Lenton type B couplers; threaded ends on reinforcing bars						
12 mm	0.09	9.34	-	25.42	nr	**34.76**
16 mm	0.09	9.34	-	29.09	nr	**38.43**
20 mm	0.09	9.34	-	32.27	nr	**41.61**
25 mm	0.09	9.34	-	37.23	nr	**46.58**
32 mm	0.09	9.34	-	50.39	nr	**59.73**
40 mm	0.09	9.34	-	73.85	nr	**83.19**
Steel fabric to BS 4483						
Fabric						
nominal mass 0.77 kg/m²; ref D49	0.02	2.08	0.23	1.37	m²	**3.68**
nominal mass 1.54 kg/m²; ref D98	0.02	2.08	0.23	1.37	m²	**3.68**
nominal mass 1.54 kg/m²; ref A98	0.03	3.11	0.35	1.25	m²	**4.72**
nominal mass 2.22 kg/m²; ref A142	0.03	3.11	0.35	1.30	m²	**4.77**
nominal mass 2.61 kg/m²; ref C283	0.03	3.11	0.35	1.53	m²	**5.00**
nominal mass 3.02 kg/m²; ref A193	0.04	4.15	0.46	1.77	m²	**6.39**
nominal mass 3.05 kg/m²; ref B196	0.04	4.15	0.46	1.78	m²	**6.40**
nominal mass 3.41 kg/m²; ref C385	0.04	4.15	0.46	2.00	m²	**6.62**
nominal mass 3.73 kg/m²; ref B283	0.04	4.15	0.46	2.20	m²	**6.81**

CLASS G: CONCRETE ANCILLARIES

Item	Gang hours	Labour £	Plant £	Material £	Unit	Total rate £
REINFORCEMENT – cont'd						
Steel fabric to BS 4483 – cont'd						
Fabric						
nominal mass 3.95 kg/m²; ref A252	0.04	4.15	0.46	2.30	m²	**6.92**
nominal mass 4.34 kg/m²; ref C503	0.05	5.19	0.58	2.54	m²	**8.32**
nominal mass 4.35 kg/m²; ref B385	0.05	5.19	0.58	2.65	m²	**8.42**
nominal mass 5.55 kg/m²; ref C636	0.05	5.19	0.58	3.25	m²	**9.03**
nominal mass 5.93 kg/m²; ref B503	0.05	5.19	0.58	3.47	m²	**9.25**
nominal mass 6.16 kg/m²; ref A393	0.07	7.27	0.82	3.60	m²	**11.68**
nominal mass 6.72 kg/m²; ref C785	0.07	7.27	0.82	3.91	m²	**11.99**
nominal mass 8.14 kg/m²; ref B785	0.08	8.30	0.93	4.76	m²	**13.99**
nominal mass 10.90 kg/m²; ref B1131	0.09	9.34	1.05	6.37	m²	**16.76**
JOINTS						
Open surface plain; average width						
ne 0.5m; scabbling concrete for subsequent pour	0.04	1.41	0.88	-	m²	**2.29**
0.5 - 1m; scabbling concrete for subsequent pour	0.03	1.06	0.67	-	m²	**1.73**
Open surface with filler; average width						
ne 0.5m; 12mm Flexcell joint filler	0.04	1.41	0.88	3.26	m²	**5.55**
0.5 - 1m; 12mm Flexcell joint filler	0.04	1.41	0.88	3.26	m²	**5.55**
ne 0.5m; 19mm Flexcell joint filler	0.05	1.77	1.08	5.61	m²	**8.47**
0.5 - 1m; 19mm Flexcell joint filler	0.05	1.77	1.08	5.61	m²	**8.47**
Formed surface plain; average width (including formwork)						
ne 0.5m	0.24	8.49	5.25	8.05	m²	**21.79**
0.5 - 1.0m	0.24	8.49	5.25	8.05	m²	**21.79**
Formed surface with filler; average width						
ne 0.5m; 10mm Flexcell joint filler	0.40	14.14	6.59	11.31	m²	**32.04**
0.5 - 1m; 10mm Flexcell joint filler	0.41	14.50	8.96	11.31	m²	**34.77**
ne 0.5m; 19mm Flexcell joint filler	0.42	14.85	9.22	13.66	m²	**37.73**
0.5 - 1m; 19mm Flexcell joint filler	0.43	15.20	9.43	13.66	m²	**38.29**
ne 0.5m; 25mm Flexcell joint filler	0.45	15.91	9.84	15.17	m²	**40.93**
0.5 - 1m; 25mm Flexcell joint filler	0.47	16.62	10.31	15.17	m²	**42.10**
PVC						
Plastics or rubber waterstops						
160 mm centre bulb	0.04	1.41	-	3.28	m	**4.70**
junction piece	0.04	1.41	-	7.14	nr	**8.56**
210 mm centre bulb	0.05	1.77	-	4.69	m	**6.46**
junction piece	0.04	1.41	-	7.29	nr	**8.71**
260 mm centre bulb	0.05	1.77	-	5.49	m	**7.25**
junction piece	0.05	1.77	-	7.78	nr	**9.55**
170 mm flat dumbell	0.04	1.41	-	4.28	m	**5.69**
junction piece	0.05	1.77	-	52.83	nr	**54.59**
210 mm flat dumbell	0.04	1.41	-	5.55	m	**6.96**
junction piece	0.07	2.48	-	59.43	nr	**61.90**
250 mm flat dumbell	0.05	1.77	-	6.96	m	**8.73**
junction piece	0.09	3.18	-	108.85	nr	**112.03**
Polysulphide sealant; gun grade						
Sealed rebates or grooves						
10 x 20 mm	0.05	1.77	-	-	m	**1.77**
20 x 20 mm	0.07	2.48	-	0.01	m	**2.48**
25 x 20 mm	0.08	2.83	-	0.01	m	**2.84**

CLASS G: CONCRETE ANCILLARIES

Item	Gang hours	Labour £	Plant £	Material £	Unit	**Total rate £**
Mild steel						
Dowels, plain or greased						
12 mm diameter x 500 mm long	0.04	1.41	-	0.51	nr	**1.92**
16 mm diameter x 750 mm long	0.05	1.59	-	1.42	nr	**3.01**
20 mm diameter x 750 mm long	0.05	1.59	-	2.22	nr	**3.82**
25 mm diameter x 750 mm long	0.05	1.59	-	3.46	nr	**5.05**
32 mm diameter x 750 mm long	0.05	1.59	-	5.69	nr	**7.28**
Dowels, sleeved or capped						
12 mm diameter x 500 mm long, debonding agent for 250 mm and capped with pvc dowel cap	0.05	1.59	-	0.57	nr	**2.16**
16 mm diameter x 750 mm long, debonding agent for 375 mm and capped with pvc dowel cap	0.05	1.87	-	1.48	nr	**3.35**
20 mm diameter x 750 mm long, debonding agent for 375 mm and capped with pvc dowel cap	0.05	1.87	-	2.29	nr	**4.16**
25 mm diameter x 750 mm long, debonding agent for 375 mm and capped with pvc dowel cap	0.06	2.12	-	3.52	nr	**5.65**
32 mm diameter x 750 mm long, debonding agent for 375 mm and capped with pvc dowel cap	0.07	2.30	-	5.75	nr	**8.05**

CLASS G: CONCRETE ANCILLARIES

Item	Gang hours	Labour £	Plant £	Material £	Unit	Total rate £
POST-TENSIONED PRESTRESSING						
The design of prestressing is based on standard patented systems, each of which has produced its own method of anchoring, joining and stressing the cables or wires. The companies marketing the systems will either supply all the materails and fittings required together with the sale or hire of suitable jacks and equipment for prestressing and grouting or they will undertake to complete the work on a sub-contract basis. The rates given below are therefore indicative only of the probable labour and plant costs and do not include for any permanent materials. The advice of specialist contractors should be sought for more accurate rates based on the design for a particular contract. Pretensioned prestressing is normally used only in the manufacture of precast units utilising special beds set up in the supplier's factory.						
Labour and plant cost in post-tensioning; material cost excluded						
form ducts to profile including supports and fixings; 50 mm internal diameter	1.00	1.03	3.05	-	m	**4.08**
Extra for grout vents	1.00	5.14	-	-	nr	**5.14**
form ducts to profile including supports and fixings; 80 mm internal diameter	1.00	1.61	3.58	-	m	**5.19**
Extra for grout vents	1.00	5.14	-	-	nr	**5.14**
form ducts to profile including supports and fixings; 100 mm internal diameter	1.00	2.20	3.77	-	m	**5.97**
Extra for grout vents	1.00	5.14	-	-	nr	**5.14**
grout ducts including provision of equipment; 50 mm internal diameter	1.00	1.53	0.53	-	m	**2.06**
grout ducts including provision of equipment; 80 mm internal diameter	1.00	1.53	0.67	-	m	**2.20**
grout ducts including provision of equipment; 100 mm internal diameter	1.00	2.09	0.79	-	m	**2.88**
form tendons including spacers etc. and pull through ducts; 7 Nr strands	0.25	5.93	3.61	-	m	**9.54**
form tendons including spacers etc. and pull through ducts; 12 Nr strands	0.45	10.67	6.50	-	m	**17.17**
form tendons including spacers etc. and pull through ducts; 19 Nr strands	0.65	15.42	9.39	-	m	**24.81**
dead end anchorage; 7 Nr strands	0.75	17.79	10.83	-	nr	**28.62**
dead end anchorage; 12 Nr strands	0.95	22.53	13.72	-	nr	**36.26**
dead end anchorage; 19 Nr strands	1.15	27.28	16.61	-	nr	**43.89**
looped buried dead end anchorage; 7 Nr strands	0.50	11.86	7.22	-	nr	**19.08**
looped buried dead end anchorage; 12 Nr strands	0.66	15.66	9.53	-	nr	**25.19**
looped buried dead end anchorage; 19 Nr strands	0.89	21.11	12.86	-	nr	**33.97**
end anchorage including reinforcement; 7 Nr strands	1.67	39.61	24.12	-	nr	**63.73**
add to last for anchorage coupling	0.46	10.91	6.64	-	nr	**17.56**
end anchorage including reinforcement; 12 Nr strands	2.04	48.39	29.47	-	nr	**77.85**
add to last for anchorage coupling	0.79	18.74	11.41	-	nr	**30.15**

CLASS G: CONCRETE ANCILLARIES

Item	Gang hours	Labour £	Plant £	Material £	Unit	**Total rate £**
end anchorage including reinforcement; 19 Nr strands	2.56	60.72	36.98	-	nr	**97.70**
add to last for anchorage coupling	1.20	28.46	17.33	-	nr	**45.80**
stress and lock off including multimatic jack; 7 Nr strands	3.21	76.14	46.37	-	nr	**122.51**
stress and lock off including multimatic jack; 12 Nr strands	4.19	99.39	60.52	-	nr	**159.91**
stress and lock off including multimatic jack; 19 Nr strands	5.58	132.36	80.60	-	nr	**212.96**
cut off and seal ends of tendons; 7 Nr strands	0.20	4.74	2.89	-	nr	**7.63**
cut off and seal ends of tendons; 12 Nr strands	0.35	8.30	5.06	-	nr	**13.36**
cut off and seal ends of tendons; 19 Nr strands	0.55	13.05	7.94	-	nr	**20.99**
CONCRETE ACCESSORIES						
Finishing of top surfaces						
wood float; level	0.02	0.71	-	-	m²	**0.71**
wood float; falls or cross-falls	0.03	1.06	-	-	m²	**1.06**
steel trowel; level	0.03	1.06	-	-	m²	**1.06**
wood float; falls or cross-falls	0.03	1.06	-	-	m²	**1.06**
steel trowel; falls or cross-falls	0.05	1.77	-	-	m²	**1.77**
granolithic finish 20 mm thick laid monolithically	0.07	2.48	0.40	7.28	m²	**10.16**
Finishing of formed surfaces						
aggregate exposure using retarder	0.05	1.77	-	0.57	m²	**2.34**
bush hammering; kango hammer	0.28	9.90	2.46	-	m²	**12.36**
rubbing down concrete surfaces after striking shutters	0.02	0.71	-	0.90	m²	**1.61**
Inserts totally within the concrete volume						
HDPE conduit 20 mm diameter	0.10	3.54	-	3.12	m	**6.66**
black enamelled steel conduit 20 mm diameter	0.10	3.54	-	6.20	m	**9.74**
galvanised steel conduit 20 mm diameter	0.10	3.54	-	9.39	m	**12.93**
Unistrut channel type P3270	0.20	7.07	-	15.26	m	**22.33**
Unistrut channel type P3370	0.20	7.07	-	11.31	m	**18.38**
Inserts projecting from surface(s) of the concrete						
expanding bolt; 10 mm diameter x 25 mm deep	0.05	1.77	-	6.35	nr	**8.11**
holding down bolt; 16 mm diameter x 250 mm deep	0.25	8.84	-	5.05	nr	**13.89**
holding down bolt; 16 mm diameter x 350 mm deep	0.25	8.84	-	5.59	nr	**14.43**
holding down bolt; 20 mm diameter x 250 mm deep	0.25	8.84	-	5.26	nr	**14.10**
holding down bolt; 20 mm diameter x 450 mm deep	0.25	8.84	-	8.25	nr	**17.09**
vitrified clay pipe to BS 65; 100 mm diameter x 1000 mm long	0.25	8.84	-	5.18	nr	**14.02**
cast iron pipe to BS 437; 100 mm diameter x 1000 mm long	0.25	8.84	-	23.85	nr	**32.69**
Grouting under plates; cement and sand (1:3)						
area ne 0.1 m²	0.10	3.54	-	0.21	nr	**3.75**
area 0.1 - 0.5 m²	0.45	15.91	-	1.01	nr	**16.92**
area 0.5 - 1.0 m²	0.78	27.58	-	2.01	nr	**29.59**
Grouting under plates; non-shrink cementitious grout						
area ne 0.1 m²	0.10	3.54	-	2.20	nr	**5.73**
area 0.1 - 0.5 m²	0.45	15.91	-	21.12	nr	**37.03**
area 0.5 - 1.0 m²	0.78	27.58	-	42.24	nr	**69.82**

CLASS H: PRECAST CONCRETE

Item	Gang hours	Labour £	Plant £	Material £	Unit	Total rate £
NOTES						
The cost of precast concrete items is very much dependant on the complexity of the moulds, the number of units to be cast from each mould and the size and weight of the unit to be handled. The unit rates below are for standard precast items that are often to be found on a civil engineering project. It would be misleading to quote for indicative costs for tailor-made precast concrete units and it is advisable to contact Specialist Manufacturers for guide prices.						
BEAMS						
Concrete mix C20						
Beams						
100 x 150 x 1050 mm long	1.00	5.17	-	11.04	nr	**16.21**
225 x 150 x 1200 mm long	1.00	7.20	1.76	31.96	nr	**40.92**
225 x 225 x 1800 mm long	1.00	9.19	3.69	56.74	nr	**69.62**
PRESTRESSED PRE-TENSIONED BEAMS						
Concrete mix C20						
Beams						
100 x 65 x 1050 mm long	1.00	2.90	-	5.41	nr	**8.30**
265 x 65 x 1800 mm long	1.00	3.63	2.53	20.41	nr	**26.57**
Bridge beams						
Inverted 'T' Beams, flange width 495 mm						
section T1; 8 m long, 380 mm deep; mass 1.88t	-	-	-	-	nr	**656.68**
section T2; 9 m long, 420 mm deep; mass 2.29t	-	-	-	-	nr	**786.66**
section T3; 11 m long, 535 mm deep; mass 3.02t	-	-	-	-	nr	**923.46**
section T4; 12 m long, 575 mm deep; mass 3.54t	-	-	-	-	nr	**1026.08**
section T5; 13 m long, 615 mm deep; mass 4.08t	-	-	-	-	nr	**1060.28**
section T6; 13 m long, 655 mm deep; mass 4.33t	-	-	-	-	nr	**1060.28**
section T7; 14 m long, 695 mm deep; mass 4.95t	-	-	-	-	nr	**1231.28**
section T8; 15 m long, 735 mm deep; mass 5.60t	-	-	-	-	nr	**1333.90**
section T9; 16 m long, 775 mm deep; mass 6.28t	-	-	-	-	nr	**1436.51**
section T10; 18 m long, 815 mm deep; mass 7.43t	-	-	-	-	nr	**1607.52**
'M' beams, flange width 970 mm						
section M2; 17 m long, 720 mm deep; mass 12.95t	-	-	-	-	nr	**3488.64**
section M3; 18 m long, 800 mm deep; mass 15.11t	-	-	-	-	nr	**3283.43**
section M6; 22 m long, 1040 mm deep; mass 20.48t	-	-	-	-	nr	**5267.17**
section M8; 25 m long, 1200 mm deep; mass 23.68t	-	-	-	-	nr	**6840.49**
'U' beams, base width 970 mm						
section U3; 16 m long, 900 mm deep; mass 19.24t	-	-	-	-	nr	**5882.81**
section U5; 20 m long, 1000 mm deep; mass 25.64t	-	-	-	-	nr	**7524.53**

CLASS H: PRECAST CONCRETE

Item	Gang hours	Labour £	Plant £	Material £	Unit	**Total rate £**
section U8; 24 m long, 1200 mm deep; mass 34.56t	-	-	-	-	nr	**10192.33**
section U12; 30 m long, 1600 mm deep; mass 52.74t	-	-	-	-	nr	**13954.59**
SLABS						
Prestressed precast concrete flooring planks; Bison 'Drycast' or similar; cement mortar grout between planks on bearings						
110 mm thick floor						
400 mm wide planks	0.21	10.05	6.06	36.77	m²	**52.88**
1200 mm wide planks	0.12	5.61	3.38	37.82	m²	**46.81**
150 mm thick floor						
400 mm wide planks	0.26	12.56	7.58	36.77	m²	**56.91**
1200 mm wide planks	0.14	7.01	4.23	37.82	m²	**49.06**
SEGMENTAL UNITS						
UNITS FOR SUBWAYS, CULVERTS AND DUCTS						
COPINGS, SILLS AND WEIR BLOCKS						
Concrete mix C30						
Coping; weathered and throated						
178 x 64 mm	1.00	6.57	4.29	20.44	m	**31.30**
305 x 76 mm	1.00	4.94	3.06	15.57	m	**23.57**

CLASS I: PIPEWORK - PIPES

Item	Gang hours	Labour £	Plant £	Material £	Unit	**Total rate £**
NOTES						
The rates assume the most efficient items of plant (excavator) and are optimum rates assuming continuous output with no delays caused by other operations or works.						
Ground conditions are assumed to be good easily worked soil with no abnormal conditions that would affect outputs and consistency of work.						
Multiplier Table for labour and plant for various site conditions for working:						
out of sequence x 2.75 (minimum)						
in hard clay x 1.75 to 2.00						
in running sand x 2.75 (minimum)						
in broken rock x 2.75 to 3.50						
below water table x 2.00 (minimum)						
Variance from CESMM3						
Fittings are included with the pipe concerned, for convenience of reference, rather than in Class J.						
RESOURCES - LABOUR						
Drainage / pipework gang (small bore)						
1 ganger/chargehand (skill rate 4) - 50% of time		6.40				
1 skilled operative (skill rate 4)		12.05				
2 unskilled operatives (general)		22.52				
1 plant operator (skill rate 3)		14.65				
1 plant operator (skill rate 3) - 50% of time		7.33				
Total Gang Rate/Hour	£	**62.95**				
Drainage / pipework gang (small bore - not in trenches)						
1 ganger/chargehand (skill rate 4) - 50% of time		6.40				
1 skilled operative (skill rate 4)		12.05				
2 unskilled operatives (general)		22.52				
Total Gang Rate/Hour	£	**40.97**				
Drainage / pipework gang (large bore)						
Note: relates to pipes exceeding 700 mm diameter.						
1 ganger/chargehand (skill rate 4) - 50% of time		6.40				
1 skilled operative (skill rate 4)		12.05				
2 unskilled operatives (general)		22.52				
1 plant operator (skill rate 3)		14.65				
Total Gang Rate/Hour	£	**55.62**				
Drainage / pipework gang (large bore - not in trenches)						
Note: relates to pipes exceeding 700 mm diameter.						
1 ganger/chargehand (skill rate 4) - 50% of time		6.40				
1 skilled operative (skill rate 4)		12.05				
2 unskilled operatives (general)		22.52				
Total Gang Rate/Hour	£	**40.97**				

CLASS I: PIPEWORK - PIPES

Item	Gang hours	Labour £	Plant £	Material £	Unit	Total rate £
RESOURCES - PLANT						
Field drains						
0.4 m3 hydraulic backacter			17.88			
2t dumper - 30% of time			1.30			
Stihl saw, 12", petrol - 30% of time			0.33			
small pump - 30% of time			0.65			
Total Rate / Hour		£	**20.15**			
Add to the above for trench supports appropriate to trench depth (see below).						
Field drains (not in trenches)						
2t dumper - 30% of time			1.30			
Stihl saw, 12", petrol - 30% of time			0.33			
Total Rate / Hour		£	**1.63**			
Drains/sewers (small bore)						
1.0 m3 hydraulic backacter			32.48			
2t dumper - 30% of time			1.30			
2.80 m3/min compressor, 2 tool - 30% of time			1.77			
disc saw - 30% of time			0.39			
extra 50ft / 15m hose - 30% of time			0.07			
small pump - 30% of time			0.65			
sundry tools - 30% of time			0.31			
Total Rate / Hour		£	**36.97**			
Add to the above for trench supports appropriate to trench depth (see below).						
Drains/sewers (small bore - not in trenches)						
2t dumper - 30% of time			1.30			
2.80 m3/min compressor, 2 tool - 30% of time			1.77			
disc saw - 30% of time			0.39			
extra 50ft / 15m hose - 30% of time			0.07			
sundry tools - 30% of time			0.31			
Total Rate / Hour		£	**3.84**			
Drains/sewers (large bore)						
1.0 m3 hydraulic backacter			32.48			
20t crawler crane - 50% of time			14.43			
2t dumper (30% of time)			1.30			
2.80 m3/min compressor, 2 tool - 30% of time			1.77			
disc saw - 30% of time			0.39			
extra 50ft / 15m hose - 30% of time			0.07			
small pump - 30% of time			0.65			
sundry tools - 30% of time			0.31			
Total Rate / Hour		£	**51.40**			
Add to the above for trench supports appropriate to trench depth (see below).						
Drains/sewers (large bore - not in trenches)						
2t dumper (30% of time)			1.30			
2.80 m3/min compressor, 2 tool - 30% of time			1.77			
disc saw - 30% of time			0.39			
extra 50ft / 15m hose - 30% of time			0.07			
sundry tools - 30% of time			0.31			
Total Rate / Hour		£	**3.84**			

CLASS I: PIPEWORK - PIPES

Item	Gang hours	Labour £	Plant £	Material £	Unit	Total rate £
RESOURCES - PLANT – cont'd						
Trench supports						
In addition to the above, the following allowances for close sheeted trench supports are included in the following unit rates, if conditions warrant it:						
ne 1.50 m deep	-	-	1.26	-	m	**1.26**
1.50 - 2.00 m deep	-	-	1.90	-	m	**1.90**
2.00 - 2.50 m deep	-	-	2.07	-	m	**2.07**
2.50 - 3.00 m deep	-	-	2.58	-	m	**2.58**
3.00 - 3.50 m deep	-	-	3.17	-	m	**3.17**
3.50 - 4.00 m deep	-	-	3.80	-	m	**3.80**
4.00 - 4.50 m deep	-	-	4.56	-	m	**4.56**
4.50 - 5.00 m deep	-	-	6.35	-	m	**6.35**
5.00 - 5.50 m deep	-	-	8.75	-	m	**8.75**
CLAY PIPES						
Field drains to BS 1196, butt joints, nominal bore; excavation and supports, backfilling						
75 mm pipes; in trench, depth						
not in trenches	0.06	2.46	0.03	3.29	m	**5.78**
ne 1.50 m deep	0.09	5.67	1.88	3.29	m	**10.84**
1.50 - 2.00 m deep	0.13	8.18	2.77	3.29	m	**14.25**
2.00 - 2.50 m deep	0.18	11.33	3.87	3.29	m	**18.49**
2.50 - 3.00 m deep	0.23	14.48	5.02	3.29	m	**22.79**
100 mm pipes; in trench, depth						
not in trenches	0.06	2.46	0.03	5.66	m	**8.15**
ne 1.50 m deep	0.10	6.29	2.09	5.66	m	**14.05**
1.50 - 2.00 m deep	0.14	8.81	2.99	5.66	m	**17.46**
2.00 - 2.50 m deep	0.19	11.96	4.07	5.66	m	**21.69**
2.50 - 3.00 m deep	0.24	15.11	5.24	5.66	m	**26.00**
150 mm pipes; in trench, depth						
not in trenches	0.06	2.46	0.03	11.59	m	**14.08**
ne 1.50 m deep	0.11	6.92	2.29	11.59	m	**20.81**
1.50 - 2.00 m deep	0.15	9.44	3.19	11.59	m	**24.23**
2.00 - 2.50 m deep	0.20	12.59	4.29	11.59	m	**28.47**
2.50 - 3.00 m deep	0.25	15.74	5.46	11.59	m	**32.79**
225 mm pipes; in trench, depth						
not in trenches	0.06	2.46	0.03	29.86	m	**32.35**
ne 1.50 m deep	0.13	8.18	2.72	29.86	m	**40.76**
1.50 - 2.00 m deep	0.17	10.70	3.62	29.86	m	**44.18**
2.00 - 2.50 m deep	0.22	13.85	4.72	29.86	m	**48.43**
2.50 - 3.00 m deep	0.27	17.00	5.89	29.86	m	**52.74**
Vitrified clay perforated field drains; BS EN295; sleeved joints; excavation and supports, backfilling						
100 mm pipes; in trench, depth						
not in trenches	0.06	2.46	0.03	7.77	m	**10.26**
ne 1.50 m deep	0.15	9.44	3.13	7.77	m	**20.34**
1.50 - 2.00 m deep	0.19	11.96	4.04	7.77	m	**23.77**
2.00 - 2.50 m deep	0.24	15.11	5.14	7.77	m	**28.02**
2.50 - 3.00 m deep	0.29	18.26	6.33	7.77	m	**32.36**
Extra for bend	0.08	5.04	-	6.25	nr	**11.28**
Extra for single junction	0.09	5.67	-	13.25	nr	**18.91**

CLASS I: PIPEWORK - PIPES

Item	Gang hours	Labour £	Plant £	Material £	Unit	**Total rate £**
150 mm pipes; in trench, depth						
not in trenches	0.06	2.46	0.03	14.11	m	**16.61**
ne 1.50 m deep	0.17	10.70	3.55	14.11	m	**28.37**
1.50 - 2.00 m deep	0.20	12.59	4.26	14.11	m	**30.96**
2.00 - 2.50 m deep	0.24	15.11	5.14	14.11	m	**34.37**
2.50 - 3.00 m deep	0.29	18.26	6.33	14.11	m	**38.70**
Extra for bend	0.13	7.87	-	12.87	nr	**20.74**
Extra for single junction	0.13	8.18	-	18.85	nr	**27.03**
225 mm pipes; in trench, depth						
not in trenches	0.08	3.28	0.03	30.26	m	**33.57**
ne 1.50 m deep	0.18	11.33	3.76	30.26	m	**45.35**
1.50 - 2.00 m deep	0.22	13.85	4.69	30.26	m	**48.80**
2.00 - 2.50 m deep	0.26	16.37	5.58	30.26	m	**52.21**
2.50 - 3.00 m deep	0.30	18.89	6.55	30.26	m	**55.70**
Extra for bend	0.12	7.55	-	62.73	nr	**70.28**
Extra for single junction	0.20	12.28	-	76.70	nr	**88.98**
300 mm pipes; in trench, depth						
not in trenches	0.10	4.10	0.16	49.44	m	**53.70**
ne 1.50 m deep	0.19	11.96	3.96	49.44	m	**65.36**
1.50 - 2.00 m deep	0.24	15.11	5.11	49.44	m	**69.66**
2.00 - 2.50 m deep	0.28	17.63	6.00	49.44	m	**73.07**
2.50 - 3.00 m deep	0.32	20.14	6.98	49.44	m	**76.57**
Extra for bend	0.24	15.11	-	121.30	nr	**136.41**
Extra for single junction	0.23	14.48	-	228.70	nr	**243.18**
Vitrified clay pipes to BS EN295, plain ends with push-fit polypropylene flexible couplings; excavation and supports, backfilling						
100 mm pipes; in trenches, depth						
not in trenches	0.06	2.46	0.24	7.72	m	**10.42**
ne 1.50 m deep	0.15	9.44	5.51	7.72	m	**22.67**
1.50 - 2.00 m deep	0.19	11.96	7.24	7.72	m	**26.92**
2.00 - 2.50 m deep	0.24	15.11	9.18	7.72	m	**32.01**
2.50 - 3.00 m deep	0.29	18.26	11.21	7.72	m	**37.19**
3.00 - 3.50 m deep	0.36	22.66	14.05	7.72	m	**44.43**
3.50 - 4.00 m deep	0.44	27.70	17.37	7.72	m	**52.78**
4.00 - 4.50 m deep	0.55	34.62	22.00	7.72	m	**64.34**
4.50 - 5.00 m deep	0.70	44.06	28.62	7.72	m	**80.41**
5.00 - 5.50 m deep	0.90	56.66	38.13	7.72	m	**102.50**
Extra for bend	0.05	3.15	-	13.68	nr	**16.83**
Extra for rest bend	0.06	3.78	-	17.81	nr	**21.59**
Extra for single junction; equal	0.09	5.67	-	24.06	nr	**29.73**
Extra for saddle; oblique	0.23	14.48	-	15.54	nr	**30.02**
150 mm pipes; in trenches, depth						
not in trenches	0.06	2.46	0.24	15.12	m	**17.82**
ne 1.50 m deep	0.17	10.70	6.42	15.12	m	**32.24**
1.50 - 2.00 m deep	0.20	12.59	7.62	15.12	m	**35.33**
2.00 - 2.50 m deep	0.24	15.11	9.18	15.12	m	**39.41**
2.50 - 3.00 m deep	0.29	18.26	11.21	15.12	m	**44.59**
3.00 - 3.50 m deep	0.39	24.55	15.21	15.12	m	**54.88**
3.50 - 4.00 m deep	0.45	28.33	17.77	15.12	m	**61.21**
4.00 - 4.50 m deep	0.58	36.51	23.22	15.12	m	**74.85**
4.50 - 5.00 m deep	0.75	47.21	30.65	15.12	m	**92.98**
5.00 - 5.50 m deep	0.96	60.43	40.65	15.12	m	**116.20**
Extra for bend	0.08	5.04	-	32.24	nr	**37.27**
Extra for rest bend	0.09	5.67	-	35.39	nr	**41.06**
Extra for single junction; equal	0.11	6.92	-	50.17	nr	**57.09**
Extra for taper reducer	0.07	4.41	-	28.45	nr	**32.85**
Extra for saddle; oblique	0.29	18.26	-	29.50	nr	**47.76**

CLASS I: PIPEWORK - PIPES

Item	Gang hours	Labour £	Plant £	Material £	Unit	**Total rate £**
CLAY PIPES – cont'd						
Vitrified clay pipes to BS EN295, plain ends with push-fit polypropylene flexible couplings; excavation and supports, backfilling – cont'd						
225 mm pipes; in trenches, depth						
not in trenches	0.08	3.28	0.31	36.51	m	**40.09**
ne 1.50 m deep	0.18	11.33	6.71	36.51	m	**54.54**
1.50 - 2.00 m deep	0.22	13.85	8.40	36.51	m	**58.75**
2.00 - 2.50 m deep	0.26	16.37	9.96	36.51	m	**62.83**
2.50 - 3.00 m deep	0.30	18.89	11.61	36.51	m	**67.00**
3.00 - 3.50 m deep	0.41	25.81	16.01	36.51	m	**78.32**
3.50 - 4.00 m deep	0.47	29.59	18.54	36.51	m	**84.63**
4.00 - 4.50 m deep	0.65	40.92	26.01	36.51	m	**103.43**
4.50 - 5.00 m deep	0.80	50.36	32.70	36.51	m	**119.56**
5.00 - 5.50 m deep	1.02	64.21	43.21	36.51	m	**143.92**
Extra for bend	0.10	6.29	-	78.13	nr	**84.43**
Extra for rest bend	0.11	6.92	-	90.68	nr	**97.60**
Extra for single junction; equal	0.16	10.07	-	124.39	nr	**134.47**
Extra for taper reducer	0.12	7.55	-	71.24	nr	**78.80**
Extra for saddle; oblique	0.36	22.66	-	70.58	nr	**93.24**
300 mm pipes; in trenches, depth						
not in trenches	0.10	4.10	0.39	64.31	m	**68.80**
ne 1.50 m deep	0.19	11.96	7.16	64.31	m	**83.43**
1.50 - 2.00 m deep	0.24	15.11	9.15	64.31	m	**88.57**
2.00 - 2.50 m deep	0.28	17.63	10.71	64.31	m	**92.65**
2.50 - 3.00 m deep	0.32	20.14	12.37	64.31	m	**96.82**
3.00 - 3.50 m deep	0.42	26.44	16.41	64.31	m	**107.16**
3.50 - 4.00 m deep	0.52	32.73	20.52	64.31	m	**117.57**
4.00 - 4.50 m deep	0.70	44.06	28.02	64.31	m	**136.40**
4.50 - 5.00 m deep	0.90	56.66	36.80	64.31	m	**157.77**
5.00 - 5.50 m deep	1.12	70.50	47.43	64.31	m	**182.25**
Extra for bend	0.15	9.44	-	195.51	nr	**204.95**
Extra for rest bend	0.16	10.07	-	241.17	nr	**251.24**
Extra for single junction; equal	0.19	11.96	-	254.43	nr	**266.39**
Extra for saddle; oblique	0.44	27.70	-	146.54	nr	**174.24**
Vitrified clay pipes to BS EN295, spigot and socket joints with sealing ring; excavation and supports, backfilling						
100 mm pipes; in trenches, depth						
not in trenches	0.06	2.46	0.26	11.13	m	**13.85**
ne 1.50 m deep	0.15	9.44	5.74	11.13	m	**26.32**
1.50 - 2.00 m deep	0.19	11.96	7.24	11.13	m	**30.33**
2.00 - 2.50 m deep	0.24	15.11	9.18	11.13	m	**35.42**
2.50 - 3.00 m deep	0.29	18.26	11.21	11.13	m	**40.60**
3.00 - 3.50 m deep	0.36	22.66	14.05	11.13	m	**47.84**
3.50 - 4.00 m deep	0.44	27.70	17.37	11.13	m	**56.19**
4.00 - 4.50 m deep	0.55	34.62	22.00	11.13	m	**67.75**
4.50 - 5.00 m deep	0.70	44.06	28.62	11.13	m	**83.82**
5.00 - 5.50 m deep	0.90	56.66	38.13	11.13	m	**105.91**
Extra for bend	0.05	3.15	-	8.80	nr	**11.95**
Extra for rest bend	0.06	3.78	-	13.60	nr	**17.38**
Extra for single junction; equal	0.09	5.67	-	19.16	nr	**24.82**
Extra fro saddle; oblique	0.23	14.48	-	12.68	nr	**27.16**

CLASS I: PIPEWORK - PIPES

Item	Gang hours	Labour £	Plant £	Material £	Unit	**Total rate £**
150 mm pipes; in trenches, depth						
not in trenches	0.06	2.46	0.24	15.29	m	**17.99**
ne 1.50 m deep	0.17	10.70	6.42	15.29	m	**32.41**
1.50 - 2.00 m deep	0.20	12.59	7.62	15.29	m	**35.51**
2.00 - 2.50 m deep	0.24	15.11	9.18	15.29	m	**39.58**
2.50 - 3.00 m deep	0.29	18.26	11.21	15.29	m	**44.76**
3.00 - 3.50 m deep	0.39	24.55	15.21	15.29	m	**55.06**
3.50 - 4.00 m deep	0.45	28.33	17.77	15.29	m	**61.39**
4.00 - 4.50 m deep	0.58	36.51	23.22	15.29	m	**75.02**
4.50 - 5.00 m deep	0.75	47.21	30.65	15.29	m	**93.15**
5.00 - 5.50 m deep	0.96	60.43	40.65	15.29	m	**116.38**
Extra for bend	0.08	5.04	-	36.30	nr	**41.33**
Extra for rest bend	0.09	5.67	-	18.64	nr	**24.30**
Extra for single junction; equal	0.11	6.92	-	37.01	nr	**43.93**
Extra for double junction; equal	0.13	8.18	-	54.97	nr	**63.15**
Extra for taper reducer	0.07	4.41	-	15.26	nr	**19.67**
Extra for saddle; oblique	0.29	18.26	-	18.04	nr	**36.29**
225 mm pipes; in trenches, depth						
not in trenches	0.08	3.28	0.31	28.85	m	**32.44**
ne 1.50 m deep	0.18	11.33	6.80	28.85	m	**46.98**
1.50 - 2.00 m deep	0.22	13.85	8.40	28.85	m	**51.10**
2.00 - 2.50 m deep	0.26	16.37	9.96	28.85	m	**55.18**
2.50 - 3.00 m deep	0.30	18.89	11.61	28.85	m	**59.35**
3.00 - 3.50 m deep	0.41	25.81	16.01	28.85	m	**70.67**
3.50 - 4.00 m deep	0.47	29.59	18.34	28.85	m	**76.78**
4.00 - 4.50 m deep	0.65	40.92	26.01	28.85	m	**95.78**
4.50 - 5.00 m deep	0.80	50.36	32.70	28.85	m	**111.91**
5.00 - 5.50 m deep	1.02	64.21	43.21	28.85	m	**136.27**
Extra for bend	0.10	6.29	-	74.03	nr	**80.33**
Extra for rest bend	0.11	6.92	-	77.01	nr	**83.94**
Extra for single junction; equal	0.16	10.07	-	78.31	nr	**88.38**
Extra for double junction; equal	0.18	11.33	-	127.92	nr	**139.25**
Extra for taper reducer	0.12	7.55	-	55.57	nr	**63.12**
Extra for saddle; oblique	0.36	22.66	-	65.63	nr	**88.29**
300 mm pipes; in trenches, depth						
not in trenches	0.10	4.10	0.39	46.44	m	**50.93**
ne 1.50 m deep	0.19	11.96	7.16	46.44	m	**65.56**
1.50 - 2.00 m deep	0.24	15.11	9.15	46.44	m	**70.70**
2.00 - 2.50 m deep	0.28	17.63	10.71	46.44	m	**74.78**
2.50 - 3.00 m deep	0.32	20.14	12.37	46.44	m	**78.95**
3.00 - 3.50 m deep	0.42	26.44	16.41	46.44	m	**89.29**
3.50 - 4.00 m deep	0.52	32.73	20.52	46.44	m	**99.70**
4.00 - 4.50 m deep	0.70	44.06	28.02	46.44	m	**118.53**
4.50 - 5.00 m deep	0.90	56.66	36.80	46.44	m	**139.90**
5.00 - 5.50 m deep	1.12	70.50	47.43	46.44	m	**164.38**
Extra for bend	0.15	9.44	-	146.10	nr	**155.54**
Extra for rest bend	0.16	10.07	-	169.74	nr	**179.82**
Extra for single junction; equal	0.19	11.96	-	164.04	nr	**176.00**
Extra for double junction; equal	0.21	13.22	-	174.89	nr	**188.11**
Extra for taper reducer	0.15	9.44	-	132.91	nr	**142.35**
Extra for saddle; oblique	0.44	27.70	-	114.22	nr	**141.92**

CLASS I: PIPEWORK - PIPES

Item	Gang hours	Labour £	Plant £	Material £	Unit	**Total rate £**
CLAY PIPES – cont'd						
Vitrified clay pipes to BS EN295, spigot and socket joints with sealing ring; excavation and supports, backfilling – cont'd						
400 mm pipes; in trenches, depth						
not in trenches	0.10	4.10	0.39	97.74	m	**102.23**
ne 1.50 m deep	0.23	14.48	8.66	97.74	m	**120.89**
1.50 - 2.00 m deep	0.29	18.26	11.06	97.74	m	**127.06**
2.00 - 2.50 m deep	0.32	20.14	12.24	97.74	m	**130.13**
2.50 - 3.00 m deep	0.38	23.92	14.70	97.74	m	**136.36**
3.00 - 3.50 m deep	0.46	28.96	17.97	97.74	m	**144.67**
3.50 - 4.00 m deep	0.58	36.51	22.90	97.74	m	**157.16**
4.00 - 4.50 m deep	0.75	47.21	30.00	97.74	m	**174.96**
4.50 - 5.00 m deep	0.95	59.80	38.82	97.74	m	**196.37**
5.00 - 5.50 m deep	1.20	75.54	50.82	97.74	m	**224.10**
Extra for bend; 90 degree	0.24	15.11	-	395.69	nr	**410.80**
Extra for bend; 45 degree	0.24	15.11	-	395.69	nr	**410.80**
Extra for bend; 22.5 degree	0.24	15.11	-	395.65	nr	**410.76**
450 mm pipes; in trenches, depth						
not in trenches	0.23	9.42	0.88	126.12	m	**136.42**
ne 1.50 m deep	0.23	14.48	8.66	126.12	m	**149.26**
1.50 - 2.00 m deep	0.30	18.89	11.45	126.12	m	**156.45**
2.00 - 2.50 m deep	0.32	20.14	12.24	126.12	m	**158.50**
2.50 - 3.00 m deep	0.38	23.92	14.70	126.12	m	**164.74**
3.00 - 3.50 m deep	0.47	29.59	18.34	126.12	m	**174.04**
3.50 - 4.00 m deep	0.60	37.77	23.68	126.12	m	**187.57**
4.00 - 4.50 m deep	0.77	48.47	30.81	126.12	m	**205.40**
4.50 - 5.00 m deep	0.97	61.06	39.65	126.12	m	**226.83**
5.00 - 5.50 m deep	1.20	75.54	50.82	126.12	m	**252.47**
Extra for bend; 90 degree	0.29	18.26	-	521.00	nr	**539.26**
Extra for bend; 45 degree	0.29	18.26	-	521.00	nr	**539.26**
Extra for bend; 22.5 degree	0.29	18.26	-	521.00	nr	**539.26**
CONCRETE PIPES						
Concrete porous pipes to BS 5911; excavation and supports, backfilling						
150 mm pipes; in trench, depth						
not in trenches	0.06	2.46	0.21	4.74	m	**7.41**
ne 1.50 m deep	0.17	10.70	3.55	4.74	m	**18.99**
1.50 - 2.00 m deep	0.20	12.59	4.26	4.74	m	**21.59**
2.00 - 2.50 m deep	0.24	15.11	5.14	4.74	m	**24.99**
2.50 - 3.00 m deep	0.29	18.26	6.33	4.74	m	**29.33**
3.00 - 3.50 m deep	0.39	24.55	8.66	4.74	m	**37.95**
3.50 - 4.00 m deep	0.45	28.33	10.20	4.74	m	**43.26**
4.00 - 4.50 m deep	0.58	36.51	13.46	4.74	m	**54.71**
4.50 - 5.00 m deep	0.75	47.21	18.04	4.74	m	**69.99**
5.00 - 5.50 m deep	0.96	60.43	24.51	4.74	m	**89.68**
225 mm pipes; in trench, depth						
not in trenches	0.08	3.28	0.13	6.04	m	**9.45**
ne 1.50 m deep	0.18	11.33	3.76	6.04	m	**21.13**
1.50 - 2.00 m deep	0.22	13.85	4.69	6.04	m	**24.58**
2.00 - 2.50 m deep	0.26	16.37	5.58	6.04	m	**27.99**
2.50 - 3.00 m deep	0.30	18.89	6.55	6.04	m	**31.48**
3.00 - 3.50 m deep	0.41	25.81	9.11	6.04	m	**40.96**

CLASS I: PIPEWORK - PIPES

Item	Gang hours	Labour £	Plant £	Material £	Unit	**Total rate £**
3.50 - 4.00 m deep	0.47	29.59	10.64	6.04	m	**46.27**
4.00 - 4.50 m deep	0.65	40.92	15.08	6.04	m	**62.03**
4.50 - 5.00 m deep	0.80	50.36	19.24	6.04	m	**75.64**
5.00 - 5.50 m deep	1.02	64.21	26.05	6.04	m	**96.29**
300 mm pipes; in trench, depth						
not in trenches	0.10	4.10	0.16	9.22	m	**13.48**
ne 1.50 m deep	0.19	11.96	3.96	9.22	m	**25.14**
1.50 - 2.00 m deep	0.24	15.11	5.11	9.22	m	**29.44**
2.00 - 2.50 m deep	0.28	17.63	6.00	9.22	m	**32.85**
2.50 - 3.00 m deep	0.32	20.14	6.98	9.22	m	**36.35**
3.00 - 3.50 m deep	0.42	26.44	9.33	9.22	m	**44.99**
3.50 - 4.00 m deep	0.52	32.73	11.78	9.22	m	**53.73**
4.00 - 4.50 m deep	0.70	44.06	16.24	9.22	m	**69.52**
4.50 - 5.00 m deep	0.90	56.66	21.66	9.22	m	**87.53**
5.00 - 5.50 m deep	1.12	70.50	28.59	9.22	m	**108.31**
Concrete pipes with rebated flexible joints to BS 5911 Class L; excavation and supports, backfilling						
300 mm pipes; in trenches, depth						
not in trenches	0.12	4.92	0.24	12.10	m	**17.26**
ne 1.50 m deep	0.22	13.85	8.31	12.10	m	**34.26**
1.50 - 2.00 m deep	0.26	16.37	9.92	12.10	m	**38.39**
2.00 - 2.50 m deep	0.30	18.89	11.49	12.10	m	**42.48**
2.50 - 3.00 m deep	0.34	21.40	13.15	12.10	m	**46.66**
3.00 - 3.50 m deep	0.44	27.70	17.17	12.10	m	**56.97**
3.50 - 4.00 m deep	0.56	35.25	22.10	12.10	m	**69.46**
4.00 - 4.50 m deep	0.73	45.95	29.21	12.10	m	**87.27**
4.50 - 5.00 m deep	0.92	57.91	37.60	12.10	m	**107.62**
5.00 - 5.50 m deep	1.19	74.91	50.39	12.10	m	**137.40**
5.50 - 6.00 m deep	1.42	89.39	61.33	12.10	m	**162.82**
Extra for bend	0.08	5.04	2.32	86.63	nr	**93.99**
375 mm pipes; in trenches, depth						
not in trenches	0.15	6.15	0.57	15.85	m	**22.56**
ne 1.50 m deep	0.24	15.11	9.05	15.85	m	**40.00**
1.50 - 2.00 m deep	0.29	18.26	11.06	15.85	m	**45.16**
2.00 - 2.50 m deep	0.33	20.77	12.63	15.85	m	**49.25**
2.50 - 3.00 m deep	0.38	23.92	14.70	15.85	m	**54.47**
3.00 - 3.50 m deep	0.46	28.96	17.97	15.85	m	**62.77**
3.50 - 4.00 m deep	0.58	36.51	22.90	15.85	m	**75.26**
4.00 - 4.50 m deep	0.75	47.21	30.00	15.85	m	**93.06**
4.50 - 5.00 m deep	0.95	59.80	38.82	15.85	m	**114.47**
5.00 - 5.50 m deep	1.20	75.54	50.82	15.85	m	**142.20**
5.50 - 6.00 m deep	1.45	91.28	62.62	15.85	m	**169.74**
Extra for bend	0.10	6.29	3.23	113.42	nr	**122.94**
450 mm pipes; in trenches, depth						
not in trenches	0.17	6.96	0.66	19.90	m	**27.53**
ne 1.50 m deep	0.25	15.74	9.43	19.90	m	**45.07**
1.50 - 2.00 m deep	0.31	19.51	11.81	19.90	m	**51.23**
2.00 - 2.50 m deep	0.35	22.03	13.39	19.90	m	**55.33**
2.50 - 3.00 m deep	0.40	25.18	15.46	19.90	m	**60.54**
3.00 - 3.50 m deep	0.48	30.22	19.20	19.90	m	**69.32**
3.50 - 4.00 m deep	0.63	39.66	24.86	19.90	m	**84.42**
4.00 - 4.50 m deep	0.80	50.36	32.01	19.90	m	**102.27**
4.50 - 5.00 m deep	1.00	62.95	40.87	19.90	m	**123.73**
5.00 - 5.50 m deep	1.25	78.69	52.94	19.90	m	**151.53**
5.50 - 6.00 m deep	1.51	95.05	65.19	19.90	m	**180.15**
Extra for bend	0.13	8.18	4.64	142.47	nr	**155.29**

CLASS I: PIPEWORK - PIPES

Item	Gang hours	Labour £	Plant £	Material £	Unit	Total rate £
CONCRETE PIPES – cont'd						
Concrete pipes with rebated flexible joints to BS 5911 Class L – cont'd						
525 mm pipes; in trenches, depth						
not in trenches	0.20	8.19	0.77	23.67	m	**32.64**
ne 1.50 m deep	0.27	17.00	10.17	23.67	m	**50.84**
1.50 - 2.00 m deep	0.33	20.77	12.59	23.67	m	**57.03**
2.00 - 2.50 m deep	0.37	23.29	14.16	23.67	m	**61.13**
2.50 - 3.00 m deep	0.42	26.44	16.24	23.67	m	**66.36**
3.00 - 3.50 m deep	0.49	30.85	19.13	23.67	m	**73.65**
3.50 - 4.00 m deep	0.65	40.92	25.66	23.67	m	**90.25**
4.00 - 4.50 m deep	0.83	52.25	33.20	23.67	m	**109.12**
4.50 - 5.00 m deep	1.03	64.84	42.09	23.67	m	**130.61**
5.00 - 5.50 m deep	1.28	80.58	54.20	23.67	m	**158.45**
5.50 - 6.00 m deep	1.55	97.57	66.92	23.67	m	**188.17**
Extra for bend	0.16	10.07	6.02	169.49	nr	**185.58**
750 mm pipes; in trenches, depth						
not in trenches	0.22	9.01	0.86	54.65	m	**64.52**
ne 1.50 m deep	0.30	16.69	15.65	54.65	m	**86.99**
1.50 - 2.00 m deep	0.36	20.02	18.92	54.65	m	**93.59**
2.00 - 2.50 m deep	0.41	22.80	21.61	54.65	m	**99.07**
2.50 - 3.00 m deep	0.47	26.14	24.94	54.65	m	**105.73**
3.00 - 3.50 m deep	0.58	32.26	31.02	54.65	m	**117.93**
3.50 - 4.00 m deep	0.80	44.50	43.12	54.65	m	**142.27**
4.00 - 4.50 m deep	1.05	58.40	57.16	54.65	m	**170.22**
4.50 - 5.00 m deep	1.30	72.31	71.90	54.65	m	**198.86**
5.00 - 5.50m deep	1.55	86.21	87.99	54.65	m	**228.86**
5.50 - 6.00 m deep	1.82	101.23	104.86	54.65	m	**260.74**
Extra for bends	0.24	13.35	15.60	391.20	nr	**420.15**
900 mm pipes; in trenches, depth						
not in trenches	0.25	10.24	0.97	74.76	m	**85.96**
ne 1.50 m deep	0.33	18.35	17.21	74.76	m	**110.32**
1.50 - 2.00 m deep	0.40	22.25	21.02	74.76	m	**118.02**
2.00 - 2.50 m deep	0.46	25.59	24.25	74.76	m	**124.59**
2.50 - 3.00 m deep	0.53	29.48	28.14	74.76	m	**132.37**
3.00 - 3.50 m deep	0.70	38.93	37.43	74.76	m	**151.12**
3.50 - 4.00 m deep	0.92	51.17	49.58	74.76	m	**175.51**
4.00 - 4.50 m deep	1.20	66.74	65.32	74.76	m	**206.82**
4.50 - 5.00 m deep	1.50	83.43	82.97	74.76	m	**241.15**
5.00 - 5.50m deep	1.80	100.12	102.19	74.76	m	**277.07**
5.50 - 6.00 m deep	2.10	116.80	120.99	74.76	m	**312.55**
Extra for bends	0.39	21.69	25.37	620.85	nr	**667.91**
1200 mm pipes; in trenches, depth						
not in trenches	0.25	10.24	0.97	115.04	m	**126.25**
ne 1.50 m deep	0.46	25.59	23.99	115.04	m	**164.62**
1.50 - 2.00 m deep	0.53	29.48	27.86	115.04	m	**172.37**
2.00 - 2.50 m deep	0.60	33.37	31.61	115.04	m	**180.03**
2.50 - 3.00 m deep	0.70	38.93	37.16	115.04	m	**191.14**
3.00 - 3.50 m deep	0.85	47.28	45.44	115.04	m	**207.76**
3.50 - 4.00 m deep	1.12	62.29	60.36	115.04	m	**237.70**
4.00 - 4.50 m deep	1.45	80.65	78.94	115.04	m	**274.63**
4.50 - 5.00 m deep	1.75	97.33	96.77	115.04	m	**309.14**
5.00 - 5.50m deep	2.05	114.02	116.40	115.04	m	**345.46**
5.50 - 6.00 m deep	2.36	131.26	135.95	115.04	m	**382.26**
Extra for bends	0.51	28.37	33.30	975.88	nr	**1037.54**

CLASS I: PIPEWORK - PIPES

Item	Gang hours	Labour £	Plant £	Material £	Unit	**Total rate £**
1500 mm pipes; in trenches, depth						
not in trenches	0.35	14.34	1.34	218.73	m	**234.41**
ne 1.50 m deep	0.60	33.37	31.28	218.73	m	**283.38**
1.50 - 2.00 m deep	0.70	38.93	36.79	218.73	m	**294.46**
2.00 - 2.50 m deep	0.81	45.05	42.69	218.73	m	**306.47**
2.50 - 3.00 m deep	0.92	51.17	48.83	218.73	m	**318.73**
3.00 - 3.50 m deep	1.05	58.40	56.14	218.73	m	**333.27**
3.50 - 4.00 m deep	1.27	70.64	68.44	218.73	m	**357.81**
4.00 - 4.50 m deep	1.70	94.55	92.55	218.73	m	**405.84**
4.50 - 5.00 m deep	2.05	114.02	113.37	218.73	m	**446.13**
5.00 - 5.50m deep	2.40	133.49	136.26	218.73	m	**488.48**
5.50 - 6.00 m deep	2.75	152.96	156.98	220.52	m	**530.46**
Extra for bends	0.63	35.04	41.13	1852.80	nr	**1928.98**
1800 mm pipes; in trenches, depth						
not in trenches	0.40	16.39	1.54	268.90	m	**286.82**
ne 1.50 m deep	0.77	42.83	40.15	268.90	m	**351.87**
1.50 - 2.00 m deep	0.91	50.61	47.81	268.90	m	**367.32**
2.00 - 2.50 m deep	1.03	57.29	54.27	268.90	m	**380.45**
2.50 - 3.00 m deep	1.12	62.29	59.44	268.90	m	**390.63**
3.00 - 3.50 m deep	1.20	66.74	64.15	268.90	m	**399.79**
3.50 - 4.00 m deep	1.52	84.54	81.92	268.90	m	**435.36**
4.00 - 4.50 m deep	2.00	111.24	108.87	268.90	m	**489.01**
4.50 - 5.00 m deep	2.40	133.49	132.72	268.90	m	**535.11**
5.00 - 5.50m deep	2.80	155.74	158.97	268.90	m	**583.60**
5.50 - 6.00 m deep	3.15	175.20	181.46	268.90	m	**625.56**
Extra for bends	0.77	42.83	50.29	2277.00	nr	**2370.11**
2100 mm pipes; in trenches, depth						
not in trenches	0.45	18.44	1.74	421.36	m	**441.53**
ne 1.50 m deep	0.98	54.51	51.10	421.36	m	**526.97**
1.50 - 2.00 m deep	1.13	62.85	59.38	421.36	m	**543.60**
2.00 - 2.50 m deep	1.23	68.41	64.01	421.36	m	**553.78**
2.50 - 3.00 m deep	1.30	72.31	69.01	421.36	m	**562.68**
3.00 - 3.50 m deep	1.50	83.43	80.20	421.36	m	**584.99**
3.50 - 4.00 m deep	1.82	101.23	98.10	421.36	m	**620.69**
4.00 - 4.50 m deep	2.35	130.71	127.92	421.36	m	**679.99**
4.50 - 5.00 m deep	2.80	155.74	154.84	421.36	m	**731.94**
5.00 - 5.50m deep	3.20	177.98	181.68	421.36	m	**781.03**
5.50 - 6.00 m deep	3.55	197.45	204.50	421.36	m	**823.31**
Extra for bends	0.89	49.50	58.19	3272.72	nr	**3380.41**
IRON PIPES						
Cast iron pipes to BS 437 plain ended pipe with "Timesaver" mechanical coupling joints; excavation and supports, backfilling						
75 mm pipes; in trenches, depth						
not in trenches	0.12	4.92	0.46	24.23	m	**29.61**
ne 1.50 m deep	0.19	11.96	6.98	24.23	m	**43.17**
1.50 - 2.00 m deep	0.21	13.22	8.01	24.23	m	**45.46**
2.00 - 2.50 m deep	0.28	17.63	10.71	24.23	m	**52.57**
2.50 - 3.00 m deep	0.34	21.40	13.15	24.23	m	**58.79**
3.00 - 3.50 m deep	0.43	27.07	16.78	24.23	m	**68.07**
3.50 - 4.00 m deep	0.55	34.62	21.70	24.23	m	**80.55**
4.00 - 4.50 m deep	0.71	44.69	28.40	24.23	m	**97.32**
4.50 - 5.00 m deep	0.89	56.03	36.38	24.23	m	**116.64**
5.00 - 5.50 m deep	1.15	72.39	48.69	24.23	m	**145.31**

CLASS I: PIPEWORK - PIPES

Item	Gang hours	Labour £	Plant £	Material £	Unit	Total rate £
IRON PIPES – cont'd						
Cast iron pipes to BS 437 plain ended pipe with "Timesaver" mechanical coupling joints – cont'd						
75 mm pipes – cont'd						
Extra for bend; 87.5 degree	0.31	19.51	0.94	39.41	nr	**59.86**
Extra for bend; 45 degree	0.31	19.51	0.94	38.33	nr	**58.79**
Extra for single junction; equal	0.48	30.22	1.86	59.77	nr	**91.85**
Extra for taper reducer	0.27	17.00	0.94	32.40	nr	**50.34**
100 mm pipes; in trenches, depth						
not in trenches	0.13	5.33	0.20	28.54	m	**34.07**
ne 1.50 m deep	0.21	13.22	7.92	28.54	m	**49.69**
1.50 - 2.00 m deep	0.23	14.48	8.76	28.54	m	**51.78**
2.00 - 2.50 m deep	0.30	18.89	11.49	28.54	m	**58.92**
2.50 - 3.00 m deep	0.37	23.29	14.31	28.54	m	**66.14**
3.00 - 3.50 m deep	0.48	30.22	18.73	28.54	m	**77.49**
3.50 - 4.00 m deep	0.60	37.77	23.68	28.54	m	**89.99**
4.00 - 4.50 m deep	0.75	47.21	30.00	28.54	m	**105.76**
4.50 - 5.00 m deep	0.95	59.80	38.82	28.54	m	**127.17**
5.00 - 5.50 m deep	1.22	76.80	51.68	28.54	m	**157.02**
Extra for bend; 87.5 degree	0.38	23.92	1.42	66.02	nr	**91.36**
Extra for bend; 45 degree	0.38	23.92	1.42	49.76	nr	**75.10**
Extra for bend; long radius	0.38	23.92	1.42	63.54	nr	**88.88**
Extra for single junction; equal	0.59	37.14	2.33	110.70	nr	**150.18**
Extra for double junction; equal	0.80	50.36	3.72	110.49	nr	**164.57**
Extra for taper reducer	0.40	25.18	2.33	46.06	nr	**73.57**
150 mm pipes; in trenches, depth						
not in trenches	0.14	5.74	0.20	49.62	m	**55.55**
ne 1.50 m deep	0.24	15.11	9.05	49.62	m	**73.78**
1.50 - 2.00 m deep	0.28	17.63	10.67	49.62	m	**77.92**
2.00 - 2.50 m deep	0.34	21.40	13.03	49.62	m	**84.05**
2.50 - 3.00 m deep	0.40	25.18	15.46	49.62	m	**90.26**
3.00 - 3.50 m deep	0.54	33.99	21.09	49.62	m	**104.70**
3.50 - 4.00 m deep	0.61	38.40	24.08	49.62	m	**112.10**
4.00 - 4.50 m deep	0.79	49.73	31.60	49.62	m	**130.95**
4.50 - 5.00 m deep	1.05	66.10	42.92	49.62	m	**158.64**
5.00 - 5.50 m deep	1.32	83.09	55.90	49.62	m	**188.61**
Extra for bend; 87.5 degree	0.56	35.25	1.85	91.19	nr	**128.29**
Ectra for bend; 45 degree	0.56	35.25	1.85	76.57	nr	**113.67**
Extra for bend; long radius	0.56	35.25	1.85	91.19	nr	**128.29**
Extra for single junction; equal	0.88	55.40	3.70	141.04	nr	**200.13**
Extra for taper reducer	0.56	35.25	3.70	104.55	nr	**143.50**
225 mm pipes; in trenches, depth						
not in trenches	0.15	6.15	0.20	144.71	m	**151.05**
ne 1.50 m deep	0.25	15.74	9.43	144.71	m	**169.87**
1.50 - 2.00 m deep	0.30	18.89	11.45	144.71	m	**175.04**
2.00 - 2.50 m deep	0.35	22.03	13.39	144.71	m	**180.12**
2.50 - 3.00 m deep	0.41	25.81	15.85	144.71	m	**186.37**
3.00 - 3.50 m deep	0.54	33.99	21.09	144.71	m	**199.79**
3.50 - 4.00 m deep	0.61	38.40	24.08	144.71	m	**207.19**
4.00 - 4.50 m deep	0.77	48.47	30.81	144.71	m	**223.99**
4.50 - 5.00 m deep	1.02	64.21	41.70	144.71	m	**250.62**
5.00 - 5.50 m deep	1.30	81.83	55.06	144.71	m	**281.60**
Extra for bend; 87.5 degree	0.77	48.47	1.85	252.37	nr	**302.69**
Extra for bend; 45 degree	0.77	48.47	1.85	252.37	nr	**302.69**
Extra for single junction; equal	1.21	76.17	4.17	394.75	nr	**475.09**
Extra for taper reducer	0.77	48.47	4.17	147.74	nr	**200.38**

CLASS I: PIPEWORK - PIPES

Item	Gang hours	Labour £	Plant £	Material £	Unit	**Total rate £**
Ductile iron pipes to BS 4772, Tyton joints; excavation and supports, backfilling						
100 mm pipes; in trenches, depth						
not in trenches	0.10	4.10	0.20	29.06	m	**33.36**
ne 1.50 m deep	0.19	11.96	6.81	29.06	m	**47.83**
1.50 - 2.00 m deep	0.20	12.59	7.62	29.06	m	**49.28**
2.00 - 2.50 m deep	0.26	16.37	9.96	29.06	m	**55.39**
2.50 - 3.00 m deep	0.32	20.14	12.37	29.06	m	**61.57**
3.00 - 3.50 m deep	0.42	26.44	16.41	29.06	m	**71.91**
3.50 - 4.00 m deep	0.53	33.36	20.92	29.06	m	**83.35**
4.00 - 4.50 m deep	0.66	41.55	26.42	29.06	m	**97.03**
4.50 - 5.00 m deep	0.84	52.88	34.33	29.06	m	**116.28**
5.00 - 5.50 m deep	1.08	67.99	45.73	29.06	m	**142.79**
Extra for bend; 90 degree	0.38	23.92	10.15	64.25	nr	**98.32**
Extra for single junction; equal	0.60	37.77	2.79	91.59	nr	**132.14**
150 mm pipes; in trenches, depth						
not in trenches	0.11	4.51	0.20	36.58	m	**41.28**
ne 1.50 m deep	0.21	13.22	7.92	36.58	m	**57.72**
1.50 - 2.00 m deep	0.25	15.74	9.54	36.58	m	**61.85**
2.00 - 2.50 m deep	0.30	18.89	11.49	36.58	m	**66.96**
2.50 - 3.00 m deep	0.37	23.29	14.31	36.58	m	**74.18**
3.00 - 3.50 m deep	0.49	30.85	19.13	36.58	m	**86.56**
3.50 - 4.00 m deep	0.55	34.62	21.70	36.58	m	**92.90**
4.00 - 4.50 m deep	0.72	45.32	28.80	36.58	m	**110.71**
4.50 - 5.00 m deep	0.94	59.17	38.43	36.58	m	**134.19**
5.00 - 5.50 m deep	1.19	74.91	50.39	36.58	m	**161.88**
Extra for bend; 90 degree	0.57	35.88	1.85	138.39	nr	**176.12**
Extra for single junction; equal	0.88	55.40	3.70	182.64	nr	**241.74**
250 mm pipes; in trenches, depth						
not in trenches	0.16	6.56	0.20	64.89	m	**71.64**
ne 1.50 m deep	0.24	15.11	9.05	64.89	m	**89.04**
1.50 - 2.00 m deep	0.30	18.89	11.45	64.89	m	**95.22**
2.00 - 2.50 m deep	0.35	22.03	13.39	64.89	m	**100.31**
2.50 - 3.00 m deep	0.42	26.44	16.24	64.89	m	**107.57**
3.00 - 3.50 m deep	0.56	35.25	21.86	64.89	m	**122.00**
3.50 - 4.00 m deep	0.64	40.29	25.26	64.89	m	**130.43**
4.00 - 4.50 m deep	0.81	50.99	32.41	64.89	m	**148.29**
4.50 - 5.00 m deep	1.02	64.21	41.70	64.89	m	**170.80**
5.00 - 5.50 m deep	1.27	79.95	53.77	64.89	m	**198.61**
Extra for bend; 90 degree	0.96	60.43	2.79	399.52	nr	**462.74**
Extra for single junction; equal	1.32	83.09	4.64	503.69	nr	**591.42**
400 mm pipes; in trenches, depth						
not in trenches	0.24	9.83	0.42	121.64	m	**131.89**
ne 1.50 m deep	0.33	20.77	12.45	121.64	m	**154.86**
1.50 - 2.00 m deep	0.43	27.07	16.38	121.64	m	**165.09**
2.00 - 2.50 m deep	0.48	30.22	18.37	121.64	m	**170.22**
2.50 - 3.00 m deep	0.57	35.88	22.03	121.64	m	**179.55**
3.00 - 3.50 m deep	0.71	44.69	27.70	121.64	m	**194.04**
3.50 - 4.00 m deep	0.91	57.28	35.91	121.64	m	**214.83**
4.00 - 4.50 m deep	1.16	73.02	46.41	121.64	m	**241.07**
4.50 - 5.00 m deep	1.47	92.54	60.08	121.64	m	**274.25**
5.00 - 5.50 m deep	1.77	111.42	74.96	121.64	m	**308.02**

CLASS I: PIPEWORK - PIPES

Item	Gang hours	Labour £	Plant £	Material £	Unit	**Total rate £**
IRON PIPES – cont'd						
Ductile iron pipes to BS 4772, Tyton joints; excavation and supports – cont'd						
600 mm pipes; in trenches, depth						
not in trenches	0.34	13.93	0.57	218.61	m	**233.12**
ne 1.50 m deep	0.47	29.59	17.71	218.61	m	**265.91**
1.50 - 2.00 m deep	0.55	34.62	20.96	218.61	m	**274.19**
2.00 - 2.50 m deep	0.66	41.55	25.27	218.61	m	**285.43**
2.50 - 3.00 m deep	0.78	49.10	30.16	218.61	m	**297.87**
3.00 - 3.50 m deep	0.89	56.03	34.74	218.61	m	**309.38**
3.50 - 4.00 m deep	1.09	68.62	43.03	218.61	m	**330.26**
4.00 - 4.50 m deep	1.37	86.24	54.82	218.61	m	**359.67**
4.50 - 5.00 m deep	1.70	107.02	69.50	218.61	m	**395.13**
5.00 - 5.50 m deep	2.03	127.79	85.96	218.61	m	**432.36**
STEEL PIPES						
Carbon steel pipes to BS 3601; welded joints;(for protection and lining refer to manufacturer); excavation and supports, backfilling						
100 mm pipes; in trenches, depth						
not in trenches	0.07	2.87	0.26	24.43	m	**27.57**
ne 1.50 m deep	0.15	9.44	5.65	24.43	m	**39.52**
1.50 - 2.00 m deep	0.17	10.70	6.49	24.43	m	**41.62**
2.00 - 2.50 m deep	0.22	13.85	8.43	24.43	m	**46.71**
2.50 - 3.00 m deep	0.27	17.00	10.43	24.43	m	**51.86**
3.00 - 3.50 m deep	0.35	22.03	13.65	24.43	m	**60.12**
3.50 - 4.00 m deep	0.44	27.70	17.37	24.43	m	**69.50**
4.00 - 4.50 m deep	0.55	34.62	22.00	24.43	m	**81.05**
4.50 - 5.00 m deep	0.70	44.06	32.24	24.43	m	**100.74**
5.00 - 5.50 m deep	0.87	54.77	36.84	24.43	m	**116.04**
Extra for bend; 45 degrees	0.07	4.41	1.38	10.18	nr	**15.97**
Extra for bend; 90 degrees	0.07	4.41	1.38	12.86	nr	**18.65**
Extra for single junction; equal	0.11	6.92	2.32	30.23	nr	**39.47**
150 mm pipes; in trenches, depth						
not in trenches	0.07	2.87	0.26	37.65	m	**40.78**
ne 1.50 m deep	0.17	10.70	6.42	37.65	m	**54.76**
1.50 - 2.00 m deep	0.20	12.59	7.62	37.65	m	**57.86**
2.00 - 2.50 m deep	0.24	15.11	9.18	37.65	m	**61.94**
2.50 - 3.00 m deep	0.29	18.26	11.21	37.65	m	**67.11**
3.00 - 3.50 m deep	0.39	24.55	15.21	37.65	m	**77.41**
3.50 - 4.00 m deep	0.44	27.70	17.37	37.65	m	**82.71**
4.00 - 4.50 m deep	0.57	35.88	22.81	37.65	m	**96.34**
4.50 - 5.00 m deep	0.72	45.32	33.15	37.65	m	**116.12**
5.00 - 5.50 m deep	0.90	56.66	38.13	37.65	m	**132.43**
Extra for bend; 45 degrees	0.09	5.67	1.85	21.39	nr	**28.90**
Extra for bend; 90 degrees	0.09	5.67	1.85	28.80	nr	**36.31**
Extra for single junction; equal	0.16	10.07	3.70	53.50	nr	**67.27**

CLASS I: PIPEWORK - PIPES

Item	Gang hours	Labour £	Plant £	Material £	Unit	**Total rate £**
200 mm pipes; in trenches, depth						
not in trenches	0.09	3.69	6.95	39.81	m	**50.45**
ne 1.50 m deep	0.18	11.33	6.80	39.81	m	**57.94**
1.50 - 2.00 m deep	0.21	13.22	8.01	39.81	m	**61.04**
2.00 - 2.50 m deep	0.25	15.74	9.58	39.81	m	**65.12**
2.50 - 3.00 m deep	0.30	18.89	11.61	39.81	m	**70.30**
3.00 - 3.50 m deep	0.40	25.18	15.61	39.81	m	**80.60**
3.50 - 4.00 m deep	0.46	28.96	18.17	39.81	m	**86.93**
4.00 - 4.50 m deep	0.59	37.14	23.60	39.81	m	**100.54**
4.50 - 5.00 m deep	0.74	46.58	30.26	39.81	m	**116.65**
5.00 - 5.50 m deep	0.92	57.91	38.96	39.81	m	**136.68**
Extra for bend; 45 degrees	0.12	7.55	1.85	45.21	nr	**54.61**
Extra for bend; 90 degrees	0.12	7.55	1.85	59.00	nr	**68.40**
Extra for single junction; equal	0.21	13.22	3.67	102.76	nr	**119.65**
250 mm pipes; in trenches, depth						
not in trenches	0.10	4.10	0.39	50.05	m	**54.54**
ne 1.50 m deep	0.18	11.33	8.34	50.05	m	**69.71**
1.50 - 2.00 m deep	0.22	13.85	10.14	50.05	m	**74.04**
2.00 - 2.50 m deep	0.26	16.37	11.87	50.05	m	**78.29**
2.50 - 3.00 m deep	0.31	19.51	14.33	50.05	m	**83.89**
3.00 - 3.50 m deep	0.41	25.81	18.97	50.05	m	**94.82**
3.50 - 4.00 m deep	0.47	29.59	21.57	50.05	m	**101.21**
4.00 - 4.50 m deep	0.60	37.77	27.39	50.05	m	**115.21**
4.50 - 5.00 m deep	0.75	47.21	34.52	50.05	m	**131.78**
5.00 - 5.50 m deep	0.94	59.17	43.29	50.05	m	**152.51**
Extra for bend; 45 degrees	0.13	8.18	1.85	79.52	nr	**89.55**
Extra for bend; 90 degrees	0.13	8.18	1.85	104.59	nr	**114.62**
Extra for single junction; equal	0.23	14.48	3.70	191.57	nr	**209.75**
300 mm pipes; in trenches, depth						
not in trenches	0.11	4.51	0.42	57.38	m	**62.31**
ne 1.50 m deep	0.20	12.59	7.54	57.38	m	**77.51**
1.50 - 2.00 m deep	0.25	15.74	9.54	57.38	m	**82.66**
2.00 - 2.50 m deep	0.28	17.63	10.71	57.38	m	**85.72**
2.50 - 3.00 m deep	0.34	21.40	13.15	57.38	m	**91.94**
3.00 - 3.50 m deep	0.44	27.70	17.17	57.38	m	**102.25**
3.50 - 4.00 m deep	0.53	33.36	20.92	57.38	m	**111.67**
4.00 - 4.50 m deep	0.68	42.81	27.20	57.38	m	**127.39**
4.50 - 5.00 m deep	0.86	54.14	35.16	57.38	m	**146.68**
5.00 - 5.50 m deep	1.06	66.73	44.90	57.38	m	**169.01**
Extra for bend; 45 degrees	0.14	8.81	2.79	106.31	nr	**117.91**
Extra for bend; 90 degrees	0.14	8.81	2.79	143.13	nr	**154.73**
Extra for single junction; equal	0.25	15.74	6.02	295.81	nr	**317.56**

CLASS I: PIPEWORK - PIPES

Item	Gang hours	Labour £	Plant £	Material £	Unit	**Total rate £**
POLYVINYL CHLORIDE PIPES						
Unplasticised pvc perforated pipes; ring seal sockets; excavation and supports, backfilling; 6 m pipe lengths unless stated otherwise						
82 mm pipes; in trench, depth						
not in trenches; 3.00 m pipe lengths	0.06	2.46	0.10	8.78	m	**11.34**
ne 1.50 m deep; 3.00 m pipe lengths	0.10	6.29	2.09	8.78	m	**17.17**
1.50 - 2.00 m deep	0.13	8.18	2.77	8.78	m	**19.74**
2.00 - 2.50 m deep	0.16	10.07	3.43	8.78	m	**22.28**
2.50 - 3.00 m deep	0.19	11.96	4.14	8.78	m	**24.89**
3.00 - 3.50 m deep	0.22	13.85	4.89	8.78	m	**27.52**
3.50 - 4.00 m deep	0.25	15.74	5.67	8.78	m	**30.18**
4.00 - 4.50 m deep	0.28	17.63	6.49	8.78	m	**32.90**
4.50 - 5.00 m deep	0.32	20.14	7.70	8.78	m	**36.62**
5.00 - 5.50 m deep	0.35	22.03	8.93	8.78	m	**39.75**
110 mm pipes; in trench, depth						
not in trenches	0.06	2.46	0.05	7.77	m	**10.28**
ne 1.50 m deep	0.10	6.29	2.09	7.77	m	**16.16**
1.50 - 2.00 m deep	0.13	8.18	2.77	7.77	m	**18.72**
2.00 - 2.50 m deep	0.16	10.07	3.43	7.77	m	**21.27**
2.50 - 3.00 m deep	0.19	11.96	4.14	7.77	m	**23.87**
3.00 - 3.50 m deep	0.22	13.85	4.89	7.77	m	**26.51**
3.50 - 4.00 m deep	0.25	15.74	5.67	7.77	m	**29.17**
4.00 - 4.50 m deep	0.28	17.63	6.49	7.77	m	**31.89**
4.50 - 5.00 m deep	0.32	20.14	7.70	7.77	m	**35.61**
5.00 - 5.50 m deep	0.36	22.66	9.19	7.77	m	**39.62**
160 mm pipes; in trench, depth						
not in trenches	0.06	2.46	0.10	15.35	m	**17.91**
ne 1.50 m deep	0.11	6.92	2.29	15.35	m	**24.57**
1.50 - 2.00 m deep	0.14	8.81	2.99	15.35	m	**27.15**
2.00 - 2.50 m deep	0.18	11.33	3.87	15.35	m	**30.55**
2.50 - 3.00 m deep	0.22	13.85	4.81	15.35	m	**34.01**
3.00 - 3.50 m deep	0.23	14.48	5.11	15.35	m	**34.94**
3.50 - 4.00 m deep	0.26	16.37	5.89	15.35	m	**37.61**
4.00 - 4.50 m deep	0.30	18.89	6.96	15.35	m	**41.20**
4.50 - 5.00 m deep	0.35	22.03	8.42	15.35	m	**45.80**
5.00 - 5.50 m deep	0.40	25.18	10.21	15.35	m	**50.74**
Unplasticised pvc pipes; ring seal sockets; excavation and supports, backfilling; 6 m pipe lengths unless stated otherwise						
82 mm pipes; in trenches, depth						
not in trenches; 3.00 m pipe lengths	0.06	2.46	0.24	8.47	m	**11.16**
ne 1.50 m deep; 3.00 m pipe lengths	0.10	6.29	3.68	8.47	m	**18.44**
1.50 - 2.00 m deep	0.13	8.18	4.96	8.47	m	**21.61**
2.00 - 2.50 m deep	0.16	10.07	6.12	8.47	m	**24.66**
2.50 - 3.00 m deep	0.19	11.96	7.34	8.47	m	**27.76**
3.00 - 3.50 m deep	0.22	13.85	8.60	8.47	m	**30.91**
3.50 - 4.00 m deep	0.25	15.74	9.87	8.47	m	**34.08**
4.00 - 4.50 m deep	0.28	17.63	11.20	8.47	m	**37.29**
4.50 - 5.00 m deep	0.32	20.14	13.08	8.47	m	**41.69**
5.00 - 5.50 m deep	0.35	22.03	14.81	8.47	m	**45.31**
Extra for bend; short radius (socket/spigot)	0.05	3.15	-	12.77	nr	**15.91**
Extra for branches; equal (socket/spigot)	0.07	4.41	-	20.15	nr	**24.56**

CLASS I: PIPEWORK - PIPES

Item	Gang hours	Labour £	Plant £	Material £	Unit	**Total rate £**
110 mm pipes; in trenches, depth						
not in trenches	0.06	2.46	0.24	7.04	m	**9.74**
ne 1.50 m deep	0.10	6.29	3.78	7.04	m	**17.12**
1.50 - 2.00 m deep	0.13	8.18	4.96	7.04	m	**20.18**
2.00 - 2.50 m deep	0.16	10.07	6.10	7.04	m	**23.21**
2.50 - 3.00 m deep	0.19	11.96	7.34	7.04	m	**26.33**
3.00 - 3.50 m deep	0.22	13.85	8.60	7.04	m	**29.49**
3.50 - 4.00 m deep	0.25	15.74	9.87	7.04	m	**32.65**
4.00 - 4.50 m deep	0.28	17.63	11.20	7.04	m	**35.87**
4.50 - 5.00 m deep	0.32	20.14	13.08	7.04	m	**40.26**
5.00 - 5.50 m deep	0.36	22.66	15.24	7.04	m	**44.94**
Extra for bend; short redius (socket/spigot)	0.05	3.15	-	13.36	nr	**16.51**
Extra for bend; adjustable (socket/spigot)	0.05	3.15	-	25.37	nr	**28.52**
Extra for reducer	0.05	3.15	-	4.60	nr	**7.75**
Extra for branches; equal (socket/spigot)	0.07	4.41	-	19.92	nr	**24.32**
160 mm pipes; in trenches, depth						
not in trenches	0.06	2.46	0.24	14.18	m	**16.88**
ne 1.50 m deep	0.11	6.92	4.14	14.18	m	**25.24**
1.50 - 2.00 m deep	0.14	8.81	5.35	14.18	m	**28.34**
2.00 - 2.50 m deep	0.18	11.33	6.90	14.18	m	**32.41**
2.50 - 3.00 m deep	0.22	13.85	8.52	14.18	m	**36.54**
3.00 - 3.50 m deep	0.23	14.48	8.97	14.18	m	**37.63**
3.50 - 4.00 m deep	0.26	16.37	10.27	14.18	m	**40.82**
4.00 - 4.50 m deep	0.30	18.89	12.02	14.18	m	**45.08**
4.50 - 5.00 m deep	0.35	22.03	14.30	14.18	m	**50.51**
5.00 - 5.50 m deep	0.40	25.18	16.94	14.18	m	**56.30**
Extra for bend; short radius (socket/spigot)	0.05	3.15	-	30.01	nr	**33.16**
Extra for branches; equal (socket/spigot)	0.07	4.41	-	38.84	nr	**43.25**
225 mm pipes; in trenches, depth						
not in trenches	0.07	2.87	0.26	7.75	m	**10.88**
ne 1.50 m deep	0.12	7.55	4.52	7.75	m	**19.83**
1.50 - 2.00 m deep	0.15	9.44	5.71	7.75	m	**22.91**
2.00 - 2.50 m deep	0.20	12.59	7.65	7.75	m	**27.99**
2.50 - 3.00 m deep	0.23	14.48	8.88	7.75	m	**31.11**
3.00 - 3.50 m deep	0.24	15.11	9.37	7.75	m	**32.23**
3.50 - 4.00 m deep	0.27	17.00	10.65	7.75	m	**35.40**
4.00 - 4.50 m deep	0.32	20.14	12.80	7.75	m	**40.70**
4.50 - 5.00 m deep	0.36	22.66	14.71	7.75	m	**45.13**
5.00 - 5.50 m deep	0.45	28.33	19.06	7.75	m	**55.14**
Extra for bend; short radius 45° (double socket)	0.07	4.41	-	52.74	nr	**57.15**
Extra for branches; equal (all socket)	0.09	5.67	-	123.87	nr	**129.54**
300 mm pipes; in trenches, depth						
not in trenches	0.08	3.28	0.31	12.85	m	**16.44**
ne 1.50 m deep	0.13	8.18	4.91	12.85	m	**25.94**
1.50 - 2.00 m deep	0.16	10.07	6.10	12.85	m	**29.02**
2.00 - 2.50 m deep	0.21	13.22	8.04	12.85	m	**34.11**
2.50 - 3.00 m deep	0.23	14.48	8.88	12.85	m	**36.21**
3.00 - 3.50 m deep	0.25	15.74	9.76	12.85	m	**38.35**
3.50 - 4.00 m deep	0.28	17.63	11.05	12.85	m	**41.53**
4.00 - 4.50 m deep	0.34	21.40	13.62	12.85	m	**47.87**
4.50 - 5.00 m deep	0.38	23.92	15.55	12.85	m	**52.32**
5.00 - 5.50 m deep	0.47	29.59	19.90	12.85	m	**62.34**
Extra for bend; short radius 45° (double socket)	0.07	4.41	-	98.08	nr	**102.49**
Extra for branches; unequal (all socket)	0.09	5.67	-	286.48	nr	**292.15**

CLASS I: PIPEWORK - PIPES

Item	Gang hours	Labour £	Plant £	Material £	Unit	**Total rate £**
POLYVINYL CHLORIDE PIPES – cont'd						
Unplasticised pvc pipes; polypropylene couplings; excavation and supports, backfilling; 6 m pipe lengths unless stated otherwise						
110 mm pipes; in trenches, depth						
not in trenches	0.06	2.46	0.24	6.13	m	**8.83**
ne 1.50 m deep	0.10	6.29	3.78	6.13	m	**16.21**
1.50 - 2.00 m deep	0.13	8.18	4.96	6.13	m	**19.27**
2.00 - 2.50 m deep	0.16	10.07	6.12	6.13	m	**22.32**
2.50 - 3.00 m deep	0.19	11.96	7.34	6.13	m	**25.43**
3.00 - 3.50 m deep	0.22	13.85	8.60	6.13	m	**28.58**
3.50 - 4.00 m deep	0.25	15.74	9.87	6.13	m	**31.74**
4.00 - 4.50 m deep	0.28	17.63	11.20	6.13	m	**34.96**
4.50 - 5.00 m deep	0.32	20.14	13.08	6.13	m	**39.35**
5.00 - 5.50 m deep	0.36	22.66	15.24	6.13	m	**44.04**
160 mm pipes; in trenches, depth						
not in trenches	0.06	2.46	0.25	19.82	m	**22.52**
ne 1.50 m deep	0.11	6.92	4.14	19.82	m	**30.88**
1.50 - 2.00 m deep	0.14	8.81	5.35	19.82	m	**33.98**
2.00 - 2.50 m deep	0.18	11.33	6.90	19.82	m	**38.05**
2.50 - 3.00 m deep	0.22	13.85	8.52	19.82	m	**42.18**
3.00 - 3.50 m deep	0.23	14.48	8.97	19.82	m	**43.26**
3.50 - 4.00 m deep	0.26	16.37	10.27	19.82	m	**46.46**
4.00 - 4.50 m deep	0.30	18.89	12.02	19.82	m	**50.72**
4.50 - 5.00 m deep	0.35	22.03	14.30	19.82	m	**56.15**
5.00 - 5.50 m deep	0.40	25.18	16.94	19.82	m	**61.93**
225 mm pipes; in trenches, depth						
not in trenches	0.07	2.87	0.26	21.65	m	**24.78**
ne 1.50 m deep	0.12	7.55	4.52	21.65	m	**33.73**
1.50 - 2.00 m deep	0.15	9.44	5.71	21.65	m	**36.80**
2.00 - 2.50 m deep	0.20	12.59	7.65	21.65	m	**41.89**
2.50 - 3.00 m deep	0.23	14.48	8.88	21.65	m	**45.01**
3.00 - 3.50 m deep	0.24	15.11	9.37	21.65	m	**46.12**
3.50 - 4.00 m deep	0.27	17.00	10.65	21.65	m	**49.29**
4.00 - 4.50 m deep	0.32	20.14	12.80	21.65	m	**54.59**
4.50 - 5.00 m deep	0.36	22.66	14.71	21.65	m	**59.03**
5.00 - 5.50 m deep	0.45	28.33	19.06	21.65	m	**69.04**
300 mm pipes; in trenches, depth						
not in trenches	0.08	3.28	0.31	42.05	m	**45.63**
ne 1.50 m deep	0.13	8.18	4.91	42.05	m	**55.14**
1.50 - 2.00 m deep	0.16	10.07	6.10	42.05	m	**58.22**
2.00 - 2.50 m deep	0.21	13.22	8.04	42.05	m	**63.31**
2.50 - 3.00 m deep	0.23	14.48	8.88	42.05	m	**65.41**
3.00 - 3.50 m deep	0.25	15.74	9.76	42.05	m	**67.55**
3.50 - 4.00 m deep	0.28	17.63	11.05	42.05	m	**70.72**
4.00 - 4.50 m deep	0.34	21.40	13.62	42.05	m	**77.07**
4.50 - 5.00 m deep	0.38	23.92	15.55	42.05	m	**81.51**
5.00 - 5.50 m deep	0.47	29.59	19.90	42.05	m	**91.53**

CLASS I: PIPEWORK - PIPES

Item	Gang hours	Labour £	Plant £	Material £	Unit	**Total rate £**
Ultrarib unplasticised pvc pipes; ring seal joints; excavation and supports, backfilling						
150 mm pipes; in trenches, depth						
not in trenches	0.13	5.33	0.50	16.93	m	**22.76**
ne 1.50 m deep	0.16	10.07	6.03	16.93	m	**33.03**
1.50 - 2.00 m deep	0.19	11.96	7.24	16.93	m	**36.13**
2.00 - 2.50 m deep	0.22	13.85	8.44	16.93	m	**39.21**
2.50 - 3.00 m deep	0.24	15.11	9.27	16.93	m	**41.31**
3.00 - 3.50 m deep	0.24	15.11	9.37	16.93	m	**41.40**
3.50 - 4.00 m deep	0.27	17.00	10.65	16.93	m	**44.58**
4.00 - 4.50 m deep	0.32	20.14	12.80	16.93	m	**49.88**
4.50 - 5.00 m deep	0.36	22.66	14.71	16.93	m	**54.31**
5.00 - 5.50 m deep	0.45	28.33	19.06	16.93	m	**64.32**
Extra for bends; short radius (socket/spigot)	0.05	3.15	-	14.93	nr	**18.08**
Extra for branches; equal (socket/spigot)	0.09	5.67	-	36.12	nr	**41.78**
225 mm pipes; in trenches, depth						
not in trenches	0.08	3.28	0.31	19.01	m	**22.59**
ne 1.50 m deep	0.13	8.18	4.91	19.01	m	**32.10**
1.50 - 2.00 m deep	0.16	10.07	6.10	19.01	m	**35.18**
2.00 - 2.50 m deep	0.21	13.22	8.04	19.01	m	**40.27**
2.50 - 3.00 m deep	0.23	14.48	8.88	19.01	m	**42.37**
3.00 - 3.50 m deep	0.25	15.74	9.76	19.01	m	**44.51**
3.50 - 4.00 m deep	0.28	17.63	11.05	19.01	m	**47.69**
4.00 - 4.50 m deep	0.34	21.40	13.62	19.01	m	**54.03**
4.50 - 5.00 m deep	0.38	23.92	15.55	19.01	m	**58.47**
5.00 - 5.50 m deep	0.47	29.59	19.90	19.01	m	**68.49**
Extra for bends; short radius (socket/spigot)	0.05	3.15	-	60.09	nr	**63.24**
Extra for branches; equal (socket/spigot)	0.09	5.67	-	107.42	nr	**113.08**
300 mm pipes; in trenches, depth						
not in trenches	0.08	3.28	0.31	30.50	m	**34.09**
ne 1.50 m deep	0.13	8.18	4.91	30.50	m	**43.59**
1.50 - 2.00 m deep	0.16	10.07	6.10	30.50	m	**46.67**
2.00 - 2.50 m deep	0.21	13.22	8.04	30.50	m	**51.76**
2.50 - 3.00 m deep	0.23	14.48	8.88	30.50	m	**53.86**
3.00 - 3.50 m deep	0.25	15.74	9.76	30.50	m	**56.00**
3.50 - 4.00 m deep	0.28	17.63	11.05	30.50	m	**59.18**
4.00 - 4.50 m deep	0.34	21.40	13.62	30.50	m	**65.52**
4.50 - 5.00 m deep	0.38	23.92	15.55	30.50	m	**69.97**
5.00 - 5.50 m deep	0.47	29.59	19.90	30.50	m	**79.99**
Extra for bends; short radius (socket/spigot)	0.07	4.41	-	97.82	nr	**102.23**
Extra for branches; equal (socket/spigot)	0.09	5.67	-	229.10	nr	**234.77**

CLASS J: PIPEWORK - FITTINGS AND VALVES

Item	Gang hours	Labour £	Plant £	Material £	Unit	**Total rate £**
NOTES						
Fittings on pipes shown with the appropriate pipe in Class I						
RESOURCES – LABOUR						
Fittings and valves gang						
1 ganger/chargehand (skill rate 4) - 50% of time		6.40				
1 skilled operative (skill rate 4)		12.05				
2 unskilled operatives (general)		22.52				
1 plant operator (skill rate 3)		14.65				
Total Gang Rate / Hour	£	**55.62**				
RESOURCES - PLANT						
Fittings and valves						
1.0 m3 hydraulic backacter			32.48			
disc saw - 30% of time			0.39			
2.80 m3/min compressor, 2 tool - 30% of time			1.77			
2t dumper - 30% of time			1.30			
compressor tools, extra 50ft / 15m hose - 30% of time			0.07			
small pump - 30% of time			0.65			
sundry tools - 30% of time			0.31			
Total Rate / Hour		£	**36.97**			
VALVES AND PENSTOCKS						
Valves						
Non-return valves; cast iron; single door; tidal flap						
250 mm	0.40	22.25	7.39	278.68	nr	**308.32**
350 mm	0.40	25.18	7.39	508.52	nr	**541.09**
450 mm	0.70	44.06	7.39	662.12	nr	**713.58**
600 mm	0.70	44.06	11.10	1053.98	nr	**1109.15**
800 mm	0.75	47.21	12.93	1966.50	nr	**2026.65**
Penstocks; cast iron; wall mounted; hand operated						
250 mm	1.66	104.50	12.93	360.80	nr	**478.23**
350 mm	2.30	144.78	14.79	509.30	nr	**668.87**
450 mm	2.90	182.56	16.64	629.20	nr	**828.40**
600 x 600 mm	5.40	339.93	29.58	1016.40	nr	**1385.91**
1000 x 1000 mm	9.00	566.55	55.47	1819.40	nr	**2441.42**

CLASS K: PIPEWORK - MANHOLES AND PIPEWORK ANCILLARIES

Item	Gang hours	Labour £	Plant £	Material £	Unit	**Total rate £**
NOTES						
The rates assume the most efficient items of plant (excavator) and are optimum rates assuming continuous output with no delays caused by other operations or works.						
Ground conditions are assumed to be good easily worked soil with no abnormal conditions that would affect outputs and consistency of work.						
Multiplier Table for labour and plant for various site conditions for working:						
out of sequence x 2.75 (minimum)						
in hard clay x 1.75 to 2.00						
in running sand x 2.75 (minimum)						
in broken rock x 2.75 to 3.50						
below water table x 2.00 (minimum)						
RESOURCES - LABOUR						
Gullies gang						
1 chargehand pipelayer (skill rate 4) - 50% of time		6.40				
1 skilled operative (skill rate 4)		12.05				
2 unskilled operatives (general)		22.52				
1 plant operator (skill rate 3)		14.65				
Total Gang Rate / Hour	£	**55.62**				
French/rubble drains, ditches and trenches gang; ducts and metal culverts gang						
1 chargehand pipelayer (skill rate 4) - 50% of time		6.40				
1 skilled operative (skill rate 4)		12.05				
2 unskilled operatives (general)		22.52				
1 plant operator (skill rate 3)		14.65				
Total Gang Rate / Hour	£	**55.62**				
RESOURCES - PLANT						
Gullies						
0.4 m^3 hydraulic excavator			17.88			
2t dumper (30% of time)			1.30			
2.80 m^3/min compressor, 2 tool (30% of time)			1.77			
compaction plate / roller (30% of time)			0.43			
2.40 m^3/min road breaker (30% of time)			0.23			
54mm poker vibrator (30% of time)			0.36			
extra 15ft / 50m hose (30% of time)			0.07			
disc saw (30% of time)			0.33			
small pump (30% of time)			0.65			
Total Rate / Hour		£	**23.02**			
French/rubble drains, ditches and trenches; ducts and metal culverts						
0.4m^3 hydraulic backacter			17.88			
2t dumper (30% of time)			1.30			
disc saw (30% of time)			0.33			
compaction plate / roller (30% of time)			0.77			
2.80 m^3/min compressor, 2 tool (30% of time)			1.77			
small pump (30% of time)			0.65			
Total Rate / Hour		£	**22.69**			

CLASS K: PIPEWORK - MANHOLES AND PIPEWORK ANCILLARIES

Item	Gang hours	Labour £	Plant £	Material £	Unit	Total rate £
MANHOLES						
Brick construction						
Design criteria used in models:						
• class A engineering bricks						
• 215 thick walls generally; 328 thick to chambers exceeding 2.5 m deep						
• 225 mm plain concrete C20/20 base slab						
• 300 mm reinforced concrete C20/20 reducing slab						
• 125 mm reinforced concrete C20/20 top slab						
• maximum height of working chamber 2.0 m above benching						
• 750 x 750 access shaft						
• plain concrete C15/20 benching, 150 mm clay main channel longitudinally and two 100 branch channels						
• step irons at 300 mm centres, doubled if depth to invert exceeds 3000 mm						
• heavy duty manhole cover and frame						
750 x 700 chamber 500 depth to invert						
excavation, support, backfilling and disposal	-	-	-	-	-	36.80
concrete base	-	-	-	-	-	98.90
brickwork chamber	-	-	-	-	-	53.13
concrete cover slab	-	-	-	-	-	120.75
concrete benching, main and branch channels	-	-	-	-	-	96.60
step irons	-	-	-	-	-	11.00
access cover and frame	-	-	-	-	-	297.00
TOTAL	-	-	-	-	£	**714.18**
750 x 700 chamber 1000 depth to invert						
excavation, support, backfilling and disposal	-	-	-	-	-	59.80
concrete base	-	-	-	-	-	98.90
brickwork chamber	-	-	-	-	-	200.97
concrete cover slab	-	-	-	-	-	120.75
concrete benching and channels	-	-	-	-	-	96.60
step irons	-	-	-	-	-	16.50
access cover and frame	-	-	-	-	-	297.00
TOTAL	-	-	-	-	£	**890.52**
750 x 700 chamber 1500 depth to invert						
excavation, support, backfilling and disposal	-	-	-	-	-	85.10
concrete base	-	-	-	-	-	98.90
brickwork chamber	-	-	-	-	-	346.50
concrete cover slab	-	-	-	-	-	120.75
concrete benching and channels	-	-	-	-	-	96.60
step irons	-	-	-	-	-	27.50
access cover and frame	-	-	-	-	-	297.00
TOTAL	-	-	-	-	£	**1072.35**
900 x 700 chamber 500 depth to invert						
excavation, support, backfilling and disposal	-	-	-	-	-	41.40
concrete base	-	-	-	-	-	102.35
brickwork chamber	-	-	-	-	-	58.91
concrete cover slab	-	-	-	-	-	133.40
concrete benching and channels	-	-	-	-	-	109.25
step irons	-	-	-	-	-	11.00
access cover and frame	-	-	-	-	-	297.00
TOTAL	-	-	-	-	£	**753.30**

CLASS K: PIPEWORK - MANHOLES AND PIPEWORK ANCILLARIES

Item	Gang hours	Labour £	Plant £	Material £	Unit	Total rate £
900 x 700 chamber 1000 depth to invert						
excavation, support, backfilling and disposal	-	-	-	-	-	69.00
concrete base	-	-	-	-	-	102.35
brickwork chamber	-	-	-	-	-	217.14
concrete cover slab	-	-	-	-	-	133.40
concrete benching and channels	-	-	-	-	-	109.25
step irons	-	-	-	-	-	16.50
access cover and frame	-	-	-	-	-	297.00
TOTAL	-	-	-	-	£	**944.64**
900 x 700 chamber 1500 depth to invert						
excavation, support, backfilling and disposal	-	-	-	-	-	96.60
concrete base	-	-	-	-	-	102.35
brickwork chamber	-	-	-	-	-	375.38
concrete cover slab	-	-	-	-	-	133.40
concrete benching and channels	-	-	-	-	-	109.25
step irons	-	-	-	-	-	27.50
access cover and frame	-	-	-	-	-	297.00
TOTAL	-	-	-	-	£	**1141.47**
1050 x 700 chamber 1500 depth to invert						
excavation, support, backfilling and disposal	-	-	-	-	-	104.65
concrete base	-	-	-	-	-	105.80
brickwork chamber	-	-	-	-	-	403.10
concrete cover slab	-	-	-	-	-	144.90
concrete benching and channels	-	-	-	-	-	123.05
step irons	-	-	-	-	-	27.50
access cover and frame	-	-	-	-	-	297.00
TOTAL	-	-	-	-	£	**1206.00**
1050 x 700 chamber 2500 depth to invert						
excavation, support, backfilling and disposal	-	-	-	-	-	177.10
concrete base	-	-	-	-	-	105.80
brickwork chamber	-	-	-	-	-	742.66
concrete cover slab	-	-	-	-	-	144.90
concrete benching and channels	-	-	-	-	-	123.05
step irons	-	-	-	-	-	44.00
access cover and frame	-	-	-	-	-	297.00
TOTAL	-	-	-	-	£	**1634.51**
1050 x 700 chamber 3500 depth to invert						
excavation, support, backfilling and disposal	-	-	-	-	-	258.75
concrete base	-	-	-	-	-	105.80
brickwork chamber	-	-	-	-	-	739.20
access shaft	-	-	-	-	-	213.68
reducing slab	-	-	-	-	-	162.15
concrete cover slab	-	-	-	-	-	97.75
concrete benching and channels	-	-	-	-	-	123.05
step irons	-	-	-	-	-	88.00
access cover and frame	-	-	-	-	-	297.00
TOTAL	-	-	-	-	£	**2085.38**
1350 x 700 chamber 2500 depth to invert						
excavation, support, backfilling and disposal	-	-	-	-	-	216.20
concrete base	-	-	-	-	-	117.30
brickwork chamber	-	-	-	-	-	863.94
concrete cover slab	-	-	-	-	-	174.80
concrete benching and channels	-	-	-	-	-	155.25
step irons	-	-	-	-	-	44.00
access cover and frame	-	-	-	-	-	297.00
TOTAL	-	-	-	-	£	**1868.49**

CLASS K: PIPEWORK - MANHOLES AND PIPEWORK ANCILLARIES

Item	Gang hours	Labour £	Plant £	Material £	Unit	Total rate £
MANHOLES – cont'd						
Brick construction – cont'd						
1350 x 700 chamber 3500 depth to invert						
excavation, support, backfilling and disposal	-	-	-	-	-	316.25
concrete base	-	-	-	-	-	117.30
brickwork chamber	-	-	-	-	-	860.48
access shaft	-	-	-	-	-	213.68
reducing slab	-	-	-	-	-	187.45
concrete cover slab	-	-	-	-	-	97.75
concrete benching and channels	-	-	-	-	-	155.25
step irons	-	-	-	-	-	88.00
access cover and frame	-	-	-	-	-	297.00
TOTAL	-	-	-	-	£	**2333.15**
1350 x 700 chamber 4500 depth to invert						
excavation, support, backfilling and disposal	-	-	-	-	-	424.35
concrete base	-	-	-	-	-	117.30
brickwork chamber	-	-	-	-	-	860.48
access shaft	-	-	-	-	-	513.98
reducing slab	-	-	-	-	-	187.45
concrete cover slab	-	-	-	-	-	97.75
concrete benching and channels	-	-	-	-	-	155.25
step irons	-	-	-	-	-	165.00
access cover and frame	-	-	-	-	-	297.00
TOTAL	-	-	-	-	£	**2818.55**
Precast concrete construction						
Design criteria used in models: • circular shafts • 150 mm plain concrete C15/20 surround • 225 mm plain concrete C20/20 base slab • precast reducing slab • precast top slab • maximum height of working chamber 2.0 m above benching • 750 mm diameter access shaft • plain concrete C15/20 benching, 150 mm clay main channel longitudinally and two 100 branch channels • step irons at 300 mm centres, doubled if depth to invert exceeds 3000 mm • heavy duty manhole cover and frame • in manholes over 6 m deep, landings at maximum intervals						
675 diameter x 500 depth to invert						
excavation, support, backfilling and disposal	-	-	-	-	-	27.60
concrete base	-	-	-	-	-	41.40
main chamber rings	-	-	-	-	-	20.70
cover slab	-	-	-	-	-	81.65
concrete benching and channels	-	-	-	-	-	54.05
concrete surround	-	-	-	-	-	42.55
step irons	-	-	-	-	-	-
access cover and frame	-	-	-	-	-	297.00
TOTAL	-	-	-	-	£	**564.95**

CLASS K: PIPEWORK - MANHOLES AND PIPEWORK ANCILLARIES

Item	Gang hours	Labour £	Plant £	Material £	Unit	**Total rate £**
675 diameter x 750 depth to invert						
excavation, support, backfilling and disposal	-	-	-	-	-	36.80
concrete base	-	-	-	-	-	41.40
main chamber rings	-	-	-	-	-	47.15
cover slab	-	-	-	-	-	81.65
concrete benching and channels	-	-	-	-	-	54.05
concrete surround	-	-	-	-	-	62.10
step irons	-	-	-	-	-	-
access cover and frame	-	-	-	-	-	297.00
TOTAL	-	-	-	-	£	**620.15**
675 diameter x 1000 depth to invert						
excavation, support, backfilling and disposal	-	-	-	-	-	44.85
concrete base	-	-	-	-	-	41.40
main chamber rings	-	-	-	-	-	74.75
cover slab	-	-	-	-	-	81.65
concrete benching and channels	-	-	-	-	-	54.05
step irons	-	-	-	-	-	-
concrete surround	-	-	-	-	-	82.80
access cover and frame	-	-	-	-	-	297.00
TOTAL	-	-	-	-	£	**676.50**
675 diameter x 1250 depth to invert						
excavation, support, backfilling and disposal	-	-	-	-	-	52.90
concrete base	-	-	-	-	-	41.40
main chamber rings	-	-	-	-	-	102.35
cover slab	-	-	-	-	-	81.65
concrete benching and channels	-	-	-	-	-	54.05
concrete surround	-	-	-	-	-	102.35
step irons	-	-	-	-	-	6.60
access cover and frame	-	-	-	-	-	297.00
TOTAL	-	-	-	-	£	**738.30**
900 diameter x 750 depth to invert						
excavation, support, backfilling and disposal	-	-	-	-	-	52.90
concrete base	-	-	-	-	-	57.50
main chamber rings	-	-	-	-	-	58.65
cover slab	-	-	-	-	-	98.90
concrete benching and channels	-	-	-	-	-	109.25
concrete surround	-	-	-	-	-	78.20
step irons	-	-	-	-	-	-
access cover and frame	-	-	-	-	-	297.00
TOTAL	-	-	-	-	£	**696.05**
900 diameter x 1000 depth to invert						
excavation, support, backfilling and disposal	-	-	-	-	-	64.40
concrete base	-	-	-	-	-	57.50
main chamber rings	-	-	-	-	-	93.15
cover slab	-	-	-	-	-	98.90
concrete benching and channels	-	-	-	-	-	109.25
concrete surround	-	-	-	-	-	102.35
step irons	-	-	-	-	-	-
access cover and frame	-	-	-	-	-	297.00
TOTAL	-	-	-	-	£	**782.30**

CLASS K: PIPEWORK - MANHOLES AND PIPEWORK ANCILLARIES

Item	Gang hours	Labour £	Plant £	Material £	Unit	Total rate £
MANHOLES – cont'd						
Precast concrete construction – cont'd						
900 diameter x 1500 depth to invert						
excavation, support, backfilling and disposal	-	-	-	-	-	92.00
concrete base	-	-	-	-	-	57.50
main chamber rings	-	-	-	-	-	161.00
cover slab	-	-	-	-	-	98.90
concrete benching and channels	-	-	-	-	-	109.25
concrete surround	-	-	-	-	-	151.80
step irons	-	-	-	-	-	13.20
access cover and frame	-	-	-	-	-	297.00
TOTAL	-	-	-	-	£	**940.40**
1200 diameter x 1500 depth to invert						
excavation, support, backfilling and disposal	-	-	-	-	-	143.75
concrete base	-	-	-	-	-	83.95
main chamber rings	-	-	-	-	-	218.50
concrete benching and channels	-	-	-	-	-	82.80
cover slab	-	-	-	-	-	140.30
concrete surround	-	-	-	-	-	198.95
step irons	-	-	-	-	-	13.20
access cover and frame	-	-	-	-	-	297.00
TOTAL	-	-	-	-	£	**1178.45**
1200 diameter x 2000 depth to invert						
excavation, support, backfilling and disposal	-	-	-	-	-	202.40
concrete base	-	-	-	-	-	83.95
main chamber rings	-	-	-	-	-	307.05
cover slab	-	-	-	-	-	140.30
concrete benching and channels	-	-	-	-	-	82.80
concrete surround	-	-	-	-	-	263.35
step irons	-	-	-	-	-	25.30
access cover and frame	-	-	-	-	-	297.00
TOTAL	-	-	-	-	£	**1402.15**
1200 diameter x 2500 depth to invert						
excavation, support, backfilling and disposal	-	-	-	-	-	248.40
concrete base	-	-	-	-	-	83.95
main chamber rings	-	-	-	-	-	399.05
cover slab	-	-	-	-	-	140.30
concrete benching and channels	-	-	-	-	-	82.80
concrete surround	-	-	-	-	-	327.75
step irons	-	-	-	-	-	31.90
access cover and frame	-	-	-	-	-	297.00
TOTAL	-	-	-	-	£	**1611.15**
1200 diameter x 3000 depth to invert						
excavation, support, backfilling and disposal	-	-	-	-	-	319.70
concrete base	-	-	-	-	-	83.95
main chamber rings	-	-	-	-	-	492.20
cover slab	-	-	-	-	-	140.30
concrete benching and channels	-	-	-	-	-	82.80
concrete surround	-	-	-	-	-	393.30
step irons	-	-	-	-	-	90.20
access cover and frame	-	-	-	-	-	297.00
TOTAL	-	-	-	-	£	**1899.45**

CLASS K: PIPEWORK - MANHOLES AND PIPEWORK ANCILLARIES

Item	Gang hours	Labour £	Plant £	Material £	Unit	Total rate £
1800 diameter x 1500 depth to invert						
excavation, support, backfilling and disposal	-	-	-	-	-	262.20
concrete base	-	-	-	-	-	147.20
main chamber rings	-	-	-	-	-	342.70
cover slab	-	-	-	-	-	242.65
concrete benching and channels	-	-	-	-	-	127.65
concrete surround	-	-	-	-	-	278.30
step irons	-	-	-	-	-	13.20
access cover and frame	-	-	-	-	-	297.00
TOTAL	-	-	-	-	£	**1710.90**
1800 diameter x 2000 depth to invert						
excavation, support, backfilling and disposal	-	-	-	-	-	368.00
concrete base	-	-	-	-	-	147.20
main chamber rings	-	-	-	-	-	491.05
cover slab	-	-	-	-	-	242.65
concrete benching and channels	-	-	-	-	-	127.65
concrete surround	-	-	-	-	-	370.30
step irons	-	-	-	-	-	25.30
access cover and frame	-	-	-	-	-	297.00
TOTAL	-	-	-	-	£	**2069.15**
1800 diameter x 2500 depth to invert						
excavation, support, backfilling and disposal	-	-	-	-	-	448.50
concrete base	-	-	-	-	-	147.20
main chamber rings	-	-	-	-	-	638.25
cover slab	-	-	-	-	-	242.65
concrete benching and channels	-	-	-	-	-	127.65
concrete surround	-	-	-	-	-	462.30
step irons	-	-	-	-	-	31.90
access cover and frame	-	-	-	-	-	297.00
TOTAL	-	-	-	-	£	**2395.45**
1800 diameter x 3000 depth to invert						
excavation, suport, backfilling and disposal	-	-	-	-	-	577.30
concrete base	-	-	-	-	-	147.20
main chamber rings	-	-	-	-	-	786.60
cover slab	-	-	-	-	-	242.65
concrete surround	-	-	-	-	-	553.15
concrete benching and channels	-	-	-	-	-	127.65
step irons	-	-	-	-	-	90.20
access cover and frame	-	-	-	-	-	297.00
TOTAL	-	-	-	-	£	**2821.75**
1800 diameter x 3500 depth to invert						
excavation, support, backfilling and disposal	-	-	-	-	-	661.25
concrete base	-	-	-	-	-	147.20
access shaft	-	-	-	-	-	93.15
main chamber rings	-	-	-	-	-	650.90
reducing slab	-	-	-	-	-	259.90
cover slab	-	-	-	-	-	98.90
concrete benching and channels	-	-	-	-	-	127.65
concrete surround	-	-	-	-	-	573.85
step irons	-	-	-	-	-	115.50
access cover and frame	-	-	-	-	-	297.00
TOTAL	-	-	-	-	£	**3025.30**

CLASS K: PIPEWORK - MANHOLES AND PIPEWORK ANCILLARIES

Item	Gang hours	Labour £	Plant £	Material £	Unit	Total rate £
MANHOLES – cont'd						
Precast concrete construction – cont'd						
1800 diameter x 4000 depth to invert						
excavation, support, backfilling and disposal	-	-	-	-	-	802.70
concrete base	-	-	-	-	-	147.20
access shaft	-	-	-	-	-	161.00
main chamber rings	-	-	-	-	-	650.90
reducing slab	-	-	-	-	-	259.90
cover slab	-	-	-	-	-	98.90
concrete benching and channels	-	-	-	-	-	127.65
concrete surround	-	-	-	-	-	625.60
step irons	-	-	-	-	-	128.70
access cover and frame	-	-	-	-	-	297.00
TOTAL	-	-	-	-	£	**3299.55**
2400 diameter x 1500 depth to invert						
excavation, support, backfilling and disposal	-	-	-	-	-	417.45
concrete base	-	-	-	-	-	227.70
main chamber rings	-	-	-	-	-	657.80
cover slab	-	-	-	-	-	709.55
concrete benching and channels	-	-	-	-	-	185.15
concrete surround	-	-	-	-	-	359.95
step irons	-	-	-	-	-	13.20
access cover and frame	-	-	-	-	-	297.00
TOTAL	-	-	-	-	£	**2867.80**
2400 diameter x 3000 depth to invert						
excavation, support, backfilling and disposal	-	-	-	-	-	906.20
concrete base	-	-	-	-	-	227.70
main chamber rings	-	-	-	-	-	1528.35
cover slab	-	-	-	-	-	709.55
concrete benching and channels	-	-	-	-	-	185.15
concrete surround	-	-	-	-	-	713.00
step irons	-	-	-	-	-	90.20
access cover and frame	-	-	-	-	-	297.00
TOTAL	-	-	-	-	£	**4657.15**
2400 diameter x 4500 depth to invert						
excavation, support, backfilling and disposal	-	-	-	-	-	1396.10
concrete base	-	-	-	-	-	227.70
access shaft	-	-	-	-	-	228.85
main chamber rings	-	-	-	-	-	1263.85
reducing slab	-	-	-	-	-	711.85
cover slab	-	-	-	-	-	98.90
concrete benching and channels	-	-	-	-	-	185.15
concrete surround	-	-	-	-	-	816.50
step irons	-	-	-	-	-	154.00
access cover and frame	-	-	-	-	-	297.00
TOTAL	-	-	-	-	£	**5379.90**
2700 diameter x 1500 depth to invert						
excavation, support, backfilling and disposal	-	-	-	-	-	509.45
concrete base	-	-	-	-	-	276.00
main chamber rings	-	-	-	-	-	754.40
cover slab	-	-	-	-	-	887.80
concrete benching and channels	-	-	-	-	-	218.50
concrete surround	-	-	-	-	-	401.35
step irons	-	-	-	-	-	13.20
access cover and frame	-	-	-	-	-	297.00
TOTAL	-	-	-	-	£	**3357.70**

CLASS K: PIPEWORK - MANHOLES AND PIPEWORK ANCILLARIES

Item	Gang hours	Labour £	Plant £	Material £	Unit	Total rate £
2700 diameter x 3000 depth to invert						
excavation, support, backfilling and disposal	-	-	-	-	-	1101.70
concrete base	-	-	-	-	-	276.00
main chamber rings	-	-	-	-	-	1774.45
cover slab	-	-	-	-	-	887.80
concrete benching and channels	-	-	-	-	-	218.50
concrete surround	-	-	-	-	-	794.65
step irons	-	-	-	-	-	90.20
access cover and frame	-	-	-	-	-	297.00
TOTAL	-	-	-	-	£	**5440.30**
2700 diameter x 4500 depth to invert						
excavation, support, backfilling and disposal	-	-	-	-	-	1692.80
concrete base	-	-	-	-	-	276.00
access shaft	-	-	-	-	-	228.85
main chamber rings	-	-	-	-	-	1467.40
reducing slab	-	-	-	-	-	882.05
cover slab	-	-	-	-	-	98.90
concrete benching and channels	-	-	-	-	-	218.50
concrete surround	-	-	-	-	-	887.80
step irons	-	-	-	-	-	154.00
access cover and frame	-	-	-	-	-	297.00
TOTAL	-	-	-	-	£	**6203.30**
3000 diameter x 3000 depth to invert						
excavation, support, backfilling and disposal	-	-	-	-	-	1328.25
concrete base	-	-	-	-	-	330.05
main chamber rings	-	-	-	-	-	2374.75
cover slab	-	-	-	-	-	1118.95
concrete benching and channels	-	-	-	-	-	235.75
concrete surround	-	-	-	-	-	882.05
step irons	-	-	-	-	-	90.20
access cover and frame	-	-	-	-	-	297.00
TOTAL	-	-	-	-	£	**6657.00**
3000 diameter x 4500 depth to invert						
excavation, support, backfilling and disposal	-	-	-	-	-	2040.10
concrete base	-	-	-	-	-	330.05
access shaft	-	-	-	-	-	228.85
main chamber rings	-	-	-	-	-	1964.20
reducing slab	-	-	-	-	-	1009.70
cover slab	-	-	-	-	-	98.90
concrete benching and channels	-	-	-	-	-	235.75
concrete surround	-	-	-	-	-	961.40
step irons	-	-	-	-	-	154.00
access cover and frame	-	-	-	-	-	297.00
TOTAL	-	-	-	-	£	**7319.95**
3000 diameter x 6000 depth to invert						
excavation, support, backfilling and disposal	-	-	-	-	-	3021.05
concrete base	-	-	-	-	-	330.05
access shaft	-	-	-	-	-	433.55
main chamber rings	-	-	-	-	-	1964.20
reducing slab	-	-	-	-	-	1009.70
cover slab	-	-	-	-	-	98.90
concrete benching and channels	-	-	-	-	-	235.75
concrete surround	-	-	-	-	-	1121.25
step irons	-	-	-	-	-	218.90
access cover and frame	-	-	-	-	-	297.00
TOTAL	-	-	-	-	£	**8730.35**

CLASS K: PIPEWORK - MANHOLES AND PIPEWORK ANCILLARIES

Item	Gang hours	Labour £	Plant £	Material £	Unit	Total rate £
MANHOLES – cont'd						
BACKDROPS TO MANHOLES						
Clayware vertical pipe complete with rest bend at base and tumbling bay junction to main drain complete with stopper; concrete grade C20 surround, 150 mm thick; additional excavation and disposal						
100 pipe						
1.15 m to invert	-	-	-	-	nr	**95.86**
2.15 m to invert	-	-	-	-	nr	**122.43**
3.15 m to invert	-	-	-	-	nr	**147.84**
4.15 m to invert	-	-	-	-	nr	**175.56**
150 pipe						
1.15 m to invert	-	-	-	-	nr	**147.84**
2.15 m to invert	-	-	-	-	nr	**177.87**
3.15 m to invert	-	-	-	-	nr	**212.52**
4.15 m to invert	-	-	-	-	nr	**247.17**
225 pipe						
1.15 m to invert	-	-	-	-	nr	**294.52**
2.15 m to invert	-	-	-	-	nr	**345.35**
3.15 m to invert	-	-	-	-	nr	**398.48**
4.15 m to invert	-	-	-	-	nr	**453.92**
GULLIES						
Vitrified clay; set in concrete grade C20, 150 mm thick; additional excavation and disposal						
Road gulley						
450 mm diameter x 900 mm deep, 100 mm or 150 mm outlet; cast iron road gulley grating and frame group 4 434 x 434 mm, on Class B engineering brick seating	0.50	27.81	0.82	271.98	nr	**300.61**
Yard gulley (mud); trapped with rodding eye; galvanised bucket; stopper						
225 mm diameter, 100 mm diameter outlet; cast iron hinged grate and frame	0.30	16.69	0.49	169.47	nr	**186.65**
Grease interceptors; internal access and bucket						
600 x 450 mm; metal tray and lid, square hopper with horizontal inlet	0.35	19.47	0.57	874.53	nr	**894.56**
Precast concrete; set in concrete grade C20, 150 mm thick; additional excavation and disposal						
Road gulley; trapped with rodding eye; galvanised bucket; stopper						
450 mm diameter x 750 mm deep; cast iron road gulley grating and frame group 4, 434 x 434 mm, on Class B engineering brick seating	0.50	27.81	0.82	182.62	nr	**211.25**
450 mm diameter x 900 mm deep; cast iron road gulley grating and frame group 4, 434 x 434 mm, on Class B engineering brick seating	0.54	30.03	0.88	183.77	nr	**214.69**

CLASS K: PIPEWORK - MANHOLES AND PIPEWORK ANCILLARIES

Item	Gang hours	Labour £	Plant £	Material £	Unit	Total rate £
450 mm diameter x 1050 mm deep;cast iron road gulley grating and frame group 4, 434 x 434 mm, on Class B engineering brick seating	0.58	32.26	0.95	186.18	nr	**219.38**
FRENCH DRAINS, RUBBLE DRAINS, DITCHES						
The rates assume the most efficient items of plant (excavator) and are optimum rates assuming continuous output with no delays caused by other operations or works.						
Ground conditions are assumed to be good easily worked soil with no abnormal conditions that would affect outputs and consistency of work.						
Multiplier Table for labour and plant for various site conditions for working:						
out of sequence x 2.75 (minimum)						
in hard clay x 1.75 to 2.00						
in running sand x 2.75 (minimum)						
in broken rock x 2.75 to 3.50						
below water table x 2.00 (minimum)						
Excavation of trenches for unpiped rubble drains (excluding trench support); cross-sectional area						
0.25 - 0.50 m²	0.10	6.29	1.96	-	m	**8.26**
0.50 - 0.75 m²	0.12	7.55	2.35	-	m	**9.91**
0.75 - 1.00 m²	0.14	8.81	2.75	-	m	**11.56**
1.00 - 1.50 m²	0.17	10.70	3.34	-	m	**14.04**
1.50 - 2.00 m²	0.20	12.59	3.92	-	m	**16.51**
Filling French and rubble drains with graded material						
graded material; 20 mm stone aggregate	0.30	18.89	5.89	19.22	m³	**43.99**
broken brick/concrete rubble	0.29	18.26	5.69	15.02	m³	**38.97**
Excavation of rectangular section ditches; unlined; cross-sectional area						
0.25 - 0.50 m²	0.11	6.92	2.11	-	m	**9.03**
0.50 - 0.75 m²	0.13	8.18	2.49	-	m	**10.68**
0.75 - 1.00 m²	0.16	10.07	3.07	-	m	**13.14**
1.00 - 1.50 m²	0.20	12.59	3.84	-	m	**16.43**
1.50 - 2.00 m²	0.25	15.74	4.79	-	m	**20.53**
Excavation of rectangular ditches; lined with precast concrete slabs; cross-sectional area						
0.25 - 0.50 m²	0.15	9.44	3.20	10.28	m	**22.92**
0.50 - 0.75 m²	0.25	15.74	5.34	16.92	m	**38.00**
0.75 - 1.00 m²	0.36	22.66	7.68	24.20	m	**54.55**
1.00 - 1.50 m²	0.40	25.18	8.54	34.30	m	**68.02**
1.50 - 2.00 m²	0.45	28.33	9.61	48.28	m	**86.22**
Excavation of vee section ditches; unlined; cross-sectional area						
0.25 - 0.50 m²	0.10	6.29	2.29	-	m	**8.58**
0.50 - 0.75 m²	0.12	7.55	2.73	-	m	**10.29**
0.75 - 1.00 m²	0.14	8.81	3.20	-	m	**12.01**
1.00 - 1.50 m²	0.18	11.33	4.11	-	m	**15.44**
1.50 - 2.00 m²	0.22	13.85	5.02	-	m	**18.87**

CLASS K: PIPEWORK - MANHOLES AND PIPEWORK ANCILLARIES

Item	Gang hours	Labour £	Plant £	Material £	Unit	Total rate £
DUCTS AND METAL CULVERTS						
Galvanised steel culverts; bitumen coated						
Sectional corrugated metal culverts, nominal internal diameter 0.5 - 1 m; 1000 mm nominal internal diameter, 1.6 mm thick						
not in trenches	0.15	8.34	0.51	79.79	m	**88.64**
in trenches, depth not exceeding 1.5 m	0.31	17.24	7.03	79.79	m	**104.06**
in trenches, depth 1.5 - 2 m	0.43	23.92	9.75	79.79	m	**113.46**
in trenches, depth 2 - 2.5 m	0.51	28.37	11.57	79.79	m	**119.72**
in trenches, depth 2.5 - 3 m	0.60	33.37	13.62	79.79	m	**126.78**
Sectional corrugated metal culverts, nominal internal diameter exceeding 1.5 m; 1600 mm nominal internal diameter, 1.6 mm thick						
not in trenches	0.21	11.68	0.72	149.39	m	**161.79**
in trenches, depth 1.5 - 2 m	0.44	24.47	9.99	149.39	m	**183.85**
in trenches, depth 2 - 2.5 m	0.56	31.15	12.71	149.39	m	**193.25**
in trenches, depth 2.5 - 3 m	0.68	37.82	15.43	149.39	m	**202.64**
in trenches, depth 3 -3.5 m	0.82	45.61	18.62	149.39	m	**213.62**
Sectional corrugated metal culverts, nominal internal diameter exceeding 1.5 m; 2000 mm nominal internal diameter, 1.6 mm thick						
not in trenches	0.26	14.46	0.89	319.47	m	**334.83**
in trenches, depth 2 - 2.5 m	0.46	25.59	10.45	319.47	m	**355.51**
in trenches, depth 2.5 - 3 m	0.60	33.37	13.62	319.47	m	**366.46**
in trenches, depth 3 - 3.5 m	0.75	41.72	17.01	319.47	m	**378.20**
in trenches, depth 3.5 - 4 m	0.93	51.73	21.11	319.47	m	**392.31**
Sectional corrugated metal culverts, nominal internal diameter exceeding 1.5 m; 2200 mm nominal internal diameter, 1.6 mm thick						
not in trenches	0.33	18.35	1.13	360.17	m	**379.65**
in trenches, depth 2.5 - 3 m	0.64	35.60	14.52	360.17	m	**410.29**
in trenches, depth 3 - 3.5 m	0.77	42.83	17.48	360.17	m	**420.47**
in trenches, depth 3.5 - 4 m	1.02	56.73	23.16	360.17	m	**440.06**
in trenches, depth exceeding 4 m	1.32	73.42	29.96	360.17	m	**463.54**
OTHER PIPEWORK ANCILLARIES						
Notes						
Refer to Section G (Concrete and concrete ancillaries) for costs relevant to the construction of Headwall Structure.						
Build Ends in						
Connections to existing manholes and other chambers, pipe bore						
150 mm diameter	0.60	37.77	2.43	9.00	nr	**49.19**
225 mm diameter	0.95	59.80	3.84	16.74	nr	**80.38**
300 mm diameter	1.25	78.69	5.07	25.40	nr	**109.16**
375 mm diameter	1.45	91.28	5.88	33.47	nr	**130.62**
450 mm diameter	1.75	97.33	39.12	42.11	nr	**178.57**

CLASS L: PIPEWORK - SUPPORTS AND PROTECTION, ANCILLARIES TO LAYING AND EXCAVATION

Item	Gang hours	Labour £	Plant £	Material £	Unit	**Total rate £**
NOTES						
The rates assume the most efficient items of plant (excavator) and are optimum rates assuming continuous output with no delays caused by other operations or works.						
Ground conditions are assumed to be good easily worked soil with no abnormal conditions that would affect outputs and consistency of work.						
Multiplier Table for labour and plant for various site conditions for working:						
out of sequence x 2.75 (minimum)						
in hard clay x 1.75 to 2.00						
in running sand x 2.75 (minimum)						
in broken rock x 2.75 to 3.50						
below water table x 2.00 (minimum)						
RESOURCES - LABOUR						
Supports and protection gang						
2 unskilled operatives (general)		22.52				
1 plant operator (skill rate 3)		14.65				
Total Gang Rate / Hour	£	**37.17**				
RESOURCES - PLANT						
Supports and protection						
0.40 m³ hydraulic backacter			17.88			
Bomag BW 65S			4.04			
Total Rate / Hour		£	**21.91**			
EXTRAS TO EXCAVATION AND BACKFILLING						
Drainage sundries						
Extra over any item of drainage for excavation in						
rock	0.65	24.16	24.00	-	m³	**48.16**
mass concrete	0.84	31.22	31.22	-	m³	**62.44**
reinforced concrete	1.18	43.86	43.77	-	m³	**87.63**
Excavation of soft spots, backfilling						
concrete grade C15P	0.30	11.15	11.08	78.33	m³	**100.57**

CLASS L: PIPEWORK - SUPPORTS AND PROTECTION, ANCILLARIES TO LAYING AND EXCAVATION

Item	Gang hours	Labour £	Plant £	Material £	Unit	Total rate £
SPECIAL PIPE LAYING METHODS						
There are many factors, apart from design consideration, which influence the cost of pipe jacking, so that it is only possible to give guide prices for a sample of the work involved. For more reliable estimates it is advisable to seek the advice of a Specialist Contractor.						
The main cost considerations are :-						
• the nature of the ground						
• length of drive						
• location						
• presence of water						
• depth below surface						
Provision of all plant, equipment and labour establishing						
thrust pit; 6 m x 4 m x 8 m deep	-	-	-	-	item	**30000.00**
reception pit; 4 m x 4 m x 8 m deep	-	-	-	-	item	**22000.00**
mobilise and set up pipe jacking equipment	-	-	-	-	item	**44000.00**
Pipe jacking, excluding the cost of non-drainage materials; concrete pipes BS 5911 Part 1 Class H with rebated joints, steel reinforcing band; length not exceeding 50 m; in sand and gravel						
900 mm nominal bore	-	-	-	-	m	**1236.00**
1200 mm nominal bore	-	-	-	-	m	**1596.00**
1500 mm nominal bore	-	-	-	-	m	**1590.00**
1800 mm nominal bore	-	-	-	-	m	**1890.00**
BEDS						
Imported sand						
100 mm deep bed for pipes nominal bore						
100 mm	0.02	0.74	0.36	1.83	m	**2.93**
150 mm	0.02	0.74	0.36	1.97	m	**3.07**
225 mm	0.03	1.12	0.54	2.30	m	**3.95**
300 mm	0.04	1.49	0.72	2.41	m	**4.61**
150 mm deep bed for pipes nominal bore						
150 mm	0.06	2.23	1.07	2.98	m	**6.28**
225 mm	0.07	2.60	1.25	3.44	m	**7.29**
300 mm	0.09	3.35	1.61	3.64	m	**8.60**
400 mm	0.12	4.46	2.15	4.08	m	**10.68**
450 mm	0.14	5.20	2.50	4.77	m	**12.48**
600 mm	0.17	6.32	3.04	5.90	m	**15.26**
750 mm	0.19	7.06	3.40	6.59	m	**17.05**
900 mm	0.21	7.81	3.75	7.72	m	**19.28**
1200 mm	0.25	9.29	4.47	9.08	m	**22.84**
Imported granular material						
100 mm deep bed for pipes nominal bore						
100 mm	0.02	0.74	0.36	1.28	m	**2.38**
150 mm	0.03	1.12	0.54	1.38	m	**3.03**
225 mm	0.04	1.49	0.72	1.59	m	**3.79**
300 mm	0.05	1.86	0.89	1.70	m	**4.45**

CLASS L: PIPEWORK - SUPPORTS AND PROTECTION, ANCILLARIES TO LAYING AND EXCAVATION

Item	Gang hours	Labour £	Plant £	Material £	Unit	**Total rate £**
150 mm deep bed for pipes nominal bore						
150 mm	0.06	2.23	1.07	2.07	m	**5.37**
225 mm	0.08	2.97	1.43	2.38	m	**6.79**
300 mm	0.10	3.72	1.79	2.54	m	**8.05**
400 mm	0.13	4.83	2.32	2.87	m	**10.02**
450 mm	0.15	5.58	2.68	3.31	m	**11.56**
600 mm	0.18	6.69	3.22	4.12	m	**14.03**
750 mm	0.20	7.43	3.58	4.59	m	**15.60**
900 mm	0.22	8.18	3.93	5.39	m	**17.50**
1200 mm	0.26	9.66	4.65	6.33	m	**20.64**
Mass concrete						
100 mm deep bed for pipes nominal bore						
100 mm	0.07	2.60	1.25	5.74	m	**9.59**
150 mm	0.08	2.97	1.43	6.22	m	**10.63**
225 mm	0.09	3.35	1.61	7.19	m	**12.14**
300 mm	0.11	4.09	1.97	7.67	m	**13.73**
150 mm deep bed for pipes nominal bore						
100 mm	0.10	3.72	1.79	8.63	m	**14.13**
150 mm	0.12	4.46	2.15	9.33	m	**15.94**
225 mm	0.14	5.20	2.50	10.78	m	**18.49**
300 mm	0.16	5.95	2.86	11.50	m	**20.31**
400 mm	0.19	7.06	3.40	12.91	m	**23.36**
450 mm	0.21	7.81	3.75	15.09	m	**26.65**
600 mm	0.24	8.92	4.29	18.66	m	**31.87**
750 mm	0.26	9.66	4.65	20.82	m	**35.13**
900 mm	0.28	10.41	5.01	24.42	m	**39.83**
1200 mm	0.32	11.89	5.72	28.71	m	**46.33**
HAUNCHES						
The following items allow for dressing the haunching material half-way up the pipe barrel for the full width of the bed and then dressing in triangular fashion to the crown of the pipe. The items exclude the drain bed.						
Mass concrete						
Haunches for pipes nominal bore						
150 mm	0.24	8.92	4.29	5.27	m	**18.48**
225 mm	0.29	10.78	5.18	8.26	m	**24.23**
300 mm	0.36	13.38	6.44	10.42	m	**30.24**
400 mm	0.43	15.98	7.69	13.78	m	**37.45**
450 mm	0.50	18.59	8.94	18.66	m	**46.18**
600 mm	0.56	20.82	10.01	28.89	m	**59.72**
750 mm	0.62	23.05	11.08	35.72	m	**69.85**
900 mm	0.69	25.65	12.34	48.89	m	**86.87**
1200 mm	0.75	27.88	13.41	63.99	m	**105.28**

CLASS L: PIPEWORK - SUPPORTS AND PROTECTION, ANCILLARIES TO LAYING AND EXCAVATION

Item	Gang hours	Labour £	Plant £	Material £	Unit	**Total rate £**
SURROUNDS						
The following items provide for dressing around the pipe above the bed. sand and granular material is quantified on the basis of the full width of the bed to the stated distance above the crown, concrete as an ellipse from the top corners of the bed to a poit at the stated distance above the crown. The items exclude the drain bed.						
Imported sand						
100 mm thick bed for pipes nominal bore						
100 mm	0.04	1.49	0.72	3.31	m	**5.51**
150 mm	0.05	1.86	0.89	4.75	m	**7.50**
225 mm	0.06	2.23	1.07	7.60	m	**10.91**
300 mm	0.08	2.97	1.43	10.60	m	**15.00**
150 mm thick bed for pipes nominal bore						
100 mm	0.10	3.72	1.79	4.02	m	**9.52**
150 mm	0.12	4.46	2.15	5.51	m	**12.12**
225 mm	0.14	5.20	2.50	8.50	m	**16.21**
300 mm	0.18	6.69	3.22	11.51	m	**21.42**
400 mm	0.24	8.92	4.29	16.80	m	**30.01**
450 mm	0.28	10.41	5.01	21.39	m	**36.81**
600 mm	0.34	12.64	6.08	34.57	m	**53.28**
750 mm	0.38	14.12	6.79	48.19	m	**69.11**
900 mm	0.42	15.61	7.51	67.10	m	**90.22**
1200 mm	0.50	18.59	8.94	106.94	m	**134.47**
Imported granular material						
100 mm thick bed for pipes nominal bore						
100 mm	0.07	4.41	1.25	2.30	m	**7.96**
150 mm	0.10	6.29	1.79	3.29	m	**11.37**
225 mm	0.13	8.18	2.32	5.27	m	**15.78**
300 mm	0.16	10.07	2.86	7.36	m	**20.29**
150 mm thick bed for pipes nominal bore						
100 mm	0.10	3.72	1.79	2.79	m	**8.29**
150 mm	0.12	4.46	2.15	3.83	m	**10.44**
225 mm	0.14	5.20	2.50	5.90	m	**13.60**
300 mm	0.18	6.69	3.22	8.02	m	**17.93**
400 mm	0.24	8.92	4.29	11.70	m	**24.91**
450 mm	0.28	10.41	5.01	14.92	m	**30.34**
600 mm	0.34	12.64	6.08	23.99	m	**42.71**
750 mm	0.38	14.12	6.79	33.56	m	**54.48**
900 mm	0.42	15.61	7.51	46.71	m	**69.83**
1200 mm	0.50	18.59	8.94	74.47	m	**101.99**
Mass concrete						
100 mm thick bed for pipes nominal bore						
100 mm	0.14	5.20	2.50	10.38	m	**18.08**
150 mm	0.16	5.95	2.86	14.90	m	**23.71**
225 mm	0.18	6.69	3.22	23.86	m	**33.77**
300 mm	0.22	8.18	3.93	33.29	m	**45.40**

CLASS L: PIPEWORK - SUPPORTS AND PROTECTION, ANCILLARIES TO LAYING AND EXCAVATION

Item	Gang hours	Labour £	Plant £	Material £	Unit	**Total rate £**
150 mm thick bed for pipes nominal bore						
150 mm	0.23	8.55	4.11	17.32	m	**29.98**
225 mm	0.26	9.66	4.65	26.66	m	**40.98**
300 mm	0.30	11.15	5.36	36.29	m	**52.80**
400 mm	0.36	13.38	6.44	52.93	m	**72.75**
450 mm	0.40	14.87	7.15	67.46	m	**89.47**
600 mm	0.45	16.73	8.04	108.44	m	**133.21**
750 mm	0.50	18.59	8.94	151.73	m	**179.25**
900 mm	0.55	20.44	9.83	211.04	m	**241.32**
1200 mm	0.61	22.67	10.91	336.37	m	**369.95**
CONCRETE STOOLS AND THRUST BLOCKS						
Mass concrete						
Concrete stools or thrust blocks (nett volume of concrete excluding volume occupied by pipes)						
0.1 m^3	0.18	6.69	3.22	8.60	nr	**18.51**
0.1 - 0.2 m^3	0.32	11.89	5.72	17.20	nr	**34.82**
0.2 - 0.5 m^3	0.62	23.05	11.08	43.08	nr	**77.21**
0.5 - 1.0 m^3	0.91	33.82	16.27	86.23	nr	**136.33**
1.0 - 2.0 m^3	1.29	47.95	23.06	172.62	nr	**243.64**
2.0 - 4.0 m^3	3.15	117.09	56.31	345.07	nr	**518.47**

CLASS M: STRUCTURAL METALWORK

Item	Gang hours	Labour £	Plant £	Material £	Unit	Total rate £
NOTES						
The following are guide prices for various structural members commonly found in a Civil Engineering contract. The list is by no means exhaustive and costs are very much dependent on the particular design and will vary greatly according to specific requirements. For more detailed prices, reference should be made to Specialist Contractors.						
FABRICATION OF MEMBERS; STEELWORK						
Columns						
universal beams; straight on plan	-	-	-	-	tonne	**1039.26**
circular hollow sections; straight on plan	-	-	-	-	tonne	**2231.43**
rectangular hollow sections; straight on plan	-	-	-	-	tonne	**2064.18**
Beams						
universal beams; straight on plan	-	-	-	-	tonne	**1008.20**
universal beams; curved on plan	-	-	-	-	tonne	**1660.42**
channels; straight on plan	-	-	-	-	tonne	**1163.48**
channels; curved on plan	-	-	-	-	tonne	**1815.72**
castellated beams; straight on plan	-	-	-	-	tonne	**1474.07**
Portal frames						
straight on plan	-	-	-	-	tonne	**1225.60**
Trestles, towers and built-up columns						
straight on plan	-	-	-	-	tonne	**1542.39**
Trusses and built-up girders						
straight on plan	-	-	-	-	tonne	**1542.39**
curved on plan	-	-	-	-	tonne	**2064.18**
Bracings						
angles; straight on plan	-	-	-	-	tonne	**1070.31**
circular hollow sections; straight on plan	-	-	-	-	tonne	**2126.30**
Purlins and cladding rails						
straight on plan	-	-	-	-	tonne	**1225.60**
Cold rolled purlins and rails						
straight on plan	-	-	-	-	tonne	**2064.18**
Anchorages and holding down bolt assemblies						
base plate and bolt assemblies complete	-	-	-	-	tonne	**2377.50**
ERECTION OF FABRICATED MEMBERS ON SITE						
Trial erection	-	-	-	-	tonne	**284.68**
Permanent erection	-	-	-	-	tonne	**216.89**
Site bolts						
black	-	-	-	-	tonne	**2819.13**
HSFG general grade	-	-	-	-	tonne	**2819.13**
HSFG higher	-	-	-	-	tonne	**3221.86**
HSFG load indicating or limit types, general grade	-	-	-	-	tonne	**3725.29**
HSFG load indicating or limit types, higher grade	-	-	-	-	tonne	**4161.58**

CLASS M: STRUCTURAL METALWORK

Item	Gang hours	Labour £	Plant £	Material £	Unit	**Total rate £**
OFF SITE SURFACE TREATMENT						
Note: The following preparation and painting systems have been calculated on the basis of 20 m^2 per tonne						
Blast cleaning	-	-	-	-	m^2	**3.73**
Galvanising	-	-	-	-	m^2	**13.00**
Painting						
one coat zinc chromate primer	-	-	-	-	m^2	**3.50**
one coat two pack epoxy zinc phosphate primer (75 microns dry film thickness)	-	-	-	-	m^2	**6.28**
two coats epoxy micaceous iron oxide (100 microns dry film thickness per coat)	-	-	-	-	m^2	**16.04**

CLASS N: MISCELLANEOUS METALWORK

Item	Gang hours	Labour £	Plant £	Material £	Unit	Total rate £
NOTES						
General The following are guide prices for various structural members commonly found in a Civil Engineering contract. The list is by no means exhaustive and costs are very much dependent on the particular design and will vary greatly according to specific requirements. For more detailed prices, reference should be made to Specialist Contractors.						
Cladding CESMM3 N.2.1 requires cladding to be measured in square metres, the item so produced being inclusive of all associated flashings at wall corners and bases, eaves, gables, ridges and around openings. As the relative quantities of these flashings will depend very much on the complexity of the building shape, the guide prices shown below for these items are shown separately to help with the accuracy of the estimate.						
Bridge bearings Bridge bearings are manufactured and installed to individual specifications. The following guide prices are for different sizes of simple bridge bearings. If requirements are known, then advice ought to be obtained from specialist manufactureres such as CCL. If there is a requirement for testing bridge bearings prior to their being installed then the tests should be enumerated separately. Specialist advice should be sought once details are known.						
RESOURCES - LABOUR						
Roofing - cladding gang						
1 ganger/chargehand (skill rate 3) - 50% of time		7.11				
2 skilled operative (skill rate 3)		26.92				
1 unskilled operative (general) - 50% of time		5.63				
Total Gang Rate / Hour	£	**39.66**				
Bridge bearing gang						
1skilled operative (skill rate 4)		12.05				
2 unskilled operatives (general)		22.52				
Total Gang Rate / Hour	£	**34.57**				
RESOURCES - PLANT						
Cladding to roofs						
15 m telescopic access platform - 50% of time			10.38			
Total Gang Rate / Hour		£	**10.38**			

CLASS N: MISCELLANEOUS METALWORK

Item	Gang hours	Labour £	Plant £	Material £	Unit	Total rate £
Cladding to walls						
15 m telescopic access platform			20.77			
Total Gang Rate / Hour		£	**20.77**			
MILD STEEL						
Mild steel						
Stairways and landings	-	-	-	-	tonne	**4030.68**
Walkways and platforms	-	-	-	-	tonne	**3665.74**
Ladders						
cat ladder; 64 x 13 mm bar strings; 19 mm rungs at 250 mm centres; 450 mm wide with safety hoops	-	-	-	-	m	**324.84**
Miscellaneous framing						
angle section; 200 x 200 x 16 mm (equal)	-	-	-	-	m	**66.78**
angle section; 150 x 150 x 10 mm (equal)	-	-	-	-	m	**31.66**
angle section; 100 x 100 x 12 mm (equal)	-	-	-	-	m	**24.53**
angle section; 200 x 150 x 15 mm (unequal)	-	-	-	-	m	**54.54**
angle section; 150 x 75 x 10 mm (unequal)	-	-	-	-	m	**23.41**
universal beams; 914 x 419 mm	-	-	-	-	m	**297.44**
universal beams; 533 x 210 mm	-	-	-	-	m	**167.99**
universal joists; 127 x 76 mm	-	-	-	-	m	**18.39**
channel section; 381 x 102 mm	-	-	-	-	m	**75.88**
channel section; 254 x 76 mm	-	-	-	-	m	**38.96**
channel section; 152 x 76 mm	-	-	-	-	m	**24.63**
tubular section; 100 x 100 x 10 mm	-	-	-	-	m	**43.23**
tubular section; 200 x 200 x 15 mm	-	-	-	-	m	**141.75**
tubular section; 76.1 x 5.0 mm	-	-	-	-	m	**13.57**
tubular section; 139.7 x 6.3 mm	-	-	-	-	m	**32.06**
Mild steel; galvanised						
Handrails						
76 mm diameter tubular handrail, 48 mm diameter standards at 750 mm centres, 48 mm diameter middle rail, 1070 mm high overall	-	-	-	-	m	**122.19**
Plate flooring						
8 mm (on plain) "Durbar" pattern floor plates, maximum weight each panel 50 kg	-	-	-	-	m²	**120.93**
Mild steel; internally and externally acid dipped, rinse and hot dip galvanised, epoxy internal paint						
Uncovered tanks						
1600 litre capacity open top water tank	-	-	-	-	nr	**929.40**
18180 litre capacity open top water tank	-	-	-	-	nr	**9485.43**
Covered tanks						
1600 litre capacity open top water tank with loose fitting lid;	-	-	-	-	nr	**1032.95**
18180 litre capacity open top water tank with loose fitting lid	-	-	-	-	nr	**10608.42**

CLASS N: MISCELLANEOUS METALWORK

Item	Gang hours	Labour £	Plant £	Material £	Unit	Total rate £
MILD STEEL – cont'd						
Corrugated steel plates to BS 1449 Pt 1, Gr H4, sealed and bolted; BS729 hot dip galvanised,epoxy internal and external paint						
Uncovered tanks						
713 m3 capacity bolted cylindrical open top tank	-	-	-	-	nr	**36172.82**
PROPRIETARY WORK						
Galvanised steel troughed sheeting; 0.70 mm metal thickness, 75 mm deep corrugations; colour coating each side; fixing with plastic capped self-tapping screws to steel purlins or rails						
Cladding						
upper surfaces inclined at an angle ne 30 degrees to the horizontal	0.15	5.95	1.56	13.89	m²	**21.39**
Extra for :						
galvanised steel inner lining sheet, 0.40 mm thick, Plastisol colour coating	0.06	2.38	1.25	5.19	m²	**8.82**
galvanised steel inner lining sheet, 0.40 mm thick, Plastisol colour coating; insulation, 80 mm thick	0.08	3.17	0.83	9.12	m²	**13.12**
surfaces inclined at an angle exceeding 60 degrees to the horizontal	0.16	6.35	3.32	11.47	m²	**21.14**
Extra for :						
galvanised steel inner lining sheet, 0.40 mm thick, Plastisol colour coating	0.11	4.36	2.28	5.19	m²	**11.84**
galvanised steel inner lining sheet, 0.40 mm thick, Plastisol colour coating; insulation, 80 mm thick	0.13	5.16	2.70	9.12	m²	**16.97**
Galvanised steel flashings; 0.90 mm metal thickness; bent to profile; fixing with plastic capped self-tapping screws to steel purlins or rails; mastic sealant						
Flashings to cladding						
250 mm girth	0.12	4.76	2.49	8.80	m	**16.05**
500 mm girth	0.18	7.14	3.74	13.68	m	**24.55**
750 mm girth	0.22	8.73	4.57	18.55	m	**31.84**
Aluminium profiled sheeting; 0.90 mm metal thickness, 75 mm deep corrugations; colour coating each side; fixing with plastic capped self-tapping screws to steel purlins or rails						
Cladding						
upper surfaces inclined at an angle exceeding 60 degrees to the horizontal	0.20	7.93	4.15	16.30	m²	**28.39**
Extra for :						
aluminium inner lining sheet, 0.70 mm thick, Plastisol colour coating	0.13	5.16	2.73	12.98	m²	**20.87**
aluminium inner lining sheet, 0.70 mm thick, Plastisol colour coating; insulation, 80 mm thick	0.15	5.95	3.12	16.91	m²	**25.97**

CLASS N: MISCELLANEOUS METALWORK

Item	Gang hours	Labour £	Plant £	Material £	Unit	**Total rate £**
Aluminium profiled sheeting; 1.00 mm metal thickness, 75 mm deep corrugations; colour coating each side; fixing with plastic capped self-tapping screws to steel purlins or rails						
Cladding						
upper surfaces inclined at an angle ne 30 degrees to the horizontal	0.17	6.74	1.77	18.72	m^2	**27.22**
Extra for :						
aluminium inner lining sheet, 0.70 mm thick, Plastisol colour coating	0.09	3.57	0.93	12.98	m^2	**17.48**
aluminium inner lining sheet, 0.70 mm thick, Plastisol colour coating; insulation, 80 mm thick	0.11	4.36	1.14	16.91	m^2	**22.41**
Aluminium flashings; 0.90 mm metal thickness; bent to profile; fixing with plastic capped self-tapping screws to steel purlins or rails; mastic sealant						
Flashings to cladding						
250 mm girth	0.12	4.76	2.49	9.77	m	**17.02**
500 mm girth	0.18	7.14	3.74	15.61	m	**26.48**
750 mm girth	0.22	8.73	4.57	21.45	m	**34.74**
Flooring; Eurogrid; galvanised mild steel						
Open grid flooring						
type 41/100; 3 x 25 mm bearer bar; 6mm diameter transverse bar	-	-	-	-	m^2	**46.87**
type 41/100; 5 x 25 mm bearer bar; 6mm diameter transverse bar	-	-	-	-	m^2	**60.63**
type 41/100; 3 x 30 mm bearer bar; 6mm diameter transverse bar	-	-	-	-	m^2	**54.40**
Duct covers; Stelduct; galvanised mild steel						
Duct covers; pedestrian duty						
225 mm clear opening	-	-	-	-	m	**91.23**
450 mm clear opening	-	-	-	-	m	**100.92**
750 mm clear opening	-	-	-	-	m	**119.72**
Duct covers; medium duty						
225 mm clear opening	-	-	-	-	m	**132.10**
450 mm clear opening	-	-	-	-	m	**149.32**
750 mm clear opening	-	-	-	-	m	**173.05**
Duct covers; heavy duty						
225 mm clear opening	-	-	-	-	m	**146.83**
450 mm clear opening	-	-	-	-	m	**218.91**
750 mm clear opening	-	-	-	-	m	**312.23**

CLASS N: MISCELLANEOUS METALWORK

Item	Gang hours	Labour £	Plant £	Material £	Unit	**Total rate £**
PROPRIETARY WORK – cont'd						
Bridge bearings						
Supply plain rubber bearings (3 m and 5 m lengths)						
150 x 20 mm	0.35	12.10	-	35.59	m	**47.69**
150 x 25 mm	0.35	12.10	-	45.76	m	**57.86**
Supply and place in position laminated elastomeric rubber bearing						
250 x 150 x 19 mm	0.25	8.64	-	12.71	nr	**21.35**
300 x 200 x 19 mm	0.25	8.64	-	19.07	nr	**27.72**
300 x 200 x 30 mm	0.27	9.33	-	30.51	nr	**39.84**
300 x 200 x 41 mm	0.27	9.33	-	40.68	nr	**50.01**
300 x 250 x 41 mm	0.30	10.37	-	50.85	nr	**61.22**
300 x 250 x 63 mm	0.30	10.37	-	78.81	nr	**89.18**
400 x 250 x 19 mm	0.32	11.06	-	31.79	nr	**42.85**
400 x 250 x 52 mm	0.32	11.06	-	86.44	nr	**97.50**
400 x 300 x 19 mm	0.32	11.06	-	38.14	nr	**49.20**
600 x 450 x 24 mm	0.35	12.10	-	106.78	nr	**118.88**
Adhesive fixings to laminated elastomeric rubber bearings						
2 mm thick epoxy adhesive	1.00	34.57	-	42.08	m²	**76.65**
15 mm thick epoxy mortar	1.50	51.85	-	252.80	m²	**304.65**
15 mm thick epoxy pourable grout	2.00	69.14	-	252.15	m²	**321.29**
Supply and install mechanical guides for laminated elastomeric rubber bearings						
500kN SLS design load; FP50 fixed pin Type 1	2.00	69.14	-	440.00	nr	**509.14**
500kN SLS design load; FP50 fixed pin Type 2	2.00	69.14	-	550.00	nr	**619.14**
750kN SLS design load; FP75 fixed pin Type 1	2.10	72.60	-	715.00	nr	**787.60**
750kN SLS design load; FP75 fixed pin Type 2	2.10	72.60	-	825.00	nr	**897.60**
300kN SLS design load; UG300 Uniguide Type 1	2.00	69.14	-	440.00	nr	**509.14**
300kN SLS design load; UG300 Uniguide Type 2	2.00	69.14	-	550.00	nr	**619.14**
Supply and install fixed pot bearings						
355 x 355; PF200	2.00	69.14	-	550.00	nr	**619.14**
425 x 425; PF300	2.10	72.60	-	605.00	nr	**677.60**
Supply and install free sliding pot bearings						
445 x 345; PS200	2.10	72.60	-	440.00	nr	**512.60**
520 x 415; PS300	2.20	76.05	-	660.00	nr	**736.05**
Supply and install guided sliding pot bearings						
455 x 375; PG200	2.20	76.05	-	715.00	nr	**791.05**
545 x 435; PG300	2.30	79.51	-	770.00	nr	**849.51**
Testing; laminated elastomeric bearings						
compression test	-	-	-	60.00	nr	**60.00**
shear test	-	-	-	75.00	nr	**75.00**
bond test (Exclusive of cost of bearings as this is a destructive test)	-	-	-	280.00	nr	**280.00**

CLASS O: TIMBER

Item	Gang hours	Labour £	Plant £	Material £	Unit	Total rate £
RESOURCES - LABOUR						
Timber gang						
1 foreman carpenter/joiner (craftsman)		18.47				
1 carpenter/joiner (craftsman)		17.35				
1 unskilled operative (general)		11.26				
1 plant operator (skill rate 3) - 50% of time		7.33				
Total Gang Rate / Hour	£	**54.41**				
Timber fixings gang						
1 carpenter/joiner (craftsman)		18.47				
1 unskilled operative (general)		11.26				
Total Gang Rate / Hour	£	**29.73**				
RESOURCES - PLANT						
Timber						
tractor / trailer			14.57			
22RB crawler crane (25% of time)			6.63			
5.6t rough terrain forklift (25% of time)			3.88			
7.5 KVA diesel generator			4.36			
two K637 rotary hammers			1.42			
two electric screwdrivers			1.11			
Total Gang Rate / Hour		£	**31.97**			
RESOURCES - MATERIALS						
The timber material prices shown below are averages, actual prices being very much affected by availability of suitably sized forest timbers capable of conversion to the sizes shown. Apart from the practicality of being able to obtain the larger sizes in one timber, normal practice and drive for economy would lead to their being built up using smaller timbers.						
HARDWOOD COMPONENTS						
Greenheart						
100 x 75 mm						
length not exceeding 1.5 m	0.15	8.16	4.82	6.25	m	**19.24**
length 1.5 - 3 m	0.13	7.07	4.18	6.25	m	**17.51**
length 3 - 5 m	0.12	6.53	3.84	6.25	m	**16.62**
length 5 - 8 m	0.11	5.99	3.54	6.41	m	**15.94**
150 x 75 mm						
length not exceeding 1.5 m	0.17	9.25	5.55	7.57	m	**22.36**
length 1.5 - 3 m	0.15	8.16	4.82	7.39	m	**20.37**
length 3 - 5 m	0.14	7.62	4.48	7.39	m	**19.48**
length 5 - 8 m	0.12	6.53	5.10	7.57	m	**19.19**
200 x 100 mm						
length not exceeding 1.5 m	0.24	13.06	7.67	15.13	m	**35.86**
length 1.5 - 3 m	0.21	11.43	6.82	15.13	m	**33.38**
length 3 - 5 m	0.20	10.61	6.25	15.13	m	**31.99**
length 5 - 8 m	0.18	9.79	5.75	15.13	m	**30.68**
length 8 - 12 m	0.16	8.71	5.12	18.47	m	**32.29**

CLASS O: TIMBER

Item	Gang hours	Labour £	Plant £	Material £	Unit	Total rate £
HARDWOOD COMPONENTS – cont'd						
Greenheart – cont'd						
200 x 200 mm						
length not exceeding 1.5 m	0.42	22.85	13.43	26.19	m	**62.46**
length 1.5 - 3 m	0.40	21.76	12.79	26.19	m	**60.74**
length 3 - 5 m	0.38	20.68	12.15	26.19	m	**59.01**
length 5 - 8 m	0.34	18.50	10.87	26.19	m	**55.55**
length 8 - 12 m	0.30	16.32	9.80	28.24	m	**54.37**
225 x 100 mm						
length not exceeding 1.5 m	0.27	14.69	8.74	15.13	m	**38.57**
length 1.5 - 3 m	0.24	13.06	7.67	15.13	m	**35.86**
length 3 - 5 m	0.22	11.97	7.24	15.13	m	**34.35**
length 5 - 8 m	0.20	10.88	8.50	15.13	m	**34.51**
length 8 - 12 m	0.18	9.79	5.96	18.47	m	**34.23**
300 x 100 mm						
length not exceeding 1.5 m	0.36	19.59	11.51	19.79	m	**50.89**
length 1.5 - 3 m	0.33	17.96	10.66	19.79	m	**48.41**
length 3 - 5 m	0.30	16.32	9.80	19.79	m	**45.92**
length 5 - 8 m	0.27	14.69	8.74	22.05	m	**45.49**
length 8 - 12 m	0.24	13.06	7.67	2.68	m	**23.41**
300 x 200 mm						
length not exceeding 1.5 m	0.50	27.20	15.99	39.57	m	**82.77**
length 1.5 - 3 m	0.45	24.48	14.50	39.57	m	**78.56**
length 3 - 5 m	0.40	21.76	12.79	39.57	m	**74.13**
length 5 - 8 m	0.35	19.04	11.30	42.01	m	**72.35**
length 8 - 12 m	0.30	16.32	9.59	42.01	m	**67.92**
300 x 300 mm						
length not exceeding 1.5 m	0.52	28.29	16.63	59.37	m	**104.29**
length 1.5 - 3 m	0.48	26.12	15.35	59.37	m	**100.83**
length 3 - 5 m	0.44	23.94	14.07	59.37	m	**97.38**
length 5 - 8 m	0.40	21.76	12.79	63.13	m	**97.69**
length 8 - 12 m	0.36	19.59	11.51	63.13	m	**94.23**
450 x 450 mm						
length not exceeding 1.5 m	0.98	53.32	31.55	131.10	m	**215.97**
length 1.5 - 3 m	0.90	48.97	28.99	131.10	m	**209.06**
length 3 - 5 m	0.83	45.16	26.65	131.10	m	**202.91**
length 5 - 8 m	0.75	40.81	24.09	134.37	m	**199.27**
length 8 - 12 m	0.68	37.00	21.74	140.93	m	**199.67**
SOFTWOOD COMPONENTS						
Softwood; stress graded SC3/4						
100 x 75 mm						
up to 3.00 m long	0.10	5.44	2.64	2.05	m	**10.14**
3.00 - 5.00 m long	0.10	5.44	2.64	2.16	m	**10.24**
150 x 75 mm						
up to 3.00 m long	0.13	7.07	3.43	3.10	m	**13.60**
3.00 - 5.00 m long	0.11	5.99	2.90	3.25	m	**12.14**
200 x 100 mm						
up to 3.00 m long	0.16	8.71	4.22	6.53	m	**19.45**
3.00 - 5.00 m long	0.14	7.62	3.70	6.90	m	**18.22**
200 x 200 mm						
up to 3.00 m long	0.25	13.60	6.59	15.44	m	**35.63**
3.00 - 5.00 m long	0.23	12.51	6.07	16.29	m	**34.87**
5.00 - 8.00 m long	0.20	10.88	5.28	17.03	m	**33.19**

CLASS O: TIMBER

Item	Gang hours	Labour £	Plant £	Material £	Unit	**Total rate £**
300 x 200 mm						
up to 3.00 m long	0.27	14.69	7.12	24.55	m	**46.37**
3.00 - 5.00 m long	0.25	13.60	6.59	25.90	m	**46.09**
5.00 - 8.00 m long	0.23	12.51	6.07	26.97	m	**45.55**
300 x 300 mm						
up to 3.00 m long	0.30	16.32	7.92	37.88	m	**62.12**
3.00 - 5.00 m long	0.27	14.69	7.12	39.97	m	**61.78**
5.00 - 8.00 m long	0.25	13.60	6.59	41.68	m	**61.87**
450 x 450 mm						
up to 3.00 m long	0.35	19.04	9.23	89.91	m	**118.19**
3.00 - 5.00 m long	0.31	16.87	8.18	94.89	m	**119.94**
5.00 - 8.00 m long	0.28	15.23	7.39	98.93	m	**121.56**
600 x 600 mm						
3.00 - 5.00 m long	0.39	21.22	8.57	166.41	m	**196.20**
5.00 - 8.00 m long	0.37	20.13	9.76	185.00	m	**214.89**
ADD to the above prices for vacuum / pressure impregnating to minimum 5.30 kg/m³ salt retention	-	-	-	29.02	m³	**29.02**
HARDWOOD DECKING						
Greenheart; wrought finish						
Thickness 25-50 mm						
150 x 50 mm	0.58	31.56	15.31	46.46	m²	**93.32**
Thickness 50-75 mm						
200 x 75 mm	0.75	40.81	19.78	59.18	m²	**119.77**
Thickness 75-100 mm						
250 x 100 mm	0.95	51.69	24.44	75.91	m²	**152.04**
SOFTWOOD DECKING						
Douglas Fir						
Thickness 25-50 mm						
150 x 50 mm	0.39	21.22	10.29	20.03	m²	**51.54**
Thickness 50-75 mm						
200 x 75 mm	0.50	27.20	13.20	27.41	m²	**67.81**
Thickness 75-100 mm						
250 x 100 mm	0.65	35.37	23.55	31.10	m²	**90.02**
FITTINGS AND FASTENINGS						
Metalwork						
Spikes; mild steel material rosehead						
14 x 14 x 275 mm long	0.13	3.86	-	0.92	nr	**4.79**
Metric mild steel bolts, nuts and washers						
M6 x 25mm long	0.05	1.49	-	0.07	nr	**1.56**
M6 x 50mm long	0.05	1.49	-	0.08	nr	**1.57**
M6 x 75mm long	0.05	1.49	-	0.11	nr	**1.59**
M6 x 100mm long	0.06	1.78	-	0.12	nr	**1.90**
M6 x 120mm long	0.06	1.78	-	0.20	nr	**1.99**
M6 x 150mm long	0.06	1.78	-	0.26	nr	**2.04**
M8 x 25mm long	0.05	1.49	-	0.09	nr	**1.58**
M8 x 50mm long	0.05	1.49	-	0.11	nr	**1.60**
M8 x 75mm long	0.06	1.78	-	0.15	nr	**1.93**
M8 x 100mm long	0.06	1.78	-	0.17	nr	**1.96**
M8 x 120mm long	0.07	2.08	-	0.30	nr	**2.38**
M8 x 150mm long	0.07	2.08	-	0.37	nr	**2.45**

CLASS O: TIMBER

Item	Gang hours	Labour £	Plant £	Material £	Unit	**Total rate £**
FITTINGS AND FASTENINGS – cont'd						
Metalwork – cont'd						
Metric mild steel bolts, nuts and washers – cont'd						
M10 x 25mm long	0.05	1.49	-	0.15	nr	**1.63**
M10 x 50mm long	0.06	1.78	-	0.17	nr	**1.96**
M10 x 75mm long	0.06	1.78	-	0.21	nr	**1.99**
M10 x 100mm long	0.07	2.08	-	0.26	nr	**2.34**
M10 x 120mm long	0.07	2.08	-	0.28	nr	**2.36**
M10 x 150mm long	0.07	2.08	-	0.53	nr	**2.61**
M10 x 200mm long	0.08	2.38	-	0.96	nr	**3.34**
M12 x 25mm long	0.06	1.78	-	0.21	nr	**1.99**
M12 x 50mm long	0.06	1.78	-	0.24	nr	**2.02**
M12 x 75mm long	0.07	2.08	-	0.29	nr	**2.37**
M12 x 100mm long	0.07	2.08	-	0.34	nr	**2.43**
M12 x 120mm long	0.08	2.38	-	0.39	nr	**2.77**
M12 x 150mm long	0.08	2.38	-	0.55	nr	**2.92**
M12 x 200mm long	0.08	2.38	-	0.81	nr	**3.19**
M12 x 240mm long	0.09	2.68	-	1.30	nr	**3.97**
M12 x 300mm long	0.10	2.97	-	1.40	nr	**4.38**
M16 x 50mm long	0.07	2.08	-	0.42	nr	**2.50**
M16 x 75mm long	0.07	2.08	-	0.49	nr	**2.57**
M16 x 100mm long	0.08	2.38	-	0.57	nr	**2.95**
M16 x 120mm long	0.08	2.38	-	0.70	nr	**3.07**
M16 x 150mm long	0.09	2.68	-	0.81	nr	**3.48**
M16 x 200mm long	0.09	2.68	-	1.19	nr	**3.86**
M16 x 240mm long	0.10	2.97	-	1.77	nr	**4.74**
M16 x 300mm long	0.10	2.97	-	1.96	nr	**4.93**
M20 x 50mm long	0.07	2.08	-	0.60	nr	**2.68**
M20 x 75mm long	0.08	2.38	-	0.79	nr	**3.17**
M20 x 100mm long	0.08	2.38	-	0.93	nr	**3.31**
M20 x 120mm long	0.09	2.68	-	1.20	nr	**3.88**
M20 x 150mm long	0.09	2.68	-	1.31	nr	**3.98**
M20 x 200mm long	0.10	2.97	-	1.78	nr	**4.75**
M20 x 240mm long	0.10	2.97	-	2.48	nr	**5.46**
M20 x 300mm long	0.11	3.27	-	2.78	nr	**6.05**
M24 x 50mm long	0.08	2.38	-	3.73	nr	**6.11**
M24 x 75mm long	0.08	2.38	-	3.90	nr	**6.28**
M24 x 100mm long	0.09	2.68	-	3.98	nr	**6.65**
M24 x 120mm long	0.09	2.68	-	4.15	nr	**6.82**
M24 x 150mm long	0.10	2.97	-	4.46	nr	**7.44**
M24 x 200mm long	0.11	3.27	-	4.91	nr	**8.18**
M24 x 240mm long	0.11	3.27	-	5.88	nr	**9.15**
M24 x 300mm long	0.12	3.57	-	6.31	nr	**9.87**
M30 x 100mm long	0.09	2.68	-	6.50	nr	**9.17**
M30 x 120mm long	0.10	2.97	-	6.73	nr	**9.70**
M30 x 150mm long	0.11	3.27	-	7.08	nr	**10.35**
M30 x 200mm long	0.11	3.27	-	7.67	nr	**10.94**
Carriage bolts, nuts and washer						
M6 x 25mm long	0.05	1.49	-	0.05	nr	**1.54**
M6 x 50mm long	0.05	1.49	-	0.06	nr	**1.55**
M6 x 75mm long	0.05	1.49	-	0.07	nr	**1.56**
M6 x 100mm long	0.06	1.78	-	0.11	nr	**1.89**
M6 x 150mm long	0.06	1.78	-	0.15	nr	**1.93**
M8 x 25mm long	0.05	1.49	-	0.07	nr	**1.56**
M8 x 50mm long	0.05	1.49	-	0.09	nr	**1.57**
M8 x 75mm long	0.06	1.78	-	0.10	nr	**1.88**
M8 x 100mm long	0.06	1.78	-	0.15	nr	**1.94**

CLASS O: TIMBER

Item	Gang hours	Labour £	Plant £	Material £	Unit	**Total rate £**
M8 x 150mm long	0.07	2.08	-	0.20	nr	**2.29**
M8 x 200mm long	0.07	2.08	-	0.50	nr	**2.58**
M10 x 25mm long	0.05	1.49	-	0.15	nr	**1.64**
M10 x 50mm long	0.06	1.78	-	0.13	nr	**1.92**
M10 x 75mm long	0.06	1.78	-	0.16	nr	**1.94**
M10 x 100mm long	0.07	2.08	-	0.21	nr	**2.30**
M10 x 150mm long	0.07	2.08	-	0.30	nr	**2.38**
M10 x 200mm long	0.08	2.38	-	0.60	nr	**2.98**
M10 x 240mm long	0.08	2.38	-	1.41	nr	**3.79**
M10 x 300mm long	0.09	2.68	-	1.55	nr	**4.22**
M12 x 25mm long	0.06	1.78	-	0.26	nr	**2.04**
M12 x 50mm long	0.06	1.78	-	0.21	nr	**1.99**
M12 x 75mm long	0.07	2.08	-	0.22	nr	**2.30**
M12 x 100mm long	0.07	2.08	-	0.30	nr	**2.38**
M12 x 150mm long	0.08	2.38	-	0.41	nr	**2.79**
M12 x 200mm long	0.08	2.38	-	0.97	nr	**3.34**
M12 x 240mm long	0.09	2.68	-	1.71	nr	**4.38**
M12 x 300mm long	0.10	2.97	-	1.88	nr	**4.85**
Galvanised steel						
Straps						
30 x 2.5 x 600 mm girth	0.13	3.86	-	1.22	nr	**5.09**
30 x 2.5 x 800 mm girth	0.13	3.86	-	1.78	nr	**5.65**
30 x 2.5 x 1000 mm girth	0.13	3.86	-	2.31	nr	**6.18**
30 x 2.5 x 1200 mm girth	0.15	4.46	-	2.81	nr	**7.27**
30 x 2.5 x 1400 mm girth	0.13	3.86	-	3.20	nr	**7.06**
30 x 2.5 x 1600 mm girth	0.15	4.46	-	3.61	nr	**8.07**
30 x 2.5 x 1800 mm girth	0.15	4.46	-	4.02	nr	**8.48**
30 x 5 x 600 mm girth	0.13	3.86	-	2.58	nr	**6.44**
30 x 5 x 800 mm girth	0.13	3.86	-	3.30	nr	**7.17**
30 x 5 x 1000 mm girth	0.13	3.86	-	4.77	nr	**8.63**
30 x 5 x 1200 mm girth	0.15	4.46	-	5.06	nr	**9.52**
30 x 5 x 1400 mm girth	0.13	3.86	-	6.00	nr	**9.86**
30 x 5 x 1600 mm girth	0.15	4.46	-	6.71	nr	**11.17**
30 x 5 x 1800 mm girth	0.15	4.46	-	7.65	nr	**12.11**
Timber connectors; round toothed plate, single sided for 10 mm or 12 mm bolts						
38 mm diameter	0.01	0.15	-	0.22	nr	**0.37**
50 mm diameter	0.01	0.15	-	0.25	nr	**0.40**
63 mm diameter	0.01	0.24	-	0.35	nr	**0.58**
75 mm diameter	0.01	0.24	-	0.77	nr	**1.01**
Timber connectors; round toothed plate, double sided for 10 mm or 12 mm bolts						
38 mm diameter	0.01	0.15	-	0.22	nr	**0.37**
50 mm diameter	0.01	0.15	-	0.25	nr	**0.40**
63 mm diameter	0.01	0.24	-	0.35	nr	**0.58**
75 mm diameter	0.01	0.24	-	0.77	nr	**1.01**
Split ring connectors						
50 mm diameter	0.06	1.78	-	1.40	nr	**3.19**
63 mm diameter	0.01	0.18	-	1.51	nr	**1.69**
101 mm diameter	0.01	0.18	-	3.15	nr	**3.33**
Shear plate connectors						
67 mm diameter	0.01	0.18	-	1.57	nr	**1.75**
101 mm diameter	0.01	0.18	-	7.88	nr	**8.05**
Flitch plates						
200 x 75 x 10 mm	0.07	2.08	-	2.08	nr	**4.16**
300 x 100 x 10 mm	0.09	2.68	-	6.02	nr	**8.69**
450 x 150 x 12 mm	0.15	4.46	-	14.18	nr	**18.63**

CLASS O: TIMBER

Item	Gang hours	Labour £	Plant £	Material £	Unit	Total rate £
FITTINGS AND FASTENINGS – cont'd						
Stainless steel						
Straps						
30 x 2.5 x 600 mm girth	0.13	3.86	-	5.04	nr	**8.90**
30 x 2.5 x 800 mm girth	0.13	3.86	-	7.01	nr	**10.87**
30 x 2.5 x 1000 mm girth	0.13	3.86	-	2.31	nr	**6.18**
30 x 2.5 x 1200 mm girth	0.15	4.46	-	2.81	nr	**7.27**
30 x 2.5 x 1400 mm girth	0.13	3.86	-	3.20	nr	**7.06**
30 x 2.5 x 1600 mm girth	0.15	4.46	-	3.61	nr	**8.07**
30 x 2.5 x 1800 mm girth	0.15	4.46	-	4.02	nr	**8.48**
30 x 5 x 600 mm girth	0.13	3.86	-	8.13	nr	**11.99**
30 x 5 x 800 mm girth	0.13	3.86	-	10.84	nr	**14.70**
30 x 5 x 1000 mm girth	0.13	3.86	-	4.77	nr	**8.63**
30 x 5 x 1200 mm girth	0.15	4.46	-	5.06	nr	**9.52**
30 x 5 x 1400 mm girth	0.13	3.86	-	6.00	nr	**9.86**
30 x 5 x 1600 mm girth	0.15	4.46	-	6.71	nr	**11.17**
30 x 5 x 1800 mm girth	0.15	4.46	-	7.65	nr	**12.11**
Coach screws						
5.0 mm diameter x 75 mm long	0.04	1.19	-	1.63	nr	**2.81**
7.0 mm diameter x 105 mm long	0.05	1.49	-	3.92	nr	**5.41**
7.0 mm diameter x 140 mm long	0.06	1.78	-	6.25	nr	**8.03**
10.0 mm diameter x 95 mm long	0.06	1.78	-	8.58	nr	**10.36**
10.0 mm diameter x 165 mm long	0.07	2.08	-	16.20	nr	**18.28**
Timber connectors; round toothed plate, single sided for 10 mm or 12 mm bolts						
38 mm diameter	0.01	0.15	-	1.97	nr	**2.12**
50 mm diameter	0.01	0.15	-	2.05	nr	**2.20**
63 mm diameter	0.01	0.24	-	2.28	nr	**2.52**
75 mm diameter	0.01	0.24	-	2.60	nr	**2.84**
Timber connectors; round toothed plate, double sided for 10 mm or 12 mm bolts						
38 mm diameter	0.01	0.15	-	1.97	nr	**2.12**
50 mm diameter	0.01	0.15	-	2.05	nr	**2.20**
63 mm diameter	0.01	0.24	-	2.28	nr	**2.52**
75 mm diameter	0.01	0.24	-	2.60	nr	**2.84**
Split ring connectors						
63 mm diameter	0.06	1.78	-	8.66	nr	**10.45**
101 mm diameter	0.06	1.78	-	17.01	nr	**18.79**
Shear plate connectors						
67 mm diameter	0.06	1.78	-	6.62	nr	**8.40**
101 mm diameter	0.06	1.78	-	38.59	nr	**40.37**

CLASS P: PILING

Item	Gang hours	Labour £	Plant £	Material £	Unit	**Total rate £**
GENERALLY						
There are a number of different types of piling which are available for use in differing situations. Selection of the most suitable type of piling for a particular site will depend on a number of factors including the physical conditions likely to be encountered during driving, the loads to be carried, the design of superstructure, etc. The most commonly used systems are included in this section						
It is essential that a thorough and adequate site investigation is carried out to ascertain details of the ground strata and bearing capacities to enable a proper assessment to be made of the most suitable and economical type of piling to be adopted.						
There are so many factors, apart from design considerations, which influence the cost of piling that it is not possible to give more than an approximate indication of costs. To obtain reliable costs for a particular contract advice should be sought from a company specialising in the particular type of piling proposed. Some Specialist Contractors will also provide a design service if required.						
BORED CAST IN PLACE CONCRETE PILES						
Generally						
The items "number of piles" are calculated based on the following:						
• allowance for provision of all plant, equipment and labour including transporting to and from site and establishing and dismantling at £5,500 in total.						
• moving the rig to and setting up at each pile position; preparing to commence driving; £35.00, 55.00 and 70.00 per 300 mm, 450 mm and 600 mm diameter piles using the tripod mounted percussion rig .						
Standing time is quoted at £90.25 per hour for tripod rig, £182.00 per hour for mobile rig and £100.00 per hour for continuous flight auger..						
Disposal of material arising from pile bores						
The disposal of excavated material is shown separately, partly as this task is generally carried out by the main contractor rather than the piling specialist, but also to allow for simple adjustment should contaminated ground be envisaged.						
Disposal of material arising from pile bores; collection from around piling operations						
storage on site; 100 m maximum distance; using 22.5 t ADT	0.05	0.82	2.09	-	m³	**2.91**

CLASS P: PILING

Item	Gang hours	Labour £	Plant £	Material £	Unit	**Total rate £**
BORED CAST IN PLACE CONCRETE PILES – cont'd						
Disposal of material arising from pile bores – cont'd						
removal; 5 km distance; using 20 t tipper	0.12	1.96	5.03	-	m³	**6.99**
removal; 15 km distance; using 20 t tipper	0.22	3.60	9.22	-	m³	**12.82**
Add to the above rates where tipping charges apply (excluding Landfill Tax):						
non-hazardous waste	-	-	-	-	m³	**26.93**
hazardous waste	-	-	-	-	m³	**32.92**
special waste	-	-	-	-	m³	**55.38**
contaminated liquid	-	-	-	-	m³	**82.31**
contaminated sludge	-	-	-	-	m³	**134.66**
Add to the above rates where Landfill Tax applies:						
exempted material	-	-	-	-	m³	**-**
inactive or inert material	-	-	-	-	m³	**3.15**
other material	-	-	-	-	m³	**22.05**
Concrete 35 N/mm², 20 mm aggregate; installed by tripod-mounted percussion rig						
The following unit costs cover the construction of small diameter bored piling using light and compact tripod rigs requiring no expensive site levelling or access ways. Piling can be constructed in very restricted headroom or on confined and difficult sites. Standard diameters are between 400 and 600 mm with a normal maximum depth of 20 m.						
The costs are based on installing 100 piles on a clear site with reasonable access.						
Diameter: 300 mm						
number of piles (see above)	-	-	-	-	nr	**109.96**
concreted length	-	-	-	-	m	**8.18**
depth bored to 10 m maximum depth	-	-	-	-	m	**45.26**
depth bored to 15 m maximum depth	-	-	-	-	m	**49.83**
depth bored to 20 m maximum depth	-	-	-	-	m	**54.41**
Diameter: 450 mm						
number of piles (see above)	-	-	-	-	nr	**133.12**
concreted length	-	-	-	-	m	**31.31**
depth bored to 10 m maximum depth	-	-	-	-	m	**84.16**
depth bored to 15 m maximum depth	-	-	-	-	m	**93.53**
depth bored to 20 m maximum depth	-	-	-	-	m	**103.03**
Diameter: 600 mm						
number of piles (see above)	-	-	-	-	nr	**150.48**
concreted length	-	-	-	-	m	**32.73**
depth bored to 10 m maximum depth	-	-	-	-	m	**150.49**
depth bored to 15 m maximum depth	-	-	-	-	m	**163.22**
depth bored to 20 m maximum depth	-	-	-	-	m	**175.96**

CLASS P: PILING

Item	Gang hours	Labour £	Plant £	Material £	Unit	**Total rate £**
Concrete 35 N/mm², 20 mm aggregate; installed by lorry/crawler-mounted rotary rig						
The following unit costs cover the construction of small diameter bored piles using lorry or crawler mounted rotary boring rigs. This type of plant is more mobile and faster in operation than the tripod rigs and is ideal for large contracts in cohesive ground. Construction of piles under bentonite suspension can be carried out to obviate the use of liners. Standard diameters of 450 - 900 mm diameter can be constructed to depths of 30 m.						
The costs are based on installing 100 piles on a clear site with reasonable access.						
Diameter: 300 mm						
number of piles (see above)	-	-	-	-	nr	**212.27**
concreted length	-	-	-	-	m	**9.26**
depth bored to 10 m maximum depth	-	-	-	-	m	**46.30**
depth bored to 15 m maximum depth	-	-	-	-	m	**46.30**
depth bored to 20 m maximum depth	-	-	-	-	m	**46.30**
Diameter: 450 mm						
number of piles (see above)	-	-	-	-	nr	**212.27**
concreted length	-	-	-	-	m	**18.52**
depth bored to 10 m maximum depth	-	-	-	-	m	**57.89**
depth bored to 15 m maximum depth	-	-	-	-	m	**57.89**
depth bored to 20 m maximum depth	-	-	-	-	m	**57.89**
Diameter: 600 mm						
number of piles (see above)	-	-	-	-	nr	**212.27**
concreted length	-	-	-	-	m	**33.57**
depth bored to 10 m maximum depth	-	-	-	-	m	**104.18**
depth bored to 15 m maximum depth	-	-	-	-	m	**104.18**
depth bored to 20 m maximum depth	-	-	-	-	m	**104.18**
Concrete 35 N/mm², 20 mm aggregate; installed by continuous flight auger						
The following unit costs cover the construction of piles by screwing a continuous flight auger into the ground to a design depth (Determined prior to commencement of piling operations and upon which the rates are based and subsequently varied to actual depths). Concrete is then pumped through the hollow stem of the auger to the bottom and the pile formed as the auger is withdrawn. Spoil is removed by the auger as it is withdrawn. This is a fast method of construction without causing disturbance or vibration to adjacent ground. No casing is required even in unsuitable soils. Reinforcement can be placed after grouting is complete.						
The costs are based on installing 100 piles on a clear site with reasonable access.						

CLASS P: PILING

Item	Gang hours	Labour £	Plant £	Material £	Unit	**Total rate £**
BORED CAST IN PLACE CONCRETE PILES – cont'd						
Concrete 35 N/mm², 20 mm aggregate; installed by continuous flight auger – cont'd						
Diameter: 300 mm						
number of piles	-	-	-	-	nr	**122.03**
concreted length	-	-	-	-	m	**8.25**
depth bored to 10 m maximum depth	-	-	-	-	nr	**12.85**
depth bored to 15 m maximum depth	-	-	-	-	nr	**12.85**
depth bored to 20 m maximum depth	-	-	-	-	nr	**12.85**
Diameter: 450 mm						
number of piles	-	-	-	-	nr	**122.03**
concreted length	-	-	-	-	m	**18.59**
depth bored to 10 m maximum depth	-	-	-	-	nr	**17.54**
depth bored to 15 m maximum depth	-	-	-	-	nr	**17.54**
depth bored to 20 m maximum depth	-	-	-	-	nr	**17.54**
Diameter: 600 mm						
number of piles	-	-	-	-	nr	**122.03**
concreted length	-	-	-	-	m	**33.08**
depth bored to 10 m maximum depth	-	-	-	-	nr	**25.71**
depth bored to 15 m maximum depth	-	-	-	-	nr	**25.71**
depth bored to 20 m maximum depth	-	-	-	-	nr	**25.71**
DRIVEN CAST IN PLACE CONCRETE PILES						
Generally						
The items "number of piles" are calculated based on the following:						
• allowance for provision of all plant, equipment and labour including transporting to and from site and establishing and dismantling at £5,676 in total for piles using the Temporary Steel Casing Method and £9,808 in total for piles using the Segmental Casing Method.						
• moving the rig to and setting up at each pile position; preparing to commence driving; £92.50 per pile.						
For the Temporary Steel Casing Method, obstructions (where within the capabilities of the normal plant) are quoted at £152.00 per hour and standing time at £145.00 per hour.						
For the Segmental Steel Casing Method, obstructions (where within the capabilities of the normal plant) are quoted at £260.00 per hour and standing time at £240.00 per hour.						

CLASS P: PILING

Item	Gang hours	Labour £	Plant £	Material £	Unit	**Total rate £**
Temporary steel casing method; concrete 35 N/mm²; reinforced for 750 kN						
The following unit costs cover the construction of piles by driving a heavy steel tube into the ground either by using an internal hammer acting on a gravel or concrete plug, as is more usual, or by using an external hammer on a driving helmet at the top of the tube. After driving to the required depth an enlarged base is formed by hammering out sucessive charges of concrete down the tube. The tube is then filled with concrete which is compacted as the tube is vibrated and withdrawn. Piles of 350 to 500 mm diameter can be constructed with rakes up to 1 in 4 to carry working loads up to 120 t per pile.						
The costs are based on installing 100 piles on a clear site with reasonable access.						
Diameter 430 mm; drive shell and form pile						
number of piles	-	-	-	-	nr	**152.73**
concreted length	-	-	-	-	m	**24.09**
depth driven; bottom-driven method	-	-	-	-	m	**7.02**
depth driven; top-driven method	-	-	-	-	m	**4.04**
Segmental casing method; concrete 35 N/mm²; nominal reinforcement						
The following unit costs cover the construction of piles by driving into hard material using a serrated thick walled tube. It is oscillated and pressed into the hard material using a hydraulic attachment to the piling rig. The hard material is broken up using chiselling methods and is then removed by mechanical grab.						
The costs are based on installing 100 piles on a clear site with reasonable access.						
Diameter 620 mm						
number of piles	-	-	-	-	nr	**203.05**
concreted length	-	-	-	-	m	**131.90**
depth bored or driven to 15 m maaximum depth	-	-	-	-	m	**10.00**
Diameter 1180 mm						
number of piles	-	-	-	-	nr	**203.05**
concreted length	-	-	-	-	m	**143.15**
depth bored or driven to 15 m maaximum depth	-	-	-	-	m	**15.00**
Diameter 1500 mm						
number of piles	-	-	-	-	nr	**203.05**
concreted length	-	-	-	-	m	**192.29**
1500 mm diameter	-	-	-	-	m	**23.76**

CLASS P: PILING

Item	Gang hours	Labour £	Plant £	Material £	Unit	**Total rate £**
PREFORMED CONCRETE PILES						
The following unit costs cover the installation of driven precast concrete piles by using a hammer acting on shoe fitted or cast into the precast concrete pile unit.						
Single pile lengths are normally a maximum of 13m long, at which point, a mechanical interlocking joint is required to extend the pile. These joints are most economically and practically formed at works.						
Lengths, sizes of sections, reinforcement details and concrete mixes vary for differing contractors, whose specialist advice should be sought for specific designs.						
The following unit costs are based on installing 100 piles on a clear site with reasonable access.						
The items "number of piles" are calculated based on the following:						
• allowance for provision of all plant, equipment and labour including transporting to and from site and establishing and dismantling at £3,000 in total for piles up to 275 x 275 mm and £3,700 in total for piles 350 x 350 mm and over.						
• moving the rig to and setting up at each pile position; preparing to commence driving; piles up to 275 x 275 mm £33.00 each; piles 350 x 350 mm and over, £53.00 each.						
• an allowance for the cost of the driving head and shoe; £35.00 for 235 x 235 mm piles, £45.00 for 275 x 275 mm and £55.00 for 350 x 350 mm.						
• cost of providing the pile of the stated length						
Typical allowances for standing time are £132.30 per hour for 235 x 235 mm piles, £150.41 for 275 x 275 mm and £180.49 for 350 x 350 mm.						
Concrete 50 N/mm²; reinforced for 600 kN						
The costs are based on installing 100 piles on a clear site with reasonable access.						
Cross-sectional area: 0.05-0.1 m²; 235 x 235 mm						
number of piles of 10 m length	-	-	-	-	nr	**303.17**
number of piles of 15 m length	-	-	-	-	nr	**395.52**
number of piles of 20 m length	-	-	-	-	nr	**487.86**
number of piles of 25 m length	-	-	-	-	nr	**580.21**
add for mechanical interlocking joint	-	-	-	-	nr	**45.32**
depth driven	-	-	-	-	m	**2.77**
Cross-sectional area: 0.05-0.1 m²; 275 x 275 mm						
number of piles of 10 m length	-	-	-	-	nr	**312.59**
number of piles of 15 m length	-	-	-	-	nr	**409.66**
number of piles of 20 m length	-	-	-	-	nr	**506.73**
number of piles of 25 m length	-	-	-	-	nr	**603.80**
add for mechanical interlocking joint	-	-	-	-	nr	**51.79**
depth driven	-	-	-	-	m	**3.23**

CLASS P: PILING

Item	Gang hours	Labour £	Plant £	Material £	Unit	**Total rate £**
Cross-sectional area: 0.1-0.15 m²; 350 x 350 mm						
number of piles of 10 m length	-	-	-	-	nr	**521.63**
number of piles of 15 m length	-	-	-	-	nr	**707.64**
number of piles of 20 m length	-	-	-	-	nr	**893.65**
number of piles of 25 m length	-	-	-	-	nr	**1079.65**
number of piles of 30 m length	-	-	-	-	nr	**1265.66**
add for mechanical interlocking joint	-	-	-	-	nr	**64.74**
depth driven	-	-	-	-	m	**3.88**
TIMBER PILES						
The items "number of piles" are calculated based on the following: • allowance for provision of all plant, equipment and labour including transporting to and from site and establishing and dismantling at £5,448 in total. • moving the rig to and setting up at each pile position and preparing to drive at £68.83 per pile allowance for the cost of the driving head and shoe at £35.00 per 225 x 225 mm pile, £45.00 for 300 x 300 mm, £55.00 for 350 x 350 mm and £65.00 for 450 x 450mm. • cost of providing the pile of the stated length A typical allowance for standing time is £245.50 per hour.						
Douglas Fir; hewn to mean pile size The costa are based on installing 100 piles on a clear site with reasonable access.						
Cross-sectional area: 0.05-0.1 m²; 225 x 225 mm						
number of piles of 10 m length	-	-	-	-	nr	**469.98**
number of piles of 15 m length	-	-	-	-	nr	**616.46**
number of piles of 20 m length	-	-	-	-	nr	**762.93**
depth driven	-	-	-	-	m	**3.11**
Cross-sectional area: 0.05-0.1 m²; 300 x 300 mm						
number of piles of 10 m length	-	-	-	-	nr	**709.51**
number of piles of 15 m length	-	-	-	-	nr	**969.96**
number of piles of 20 m length	-	-	-	-	nr	**1230.41**
depth driven	-	-	-	-	m	**3.72**
Cross-sectional area: 0.1-0.15 m²; 350 x 350 mm						
number of piles of 10 m length	-	-	-	-	nr	**909.25**
number of piles of 15 m length	-	-	-	-	nr	**1263.78**
number of piles of 20 m length	-	-	-	-	nr	**1618.31**
depth driven	-	-	-	-	m	**4.26**
Cross-sectional area: 0.15-0.25 m²; 450 x 450 mm						
number of piles of 10 m length	-	-	-	-	nr	**1383.87**
number of piles of 15 m length	-	-	-	-	nr	**1969.93**
number of piles of 20 m length	-	-	-	-	nr	**2555.99**
number of piles of 25 m length	-	-	-	-	nr	**3142.04**
depth driven	-	-	-	-	m	**5.59**

CLASS P: PILING

Item	Gang hours	Labour £	Plant £	Material £	Unit	Total rate £
TIMBER PILES – cont'd						
Greenheart; hewn to mean pile size						
The costs are based on installing 100 piles on a clear site with reasonable access.						
Cross-sectional area: 0.05-0.1 m²; 225 x 225 mm						
number of piles of 10 m length	-	-	-	-	nr	**528.68**
number of piles of 15 m length	-	-	-	-	nr	**704.50**
number of piles of 20 m length	-	-	-	-	nr	**880.32**
depth driven	-	-	-	-	m	**3.11**
Cross-sectional area: 0.05-0.1 m²; 300 x 300 mm						
number of piles of 10 m length	-	-	-	-	nr	**813.77**
number of piles of 15 m length	-	-	-	-	nr	**1126.36**
number of piles of 20 m length	-	-	-	-	nr	**1438.94**
depth driven	-	-	-	-	m	**3.72**
Cross-sectional area: 0.1-0.15 m²; 350 x 350 mm						
number of piles of 10 m length	-	-	-	-	nr	**1051.00**
number of piles of 15 m length	-	-	-	-	nr	**1476.41**
number of piles of 20 m length	-	-	-	-	nr	**1901.81**
depth driven	-	-	-	-	m	**4.26**
Cross-sectional area: 0.15-0.25 m²; 450 x 450 mm						
number of piles of 10 m length	-	-	-	-	nr	**1622.64**
number of piles of 15 m length	-	-	-	-	nr	**2328.08**
number of piles of 20 m length	-	-	-	-	nr	**3033.53**
number of piles of 25 m length	-	-	-	-	nr	**3738.97**
depth driven	-	-	-	-	m	**5.59**
ISOLATED STEEL PILES						
Steel bearing piles are commonly carried out by a Specialist Contractor and whose advice should be sought to arrive at accurate costing. However the following items can be used to assess a budget cost for such work.						
The following unit costs are based upon driving 100 nr steel bearing piles on a clear site with reasonable access.						
The items "number of piles" are calculated based on the following:						
• allowance for provision of all plant, equipment and labour including transporting to and from site and establishing and dismantling at £6,289 in total up to a maximum 100 miles radius from base and £15,724 up to a maximum 250 miles radius from base.						
• moving the rig to and setting up at each pile position; preparing to commence driving; £183.96 per pile.						
• cost of providing the pile of the stated length						
A typical allowance for standing time is £283.90 per hour.						

CLASS P: PILING

Item	Gang hours	Labour £	Plant £	Material £	Unit	**Total rate £**
Steel EN 10025 grade S275; within 100 miles of steel plant						
The costs are based upon installing 100 nr on a clear site with reasonable access.						
Mass 45 kg/m; 203 x 203 mm						
number of piles: length 10 m	-	-	-	580.94	nr	**580.94**
number of piles: length 15 m	-	-	-	724.57	nr	**724.57**
number of piles: length 20 m	-	-	-	865.73	nr	**865.73**
depth driven; vertical	-	-	-	-	m	**8.80**
depth driven; raking	-	-	-	-	m	**10.64**
Mass 54 kg/m; 203 x 203 mm						
number of piles; length 10 m	-	-	-	638.57	nr	**638.57**
number of piles; length 15 m	-	-	-	811.76	nr	**811.76**
number of piles; length 20 m	-	-	-	981.98	nr	**981.98**
depth driven; vertical	-	-	-	-	m	**8.80**
depth driven; raking	-	-	-	-	m	**10.64**
Mass 63 kg/m; 254 x 254 mm						
number of piles; length 10 m	-	-	-	697.72	nr	**697.72**
number of piles; length 15 m	-	-	-	901.23	nr	**901.23**
number of piles; length 20 m	-	-	-	1101.28	nr	**1101.28**
depth driven; vertical	-	-	-	-	m	**9.42**
depth driven; raking	-	-	-	-	m	**11.40**
Mass 71 kg/m; 254 x 254 mm						
number of piles; length 10 m	-	-	-	748.09	nr	**748.09**
number of piles; length 15 m	-	-	-	977.44	nr	**977.44**
number of piles; length 20 m	-	-	-	1202.89	nr	**1202.89**
depth driven; vertical	-	-	-	-	m	**9.42**
depth driven; raking	-	-	-	-	m	**11.40**
Mass 85 kg/m; 254 x 254 mm						
number of piles; length 10 m	-	-	-	836.23	nr	**836.23**
number of piles; length 15 m	-	-	-	1110.81	nr	**1110.81**
number of piles; length 20 m	-	-	-	1380.71	nr	**1380.71**
depth driven; vertical	-	-	-	-	m	**9.42**
depth driven; raking	-	-	-	-	m	**11.40**
Mass 79 kg/m; 305 x 305 mm						
number of piles; length 10 m	-	-	-	810.62	nr	**810.62**
number of piles; length 15 m	-	-	-	1071.90	nr	**1071.90**
number of piles; length 20 m	-	-	-	1328.83	nr	**1328.83**
depth driven; vertical	-	-	-	-	m	**9.42**
depth driven; raking	-	-	-	-	m	**11.40**
Mass 95 kg/m; 305 x 305 mm						
number of piles; length 10 m	-	-	-	913.81	nr	**913.81**
number of piles; length 15 m	-	-	-	1228.01	nr	**1228.01**
number of piles; length 20 m	-	-	-	1536.98	nr	**1536.98**
depth driven; vertical	-	-	-	-	m	**10.04**
depth driven; raking	-	-	-	-	m	**12.17**
Mass 110 kg/m; 305 x 305 mm						
number of piles; length 10 m	-	-	-	1010.56	nr	**1010.56**
number of piles; length 15 m	-	-	-	1374.36	nr	**1374.36**
number of piles; length 20 m	-	-	-	1732.12	nr	**1732.12**
depth driven; vertical	-	-	-	-	m	**10.04**
depth driven; raking	-	-	-	-	m	**12.17**
Mass 109 kg/m; 356 x 368 mm						
number of piles; length 10 m	-	-	-	1020.90	nr	**1020.90**
number of piles; length 15 m	-	-	-	1389.79	nr	**1389.79**
number of piles; length 20 m	-	-	-	1752.68	nr	**1752.68**
depth driven; vertical	-	-	-	-	m	**10.04**
depth driven; raking	-	-	-	-	m	**12.17**

CLASS P: PILING

Item	Gang hours	Labour £	Plant £	Material £	Unit	Total rate £
ISOLATED STEEL PILES – cont'd						
Steel EN 10025 grade S275; within 100 miles of steel plant – cont'd						
Mass 126 kg/m; 305 x 305 mm						
number of piles; length 10 m	-	-	-	1113.75	nr	**1113.75**
number of piles; length 15 m	-	-	-	1530.48	nr	**1530.47**
number of piles; length 20 m	-	-	-	1940.27	nr	**1940.27**
driving piles; vertical	-	-	-	-	m	**10.67**
driving piles; raking	-	-	-	-	m	**12.91**
Mass 149 kg/m; 305 x 305 mm						
number of piles; length 10 m	-	-	-	1262.09	nr	**1262.09**
number of piles; length 15 m	-	-	-	1754.88	nr	**1754.88**
number of piles; length 20 m	-	-	-	2239.48	nr	**2239.48**
driving piles; vertical	-	-	-	-	m	**10.67**
driving piles; raking	-	-	-	-	m	**12.91**
Mass 186 kg/m; 305 x 305 mm						
number of piles; length 10 m	-	-	-	1500.73	nr	**1500.73**
number of piles; length 15 m	-	-	-	2115.89	nr	**2115.89**
number of piles; length 20 m	-	-	-	2720.82	nr	**2720.82**
driving piles; vertical	-	-	-	-	m	**11.30**
drive piles; raking	-	-	-	-	m	**14.44**
Mass 223 kg/m; 305 x 305 mm						
number of piles; length 10 m	-	-	-	1739.37	nr	**1739.37**
number of piles; length 15 m	-	-	-	2476.90	nr	**2476.90**
number of piles; length 20 m	-	-	-	3202.17	nr	**3202.16**
driving piles; vertical	-	-	-	-	m	**12.56**
driving piles; raking	-	-	-	-	m	**15.20**
Mass 133 kg/m; 356 x 368 mm						
number of piles; length 10 m	-	-	-	1179.38	nr	**1179.38**
number of piles; length 15 m	-	-	-	1629.50	nr	**1629.50**
number of piles; length 20 m	-	-	-	2072.30	nr	**2072.30**
driving piles; vertical	-	-	-	-	m	**10.67**
driving piles; raking	-	-	-	-	m	**12.91**
Mass 152 kg/m; 356 x 368 mm						
number of piles; length 10 m	-	-	-	1304.85	nr	**1304.85**
number of piles; length 15 m	-	-	-	1819.27	nr	**1819.27**
number of piles; length 20 m	-	-	-	2325.32	nr	**2325.32**
driving piles; vertical	-	-	-	-	m	**11.30**
driving piles; raking	-	-	-	-	m	**13.69**
Mass 174 kg/m; 356 x 368 mm						
number of piles; length 10 m	-	-	-	1450.13	nr	**1450.13**
number of piles; length 15 m	-	-	-	2039.00	nr	**2039.00**
number of piles; length 20 m	-	-	-	2618.30	nr	**2618.30**
driving piles; vertical	-	-	-	-	m	**11.30**
driving piles; raking	-	-	-	-	m	**13.69**

CLASS P: PILING

Item	Gang hours	Labour £	Plant £	Material £	Unit	**Total rate £**
INTERLOCKING STEEL PILES						
Sheet steel piling is commonly carried out by a Specialist Contractor, whose advice should be sought to arrive at accurate costings. However the following items can be used to assess a budget for such work.						
Note: area of driven piles will vary from area supplied dependant upon pitch line of piling and provision for such allowance has been made in PC for supply.						
The materials cost below includes the manufacturers tarrifs for a 200 mile delivery radius from works, delivery 10 - 20t loads and with an allowance of 10% to cover waste / projecting piles etc.						
ProfilARBED Z section steel piles; EN 10248 grade S270GP steel						
The following unit costs are based on driving/extracting 1,500 m² of sheet piling on a clear site with reasonable access.						
Provision of all plant, equipment and labour including transport to and from the site and establishing and dismantling for						
driving of sheet piling	-	-	-	-	sum	**5660.40**
extraction of sheet piling	-	-	-	-	sum	**5030.40**
Standing time	-	-	-	-	hr	**272.54**
Section modulus 800 - 1200 cm³/m; section reference AZ 12; mass 98.7 kg/m², sectional modulus 1200 cm³/m; EN 10248 grade S270GP steel						
length of welded corner piles	-	-	-	-	m	**65.83**
length of welded junction piles	-	-	-	-	m	**92.16**
driven area	-	-	-	-	m²	**34.39**
area of piles of length not exceeding 14 m	-	-	-	69.95	m²	**69.95**
area of piles of length 14 - 24 m	-	-	-	75.19	m²	**75.19**
area of piles of length exceeding 24 m	-	-	-	77.70	m²	**77.70**
Section modulus 1200 - 20000 cm³/m; section reference AZ 17; mass 108.6 kg/m²; sectional modulus 1665 cm³/m; EN 10248 grade S270GP steel						
length of welded corner piles	-	-	-	-	m	**65.83**
length of welded junction piles	-	-	-	-	m	**92.16**
driven area	-	-	-	-	m²	**34.39**
area of piles of length not exceeding 14 m	-	-	-	73.94	m²	**73.94**
area of piles of length 14 - 24 m	-	-	-	75.19	m²	**75.19**
area of piles of length exceeding 24 m	-	-	-	77.70	m²	**77.70**

CLASS P: PILING

Item	Gang hours	Labour £	Plant £	Material £	Unit	Total rate £
INTERLOCKING STEEL PILES – cont'd						
ProfilARBED Z section steel piles; EN 10248 grade S270GP steel – cont'd						
Section modulus 2000 - 3000 cm³/m; section reference AZ 26; mass 155.2 kg/m²; sectional modulus 2600 cm³/m; EN 10248 grade S270GP steel						
driven area	-	-	-	-	m²	**32.10**
area of piles of length 6 - 18 m	-	-	-	83.78	m²	**83.78**
area of piles of length 18 - 24 m	-	-	-	85.20	m²	**85.20**
Section modulus 3000 - 4000 cm³/m; section reference AZ 36; mass 194.0 kg/m²; sectional modulus 3600 cm³/m; EN 10248 grade S270GP steel						
driven area	-	-	-	-	m²	**29.16**
area of piles of length 6 - 18 m	-	-	-	102.29	m²	**102.29**
area of piles of length 18 - 24 m	-	-	-	104.03	m²	**104.03**
Straight section modulus ne 500 cm³/m; section reference AS 500-12 mass 149 kg/m²; sectional modulus 51 cm³/m; EN 10248 grade S270GP steel						
driven area	-	-	-	-	m²	**26.95**
area of piles of length 6 - 18 m	-	-	-	138.58	m²	**138.58**
area of piles of length 18 - 24 m	-	-	-	141.82	m²	**141.82**
One coat black tar vinyl (PC1) protective treatment applied all surfaces at shop to minimum dry film thickness up to 150 microns to steel piles						
section reference AZ 12; pile area	-	-	-	7.56	m²	**7.56**
section reference AZ 17; pile area	-	-	-	7.73	m²	**7.73**
section reference AZ 26; pile area	-	-	-	8.45	m²	**8.45**
section reference AZ 36; pile area	-	-	-	9.00	m²	**9.00**
section reference AS 500 - 12; pile area	-	-	-	9.40	m²	**9.40**
One coat black high build isocyanate cured epoxy pitch (PC2) protective treatment applied all surfaces at shop to minimum dry film thickness up to 450 microns to steel piles						
section reference AZ 12; pile area	-	-	-	15.12	m²	**15.12**
section reference AZ 17; pile area	-	-	-	15.47	m²	**15.47**
section reference AZ 26; pile area	-	-	-	16.91	m²	**16.91**
section reference AZ 36; pile area	-	-	-	18.00	m²	**18.00**
section reference AS 500 - 12; pile area	-	-	-	18.80	m²	**18.80**

CLASS P: PILING

Item	Gang hours	Labour £	Plant £	Material £	Unit	**Total rate £**
ProfilARBED U section steel piles; EN 10248 grade S270GP steel						
The following unit costs are based on driving/extracting 1,500 m² of sheet piling on a clear site with reasonable access.						
Provision of plant, equipment and labour including transport to and from the site and establishing and dismantling						
driving of sheet piling	-	-	-	-	sum	**5445.25**
extraction of sheet piling	-	-	-	-	sum	**4821.95**
Standing time	-	-	-	-	hr	**261.19**
Section modulus 500 - 800 cm³/m; section reference PU 6; mass 76.0 kg/m²; sectional modulus 600 cm³/m						
driven area	-	-	-	-	m²	**33.90**
area of piles of length 6 - 18 m	-	-	-	58.51	m²	**58.51**
area of piles of length 18 - 24 m	-	-	-	59.50	m²	**59.50**
Section modulus 800 - 1200 cm³/m; section reference PU 8; mass 90.9 kg/m²; sectional modulus 830 cm³/m						
driven area	-	-	-	-	m²	**28.89**
area of piles of length 6 - 18 m	-	-	-	58.51	m²	**58.51**
area of piles of length 18 - 24 m	-	-	-	59.56	m²	**59.56**
Section modulus 1200 - 2000 cm³/m; section reference PU 12; mass 110.1 kg/m²; sectional modulus 1200 cm³/m						
driven area	-	-	-	-	m²	**25.75**
area of piles of length 6 - 18 m	-	-	-	71.09	m²	**71.09**
area of piles of length 18 - 24 m	-	-	-	72.32	m²	**72.32**
Section modulus 1200 - 2000 cm³/m; section reference PU 18; mass 128.2 kg/m²; sectional modulus 1800 cm³/m						
driven area	-	-	-	-	m²	**22.91**
area of piles of length 6 - 18 m	-	-	-	82.57	m²	**82.57**
area of piles of length 18 - 24 m	-	-	-	84.00	m²	**84.00**
Section modulus 2000 - 3000 cm³/m; section reference PU 22; mass 143.6 kg/m²; sectional modulus 2200 cm³/m						
driven area	-	-	-	-	m²	**22.91**
area of piles of length 6 - 18 m	-	-	-	99.94	m²	**99.94**
area of piles of length 18 - 24 m	-	-	-	102.35	m²	**102.35**
Section modulus 3000 - 4000 cm³/m; section reference PU 32; mass 190.2 kg/m²; sectional modulus 3200 cm³/m						
driven area	-	-	-	-	m²	**18.20**
area of piles of length 6 - 18 m	-	-	-	112.98	m²	**112.98**
area of piles of length 18 - 24 m	-	-	-	115.70	m²	**115.70**

CLASS P: PILING

Item	Gang hours	Labour £	Plant £	Material £	Unit	**Total rate £**
INTERLOCKING STEEL PILES – cont'd						
ProfilARBED U section steel piles; EN 10248 grade S270GP steel – cont'd						
One coat black tar vinyl (PC1) protective treatment applied all surfaces at shop to minimum dry film thickness up to 150 microns to steel piles						
section reference PU 6; pile area	-	-	-	6.47	m²	**6.47**
section reference PU 8; pile area	-	-	-	6.36	m²	**6.36**
section reference PU 12; pile area	-	-	-	6.76	m²	**6.76**
section reference PU 18; pile area	-	-	-	7.25	m²	**7.25**
section reference PU 22; pile area	-	-	-	7.52	m²	**7.52**
section reference PU 32; pile area	-	-	-	7.63	m²	**7.63**
One coat black high build isocyanate cured epoxy pitch (PC2) protective treatment applied all surfaces at shop to minimum dry film thickness up to 450 microns to steel piles						
section reference PU 6; pile area	-	-	-	12.94	m²	**12.94**
section reference PU 8; pile area	-	-	-	12.72	m²	**12.72**
section reference PU 12; pile area	-	-	-	13.52	m²	**13.52**
section reference PU 18; pile area	-	-	-	14.50	m²	**14.50**
section reference PU 22; pile area	-	-	-	15.04	m²	**15.04**
section reference PU 32; pile area	-	-	-	15.26	m²	**15.26**

CLASS Q: PILING ANCILLARIES

Item	Gang hours	Labour £	Plant £	Material £	Unit	**Total rate £**
CAST IN PLACE CONCRETE PILES						
Bored; tripod-mounted percussion rig						
Backfilling empty bore with selected excavated material						
diameter 500mm	-	-	-	-	m	**3.24**
Permanent casings; each length not exceeding 13 m						
diameter 500mm	-	-	-	-	m	**10.06**
Permanent casings; each length exceeding 13 m						
diameter 500mm	-	-	-	-	m	**10.74**
Enlarged bases						
diameter 1500 mm; to 500mm diameter pile	-	-	-	-	nr	**239.27**
Cutting off surplus lengths						
diameter 500mm	-	-	-	-	m	**25.97**
Preparing heads						
500mm diameter	-	-	-	-	nr	**38.95**
Bored; lorry/crawler mounted rotary rig						
Backfilling empty bore with selected excavated material						
diameter 500mm	-	-	-	-	m	**3.24**
Permanent casings; each length not exceeding 13 m						
diameter 500mm	-	-	-	-	m	**10.06**
Permanent casings; each length exceeding 13 m						
diameter 500mm	-	-	-	-	m	**10.74**
Enlarged bases						
diameter 1500 mm; to 500mm diameter pile	-	-	-	-	nr	**239.27**
Cutting off surplus lengths						
diameter 500mm	-	-	-	-	m	**25.97**
Preparing heads						
500mm diameter	-	-	-	-	nr	**38.95**
Bored; continuous flight auger						
Backfilling empty bore with selected excavated material						
450 mm diameter piles	-	-	-	-	m	**2.60**
600 mm diameter piles	-	-	-	-	m	**3.56**
750 mm diameter piles	-	-	-	-	m	**3.89**
Permanent casings; each length not exceeding 13 m						
450 mm diameter piles	-	-	-	-	m	**14.76**
600 mm diameter piles	-	-	-	-	m	**17.44**
750 mm diameter piles	-	-	-	-	m	**20.14**
Permanent casings ;each length exceeding 13 m						
450 mm diameter piles	-	-	-	-	m	**9.41**
600 mm diameter piles	-	-	-	-	m	**10.74**
750 mm diameter piles	-	-	-	-	m	**218.76**
Enlarged bases						
diameter 1400 mm; to 450 mm diameter piles	-	-	-	-	nr	**218.76**
diameter 1800 mm; to 600 mm diameter piles	-	-	-	-	nr	**266.62**
diameter 2100 mm; to 750 mm diameter piles	-	-	-	-	nr	**161.83**
Cutting off surplus lengths						
450 mm diameter piles	-	-	-	-	m	**22.71**
600 mm diameter piles	-	-	-	-	m	**32.45**
750 mm diameter piles	-	-	-	-	m	**38.95**

CLASS Q: PILING ANCILLARIES

Item	Gang hours	Labour £	Plant £	Material £	Unit	**Total rate £**
CAST IN PLACE CONCRETE PILES – cont'd						
Bored; continuous flight auger – cont'd						
Preparing heads						
450 mm diameter piles	-	-	-	-	nr	**24.31**
600 mm diameter piles	-	-	-	-	nr	**38.95**
750 mm diameter piles	-	-	-	-	nr	**58.41**
Collection from around pile heads of spoil accruing from piling operations and depositing in spoil heaps (For final disposal see Section F2 - Earthworks Disposal)	-	-	-	-	m^3	**2.82**
Reinforcement; mild steel						
Straight bars, nominal size						
6 mm	-	-	-	-	tonne	**892.01**
8 mm	-	-	-	-	tonne	**788.70**
10 mm	-	-	-	-	tonne	**777.94**
12 mm	-	-	-	-	tonne	**769.33**
16 mm	-	-	-	-	tonne	**733.83**
25 mm	-	-	-	-	tonne	**638.05**
32 mm	-	-	-	-	tonne	**639.12**
50 mm	-	-	-	-	tonne	**659.57**
Helical bars, nominal size						
6 mm	-	-	-	-	tonne	**971.67**
8 mm	-	-	-	-	tonne	**849.48**
10 mm	-	-	-	-	tonne	**847.88**
12 mm	-	-	-	-	tonne	**822.67**
Reinforcement; high tensile steel						
Straight bars, nominal size						
6 mm	-	-	-	-	tonne	**907.08**
8 mm	-	-	-	-	tonne	**793.50**
10 mm	-	-	-	-	tonne	**772.56**
12 mm	-	-	-	-	tonne	**761.80**
16 mm	-	-	-	-	tonne	**729.51**
25 mm	-	-	-	-	tonne	**632.66**
32 mm	-	-	-	-	tonne	**633.73**
50 mm	-	-	-	-	tonne	**654.19**
Helical bars, nominal size						
6 mm	-	-	-	-	tonne	**987.98**
8 mm	-	-	-	-	tonne	**864.03**
10 mm	-	-	-	-	tonne	**837.12**
12 mm	-	-	-	-	tonne	**826.36**
Couplers; Lenton type A; threaded ends on reinforcing bars						
12 mm	0.09	9.34	-	8.96	nr	**18.30**
16 mm	0.09	9.34	-	10.31	nr	**19.65**
20 mm	0.09	9.34	-	14.68	nr	**24.02**
25 mm	0.09	9.34	-	19.89	nr	**29.24**
32 mm	0.09	9.34	-	27.39	nr	**36.73**
40 mm	0.09	9.34	-	37.78	nr	**47.13**

CLASS Q: PILING ANCILLARIES

Item	Gang hours	Labour £	Plant £	Material £	Unit	**Total rate £**
Couplers; Lenton type B; threaded ends on reinforcing bars						
12 mm	0.09	9.34	-	25.42	nr	**34.76**
16 mm	0.09	9.34	-	29.09	nr	**38.43**
20 mm	0.09	9.34	-	32.27	nr	**41.61**
25 mm	0.09	9.34	-	37.23	nr	**46.58**
32 mm	0.09	9.34	-	50.39	nr	**59.73**
40 mm	0.09	9.34	-	73.85	nr	**83.19**
PREFORMED CONCRETE PILES						
General						
Preparing heads						
235 x 235 mm piles	-	-	-	-	nr	**26.77**
275 x 275 mm piles	-	-	-	-	nr	**36.57**
350 x 350 mm piles	-	-	-	-	nr	**56.24**
TIMBER PILES						
Douglas Fir						
Cutting off surplus lengths						
cross-sectional area: 0.025-0.05 m^2	-	-	-	-	nr	**2.67**
cross-sectional area: 0.05-0.1 m^2	-	-	-	-	nr	**4.77**
cross-sectional area: 0.1-0.15 m^2	-	-	-	-	nr	**6.12**
cross-sectional area: 0.15-0.25 m^2	-	-	-	-	nr	**11.71**
Preparing heads						
cross-sectional area: 0.025-0.05 m^2	-	-	-	-	nr	**2.67**
cross-sectional area: 0.05-0.1 m^2	-	-	-	-	nr	**4.77**
cross-sectional area: 0.1-0.15 m^2	-	-	-	-	nr	**6.12**
cross-sectional area: 0.15-0.25 m^2	-	-	-	-	nr	**11.71**
Greenheart						
Cutting off surplus lengths						
cross-sectional area: 0.025-0.05 m^2	-	-	-	-	nr	**5.28**
cross-sectional area: 0.05-0.1 m^2	-	-	-	-	nr	**9.49**
cross-sectional area: 0.1-0.15 m^2	-	-	-	-	nr	**12.29**
cross-sectional area: 0.15-0.25 m^2	-	-	-	-	nr	**22.77**
Preparing heads						
cross-sectional area: 0.025-0.05 m^2	-	-	-	-	nr	**5.28**
cross-sectional area: 0.05-0.1 m^2	-	-	-	-	nr	**9.49**
cross-sectional area: 0.1-0.15 m^2	-	-	-	-	nr	**12.29**
cross-sectional area: 0.15-0.25 m^2	-	-	-	-	nr	**22.77**

CLASS Q: PILING ANCILLARIES

Item	Gang hours	Labour £	Plant £	Material £	Unit	Total rate £
ISOLATED STEEL PILES						
Steel bearing piles						
Steel bearing piles are commonly carried out by a Specialist Contractor and whose advice should be sought to arrive at accurate costing. However the following items can be used to assess a budget cost for such work. The item for number of pile extensions includes for the cost of setting up the rig at the pile position, together with welding the extension to the top of the steel bearing pile. The items for length of pile extension cover the material only, the driving cost being included in Class P.						
Number of pile extensions						
at each position	-	-	-	-	nr	**235.06**
Length of pile extensions, each length not exceeding 3 m; steel EN 10025 grade S275						
mass 45 kg/m	-	-	-	28.23	m	**28.23**
mass 54 kg/m	-	-	-	34.04	m	**34.04**
mass 63 kg/m	-	-	-	40.01	m	**40.01**
mass 71 kg/m	-	-	-	45.09	m	**45.09**
mass 79 kg/m	-	-	-	51.39	m	**51.39**
mass 85 kg/m	-	-	-	53.98	m	**53.98**
mass 95 kg/m	-	-	-	61.79	m	**61.79**
mass 109 kg/m	-	-	-	72.58	m	**72.58**
mass 110 kg/m	-	-	-	71.55	m	**71.55**
mass 126 kg/m	-	-	-	81.96	m	**81.96**
mass 149 kg/m	-	-	-	96.92	m	**96.92**
mass 133 kg/m	-	-	-	88.56	m	**88.56**
mass 152 kg/m	-	-	-	101.21	m	**101.21**
mass 174 kg/m	-	-	-	115.86	m	**115.86**
mass 186 kg/m	-	-	-	120.99	m	**120.99**
mass 223 kg/m	-	-	-	145.05	m	**145.05**
Length of pile extensions, each length exceeding 3 m; steel EN 10025 grade S275						
mass 45 kg/m	-	-	-	28.23	m	**28.23**
mass 54 kg/m	-	-	-	34.04	m	**34.04**
mass 63 kg/m	-	-	-	40.01	m	**40.01**
mass 71 kg/m	-	-	-	45.09	m	**45.09**
mass 79 kg/m	-	-	-	51.39	m	**51.39**
mass 85 kg/m	-	-	-	53.98	m	**53.98**
mass 95 kg/m	-	-	-	61.79	m	**61.79**
mass 109 kg/m	-	-	-	72.58	m	**72.58**
mass 110 kg/m	-	-	-	71.55	m	**71.55**
mass 126 kg/m	-	-	-	81.96	m	**81.96**
mass 149 kg/m	-	-	-	96.92	m	**96.92**
mass 133 kg/m	-	-	-	88.56	m	**88.56**
mass 152 kg/m	-	-	-	101.21	m	**101.21**
mass 174 kg/m	-	-	-	115.86	m	**115.86**
mass 186 kg/m	-	-	-	120.99	m	**120.99**
mass 223 kg/m	-	-	-	145.05	m	**145.05**
Number of pile extensions						
section size 203 x 203 x any kg/m	-	-	-	-	nr	**100.45**
section size 254 x 254 x any kg/m	-	-	-	-	nr	**125.57**
section size 305 x 305 x any kg/m	-	-	-	-	nr	**150.68**
section size 356 x 368 x any kg/m	-	-	-	-	nr	**175.79**

CLASS Q: PILING ANCILLARIES

Item	Gang hours	Labour £	Plant £	Material £	Unit	**Total rate £**
Cutting off surplus lengths						
mass 30-60 kg/m	-	-	-	-	nr	**5.21**
mass 60-120 kg/m	-	-	-	-	nr	**7.83**
mass 120-250 kg/m	-	-	-	-	nr	**10.44**
Burning off tops of piles to level						
mass 30-60 kg/m	-	-	-	-	nr	**5.21**
mass 60-120 kg/m	-	-	-	-	nr	**7.83**
mass 120-250 kg/m	-	-	-	-	nr	**10.44**
INTERLOCKING STEEL PILES						
Steel sheet piling is commonly carried out by a Specialist Contractor and whose advice should be sought to arrive at accurate costing. However the following items can be used to assess a budget cost for such work.						
ProfilARBED Z section steel piles; EN 10248 grade S270GP steel						
Cutting off surplus lengths						
Section modulus 500 - 800 cm³/m	-	-	-	-	m	**9.95**
Section modulus 800 - 1200 cm³/m	-	-	-	-	m	**9.95**
Section modulus 1200 - 2000 cm³/m	-	-	-	-	m	**9.95**
Section modulus 2000 - 3000 cm³/m	-	-	-	-	m	**10.65**
Section modulus 3000 - 4000 cm³/m	-	-	-	-	m	**12.78**
Extract piling and stacking on site						
Section modulus 500 - 800 cm³/m	-	-	-	-	m²	**24.82**
Section modulus 800 - 1200 cm³/m	-	-	-	-	m²	**24.82**
Section modulus 1200 - 2000 cm³/m	-	-	-	-	m²	**22.00**
Section modulus 2000 - 3000 cm³/m	-	-	-	-	m²	**17.02**
Section modulus 3000 - 4000 cm³/m	-	-	-	-	m²	**17.40**
ProfilARBED Z section steel piles; EN 10248 grade S270GP steel						
Cutting off surplus lengths						
section modulus 500 - 800 cm³/m	-	-	-	-	m	**8.51**
section modulus 800 - 1200 cm³/m	-	-	-	-	m	**9.95**
section modulus 1200 - 2000 cm³/m	-	-	-	-	m	**9.95**
section modulus 2000 - 3000 cm³/m	-	-	-	-	m	**12.06**
section modulus 3000 - 4000 cm³/m	-	-	-	-	m	**12.78**
Extract piling and stack on site						
Section modulus 500 - 800 cm³/m	-	-	-	-	m²	**24.14**
Section modulus 800 - 1200 cm³/m	-	-	-	-	m²	**21.65**
Section modulus 1200 - 2000 cm³/m PU 12	-	-	-	-	m²	**17.15**
Section modulus 1200 - 2000 cm³/m PU 18	-	-	-	-	m²	**17.02**
Section modulus 2000 - 3000 cm³/m	-	-	-	-	m²	**15.63**
Section modulus 3000 - 4000 cm³/m	-	-	-	-	m²	**16.63**

CLASS Q: PILING ANCILLARIES

Item	Gang hours	Labour £	Plant £	Material £	Unit	Total rate £
OBSTRUCTIONS						
General						
Obstructions	-	-	-	-	hr	**99.28**
PILE TESTS						
Cast in place						
Pile tests; 500 mm diameter working pile; maximum test load of 600kN using non-working tension piles as reaction tripod						
first pile	-	-	-	-	nr	**4340.10**
subsequent pile	-	-	-	-	nr	**3100.40**
Take and test undisturbed soil samples; tripod	-	-	-	-	nr	**156.12**
Make, cure and test concrete cubes; tripod	-	-	-	-	nr	**8.98**
Pile tests; working pile; maximum test load of 1½ times working load; first pile						
450 mm / 650kN	-	-	-	-	nr	**2497.40**
600 mm / 1400kN	-	-	-	-	nr	**3121.74**
750 mm / 2200kN	-	-	-	-	nr	**3746.09**
Pile tests; working pile; maximum test load of 1½ times working load; second and subsequent piles						
450 mm / 650kN	-	-	-	-	nr	**1248.70**
600 mm / 1400kN	-	-	-	-	nr	**1873.05**
750 mm / 2200kN	-	-	-	-	nr	**2497.40**
Pile tests; working pile; electronic integrity testing; each pile (minimum 40 piles per visit)	-	-	-	-	nr	**15.60**
Make, cure and test concrete cubes	-	-	-	-	nr	**10.50**
Preformed						
Pile tests; working pile; maximum test load of 1.5 times working load	-	-	-	-	nr	**2402.96**
Pile tests; working pile; dynamic testing with piling hammer	-	-	-	-	nr	**600.74**
Steel bearing piles						
Steel bearing piles are commonly carried out by a Specialist Contractor and whose advice should be sought to arrive at accurate costing. However the following items can be used to assess a budget cost for such work.						
The following unit costs are based upon driving 100 nr steel bearing piles 15-24m long on a clear site with reasonable access. Supply is based on delivery 75 miles from works, in loads over 20t.						
Establishment of pile testing equipment on site preliminary to any piling operation	-	-	-	-	sum	**19890.40**
Carry out pile test on bearing piles irrespective of section using pile testing equipment on site up to 108 t load	-	-	-	-	nr	**7232.35**

CLASS Q: PILING ANCILLARIES

Item	Gang hours	Labour £	Plant £	Material £	Unit	**Total rate £**
Driven-temporary casing						
Pile tests; 430 mm diameter working pile; maximum test load of 1125kN using non-working tension piles as reaction; first piles						
bottom driven	-	-	-	-	nr	**3121.74**
top driven	-	-	-	-	nr	**2497.40**
Pile tests; 430 mm diameter working pile; maximum test load of 1125kN using non-working tension piles as reaction; subsequent piles						
bottom driven	-	-	-	-	nr	**1248.70**
top driven	-	-	-	-	nr	**1248.70**
Pile tests; working pile; electronic integrity testing; each pile (minimum 40 piles per visit)	-	-	-	-	nr	**15.62**
Make cure and test concrete cubes	-	-	-	-	nr	**8.11**
Driven - segmental casing						
Pile tests; 500 mm diameter working pile; maximum test load of 600kN using non-working tension piles as reaction						
first pile	-	-	-	-	nr	**4515.00**
subsequent piles	-	-	-	-	nr	**3412.50**

CLASS R: ROADS AND PAVINGS

Item	Gang hours	Labour £	Plant £	Material £	Unit	**Total rate £**
GENERAL						
Notes - Labour and Plant						
All outputs are based on clear runs without undue delay to two pavers with 75% utilisation.						
The outputs can be adjusted as follows to take account of space or time influences on the utilisation.						
Factors for varying utilisation of Labour and Plant:						
1 paver @ 75 % utilisation = x 2.00						
1 paver @ 100 % utilisation = x 1.50						
2 paver @ 100 % utilisation = x 0.75						
RESOURCES - LABOUR						
Sub-base laying gang						
1 ganger/chargehand (skill rate 4)		12.80				
1 skilled operative (skill rate 4)		12.05				
2 unskilled operatives (general)		22.52				
1 plant operator (skill rate 2)		16.35				
1 plant operator (skill rate 3)		14.65				
Total Gang Rate / Hour	£	**78.37**				
Flexible paving gang						
1 ganger/chargehand (skill rate 4)		12.80				
2 skilled operatives (skill rate 4)		24.10				
4 unskilled operatives (general)		45.04				
4 plant operators (skill rate 3)		58.60				
Total Gang Rate / Hour	£	**140.54**				
Concrete paving gang						
1 ganger/chargehand (skill rate 4)		12.80				
2 skilled operatives (skill rate 4)		24.10				
4 unskilled operatives (general)		45.04				
1 plant operator (skill rate 2)		16.35				
1 plant operator (skill rate 3)		14.65				
Total Gang Rate / Hour	£	**112.94**				
Road surface spraying gang						
1 plant operator (skill rate 3)		14.65				
Total Gang Rate / Hour	£	**14.65**				
Road chippings gang						
1 ganger/chargehand (skill rate 4) - 50% of time		6.40				
1 skilled operative (skill rate 4)		12.05				
2 unskilled operatives (general)		22.52				
3 plant operators (skill rate 3)		43.95				
Total Gang Rate / Hour	£	**84.92**				
Cutting slabs gang						
1 unskilled operative (generally)		11.26				
Total Gang Rate / Hour	£	**11.26**				
Concrete filled joints gang						
1 ganger/chargehand (skill rate 4) - 50% of time		6.40				
1 skilled operatives (skill rate 4)		12.05				
2 unskilled operatives (general)		22.52				
Total Gang Rate / Hour	£	**40.97**				

CLASS R: ROADS AND PAVINGS

Item	Gang hours	Labour £	Plant £	Material £	Unit	Total rate £
Milling gang						
1 ganger/chargehand (skill rate 4)		12.80				
2 skilled operatives (skill rate 4)		24.10				
4 unskilled operatives (general)		45.04				
1 plant operators (skill rate 3)		14.65				
1 plant operator (skill rate 2)		16.35				
Total Gang Rate / Hour	£	**112.94**				
Rake and compact planed material gang						
1 ganger/chargehand (skill rate 4)		12.80				
1 skilled operatives (skill rate 4)		12.05				
3 unskilled operatives (general)		33.78				
1 plant operator (skill rate 3)		14.65				
1 plant operator (skill rate 4)		13.17				
Total Gang Rate / Hour	£	**86.45**				
Kerb laying gang						
3 skilled operatives (skill rate 4)		36.15				
1 unskilled operative (general)		11.26				
1 plant operator (skill rate 3) - 25% of time		3.66				
Total Gang Rate / Hour	£	**51.07**				
Path sub-base, bitmac and gravel laying gang						
1 ganger/chargehand (skill rate 4)		12.80				
2 unskilled operatives (general)		22.52				
1 plant operator (skill rate 3)		14.65				
Total Gang Rate / Hour	£	**49.97**				
Paviors and flagging gang						
1 skilled operative (skill rate 4)		12.05				
1 unskilled operative (general)		11.26				
Total Gang Rate / Hour	£	**23.31**				
Traffic signs gang						
1 ganger/chargehand (skill rate 3)		14.21				
1 skilled operative (skill rate 3)		13.46				
2 unskilled operatives (general)		22.52				
1 plant operator (skill rate 3) - 25% of time		3.66				
Total Gang Rate / Hour	£	**53.85**				
RESOURCES - PLANT						
Sub-base laying						
93 KW motor grader			24.27			
0.80 m³ tractor loader			19.79			
6t towed roller			9.72			
Total Rate / Hour		£	**53.78**			
Flexible paving						
2 asphalt pavers, 35 kW, 4.0 m			86.36			
2 deadweight rollers, 3 point, 10 t			30.41			
tractor with front bucket and integral 2 tool compressor			17.04			
compressor tools: scabbler			1.73			
tar sprayer, 100 litre			4.17			
self propelled chip spreader			8.63			
channel (heat) iron			1.39			
Total Rate / Hour		£	**149.74**			
Concrete paving						
wheeled loader, 2.60 m²			52.32			
concrete paver, 6.0 m			70.57			
compaction slipform finisher			16.78			
Total Rate / Hour		£	**139.67**			

CLASS R: ROADS AND PAVINGS

Item	Gang hours	Labour £	Plant £	Material £	Unit	Total rate £
RESOURCES – PLANT – cont'd						
Road surface spraying						
tar sprayer, 100 litre			4.17			
Total Rate / Hour		£	**4.17**			
Road chippings						
deadweight rollers, 3 point, 10 t			15.20			
tar sprayer, 100 litre			4.17			
self propelled chip spreader			8.63			
channel (heat) iron			1.39			
Total Rate / Hour		£	**29.11**			
Cutting slabs						
compressor, 65 cfm			4.63			
12" disc cutter			1.30			
Total Rate / Hour		£	**5.93**			
Milling						
cold planer, 2.10 m			50.38			
wheeled loader, 2.60m#			52.32			
Total Rate/ Hour		£	**102.70**			
Heat planing						
heat planer, 4.5 m			72.82			
wheeled loader, 2.60 m#			52.32			
Total Rate/ Hour		£	**125.14**			
Rake and compact planed material						
deadweight roller, 3 point, 10t			15.20			
tractor with front bucket and integral 2 tool compressor			17.04			
channel (heat) iron			1.10			
Total Rate/ Hour		£	**33.35**			
Kerb laying						
backhoe JCB 3CX (25% of time)JCB 3CX (25% of time)			4.34			
12" stihl saw			1.08			
road forms			1.78			
TOTAL		£	**7.28**			
Path sub-base, bitmac and gravel laying						
backhoe JCB 3CX			17.35			
2 t dumper			4.34			
pedestrian roller Bomag BW 90S			4.31			
TOTAL		£	**26.00**			
Paviors and flagging						
2 t dumper (33% of time)			1.45			
Total Rate / Hour		£	**1.45**			
Traffic signs						
JCB 3CX backhoe - 50% of time			8.68			
125 cfm compressor - 50% of time			3.41			
compressor tools: hand held hammer drill - 50% of time			0.43			
compressor tools: clay spade - 50% of time			0.20			
compressor tools: extra 15 m hose - 50% of time			0.12			
8 t lorry with hiab lift - 50% of time			12.31			
Total Rate / Hour		£	**25.14**			

CLASS R: ROADS AND PAVINGS

Item	Gang hours	Labour £	Plant £	Material £	Unit	**Total rate £**
SUB-BASES, FLEXIBLE ROAD BASES AND SURFACING						
Granular material DTp specified type 1						
Sub-base; spread and graded						
75 mm deep	0.04	2.74	1.88	21.47	m³	**26.09**
100 mm deep	0.04	3.13	2.15	21.47	m³	**26.75**
150 mm deep	0.05	3.53	2.42	21.47	m³	**27.41**
200 mm deep	0.05	3.92	2.69	21.47	m³	**28.07**
Lean concrete DTp specified strength mix C20P/20 mm aggregate						
Sub-base; spread and graded						
100 mm deep	0.05	3.53	2.42	67.86	m³	**73.80**
200 mm deep	0.05	3.92	2.69	67.86	m³	**74.46**
Hardcore						
Sub-base; spread and graded						
100 mm deep	0.04	3.13	2.15	13.61	m³	**18.90**
150 mm deep	0.05	3.53	2.42	13.61	m³	**19.56**
200 mm deep	0.05	3.92	2.69	13.61	m³	**20.22**
Geotextiles						
refer to Class E						
Wet mix macadam; DTp clause 808						
Sub-base; spread and graded						
75 mm deep	0.04	2.74	2.85	20.96	m³	**26.55**
100 mm deep	0.04	3.13	2.96	20.96	m³	**27.05**
200 mm deep	0.05	3.92	3.12	20.96	m³	**27.99**
Dense Bitumen Macadam						
Road Base to DTp clause 903						
100 mm deep	0.02	2.81	2.99	6.05	m²	**11.85**
150 mm deep	0.03	3.51	3.74	9.07	m²	**16.33**
200 mm deep	0.03	4.22	4.50	12.09	m²	**20.81**
Base Course to DTp clause 906						
50 mm deep	0.02	2.11	2.25	2.55	m²	**6.91**
100 mm deep	0.02	2.81	2.99	5.10	m²	**10.91**
Wearing Course to DTp clause 912						
30 mm deep	0.01	1.41	1.50	2.17	m²	**5.08**
50 mm deep	0.02	2.11	2.25	3.62	m²	**7.98**
Bitumen Macadam						
Base Course to DTp clause 908						
35 mm deep	0.01	1.41	1.50	2.31	m²	**5.22**
70 mm deep	0.02	2.11	2.25	4.62	m²	**8.98**
Dense Tarmacadam						
Road Base to DTp clause 902						
50 mm deep	0.02	2.11	2.23	3.02	m²	**7.36**
100 mm deep	0.02	2.11	2.23	6.05	m²	**10.39**
Base Course to DTp clause 907						
60 mm deep	0.02	2.11	2.25	3.94	m²	**8.30**
80 mm deep	0.02	2.11	2.25	5.25	m²	**9.61**

CLASS R: ROADS AND PAVINGS

Item	Gang hours	Labour £	Plant £	Material £	Unit	**Total rate £**
SUB-BASES, FLEXIBLE ROAD BASES AND SURFACING – cont'd						
Dense Tar Surfacing						
Wearing Course to DTp clause 913						
30 mm deep	0.01	1.41	1.50	2.24	m²	**5.15**
50 mm deep	0.02	2.11	2.22	3.74	m²	**8.07**
Cold Asphalt						
Wearing Course to DTp clause 914						
15 mm deep	0.01	1.41	1.50	1.14	m²	**4.05**
30 mm deep	0.01	1.41	1.50	2.29	m²	**5.20**
Rolled Asphalt						
Base Course to DTp clause 905						
60 mm deep	0.02	2.11	2.25	3.96	m²	**8.31**
80 mm deep	0.02	2.11	2.25	5.28	m²	**9.63**
Wearing Course to DTp clause 911						
40 mm deep	0.02	2.11	2.25	3.59	m²	**7.94**
60 mm deep	0.02	2.11	2.25	5.38	m²	**9.73**
Slurry sealing; BS 434 class K3						
Sealing to DTp clause 918						
3 mm deep	0.02	0.22	0.06	1.10	m²	**1.39**
4 mm deep	0.02	0.22	0.06	1.52	m²	**1.80**
Coated chippings, 9 - 11 kg/m²						
Surface dressing to DTp clause 915						
6 mm nominal size	0.01	0.85	0.29	0.49	m²	**1.63**
8 mm nominal size	0.01	0.85	0.29	0.51	m²	**1.66**
10 mm nominal size	0.01	0.85	0.29	0.53	m²	**1.67**
12 mm nominal size	0.01	0.85	0.29	0.58	m²	**1.72**
Bituminous spray; BS 434 K1 - 40						
Tack coat to DTp clause 920						
large areas; over 20 m²	0.02	0.22	0.06	0.27	m²	**0.55**
small areas; under 20 m²	0.02	0.22	0.06	0.27	m²	**0.55**
Removal of flexible surface						
Trimming edges only of existing slabs, floors or similar surfaces (wet or dry); 6 mm cutting width						
50 mm deep	0.02	0.23	0.12	3.00	m	**3.34**
100 mm deep	0.03	0.34	0.19	3.14	m	**3.66**
Cutting existing slabs, floors or similar surfaces (wet or dry); 8 mm cutting width						
50 mm deep	0.03	0.28	0.14	3.00	m	**3.43**
100 mm deep	0.06	0.68	0.36	3.14	m	**4.17**
150 mm deep	0.08	0.90	0.47	3.32	m	**4.70**
Milling pavement (assumes disposal on site or re-use as fill but excludes transport if required)						
75 mm deep	0.03	3.05	2.77	-	m²	**5.82**
100 mm deep	0.04	4.07	3.70	-	m²	**7.76**
50 mm deep; scarifying surface	0.02	2.48	2.26	-	m²	**4.74**
75 mm deep; scarifying surface	0.04	4.18	3.80	-	m²	**7.98**
25 mm deep; heat planing for re-use	0.03	3.61	4.00	-	m²	**7.62**
50 mm deep; heat planing for re-use	0.06	6.32	7.01	-	m²	**13.33**

CLASS R: ROADS AND PAVINGS

Item	Gang hours	Labour £	Plant £	Material £	Unit	**Total rate £**
Raking over scarified or heat planed material; compacting with 10 t roller						
ne 50 mm deep	0.01	0.86	0.34	-	m²	**1.20**
CONCRETE PAVEMENTS						
The following unit costs are for jointed reinforced concrete slabs, laid in reasonable areas (over 200m²) by paver train/slipformer.						
Designed mix; cement to BS 12; grade C30, 20 mm aggregate						
Carriageway slabs of DTp Specified paving quality						
180 mm deep	0.02	1.69	2.10	13.79	m²	**17.58**
220 mm deep	0.02	2.03	2.51	16.86	m²	**21.40**
260 mm deep	0.02	2.48	3.07	19.92	m²	**25.48**
300 mm deep	0.03	2.82	3.49	22.99	m²	**29.30**
Fabric						
Steel fabric reinforcement to BS 4483						
Ref A142 nominal mass 2.22 kg	0.03	3.39	-	1.30	m²	**4.69**
Ref A252 nominal mass 3.95 kg	0.04	4.52	-	2.30	m²	**6.82**
Ref B385 nominal mass 4.53 kg	0.04	4.52	-	2.65	m²	**7.16**
Ref C636 nominal mass 5.55 kg	0.05	5.65	-	3.25	m²	**8.90**
Ref B503 nominal mass 5.93 kg	0.05	5.65	-	3.47	m²	**9.12**
Mild Steel bar reinforcement BS 4449						
Bars; supplied in bent and cut lengths						
6 mm nominal size	8.00	903.52	-	467.63	tonne	**1371.15**
8 mm nominal size	6.74	761.22	-	456.52	tonne	**1217.74**
10 mm nominal size	6.74	761.22	-	446.42	tonne	**1207.64**
12 mm nominal size	6.74	761.22	-	444.40	tonne	**1205.62**
16 mm nominal size	6.15	694.58	-	441.37	tonne	**1135.95**
High yield steel bar reinforcement BS 4449 or 4461						
Bars; supplied in bent and cut lengths						
6 mm nominal size	8.00	903.52	-	438.34	tonne	**1341.86**
8 mm nominal size	6.74	761.22	-	428.24	tonne	**1189.46**
10 mm nominal size	6.74	761.22	-	417.13	tonne	**1178.35**
12 mm nominal size	6.74	761.22	-	419.15	tonne	**1180.37**
16 mm nominal size	6.15	694.58	-	414.10	tonne	**1108.68**
Sheeting to prevent moisture loss						
Polyethylene sheeting; lapped joints; horizontal below concrete pavements						
250 micron	0.01	1.13	-	0.40	m²	**1.53**
500 micron	0.01	1.13	-	0.72	m²	**1.85**

CLASS R: ROADS AND PAVINGS

Item	Gang hours	Labour £	Plant £	Material £	Unit	Total rate £
JOINTS IN CONCRETE PAVEMENTS						
General						
Longitudinal joints						
180 mm deep	0.01	1.36	1.68	18.77	m	**21.81**
220 mm deep	0.01	1.36	1.68	19.79	m	**22.82**
260 mm deep	0.01	1.36	1.68	23.64	m	**26.67**
300 mm deep	0.01	1.36	1.68	25.01	m	**28.04**
Expansion joints						
180 mm deep	0.01	1.36	1.68	28.83	m	**31.86**
220 mm deep	0.01	1.36	1.68	34.05	m	**37.08**
260 mm deep	0.01	1.36	1.68	39.26	m	**42.29**
300 mm deep	0.01	1.36	1.68	40.29	m	**43.32**
Contraction joints						
180 mm deep	0.01	1.36	1.68	15.61	m	**18.64**
220 mm deep	0.01	1.36	1.68	16.67	m	**19.71**
260 mm deep	0.01	1.36	1.68	19.59	m	**22.62**
300 mm deep	0.01	1.36	1.68	21.88	m	**24.91**
Construction joints						
180 mm deep	0.01	1.36	1.68	9.32	m	**12.36**
220 mm deep	0.01	1.36	1.68	10.43	m	**13.46**
260 mm deep	0.01	1.36	1.68	11.47	m	**14.50**
300 mm deep	0.01	1.36	1.68	12.49	m	**15.53**
Open joints with filler						
ne 0.5 m; 10 mm flexcell joint filler	0.11	4.51	-	3.22	m	**7.73**
0.5 - 1.00 m; 10 mm flexcell joint filler	0.11	4.51	-	4.63	m	**9.14**
Joint sealants						
10 x 20 mm cold polysulphide sealant	0.14	5.74	-	2.90	m	**8.63**
20 x 20 mm cold polysulphide sealant	0.18	7.37	-	5.76	m	**13.14**
KERBS, CHANNELS AND EDGINGS						
Foundations to kerbs etc.						
Measurement Note: the following are shown separate from their associated kerb etc. to simplify the presentation of cost alternatives.						
Mass concrete						
200 x 100 mm	0.01	0.51	0.07	1.53	m	**2.10**
300 x 150 mm	0.02	0.77	0.10	3.49	m	**4.36**
450 x 150 mm	0.02	1.02	0.15	5.16	m	**6.33**
100 x 100 mm haunching, one side	0.01	0.26	0.04	0.37	m	**0.66**
Precast concrete kerbs; BS 7263:bedded, jointed and pointed in cement mortar						
Kerbs; bullnosed; splayed or half battered; straight or curved over 12 m radius						
125 x 150 mm	0.06	3.06	0.44	3.39	m	**6.90**
125 x 255 mm	0.07	3.57	0.51	4.33	m	**8.41**
150 x 305 mm	0.07	3.57	0.52	7.08	m	**11.18**
Kerbs; bullnosed; splayed or half battered; curved ne 12 m radius						
125 x 150 mm	0.07	3.32	0.47	3.39	m	**7.19**
125 x 255 mm	0.08	3.83	0.55	4.33	m	**8.70**
150 x 305 mm	0.08	3.83	0.56	7.08	m	**11.47**

CLASS R: ROADS AND PAVINGS

Item	Gang hours	Labour £	Plant £	Material £	Unit	**Total rate £**
Quadrants						
305 x 305 x 150 mm	0.08	4.09	0.59	6.95	nr	**11.63**
455 x 455 x 255 mm	0.10	5.11	0.74	9.39	nr	**15.24**
Drop kerbs						
125 x 255 mm	0.07	3.57	0.52	6.35	m	**10.44**
150 x 305 mm	0.07	3.57	0.52	13.37	m	**17.46**
Channel; straight or curved over 12 m radius						
255 x 125 mm	0.07	3.57	0.52	5.74	m	**9.84**
Channel; curved radius ne 12 m						
255 x 125 mm	0.07	3.57	0.52	5.74	m	**9.84**
Edging; straight or curved over 12 m radius						
50 x 150 mm	0.04	2.04	0.30	2.00	m	**4.34**
Edging; curved ne 12 m radius						
50 x 150 mm	0.05	2.30	0.33	2.00	m	**4.63**
Precast concrete drainage channels; Charcon Safeticurb; channels jointed with plastic rings and bedded; jointed and pointed in cement mortar						
Channel unit; straight; Type DBA/3						
250 x 254 mm; medium duty	0.08	3.83	0.56	32.97	m	**37.36**
305 x 305 mm; heavy duty	0.10	4.85	0.71	70.08	m	**75.64**
Precast concrete Ellis Trief safety kerb; bedded jointed and pointed in cement mortar						
Kerb; straight or curved over 12 m radius						
415 x 380 mm	0.23	11.49	1.61	57.65	m	**70.75**
Kerb; curved ne 12 m radius						
415 x 380 mm	0.25	12.77	1.80	57.65	m	**72.21**
Precast concrete combined kerb and drainage block Beany block system; bedded jointed and pointed in cement mortar						
Kerb; top block, shallow base unit, standard cover plate and frame						
straight or curved over 12 m radius	0.15	7.66	1.08	100.82	m	**109.56**
curved ne 12 m radius	0.20	10.21	1.43	140.86	m	**152.50**
Kerb; top block, standard base unit, standard cover plate and frame						
straight or curved over 12 m radius	0.15	7.66	1.09	106.32	m	**115.07**
curved ne 12 m radius	0.20	10.21	1.43	148.56	m	**160.20**
Kerb; top block, deep base unit, standard cover plate and frame						
straight or curved over 12 m radius	0.15	7.66	1.07	132.72	m	**141.45**
curved ne 12 m radius	0.20	10.21	1.41	185.52	m	**197.15**
base block depth tapers	0.10	5.30	0.72	55.72	m	**61.74**

CLASS R: ROADS AND PAVINGS

Item	Gang hours	Labour £	Plant £	Material £	Unit	Total rate £
KERBS, CHANNELS AND EDGINGS – cont'd						
Extruded asphalt kerbs to BS 5931; extruded and slip formed						
Kerb; straight or curved over 12 m radius						
75 mm kerb height	-	-	-	5.41	m	**5.41**
100 mm kerb height	-	-	-	7.91	m	**7.91**
125 mm kerb height	-	-	-	15.40	m	**15.40**
Channel; straight or curved over 12 m radius						
300 mm channel width	-	-	-	11.64	m	**11.64**
250 mm channel width	-	-	-	11.64	m	**11.64**
Kerb; curved to radius ne 12 m						
75 mm kerb height	-	-	-	10.81	m	**10.81**
100 mm kerb height	-	-	-	7.93	m	**7.93**
125 mm kerb height	-	-	-	10.39	m	**10.39**
Channel; curved to radius ne 12 m						
300 mm channel width	-	-	-	15.80	m	**15.80**
250 mm channel width	-	-	-	12.05	m	**12.05**
Extruded concrete; slip formed						
Kerb; straight or curved over 12 m radius						
100 mm kerb height	-	-	-	10.02	m	**10.02**
125 mm kerb height	-	-	-	11.86	m	**11.86**
Kerb; curved to radius ne 12 m						
100 mm kerb height	-	-	-	10.02	-	**10.02**
125 mm kerb height	-	-	-	10.48	-	**10.48**
LIGHT DUTY PAVEMENTS						
Sub-bases						
Measurement Note: the following are shown separate from their associated paving to simplify the presentation of cost alternatives.						
To paved area; sloping not exceeding 10 degrees to the horizontal						
100 mm thick sand	0.01	0.45	0.23	3.10	m²	**3.79**
150 mm thick sand	0.01	0.60	0.31	4.65	m²	**5.57**
100 mm thick gravel	0.01	0.45	0.23	2.45	m²	**3.14**
150 mm thick gravel	0.01	0.60	0.31	3.68	m²	**4.59**
100 mm thick hardcore	0.01	0.45	0.23	2.27	m²	**2.95**
150 mm thick hardcore	0.01	0.60	0.31	3.40	m²	**4.32**
100 mm thick concrete grade 20/20	0.02	1.05	0.55	7.59	m²	**9.19**
150 mm thick concrete grade 20/20	0.03	1.60	0.83	11.39	m²	**13.82**
Bitumen macadam surfacing; BS 4987; base course of 20 mm open graded aggregate to clause 2.6.1 tables 5-7; wearing course of 6 mm medium graded aggregate to clause 2.7.6 tables 32-33						
Paved area comprising base course 40 mm thick wearing course 20 mm thick						
sloping not exceeding 10 degrees to the horizontal	0.09	4.25	2.21	9.37	m²	**15.82**
sloping not exceeding 10 degrees to the horizontal; red additives	0.09	4.25	2.21	11.14	m²	**17.60**

CLASS R: ROADS AND PAVINGS

Item	Gang hours	Labour £	Plant £	Material £	Unit	**Total rate £**
sloping not exceeding 10 degrees to the horizontal; green additives	0.09	4.25	2.21	12.37	m²	**18.82**
sloping exceeding 10 degrees to the horizontal	0.10	4.75	2.47	9.37	m²	**16.59**
sloping exceeding 10 degrees to the horizontal; red additives	0.10	4.75	2.47	11.14	m²	**18.37**
sloping exceeding 10 degrees to the horizontal; green additives	0.10	4.75	2.47	12.37	m²	**19.59**
Granular base surfacing; Central Reserve Treatments Limestone, graded 10 mm down; laid and compacted						
Paved area 100 mm thick; surface sprayed twice with two coats of cold bituminous emulsion; blinded with 6 mm quartizite fine gravel						
sloping not exceeding 10 degrees to the horizontal	0.02	1.00	0.52	5.38	m²	**6.90**
Breedon plc Golden gravel; graded 13 mm to fines; rolled wet						
Paved area 50 mm thick; single layer						
sloping not exceeding 10 degrees to the horizontal	0.03	1.50	0.78	6.85	m²	**9.13**
Precast concrete flags; BS 7263; grey; bedding in cement mortar						
Paved area; sloping not exceeding 10 degrees to the horizontal						
900 x 600 x 63 mm	0.21	4.90	0.30	7.62	m²	**12.82**
900 x 600 x 50 mm	0.20	4.66	0.29	6.73	m²	**11.69**
600 x 600 x 63 mm	0.25	5.83	0.36	9.31	m²	**15.50**
600 x 600 x 50 mm	0.24	5.59	0.35	7.82	m²	**13.76**
600 x 450 x 50 mm	0.28	6.53	0.40	9.70	m²	**16.63**
Extra for coloured, 50 mm thick	-	-	-	3.84	m²	**3.84**
Precast concrete rectangular paving blocks; BS 6717; grey; bedding on 50 mm thick dry sharp sand; filling joints; excluding sub-base						
Paved area; sloping not exceeding 10 degrees to the horizontal						
200 x 100 x 80 mm thick	0.30	6.99	0.43	11.77	m²	**19.20**
200 x 100 x 80 mm thick; coloured blocks	0.30	6.99	0.43	12.49	m²	**19.91**
Brick paviors; bedding on 20 mm thick mortar; excluding sub-base						
Paved area; sloping not exceeding 10 degrees to the horizontal						
215 x 103 x 65 mm	0.30	6.99	0.43	19.95	m²	**27.37**
Granite setts; bedding on 25 mm cement mortar; excluding sub-base						
Paved area; sloping not exceeding 10 degrees to the horizontal						
to random pattern	0.90	20.98	1.30	49.27	m²	**71.55**
to specific pattern	1.20	27.97	1.74	49.27	m²	**78.97**

CLASS R: ROADS AND PAVINGS

Item	Gang hours	Labour £	Plant £	Material £	Unit	Total rate £
LIGHT DUTY PAVEMENTS – cont'd						
Cobble paving; 50 - 75 mm; bedding oin 25 mm cement mortar; filling joints; excluding sub-base						
Paved area; sloping not exceeding 10 degrees to the horizontal						
50 - 75 mm diameter cobbles	1.00	23.31	1.43	11.97	m²	**36.71**
ANCILLARIES						
Traffic signs						
In this section prices will vary depending upon the diagram configurations. The following are average costs of signs and bollards. Diagram numbers refer to the Traffic Signs Regulations and General Directions 1989 and the figure numbers refer to the Traffic Sign Manual.						
Examples of Prime Costs for Class 1 (High Intensity) traffic and road signs (ex works) for orders exceeding £600.						
600 x 450 mm	-	-	-	78.53	nr	**78.53**
600 mm diameter	-	-	-	98.62	nr	**98.62**
600 mm triangular	-	-	-	82.27	nr	**82.27**
500 x 500 mm	-	-	-	69.34	nr	**69.34**
450 x 450 mm	-	-	-	59.21	nr	**59.21**
450 x 300 mm	-	-	-	49.00	nr	**49.00**
1200 x 400 mm (CHEVRONS)	-	-	-	125.95	nr	**125.95**
Examples of Prime Costs for Class 21 (Engineering Grade) traffic and road signs (ex works) for orders exceeding £600.						
600 x 450 mm	-	-	-	63.61	nr	**63.61**
600 mm diameter	-	-	-	107.39	nr	**107.39**
600 mm triangular	-	-	-	89.59	nr	**89.59**
500 x 500 mm	-	-	-	56.25	nr	**56.25**
450 x 450 mm	-	-	-	47.73	nr	**47.73**
450 x 300 mm	-	-	-	41.12	nr	**41.12**
1200 x 400 mm (CHEVRONS)	-	-	-	137.16	nr	**137.16**
Standard reflectorised traffic signs						
Note: Unit costs do not include concrete foundations						
Standard one post signs; 600 x 450 mm type C1 signs						
fixed back to back to another sign (measured separately) with aluminium clips to existing post (measured separately)	0.04	2.15	1.01	82.46	-	**85.62**
Extra for fixing singly with aluminium clips	0.01	0.54	0.17	1.53	-	**2.24**
Extra for fixing singly with stainless steel clips	0.01	0.54	0.40	10.75	-	**11.69**
fixed back to back to another sign (measured separately) with stainless steel clips to one new						

CLASS R: ROADS AND PAVINGS

Item	Gang hours	Labour £	Plant £	Material £	Unit	**Total rate £**
76 mm diameter plastic coated steel posts						
1.75 m long	0.27	14.54	6.79	121.17	nr	**142.51**
Extra for fixing singly to one face only	0.01	0.54	0.17	-	nr	**0.71**
Extra for 76 mm diameter 1.75 m long aluminium post	0.02	1.08	0.37	11.44	nr	**12.89**
Extra for 76 mm diameter 3.5 m long plastic coated steel post	0.02	1.08	0.37	25.85	nr	**27.30**
Extra for 76 mm diameter 3.5 m long aluminium post	0.02	1.08	0.37	48.99	nr	**50.43**
Extra for excavation for post, in hard material	1.10	59.23	19.79	-	nr	**79.02**
Extra for single external illumination unit with fitted photo cell (excluding trenching and cabling); unit cost per face illuminated	0.33	17.77	5.93	62.26	nr	**85.96**
Standard two post signs; 1200 x 400 mm; signs fixed back to back to another sign (measured separately) with stainless steel clips to two new 76 mm diameter plastic coated steel posts	0.51	27.46	12.83	214.13	nr	**254.42**
Extra for fixing singly to one face only	0.02	1.08	0.37	-	-	**1.45**
Extra for two 76 mm diameter 1.75 m long aluminium posts	0.04	2.15	0.72	22.89	-	**25.76**
Extra for two 76 mm diameter 1.75 m long Plastic coated steel posts	0.04	2.15	0.72	51.70	-	**54.57**
Extra for two 76 mm diameter 3.5 m long aluminium posts	0.04	2.15	0.72	97.97	-	**100.84**
Extra for excavation for post, in hard material	1.10	59.23	19.79	-	-	**79.02**
Extra for single external illumination unit with fitted photo cell (including trenching and cabling); unit cost per face illuminated	0.58	31.23	10.44	87.21	-	**128.88**
Standard internally illuminated traffic signs						
Bollard with integral mould-in translucent graphics (excluding trenching and cabling)						
fixing to concrete base	0.48	25.85	12.07	146.00	nr	**183.92**
Special traffic signs						
Note: Unit costs do not include concrete foundations or trenching and cabling						
Externally illuminated reflectorised traffic signs manufactured to order						
special signs, surface area 1.50 m² on two 100 mm diameter steel posts	-	-	-	-	nr	**559.57**
special signs, surface area 4.00 m² on three 100 mm diameter steel posts	-	-	-	-	nr	**921.22**
Internally illuminated traffic signs manufactured to order						
special signs, surface area 0.25 m² on one new 76 mm diameter post	-	-	-	-	nr	**201.41**
special signs, surface area 0.75 m² on one new 100 mm diameter steel posts	-	-	-	-	nr	**282.24**
special signs, surface area 4.00 m² on four new 120 mm diameter steel posts	-	-	-	-	nr	**718.41**

CLASS R: ROADS AND PAVINGS

Item	Gang hours	Labour £	Plant £	Material £	Unit	**Total rate £**
ANCILLARIES – cont'd						
Signs on gantries						
Externally illuminated reflectorised signs						
1.50 m^2	1.78	96.12	56.00	186.88	nr	**339.00**
2.50 m^2	2.15	115.78	67.45	193.70	nr	**376.93**
3.00 m^2	3.07	165.32	96.31	204.62	nr	**466.25**
Internally illuminated sign with translucent optical reflective sheeting and remote light source						
0.75 m^2	1.56	84.01	48.94	1148.76	nr	**1281.71**
1.00 m^2	1.70	91.55	53.33	1531.67	nr	**1676.55**
1.50 m^2	2.41	129.78	75.61	2297.50	nr	**2502.89**
Existing signs						
Take from store and re-erect						
3.0 m high road sign	0.28	15.08	8.78	60.10	nr	**83.96**
road sign on two posts	0.50	26.93	15.69	120.20	nr	**162.81**
ROAD MARKINGS						
Surface markings; reflectorised white						
Letters and shapes						
triangles; 1.6 m high	-	-	-	-	nr	**7.57**
triangles; 2.0 m high	-	-	-	-	nr	**10.28**
triangles; 3.75 m high	-	-	-	-	nr	**13.52**
circles with enclosing arrows; 1.6 m diameter	-	-	-	-	nr	**54.08**
arrows; 4.0 m long; straight	-	-	-	-	nr	**21.63**
arrows; 4.0 m long; turning	-	-	-	-	nr	**21.63**
arrows; 6.0 m long; straight	-	-	-	-	nr	**27.04**
arrows; 6.0 m long; turning	-	-	-	-	nr	**27.04**
arrows; 6.0 m long; curved	-	-	-	-	nr	**27.04**
arrows; 6.0 m long; double headed	-	-	-	-	nr	**37.85**
arrows; 8.0 m long; double headed	-	-	-	-	nr	**54.08**
arrows; 16.0 m long; double headed	-	-	-	-	nr	**81.11**
arrows; 32.0 m long; double headed	-	-	-	-	nr	**108.15**
letters or numerals; 1.6 m high	-	-	-	-	nr	**7.04**
letters or numerals; 2.0 m high	-	-	-	-	nr	**10.28**
letters or numerals; 3.75 m high	-	-	-	-	nr	**17.85**
Continuous lines						
150 mm wide	-	-	-	-	m	**0.81**
200 mm wide	-	-	-	-	m	**1.08**
Intermittent lines						
60 mm wide; 0.60 m line and 0.60 m gap	-	-	-	-	m	**0.65**
100 mm wide; 1.0 m line and 5.0 m gap	-	-	-	-	m	**0.65**
100 mm wide; 2.0 m line and 7.0 m gap	-	-	-	-	m	**0.65**
100 mm wide; 4.0 m line and 2.0 m gap	-	-	-	-	m	**0.65**
100 mm wide; 6.0 m line and 3.0 m gap	-	-	-	-	m	**0.65**
150 mm wide; 1.0 m line and 5.0 m gap	-	-	-	-	m	**0.98**
150 mm wide; 6.0 m line and 3.0 m gap	-	-	-	-	m	**0.98**
150 mm wide; 0.60 m line and 0.30 m gap	-	-	-	-	m	**0.98**
200 mm wide; 0.60 m line and 0.30 m gap	-	-	-	-	m	**1.30**
200 mm wide; 1.0 m line and 1.0 m gap	-	-	-	-	m	**1.30**

CLASS R: ROADS AND PAVINGS

Item	Gang hours	Labour £	Plant £	Material £	Unit	Total rate £
Surface markings; reflectorised yellow						
Continuous lines						
100 mm wide	-	-	-	-	m	**0.55**
150 mm wide	-	-	-	-	m	**0.81**
Intermittent lines						
kerb marking; 0.25 m long	-	-	-	-	nr	**0.55**
Surface markings; thermoplastic screed or spray						
Note: Unit costs based upon new road with clean surface closed to traffic.						
Continuous line in reflectorised white						
150 mm wide	-	-	-	-	m	**0.81**
200 mm wide	-	-	-	-	m	**1.08**
Continuous line in reflectorised yellow						
100 mm wide	-	-	-	-	m	**0.55**
150 mm wide	-	-	-	-	m	**0.81**
Intermittent line in reflectorised white						
60 mm wide with 0.60 m line and 0.60 m gap	-	-	-	-	m	**0.65**
100 mm wide with 1.0 m line and 5.0 m gap	-	-	-	-	m	**0.65**
100 mm wide with 2.0 m line and 7.0 m gap	-	-	-	-	m	**0.65**
100 mm wide with 4.0 m line and 2.0 m gap	-	-	-	-	m	**0.65**
100 mm wide with 6.0 m line and 3.0 m gap	-	-	-	-	m	**0.65**
150 mm wide with 1.0 m line and 5.0 m gap	-	-	-	-	m	**0.98**
150 mm wide with 6.0 m line and 3.0 m gap	-	-	-	-	m	**0.98**
150 mm wide with 0.6 m line and 0.3 m gap	-	-	-	-	m	**0.98**
200 mm wide with 0.6 m line and 0.3 m gap	-	-	-	-	m	**1.30**
200 mm wide with 1.0 m line and 1.0 m gap	-	-	-	-	m	**1.30**
Ancillary line in reflectorised white						
150 mm wide in hatched areas	-	-	-	-	m	**0.81**
200 mm wide in hatched areas	-	-	-	-	m	**1.30**
Ancillary line in reflectorised yellow						
150 mm wide in hatched areas	-	-	-	-	m	**0.81**
Triangles in reflectorised white						
1.6 m high	-	-	-	-	nr	**7.57**
2.0 m high	-	-	-	-	nr	**10.28**
3.75 m high	-	-	-	-	nr	**13.52**
Circles with enclosing arrows in reflectorised white						
1.6 m diameter	-	-	-	-	nr	**54.08**
Arrows in reflectorised white						
4.0 m long straight or turning	-	-	-	-	nr	**21.63**
6.0 m long straight or turning	-	-	-	-	nr	**27.04**
6.0 m long curved	-	-	-	-	nr	**27.04**
6.0 m long double headed	-	-	-	-	nr	**37.85**
8.0 m long double headed	-	-	-	-	nr	**54.08**
16.0 m long double headed	-	-	-	-	nr	**81.11**
32.0 m long double headed	-	-	-	-	nr	**108.15**
Kerb markings in yellow						
250 mm long	-	-	-	-	nr	**0.55**
Letters or numerals in reflectorised white						
1.6 m high	-	-	-	-	nr	**7.04**
2.0 m high	-	-	-	-	nr	**10.28**
3.75 m high	-	-	-	-	nr	**17.85**

CLASS R: ROADS AND PAVINGS

Item	Gang hours	Labour £	Plant £	Material £	Unit	Total rate £
ROAD MARKINGS – cont'd						
Surface markings; Verynyl strip markings Note: Unit costs based upon new road with clean surface closed to traffic.						
Verynyl' strip markings (pedestrian crossings and similar locations)						
200 mm wide line	-	-	-	-	m	**7.09**
600 x 300 mm single stud tile	-	-	-	-	nr	**12.11**
Removal of thermoplastic screed or spray markings						
Removal of existing reflectorised thermoplastic markings						
100 mm wide line	-	-	-	-	m	**1.05**
150 mm wide line	-	-	-	-	m	**1.57**
200 mm wide line	-	-	-	-	m	**2.10**
arrow or letter ne 6.0 m long	-	-	-	-	nr	**16.27**
arrow or letter 6.0 - 16.00 m long	-	-	-	-	nr	**68.25**
REFLECTING ROAD STUDS						
100 x 100 mm square bi-directional reflecting road studs with amber corner cube reflectors	-	-	-	-	nr	**5.25**
140 x 254 mm rectangular one way reflecting road studs with red catseye reflectors	-	-	-	-	nr	**11.50**
140 x 254 mm rectangular one way reflecting road studs with green catseye reflectors	-	-	-	-	nr	**11.50**
140 x 254 mm rectangular bi-directional reflecting road studs with white catseye reflectors	-	-	-	-	nr	**12.07**
140 x 254 mm rectangular bi-directional reflecting road studs with amber catseye reflectors	-	-	-	-	nr	**12.07**
140 x 254 mm rectangular bi-directional reflecting road stud without catseye reflectors	-	-	-	-	nr	**7.88**
REMOVAL OF ROAD STUDS						
Removal of road studs						
100 x 100 mm corner cube type	-	-	-	-	nr	**1.05**
140 x 254 mm cateye type	-	-	-	-	nr	**3.36**
REMOVAL FROM STORE AND REFIX ROAD STUD						
General						
Remove from store and re-install 100 x 100 mm square bi-directional reflecting road stud with corner cube reflectors	-	-	-	-	nr	**2.63**
Remove from store and re-install 140 x 254 mm rectangular one way reflecting road stud with catseye reflectors	-	-	-	-	nr	**6.30**

CLASS R: ROADS AND PAVINGS

Item	Gang hours	Labour £	Plant £	Material £	Unit	**Total rate £**
TRAFFIC SIGNAL INSTALLATIONS						
Traffic signal installation is carried out exclusively by specialist contractors, although certain items are dealt with by the main contractor or a sub-contractor. The following of signal pedestals, loop detector unit pedestals, controller unit boxes and cable connection pillars						
Installation of signal pedestals, loop detector unit pedestals, controller unit boxes and cable connection pillars						
signal pedestal	-	-	-	-	nr	**27.17**
loop detector unit pedestral	-	-	-	-	nr	**16.82**
Excavate trench for traffic signal cable, depth ne 1.50 m; supports, backfilling						
450 mm wide	-	-	-	-	m	**5.17**
Extra for excavating in hard material	-	-	-	-	m^3	**28.46**
Saw cutting grooves in pavement for detector loops and feeder cables; seal with hot bitumen sealant after installation						
25 mm deep	-	-	-	-	m	**4.52**

CLASS S: RAIL TRACK

Item	Gang hours	Labour £	Plant £	Material £	Unit	Total rate £
NOTES						
Generally The following unit costs are for guidance only. For more reliable estimates it is advisable to seek the advice of a Specialist Contractor. These rates are for the supply and laying of track other than in connection with the Permanent Way.						
Permanent Way The following rates would not reflect work carried out on the existing public track infrastructure (Permanent Way), which tends to be more costly not merely due to differences in technology and the level of specification and control standards, but also due to a number of logistical factors, such as: • access to the works for personnel, plant and machinery would be via approved access points to the rail followed by travel along the rail to the work area; this calls for the use of additional and expensive transport plant as well as reducing the effective shift time of the works gang • effect of track possession periods will dictate when the work can be carried out and could well force night-time or weekend working and perhaps severely reducing the effective shift hours where couipled to long travel to work distances and the need to clear away before the resumption of traffic. • the labour gang will be composed of more highly paid personnel, reflecting the additional training received; in addition there may well be additional gang members acting as look-outs; this could add 30% to the gang rates shown below • plant will tend to cost more, especially if the circumstances of the work call for rail/road plant; this could add 20 % to the gang rates shown						
Possession costs Where the contractor's work is on, over or poses a risk to the safety of the railway, then the contractor normally applies for possession of the track. During the period for which the contractor is given possession, rail traffic stops . Possessions of the Operational Safety Zone may well be fragmented rather than a single continuous period, dependant upon windows in the pattern of rail traffic Costs for working alongside an operational rail system are high, the need for safety demanding a high degree of supervision, look-outs, the induction of labour gangs and may involve temporary works such as safety barriers.						

CLASS S: RAIL TRACK

Item	Gang hours	Labour £	Plant £	Material £	Unit	Total rate £
TRACK FOUNDATIONS						
Imported crushed granite						
Bottom ballast	-	-	-	-	m³	**47.44**
Top ballast	-	-	-	-	m³	**60.25**
Imported granular material						
Blankets; 150 mm thick	-	-	-	-	m²	**6.09**
Imported sand						
Blinding; 100 mm thick	-	-	-	-	m²	**3.44**
Polythene sheeting						
Waterproof membrane; 1200 gauge	-	-	-	-	m²	**2.00**
LIFTING, PACKING AND SLEWING						
Maximum distance of slew 300 mm; maximum lift 100 mm; no extra ballast allowed						
Bullhead rail track; fishplated; timber sleepers	-	-	-	-	m	**16.91**
Bullhead rail track; fishplated; concrete sleepers	-	-	-	-	m	**18.57**
Bullhead rail track with turnout; timber sleepers	-	-	-	-	nr	**532.46**
Flat bottom rail track; welded; timber sleepers	-	-	-	-	m	**16.91**
Flat bottom rail track; welded; concrete sleepers	-	-	-	-	m	**15.38**
Flat bottom rail track with turnout; concrete sleepers	-	-	-	-	nr	**532.46**
Buffer stops	-	-	-	-	nr	**187.48**
TAKING UP						
Taking up; dismantling into individual components; sorting; storing on site where directed						
Bullhead or flat bottom rails						
plain track; fishplated; timber sleepers	-	-	-	-	m	**6.98**
plain track; fishplated; concrete sleepers	-	-	-	-	m	**9.10**
plain track; welded; timber sleepers	-	-	-	-	m	**8.60**
plain track; welded; concrete sleepers	-	-	-	-	m	**11.32**
turnouts; fishplated; concrete sleepers	-	-	-	-	nr	**444.90**
diamond crossings; fishplated; timber sleepers	-	-	-	-	nr	**396.86**
Dock and crane rails						
plain track; welded; base plates	-	-	-	-	m	**9.36**
turnouts; welded; base plates	-	-	-	-	nr	**257.58**
diamonds; welded; base plates	-	-	-	-	nr	**229.98**
Check and guard rails						
plain track; fishplated	-	-	-	-	m	**3.32**
Conductor rails						
plain track; fishplated	-	-	-	-	m	**3.46**
Sundries						
buffer stops	-	-	-	-	nr	**95.28**
retarders	-	-	-	-	nr	**29.58**
wheel stops	-	-	-	-	nr	**19.72**
lubricators	-	-	-	-	nr	**36.14**
switch heaters	-	-	-	-	nr	**23.00**
switch levers	-	-	-	-	nr	**9.86**

CLASS S: RAIL TRACK

Item	Gang hours	Labour £	Plant £	Material £	Unit	Total rate £
SUPPLYING (STANDARD GAUGE TRACK)						
Supplying						
Bullhead rails; BS 11; delivered in standard 18.288 m lengths						
BS95R section, 47 kg/m; for jointed track	-	-	-	-	tonne	**864.42**
BS95R section, 47 kg/m; for welded track	-	-	-	-	tonne	**864.42**
Flat bottom rails; BS 11; delivered in standard 18.288 m lengths						
BS113'A' section, 56 kg/m; for jointed track	-	-	-	-	tonne	**839.75**
BS113'A' section, 56 kg/m; for welded track	-	-	-	-	tonne	**839.75**
Extra for curved rails to form super elevation; radius over 600 m	-	-	-	-	%	**18.00**
Check and guard rails; BS 11; delivered in standard 18.288 m lengths; flange planed to allow 50 mm free wheel clearance						
BS113'A' section, 56 kg/m; for bolting	-	-	-	-	tonne	**1041.13**
Conductor rails; BS 11; delivered in standard 18.288 m lengths						
BS113'A' section, 56 kg/m; for bolting	-	-	-	-	tonne	**1041.13**
Twist rails; BS 11; delivered in standard 18.288 m lengths						
BS113'A' section, 56 kg/m; for bolting	-	-	-	-	tonne	**1041.13**
Sleepers; bitumen saturated French Maritime pine						
2600 x 250 x 130 mm	-	-	-	-	nr	**29.84**
Sleepers; bitumen saturated Douglas fir						
2600 x 250 x 130 mm	-	-	-	-	nr	**46.15**
Sleepers; prestressed concrete						
2525 x 264 x 204 mm; BR type F27, Pandrol inserts	-	-	-	-	nr	**30.52**
Fittings						
Cast iron chairs complete with chair screws, plastic ferrules, spring steels and keys; BR type S1	-	-	-	-	nr	**27.77**
Cast iron chairs complete with chair screws, plastic ferrules, spring steels and keys; BR type CC	-	-	-	-	nr	**51.38**
Cast iron chairs complete with resilient pad, chair screws, ferrules, rail clips and nylon insulators; BR type PAN 6	-	-	-	-	nr	**37.81**
Cast iron chairs complete with resilient pad, chair screws, ferrules, rail clips and nylon insulators; BR type VN	-	-	-	-	nr	**38.52**
Cast iron chairs complete with resilient pad, chair screws, ferrules, rail clips and nylon insulators; BR type C	-	-	-	-	nr	**49.03**
Pandrol rail clips and nylon insulator	-	-	-	-	nr	**2.81**
plain fishplates; for BS95R section rail, skirted pattern; sets of two; complete with fishbolts, nuts and washers	-	-	-	-	nr	**44.21**
plain fishplates; for BS95R section rail, joggled pattern; sets of two; complete with fishbolts, nuts and washers	-	-	-	-	nr	**104.30**
plain fishplates; for BS 113 'A' section rail, shallow section; sets of two; complete with fishbolts, nuts and washers	-	-	-	-	nr	**50.39**

CLASS S: RAIL TRACK

Item	Gang hours	Labour £	Plant £	Material £	Unit	**Total rate £**
insulated fishplates; for BS95R section rail, steel billet pattern; sets of two; complete with high tensile steel bolts, nuts and washers	-	-	-	-	nr	**139.61**
insulated fishplates; BS95R section rail, steel billet pattern; sets of two; complete with high tensile steel bolts, nuts and washers	-	-	-	-	nr	**127.56**
cast iron spacer block between running and guard rails; for BS95R section rail; M25 x 220 mm bolt, nut and washers	-	-	-	-	nr	**12.94**
cast iron spacer block between running and guard rails; for BS 113 'A' section rail; M25 x 220 mm bolt, nut and washers	-	-	-	-	nr	**24.90**
Turnouts; complete with closures, check rails, fittings, timber sleepers						
Type B8; BS 95R bullhead rail	-	-	-	-	nr	**14209.90**
Type C10; BS 95R bullhead rail	-	-	-	-	nr	**15776.59**
Type Bx8; BS 113 'A' section flat bottom rail	-	-	-	-	nr	**23633.11**
Type Cv9.25; BS 113 'A' section flat bottom rail	-	-	-	-	nr	**25876.33**
Diamond crossings; complete with closures, check rails, fittings, timber sleepers						
RT standard design, angle 1 in 4; BS95R bullhead rail	-	-	-	-	nr	**90215.33**
RT standard design, angle 1 in 4; BS 113 'A' section flat bottom rail	-	-	-	-	nr	**97519.36**
Sundries						
buffer stops; single raker, steel rail and timber; 2 tonnes approximate weight	-	-	-	-	nr	**2234.58**
buffer stops; double raker, steel rail and timber; 2.5 tonnes approximate weight	-	-	-	-	nr	**2454.38**
wheel stops; steel; 100 kg approximate weight	-	-	-	-	nr	**124.13**
lubricators; single rail	-	-	-	-	nr	**1633.86**
lubricators; double rail	-	-	-	-	nr	**1738.84**
switch levers; upright pattern	-	-	-	-	nr	**330.20**
switch levers; flush type	-	-	-	-	nr	**627.85**
LAYING (STANDARD GAUGE TRACK)						
Laying						
Bullhead rails; jointed with fishplates; softwood sleepers						
plain track	-	-	-	-	m	**31.93**
form curve in plain track, radius ne 300 m	-	-	-	-	m	**9.05**
form curve in plain track, radius over 300 m	-	-	-	-	m	**12.01**
turnouts; standard, type B8	-	-	-	-	nr	**2254.50**
turnouts; standard, type C10	-	-	-	-	nr	**2724.95**
diamond crossings; standard	-	-	-	-	nr	**1968.61**
welded joints; alumino-thermic welding including refractory mould	-	-	-	-	nr	**163.92**
spot re-sleepering	-	-	-	-	nr	**81.96**
Bullhead rails; jointed with fishplates; concrete sleepers						
plain track	-	-	-	-	m	**28.40**
form curve in plain track, radius ne 300 m	-	-	-	-	m	**8.45**
form curve in plain track, radius over 300 m	-	-	-	-	m	**11.03**
turnouts; standard, type B8	-	-	-	-	nr	**2254.50**
turnouts; standard, type C10	-	-	-	-	nr	**2724.95**
diamond crossings; standard	-	-	-	-	nr	**1968.61**
welded joints; alumino-thermic welding	-	-	-	-	nr	**163.92**
spot re-sleepering	-	-	-	-	nr	**81.96**

CLASS S: RAIL TRACK

Item	Gang hours	Labour £	Plant £	Material £	Unit	Total rate £
LAYING (STANDARD GAUGE TRACK) – cont'd						
Laying – cont'd						
Bullhead rails; welded joints; softwood sleepers						
plain track	-	-	-	-	m	**49.75**
form curve in plain track, radius ne 300 m	-	-	-	-	m	**9.41**
form curve in plain track, radius over 300 m	-	-	-	-	m	**8.86**
turnouts; standard, type B8	-	-	-	-	nr	**2254.50**
turnouts; standard, type C10	-	-	-	-	nr	**2724.95**
diamond crossings; standard	-	-	-	-	nr	**1968.61**
spot re-sleepering	-	-	-	-	nr	**81.96**
Bullhead rails; welded joints; concrete sleepers						
plain track	-	-	-	-	m	**39.98**
form curve in plain track, radius ne 300 m	-	-	-	-	m	**9.05**
form curve in plain track, radius over 300 m	-	-	-	-	m	**11.58**
turnouts; standard, type B8	-	-	-	-	nr	**2254.50**
turnouts; standard, type C10	-	-	-	-	nr	**2724.95**
diamond crossings; standard	-	-	-	-	nr	**1968.61**
spot re-sleepering	-	-	-	-	nr	**81.96**
Flat bottom rails; jointed with fishplates; softwood sleepers						
plain track	-	-	-	-	m	**45.05**
form curve in plain track, radius not exceeding 300 m	-	-	-	-	m	**10.49**
form curve in plain track, radius exceeding 300 m	-	-	-	-	m	**9.95**
turnouts; standard, type Bv8	-	-	-	-	nr	**3090.41**
turnouts; standard, type Cv9.25	-	-	-	-	nr	**3517.44**
diamond crossings; standard	-	-	-	-	nr	**2337.71**
welded joints; alumino-thermic welding including refractory mould	-	-	-	-	nr	**174.23**
spot re-sleepering	-	-	-	-	nr	**81.96**
Flat bottom rails; jointed with fishplates; concrete sleepers						
plain track	-	-	-	-	m	**47.05**
form curve in plain track, radius ne 300 m	-	-	-	-	m	**10.49**
form curve in plain track, radius over 300 m	-	-	-	-	m	**9.95**
turnouts; standard, type Bv8	-	-	-	-	nr	**3090.41**
turnouts; standard, type Cv9.25	-	-	-	-	nr	**3517.44**
diamond crossings; standard	-	-	-	-	nr	**2337.71**
welded joints; alumino-thermic welding including refractory mould	-	-	-	-	nr	**168.23**
spot re-sleepering	-	-	-	-	nr	**81.96**
Flat bottom rails; welded joints; softwood sleepers						
plain track	-	-	-	-	m	**63.32**
form curve in plain track, radius ne 300 m	-	-	-	-	m	**11.23**
form curve in plain track, radius over 300 m	-	-	-	-	m	**10.32**
turnouts; standard, type Bv8	-	-	-	-	nr	**3090.41**
turnouts; standard, type Cv9.25	-	-	-	-	nr	**3517.44**
diamond crossings; standard	-	-	-	-	nr	**2337.71**
spot re-sleepering	-	-	-	-	nr	**81.96**

CLASS S: RAIL TRACK

Item	Gang hours	Labour £	Plant £	Material £	Unit	**Total rate £**
Flat bottom rails; welded joints; concrete sleepers						
plain track	-	-	-	-	m	**58.93**
form curve in plain track, radius ne 300 m	-	-	-	-	m	**10.99**
form curve in plain track, radius over 300 m	-	-	-	-	m	**10.07**
turnouts; standard, type Bv8	-	-	-	-	nr	**3236.20**
turnouts; standard, type Cv9.25	-	-	-	-	nr	**3683.35**
diamond crossings; standard	-	-	-	-	nr	**2447.99**
spot re-sleepering	-	-	-	-	nr	**81.96**
Check rails, flat bottom; jointed with fishplates						
rail	-	-	-	-	m	**8.34**
welded joints; alumino-thermic welding including refractory mould	-	-	-	-	nr	**182.46**
Guard rails, bullhead; jointed with fishplates						
rail	-	-	-	-	m	**9.10**
welded joints; alumino-thermic welding including refractory mould	-	-	-	-	nr	**182.46**
Guard rails, flat bottom; jointed with fishplates						
rail	-	-	-	-	m	**8.34**
welded joints; alumino-thermic welding including refractory mould	-	-	-	-	nr	**182.46**
Conductor rails, bullhead; jointed with fishplates						
rail	-	-	-	-	m	**6.65**
welded joints; alumino-thermic welding including refractory mould	-	-	-	-	nr	**182.46**
Sundries						
buffer stops; single raker, steel rail and timber; 2 tonnes approximate weight	-	-	-	-	nr	**263.35**
buffer stops; double raker, steel rail and timber; 2.5 tonnes approximate weight	-	-	-	-	nr	**322.10**
wheel stops; steel; 100 kg approximate weight	-	-	-	-	nr	**98.52**
lubricators; single rail	-	-	-	-	nr	**227.36**
lubricators; double rail	-	-	-	-	nr	**303.16**
switch levers; upright pattern	-	-	-	-	nr	**94.73**
switch levers; flush type	-	-	-	-	nr	**94.73**
conductor rail guard boards	-	-	-	-	m	**13.27**
DECAUVILLE TRACK						
Supplying						
Dock and crane rails; for welded track						
section 56 crane rail; 12.2 m lengths	-	-	-	-	nr	**1293.10**
section 101 crane rail; 9.144 m lengths	-	-	-	-	nr	**1119.14**
Fittings						
20 mm mild steel sole plate 400 mm wide; drilled with two bolt holes at 1200 mm centres	-	-	-	-	m	**59.78**
M20 x 250 mm holding down bolt, nut and washers	-	-	-	-	nr	**2.54**
rail clips, spring type, adjustable; complete with M20 x 60 mm stud welded to sole plate	-	-	-	-	nr	**14.66**
Sundries						
wheel stops; 200 kg each	-	-	-	-	nr	**170.51**
Laying						
Crane rails, section 56; continuous sole plate; welded						
plain track	-	-	-	-	m	**50.23**
form curve in plain track, radius ne 300 m	-	-	-	-	m	**9.32**

CLASS S: RAIL TRACK

Item	Gang hours	Labour £	Plant £	Material £	Unit	**Total rate £**
DECAUVILLE TRACK – cont'd						
Laying – cont'd						
Crane rails, section 101; continuous sole plate; welded						
plain track	-	-	-	-	m	**63.61**
form curve in plain track, radius ne 300 m	-	-	-	-	m	**9.32**
Sundries						
wheel stops	-	-	-	-	nr	**93.30**

CLASS T: TUNNELS

Item	Gang hours	Labour £	Plant £	Material £	Unit	Total rate £
NOTES						

Notes

There are so many factors, apart from design considerations, which influence the cost of tunnelling that it is only possible to give guide prices for a sample of the work involved. For more reliable estimates it is advisable to seek the advice of a Specialist Contractor.

The main cost considerations are :

- the nature of the ground
- size of tunnel
- length of drive
- depth below surface
- anticipated overbreak
- support of face and roof of tunnel (rock bolting etc.)
- necessity for pre-grouting
- ventilation
- presence of water
- use of compressed air working

The following rates for mass concrete work cast in primary and secondary linings to tunnels and access shafts are based on a 5.0 m depth of shaft and 15.0 m head of tunnel

Apply the following factors for differing depths and lengths :

HEAD LENGTH :	15 m	30 m	60 m	90 m
Shaft depth 5m	+0%	+10%	+20%	+32½%
Shaft depth 10m	+5%	+12½%	+27½%	+35%
Shaft depth 15m	+10%	+17½%	+32½%	+40%
Shaft depth 20m	+15%	+20%	+37½%	+42½%

Item	Gang hours	Labour £	Plant £	Material £	Unit	Total rate £
EXCAVATION						
Excavating tunnels in rock						
1.5 m diameter	-	-	-	-	m³	**449.56**
3.0 m diameter	-	-	-	-	m³	**277.57**
Excavating tunnels in soft material						
1.5 m diameter	-	-	-	-	m³	**204.59**
3.0 m diameter	-	-	-	-	m³	**111.97**
Excavating shafts in rock						
3.0 m diameter	-	-	-	-	m³	**163.21**
4.5 m diameter	-	-	-	-	m³	**137.63**
Excavating shafts in soft material						
3.0 m diameter	-	-	-	-	m³	**96.10**
4.5 m diameter	-	-	-	-	m³	**81.83**
Excavating other cavities in rock						
1.5 m diameter	-	-	-	-	m³	**449.56**
3.0 m diameter	-	-	-	-	m³	**277.57**
Excavating other cavities in soft material						
1.5 m diameter	-	-	-	-	m³	**204.59**
3.0 m diameter	-	-	-	-	m³	**111.97**
Excavating surfaces in rock	-	-	-	-	m²	**16.09**
Excavating surfaces in soft material	-	-	-	-	m²	**16.09**

CLASS T: TUNNELS

Item	Gang hours	Labour £	Plant £	Material £	Unit	Total rate £
IN-SITU LINING TO TUNNELS						
Notes						
The following rates for mass concrete work cast in primary and secondary linings to tunnels and access shafts are based on a 5.0 m depth of shaft and 15.0 m head of tunnel						
See above for additions for differing shaft depths and tunnel lengths.						
Mass concrete; grade C30, 20 mm aggregate						
Cast primary lining to tunnels						
1.5 m diameter	-	-	-	-	m³	**258.92**
3.0 m diameter	-	-	-	-	m³	**219.67**
Secondary lining to tunnels						
1.5 m diameter	-	-	-	-	m³	**299.92**
3.0 m diameter	-	-	-	-	m³	**235.15**
Formwork; rough finish						
Tunnel lining						
1.5 m diameter	-	-	-	-	m²	**43.52**
3.0 m diameter	-	-	-	-	m²	**43.52**
IN-SITU LINING TO ACCESS SHAFTS						
Notes						
The following rates for mass concrete work cast in primary and secondary linings to tunnels and access shafts are based on a 5.0 m depth of shaft and 15.0 m head of tunnel						
See above for additions for differing shaft depths and tunnel lengths.						
Mass concrete; grade C30, 20 mm aggregate						
Secondary linings to shafts						
3.0 m int diameter	-	-	-	-	m³	**234.84**
4.5 m int diameter	-	-	-	-	m³	**229.19**
Cast primary lining to shafts						
3.0 m int diameter	-	-	-	-	m³	**251.91**
4.5 m int diameter	-	-	-	-	m³	**239.96**
Formwork; rough finish						
Shaft lining						
3.0 m int diameter	-	-	-	-	m²	**66.16**
4.5 m int diameter	-	-	-	-	m²	**45.10**
IN-SITU LINING TO OTHER CAVITIES						
Notes						
The following rates for mass concrete work cast in primary and secondary linings to tunnels and access shafts are based on a 5.0 m depth of shaft and 15.0 m head of tunnel						
See above for additions for differing shaft depths and tunnel lengths.						

CLASS T: TUNNELS

Item	Gang hours	Labour £	Plant £	Material £	Unit	Total rate £
Mass concrete; grade C30, 20 mm aggregate						
Cast primary lining to other cavities						
1.5 m int diameter	-	-	-	-	m³	**258.92**
3.0 m int diameter	-	-	-	-	m³	**219.67**
Secondary linings to other cavities						
1.5 m int diameter	-	-	-	-	m³	**299.92**
3.0 m int diameter	-	-	-	-	m³	**235.15**
Formwork; rough finish						
Other cavities lining						
1.5 m int diameter	-	-	-	-	m²	**43.52**
3.0 m int diameter	-	-	-	-	m²	**43.52**
PREFORMED SEGMENTAL LININGS TO TUNNELS						
Precast concrete bolted rings; flanged; including packing; guide price/ring based upon standard bolted concrete segmental rings; ring width 610mm						
Linings to tunnels						
1.5 m int diameter; 6 segments , maximum piece weight 1139 kg	-	-	-	-	nr	**483.17**
3.0 m int diameter; 7 segments, maximum piece weight 247 kg	-	-	-	-	nr	**867.75**
Lining ancillaries; bitumen impregnated fibreboard						
Parallel circumferential packing						
1.5 m int diameter	-	-	-	-	nr	**4.85**
3.0 m int diameter	-	-	-	-	nr	**9.70**
Lining ancillaries; PC4AF caulking compound						
Caulking						
1.5 m int diameter	-	-	-	-	m	**7.99**
3.0 m int diameter	-	-	-	-	m	**7.99**
PREFORMED SEGMENTAL LININGS TO SHAFTS						
Precast concrete bolted rings; flanged; including packing; guide price/ring based upon standard bolted concrete segmental rings; ring width 610mm						
Linings to shafts						
3.0 m int diameter; 7 segments, maximum piece weight 247 kg	-	-	-	-	nr	**695.51**
Lining ancillaries; bitumen impregnated fibreboard						
Parallel circumferential packing						
3.0 m int diameter	-	-	-	-	nr	**9.70**
Lining ancillaries; PC4AF caulking compound						
Caulking						
3.0 m int diameter	-	-	-	-	m	**7.99**

CLASS T: TUNNELS

Item	Gang hours	Labour £	Plant £	Material £	Unit	**Total rate £**
PREFORMED SEGMENTAL LININGS TO OTHER CAVITIES						
Precast concrete bolted rings; flanged; including packing; guide price/ring based upon standard bolted concrete segmental rings; ring width 610mm						
Linings to tunnels						
1.5 m int diameter; 6 segments, maximum piece weigth 139 kg	-	-	-	-	nr	**483.17**
3.0 m int diameter; 7 segments , maximum piece weight 247 kg	-	-	-	-	nr	**867.75**
Linings to shafts						
3.0 m int diameter; 7 segments, maximum piece weight 247 kg	-	-	-	-	nr	**695.51**
Lining ancillaries; bitumen impregnated fibreboard						
Parallel circumferential packing						
1.5 m int diameter	-	-	-	-	nr	**4.85**
3.0 m int diameter	-	-	-	-	nr	**9.70**
Lining ancillaries; PC4AF caulking compound						
Caulking						
1.5 m int diameter	-	-	-	-	m	**7.99**
3.0 m int diameter	-	-	-	-	m	**7.99**
SUPPORT AND STABILIZATION						
Rock bolts						
mechanical	-	-	-	-	m	**24.35**
mechanical grouted	-	-	-	-	m	**38.03**
pre-grouted impacted	-	-	-	-	m	**36.58**
chemical end anchor	-	-	-	-	m	**36.58**
chemical grouted	-	-	-	-	m	**26.02**
chemically filled	-	-	-	-	m	**40.93**
Internal support						
steel arches; supply	-	-	-	-	tonne	**1070.22**
steel arches; erection	-	-	-	-	tonne	**422.25**
timber supports; supply	-	-	-	-	m^3	**311.66**
timber supports; erection	-	-	-	-	m^3	**268.72**
lagging	-	-	-	-	m^2	**23.04**
sprayed concrete	-	-	-	-	m^2	**26.36**
mesh or link	-	-	-	-	m^2	**8.51**
Pressure grouting						
sets of drilling and grouting plant	-	-	-	-	nr	**1155.27**
face packers	-	-	-	-	nr	**51.17**
deep packers	-	-	-	-	nr	**87.74**
drilling and flushing to 40 mm diameter	-	-	-	-	m	**19.38**
re-drilling and flushing	-	-	-	-	m	**15.37**
injection of grout materials; chemical grout	-	-	-	-	tonne	**614.20**
Forward probing	-	-	-	-	m	**19.97**

CLASS U: BRICKWORK, BLOCKWORK, MASONRY

Item	Gang hours	Labour £	Plant £	Material £	Unit	**Total rate £**
NOTES						
Apply the following multipliers to both labour and plant for rubble walls:-						
height 2 to 5 m 1.21						
height 5 to 10 m 1.37						
wall to radius small 1.75						
wall to radius large 1.50						
wall to rake or batter 1.15						
wall in piers or stanchion 1.50						
wall in butresses 1.15						
RESOURCES - LABOUR						
Brickwork, blockwork and masonry gang						
1 foreman bricklayer (craftsman)		18.47				
4 bricklayers (craftsman)		69.40				
1 unskilled operative (general)		11.26				
Total Gang Rate / Hour	£	**99.13**				
RESOURCES - PLANT						
Brickwork, blockwork and masonry						
2t dumper (50% of time)			2.17			
mixer (50% of time)			1.02			
small power tools			1.78			
scaffold, etc.			1.27			
Total Rate / Hour		£	**6.24**			
COMMON BRICKWORK						
Common bricks in cement mortar designation (ii)						
Thickness 103 mm						
vertical straight walls	0.23	22.60	1.42	10.71	m²	**34.74**
vertical curved walls	0.30	29.34	1.83	10.71	m²	**41.89**
battered straight walls	0.33	32.61	2.05	10.71	m²	**45.38**
battered curved walls	0.37	36.68	2.31	10.71	m²	**49.70**
vertical facing to concrete	0.25	24.49	1.54	18.18	m²	**44.21**
battered facing to concrete	0.37	36.68	2.31	18.19	m²	**57.17**
casing to metal sections	0.32	31.23	1.96	18.19	m²	**51.38**
Thickness 215 mm						
vertical straight walls	0.44	43.82	2.76	21.90	m²	**68.48**
vertical curved walls	0.57	56.50	3.56	21.90	m²	**81.96**
battered straight walls	0.63	62.45	3.93	21.90	m²	**88.29**
battered curved walls	0.71	70.38	4.40	21.90	m²	**96.69**
vertical facing to concrete	0.48	47.58	2.98	29.37	m²	**79.94**
battered facing to concrete	0.71	70.38	4.40	29.37	m²	**104.16**
casing to metal sections	0.61	60.47	3.78	29.37	m²	**93.62**
Thickness 328 mm						
vertical straight walls	0.64	63.44	4.02	33.45	m²	**100.92**
vertical curved walls	0.82	81.29	5.14	33.45	m²	**119.88**
battered straight walls	0.91	90.21	5.66	33.45	m²	**129.32**
battered curved walls	1.01	100.12	6.30	33.45	m²	**139.88**
vertical facing to concrete	0.69	68.40	4.33	40.92	m²	**113.66**
battered facing to concrete	1.01	100.12	6.30	40.92	m²	**147.35**
casing to metal sections	0.87	86.24	5.43	40.92	m²	**132.60**

CLASS U: BRICKWORK, BLOCKWORK, MASONRY

Item	Gang hours	Labour £	Plant £	Material £	Unit	**Total rate £**
COMMON BRICKWORK – cont'd						
Common bricks in cement mortar designation (ii) – cont'd						
Thickness 440 mm						
vertical straight walls	0.84	83.27	5.21	44.86	m²	**133.33**
vertical curved walls	1.06	105.08	6.60	44.86	m²	**156.54**
battered straight walls	1.16	114.99	7.25	44.86	m²	**167.10**
battered curved walls	1.29	127.88	8.04	44.86	m²	**180.77**
vertical facing to concrete	0.90	89.22	5.60	52.24	m²	**147.07**
battered facing to concrete	1.29	127.88	8.04	52.33	m²	**188.24**
casing to metal sections	1.12	111.03	6.98	52.33	m²	**170.33**
Thickness 890 mm						
vertical straight walls	1.50	148.69	9.36	90.53	m²	**248.59**
vertical curved walls	1.85	183.39	11.56	90.53	m²	**285.49**
battered straight walls	2.00	198.26	12.54	90.53	m²	**301.33**
battered curved walls	2.20	218.09	13.70	90.53	m²	**322.32**
vertical facing to concrete	1.60	158.61	10.00	98.00	m²	**266.61**
battered facing to concrete	2.20	218.09	13.70	98.00	m²	**329.79**
casing to metal sections	1.94	192.31	12.13	98.00	m²	**302.44**
Thickness exceeding 1 m						
vertical straight walls	1.64	162.57	10.21	101.52	m²	**274.31**
vertical curved walls	2.00	198.26	12.53	101.52	m²	**312.31**
battered straight walls	2.17	215.11	13.56	101.52	m²	**330.19**
battered curved walls	2.37	234.94	14.77	101.52	m²	**351.23**
vertical facing to concrete	1.74	172.49	10.88	108.99	m²	**292.36**
battered facing to concrete	2.37	234.94	14.77	108.99	m²	**358.70**
casing to metal sections	2.10	208.17	13.13	108.99	m²	**330.29**
Columns and piers						
215 x 215 mm	0.13	12.89	0.81	4.73	m	**18.42**
440 x 215 mm	0.24	23.79	1.50	9.66	m	**34.95**
665 x 328 mm	0.44	43.62	2.75	21.98	m	**68.35**
890 x 890 mm	1.10	109.04	6.87	79.38	m	**195.29**
Surface features						
copings; standard header-on-edge; 215 mm wide x 103 mm high	0.10	9.91	0.66	2.40	m	**12.97**
sills; standard header-on-edge; 215 mm wide x 103 mm high	0.13	12.89	0.78	2.26	m	**15.93**
rebates	0.30	29.74	1.87	-	m	**31.61**
chases	0.35	34.70	2.19	-	m	**36.88**
band courses;flush;215 mm wide	0.05	4.96	0.28	-	m	**5.24**
band courses; projection 103mm;215 mm wide	0.05	4.96	0.31	-	m	**5.27**
corbels; maximum projection 103mm;215 mm wide	0.15	14.87	0.94	2.19	m	**17.99**
pilasters;328 mm wide x 103 mm projection	-	-	0.44	4.73	m	**5.16**
pilasters;440 mm wide x 215 mm projection	0.12	11.90	0.75	10.68	m	**23.32**
plinths; projection 103 mm x 900 mm wide	0.19	18.83	1.19	10.00	m	**30.02**
fair facing	0.06	5.95	0.36	-	m²	**6.30**
Ancillaries						
bonds to existing work; to brickwork	1.50	148.69	9.36	11.54	m²	**169.60**
built-in pipes and ducts, cross-sectional area not exceeding 0.05 m²; excluding supply; brickwork 103 mm thick	0.06	5.95	0.38	0.49	nr	**6.82**

CLASS U: BRICKWORK, BLOCKWORK, MASONRY

Item	Gang hours	Labour £	Plant £	Material £	Unit	**Total rate £**
built-in pipes and ducts, cross-sectional area not exceeding 0.05 m²; excluding supply; brickwork 215 mm thick	0.12	11.90	0.72	0.78	nr	**13.39**
built-in pipes and ducts, cross-sectional area 0.05 - 0.25 m²; excluding supply; brickwork 103 mm thick	0.15	14.87	0.94	0.91	nr	**16.71**
built-in pipes and ducts, cross-sectional area 0.05 - 0.25 m²; excluding supply; brickwork 215 mm thick	0.29	28.75	1.81	1.40	nr	**31.95**
built-in pipes and ducts, cross-sectional area 0.25-0.5 m²; excluding supply; brickwork 103 mm thick	0.18	17.84	1.09	1.12	nr	**20.06**
built-in pipes and ducts, cross-sectional area 0.25-0.5 m²; excluding supply; brickwork 215 mm thick	0.34	33.70	2.12	1.67	nr	**37.50**
FACING BRICKWORK						
Facing bricks; in plasticised cement mortar designation (iii)						
Thickness 103 mm						
vertical straight walls	0.34	33.70	2.12	29.45	m²	**65.28**
vertical curved walls	0.45	44.61	2.81	29.45	m²	**76.87**
battered straight walls	0.45	44.61	2.81	29.45	m²	**76.87**
battered curved walls	0.56	55.51	3.51	29.45	m²	**88.48**
vertical facing to concrete	0.37	36.68	2.22	36.92	m²	**75.82**
battered facing to concrete	0.56	55.51	3.51	36.92	m²	**95.95**
casing to metal sections	0.49	48.57	3.06	36.92	m²	**88.55**
Thickness 215 mm						
vertical straight walls	0.57	56.50	3.95	59.38	m²	**119.84**
vertical curved walls	0.84	83.27	5.24	59.38	m²	**147.90**
battered straight walls	0.84	83.27	5.24	59.38	m²	**147.90**
battered curved walls	1.02	101.11	6.39	59.38	m²	**166.88**
vertical facing to concrete	0.66	65.43	4.12	66.85	m²	**136.40**
battered facing to concrete	1.02	101.11	6.39	66.85	m²	**174.35**
casing to metal sections	0.82	81.29	5.12	66.85	m²	**153.26**
Thickness 328 mm						
vertical straight walls	0.83	82.28	5.16	89.67	m²	**177.11**
vertical curved walls	1.12	111.03	6.96	89.67	m²	**207.66**
battered straight walls	1.12	111.03	6.96	89.67	m²	**207.66**
battered curved walls	1.36	134.82	8.49	89.67	m²	**232.98**
vertical facing to concrete	0.88	87.23	5.47	97.14	m²	**189.85**
battered facing to concrete	1.36	134.82	8.49	97.14	m²	**240.45**
casing to metal sections	1.18	116.97	7.37	97.14	m²	**221.48**
Thickness 440 mm						
vertical straight walls	1.08	107.06	6.74	119.82	m²	**233.62**
vertical curved walls	1.39	137.79	8.68	119.82	m²	**266.29**
battered straight walls	1.39	137.79	8.68	119.82	m²	**266.29**
vertical facing to concrete	1.09	108.05	6.82	127.29	m²	**242.16**
battered facing to concrete	1.39	137.79	8.68	127.29	m²	**273.76**
casing to metal sections	1.39	137.79	8.68	127.29	m²	**273.76**
Columns and piers						
215 x 215 mm	0.17	16.85	1.04	13.17	m	**31.06**
440 x 215 mm	0.29	28.75	1.81	26.52	m	**57.08**
665 x 328 mm	0.58	57.50	3.60	59.93	m	**121.03**

CLASS U: BRICKWORK, BLOCKWORK, MASONRY

Item	Gang hours	Labour £	Plant £	Material £	Unit	Total rate £
FACING BRICKWORK – cont'd						
Facing bricks; in plasticised cement mortar designation (iii) – cont'd						
Surface features						
copings; standard header-on-edge; standard bricks; 215 mm wide x 103 mm high	0.13	12.89	0.84	6.19	m	**19.91**
flat arches; standard stretcher-on-end; 215 mm wide x 103 mm high	0.21	20.82	1.31	5.98	m	**28.10**
flat arches; standard stretcher-on-end; bullnosed special bricks;103 mm x 215 mm high	0.22	21.81	1.38	42.59	m	**65.77**
segmental arches; single ring; standard bricks; 103 mm wide x 215 mm high	0.37	36.68	2.31	5.98	m	**44.96**
segmental arches; two ring; standard bricks; 103 mm wide x 440 mm high	0.49	48.57	3.06	12.09	m	**63.73**
segmental arches; cut voussoirs; 103 mm wide x 215 mm high	0.39	38.66	2.44	190.01	m	**231.11**
rebates	0.33	32.71	2.06	-	m	**34.77**
chases	0.37	36.68	2.31	-	m	**38.99**
cornices; maximum projection 103mm; 215 mm wide	0.37	36.68	2.31	6.30	m	**45.28**
band courses; projection 113mm; 215 mm wide	0.06	5.95	0.34	-	m	**6.29**
corbels; maximum projection 113mm; 215 mm wide	0.37	36.68	2.31	6.30	m	**45.28**
pilasters;328 mm wide x 113 mm projection	0.05	4.96	0.31	13.16	m	**18.43**
pilasters;440 mm wide x 215 mm projection	0.06	5.95	0.34	29.65	m	**35.94**
plinths; projection 113 mm x 900 mm wide	0.24	23.79	1.50	26.86	m	**52.15**
fair facing; pointing as work proceeds	0.06	5.95	0.38	-	m^2	**6.32**
Ancillaries						
bonds to existing work; to brickwork	1.50	148.69	9.36	10.88	m	**168.94**
built-in pipes and ducts, cross-sectional area not exceeding 0.05 m^2; excluding supply; brickwork half brick thick	0.10	9.91	0.59	0.73	nr	**11.24**
built-in pipes and ducts, cross-sectional area not exceeding 0.05 m^2; excluding supply; brickwork one brick thick	0.15	14.87	0.69	1.25	nr	**16.81**
built-in pipes and ducts, cross-sectional area 0.05-0.25 m^2; excluding supply; brickwork half brick thick	0.19	18.83	1.19	1.17	nr	**21.19**
built-in pipes and ducts, cross-sectional area 0.05-0.25 m^2; excluding supply; brickwork one brick thick	0.33	32.71	2.06	1.95	nr	**36.72**
built-in pipes and ducts, cross-sectional area 0.25-0.5 m^2; excluding supply; brickwork half brick thick	0.23	22.80	1.44	1.37	nr	**25.61**
built-in pipes and ducts, cross-sectional area 0.25-0.5 m^2; excluding supply; brickwork one brick thick	0.40	39.65	2.46	2.29	nr	**44.41**

CLASS U: BRICKWORK, BLOCKWORK, MASONRY

Item	Gang hours	Labour £	Plant £	Material £	Unit	**Total rate £**
ENGINEERING BRICKWORK						
Class A engineering bricks, solid; in cement mortar designation (ii)						
Thickness 103 mm						
vertical straight walls	0.27	26.77	1.68	20.51	m²	**48.96**
vertical curved walls	0.37	36.68	2.31	20.51	m²	**59.50**
battered straight walls	0.37	36.68	2.31	20.51	m²	**59.50**
battered curved walls	0.41	40.64	2.57	20.51	m²	**63.73**
vertical facing to concrete	0.32	31.72	1.96	27.98	m²	**61.67**
battered facing to concrete	0.46	45.60	2.89	27.98	m²	**76.47**
casing to metal sections	0.41	40.64	2.57	27.98	m²	**71.20**
Thickness 215 mm						
vertical straight walls	0.52	51.55	3.25	41.50	m²	**96.30**
vertical curved walls	0.71	70.38	4.40	41.50	m²	**116.29**
battered straight walls	0.71	70.38	4.40	41.50	m²	**116.29**
battered curved walls	0.78	77.32	4.87	41.50	m²	**123.70**
vertical facing to concrete	0.61	60.47	3.78	48.97	m²	**113.22**
battered facing to concrete	0.87	86.24	5.44	48.97	m²	**140.65**
casing to metal sections	0.78	77.32	4.87	48.97	m²	**131.17**
Thickness 328 mm						
vertical straight walls	0.75	74.35	4.70	62.85	m²	**141.90**
vertical curved walls	1.01	100.12	6.30	62.85	m²	**169.28**
battered straight walls	1.01	100.12	6.30	62.85	m²	**169.28**
battered curved walls	1.11	110.03	6.94	62.85	m²	**179.82**
vertical facing to concrete	0.87	86.24	5.44	70.32	m²	**162.00**
battered facing to concrete	1.24	122.92	7.71	70.32	m²	**200.95**
casing to metal sections	1.11	110.03	6.94	70.32	m²	**187.29**
Thickness 440 mm						
vertical straight walls	0.97	96.16	6.06	84.06	m²	**186.27**
vertical curved walls	1.29	127.88	8.04	84.06	m²	**219.97**
battered straight walls	1.29	127.88	8.04	84.06	m²	**219.97**
battered curved walls	1.41	139.77	8.81	84.06	m²	**232.64**
vertical facing to concrete	1.12	111.03	6.98	91.53	m²	**209.53**
battered facing to concrete	1.56	154.64	9.73	91.53	m²	**255.90**
casing to metal sections	1.41	139.77	8.81	91.53	m²	**240.11**
Thickness 890 mm						
vertical straight walls	1.72	170.50	10.72	168.87	m²	**350.10**
vertical curved walls	2.20	218.09	13.70	168.87	m²	**400.66**
battered straight walls	2.20	218.09	13.70	168.87	m²	**400.66**
battered curved walls	2.37	234.94	14.79	168.87	m²	**418.60**
vertical facing to concrete	1.94	192.31	12.13	176.34	m²	**380.78**
battered facing to concrete	2.58	255.76	16.08	176.34	m²	**448.18**
casing to metal sections	2.37	234.94	14.79	176.34	m²	**426.07**
Thickness exceeding 1 m						
vertical straight walls	1.87	185.37	11.65	189.39	m³	**386.41**
vertical curved walls	2.37	234.94	14.77	189.39	m³	**439.10**
battered straight walls	2.37	234.94	14.77	189.39	m³	**439.10**
battered curved walls	2.55	252.78	15.91	189.39	m³	**458.08**
vertical facing to concrete	2.10	208.17	13.13	196.79	m³	**418.09**
battered facing to concrete	2.76	273.60	17.23	196.79	m³	**487.62**
casing to metal sections	2.55	252.78	15.91	196.79	m³	**465.48**
Columns and piers						
215 x 215 mm	0.16	15.86	0.99	9.14	m	**25.99**
440 x 215 mm	0.28	27.76	1.74	18.48	m	**47.97**
665 x 328 mm	0.56	55.51	3.47	41.83	m	**100.81**
890 x 890 mm	1.78	176.45	11.10	149.94	m	**337.49**

CLASS U: BRICKWORK, BLOCKWORK, MASONRY

Item	Gang hours	Labour £	Plant £	Material £	Unit	Total rate £
ENGINEERING BRICKWORK – cont'd						
Class A engineering bricks, solid; in cement mortar designation (ii) – cont'd						
Surface features						
copings; standard header-on-edge; 215 mm wide x 103 mm high	0.11	10.90	0.66	4.60	m	**16.16**
sills; standard header-on-edge; 215 mm wide x 103 mm high	0.13	12.89	0.78	4.47	m	**18.13**
rebates	0.33	32.71	2.06	-	m	**34.77**
chases	0.37	36.68	2.31	-	m	**38.99**
band courses;flush;215 mm wide	0.05	4.96	0.28	-	m	**5.24**
band courses; projection 103mm; 215 mm wide	0.05	4.96	0.31	-	m	**5.27**
corbels; maximum projection 103mm; 215 mm wide	0.15	14.87	0.94	4.39	m	**20.20**
pilasters;328 mm wide x 103 mm projection	0.07	6.94	0.44	9.14	m	**16.51**
pilasters;440 mm wide x 215 mm projection	0.12	11.90	0.75	20.60	m	**33.25**
plinths; projection 103 mm x 900 mm wide	0.19	18.83	1.19	18.82	m	**38.84**
fair facing	0.06	5.95	0.38	-	m²	**6.32**
Ancillaries						
bonds to existing brickwork; to brickwork	0.29	28.75	1.81	14.94	m²	**45.50**
built-in pipes and ducts, cross-sectional area not exceeding 0.05 m²; excluding supply; brickwork half brick thick	0.10	9.91	0.59	0.63	nr	**11.14**
built-in pipes and ducts, cross-sectional area not exceeding 0.05 m²; excluding supply; brickwork one brick thick	0.15	14.87	0.94	0.89	nr	**16.70**
built-in pipes and ducts, cross-sectional area 0.05-0.25 m²; excluding supply; brickwork half brick thick	0.19	18.83	1.19	1.10	nr	**21.12**
built-in pipes and ducts, cross-sectional area 0.05-0.25 m²; excluding supply; brickwork one brick thick	0.33	32.71	2.06	1.68	nr	**36.45**
built-in pipes and ducts, cross-sectional area 0.025-0.5 m²; excluding supply; brickwork half brick thick	0.23	22.80	1.44	1.33	nr	**25.56**
built-in pipes and ducts, cross-sectional area 0.025-0.5 m²; excluding supply; brickwork one brick thick	0.40	39.65	2.46	2.02	nr	**44.13**
Class B engineering bricks, perforated; in cement mortar designation (ii)						
Thickness 103 mm						
vertical straight walls	0.27	26.77	1.68	13.54	m²	**41.99**
vertical curved walls	0.37	36.68	2.31	13.54	m²	**52.53**
battered straight walls	0.37	36.68	2.31	13.54	m²	**52.53**
vertical facing to concrete	0.32	31.72	1.96	21.01	m²	**54.70**
casings to metal sections	0.41	40.64	2.57	21.01	m²	**64.23**
Thickness 215 mm						
vertical straight walls	0.52	51.55	3.25	27.57	m²	**82.36**
vertical curved walls	0.71	70.38	4.40	27.57	m²	**102.35**
battered straight walls	0.71	70.38	4.40	27.57	m²	**102.35**
vertical facing to concrete	0.61	60.47	3.78	35.04	m²	**99.28**
casings to metal sections	0.78	77.32	4.87	35.04	m²	**117.23**

CLASS U: BRICKWORK, BLOCKWORK, MASONRY

Item	Gang hours	Labour £	Plant £	Material £	Unit	**Total rate £**
Thickness 328 mm						
vertical straight walls	0.76	75.34	4.72	41.95	m²	**122.01**
vertical curved walls	1.01	100.12	6.30	41.95	m²	**148.37**
battered straight walls	1.01	100.12	6.30	41.95	m²	**148.37**
vertical facing to concrete	0.87	86.24	5.44	49.42	m²	**141.10**
casings to metal sections	1.11	110.03	6.94	49.42	m²	**166.39**
Thickness 440 mm						
vertical straight walls	0.97	96.16	6.06	56.18	m²	**158.40**
vertical curved walls	1.29	127.88	8.04	56.18	m²	**192.10**
battered straight walls	1.29	127.88	8.04	56.18	m²	**192.10**
vertical facing to concrete	1.12	111.03	6.98	63.65	m²	**181.66**
casings to metal sections	1.41	139.77	8.81	63.65	m²	**212.23**
Thickness 890 mm						
vertical straight walls	1.72	170.50	10.72	113.12	m²	**294.35**
vertical curved walls	2.20	218.09	13.70	113.12	m²	**344.91**
battered straight walls	2.20	218.09	13.70	113.12	m²	**344.91**
battered curved walls	2.37	234.94	14.79	113.12	m²	**362.85**
vertical facing to concrete	1.94	192.31	12.13	120.59	m²	**325.03**
battered facing to concrete	2.58	255.76	16.08	120.59	m²	**392.43**
casing to metal sections	2.37	234.94	14.79	120.59	m²	**370.32**
Thickness exceeding 1 m						
vertical straight walls	1.87	185.37	11.65	127.06	m³	**324.08**
vertical curved walls	2.37	234.94	14.77	127.06	m³	**376.77**
battered straight walls	2.37	234.94	14.77	127.06	m³	**376.77**
battered curved walls	2.55	252.78	15.91	127.06	m³	**395.75**
vertical facing to concrete	2.10	208.17	13.13	134.47	m³	**355.77**
battered facing to concrete	2.76	273.60	17.23	134.47	m³	**425.29**
casing to metal sections	2.55	252.78	15.91	134.47	m³	**403.15**
Columns and piers						
215 x 215 mm	0.16	15.86	0.99	6.00	m	**22.85**
440 x 215 mm	0.28	27.76	1.74	12.20	m	**41.70**
665 x 328 mm	0.56	55.51	3.47	27.72	m	**86.70**
890 x 890 mm	1.78	176.45	11.10	99.77	m	**287.31**
Surface features						
copings; standard header-on-edge; 215 mm wide x 103 mm high	0.11	10.90	0.66	3.04	m	**14.60**
sills; standard header-on-edge; 215 mm wide x 103 mm high	0.13	12.89	0.78	2.90	m	**16.57**
rebates	0.33	32.71	2.06	-	m	**34.77**
chases	0.37	36.68	2.31	-	m	**38.99**
band courses;flush; 215 mm wide	0.05	4.96	0.28	-	m	**5.24**
band courses; projection 103mm; 215 mm wide	0.05	4.96	0.31	-	m	**5.27**
corbels; maximum projection 103mm; 215 mm wide	0.15	14.87	0.94	2.82	m	**18.63**
pilasters;328 mm wide x 103 mm projection	0.07	6.94	0.44	6.00	m	**13.38**
pilasters;440 mm wide x 215 mm projection	0.12	11.90	0.75	13.54	m	**26.19**
plinths; projection 103 mm x 900 mm wide	0.19	18.83	1.19	12.55	m	**32.57**
fair facing	0.06	5.95	0.38	-	m²	**6.32**
Ancillaries						
bonds to existing brickwork; to brickwork	0.29	28.75	1.81	14.82	m²	**45.38**
built-in pipes and ducts, cross-sectional area not exceeding 0.05 m²; excluding supply; brickwork half brick thick	0.10	9.91	0.59	0.63	nr	**11.14**

CLASS U: BRICKWORK, BLOCKWORK, MASONRY

Item	Gang hours	Labour £	Plant £	Material £	Unit	Total rate £
ENGINEERING BRICKWORK – cont'd						
Class B engineering bricks, perforated; in cement mortar designation (ii) – cont'd						
Ancillaries – cont'd						
built-in pipes and ducts, cross-sectional area not exceeding 0.05 m²; excluding supply; brickwork one brick thick	0.15	14.87	0.94	0.89	nr	**16.70**
built-in pipes and ducts, cross-sectional area 0.05-0.25 m²; excluding supply; brickwork half brick thick	0.19	18.83	1.19	1.10	nr	**21.12**
built-in pipes and ducts, cross-sectional area 0.05-0.25 m²; excluding supply; brickwork one brick thick	0.33	32.71	2.06	1.68	nr	**36.45**
built-in pipes and ducts, cross-sectional area 0.025-0.5 m²; excluding supply; brickwork half brick thick	0.23	22.80	1.44	1.33	nr	**25.56**
built-in pipes and ducts, cross-sectional area 0.025-0.5 m²; excluding supply; brickwork one brick thick	0.40	39.65	2.46	2.00	nr	**44.12**
LIGHTWEIGHT BLOCKWORK						
Lightweight concrete blocks; 3.5 N/mm²; in cement-lime mortar designation (iii)						
Thickness 100 mm; PC £ 6.20/m²						
vertical straight walls	0.17	16.85	1.09	6.78	m²	**24.72**
vertical curved walls	0.23	22.80	1.44	6.78	m²	**31.02**
vertical facework to concrete	0.18	17.84	1.12	8.40	m²	**27.36**
casing to metal sections	0.21	20.82	1.31	8.40	m²	**30.53**
Thickness 140 mm; PC £ 8.20/m²						
vertical straight walls	0.23	22.80	1.40	8.95	m²	**33.15**
vertical curved walls	0.30	29.74	1.86	8.95	m²	**40.55**
vertical facework to concrete	0.23	22.80	1.45	10.49	m²	**34.74**
casing to metal sections	0.27	26.77	1.69	10.49	m²	**38.94**
Thickness 215 mm; PC £ 13.13/m²						
vertical straight walls	0.28	27.76	1.72	14.34	m²	**43.82**
vertical curved walls	0.37	36.68	2.29	14.34	m²	**53.31**
vertical facework to concrete	0.28	27.76	1.77	15.93	m²	**45.45**
casing to metal sections	0.31	30.73	1.94	15.93	m²	**48.60**
Columns and piers						
440 x 100 mm	0.08	7.93	0.50	3.00	m	**11.43**
890 x 140 mm	0.22	21.81	1.38	8.02	m	**31.20**
Surface features						
fair facing	0.06	5.95	0.38	-	m²	**6.32**
Ancillaries						
built-in pipes and ducts, cross-sectional area not exceeding 0.05 m²; excluding supply; blockwork 100 mm thick	0.04	3.97	0.25	0.26	nr	**4.48**
built-in pipes and ducts, cross-sectional area not exceeding 0.05 m²; excluding supply; blockwork 140 mm thick	0.09	8.92	0.56	0.67	nr	**10.15**
built-in pipes and ducts, cross-sectional area not exceeding 0.05 m²; excluding supply; blockwork 215 mm thick	0.13	12.89	0.79	0.87	nr	**14.55**

CLASS U: BRICKWORK, BLOCKWORK, MASONRY

Item	Gang hours	Labour £	Plant £	Material £	Unit	**Total rate £**
built-in pipes and ducts, cross-sectional area 0.05-0.25 m²; excluding supply; blockwork 100 mm thick	0.13	12.89	0.79	0.89	nr	**14.57**
built-in pipes and ducts, cross-sectional area 0.05-0.25 m²; excluding supply; blockwork 140 mm thick	0.18	17.84	1.10	1.34	nr	**20.28**
built-in pipes and ducts, cross-sectional area 0.05-0.25 m²; excluding supply; blockwork 215 mm thick	0.25	24.78	1.56	1.80	nr	**28.14**
built-in pipes and ducts, cross-sectional area 0.25-0.5 m²; excluding supply; blockwork 100 mm thick	0.15	14.87	0.94	1.14	nr	**16.94**
built-in pipes and ducts, cross-sectional area 0.25-0.5 m²; excluding supply; blockwork 140 mm thick	0.21	20.82	1.31	1.34	nr	**23.47**
built-in pipes and ducts, cross-sectional area 0.25-0.5 m²; excluding supply; blockwork 215 mm thick	0.30	29.74	1.84	2.15	nr	**33.73**
DENSE CONCRETE BLOCKWORK						
Dense concrete blocks; 7 N/mm²; in cement mortar designation (iii)						
Walls, built vertical and straight						
100 mm thick; PC £5.53/m²	0.19	18.83	1.19	6.07	m²	**26.10**
140 mm thick; PC £8.01/m²	0.25	24.78	1.55	8.73	m²	**35.06**
215 mm thick; PC £14.60/m²	0.34	33.70	2.12	15.85	m²	**51.68**
Walls, built vertical and curved						
100 mm thick	0.25	24.78	1.59	6.10	m²	**32.47**
140 mm thick	0.33	32.71	2.06	8.73	m²	**43.50**
215 mm thick	0.45	44.61	2.82	15.88	m²	**63.31**
Walls, built vertical in facework to concrete						
100 mm thick	0.20	19.83	1.23	7.54	m²	**28.60**
140 mm thick	0.26	25.77	1.59	10.26	m²	**37.63**
215 mm thick	0.35	34.70	2.19	17.62	m²	**54.50**
Walls, as casings to metal sections, built vertical and straight						
100 mm thick	0.23	22.80	1.44	7.54	m²	**31.78**
140 mm thick	0.30	29.74	1.84	10.26	m²	**41.84**
Columns and piers						
440 x 100 mm	0.08	7.93	0.50	2.53	m	**10.95**
890 x 140 mm thick	0.22	21.81	1.38	7.85	m	**31.03**
Ancillaries						
built-in pipes and ducts, cross-sectional area not exceeding 0.05 m²; excluding supply; blockwork 100 mm thick	0.05	4.96	0.28	0.26	nr	**5.50**
built-in pipes and ducts, cross-sectional area not exceeding 0.05 m²; excluding supply; blockwork 140 mm thick	0.10	9.91	0.63	0.64	nr	**11.18**
built-in pipes and ducts, cross-sectional area not exceeding 0.05 m²; excluding supply; blockwork 215 mm thick	0.14	13.88	0.84	0.98	nr	**15.70**
built-in pipes and ducts, cross-sectional area 0.05-0.25 m²; excluding supply; blockwork 100 mm thick	0.14	13.88	0.84	0.86	nr	**15.58**

CLASS U: BRICKWORK, BLOCKWORK, MASONRY

Item	Gang hours	Labour £	Plant £	Material £	Unit	Total rate £
DENSE CONCRETE BLOCKWORK – cont'd						
Dense concrete blocks; 7 N/mm²; in cement mortar designation (iii) – cont'd						
built-in pipes and ducts, cross-sectional area 0.05-0.25 m²; excluding supply; blockwork 140 mm thick	0.19	18.83	1.19	1.31	nr	**21.34**
built-in pipes and ducts, cross-sectional area 0.05-0.25 m²; excluding supply; blockwork 215 mm thick	0.27	26.77	1.69	2.00	nr	**30.45**
built-in pipes and ducts, cross-sectional area 0.25-0.5 m²; excluding supply; blockwork 100 mm thick	0.16	15.86	1.02	1.05	nr	**17.93**
built-in pipes and ducts, cross-sectional area 0.25-0.5 m²; excluding supply; blockwork 140 mm thick	0.23	22.80	1.44	1.34	nr	**25.58**
built-in pipes and ducts, cross-sectional area 0.25-0.5 m²; excluding supply; blockwork 215 mm thick	0.32	31.72	2.00	2.38	nr	**36.09**
ARTIFICIAL STONE BLOCKWORK						
Reconstituted stone masonry blocks; Bradstone 100 bed weathered Cotswold or North Cearney rough hewn rockfaced blocks; in coloured cement-lime mortar designation (iii)						
Thickness 100 mm vertical facing						
vertical straight walls	0.30	29.74	1.87	31.82	m²	**63.44**
vertical curved walls	0.39	38.66	2.44	31.82	m²	**72.92**
vertical facework to concrete	0.31	30.73	1.95	31.85	m²	**64.53**
casing to metal sections	0.40	39.65	2.50	31.85	m²	**74.00**
Ancillaries						
built-in pipes and ducts, cross-sectional area not exceeding 0.05 m²; excluding supply	0.07	6.94	0.44	0.40	nr	**7.78**
built-in pipes and ducts, cross-sectional area 0.05-0.25 m²; excluding supply	0.22	21.81	1.38	3.40	nr	**26.58**
built-in pipes and ducts, cross-sectional area 0.25-0.5 m²; excluding supply	0.26	25.77	1.62	4.27	nr	**31.67**
Reconstituted stone masonry blocks; Bradstone Architectural dressing in weathered Cotswold or North Cerney shades; in coloured cement-lime mortar designation (iii)						
Surface features; Pier Caps						
305 x 305 mm, weathered and throated	0.09	8.92	0.56	13.03	nr	**22.51**
381 x 381 mm, weathered and throated	0.11	10.90	0.69	18.42	nr	**30.01**
457 x 457 mm, weathered and throated	0.13	12.89	0.81	25.20	nr	**38.90**
533 x 533 mm, weathered and throated	0.15	14.87	0.94	35.09	nr	**50.90**
Surface features; Copings						
152 x 76 mm, twice weathered and throated	0.08	7.93	0.50	7.18	m	**15.61**
152 x 76 mm, curved on plan, twice weathered and throated	0.11	10.90	0.66	47.45	m	**59.02**
305 x 76 mm, twice weathered and throated	0.10	9.91	2.50	17.66	m	**30.07**
305 x 76 mm, curved on plan, twice weathered and throated	0.13	12.89	0.83	71.14	m	**84.86**

CLASS U: BRICKWORK, BLOCKWORK, MASONRY

Item	Gang hours	Labour £	Plant £	Material £	Unit	Total rate £
Surface features; Pilasters						
440 x 100 mm	0.14	13.88	0.88	14.02	m	**28.77**
Surface features; Corbels						
479 x 100 x 215 mm, splayed	0.49	48.57	3.06	42.75	nr	**94.38**
665 x 100 x 215 mm, splayed	0.55	54.52	3.43	42.64	nr	**100.59**
Surface features; Lintels						
100 x 140 mm	0.11	10.90	0.69	27.49	m	**39.09**
100 x 215 mm	0.16	15.86	1.00	35.93	m	**52.79**
ASHLAR MASONRY						
Portland Whitbed limestone; in cement-lime mortar designation (iv); pointed one side cement-lime mortar designation (iii) incorporating stone dust						
Thickness 50 mm						
vertical facing to concrete	0.85	84.26	5.30	110.00	m²	**199.57**
Thickness 75 mm						
vertical straight walls	0.95	94.17	5.93	195.00	m²	**295.10**
vertical curved walls	1.45	143.74	9.05	346.17	m²	**498.96**
Surface features						
copings; weathered and twice throated; 250 x 150 mm	0.45	44.61	2.81	142.00	m	**189.42**
copings; weathered and twice throated; 250 x 150 mm; curved on plan	0.45	44.61	2.81	170.37	m	**217.79**
copings; weathered and twice throated; 400 x 150 mm	0.49	48.57	3.06	208.74	m	**260.37**
copings; weathered and twice throated; 400 x 150 mm; curved on plan	0.49	48.57	3.06	250.47	m	**302.10**
string courses; shaped and dressed; 75 mm projection x 150 mm high	0.45	44.61	2.81	112.27	m	**159.68**
corbel; shaped and dressed; 500 x 450 x 300 mm	0.55	54.52	3.43	160.59	nr	**218.54**
keystone; shaped and dressed; 750 x 900 x 300 mm (extreme)	1.30	128.87	8.12	627.64	nr	**764.63**
RUBBLE MASONRY						
Rubble masonry; random stones; in cement-lime mortar designation (iii)						
Walls, built vertical and straight; not exceeding 2m high						
300 mm thick	1.25	123.91	7.80	283.42	m²	**415.13**
450 mm thick	1.80	178.43	11.23	489.59	m²	**679.26**
600 mm thick	2.40	237.91	14.98	566.79	m²	**819.69**
Walls, built vertical, curved on plan; not exceeding 2 m high						
300 mm thick	1.40	138.78	4.72	290.04	m²	**433.54**
450 mm thick	2.00	198.26	12.48	425.14	m²	**635.88**
600 mm thick	2.65	262.69	16.54	566.79	m²	**846.03**
Walls, built with battered face; not exceeding 2 m high						
300 mm thick	1.40	138.78	8.74	283.42	m²	**430.94**
450 mm thick	2.00	198.26	12.48	425.14	m²	**635.88**
600 mm thick	2.65	262.69	16.54	566.79	m²	**846.03**

CLASS U: BRICKWORK, BLOCKWORK, MASONRY

Item	Gang hours	Labour £	Plant £	Material £	Unit	**Total rate £**
RUBBLE MASONRY – cont'd						
Rubble masonry; squared stones in cement-lime mortar designation (iii)						
Walls, built vertical and straight; not exceeding 2 m high						
300 mm thick	1.25	123.91	7.80	512.19	m^2	**643.90**
450 mm thick	1.80	178.43	11.23	768.27	m^2	**957.94**
600 mm thick	2.40	237.91	14.98	1024.38	m^2	**1277.27**
Walls, built vertical, curved on plan; not exceeding 2 m high						
300 mm thick	1.40	138.78	8.74	512.19	m^2	**659.71**
450 mm thick	2.00	198.26	12.48	768.27	m^2	**979.01**
600 mm thick	2.40	237.91	14.98	1078.57	m^2	**1331.46**
Walls, built with battered face; not exceeding 2 m high						
300 mm thick	1.40	138.78	8.74	512.19	m^2	**659.71**
450 mm thick	2.00	198.26	12.48	768.27	m^2	**979.01**
600 mm thick	2.40	237.91	14.98	1078.57	m^2	**1331.46**
Dry stone walling; random stones						
Average thickness 300 mm						
battered straight walls	1.15	114.00	2.50	261.99	m^2	**378.49**
Average thickness 450 mm						
battered straight walls	1.65	163.56	3.58	392.93	m^2	**560.08**
Average thickness 600 mm						
battered straight walls	2.15	213.13	4.66	523.94	m^2	**741.74**
Surface features						
copings; formed of rough stones 275 x 200 mm (average) high	0.45	44.61	2.81	67.56	m	**114.98**
copings; formed of rough stones 500 x 200 mm (average) high	0.55	54.52	3.43	122.85	m	**180.80**
ANCILLARIES COMMON TO ALL DIVISIONS						
Expamet joint reinforcement						
Ancillaries						
joint reinforcement; 65 mm wide	0.01	0.99	0.04	0.58	m	**1.61**
joint reinforcement; 115 mm wide	0.01	0.99	0.05	1.01	m	**2.05**
joint reinforcement; 175 mm wide	0.01	0.99	0.07	1.58	m	**2.64**
joint reinforcement; 225 mm wide	0.01	0.99	0.09	2.03	m	**3.11**
Hyload pitch polymer damp proof course; lapped joints; in cement mortar						
Ancillaries						
103 mm wide; horizontal	0.01	0.99	0.07	1.11	m	**2.17**
103 mm wide; vertical	0.02	1.98	0.11	1.11	m	**3.20**
215 mm wide; horizontal	0.03	2.97	0.16	2.35	m	**5.49**
215 mm wide; vertical	0.04	3.97	0.25	2.35	m	**6.57**
328 mm wide; horizontal	0.04	3.97	0.24	3.96	m	**8.16**
328 mm wide; vertical	0.06	5.95	0.36	3.96	m	**10.26**

CLASS U: BRICKWORK, BLOCKWORK, MASONRY

Item	Gang hours	Labour £	Plant £	Material £	Unit	Total rate £
Pre-formed closed cell joint filler; pointing with polysulphide sealant						
Ancillaries						
movement joints; 12 mm filler 90 wide; 12 x 12 sealant one side	0.06	5.95	0.34	0.67	m	**6.96**
movement joints; 12 mm filler 200 wide; 12 x 12 sealant one side	0.07	6.94	0.41	1.00	m	**8.35**
Dritherm cavity insulation						
Infills						
50 mm thick	0.04	3.97	0.25	3.94	m²	**8.15**
75 mm thick	0.05	4.96	0.28	5.23	m²	**10.47**
Concrete						
Infills						
50 mm thick	0.06	5.95	0.36	3.99	m²	**10.29**
Galvanised steel wall ties						
Fixings and ties						
vertical twist strip type; 900 mm horizontal and 450 mm vertical staggered spacings	0.02	1.98	0.09	1.83	m²	**3.91**
Stainless steel wall ties						
Fixings and ties						
vertical twist strip type; 900 mm horizontal and 450 mm vertical staggered spacings	0.02	1.98	0.09	4.04	m²	**6.11**

CLASS V: PAINTING

Item	Gang hours	Labour £	Plant £	Material £	Unit	Total rate £
RESOURCES - LABOUR						
Painting gang						
1 ganger (skill rate 3)		14.21				
3 painters (skill rate 3)		40.38				
1 unskilled operative (general)		11.26				
Total Gang Rate / Hour	£	**65.85**				
RESOURCES - PLANT						
Painting						
1.5KVA diesel generator			2.26			
transformers/cables; junction box			0.24			
4.5" electric grinder			0.36			
transit van (50% of time)			4.26			
ladders			1.33			
Total Rate / Hour		£	**8.46**			
LEAD BASED PRIMER PAINT						
One coat calcium plumbate primer						
Metal						
upper surfaces inclined at an angle ne 30 degrees to the horizontal	0.03	1.65	0.21	0.54	m²	**2.40**
upper surfaces inclined at an angle 30 - 60 degrees to the horizontal	0.03	1.65	0.21	0.54	m²	**2.40**
surfaces inclined at an angle exceeding 60 degrees to the horizontal	0.03	1.65	0.21	0.54	m²	**2.40**
soffit surfaces and lower surfaces inclined at an angle ne 60 degrees to the horizontal	0.03	1.98	0.26	0.54	m²	**2.77**
surfaces of width ne 300 mm	0.01	0.66	0.09	0.16	m	**0.91**
surfaces of width 300 mm - 1 m	0.03	1.65	0.21	0.35	m	**2.21**
Metal sections	0.03	1.98	0.25	0.54	m²	**2.76**
Pipework	0.03	1.98	0.25	0.54	m²	**2.76**
IRON BASED PRIMER PAINT						
One coat iron oxide primer						
Metal						
upper surfaces inclined at an angle ne 30 degrees to the horizontal	0.03	1.65	0.21	0.63	m²	**2.49**
upper surfaces inclined at an angle 30 - 60 degrees to the horizontal	0.03	1.65	0.21	0.63	m²	**2.49**
surfaces inclined at an angle exceeding 60 degrees to the horizontal	0.03	1.65	0.21	0.63	m²	**2.49**
soffit surfaces and lower surfaces inclined at an angle ne 60 degrees to the horizontal	0.03	1.98	0.26	0.63	m²	**2.87**
surfaces of width ne 300 mm	0.01	0.66	0.09	0.19	m	**0.94**
surfaces of width 300 mm - 1 m	0.03	1.65	0.21	0.42	m	**2.28**
Metal sections	0.03	1.98	0.25	0.63	m²	**2.85**
Pipework	0.03	1.98	0.25	0.63	m²	**2.85**

CLASS V: PAINTING

Item	Gang hours	Labour £	Plant £	Material £	Unit	Total rate £
ACRYLIC PRIMER PAINT						
One coat acrylic wood primer						
Timber						
upper surfaces inclined at an angle ne 30 degrees to the horizontal	0.02	1.32	0.17	0.38	m²	**1.87**
upper surfaces inclined at an angle 30 - 60 degrees to the horizontal	0.02	1.32	0.17	0.38	m²	**1.87**
surfaces inclined at an angle exceeding 60 degrees to the horizontal	0.02	1.32	0.16	0.38	m²	**1.86**
soffit surfaces and lower surfaces inclined at an angle ne 60 degrees to the horizontal	0.03	1.65	0.21	0.38	m²	**2.24**
surfaces of width ne 300 mm	0.01	0.66	0.08	0.11	m	**0.85**
surfaces of width 300 mm - 1 m	0.03	1.65	0.21	0.25	m	**2.10**
GLOSS PAINT						
One coat calcium plumbate primer; one undercoat and one finishing coat of gloss paint						
Metal						
upper surfaces inclined at an angle ne 30 degrees to the horizontal	0.07	4.61	0.60	1.18	m²	**6.39**
upper surfaces inclined at an angle 30 - 60 degrees to the horizontal	0.08	4.94	0.64	1.18	m²	**6.76**
surfaces inclined at an angle exceeding 60 degrees to the horizontal	0.08	5.27	0.68	1.18	m²	**7.13**
soffit surfaces and lower surfaces inclined at an angle ne 60 degrees to the horizontal	0.09	5.60	0.72	1.18	m²	**7.50**
surfaces of width ne 300 mm	0.03	1.65	0.21	0.35	m	**2.21**
surfaces of width 300 mm - 1 m	0.05	3.29	0.43	0.77	m	**4.49**
Metal sections	0.09	5.93	0.73	1.18	m²	**7.84**
Pipework	0.09	5.93	0.73	1.18	m²	**7.84**
One coat acrylic wood primer; one undercoat and one finishing coat of gloss paint						
Timber						
upper surfaces inclined at an angle ne 30 degrees to the horizontal	0.07	4.28	0.49	1.02	m²	**5.80**
upper surfaces inclined at an angle 30 - 60 degrees to the horizontal	0.07	4.61	0.60	1.02	m²	**6.23**
surfaces inclined at an angle exceeding 60 degrees to the horizontal	0.08	4.94	0.64	1.03	m²	**6.60**
soffit surfaces and lower surfaces inclined at an angle ne 60 degrees to the horizontal	0.08	5.27	0.68	1.02	m²	**6.97**
surfaces of width ne 300 mm	0.03	1.65	0.14	0.31	m	**2.09**
surfaces of width 300 mm - 1 m	0.05	3.29	0.43	0.66	m	**4.38**
One coat calcium plumbate primer; two undercoats and one finishing coat of gloss paint						
Metal						
upper surfaces inclined at an angle ne 30 degrees to the horizontal	0.06	3.95	0.51	1.50	m²	**5.95**
upper surfaces inclined at an angle 30 - 60 degrees to the horizontal	0.07	4.28	0.55	1.50	m²	**6.32**

CLASS V: PAINTING

Item	Gang hours	Labour £	Plant £	Material £	Unit	**Total rate £**
GLOSS PAINT – cont'd						
One coat calcium plumbate primer; two undercoats and one finishing coat of gloss paint – cont'd						
surfaces inclined at an angle exceeding 60 degrees to the horizontal	0.07	4.61	0.60	1.50	m²	**6.70**
soffit surfaces and lower surfaces inclined at an angle ne 60 degrees to the horizontal	0.08	4.94	0.64	1.46	m²	**7.03**
surfaces of width ne 300 mm	0.03	1.98	0.26	0.45	m	**2.68**
surfaces of width 300 mm - 1 m	0.06	3.95	0.51	0.97	m	**5.43**
One coat acrylic wood primer; two undercoats and one finishing coat of gloss paint						
Timber						
upper surfaces inclined at an angle ne 30 degrees to the horizontal	0.07	4.28	0.55	1.33	m²	**6.16**
upper surfaces inclined at an angle 30 - 60 degrees to the horizontal	0.07	4.61	0.60	1.33	m²	**6.54**
surfaces inclined at an angle exceeding 60 degrees to the horizontal	0.08	4.94	0.64	1.34	m²	**6.91**
soffit surfaces and lower surfaces inclined at an angle ne 60 degrees to the horizontal	0.08	5.27	0.63	1.33	m²	**7.23**
surfaces of width ne 300 mm	0.03	1.98	0.60	0.40	m	**2.97**
surfaces of width 300 mm - 1 m	0.06	3.95	0.51	0.87	m	**5.32**
One coat alkali resisting primer; two undercoats and one finishing coat of gloss paint						
Smooth concrete						
upper surfaces inclined at an angle ne 30 degrees to the horizontal	0.04	2.30	0.30	1.33	m²	**3.93**
upper surfaces inclined at an angle 30 - 60 degrees to the horizontal	0.04	2.63	0.34	1.33	m²	**4.30**
surfaces inclined at an angle exceeding 60 degrees to the horizontal	0.05	2.96	0.38	1.33	m²	**4.68**
soffit surfaces and lower surfaces inclined at an angle ne 60 degrees to the horizontal	0.05	3.29	0.43	1.33	m²	**5.05**
surfaces of width ne 300 mm	0.02	0.99	0.13	0.40	m	**1.52**
surfaces of width 300 mm - 1 m	0.03	1.98	0.26	0.86	m	**3.10**
Brickwork and rough concrete						
upper surfaces inclined at an angle ne 30 degrees to the horizontal	0.04	2.63	0.34	1.35	m²	**4.32**
upper surfaces inclined at an angle 30 - 60 degrees to the horizontal	0.05	2.96	0.38	1.36	m²	**4.70**
surfaces inclined at an angle exceeding 60 degrees to the horizontal	0.05	3.29	0.43	1.39	m²	**5.11**
soffit surfaces and lower surfaces inclined at an angle ne 60 degrees to the horizontal	0.06	3.62	0.47	1.39	m²	**5.48**
surfaces of width ne 300 mm	0.02	1.32	0.17	0.42	m	**1.90**
surfaces of width 300 mm - 1 m	0.04	2.30	0.30	0.89	m	**3.50**
Blockwork						
upper surfaces inclined at an angle ne 30 degrees to the horizontal	0.06	3.62	0.47	1.39	m²	**5.48**
soffit surfaces and lower surfaces inclined at an angle ne 60 degrees to the horizontal	0.06	3.95	0.51	1.39	m²	**5.84**
surfaces of width ne 300 mm	0.03	1.65	0.21	0.42	m	**2.27**
surfaces of width 300 mm - 1 m	0.04	2.63	0.34	0.90	m	**3.87**

CLASS V: PAINTING

Item	Gang hours	Labour £	Plant £	Material £	Unit	**Total rate £**
Two coats anti-condensation paint						
Metal						
upper surfaces inclined at an angle ne 30 degrees to the horizontal	0.08	4.94	0.64	3.25	m²	**8.82**
upper surfaces inclined at an angle 30 - 60 degrees to the horizontal	0.08	4.94	0.64	3.25	m²	**8.82**
surfaces inclined at an angle exceeding 60 degrees to the horizontal	0.08	4.94	0.64	3.25	m²	**8.82**
soffit surfaces and lower surfaces inclined at an angle ne 60 degrees to the horizontal	0.08	5.27	0.68	3.25	m²	**9.19**
surfaces of width ne 300 mm	0.03	1.98	0.26	0.97	m	**3.21**
surfaces of width 300 mm - 1 m	0.05	3.29	0.43	2.11	m	**5.83**
Metal sections	0.09	5.93	0.77	3.25	m²	**9.94**
EMULSION PAINT						
One thinned coat, two coats vinyl emulsion paint						
Smooth concrete						
upper surfaces inclined at an angle ne 30 degrees to the horizontal	0.07	4.28	0.55	0.87	m²	**5.70**
upper surfaces inclined at an angle 30 - 60 degrees to the horizontal	0.07	4.61	0.60	0.87	m²	**6.08**
surfaces inclined at an angle exceeding 60 degrees to the horizontal	0.08	4.94	0.64	0.87	m²	**6.45**
soffit surfaces and lower surfaces inclined at an angle ne 60 degrees to the horizontal	0.08	5.27	0.68	0.87	m²	**6.82**
surfaces of width ne 300 mm	0.03	1.98	0.26	0.26	m	**2.50**
surfaces of width 300 mm - 1 m	0.07	4.61	0.60	0.57	m	**5.77**
Brickwork and rough concrete						
upper surfaces inclined at an angle ne 30 degrees to the horizontal	0.08	4.94	0.64	0.90	m²	**6.48**
upper surfaces inclined at an angle 30 - 60 degrees to the horizontal	0.08	5.27	0.68	0.90	m²	**6.85**
surfaces inclined at an angle exceeding 60 degrees to the horizontal	0.09	5.60	0.72	0.90	m²	**7.21**
soffit surfaces and lower surfaces inclined at an angle ne 60 degrees to the horizontal	0.09	5.93	0.77	0.90	m²	**7.59**
surfaces of width ne 300 mm	0.03	1.98	0.26	0.27	m	**2.50**
surfaces of width 300 mm - 1 m	0.08	5.27	0.68	0.62	m	**6.56**
Blockwork						
surfaces inclined at an angle exceeding 60 degrees to the horizontal	0.09	5.93	0.77	0.92	m²	**7.61**
soffit surfaces and lower surfaces inclined at an angle ne 60 degrees to the horizontal	0.10	6.58	0.85	0.92	m²	**8.35**
surfaces of width ne 300 mm	0.04	2.63	0.34	0.28	m	**3.25**
surfaces of width 300 mm - 1 m	0.08	5.27	0.68	0.60	m	**6.54**

CLASS V: PAINTING

Item	Gang hours	Labour £	Plant £	Material £	Unit	**Total rate £**
CEMENT PAINT						
Two coats masonry paint						
Smooth concrete						
upper surfaces inclined at an angle ne 30 degrees to the horizontal	0.07	4.28	0.55	0.69	m²	**5.52**
upper surfaces inclined at an angle 30 - 60 degrees to the horizontal	0.07	4.61	0.60	0.69	m²	**5.90**
surfaces inclined at an angle exceeding 60 degrees to the horizontal	0.08	4.94	0.64	0.69	m²	**6.26**
soffit surfaces and lower surfaces inclined at an angle ne 60 degrees to the horizontal	0.08	5.27	0.68	0.69	m²	**6.63**
surfaces of width ne 300 mm	0.03	1.98	0.26	0.21	m	**2.44**
surfaces of width 300 mm - 1 m	0.07	4.61	0.60	0.45	m	**5.65**
Brickwork and rough concrete						
upper surfaces inclined at an angle ne 30 degrees to the horizontal	0.08	4.94	0.64	0.79	m²	**6.37**
upper surfaces inclined at an angle 30 - 60 degrees to the horizontal	0.08	5.27	0.68	0.79	m²	**6.74**
surfaces inclined at an angle exceeding 60 degrees to the horizontal	0.09	5.60	0.72	0.79	m²	**7.11**
soffit surfaces and lower surfaces inclined at an angle ne 60 degrees to the horizontal	0.09	5.93	0.77	0.79	m²	**7.49**
surfaces of width ne 300 mm	0.03	1.98	0.26	0.24	m	**2.47**
surfaces of width 300 mm - 1 m	0.08	5.27	0.68	0.52	m	**6.46**
Blockwork						
surfaces inclined at an angle exceeding 60 degrees to the horizontal	0.09	5.93	0.77	1.03	m²	**7.73**
soffit surfaces and lower surfaces inclined at an angle ne 60 degrees to the horizontal	0.10	6.58	0.85	1.03	m²	**8.46**
surfaces of width ne 300 mm	0.04	2.63	0.34	0.31	m	**3.28**
surfaces of width 300 mm - 1 m	0.08	5.27	0.68	0.67	m	**6.62**
One thinned coat, two coats concrete floor paint						
Smooth concrete						
upper surfaces inclined at an angle ne 30 degrees to the horizontal	0.07	4.28	0.55	3.27	m²	**8.10**
upper surfaces inclined at an angle 30 - 60 degrees to the horizontal	0.07	4.61	0.60	3.27	m²	**8.47**
surfaces inclined at an angle exceeding 60 degrees to the horizontal	0.08	4.94	0.64	3.27	m²	**8.84**
soffit surfaces and lower surfaces inclined at an angle ne 60 degrees to the horizontal	0.08	5.27	0.68	2.12	m²	**8.06**
surfaces of width ne 300 mm	0.03	1.98	0.26	1.01	m	**3.25**
surfaces of width 300 mm - 1 m	0.07	4.61	0.60	2.12	m	**7.33**
Additional coats						
width exceeding 1 m	0.03	1.98	0.26	1.31	m²	**3.54**
surfaces of width ne 300 mm	0.01	0.66	0.09	0.39	m	**1.14**
surfaces of width 300 mm - 1 m	0.02	0.99	0.21	0.85	m	**2.05**

CLASS V: PAINTING

Item	Gang hours	Labour £	Plant £	Material £	Unit	**Total rate £**
EPOXY OR POLYURETHANE PAINT						
Blast clean to BS 7079; one coat zinc chromate etch primer, two coats zinc phosphate CR/alkyd undercoat off site; one coat MIO CR undercoat and one coat CR finish on site						
Metal sections	-	-	-	-	m²	**24.56**
Blast clean to BS 7079; one coat zinc rich 2 pack primer, one coat MIO high build epoxy 2 pack paint off site; one coat polyurethane 2 pack undercoat and one coat polyurethane 2 pack finish on site						
Metal sections	-	-	-	-	m²	**26.78**
Two coats clear polyurethane varnish						
Timber						
upper surfaces inclined at an angle ne 30 degrees to the horizontal	0.04	2.63	0.34	1.37	m²	**4.34**
upper surfaces inclined at an angle 30 - 60 degrees to the horizontal	0.05	2.96	0.38	1.37	m²	**4.71**
surfaces inclined at an angle exceeding 60 degrees to the horizontal	0.05	3.29	0.43	1.37	m²	**5.09**
soffit surfaces and lower surfaces inclined at an angle ne 60 degrees to the horizontal	0.06	3.95	0.51	1.37	m²	**5.83**
surfaces of width ne 300 mm	0.03	1.65	0.21	0.41	m	**2.27**
surfaces of width 300 mm - 1 m	0.05	3.29	0.43	0.89	m	**4.61**
Two coats colour stained polyurethane varnish						
Timber						
upper surfaces inclined at an angle ne 30 degrees to the horizontal	0.04	2.63	0.34	1.04	m²	**4.02**
upper surfaces inclined at an angle 30 - 60 degrees to the horizontal	0.05	2.96	0.38	1.04	m²	**4.38**
surfaces inclined at an angle exceeding 60 degrees to the horizontal	0.05	3.29	0.43	1.04	m²	**4.76**
soffit surfaces and lower surfaces inclined at an angle ne 60 degrees to the horizontal	0.06	3.95	0.51	1.04	m²	**5.50**
surfaces of width ne 300 mm	0.03	1.65	0.21	0.31	m	**2.17**
surfaces of width 300 mm - 1 m	0.05	3.29	0.43	0.68	m	**4.40**
Three coats colour stained polyurethane varnish						
Timber						
upper surfaces inclined at an angle ne 30 degrees to the horizontal	0.06	3.62	0.47	1.57	m²	**5.65**
upper surfaces inclined at an angle 30 - 60 degrees to the horizontal	0.07	4.28	0.55	1.57	m²	**6.39**
surfaces inclined at an angle exceeding 60 degrees to the horizontal	0.07	4.61	0.60	1.57	m²	**6.77**
soffit surfaces and lower surfaces inclined at an angle ne 60 degrees to the horizontal	0.09	5.60	0.72	1.57	m²	**7.88**
surfaces of width ne 300 mm	0.04	2.30	0.30	0.47	m	**3.07**
surfaces of width 300 mm - 1 m	0.07	4.61	0.60	1.02	m	**6.22**

CLASS V: PAINTING

Item	Gang hours	Labour £	Plant £	Material £	Unit	**Total rate £**
EPOXY OR POLYURETHANE PAINT – cont'd						
One coat hardwood stain basecoat and two coats hardwood woodstain						
Timber						
upper surfaces inclined at an angle ne 30 degrees to the horizontal	0.06	3.95	0.51	1.40	m²	**5.86**
upper surfaces inclined at an angle 30 - 60 degrees to the horizontal	0.07	4.28	0.55	1.40	m²	**6.23**
surfaces inclined at an angle exceeding 60 degrees to the horizontal	0.07	4.61	0.60	1.40	m²	**6.60**
soffit surfaces and lower surfaces inclined at an angle ne 60 degrees to the horizontal	0.08	5.27	0.68	1.50	m²	**7.45**
surfaces of width ne 300 mm	0.04	2.30	0.30	0.42	m	**3.02**
surfaces of width 300 mm - 1 m	0.07	4.61	0.60	0.91	m	**6.11**
BITUMINOUS OR COAL TAR PAINT						
Two coats golden brown creosote wood preservative						
Timber						
upper surfaces inclined at an angle ne 30 degrees to the horizontal	0.05	2.96	0.38	0.10	m²	**3.44**
upper surfaces inclined at an angle 30 - 60 degrees to the horizontal	0.05	3.29	0.43	0.10	m²	**3.82**
surfaces inclined at an angle exceeding 60 degrees to the horizontal	0.06	3.62	0.47	0.10	m²	**4.19**
soffit surfaces and lower surfaces inclined at an angle ne 60 degrees to the horizontal	0.07	4.28	0.55	0.10	m²	**4.93**
surfaces of width ne 300 mm	0.03	1.98	0.26	0.03	m	**2.26**
surfaces of width 300 mm - 1 m	0.06	3.62	0.47	0.06	m	**4.15**
Two coats bituminous paint						
Metal sections	0.08	5.27	0.68	1.33	m²	**7.28**
Pipework	0.09	5.93	0.81	1.33	m²	**8.07**

CLASS W: WATERPROOFING

Item	Gang hours	Labour £	Plant £	Material £	Unit	Total rate £
NOTES						
Asphalt roofing						
Work has been presented in more detail than CESMM3 to allow the user greater freedom to access an appropriate rate to suit the complexity of his work.						
RESOURCES - LABOUR						
Roofing - cladding gang						
1 ganger/chargehand (skill rate 3) - 50% of time		7.11				
2 skilled operative (skill rate 3)		26.92				
1 unskilled operative (general) - 50% of time		5.63				
Total Gang Rate / Hour	£	**39.66**				
Damp proofing gang						
1 ganger (skill rate 4)		12.80				
1 skilled operative (skill rate 4)		12.05				
1 unskilled labour (general)		11.26				
Total Gang Rate / Hour	£	**36.11**				
Roofing - asphalt gang						
1 ganger/chargehand (skill rate 4) - 50% of time		6.40				
1 skilled operative (skill rate 4)		12.05				
1 unskilled operative (general)		11.26				
Total Gang Rate / Hour	£	**29.71**				
Tanking - asphalt gang						
1 ganger/chargehand (skill rate 4) - 50% of time		6.40				
1 skilled operative (skill rate 4)		12.05				
1 unskilled operative (general)		11.26				
Total Gang Rate / Hour	£	**29.71**				
Tanking - waterproof sheeting gang						
1 skilled operative (skill rate 4)		12.05				
Total Gang Rate / Hour	£	**12.05**				
Tanking - rendering gang						
1 ganger/chargehand (skill rate 4)		12.80				
1 skilled operative (skill rate 4)		12.05				
1 unskilled operative (generally)		11.26				
Total Gang Rate / Hour	£	**36.11**				
Protective layers - flexible sheeting, sand and pea gravel coverings gang						
1 unskilled operative (generally)		11.26				
Total Gang Rate / Hour	£	**11.26**				
Protective layers - screed gang						
1 ganger/chargehand (skill rate 4)		12.80				
1 skilled operative (skill rate 4)		12.05				
1 unskilled operative (general)		11.26				
Total Gang Rate / Hour	£	**36.11**				
Sprayed or brushed waterproofing gang						
1 ganger/chargehand (skill rate 4) - 30% of time		3.84				
1 skilled operative (skill rate 4)		12.05				
Total Gang Rate / Hour	£	**15.89**				

CLASS W: WATERPROOFING

Item	Gang hours	Labour £	Plant £	Material £	Unit	**Total rate £**
RESOURCES - PLANT						
Damp proofing						
2t dumper (50% of time)			2.17			
Total Rate / Hour		**£**	**2.17**			
Tanking - asphalt						
tar boiler (50% of time)			7.39			
2t dumper (50% of time)			2.17			
Total Rate / Hour		**£**	**9.56**			
Tanking - waterproof sheeting						
2t dumper (50% of time)			2.17			
Total Rate / Hour		**£**	**2.17**			
Tanking - rendering						
mixer			0.84			
2t dumper (50% time)			2.17			
Total Rate / Hour		**£**	**3.01**			
Protective layers - flexible sheeting, sand and pea gravel coverings						
2t dumper (50% of time)			2.17			
Total Rate / Hour		**£**	**2.17**			
Protective layers - screed						
mixer			0.84			
2t dumper (50% time)			2.17			
Total Rate / Hour		**£**	**3.01**			
Sprayed or brushed waterproofing						
2t dumper (50% of time)			2.17			
Total Rate / Hour		**£**	**2.17**			
DAMP PROOFING						
Waterproof sheeting						
0.3 mm polythene sheet						
ne 300 mm wide	0.01	0.11	0.02	0.54	m	**0.67**
300 mm - 1 m wide	0.01	0.12	0.02	1.08	m	**1.22**
on horizontal or included surfaces	0.01	0.14	0.03	1.64	m²	**1.82**
on vertical surfaces	0.01	0.12	0.02	1.64	m²	**1.79**
TANKING						
Asphalt						
13 mm Mastic asphalt to BS 6925, Type T 1097; two coats; on concrete surface						
upper surfaces inclined at an angle ne 30 degrees to the horizontal	0.45	13.37	4.72	5.30	m²	**23.39**
upper surfaces inclined at an angle 30 - 60 degrees to the horizontal	0.60	17.83	6.29	5.30	m²	**29.42**
upper surfaces inclined at an angle exceeding 60 degrees to the horizontal	1.00	29.71	10.49	5.30	m²	**45.50**
curved surfaces	1.20	35.65	12.59	5.30	m²	**53.54**
domed surfaces	1.50	44.56	15.74	5.30	m²	**65.60**
ne 300 mm wide	0.20	5.94	2.10	1.60	m	**9.64**
300 mm - 1 m wide	0.45	13.37	4.72	5.30	m	**23.39**

CLASS W: WATERPROOFING

Item	Gang hours	Labour £	Plant £	Material £	Unit	**Total rate £**
20 mm Mastic asphalt to BS 6925, Type T 1097; two coats; on concrete surface						
upper surfaces inclined at an angle ne 30 degrees to the horizontal	0.50	14.86	5.25	8.14	m²	**28.24**
upper surfaces inclined at an angle 30 - 60 degrees to the horizontal	0.70	20.80	7.34	8.14	m²	**36.28**
upper surfaces inclined at an angle exceeding 60 degrees to the horizontal	1.20	35.65	12.59	8.14	m²	**56.38**
curved surfaces	1.40	41.59	14.69	8.14	m²	**64.42**
domed surfaces	1.75	51.99	18.36	8.14	m²	**78.49**
ne 300 mm wide	0.23	6.83	2.41	2.44	m	**11.68**
300 mm - 1 m wide	0.50	14.86	5.25	8.14	m	**28.24**
13 mm Mastic asphalt to BS 6925, Type T 1097; two coats; on brickwork surface; raking joints to form key						
upper surfaces inclined at an angle 30 - 60 degrees to the horizontal	0.90	26.74	6.32	8.17	m²	**41.23**
upper surfaces inclined at an angle exceeding 60 degrees to the horizontal	1.30	38.62	13.64	8.17	m²	**60.43**
curved surfaces	1.50	44.56	15.74	8.17	m²	**68.47**
domed surfaces	1.80	53.48	18.88	8.17	m²	**80.53**
ne 300 mm wide	0.30	8.91	2.10	2.45	m	**13.47**
300 mm - 1 m wide	0.75	22.28	7.97	8.17	m	**38.42**
Waterproof sheeting						
Bituthene 3000; lapped joints						
upper surfaces inclined at an angle ne 30 degrees to the horizontal	0.05	1.81	0.11	5.23	m²	**7.14**
upper surfaces inclined at an angle 30 - 60 degrees to the horizontal	0.05	1.81	0.11	5.23	m²	**7.14**
ne 300 mm wide	0.03	0.90	0.05	1.38	m	**2.33**
300 mm - 1 m wide	0.04	1.44	0.09	2.94	m	**4.47**
Bituthene 3000; lapped joints; primer coat						
upper surfaces inclined at an angle exceeding 60 degrees to the horizontal	0.06	2.17	0.13	5.55	m²	**7.85**
Rendering in waterproof cement mortar						
19 mm render in waterproof cement mortar (1:3); two coat work						
upper surfaces inclined at an angle exceeding 60 degrees to the horizontal	0.11	3.97	0.33	9.04	m²	**13.34**
ne 300 mm wide	0.07	2.53	0.21	3.35	m	**6.09**
300 mm - 1 m wide	0.11	3.97	0.33	6.21	m	**10.52**
32 mm render in waterproof cement mortar (1:3); one coat work						
upper surfaces inclined at an angle ne 30 degrees to the horizontal	0.11	3.97	0.33	11.36	m²	**15.66**
ne 300 mm wide	0.07	2.53	0.21	3.30	m	**6.04**
300 mm - 1 m wide	0.11	3.97	0.33	5.76	m	**10.06**

CLASS W: WATERPROOFING

Item	Gang hours	Labour £	Plant £	Material £	Unit	**Total rate £**
ROOFING						
Asphalt						
13 mm Mastic asphalt to BS 6925 Type R 988; two coats; on concrete surface						
ne 300 mm wide	0.20	5.94	2.10	1.93	m	**9.97**
20 mm Mastic asphalt to BS 6925 Type R 988; two coats; on concrete surface						
upper surfaces inclined at an angle ne 30 degrees to the horizontal	0.50	14.86	5.25	10.28	m²	**30.38**
upper surfaces inclined at an angle 30 - 60 degrees to the horizontal	0.70	20.80	7.34	10.28	m²	**38.42**
surfaces inclined at an angle exceeding 60 degrees to the horizontal	1.20	35.65	12.59	10.28	m²	**58.52**
curved surfaces	1.50	44.56	15.74	10.28	m²	**70.58**
domed surfaces	1.75	51.99	18.36	10.28	m²	**80.63**
ne 300 mm wide	0.23	6.83	2.41	3.08	m	**12.33**
300 mm - 1 m wide	0.50	14.86	5.25	10.28	m	**30.38**
Extra for :						
10 mm thick limestone chippings bedded in hot bitumen	0.05	1.49	0.28	1.04	m²	**2.80**
dressing with solar reflective paint	0.05	1.49	0.28	2.54	m²	**4.30**
300 x 300 x 8 mm GRP tiles bedded in hot bitumen	0.30	8.91	1.65	22.71	m²	**33.28**
PROTECTIVE LAYERS						
Flexible sheeting						
3 mm Servi-pak protection board to Bituthene						
upper surfaces inclined at an angle ne 30 degrees to the horizontal	0.20	2.25	0.43	14.17	m²	**16.86**
ne 300 mm wide	0.10	1.13	0.22	6.05	m	**7.39**
300 mm - 1 m wide	0.20	2.25	0.43	14.17	m	**16.86**
3 mm Servi-pak protection board to Bituthene; fixing with adhesive dabs						
upper surfaces inclined at an angle exceeding 60 degrees to the horizontal	0.25	2.81	0.54	15.01	m²	**18.37**
ne 300 mm wide	0.12	1.35	0.26	6.37	m	**7.98**
300 mm - 1 m wide	0.25	2.81	0.54	15.01	m	**18.37**
6 mm Servi-pak protection board to Bituthene						
upper surfaces inclined at an angle ne 30 degrees to the horizontal	0.20	2.25	0.43	22.25	m²	**24.93**
ne 300 mm wide	0.10	1.13	0.22	9.16	m	**10.50**
300 mm 1 m wide	0.20	2.25	0.43	22.25	m	**24.93**
6 mm Servi-pak protection board to Bituthene; fixing with adhesive dabs						
upper surfaces inclined at an angle exceeding 60 degrees to the horizontal	0.25	2.81	0.54	23.33	m²	**26.69**
ne 300 mm wide	0.12	1.35	0.26	9.82	m	**11.44**
300 mm 1 m wide	0.25	2.81	0.54	23.33	m	**26.69**
Sand covering						
25 mm thick						
upper surfaces inclined at an angle ne 30 degrees to the horizontal	0.02	0.23	0.02	0.53	m²	**0.77**

CLASS W: WATERPROOFING

Item	Gang hours	Labour £	Plant £	Material £	Unit	**Total rate £**
Pea gravel covering						
50 mm thick						
upper surfaces inclined at an angle ne 30 degrees to the horizontal	0.02	0.23	0.02	1.61	m²	**1.85**
Sand and cement screed						
50 mm screed in cement mortar (1:4); one coat work						
upper surfaces inclined at an angle ne 30 degrees to the horizontal	0.13	4.69	0.29	4.02	m²	**9.00**
SPRAYED OR BRUSHED WATERPROOFING						
Two coats RIW liquid asphaltic composition						
on horizontal or vertical surfaces	0.06	0.95	0.13	4.10	m²	**5.18**
Two coats Aquaseal						
on horizontal or vertical surfaces	0.06	0.95	0.13	2.59	m²	**3.67**
One coat Ventrot primer; one coat Ventrot hot applied damp proof membrane						
on horizontal or vertical surfaces	0.05	0.79	0.11	5.82	m²	**6.72**

CLASS X: MISCELLANEOUS WORK

Item	Gang hours	Labour £	Plant £	Material £	Unit	**Total rate £**
RESOURCES - LABOUR						
Fencing / barrier gang						
1 ganger/chargehand (skill rate 4) - 50% of time		6.40				
1 skilled operative (skill rate 4) - 50% of time		6.03				
1 unskilled operative (general)		11.26				
1 plant operator (skill rate 4)		13.17				
Total Gang Rate / Hour	£	**36.86**				
Safety fencing gang						
1 ganger/chargehand (skill rate 4)		12.80				
1 skilled operative (skill rate 4)		12.05				
2 unskilled operatives (general)		22.52				
1 plant operator (skill rate 4)		13.17				
Total Gang Rate / Hour	£	**60.54**				
Guttering gang						
2 skilled operatives (skill rate 4)		24.10				
Total Gang Rate / Hour	£	**24.10**				
Rock filled gabions gang						
1 ganger/chargehand (skill rate 4)		12.80				
4 unskilled operatives (general)		45.04				
1 plant operator (skill rate 3) - 50% of time		7.33				
Total Gang Rate / Hour	£	**65.17**				
RESOURCES - PLANT						
Fencing/barriers						
agricultural type tractor; fencing auger			12.69			
gas oil for ditto			2.63			
drop sided trailer; two axles			0.31			
power tools (fencing)			2.80			
Total Rate / Hour		£	**18.42**			
Guttering						
ladders			1.66			
Total Rate / Hour		£	**1.66**			
Rock filled gabions						
16 tonne crawler backacter (50% of time)			11.64			
Total Rate / Hour		£	**11.64**			
FENCES						
Timber fencing						
Timber post and wire						
1.20 m high; DTp type 3; timber posts, driven; cleft chestnut paling	0.07	2.58	1.29	4.27	m	**8.14**
0.90 m high; DTp type 4; galvanised rectangular wire mesh	0.13	4.79	2.39	4.39	m	**11.58**
1.275 m high; DTp type 1; galvanised wire, 2 barbed, 4 plain	0.06	2.21	1.11	4.77	m	**8.09**
1.275 m high; DTp type 2; galvanised wire, 2 barbed, 4 plain	0.06	2.21	1.11	4.39	m	**7.71**

CLASS X: MISCELLANEOUS WORK

Item	Gang hours	Labour £	Plant £	Material £	Unit	**Total rate £**
Concrete post and wire						
1.20 m high; DTp type 3; timber posts, driven; cleft chestnut paling	0.07	2.58	1.29	4.27	m	**8.14**
0.90 m high; DTp type 4; galvanised rectangular wire mesh	0.13	4.79	2.39	4.39	m	**11.58**
1.275 m high; DTp type 1; galvanised wire, 2 barbed, 4 plain	0.06	2.21	1.11	4.77	m	**8.09**
1.275 m high; DTp type 2; galvanised wire, 2 barbed, 4 plain	0.06	2.21	1.11	4.39	m	**7.71**
Metal post and wire						
1.20 m high; DTp type 3; timber posts, driven; cleft chestnut paling	0.07	2.58	1.29	4.27	m	**8.14**
0.90 m high; DTp type 4; galvanised rectangular wire mesh	0.13	4.79	2.39	4.39	m	**11.58**
1.275 m high; DTp type 1; galvanised wire, 2 barbed, 4 plain	0.06	2.21	1.11	4.77	m	**8.09**
1.275 m high; DTp type 2; galvanised wire, 2 barbed, 4 plain	0.06	2.21	1.11	4.39	m	**7.71**
Timber post and wire						
1.20 m high; DTp type 3; timber posts, driven; cleft chestnut paling	0.07	2.58	1.29	4.27	m	**8.14**
0.90 m high; DTp type 4; galvanised rectangular wire mesh	0.13	4.79	2.39	4.39	m	**11.58**
1.275 m high; DTp type 1; galvanised wire, 2 barbed, 4 plain	0.06	2.21	1.11	4.77	m	**8.09**
1.275 m high; DTp type 2; galvanised wire, 2 barbed, 4 plain	0.06	2.21	1.11	4.39	m	**7.71**
Timber close boarded; concrete posts						
Timber close boarded						
1.80 m high; 125 x 125 mm posts	0.30	11.06	5.53	36.18	m	**52.77**
Wire rope safety fencing to BS 5750; based on 600 m lengths						
Metal crash barriers						
600 mm high; 4 wire ropes; long line posts at 2.40 m general spacings, driven	0.16	9.69	2.95	63.62	m	**76.26**
600 mm high; 4 wire ropes; short line posts at 2.40 m general spacings, 400 x 400 x 600 mm concrete footing	0.27	16.35	4.97	66.06	m	**87.38**
600 mm high; 4 wire ropes; short line posts at 2.40 m general spacings, 400 x 400 x 600 mm concrete footing, socketed	0.32	19.37	5.90	66.06	m	**91.33**
600 mm high; 4 wire ropes; short line posts at 2.40 m general spacings, bolted to structure	0.20	12.11	3.69	62.86	m	**78.65**
Pedestrian guard rails						
Metal guard rails						
1000 mm high; tubular galvanised mild steel to BS 3049, mesh infill (105 swg, 50 x 50 mm mesh; steel posts with concrete footing	0.80	9.57	2.95	165.98	m	**178.49**

CLASS X: MISCELLANEOUS WORK

Item	Gang hours	Labour £	Plant £	Material £	Unit	Total rate £
FENCES – cont'd						
Beam safety fencing; based on 600 m lengths						
Metal crash barriers						
600 mm high; untensioned corrugated beam, single sided; long posts at 3.20 m general spacings, driven	0.10	6.05	1.84	29.92	m	**37.82**
600 mm high; untensioned corrugated beam, double sided; long posts at 3.20 m general spacings, driven	0.26	15.74	4.79	47.12	m	**67.65**
600 mm high; untensioned open box beam, single sided; long posts at 3.20 m general spacings, driven	0.10	6.05	1.84	53.62	m	**61.51**
600 mm high; untensioned open box beam, double sided; long posts at 3.20 m general spacings, driven	0.26	15.74	4.79	94.86	m	**115.39**
600 mm high; untensioned open box beam, double height; long posts at 3.20 m general spacings, driven	0.30	18.16	5.53	118.11	m	**141.80**
600 mm high; tensioned corrugated beam, single sided; long posts at 3.20 m general spacings, driven	0.13	7.87	2.39	33.62	m	**43.88**
600 mm high; tensioned corrugated beam, double sided; long posts at 3.20 m general spacings, driven	0.37	22.40	6.82	54.51	m	**83.73**
Gates and stiles						
Timber field gates						
single; 3.00 m wide x 1.27 m high	1.84	67.82	33.34	145.40	nr	**246.56**
single; 3.60 m wide x 1.27 m high	1.90	70.03	34.43	168.37	nr	**272.83**
single; 4.10 m wide x 1.27 m high	2.00	73.72	36.24	203.59	nr	**313.54**
single; 4.71 m wide x 1.27 m high	2.00	73.72	36.24	209.68	nr	**319.64**
Timber wicket gates						
single;1.20 m wide x 1.20 m high; DTp Type 1	1.20	44.23	21.74	45.92	nr	**111.89**
single; 1.20 m wide x 1.02 m high; DTp Type 2	1.20	44.23	21.74	52.37	nr	**118.34**
Metal field gates						
single; steel tubular; 3.60 m wide x 1.175 m high	0.30	11.06	5.44	58.29	nr	**74.79**
single; steel tubular; 4.50 m wide x 1.175 m high	0.30	11.06	5.44	74.94	nr	**91.43**
single; steel tubular; half mesh; 3.60 m wide x 1.175 m high	0.30	11.06	5.44	74.94	nr	**91.43**
single; steel tubular; half mesh; 4.50 m wide x 1.175 m high	0.30	11.06	5.44	91.58	nr	**108.08**
single; steel tubular; extra wide; 4.88 m wide x 1.175 m high	0.30	11.06	5.44	83.26	nr	**99.75**
double; steel tubular; 5.02 m wide x 1.175 m high	0.60	22.12	10.87	124.92	nr	**157.90**
Stiles						
1.00 m wide x 1.45 m high; DTp Type 1	1.50	55.29	27.18	106.98	nr	**189.45**
1.00 m wide x 1.45 m high; DTp Type 2	1.40	51.60	25.37	80.40	nr	**157.37**
GATES AND STILES						
Notes						
Refer to Part 5 Series 300.						

CLASS X: MISCELLANEOUS WORK

Item	Gang hours	Labour £	Plant £	Material £	Unit	**Total rate £**
DRAINAGE TO STRUCTURES ABOVE GROUND						
Note						
Outputs are based on heights up to 3 m above ground and exclude time spent on erecting access equipment, but include marking, cutting, drilling to wood, brick or concrete and all fixings. testing of finished work is not included.						
Output multipliers for labour and plant for heights over 3 m :						
3 - 6 m x 1.25						
6 - 9 m x 1.50						
9 - 12 m x 1.75						
12 - 15 m x 2.00						
Cast iron gutters and fittings; BS 460						
100 x 75 mm gutters; support brackets	0.25	6.03	0.41	30.33	m	**36.77**
stop end	0.03	0.65	0.04	23.78	nr	**24.47**
running outlet	0.04	0.84	0.06	25.03	nr	**25.93**
angle	0.04	0.84	0.06	37.85	nr	**38.75**
125 x 75 mm gutters; support brackets	0.30	7.23	0.50	37.60	m	**45.33**
stop end	0.03	0.77	0.05	27.80	nr	**28.63**
running outlet	0.04	0.96	0.07	28.98	nr	**30.01**
angle	0.04	0.96	0.07	45.95	nr	**46.98**
Cast iron rainwater pipes and fittings; BS 460						
65 mm diameter; support brackets	0.26	6.27	0.43	19.65	m	**26.35**
bend	0.28	6.75	0.46	11.19	nr	**18.40**
offset, 75 mm projection	0.30	7.23	0.50	16.81	nr	**24.53**
offset, 225 mm projection	0.30	7.23	0.50	19.56	nr	**27.29**
offset, 455 mm projection	0.30	7.23	0.50	53.53	nr	**61.26**
shoe	0.18	4.34	0.30	15.55	nr	**20.18**
75 mm diameter; support brackets	0.28	6.75	0.46	19.65	m	**26.86**
bend	0.28	6.75	0.46	13.32	nr	**20.54**
offset, 75 mm projection	0.30	7.23	0.50	16.81	nr	**24.53**
offset, 225 mm projection	0.30	7.23	0.50	19.56	nr	**27.29**
offset, 455 mm projection	0.30	7.23	0.50	53.53	nr	**61.26**
shoe	0.18	4.34	0.30	15.55	nr	**20.18**
PVC-U gutters and fittings; Marley						
116 x 75 mm gutters; support brackets	0.18	4.34	0.30	7.26	m	**11.90**
stop end	0.05	1.21	0.08	2.89	nr	**4.18**
running outlet	0.05	1.21	0.08	10.36	nr	**11.65**
angle	0.14	3.37	0.23	12.64	nr	**16.25**
150 mm half round gutters; support brackets	0.18	4.34	0.30	10.59	m	**15.22**
stop end	0.05	1.21	0.08	3.50	nr	**4.79**
running outlet	0.05	1.21	0.08	12.96	nr	**14.24**
angle	0.14	3.37	0.23	11.84	nr	**15.44**
PVC-U external rainwater pipes and fittings; Marley						
68 mm diameter; support brackets	0.15	3.62	0.25	3.29	m	**7.16**
bend	0.16	3.86	0.27	7.08	nr	**11.20**
offset bend	0.18	4.34	0.30	3.92	nr	**8.55**
shoe	0.10	2.41	0.17	6.12	nr	**8.69**
110 mm diameter; support brackets	0.15	3.62	0.25	8.31	m	**12.17**
bend	0.16	3.86	0.27	18.75	nr	**22.87**
offset bend	0.18	4.34	0.30	16.71	nr	**21.34**
shoe	0.10	2.41	0.17	14.38	nr	**16.96**

CLASS X: MISCELLANEOUS WORK

Item	Gang hours	Labour £	Plant £	Material £	Unit	**Total rate £**
ROCK FILLED GABIONS						
Gabions						
PVC coated galvanised wire mesh box gabions, wire laced; graded broken stone filling						
1.0 x 1.0 m module sizes	0.65	42.36	7.57	49.69	m³	**99.61**
1.0 x 0.5 m module sizes	0.80	52.14	9.31	61.62	m³	**123.07**
Heavily galvanised woven wire mesh box gabions, wire laced; graded broken stone filling						
1.0 x 1.0 m module sizes	0.65	42.36	7.57	45.40	m³	**95.33**
1.0 x 0.5 m module sizes	0.80	52.14	9.31	54.84	m³	**116.29**
Reno mattresses						
PVC coated woven wire mesh mattresses, wire tied; graded broken stone filling						
230 mm deep	0.15	9.78	1.75	15.41	m²	**26.93**
Heavily galvanised woven wire mesh, wire tied; graded broken stone filling						
300 mm deep	0.15	9.78	1.75	18.58	m²	**30.10**

CLASS Y: SEWER AND WATER MAIN RENOVATION AND ANCILLARY WORKS

Item	Gang hours	Labour £	Plant £	Material £	Unit	**Total rate £**
RESOURCES - LABOUR						
Drain repair gang						
1 chargehand pipelayer (skill rate 4) - 50% of time		6.40				
1 skilled operative (skill rate 4)		12.05				
2 unskilled operatives (generally)		22.52				
1 plant operator (skill rate 3)		14.65				
Total Gang Rate / Hour	£	**55.62**				
RESOURCES - PLANT						
Drainage						
0.40 m³ hydraulic excavator			17.88			
trench sheets, shores, props etc.			9.27			
2t dumper (30% of time)			1.30			
compaction / plate roller (30% of time)			0.36			
7.30 m³/min compressor			3.00			
small pump			0.65			
Total Rate / Hour		£	**32.45**			
PREPARATION OF EXISTING SEWERS						
Cleaning						
eggshape sewer 1300 mm high	-	-	-	-	m	**13.40**
Removing intrusions						
brickwork	-	-	-	-	m³	**70.20**
concrete	-	-	-	-	m³	**121.47**
reinforced concrete	-	-	-	-	m³	**154.38**
Plugging laterals with concrete plug						
bore not exceeding 300 mm	-	-	-	-	nr	**67.74**
bore 450 mm	-	-	-	-	nr	**104.55**
Plugging laterals with brickwork plug						
bore 750 mm	-	-	-	-	nr	**314.78**
Local internal repairs to brickwork						
area not exceeding 0.1 m²	-	-	-	-	nr	**13.95**
area 0.1 - 0.25 m²	-	-	-	-	nr	**36.41**
area 1m²	-	-	-	-	nr	**88.60**
area 10 m²	-	-	-	-	nr	**408.48**
Grouting ends of redundant drains and sewers						
100 mm diameter	0.03	1.67	0.97	4.48	nr	**7.11**
300 mm diameter	0.13	6.95	4.04	17.92	nr	**28.91**
450 mm diameter	0.26	14.46	8.45	39.10	nr	**62.02**
600 mm diameter	0.50	27.81	16.25	79.13	nr	**123.19**
1200 mm diameter	1.70	94.55	55.18	223.97	nr	**373.70**
STABILIZATION OF EXISTING SEWERS						
Pointing, cement mortar (1:3)						
faces of brickwork	-	-	-	-	m²	**29.77**

CLASS Y: SEWER AND WATER MAIN RENOVATION AND ANCILLARY WORKS

Item	Gang hours	Labour £	Plant £	Material £	Unit	**Total rate £**
RENOVATION OF EXISTING SEWERS						
Sliplining						
GRP one piece unit; eggshape sewer 21300 mm high	-	-	-	-	m	**316.30**
Renovation of existing sewers						
sliplining	-	-	-	-	hr	**76.70**
LATERALS TO RENOVATED SEWERS						
Jointing						
bore not exceeding 150 mm	-	-	-	-	nr	**49.02**
bore 150 - 300 mm	-	-	-	-	nr	**86.98**
bore 450 mm	-	-	-	-	nr	**121.77**
EXISTING MANHOLES						
Abandonment						
sealing redundant road gullies with grade C15 concrete	0.02	1.28	0.76	14.18	nr	**16.22**
sealing redundant chambers with grade C15 concrete						
ne 1.0 m deep to invert	0.09	5.01	2.93	57.11	nr	**65.05**
1.0 - 2.0 m deep to invert	0.21	11.68	6.82	92.20	nr	**110.71**
2.0 - 3.0 m deep to invert	0.55	30.59	17.84	153.42	nr	**201.85**
Alteration						
100 x 100 mm water stop tap boxes on 100 x 100 mm brick chambers						
raising the level by 150 mm or less	0.06	3.34	1.96	17.11	nr	**22.41**
lowering the level by 150 mm or less	0.04	2.22	1.30	9.64	nr	**13.16**
420 x 420 mm cover and frame on 420 x 420 mm in-situ concrete chamber						
raising the level by 150 mm or less	0.10	5.56	3.26	30.08	nr	**38.90**
lowering the level by 150 mm or less	0.06	3.34	1.96	18.97	nr	**24.27**
Raising the level of 700 x 700 mm cover and frame on 700 x 500 mm in-situ concrete chamber by 150 mm or less	0.17	9.46	5.53	43.85	nr	**58.84**
600 x 600 mm grade "A" heavy duty manhole cover and frame on 600 x 600 mm brick chamber						
raising the level by 150 mm or less	0.17	9.46	5.53	43.85	nr	**58.84**
raising the level by 150 - 300 mm	0.21	11.68	6.82	55.05	nr	**73.56**
lowering the level by 150 mm or less	0.10	5.56	3.26	27.06	nr	**35.88**
INTERRUPTIONS						
Preparation of existing sewers						
cleaning	-	-	-	-	hr	**324.78**
Stabilization of existing sewers						
pointing	-	-	-	-	hr	**65.83**

CLASS Z: SIMPLE BUILDING WORKS INCIDENTAL TO CIVIL ENGINEERING WORKS

Item	Gang hours	Labour £	Plant £	Material £	Unit	**Total rate £**
NOTES						
The user should refer to Part 6: Unit Costs (Ancillary Building Works for cost guidance on the matters listed in this section.						
RESOURCES - LABOUR						
Carpentry and joinery gang						
1 foreman carpenter (craftsman)		18.47				
5 carpenters (craftsmen)		86.75				
1 unskilled operative (general)		11.26				
Total Gang Rate/Hour	£	**116.48**				
Ironmongery gang						
5 carpenters (craftsmen)		17.35				
Total Gang Rate/Hour	£	**17.35**				
Glazing gang						
1 glazier (craftsman)		17.35				
Total Gang Rate/Hour	£	**17.35**				
Finishings gang						
2 plasterers / ceramic tilers (craftsmen)		34.70				
1 unskilled operative (general)		11.26				
Total Gang Rate/Hour	£	**45.96**				
Vinyl tiling gang						
1 tiler (craftsman)		17.35				
1 unskilled operative (general)		11.26				
Total Gang Rate/Hour	£	**28.61**				
Plumbing gang						
1 plumber (craftsman)		17.35				
1 unskilled operative (general)		11.26				
Total Gang Rate/Hour	£	**28.61**				
CARPENTRY AND JOINERY						
Softwood; structural grade SC3; sawn; tanalised						
Structural and carcassing timber; floors						
38 x 100	0.02	2.56	-	1.32	m	**3.89**
38 x 125	0.02	2.68	-	1.72	m	**4.40**
38 x 150	0.03	2.91	-	2.15	m	**5.06**
50 x 100	0.03	2.91	-	1.52	m	**4.43**
50 x 125	0.03	3.03	-	1.79	m	**4.82**
50 x 150	0.03	3.15	-	2.29	m	**5.43**
50 x 175	0.03	3.15	-	2.52	m	**5.67**
50 x 200	0.03	3.38	-	3.09	m	**6.47**
50 x 225	0.03	3.38	-	3.47	m	**6.85**
75 x 200	0.03	3.61	-	5.21	m	**8.82**
75 x 225	0.03	3.61	-	6.24	m	**9.85**
100 x 200	0.04	4.89	-	7.37	m	**12.27**
100 x 225	0.05	5.24	-	8.84	m	**14.08**

CLASS Z: SIMPLE BUILDING WORKS INCIDENTAL TO CIVIL ENGINEERING WORKS

Item	Gang hours	Labour £	Plant £	Material £	Unit	Total rate £
CARPENTRY AND JOINERY – cont'd						
Softwood; structural grade SC3; sawn; tanalised – cont'd						
Structural and carcassing timber; walls and partitions						
38 x 100	0.03	3.15	-	1.32	m	**4.47**
38 x 125	0.03	3.49	-	1.72	m	**5.21**
38 x 150	0.03	3.73	-	2.15	m	**5.88**
50 x 100	0.03	3.84	-	1.52	m	**5.36**
50 x 125	0.04	4.08	-	1.79	m	**5.87**
Structural and carcassing timber; flat roofs						
38 x 100	0.03	2.91	-	1.32	m	**4.23**
38 x 125	0.02	2.68	-	1.72	m	**4.40**
38 x 150	0.03	2.91	-	2.15	m	**5.06**
50 x 100	0.03	2.91	-	1.52	m	**4.43**
50 x 125	0.03	3.03	-	1.79	m	**4.82**
50 x 150	0.03	3.15	-	2.29	m	**5.43**
50 x 175	0.03	3.15	-	2.52	m	**5.67**
50 x 200	0.03	3.38	-	3.09	m	**6.47**
50 x 225	0.03	3.38	-	3.47	m	**6.85**
75 x 200	0.03	3.61	-	5.21	m	**8.82**
75 x 225	0.03	3.61	-	6.24	m	**9.85**
100 x 200	0.04	4.89	-	7.37	m	**12.27**
100 x 225	0.05	5.24	-	8.84	m	**14.08**
Structural and carcassing timber; pitched roofs						
38 x 100	0.03	3.15	-	1.32	m	**4.47**
38 x 125	0.02	2.68	-	1.72	m	**4.40**
38 x 150	0.03	3.15	-	2.15	m	**5.30**
50 x 100	0.03	3.84	-	1.52	m	**5.36**
50 x 125	0.03	3.03	-	1.79	m	**4.82**
50 x 150	0.04	4.43	-	2.29	m	**6.71**
50 x 175	0.04	4.43	-	2.52	m	**6.95**
50 x 200	0.04	4.43	-	3.09	m	**7.52**
50 x 225	0.04	4.43	-	3.47	m	**7.90**
75 x 125	0.05	5.24	-	3.47	m	**8.71**
75 x 150	0.05	5.24	-	3.89	m	**9.14**
Structural and carcassing timber; plates and bearers						
38 x 100	0.01	1.28	-	1.32	m	**2.60**
50 x 75	0.01	1.28	-	1.15	m	**2.43**
50 x 100	0.01	1.51	-	1.52	m	**3.03**
75 x 100	0.01	1.51	-	2.60	m	**4.11**
75 x 125	0.02	1.75	-	3.47	m	**5.21**
75 x 150	0.02	1.75	-	3.89	m	**5.64**
Structural and carcassing timber; struts						
38 x 100	0.06	6.41	-	1.32	m	**7.73**
50 x 75	0.06	6.41	-	1.15	m	**7.56**
50 x 100	0.06	6.41	-	1.52	m	**7.93**
Structural and carcassing timber; cleats						
225 mm x 100 mm x 75 mm	0.04	4.19	-	0.60	nr	**4.79**
Structural and carcassing timber; trussed rafters and roof trusses						

CLASS Z: SIMPLE BUILDING WORKS INCIDENTAL TO CIVIL ENGINEERING WORKS

Item	Gang hours	Labour £	Plant £	Material £	Unit	Total rate £
Softwood; joinery quality; wrought; tanalised						
Strip boarding; walls						
18 mm nominal thick	0.14	16.31	-	11.68	m²	**27.99**
18 mm nominal thick; not exceeding 100 mm wide	0.05	5.24	-	1.17	m²	**6.41**
18 mm nominal thick; 100 - 200 mm wide	0.05	5.82	-	2.34	m²	**8.16**
18 mm nominal thick; 200 - 300 mm wide	0.06	6.99	-	3.50	m²	**10.49**
Softwood; joinery quality; wrought						
Strip boarding; walls						
18 mm nominal thick	0.14	16.31	-	11.68	m²	**27.99**
18 mm nominal thick; not exceeding 100 mm wide	0.05	5.24	-	1.17	m²	**6.41**
18 mm nominal thick; 100 - 200 mm wide	0.05	5.82	-	2.34	m²	**8.16**
18 mm nominal thick; 200 - 300 mm wide	0.06	6.99	-	3.50	m²	**10.49**
Miscellaneous joinery; skirtings						
19 x 100	0.02	2.33	-	2.35	m	**4.68**
25 x 150	0.03	2.91	-	3.61	m	**6.53**
Miscellaneous joinery; architraves						
12 x 50	0.02	2.80	-	1.69	m	**4.48**
19 x 63	0.02	2.80	-	1.93	m	**4.72**
Miscellaneous joinery; trims						
12 x 25	0.02	2.80	-	1.32	m	**4.12**
12 x 50	0.02	2.80	-	1.69	m	**4.48**
16 x 38	0.02	2.80	-	1.30	m	**4.09**
19 x 19	0.02	2.80	-	1.24	m	**4.04**
Plywood; marine quality						
Sheet boarding; walls						
18 mm nominal tihck	0.14	16.31	-	14.69	m²	**31.00**
Plywood; external quality						
Sheet boarding; floors						
18 mm nominal thick	0.11	12.81	-	11.68	m²	**24.50**
18 mm nominal thick; not exceeding 100 mm wide	0.04	4.08	-	1.17	m²	**5.25**
18 mm nominal thick; 100 - 200 mm wide	0.04	4.43	-	2.34	m²	**6.76**
18 mm nominal thick; 200 - 300 mm wide	0.05	5.24	-	3.50	m²	**8.75**
Sheet boarding; walls						
18 mm nominal thick	0.14	16.31	-	11.68	m²	**27.99**
18 mm nominal thick; not exceeding 100 mm wide	0.05	5.24	-	1.17	m²	**6.41**
18 mm nominal thick; 100 - 200 mm wide	0.05	5.82	-	2.34	m²	**8.16**
18 mm nominal thick; 200 - 300 mm wide	0.06	6.99	-	3.50	m²	**10.49**
Sheet boarding; soffits						
18 mm nominal thick	0.16	18.05	-	11.68	m²	**29.74**
18 mm nominal thick; not exceeding 100 mm wide	0.05	5.82	-	1.17	m²	**6.99**
18 mm nominal thick; 100 - 200 mm wide	0.06	6.41	-	2.34	m²	**8.74**
18 mm nominal thick; 200 - 300 mm wide	0.07	7.69	-	3.50	m²	**11.19**

CLASS Z: SIMPLE BUILDING WORKS INCIDENTAL TO CIVIL ENGINEERING WORKS

Item	Gang hours	Labour £	Plant £	Material £	Unit	Total rate £
INSULATION						
Vapour barrier; Sisalkraft Moistop building paper grade 728 (Class A1F); 150 mm laps; fixed to softwood						
Sheets						
floors	0.01	1.51	-	1.18	m²	**2.69**
sloping upper surfaces	0.01	1.63	-	1.18	m²	**2.81**
walls	0.02	1.98	-	1.18	m²	**3.16**
soffits	0.02	2.33	-	1.18	m²	**3.51**
Insulation quilt; Gypglas 1000 glass fibre; laid loose between members at 600 mm centres						
Quilts; floors						
60 mm thick	0.03	3.26	-	2.55	m²	**5.82**
80 mm thick	0.03	3.26	-	3.35	m²	**6.61**
100 mm thick	0.03	3.26	-	3.89	m²	**7.15**
150 mm thick	0.03	3.26	-	6.08	m²	**9.34**
Quilts; sloping upper surfaces						
60 mm thick	0.03	3.61	-	2.55	m²	**6.17**
80 mm thick	0.03	3.96	-	3.35	m²	**7.31**
100 mm thick	0.04	4.19	-	3.89	m²	**8.08**
150 mm thick	0.04	4.66	-	6.08	m²	**10.74**
Insulation quilt; Gypglas 1000 glass fibre; laid between members at 600 mm centres; fixing with staples						
Quilts; walls						
60 mm thick	0.04	4.08	-	2.78	m²	**6.86**
80 mm thick	0.04	4.43	-	3.57	m²	**8.00**
100 mm thick	0.04	4.89	-	4.11	m²	**9.00**
150 mm thick	0.05	5.82	-	6.30	m²	**12.13**
Quilts; soffits						
60 mm thick	0.04	4.89	-	2.78	m²	**7.67**
80 mm thick	0.05	5.47	-	3.57	m²	**9.04**
100 mm thick	0.05	5.82	-	4.11	m²	**9.94**
150 mm thick	0.05	6.29	-	6.30	m²	**12.59**
Insulation board; Jablite expanded polystyrene standard grade; fixing with adhesive						
Boards; floors						
25 mm thick	0.06	6.52	-	4.49	m²	**11.01**
40 mm thick	0.06	6.52	-	5.53	m²	**12.06**
50 mm thick	0.06	6.52	-	6.34	m²	**12.86**
Boards; sloping upper surfaces						
25 mm thick	0.07	7.57	-	4.49	m²	**12.06**
40 mm thick	0.07	7.57	-	5.53	m²	**13.11**
50 mm thick	0.07	7.57	-	6.34	m²	**13.91**
Boards; walls						
25 mm thick	0.08	8.74	-	4.49	m²	**13.22**
40 mm thick	0.08	8.74	-	5.53	m²	**14.27**
50 mm thick	0.08	8.74	-	6.34	m²	**15.08**
Boards; soffits						
25 mm thick	0.08	9.67	-	4.49	m²	**14.15**
40 mm thick	0.08	9.67	-	5.53	m²	**15.20**
50 mm thick	0.08	9.67	-	6.34	m²	**16.01**

CLASS Z: SIMPLE BUILDING WORKS INCIDENTAL TO CIVIL ENGINEERING WORKS

Item	Gang hours	Labour £	Plant £	Material £	Unit	Total rate £
WINDOWS, DOORS AND GLAZING						
Timber windows; treated planed softwood; Boulton & Paul; plugged and screwed to masonry						
High performance top hung reversible windows; ventilators; weather stripping; opening sashes and fanlights hung on rustproof hinges with aluminized lacquered espagnolette bolts; glazing 4 mm OQ glass						
600 mm x 900 mm; ref R0609	0.16	19.10	-	153.68	nr	**172.78**
915 mm x 1050 mm; ref R0910	0.21	24.34	-	179.02	nr	**203.36**
1200 mm x 1050 mm; ref R1210	0.25	28.54	-	199.57	nr	**228.11**
1800 mm x 1050 mm; ref R1810	0.31	35.99	-	247.51	nr	**283.51**
Metal windows; steel fixed light; factory finished polyester powder coating; Crittal Homelight range; fixing lugs plugged and screwed to masonry						
One piece composites; glazing 4 mm OQ glass; easy-glaze beads and weather stripping						
628 mm x 923 mm; ref ZNC5	0.22	25.39	-	45.31	nr	**70.70**
1237 mm x 923 mm; ref ZNC13	0.32	37.04	-	70.82	nr	**107.86**
1237 mm x 1218 mm; ref ZND13	0.32	37.04	-	88.25	nr	**125.29**
1846 mm x 1513 mm; ref ZNDV14	0.32	37.04	-	129.48	nr	**166.52**
Plastics windows; PVC-U, reinforced where appropriate with aluminium alloy; standard ironmongery; fixing lugs plugged and screwed to masonry						
Casement / fixed light; glazing 4 mm OQ glass; e.p.d.m. glazing gaskets and weather seals						
600 mm x 1200 mm; single glazed	0.32	37.04	-	175.09	nr	**212.13**
600 mm x 1200 mm; double glazed	0.32	37.04	-	181.13	nr	**218.17**
1200 mm x 1200 mm; single glazed	0.36	42.40	-	265.65	nr	**308.05**
1200 mm x 1200 mm; double glazed	0.36	42.40	-	277.73	nr	**320.12**
1800 mm x 1200 mm; single glazed	0.41	47.99	-	392.44	nr	**440.43**
1800 mm x 1200 mm; double glazed	0.41	47.99	-	422.63	nr	**470.61**
Timber doors; treated planed softwood						
Matchboarded, ledged and braced doors; 25 mm thick ledges and braces; 19 mm thick tongued, grooved and v-jointed one side vertical boarding						
762 mm x 1981 mm	0.27	31.80	-	45.40	nr	**77.20**
838 mm x 1981 mm	0.27	31.80	-	45.40	nr	**77.20**
Panelled doors; one open panel for glass; including beads						
762 x 1981 x 44 mm	0.32	37.27	-	78.13	nr	**115.40**
Panelled doors; two open panels for glass; including beads						
762 x 1981 x 44 mm	0.32	37.27	-	110.49	nr	**147.76**
838 x 1981 x 44 mm	0.32	37.27	-	115.92	nr	**153.19**

CLASS Z: SIMPLE BUILDING WORKS INCIDENTAL TO CIVIL ENGINEERING WORKS

Item	Gang hours	Labour £	Plant £	Material £	Unit	Total rate £
WINDOWS, DOORS AND GLAZING – cont'd						
Timber doors;standard flush pattern						
Flush door; internal quality; skeleton or cellular core; hardboard faced both sides; lipped on two long edges; primed						
626 x 2040 x 40 mm	0.22	25.98	-	20.89	nr	**46.86**
726 x 2040 x 40 mm	0.22	25.98	-	20.89	nr	**46.86**
826 x 2040 x 40 mm	0.22	25.98	-	20.89	nr	**46.86**
Flush door; internal quality; skeleton or cellular core; chipboard faced both sides; lipped all edges; primed						
626 x 2040 x 40 mm	0.22	25.98	-	30.79	nr	**56.77**
726 x 2040 x 40 mm	0.22	25.98	-	30.79	nr	**56.77**
826 x 2040 x 40 mm	0.22	25.98	-	30.79	nr	**56.77**
Flush door; internal quality; skeleton or cellular core; Sapele veneered both sides; lipped all edges; primed						
626 x 2040 x 40 mm	0.34	39.60	-	34.05	nr	**73.65**
726 x 2040 x 40 mm	0.34	39.60	-	34.05	nr	**73.65**
826 x 2040 x 40 mm	0.34	39.60	-	34.05	nr	**73.65**
Flush door; half-hour fire check (30/20); solid core; chipboard faced both sides; lipped all edges; primed						
626 x 2040 x 44 mm	0.32	37.27	-	47.73	nr	**85.00**
726 x 2040 x 44 mm	0.32	37.27	-	47.73	nr	**85.00**
826 x 2040 x 44 mm	0.32	37.27	-	47.73	nr	**85.00**
Timber frames or lining sets; treated planed softwood						
Internal door frame or lining; 30 x 107 mm lining with 12 x 38 mm door stop						
for 726 x 2040 mm door	0.14	16.89	-	36.31	nr	**53.20**
for 826 x 2040 mm door	0.14	16.89	-	36.31	nr	**53.20**
Internal door frame or lining; 30 x 133 mm lining with 12 x 38 mm door stop						
for 726 x 2040 mm door	0.14	16.89	-	40.65	nr	**57.54**
for 826 x 2040 mm door	0.14	16.89	-	40.65	nr	**57.54**
Ironmongery						
Hinges						
100 mm; light steel	0.17	3.04	-	0.68	nr	**3.72**
S/D rising butts; R/L hand; 102 x 67 mm; BRS	0.17	3.04	-	22.65	nr	**25.68**
Door closers						
light duty surface fixed door closer; L/R hand; SIL	1.30	22.55	-	93.01	nr	**115.56**
door selector; face fixing; SAA	0.80	13.88	-	79.60	nr	**93.48**
floor spring; single and double action; ZP	3.33	57.78	-	195.47	nr	**253.25**
Locks						
mortice dead lock; 63 x 108 mm; SSS	1.00	17.35	-	12.88	nr	**30.23**
rim lock; 140 x 73 mm; GYE	0.56	9.72	-	9.79	nr	**19.51**
upright mortice lock; 103 x 82 mm; 3 lever	1.11	19.26	-	13.27	nr	**32.53**
Bolts						
flush; 152 x 25 mm; SCP	0.80	13.88	-	15.84	nr	**29.72**
indicating; 76 x 41 mm; SAA	0.89	15.44	-	11.58	nr	**27.02**
panic; single; SVE	3.33	57.78	-	73.08	nr	**130.86**
panic; double; SVE	4.67	81.02	-	99.92	nr	**180.94**
necked tower; 203 mm; BJ	0.40	6.94	-	4.33	nr	**11.27**

CLASS Z: SIMPLE BUILDING WORKS INCIDENTAL TO CIVIL ENGINEERING WORKS

Item	Gang hours	Labour £	Plant £	Material £	Unit	**Total rate £**
Handles						
pull; 225 mm; back fixing; PAA	0.23	3.99	-	7.01	nr	**11.00**
pull; 225 mm; face fixing with cover rose; PAA	0.44	7.63	-	37.68	nr	**45.32**
lever; PAA	0.44	7.63	-	26.26	nr	**33.90**
Plates						
finger plate; 300 x 75 x 3 mm; SAA	0.23	3.99	-	4.59	nr	**8.58**
kicking plate; 1000 x 150 x 3 mm; PAA	0.44	7.63	-	13.33	nr	**20.96**
letter plate; 330 x 76 mm; aluminium finish	1.77	30.71	-	9.55	nr	**40.26**
Brackets						
head bracket; open; side fixing; bolting to masonry	0.50	8.68	-	7.90	nr	**16.58**
head bracket; open; soffit fixing; bolting to masonry	0.50	8.68	-	3.88	nr	**12.56**
Sundries						
rubber door stop; SAA	0.11	1.91	-	1.33	nr	**3.24**
Glazing; standard plain glass to BS 952, clear float; glazing with putty or bradded beads						
Glass						
3 mm thick	0.95	16.48	-	9.85	m²	**26.33**
4 mm thick	0.95	16.48	-	10.67	m²	**27.15**
5 mm thick	0.95	16.48	-	13.96	m²	**30.44**
6 mm thick	0.95	16.48	-	14.78	m²	**31.26**
Hermetically sealed units, factory made						
two panes 4 mm thick, 6 mm air space	0.95	16.48	-	49.61	m²	**66.09**
Glazing; standard plain glass to BS 952, rough cast; glazing with putty or bradded beads						
Glass						
6 mm thick	0.95	16.48	-	22.76	m²	**39.24**
Glazing; standard plain glass to BS 952, Georgian wired cast; glazing with putty or bradded beads						
Glass						
7 mm thick	0.95	16.48	-	28.11	m²	**44.60**
Glazing; standard plain glass to BS 952, Georgian wired polished; glazing with putty or bradded beads						
Glass						
6 mm thick	0.95	16.48	-	60.91	m²	**77.40**
Glazing; special glass to BS 952, toughened clear float; glazing with putty or bradded beads						
Glass						
4 mm thick	0.95	16.48	-	45.24	m²	**61.72**
5 mm thick	0.95	16.48	-	53.45	m²	**69.94**
6 mm thick	0.95	16.48	-	59.62	m²	**76.10**
10 mm thick	1.05	18.22	-	116.50	m²	**134.72**

CLASS Z: SIMPLE BUILDING WORKS INCIDENTAL TO CIVIL ENGINEERING WORKS

Item	Gang hours	Labour £	Plant £	Material £	Unit	Total rate £
WINDOWS, DOORS AND GLAZING – cont'd						
Glazing; special glass to BS 952, clear laminated safety; glazing with putty or bradded beads						
Glass						
4.4 mm thick	0.95	16.48	-	44.98	m²	**61.46**
5.4 mm thick	0.95	16.48	-	56.23	m²	**72.71**
6.4 mm thick	0.95	16.48	-	65.60	m²	**82.08**
Glazing; standard plain glass to BS 952, clear float; glazing with putty or bradded beads						
Glass						
4.4 mm thick	0.95	16.48	-	44.98	m²	**61.46**
5.4 mm thick	0.95	16.48	-	56.23	m²	**72.71**
6.4 mm thick	0.95	16.48	-	65.60	m²	**82.08**
Patent glazing; aluminium alloy bars 2.55 m long at 622 mm centres; fixed to supports						
Patent glazing						
roofs	-	-	-	-	m²	**127.00**
SURFACE FINISHES, LININGS AND PARTITIONS						
In situ finishes; cement and sand (1:3); steel trowelled						
Floors						
30 mm thick	0.12	5.52	-	2.25	m²	**7.76**
40 mm thick	0.12	5.52	-	2.99	m²	**8.51**
50 mm thick	0.14	6.43	-	3.74	m²	**10.18**
60 mm thick	0.15	6.89	-	4.49	m²	**11.38**
70 mm thick	0.16	7.35	-	5.24	m²	**12.59**
Sloping upper surfaces						
30 mm thick	0.16	7.35	-	2.25	m²	**9.60**
40 mm thick	0.16	7.35	-	2.99	m²	**10.35**
50 mm thick	0.18	8.27	-	3.74	m²	**12.02**
60 mm thick	0.19	8.73	-	4.49	m²	**13.22**
70 mm thick	0.20	9.19	-	5.24	m²	**14.43**
Walls						
12 mm thick	0.31	14.25	-	0.90	m²	**15.15**
15 mm thick	0.34	15.63	-	1.09	m²	**16.72**
20 mm thick	0.38	17.46	-	1.50	m²	**18.96**
Surfaces of width not exceeding 300 mm						
12 mm thick	0.15	6.89	-	0.90	m²	**7.79**
15 mm thick	0.13	5.97	-	1.12	m²	**7.10**
20 mm thick	0.12	5.52	-	1.50	m²	**7.01**
30 mm thick	0.08	3.68	-	2.25	m²	**5.93**
40 mm thick	0.08	3.68	-	2.99	m²	**6.67**
50 mm thick	0.09	4.14	-	3.74	m²	**7.88**
60 mm thick	0.09	4.14	-	4.49	m²	**8.63**
70 mm thick	0.10	4.60	-	5.24	m²	**9.84**

CLASS Z: SIMPLE BUILDING WORKS INCIDENTAL TO CIVIL ENGINEERING WORKS

Item	Gang hours	Labour £	Plant £	Material £	Unit	**Total rate £**
Surfaces of width 300 mm - 1 m						
12 mm thick	0.23	10.57	-	0.90	m²	**11.47**
15 mm thick	0.26	11.95	-	1.12	m²	**13.07**
20 mm thick	0.26	11.95	-	1.50	m²	**13.45**
30 mm thick	0.12	5.52	-	2.25	m²	**7.76**
40 mm thick	0.12	5.52	-	2.99	m²	**8.51**
50 mm thick	0.13	5.97	-	3.74	m²	**9.72**
60 mm thick	0.14	6.43	-	4.49	m²	**10.92**
70 mm thick	0.15	6.89	-	5.24	m²	**12.14**
In situ finishes; lightweight plaster; Carlite; steel trowelled						
Walls						
12 mm thick	0.22	10.11	-	0.83	m²	**10.94**
Surfaces of width not exceeding 300 mm						
12 mm thick	0.22	10.11	-	0.25	m²	**10.36**
Surfaces of width 300 mm - 1 m						
12 mm thick	0.16	7.35	-	0.83	m²	**8.18**
Beds and backings; cement and sand (1:3)						
Floors						
50 mm thick	0.14	6.43	-	3.74	m²	**10.18**
70 mm thick	0.16	7.35	-	5.24	m²	**12.59**
Sloping upper surfaces						
30 mm thick	0.16	7.35	-	2.25	m²	**9.60**
40 mm thick	0.16	7.35	-	2.99	m²	**10.35**
Walls						
12 mm thick	0.31	14.25	-	0.90	m²	**15.15**
20 mm thick	0.38	17.46	-	1.50	m²	**18.96**
Surfaces of width not exceeding 300 mm						
12 mm thick	0.15	6.89	-	0.90	m²	**7.79**
20 mm thick	0.12	5.52	-	1.50	m²	**7.01**
30 mm thick	0.08	3.68	-	2.25	m²	**5.93**
50 mm thick	0.09	4.14	-	3.74	m²	**7.88**
70 mm thick	0.10	4.60	-	5.24	m²	**9.84**
Surfaces of width 300 mm - 1 m						
12 mm thick	0.23	10.57	-	0.90	m²	**11.47**
20 mm thick	0.26	11.95	-	1.50	m²	**13.45**
30 mm thick	0.12	5.52	-	2.25	m²	**7.76**
50 mm thick	0.13	5.97	-	3.74	m²	**9.72**
70 mm thick	0.15	6.89	-	5.24	m²	**12.14**
Tiles; Daniel Platt Crown; red; bedding 10 mm thick and jointing in cement mortar (1:3); grouting with cement mortar (1:1)						
Floors						
150 x 150 x 12.5 mm	0.32	14.71	-	24.27	m²	**38.98**
200 x 200 x 19 mm	0.32	14.71	-	33.78	m²	**48.49**
Surfaces of width not exceeding 300 mm						
150 x 150 x 12.5 mm	0.16	7.35	-	7.28	m²	**14.63**
200 x 200 x 19 mm	0.13	5.97	-	10.14	m²	**16.12**
Surfaces of width 300 mm - 1 m						
150 x 150 x 12.5 mm	0.24	11.03	-	24.27	m²	**35.30**
200 x 200 x 19 mm	0.24	11.03	-	33.80	m²	**44.83**

CLASS Z: SIMPLE BUILDING WORKS INCIDENTAL TO CIVIL ENGINEERING WORKS

Item	Gang hours	Labour £	Plant £	Material £	Unit	**Total rate £**
SURFACE FINISHES, LININGS AND PARTITIONS – cont'd						
Tiles; Daniel Platt Crown; brown; bedding 10 mm thick and jointing in cement mortar (1:3); grouting with cement mortar (1:1)						
Floors						
150 x 150 x 12.5 mm	0.32	14.71	-	28.35	m²	**43.06**
200 x 200 x 19 mm	0.32	14.71	-	39.60	m²	**54.31**
Surfaces of width not exceeding 300 mm						
150 x 150 x 12.5 mm	0.16	7.35	-	8.51	m²	**15.86**
200 x 200 x 19 mm	0.13	5.97	-	11.89	m²	**17.86**
Surfaces of width 300 mm - 1 m						
150 x 150 x 12.5 mm	0.24	11.03	-	28.35	m²	**39.38**
200 x 200 x 19 mm	0.24	11.03	-	39.62	m²	**50.65**
Tiles; glazed ceramic wall tiles; BS 6431; white; fixing with adhesive; pointing joints with grout						
Sloping upper surfaces						
152 x 152 x 5.5 mm thick	0.28	12.87	-	23.46	m²	**36.33**
200 x 100 x 6.5 mm thick	0.28	12.87	-	23.46	m²	**36.33**
Walls						
152 x 152 x 5.5 mm thick	0.24	11.03	-	23.46	m²	**34.49**
200 x 100 x 6.5 mm thick	0.24	11.03	-	23.46	m²	**34.49**
Soffit						
152 x 152 x 5.5 mm thick	0.28	12.87	-	23.46	m²	**36.33**
200 x 100 x 6.5 mm thick	0.28	12.87	-	23.46	m²	**36.33**
Surfaces of width not exceeding 300 mm						
152 x 152 x 5.5 mm thick	0.12	5.52	-	7.07	m²	**12.58**
200 x 100 x 6.5 mm thick	0.12	5.52	-	7.07	m²	**12.58**
Surfaces of width 300 mm - 1 m						
152 x 152 x 5.5 mm thick	0.18	8.27	-	23.46	m²	**31.73**
200 x 100 x 6.5 mm thick	0.18	8.27	-	23.46	m²	**31.73**
Tiles; slate; Riven Welsh; bedding 10 mm thick and jointing in cement mortar (1:3); grouting with cement mortar (1:1)						
Floors						
250 x 250 x 12 - 15 mm	0.24	11.03	-	33.12	m²	**44.15**
Surfaces of width not exceeding 300 mm						
250 x 250 x 12 - 15 mm	0.12	5.52	-	9.94	m²	**15.45**
Surfaces of width 300 mm - 1 m						
250 x 250 x 12 - 15 mm	0.18	8.27	-	33.12	m²	**41.40**
Tiles; vinyl; Accoflex; fixing with adhesive						
Floors						
300 x 300 x 2 mm	0.17	4.78	-	8.65	m²	**13.43**
300 x 300 x 2.5 mm	0.17	4.78	-	8.58	m²	**13.36**
Surfaces of width not exceeding 300 mm						
300 x 300 x 2 mm	0.08	2.37	-	2.62	m²	**4.99**
300 x 300 x 2.5 mm	0.08	2.37	-	2.60	m²	**4.97**
Surfaces of width 300 mm - 1 m						
300 x 300 x 2 mm	0.13	3.58	-	8.65	m²	**12.23**
300 x 300 x 2.5 mm	0.13	3.58	-	8.58	m²	**12.16**

CLASS Z: SIMPLE BUILDING WORKS INCIDENTAL TO CIVIL ENGINEERING WORKS

Item	Gang hours	Labour £	Plant £	Material £	Unit	Total rate £
Tiles; vinyl; Marley HD; fixing with adhesive						
Floors						
300 x 300 x 2 mm	0.23	6.67	-	10.75	m²	**17.41**
Surfaces of width not exceeding 300 mm						
300 x 300 x 2 mm	0.12	3.35	-	3.22	m²	**6.57**
Surfaces of width 300 mm - 1 m						
300 x 300 x 2 mm	0.17	5.01	-	10.75	m²	**15.76**
Tiles; rubber studded; Altro Mondopave; type MRB; black; fixing with adhesive						
Floors						
500 x 500 x 2.5 mm	0.40	11.44	-	30.69	m²	**42.14**
500 x 500 x 4.0 mm	0.40	11.44	-	29.50	m²	**40.94**
Surfaces of width not exceeding 300 mm						
500 x 500 x 2.5 mm	0.20	5.72	-	9.14	m²	**14.86**
500 x 500 x 4.0 mm	0.20	5.72	-	8.85	m²	**14.57**
Surfaces of width 300 mm - 1 m						
500 x 500 x 2.5 mm	0.30	8.58	-	30.69	m²	**39.28**
500 x 500 x 4.0 mm	0.30	8.58	-	29.50	m²	**38.08**
Tiles; rubber studded; Altro Mondopave; type MRB; colour; fixing with adhesive						
Floors						
500 x 500 x 2.5 mm	0.40	11.44	-	29.59	m²	**41.03**
500 x 500 x 4.0 mm	0.40	11.44	-	29.36	m²	**40.80**
Surfaces of width not exceeding 300 mm						
500 x 500 x 2.5 mm	0.20	5.72	-	8.81	m²	**14.53**
500 x 500 x 4.0 mm	0.20	5.72	-	8.81	m²	**14.53**
Surfaces of width 300 mm - 1 m						
500 x 500 x 2.5 mm	0.30	8.58	-	29.59	m²	**38.17**
500 x 500 x 4.0 mm	0.30	8.58	-	29.36	m²	**37.94**
Tiles; linoleum; Forbo Nairn; Marmoleum Dual; level; fixing with adhesive						
Floors						
2.50 mm thick; marbled	0.27	7.64	-	13.93	m²	**21.56**
Surfaces of width not exceeding 300 mm						
2.50 mm thick; marbled	0.13	3.81	-	4.18	m²	**7.98**
Surfaces of width 300 mm - 1 m						
2.50 mm thick; marbled	0.20	5.72	-	2.35	m²	**8.07**
Flexible sheer; linoleum; Forbo Nairn; Marmoleum Real; level; fixing with adhesive						
Floors						
2.00 mm thick; marbled	0.33	9.54	-	10.62	m²	**20.16**
2.50 mm thick; marbled	0.33	9.54	-	12.92	m²	**22.46**
3.20 mm thick; marbled	0.33	9.54	-	13.40	m²	**22.94**
4.00 mm thick; marbled	0.33	9.54	-	16.46	m²	**25.99**
Surfaces of width not exceeding 300 mm						
2.00 mm thick; marbled	0.17	4.78	-	3.19	m²	**7.96**
2.50 mm thick; marbled	0.17	4.78	-	3.88	m²	**8.65**
3.20 mm thick; marbled	0.17	4.78	-	4.02	m²	**8.80**
4.00 mm thick; marbled	0.17	4.78	-	4.82	m²	**9.60**
Surfaces of width 300 mm - 1m						
2.00 mm thick; marbled	0.25	7.15	-	10.62	m²	**17.77**
2.50 mm thick; marbled	0.25	7.15	-	12.92	m²	**20.08**
3.20 mm thick; marbled	0.25	7.15	-	13.40	m²	**20.55**
4.00 mm thick; marbled	0.25	7.15	-	16.46	m²	**23.61**

CLASS Z: SIMPLE BUILDING WORKS INCIDENTAL TO CIVIL ENGINEERING WORKS

Item	Gang hours	Labour £	Plant £	Material £	Unit	Total rate £
SURFACE FINISHES, LININGS AND PARTITIONS – cont'd						
Flexible sheet; vinyl; slip resistant; Forbo Nairn; Surestep; fixing with adhesive						
Floors						
2.00 mm thick	0.33	9.53	-	15.91	m²	**25.44**
Surfaces of width not exceeding 300 mm						
2.00 mm thick	0.17	4.78	-	4.77	m²	**9.55**
Surfaces of width 300 mm - 1 m						
2.00 mm thick	0.25	7.15	-	15.91	m²	**23.06**
Flexible sheet; vinyl; Marley HD; fixing with adhesive						
Floors						
2.00 mm thick	0.30	8.58	-	10.75	m²	**19.33**
2.50 mm thick	0.33	9.54	-	11.98	m²	**21.51**
Surfaces of width not exceeding 300 mm						
2.00 mm thick	0.15	4.29	-	3.22	m²	**7.52**
2.50 mm thick	0.17	4.78	-	3.59	m²	**8.37**
Surfaces of width 300 mm - 1 m						
2.00 mm thick	0.23	6.44	-	10.75	m²	**17.19**
2.50 mm thick	0.25	7.15	-	11.98	m²	**19.13**
Flexible sheet; vinyl; Armstrong Rhinocontract Interior; fixing with adhesive						
Floors						
2.00 mm thick	0.33	9.54	-	14.40	m²	**23.93**
Surfaces of width not exceeding 300 mm						
2.00 mm thick	0.17	4.78	-	4.34	m²	**9.12**
Surfaces of width 300 mm - 1 m						
2.00 mm thick	0.25	7.15	-	14.40	m²	**21.55**
Flexible sheet; carpet; Marleytex sheet; fixing with adhesive						
Floors						
5.00 mm thick	0.17	4.78	-	7.41	m²	**12.19**
Surfaces of width not exceeding 300 mm						
5.00 mm thick	0.08	2.37	-	2.22	m²	**4.60**
Surfaces of width 300 mm - 1 m						
5.00 mm thick	0.13	3.58	-	7.41	m²	**10.99**
Suspended ceilings; mineral fibre tiles in exposed grid; suspension system and wire hangers to structural soffit						
Suspended ceiling						
150 - 500 mm depth of suspension	-	-	-	-	m²	**18.38**

CLASS Z: SIMPLE BUILDING WORKS INCIDENTAL TO CIVIL ENGINEERING WORKS

Item	Gang hours	Labour £	Plant £	Material £	Unit	**Total rate £**
Suspended ceilings; mineral fibre tiles in concealed grid; suspension system and wire hangers to structural soffit						
Suspended ceiling						
150 - 500 mm depth of suspension	-	-	-	-	m²	**28.92**
Bulkheads						
250 mm girth	-	-	-	-	m	**10.44**
500 mm girth	-	-	-	-	m	**20.88**
PIPED BUILDING SERVICES						
Pipework; copper pipes EN 1057; capillary fittings; pipe clips screwed to background						
Pipes						
15 mm	0.13	3.58	-	1.85	m	**5.43**
22 mm	0.13	3.72	-	2.35	m	**6.07**
28 mm	0.14	4.15	-	4.63	m	**8.78**
35 mm	0.17	4.72	-	11.54	m	**16.26**
42 mm	0.19	5.44	-	13.71	m	**19.15**
Fittings						
15 mm made bend	0.08	2.15	-	-	nr	**2.15**
15 mm elbow	0.08	2.15	-	0.70	nr	**2.85**
15 mm equal tee	0.13	3.58	-	1.32	nr	**4.90**
15 mm straight coupling	0.09	2.43	-	0.51	nr	**2.94**
22 mm made bend	0.10	2.86	-	-	nr	**2.86**
22 mm elbow	0.11	3.15	-	1.60	nr	**4.75**
22 mm equal tee	0.17	4.72	-	3.57	nr	**8.29**
22 mm straight coupling	0.11	3.15	-	1.17	nr	**4.32**
28 mm made bend	0.13	3.58	-	-	nr	**3.58**
28 mm elbow	0.14	4.01	-	2.62	nr	**6.62**
28 mm equal tee	0.20	5.87	-	6.67	nr	**12.54**
28 mm straight coupling	0.14	4.01	-	1.94	nr	**5.95**
35 mm made bend	0.15	4.29	-	-	nr	**4.29**
35 mm elbow	0.17	4.72	-	9.87	nr	**14.59**
35 mm equal tee	0.23	6.58	-	16.47	nr	**23.05**
35 mm straight coupling	0.17	4.72	-	5.61	nr	**10.33**
42 mm made bend	0.20	5.72	-	-	nr	**5.72**
42 mm elbow	0.20	5.72	-	1.12	nr	**6.84**
42 mm equal tee	0.26	7.44	-	25.02	nr	**32.46**
42 mm straight coupling	0.20	5.72	-	8.41	nr	**14.13**
Pipework; 19 mm thick rigid mineral glass fibre sectional pipe lagging; plain finish; fixing with aluminium bands						
Insulation						
around 15 mm pipes	0.04	1.00	-	5.90	m	**6.90**
around 22 mm pipes	0.05	1.43	-	6.36	m	**7.79**
around 28 mm pipes	0.06	1.57	-	6.95	m	**8.52**
around 35 mm pipes	0.06	1.72	-	9.15	m	**10.86**
around 42 mm pipes	0.07	1.86	-	9.74	m	**11.60**

CLASS Z: SIMPLE BUILDING WORKS INCIDENTAL TO CIVIL ENGINEERING WORKS

Item	Gang hours	Labour £	Plant £	Material £	Unit	Total rate £
PIPED BUILDING SERVICES – cont'd						
Equipment; polyethylene cold water feed and expansion cistern to BS 4213, with cover; placing in position						
Cisterns and tanks						
68 litres; ref SC15	0.63	17.88	-	42.40	nr	**60.28**
114 litres; ref SC25	0.72	20.74	-	49.88	nr	**70.62**
Equipment; grp cold water storage cistern, with cover; placing in position						
Cisterns and tanks						
68 litres	0.63	17.88	-	108.53	nr	**126.41**
114 litres	0.72	20.74	-	134.82	nr	**155.57**
Equipment; copper single feed coil indirect cylinder to BS 1566 Part 2 grade 3; placing in position						
Cisterns and tanks						
96 litres; ref 2	1.00	28.61	-	143.78	nr	**172.39**
114 litres; ref 3	1.13	32.19	-	115.48	nr	**147.67**
Equipment; combination copper coil direct hot water storage units to BS 3198; placing in position						
Cisterns and tanks						
400 x 900 mm; 65/20 litres	1.40	40.05	-	144.99	nr	**185.04**
450 x 1075 mm; 115/25 litres	2.45	70.09	-	168.67	nr	**238.77**
Sanitary appliances and fittings						
Sink; white glazed fireclay to BS 1206 with pair of cast iron cantilever brackets						
610 x 455 x 205 mm	1.50	42.91	-	187.16	nr	**230.08**
610 x 455 x 205 mm	1.50	42.91	-	265.65	nr	**308.56**
Sink; stainless steel combined bowl and drainer; pair 19 mm chromium plated high neck pillar taps; chain and self colour plug to BS 3380; setting on base unit						
1050 x 500, single drainer, single bowl 420 x 350 x 175 mm	0.88	25.03	-	136.85	nr	**161.88**
1550 x 500, double drainer, single bowl 420 x 350 x 200 mm	0.88	25.03	-	156.40	nr	**181.43**
Lavatory basin; white vitreous china to BS 1188; pair 12 mm chromium plated pillar taps; chain and self colour plug to BS 3380; trap; painted cast iron brackets plugged and screwed to masonry						
560 x 405 mm	1.15	32.90	-	120.75	nr	**153.65**
635 x 455 mm	1.15	32.90	-	217.35	nr	**250.25**
Add for coloured	-	-	-	20.53	nr	**20.53**
Add for pedestal in lieu of brackets	0.10	2.86	-	25.66	nr	**28.52**

CLASS Z: SIMPLE BUILDING WORKS INCIDENTAL TO CIVIL ENGINEERING WORKS

Item	Gang hours	Labour £	Plant £	Material £	Unit	**Total rate £**
WC suite; low level; white glazed vitreous china pan; black plastic seat; 9 litre white glazed vitreous china cistern and brackets; low pressure ball valve; plastic flush pipe; building bracket into masonry; plugging and screwing pipe brackets and pan; bedding pan in mastic						
P trap outlet	1.50	42.91	-	193.20	nr	**236.11**
Add for coloured	-	-	-	35.92	nr	**35.92**
Bowl type wall urinal; white glazed vitreous china; white glazed vitreous china automatic flushing cistern and brackets; chromium plated flush pipe and spreaders; building brackets into masonry; plugging and screwing pipe brackets						
single; 455 x 380 x 330 mm	2.00	57.22	-	-	nr	**57.22**
range of two; 455 x 380 x 330 mm	5.50	157.35	-	452.81	nr	**610.17**
Add for each additional urinal	1.60	45.78	-	150.94	nr	**196.71**
Add for division between urinals	0.38	10.73	-	72.45	nr	**83.18**

DAVIS LANGDON

EUROPE & MIDDLE EAST
office locations

EUROPE & MIDDLE EAST

ENGLAND

DAVIS LANGDON
LONDON
MidCity Place
71 High Holborn
London WC1V 6QS
Tel: (020) 7061 7000
Fax: (020) 7061 7061
Email: simon.johnson@davislangdon.com

BIRMINGHAM
75-77 Colmore Row
Birmingham B3 2HD
Tel: (0121) 710 1100
Fax: (0121) 710 1399
Email: david.daly@davislangdon.com

BRISTOL
St Lawrence House
29/31 Broad Street
Bristol BS1 2HF
Tel: (0117) 927 7832
Fax: (0117) 925 1350
Email: alan.francis@davislangdon.com

CAMBRIDGE
36 Storey's Way
Cambridge CB3 ODT
Tel: (01223) 351 258
Fax: (01223) 321 002
Email: laurence.brett@davislangdon.com

LEEDS
No 4 The Embankment
Victoria Wharf
Sovereign Street
Leeds LS1 4BA
Tel: (0113) 243 2481
Fax: (0113) 242 4601
Email: duncan.sissons@davislangdon.com

LIVERPOOL
Cunard Building
Water Street
Liverpool
L3 1JR
Tel: (0151) 236 1992
Fax: (0151) 227 5401
Email: andrew.stevenson@davislangdon.com

MAIDSTONE
11 Tower View
Kings Hill
West Malling
Kent ME19 4UY
Tel: (01732) 840 429
Fax: (01732) 842 305
Email: nick.leggett@davislangdon.com

MANCHESTER
Cloister House
Riverside
New Bailey Street
Manchester
M3 5AG
Tel: (0161) 819 7600
Fax: (0161) 819 1818
Email: paul.stanion@davislangdon.com

MILTON KEYNES
Everest House
Rockingham Drive
Linford Wood
Milton Keynes
MK14 6LY
Tel: (01908) 304 700
Fax: (01908) 660 059
Email: kevin.sims@davislangdon.com

NORWICH
63 Thorpe Road
Norwich NR1 1UD
Tel: (01603) 628 194
Fax: (01603) 615 928
Email: michael.ladbrook@davislangdon.com

OXFORD
Avalon House
Marcham Road
Abingdon
Oxford OX14 1TZ
Tel: (01235) 555 025
Fax: (01235) 554 909
Email: paul.coomber@davislangdon.com

PETERBOROUGH
Clarence House
Minerva Business Park
Lynchwood
Peterborough PE2 6FT
Tel: (01733) 362 000
Fax: (01733) 230 875
Email: stuart.bremner@davislangdon.com

PLYMOUTH
1 Ensign House
Parkway Court
Longbridge Road
Plymouth PL6 8LR
Tel: (01752) 827 444
Fax: (01752) 221 219
Email: gareth.steventon@davislangdon.com

SOUTHAMPTON
Brunswick House
Brunswick Place
Southampton SO15 2AP
Tel: (023) 8033 3438
Fax: (023) 8022 6099
Email: chris.tremellen@davislangdon.com

LEGAL SUPPORT
LONDON
MidCity Place
71 High Holborn
London WC1V 6QS
Tel: (020) 7061 7000
Fax: (020) 7061 7061
Email: mark.hackett@davislangdon.com

MANAGEMENT CONSULTING
LONDON
MidCity Place
71 High Holborn
London WC1V 6QS
Tel: (020) 7061 7000
Fax: (020) 7061 7005
Email:john.connaughton@davislangdon.com

SPECIFICATION CONSULTING
Davis Langdon Schumann Smith
STEVENAGE
Southgate House
Southgate
Stevenage
SG1 1HG
Tel: (01438) 742 642
Fax: (01438) 742 632
Email: nick.schumann@schumannsmith.com

MANCHESTER
Cloister House
Riverside, New Bailey Street
Manchester M3 5AG
Tel: (0161) 819 7600
Fax: (0161) 819 1818
Email: richard.jackson@davislangdon.com

ENGINEERING SERVICES
Davis Langdon Mott Green Wall
MidCity Place
71 High Holborn
London WC1V 6QS
Tel: (020) 7061 7777
Fax: (020) 7061 7009
Email: barry.nugent@mottgreenwall.co.uk

PROPERTY TAX & FINANCE
LONDON
Davis Langdon Crosher & James
MidCity Place
71 High Holborn
London WC1V 6QS
Tel: (020) 7061 7077
Fax: (020) 7061 7078
Email: tony.llewellyn@crosherjames.com

BIRMINGHAM
102 New Street
Birmingham
B2 4HQ
Tel: (0121) 632 3600
Fax: (0121) 632 3601
Email: clive.searle@crosherjames.com

SOUTHAMPTON
Brunswick House
Brunswick Place
Southampton SO15 2AP
Tel: (023) 8068 2800
Fax: (0870) 048 8141
Email: david.rees@crosherjames.com

SCOTLAND

DAVIS LANGDON
EDINBURGH
39 Melville Street
Edinburgh EH3 7JF
Tel: (0131) 240 1350
Fax: (0131) 240 1399
Email: sam.mackenziel@davislangdon.com

GLASGOW
Monteith House
11 George Square
Glasgow G2 1DY
Tel: (0141) 248 0300
Fax: (0141) 248 0303
Email: sam.mackenzie@davislangdon.com

PROPERTY TAX & FINANCE
Davis Langdon Crosher & James
EDINBURGH
39 Melville Street
Edinburgh EH3 7JF
Tel: (0131) 220 4225
Fax: (0131) 220 4226
Email: ian.mcfarlane@crosherjames.com

GLASGOW
Monteith House
11 George Square
Glasgow G2 1DY
Tel: (0141) 248 0333
Fax: (0141) 248 0313
Email: ken.fraser@crosherjames.com

WALES

DAVIS LANGDON
CARDIFF
4 Pierhead Street
Capital Waterside
Cardiff CF10 4QP
Tel: (029) 2049 7497
Fax: (029) 2049 7111
Email: paul.edwards@davislangdon.com

PROPERTY TAX & FINANCE
Davis Langdon Crosher & James
CARDIFF
4 Pierhead Street
Capital Waterside
Cardiff CF10 4QP
Tel: (029) 2049 7497
Fax: (029) 2049 7111
Email: michael.murray@crosherjames.com

IRELAND

DAVIS LANGDON PKS
CORK
Hibernian House
80A South Mall
Cork. Ireland
Tel: (00 353 21) 4222 800
Fax: (00 353 21) 4222 801
Email: alangmaid@dlpks.ie

DUBLIN
24 Lower Hatch Street
Dublin 2, Ireland
Tel: (00 353 1) 676 3671
Fax: (00 353 1) 676 3672
Email: mwebb@dlpks.ie

GALWAY
Heritage Hall
Kirwan's Lane
Galway
Ireland
Tel: (00 353 91) 530 199
Fax: (00 353 91) 530 198
Email: joregan@dlpks.ie

LIMERICK
8 The Crescent
Limerick, Ireland
Tel: (00 353 61) 318 870
Fax: (00 353 61) 318 871
Email: cbarry@dlpks.ie

SPECIFICATION CONSULTING
DUBLIN
24 Lower Hatch Street
Dublin 2, Ireland
Tel: (00 353 1) 676 3671
Fax: (00 353 1) 676 3672
Email: jhartnett@dlpks.ie

SPAIN

DAVIS LANGDON EDETCO
BARCELONA
C/Muntaner, 479, 1-2
Barcelona 08021
Spain
Tel: (00 34 93) 418 6899
Fax: (00 34 93) 211 0003
Contact: Francesc Monells
Email: barcelona@edetco.com

GIRONA
C/Salt 10 2on
Girona 17005
Spain
Tel: (00 34 97) 223 8000
Fax: (00 34 97) 224 2661
Contact: Francesc Monells
Email: girona@edetco.com

RUSSIA

RUPERTI PROJECT SERVICES
INTERNATIONAL
MOSCOW
Office 5, 15 Myasnitskaya ul
Moscow 101000
Russia
Tel: (00 7 495) 933 7810
Fax: (00 7 495) 933 7811
Email: anthony.ruperti@davislangdon.co

LEBANON

DAVIS LANGDON
BEIRUT
PO Box 13-5422-Shouran
Beirut
Lebanon
Tel: (00 9611) 780 111
Fax: (00 9611) 809 045
Contact: Muhyidden Itani
Email: DLL.MI@cyberia.net.lb

BAHRAIN

DAVIS LANGDON
MANAMA
3rd Floor Building 256
Road No 3605
Area No 336
PO Box 640
Manama
Bahrain
Tel: (00 973) 1782 7567
Fax: (00 973) 1772 7210
Email: steven.coates@davislangdon-bahrain.com

UNITED ARAB EMIRATES

DAVIS LANGDON
DUBAI
PO Box 7856
No. 410
Oud Metha Office Building
Dubai
UAE
Tel: (00 9714) 32 42 919
Fax: (00 9714) 32 42 838
Email: neil.taylor@davislangdon-dubai.com

QATAR

DAVIS LANGDON
DOHA
PO Box 3206
Doha
State of Qatar
Tel: (00 974) 4580 150
Fax: (00 974) 4697 905
Email:steven.humphrey@davislangdon-qatar.com

EGYPT

DAVIS LANGDON
CAIRO
35 Misr Helwan Road
Maadi 11431
Cairo
Egypt
Tel: (00 20 2) 526 2319
Fax: (00 20 2) 527 1338
Email: bob.ames@dlegypt.com

Specialist Service Lines
Project Management | Cost Management | Management Consulting | Legal Support | Specification Consulting | Engineering Services | Property Tax & Finance

Specialist Sectors
Arts | Commercial Offices | Distribution | Education | Food Processing | Health | Heritage | Hotels & Leisure | Industrial | Infrastructure | Public Buildings | Regeneration | Residential | Retail | Sports | Transportation

Davis Langdon LLP is a member firm of Davis Langdon & Seah International, with offices throughout Europe and the Middle East, Asia, Australasia, Africa and the USA

PART 5

Unit Costs (Highway Works)

INTRODUCTORY NOTES

The Unit Costs in this part represent the net cost to the Contractor of executing the work on site; they are not the prices which would be entered in a tender Bill of Quantities.

It must be emphasised that the unit rates are averages calculated on unit outputs for typical site conditions. Costs can vary considerably from contract to contract depending on individual Contractors, site conditions, working methods and other factors. Reference should be made to Part 1 for a general discussion on Civil Engineering Estimating.

Guidance prices are included for work normally executed by specialists, with a brief description where necessary of the assumptions upon which the costs have been based. Should the actual circumstances differ, it would be prudent to obtain check prices from the specialists concerned, on the basis of actual / likely quantity of the work, nature of site conditions, geographical location, time constraints, etc.

The method of measurement adopted in this section is the **Method of Measurement for Highway Works**, *subject to variances where this has been felt to be of advantage to produce more helpful price guidance.*

We have structured this Unit Cost section to cover as many aspects of Civil and Highway works as possible.

The Gang hours column shows the output per measured unit in actual time, not the total labour hours; thus for an item involving a gang of 5 men each for 0.3 hours, the total labour hours would naturally be 1.5, whereas the Gang hours shown would be 0.3.

This section is structured in such a manner as to provide the User with adequate background information on how the rates have been calculated, so as to allow them to be readily adjusted to suit other conditions to the example presented:

- *alternative gang structures as well as the effect of varying bonus levels, travelling costs etc.*
- *Other types of plant or else different running costs from the medium usage presumed*
- *Other types of materials or else different discount / waste allowances from the levels presumed*

Reference to Part 3 giving basic costs of labour, materials and plant together with Parts 13 and 14 will assist the reader in making adjustments to the unit costs.

GUIDANCE NOTES

Generally

Adjustments should be made to the rates shown for time, location, local conditions, site constraints and any other factors likely to affect the costs of a specific scheme.

Materials cost

Materials costs within the rates have been calculated using the 'list prices' contained in Part 3: Resources (pages 40 to 118), with an index appearing on page 39), adjusted to allow for delivery charges (if any) and a 'reasonable' level of discount obtainable by the contractor, this will vary very much depending on the contractor's standing, the potential size of the order and the supplier's eagerness and will vary also between raw traded goods such as timber which will attract a low discount of perhaps 3%, if at all, and manufactured goods where the room for bargaining is much greater and can reach levels of 30% to 40%. High demand for a product at the time of pricing can dramatically reduce the potential discount, as can the world economy in the case of imported goods such as timber and copper. Allowance has also been made for wastage on site (generally 2½% to 5%) dependent upon the risk of damage, the actual level should take account of the nature of the material and its method of storage and distribution about the site.

Labour cost

The composition of the labour and type of plant is generally stated at the beginning of each section, more detailed information on the calculation of the labour rates is given in Part 3: Resources, pages 33 to 37. In addition on pages 37 and 38 is a summary of labour grades and responsibilities extracted from the Working Rule Agreement. Within Parts 4 and 5, each section is prefaces by a detailed build-up of the labour gang assumed for each type of work. This should allow the user to see the cost impact of a different gang as well as different levels of bonus payments, allowances for skills and conditions, travelling allowances etc. The user should be aware that the output constants are based on the gangs shown and would also need to be changed.

Plant cost

A rate build-up of suitable plant is generally stated at the beginning of each section, with more detailed information on alternative machines and their average fuel costs being given in Part 3: Resources, pages 117 to 146. Within Parts 4 and 5, each section is prefaces by a detailed build-up of plant assumed for each type of work This should allow the user to see the cost impact of using alternative plant as well as different levels of usage (see note on pages 119 and 120). The user should be aware that the output constants are based on the plant shown and would also need to be changed.

Outputs

The user is directed to Part 13: Outputs (pages 589 to 698) which contains a selection of output constants and in particular a chart of haulage times for various capacities of Tippers on page 591.

Method of Measurement

A keynote to bills of quantities for highway works is the brevity of descriptions due to a strong emphasis being placed on the estimator pricing the work described in the Specification and shown on the Drawings.

Although this part of the book is primarily based on MMHW, the specific rules have been varied from in cases where it has been felt that an alternative presentation would be of value to the book's main purpose of providing guidance on prices. This is especially so with a number of specialist contractors but also in the cases of work where a more detailed presentation will enable the user to allow for ancillary items.

LEVEL 1 DIVISION		LEVEL 2 CONSTRUCTION HEADING	LEVEL 3 MMHW SERIES HEADINGS
(i) Preliminaries		Preliminaries	Series 100
(ii) Roadworks		Roadworks General	Series 200 Series 300 Series 400 Series 600
		Main Carriageway	Series 500 Series 700 Series 1100
		Interchanges	Series 500 Series 700 Series 1100
		Side Roads	Series 500 Series 700 Series 1100
		Signs, Motorway Communications and Lighting	Series 1200 Series 1300 Series 1400 Series 1500
		Landscape and Ecology	Series 3000
(iii) Structures	Structure in form of Bridge or Viaduct; Name or Reference	Special Preliminaries	
		Piling	Series 1600
		Substructure – End Supports	Series 500 Series 600 Series 1100 Series 1700 Series 1800
			Series 1900 Series 2300 Series 2400
		Substructure – Intermediate Supports Substructure – Main Span Substructure – Approach Spans	As for End Supports
		Superstructure – Main Span Superstructure – Approach Spans Superstructure – Arch Ribs	Series 500 Series 1700 Series 1800 Series 1900 Series 2100 Series 2300 Series 2400
		Finishings	Series 400 Series 600 Series 700 Series 1100 Series 2000 Series 2200 Series 2400

LEVEL 1 DIVISION	LEVEL 2 CONSTRUCTION HEADING	LEVEL 3 MMHW SERIES HEADINGS	LEVEL 1 DIVISION
	Retaining wall, Culvert, Subway, Gantry, Large Headwall, Gabion Wall, Diaphragm wall, Pocket Type Reinforced Brickwork Retaining Wall and the like; Name or Reference	Special Preliminaries	
		Main Construction	Series 500 Series 600 Series 1100 Series 1600 Series 1700 Series 1800 Series 1900 Series 2300 Series 2400
		Finishings	Series 400 Series 600 Series 700 Series 1100 Series 2000 Series 2200 Series 2400
(iv) Structures where a choice of designs is offered	Structure Designed by the Overseeing Organisation; Name or Reference	To comply with the principles set down above for Structures	
	Structure Designed by the Contractor; Name or Reference		
(v) Structures Designed by the Contractor	Structure; Name or Reference		
(vi) Service Areas		Roadworks Structures	To comply with the principles set down above for Roadworks and Structures
(vii) Maintenance Compounds		Roadworks Structures	
(viii) Accommodation Works		Interest; Name or Reference	
(ix) Works for Statutory or Other Bodies		Body; Name or Reference	To comply with the principles set down above for Roadworks and Structures
(x) Daywork		Daywork	
(xi) PC & Provisional Sum		PC & Provisional Sum	

SERIES 100: PRELIMINARIES

Item	Gang hours	Labour £	Plant £	Material £	Unit	Total rate £
NOTES						
General						
Refer also to the example calculation of Preliminaries in Part 2 and also to Part 8 Oncosts and profit.						
TEMPORARY ACCOMMODATION						
Erection of principal offices for the Overseeing Organisation						
prefabricated unit; connect to services	-	-	480.63	-	nr	**480.63**
Erection of offices and messes for the Contractor						
prefabricated unit; connect to services	-	-	480.63	-	nr	**480.63**
Erection of stores and workshops for the Contractor						
prefabricated unit; connect to services	-	-	480.63	-	nr	**480.63**
Servicing of principal offices for the Overseeing Organisation						
jack leg hutment; 3.7 m x 2.6 m	-	-	29.77	-	week	**29.77**
jack leg hutment; 7.35 m x 3.1 m	-	-	45.81	-	week	**45.81**
jack leg hutment; 14.7 m x 3.7 m	-	-	80.18	-	week	**80.18**
jack leg toilet unit; 4.9 m x 2.6 m	-	-	28.79	-	week	**28.79**
Servicing of portable offices for the Overseeing Organisation						
wheeled cabin; 3.7 m x 2.3 m	-	-	22.26	-	week	**22.26**
wheeled cabin; 6.7 m x 2.3 m	-	-	44.54	-	week	**44.54**
Servicing of offices and messes for the Overseeing Organisation						
jack leg hutment; 3.7 m x 2.6 m	-	-	29.77	-	week	**29.77**
jack leg hutment; 7.35 m x 3.1 m	-	-	45.81	-	week	**45.81**
jack leg hutment; 14.7 m x 3.7 m	-	-	80.18	-	week	**80.18**
wheeled cabin; 3.7 m x 2.3 m	-	-	22.26	-	week	**22.26**
wheeled cabin; 6.7 m x 2.3 m	-	-	44.54	-	week	**44.54**
jack leg toilet unit; 4.9 m x 2.6 m	-	-	28.79	-	week	**28.79**
canteen unit; 9.8 m x 2.6 m	-	-	68.72	-	week	**68.72**
Servicing of stores and workshops for the Contractor						
jack leg hutment; 3.7 m x 2.6 m	-	-	29.77	-	week	**29.77**
jack leg hutment; 7.35 m x 3.1 m	-	-	45.81	-	week	**45.81**
wheeled cabin; 3.7 m x 2.3 m	-	-	22.26	-	week	**22.26**
wheeled cabin; 6.7 m x 2.3 m	-	-	44.54	-	week	**44.54**
pollution decontamination unit; 6.7 m x 2.3 m	-	-	82.47	-	week	**82.47**
Dismantling of principal offices for the Overseeing Organisation						
prefabricated unit; disconnect from services; removing	-	-	480.63	-	nr	**480.63**
Dismantling of offices and messes for the Contractor						
prefabricated unit; disconnect from services; removing	-	-	480.63	-	nr	**480.63**
Dismantling of stores and workshops for the Contractor						
prefabricated unit; disconnect from services; removing	-	-	480.63	-	nr	**480.63**

SERIES 100: PRELIMINARIES

Item	Gang hours	Labour £	Plant £	Material £	Unit	Total rate £
VEHICLES FOR THE OVERSEEING ORGANISATION						
Vehicles for the Overseeing Organisation						
Land Rover short wheelbase	-	-	380.53	-	week	**380.53**
Land Rover long wheelbase	-	-	492.93	-	week	**492.93**
For other types of transport vehicles refer to Resources-Plant page *138.	-	-	-	-	-	
OPERATIVES FOR THE ENGINEER						
Operatives for the Overseeing Organisation						
Chainman for the Overseeing Organisation	40.00	380.00	-	-	week	**380.00**
Driver for the Overseeing Organisation	40.00	380.00	-	-	week	**380.00**
Laboratory assistant for the Overseeing Organisation	40.00	306.00	-	-	week	**306.00**
Example	1.00	10.35	2.27	1135.63	week	**1148.24**

SERIES 200: SITE CLEARANCE

Item	Gang hours	Labour £	Plant £	Material £	Unit	**Total rate £**
NOTES						
General						
The prices in this section are to include for the removal of superficial obstructions down to existing ground level.						
Demolition of individual or groups of buildings or structures						
MMHW states that individual structures should be itemised. The following rates are given as £ per m³ to simplify the pricing of different sized structures. (Refer also to Part 4 of this book Section D)						
RESOURCES - LABOUR						
Clearance gang						
1 ganger/chargehand (skill rate 4)		12.80				
1 skilled operative (skill rate 4)		12.05				
2 unskilled operatives (general)		22.52				
1 plant operator (skill rate 3)		14.65				
Total Gang Rate / Hour	£	**62.02**				
RESOURCES - PLANT						
Clearance						
0.8 m³ tractor loader - 50% of time			9.14			
8 t lorry with hiab - 25% of time			5.05			
4 t dumper - 50% of time			2.98			
20 t mobile crane - 25% of time			8.45			
compressor, 11.3 m³/min (450 cfm), 4 tool			6.33			
compressor tools: two brick hammers / picks - 50% of time			0.31			
compressor tools: chipping hammer			0.42			
compressor tools: medium rock drill 30			0.08			
compressor tools: road breaker			0.39			
compressor tools: two 15 m lengths hose			0.26			
Total Gang Rate / Hour		£	**33.41**			
SITE CLEARANCE						
General site clearance						
open field site	9.91	614.62	331.15	-	ha	**945.77**
medium density wooded	20.61	1278.23	688.62	-	ha	**1966.85**
heavy density wooded	32.09	1990.22	1072.19	-	ha	**3062.41**
urban areas (town centre)	30.60	1897.81	1022.50	-	ha	**2920.31**
live dual carriageway	30.60	1897.81	1022.50	-	ha	**2920.31**
Demolition of building or structure						
building; brick construction with timber floor and roof	-	-	-	-	m³	**5.34**

SERIES 200: SITE CLEARANCE

Item	Gang hours	Labour £	Plant £	Material £	Unit	Total rate £
SITE CLEARANCE – cont'd						
Demolition of building or structure – cont'd						
building; brick construction with concrete floor and roof	-	-	-	-	m³	**8.82**
building; masonry construction with timber floor and roof	-	-	-	-	m³	**6.90**
building; reinforced concrete frame construction with brick infill	-	-	-	-	m³	**9.19**
building; steel frame construction with brick infill	-	-	-	-	m³	**5.00**
building; steel frame construction with cladding	-	-	-	-	m³	**4.74**
building; timber	-	-	-	-	m³	**4.21**
reinforced concrete bridge deck or superstructure	-	-	-	-	m³	**8.10**
reinforced concrete bridge abutment or bank seat	-	-	-	-	m³	**31.86**
reinforced concrete retaining wall	-	-	-	-	m³	**118.07**
brick or masonry retaining wall	-	-	-	-	m³	**64.94**
brick or masonry boundary wall	-	-	-	-	m³	**53.13**
dry stone boundary wall	-	-	-	-	m³	**64.94**
TAKE UP OR DOWN AND SET ASIDE FOR RE-USE OR REMOVE TO STORE OR TIP OFF SITE						
Take up or down and set aside for re-use						
precast concrete kerbs and channels	0.02	1.05	1.65	-	m	**2.70**
precast concrete edgings	0.01	0.79	1.26	-	m	**2.06**
precast concrete drainage and kerb blocks	0.02	1.32	2.03	-	m	**3.35**
precast concrete drainage channel systems	0.02	1.32	2.03	-	m	**3.35**
tensioned single sided corrugated beam safety fence	0.14	8.68	4.70	-	m	**13.38**
timber post and 4 rail fence	0.08	4.96	2.67	-	m	**7.63**
bench seat	0.13	8.06	5.25	-	nr	**13.31**
cattle trough	0.16	9.68	6.28	-	nr	**15.95**
permanent bollard	0.13	8.06	5.25	-	nr	**13.31**
pedestrian crossing lights; pair	0.26	16.13	8.71	-	nr	**24.83**
lighting column including bracket arm and lantern; 5m high	0.58	35.97	19.39	-	nr	**55.36**
lighting column including bracket arm and lantern; 10 m high	0.61	37.83	20.36	-	nr	**58.19**
traffic sign	0.26	16.13	8.71	-	nr	**24.83**
timber gate	0.13	8.06	5.25	-	nr	**13.31**
timber gate	0.13	8.06	5.25	-	nr	**13.31**
stile	0.13	8.06	5.25	-	nr	**13.31**
road stud	0.03	1.61	1.03	-	nr	**2.64**
chamber cover and frame	0.03	1.61	1.03	-	nr	**2.64**
gully grating and frame	0.03	1.61	1.03	-	nr	**2.64**
feeder pillars	0.03	1.61	1.03	-	nr	**2.64**
Take up or down and remove to store off site						
precast concrete kerbs and channels	0.02	1.05	3.73	-	m	**4.78**
precast concrete edgings	0.01	0.79	2.82	-	m	**3.62**
precast concrete drainage and kerb blocks	0.02	1.32	4.63	-	m	**5.95**
precast concrete drainage channel systems	0.02	1.32	4.63	-	m	**5.95**
tensioned single sided corrugated beam safety fence	0.14	8.68	6.41	-	m	**15.10**

SERIES 200: SITE CLEARANCE

Item	Gang hours	Labour £	Plant £	Material £	Unit	**Total rate £**
timber post and 4 rail fence	0.08	4.96	3.65	-	m	**8.61**
bench seat	0.13	8.06	11.62	-	nr	**19.68**
cattle trough	0.16	9.68	13.92	-	nr	**23.60**
permanent bollard	0.13	8.06	11.62	-	nr	**19.68**
pedestrian crossing lights; pair	0.26	16.13	11.89	-	nr	**28.02**
lighting column including bracket arm and lantern; 5m high	0.58	35.97	26.50	-	nr	**62.47**
lighting column including bracket arm and lantern; 10 m high	0.61	37.83	27.83	-	nr	**65.66**
traffic sign	0.26	16.13	11.89	-	nr	**28.01**
timber gate	0.13	8.06	11.62	-	nr	**19.68**
timber gate	0.13	8.06	11.62	-	nr	**19.68**
stile	0.13	8.06	11.62	-	nr	**19.68**
road stud	0.03	1.61	2.31	-	nr	**3.92**
chamber cover and frame	0.03	1.61	2.31	-	nr	**3.92**
gully grating and frame	0.03	1.61	2.31	-	nr	**3.92**
feeder pillars	0.03	1.61	2.31	-	nr	**3.92**
Take up or down and remove to tip off site						
tensioned single sided corrugated beam safety fence	0.17	10.54	5.79	-	m	**16.33**
timber post and 4 rail fence	0.09	5.58	3.06	-	m	**8.64**
low pressure gas mains up to 150 mm diameter	0.04	2.48	1.36	-	m	**3.85**
low pressure water mains up to 75 mm diameter	0.03	1.86	1.04	-	m	**2.90**
power cable laid singly	0.03	1.86	1.04	-	m	**2.90**
lighting column including bracket arm and lantern; 5m high	0.82	50.86	28.00	-	nr	**78.85**
lighting column including bracket arm and lantern; 10 m high	0.85	52.72	28.99	-	nr	**81.70**
traffic sign including posts	0.38	23.57	12.59	-	nr	**36.16**
Removal of existing reflectorised thermoplastic road markings						
100 mm wide line	-	-	-	-	m	**1.10**
150 mm wide line	-	-	-	-	m	**1.65**
200 mm wide line	-	-	-	-	m	**2.21**
arrow or letter ne 6.0 m long	-	-	-	-	nr	**17.08**
arrow or letter 6.0 - 16.0 m long	-	-	-	-	nr	**71.66**

SERIES 300: FENCING

Item	Gang hours	Labour £	Plant £	Material £	Unit	Total rate £
NOTES						
General						
This section is restricted to those fences and barriers which are most commonly found on Highway Works. Hedges have been included, despite not being specifically catered for in the MMHW.						
The re-erection cost for fencing taken from store assumes that major components are in good condition; the prices below allow a sum of 20 % of the value of new materials to cover minor repairs, new fixings and touching up any coatings.						
RESOURCES - LABOUR						
Fencing/barrier gang						
1 ganger/chargehand (skill rate 4) - 50% of time		6.40				
1 skilled operative (skill rate 4) - 50% of time		6.03				
1 unskilled operative (general)		11.26				
1 plant operator (skill rate 4)		13.17				
Total Gang Rate / Hour	£	**36.86**				
Horticultural works gang						
1 skilled operative (skill rate 4)		12.05				
1 unskilled operative (general)		11.26				
Total Gang Rate / Hour	£	**23.31**				
RESOURCES - PLANT						
Fencing/Barriers						
agricultural type tractor; fencing auger			12.93			
gas oil for ditto			2.09			
drop sided trailer, two axles			0.32			
power tools etc. (fencing)			2.75			
Total Gang Rate / Hour		£	**18.09**			
RESOURCES - MATERIALS						
All rates for materials are based on the most economically available materials with a minimum waste allowance of 2.5% and supplier's discount of 15%.						
ENVIRONMENTAL BARRIERS						
Note: this section has been retained in the book for convenience in pricing although it is noted that it has been removed from the MMHW.						

SERIES 300: FENCING

Item	Gang hours	Labour £	Plant £	Material £	Unit	**Total rate £**
Hedges						
Set out, nick out and excavate trench minimum 400 mm deep and break up subsoil to minimum depth 300 mm	0.15	3.50	-	-	m	**3.50**
Supply and plant hedging plants ; backfilling with excavated topsoil						
single row plants at 200 mm centres	0.25	5.83	-	5.01	m	**10.84**
single row plants at 300 mm centres	0.17	3.96	-	3.34	m	**7.30**
single row plants at 400 mm centres	0.13	2.91	-	2.51	m	**5.42**
single row plants at 500 mm centres	0.10	2.33	-	2.00	m	**4.34**
single row plants at 600 mm centres	0.08	1.86	-	1.66	m	**3.53**
double row plants at 200 mm centres	0.50	11.65	-	10.02	m	**21.68**
double row plants at 300 mm centres	0.34	7.93	-	6.68	m	**14.60**
double row plants at 400 mm centres	0.25	5.83	-	5.01	m	**10.84**
double row plants at 500 mm centres	0.20	4.66	-	4.01	m	**8.67**
double row plants at 600 mm centres	0.16	3.73	-	3.33	m	**7.06**
Extra for incorporating manure at 1 m^3 / 30m^3	0.60	6.76	-	0.22	m^3	**6.97**
Noise barriers						
Noise barriers consist of the erection of reflective or absorbent acoustical screening to reduce nuisance from noise. Due to the divergence in performance requirements and specification for various locations it is not practical to state all inclusive unit costs. Therefore advice should be sought from Specialist Contractors in order to obtain accurate costings. However listed below are examples of sample specification together with approximate costings in order to obtain budget prices.						
NB:- The following unit costs are based upon a 2.0 m high barrier						
Noise reflective barriers						
Barrier with architectural precast concrete panels and integral posts	-	-	-	-	m^2	**138.72**
Barrier with acoustical timber planks post support system	-	-	-	-	m^2	**125.23**
Sound Absorptive barriers						
Barrier with architectural precast wood fibre concrete panels and integral posts	-	-	-	-	m^2	**165.86**
Barrier with perforated steel and mineral wool blankets in self-supporting system	-	-	-	-	m^2	**187.83**

SERIES 300: FENCING

Item	Gang hours	Labour £	Plant £	Material £	Unit	Total rate £
FENCING, GATES AND STILES						
Temporary fencing						
Type 1; 1.275 m high, timber posts and two strands of galvanised barbed wire and four strands of galvanised plain wire	0.06	2.21	1.09	4.77	m	**8.07**
Type 2; 1.275 m high, timber posts and two strands of galvanised barbed wire and four strands of galvanised plain wire	0.06	2.21	1.09	4.39	m	**7.69**
Type 3; 1.2 m high, timber posts and cleft chestnut paling	0.07	2.58	1.26	4.27	m	**8.12**
Type 4; 0.9 m high, timber posts and galvanised rectangular wire mesh	0.13	4.79	2.35	4.39	m	**11.53**
Timber rail fencing						
1.4 m high, timber posts and four rails	0.13	4.79	2.35	13.11	m	**20.25**
Plastic coated heavy pattern chain link fencing						
1.40 m high with 125 x 125 mm concrete posts	0.05	1.84	0.90	8.07	m	**10.82**
1.80 m high with 125 x 125 mm concrete posts	0.06	2.21	1.09	10.62	m	**13.92**
Plastic coated strained wire fencing						
1.35 m high, nine strand with 40 x 40 x 3 mm plastic coated RHS posts	0.16	5.90	2.89	15.01	m	**23.81**
1.80 m high, eleven strand with 50 x 50 x 3 mm plastic coated RHS posts	0.20	7.37	3.62	20.00	m	**30.99**
2.10 m high, fifteen strand with 50 x 50 x 3 mm plastic coated RHS posts	0.22	8.11	3.98	23.34	m	**35.43**
Woven wire fencing						
1.23 m high, galvanised wire with 75 x 150 mm timber posts	0.06	2.21	1.09	6.62	m	**9.92**
Close boarded fencing						
1.80 m high with 125 x 125 mm concrete posts	0.30	11.06	5.43	37.99	m	**54.47**
Concrete foundation						
to main posts	0.09	3.32	0.28	4.37	nr	**7.96**
to straining posts	0.09	3.32	0.28	4.37	nr	**7.96**
to struts	0.09	3.32	0.28	4.37	nr	**7.96**
to intermediate posts	0.09	3.32	0.28	4.37	nr	**7.96**
Steel tubular frame single field gates						
1.175 m high 3.60 m wide	0.30	11.06	5.43	58.29	nr	**74.78**
1.175 m high 4.50 m wide	0.30	11.06	5.43	74.94	nr	**91.42**
Steel tubular frame half mesh single field gates						
1.175 m high 3.60 m wide	0.30	11.06	5.43	74.94	nr	**91.42**
1.175 m high 4.50 m wide	0.30	11.06	5.43	91.58	nr	**108.07**
Steel tubular frame extra wide single field gates						
1.175 m high 4.88 m wide	0.30	11.06	5.43	83.26	nr	**99.75**
Steel tubular frame double field gates						
1.175 m high 5.02 m wide	0.60	22.12	10.86	124.92	nr	**157.89**
Timber single field gates						
1.27 m high 3.00 m wide	1.84	67.82	33.29	145.40	nr	**246.50**
1.27 m high 3.60 m wide	1.90	70.03	34.37	168.37	nr	**272.77**
1.27 m high 4.10 m wide	2.00	73.72	36.18	203.59	nr	**313.49**
1.27 m high 4.71 m wide	2.00	73.72	36.18	209.68	nr	**319.58**
Timber Type 1 wicket gates						
1.27 m high 1.20 m wide	1.20	44.23	21.71	45.92	nr	**111.86**
Timber Type 2 wicket gates						
1.27 m high 1.02 m wide	1.20	44.23	21.71	52.37	nr	**118.31**
Timber kissing gates						
1.27 m high 1.77 m wide	2.00	73.72	36.18	145.40	nr	**255.30**

SERIES 300: FENCING

Item	Gang hours	Labour £	Plant £	Material £	Unit	**Total rate £**
Timber stiles Type 1						
1.45 m high 1.00 m wide	1.50	55.29	27.14	112.05	nr	**194.48**
Timber stiles Type 2						
1.45 m high 1.00 m wide	1.40	51.60	25.32	84.21	nr	**161.14**
Extra for						
sheep netting on post and wire	0.04	1.29	0.11	0.51	m	**1.91**
pig netting on post and wire	0.04	1.29	0.11	0.55	m	**1.95**
REMOVE FROM STORE AND RE-ERECT FENCING, GATES AND STILES						
Timber rail fencing						
1.4 m high, timber posts and four rails	0.13	4.79	2.35	2.32	m	**9.46**
Plastic coated heavy pattern chain link fencing						
1.40 m high with 125 x 125 mm concrete posts	0.05	1.84	0.90	1.45	m	**4.20**
1.80 m high with 125 x 125 mm concrete posts	0.06	2.21	1.09	1.92	m	**5.22**
Plastic coated strained wire fencing						
1.35 m high, nine strand with 40 x 40 x 3 mm plastic coated RHS posts	0.16	5.90	2.89	2.71	m	**11.50**
1.80 m high, eleven strand with 50 x 50 x 3 mm plastic coated RHS posts	0.20	7.37	3.62	4.03	m	**15.02**
2.10 m high, fifteen strand with 50 x 50 x 3 mm plastic coated RHS posts	0.22	8.11	3.98	4.27	m	**16.36**
Woven wire fencing						
1.23 m high, galvanised wire with 75 x 150 mm timber posts	0.06	2.21	1.09	1.34	m	**4.63**
Close boarded fencing						
1.80 m high with 125 x 125 mm concrete posts	0.30	11.06	5.43	7.21	m	**23.70**
Steel tubular frame single field gates						
1.175 m high 3.60 m wide	0.30	11.06	5.43	12.09	nr	**28.58**
1.175 m high 4.50 m wide	0.30	11.06	5.43	15.53	nr	**32.02**
Steel tubular frame half mesh single field gates						
1.175 m high 3.60 m wide	0.30	11.06	5.43	15.53	nr	**32.02**
1.175 m high 4.50 m wide	0.30	11.06	5.43	19.00	nr	**35.48**
Steel tubular frame extra wide single field gates						
1.175 m high 4.88 m wide	0.30	11.06	5.43	17.28	nr	**33.76**
Steel tubular frame double field gates						
1.175 m high 5.02 m wide	0.60	22.12	10.86	25.93	nr	**58.90**
Timber single field gates						
1.27 m high 3.00 m wide	1.84	67.82	33.29	30.45	nr	**131.56**
1.27 m high 3.60 m wide	1.90	70.03	34.37	35.27	nr	**139.68**
1.27 m high 4.10 m wide	2.00	73.72	36.18	42.66	nr	**152.55**
1.27 m high 4.71 m wide	2.00	73.72	36.18	43.93	nr	**153.83**
Timber Type 1 wicket gates						
1.27 m high 1.20 m wide	1.20	44.23	21.71	9.62	nr	**75.56**
Timber Type 2 wicket gates						
1.27 m high 1.02 m wide	1.20	44.23	21.71	10.98	nr	**76.92**
Timber kissing gates						
1.27 m high 1.77 m wide	2.00	73.72	36.18	30.45	nr	**140.35**
Timber stiles Type 1						
1.45 m high 1.00 m wide	1.50	55.29	27.14	25.17	nr	**107.59**
Timber stiles Type 2						
1.45 m high 1.00 m wide	1.40	51.60	25.32	18.93	nr	**95.86**

SERIES 400: SAFETY FENCES/BARRIERS AND PEDESTRIAN GUARDRAILS

Item	Gang hours	Labour £	Plant £	Material £	Unit	Total rate £
NOTES						
General						
The re-erection cost for safety fencing taken from store assumes that major components are in good condition; the prices below allow a sum of 20 % of the value of new materials to cover minor repairs, new fixings and touching up any coatings.						
RESOURCES - LABOUR						
Safety fencing gang						
1 ganger/chargehand (skill rate 4)		12.80				
1 skilled operative (skill rate 4)		12.05				
2 unskilled operatives (general)		22.52				
1 plant operator (skill rate 4)		13.17				
Total Gang Rate / Hour	£	**60.54**				
RESOURCES - PLANT						
Safety fencing						
agricultural type tractor; fencing auger			12.93			
gas oil for ditto			2.09			
drop sided trailer, two axles			0.32			
small tools (part time)			1.92			
Total Gang Rate / Hour		£	**17.26**			
BEAM SAFETY FENCES						
Prices generally are for beams "straight or curved exceeding 120m radius", for work to a tighter radius						
50 - 120 m radius — Add 15 %						
not exceeding 50 m radius — Add 40 %						
Untensioned beams						
single sided corrugated beam	0.07	4.24	1.21	16.59	m	**22.03**
double sided corrugated beam	0.22	13.32	3.80	33.15	m	**50.27**
single sided open box beam	0.07	4.24	1.21	33.65	m	**39.09**
single sided double rail open box beam	0.22	13.32	3.80	67.30	m	**84.42**
double height open box beam	0.24	14.53	4.14	69.40	m	**88.07**
Tensioned beams						
single sided corrugated beam	0.10	6.05	1.72	17.53	m	**25.31**
double sided corrugated beam	0.12	7.26	2.07	36.14	m	**45.48**
Long driven post						
for single sided tensioned corrugated beam	0.06	3.33	0.95	39.29	nr	**43.57**
for double sided tensioned corrugated beam	0.06	3.33	0.95	39.29	nr	**43.57**
for single sided open box beam	0.06	3.33	0.95	39.29	nr	**43.57**
for double sided open box beam	0.06	3.33	0.95	39.29	nr	**43.57**
for double height open box beam	0.06	3.33	0.95	39.29	nr	**43.57**
Short post for setting in concrete or socket						
for single sided tensioned corrugated beam	0.06	3.63	1.04	15.15	nr	**19.82**
for double sided tensioned corrugated beam	0.06	3.63	1.04	15.43	nr	**20.10**
for single sided open box beam	0.06	3.63	1.04	18.39	nr	**23.06**
for double sided open box beam	0.06	3.63	1.04	21.92	nr	**26.59**

SERIES 400: SAFETY FENCES/BARRIERS AND PEDESTRIAN GUARDRAILS

Item	Gang hours	Labour £	Plant £	Material £	Unit	**Total rate £**
Mounting bracket fixed to structure						
for single sided open box beam	0.16	9.69	2.76	59.04	nr	**71.49**
Terminal section						
for untensioned single sided corrugated beam	0.71	42.98	12.26	425.00	nr	**480.24**
for untensioned double sided corrugated beam	1.25	75.67	21.58	450.00	nr	**547.25**
for untensioned single sided open box beam	1.01	61.15	17.43	455.00	nr	**533.58**
for untensioned double sided open box beam	1.78	107.76	30.72	765.00	nr	**903.48**
for tensioned single sided corrugated beam	0.96	58.12	16.57	296.50	nr	**371.19**
for tensioned double sided corrugated beam	1.70	102.92	29.34	453.75	nr	**586.01**
Full height anchorage						
for single sided tensioned corrugated beam	3.95	239.13	68.18	655.25	nr	**962.56**
for double sided tensioned corrugated beam	4.35	263.35	75.08	764.00	nr	**1102.43**
for single sided open box beam	3.80	230.05	65.59	564.50	nr	**860.14**
for double sided open box beam	4.20	254.27	72.50	665.00	nr	**991.76**
Expansion joint anchorage						
for single sided open box beam	4.52	273.64	78.02	1163.50	nr	**1515.16**
for double sided open box beam	5.15	311.78	88.89	1409.50	nr	**1810.17**
Type 048 connection to bridge parapet						
for single sided open box beam	0.70	42.38	12.08	148.91	nr	**203.37**
Connection piece for single sided open box beam to single sided corrugated beam	0.78	47.22	13.46	152.75	nr	**213.43**
Standard concrete foundation						
for post for corrugated beam	0.23	13.92	3.90	6.15	nr	**23.97**
for post for open box beam	0.23	13.92	3.97	7.66	nr	**25.55**
Concrete foundation Type 1 spanning filter drain						
for post for corrugated beam	0.25	15.13	4.32	9.98	nr	**29.44**
for post for open box beam	0.25	15.13	4.32	8.84	nr	**28.29**
Standard socketed foundation for post for open box beam	0.30	18.16	5.18	8.86	nr	**32.20**
REMOVE FROM STORE AND RE-ERECT BEAM SAFETY FENCES						
Prices generally are for beams "straight or curved exceeding 120m radius", for work to a tighter radius						
50 - 120 m radius — Add 15 %						
not exceeding 50 m radius — Add 25 %						
Untensioned beams						
single sided corrugated beam	0.07	4.24	1.21	3.31	m	**8.75**
double sided corrugated beam	0.22	13.32	3.80	6.64	m	**23.76**
single sided open box beam	0.07	4.24	1.21	6.74	m	**12.18**
single sided double rail open box beam	0.22	13.32	3.80	13.69	m	**30.81**
double height open box beam	0.24	14.53	4.14	13.88	m	**32.55**
Tensioned beams						
single sided corrugated beam	0.10	6.05	1.72	3.50	m	**11.28**
double sided corrugated beam	0.12	7.26	2.07	7.23	m	**16.57**
Long driven post						
for single sided tensioned corrugated beam	0.06	3.33	0.95	7.85	nr	**12.13**
for double sided tensioned corrugated beam	0.06	3.33	0.95	7.86	nr	**12.14**
for single sided open box beam	0.06	3.33	0.95	7.92	nr	**12.20**
for double sided open box beam	0.06	3.33	0.95	9.21	nr	**13.49**
for double height open box beam	0.06	3.33	0.95	23.44	nr	**27.72**

SERIES 400: SAFETY FENCES/BARRIERS AND PEDESTRIAN GUARDRAILS

Item	Gang hours	Labour £	Plant £	Material £	Unit	Total rate £
REMOVE FROM STORE AND RE-ERECT BEAM SAFETY FENCES – cont'd						
Short post for setting in concrete or socket						
for single sided tensioned corrugated beam	0.06	3.63	1.04	15.28	nr	**19.95**
for double sided tensioned corrugated beam	0.06	3.63	1.04	15.58	nr	**20.25**
for single sided open box beam	0.06	3.63	1.04	18.56	nr	**23.23**
for double sided open box beam	0.06	3.63	1.04	22.12	nr	**26.79**
Mounting bracket fixed to structure						
for single sided open box beam	0.16	9.69	2.76	5.85	nr	**18.30**
Terminal section						
for untensioned single sided corrugated beam	0.71	42.98	12.26	85.00	nr	**140.24**
for untensioned double sided corrugated beam	1.25	75.67	21.58	90.10	nr	**187.35**
for untensioned single sided open box beam	1.01	61.15	17.43	91.00	nr	**169.58**
for untensioned double sided open box beam	1.78	107.76	30.72	148.50	nr	**286.98**
for tensioned single sided corrugated beam	0.96	58.12	16.57	56.75	nr	**131.44**
for tensioned double sided corrugated beam	1.70	102.92	29.34	88.45	nr	**220.71**
Full height anchorage						
for single sided tensioned corrugated beam	3.95	239.13	68.18	112.25	nr	**419.56**
for double sided tensioned corrugated beam	4.35	263.35	75.08	143.25	nr	**481.68**
for single sided open box beam	3.80	230.05	65.59	112.90	nr	**408.54**
for double sided open box beam	4.20	254.27	72.50	133.10	nr	**459.86**
Expansion joint anchorage						
for single sided open box beam	4.52	273.64	78.02	232.75	nr	**584.41**
for double sided open box beam	5.15	311.78	88.89	282.00	nr	**682.67**
Type 048 connection to bridge parapet						
for single sided open box beam	0.70	42.38	12.08	6.52	nr	**60.98**
Connection piece for single sided open box beam to single sided corrugated beam	0.78	47.22	13.46	152.75	nr	**213.43**
Standard concrete foundation						
for post for corrugated beam	0.23	13.92	3.90	6.15	nr	**23.97**
for post for open box beam	0.23	13.92	3.97	7.66	nr	**25.55**
Concrete foundation Type 1 spanning filter drain						
for post for corrugated beam	0.25	15.13	4.32	9.98	nr	**29.44**
for post for open box beam	0.25	15.13	4.32	8.84	nr	**28.29**
Standard socketed foundation for post for open box beam	0.30	18.16	5.18	8.86	nr	**32.20**
WIRE ROPE SAFETY FENCES						
Wire rope safety fencing to BS 5750; based on 600 m lengths; 4 rope system; posts at 2.40 m general spacings.						
Wire rope	0.03	1.82	0.52	13.34	m	**15.67**
Long driven line posts	0.05	3.03	0.86	24.63	nr	**28.52**
Long driven deflection posts	0.05	3.03	0.86	24.63	nr	**28.52**
Long driven height restraining posts	0.05	3.03	0.86	24.63	nr	**28.52**
Short line post for setting in concrete or socket	0.06	3.63	1.04	22.79	nr	**27.46**
Short deflection post for setting in concrete or socket	0.06	3.63	1.04	22.79	nr	**27.46**
Short height restraining post for setting in concrete or socket	0.06	3.63	1.04	22.79	nr	**27.46**
Fixed height surface mounted post fixed to structure or foundation	0.09	5.45	1.55	25.88	nr	**32.88**
Standard intermediate anchorage	2.00	121.08	34.52	299.00	nr	**454.60**
Standard end anchorage	2.00	121.08	34.52	107.78	nr	**263.38**

SERIES 400: SAFETY FENCES/BARRIERS AND PEDESTRIAN GUARDRAILS

Item	Gang hours	Labour £	Plant £	Material £	Unit	Total rate £
In situ standard concrete foundation for post	0.23	13.92	3.97	7.69	nr	**25.59**
In situ standard socketed foundation for post	0.33	19.98	5.70	7.69	nr	**33.37**
Concrete foundation Type 1 spanning filter drain for post	0.37	22.40	6.38	9.25	nr	**38.03**
CONCRETE SAFETY BARRIERS						
Permanent vertical concrete safety barrier; TRRL Design - D.Tp Approved						
Intermediate units Type V01 & V02; 3 m long						
straight or curved exceeding 50 m radius	0.16	9.69	2.76	375.73	nr	**388.18**
curved not exceeding 50 m radius	0.20	12.11	3.45	375.73	nr	**391.29**
Make up units Type V05 & V06; 1 m long						
straight or curved exceeding 50 m radius	0.30	18.16	5.18	485.85	nr	**509.20**
Termination units Type V03 & V04; 3 m long	0.50	30.27	8.63	465.98	nr	**504.88**
Transition to single sided open box beam unit Type V08 & V09; 1.5 m long	0.37	22.40	6.38	553.98	nr	**582.76**
Transition to rectangular hollow section beam unit Type V10, V11 & V12; 1.5 m long	0.37	22.40	6.38	476.98	nr	**505.76**
Transition to double sided open box beam; 1.5 m long unit Type V07	0.37	22.40	6.38	532.98	nr	**561.76**
Anchor plate sets (normally two plates per first and last three units in any run)	0.15	9.08	5.45	36.92	nr	**51.45**
Temporary concrete safety barrier; TRRL Design - D.Tp Approved						
Intermediate units Type V28; 3 m long						
straight or curved exceeding 50 m radius	0.16	9.69	2.76	335.38	nr	**347.83**
curved not exceeding 50 m radius	0.20	12.11	3.45	335.38	nr	**350.94**
Termination units Type V29; 3 m long	0.50	30.27	8.63	375.88	nr	**414.78**
PEDESTRIAN GUARD RAILS AND HANDRAILS						
New work						
Tubular galvanised mild steel pedestrian guard rails to BS 3049 with mesh infill (105 swg, 50 x 50 mm mesh); 1.0 m high						
mounted on posts with concrete footing	0.16	9.69	2.76	157.23	m	**169.67**
mounted on posts bolted to structure or ground beam	0.14	8.48	2.42	157.23	m	**168.12**
Solid section galvanised steel pedestrian guard rails with vertical rails (group P4 parapet); 1.0 m high						
mounted on posts with concrete footing	0.19	11.50	3.28	19.71	m	**34.49**
mounted on posts bolted to structure or ground beam	0.17	10.29	2.94	136.35	m	**149.58**
Tubular double ball galvanised steel handrail						
50 mm diameter; 1.20 m high posts	0.15	9.08	2.59	123.04	m	**134.71**
63 mm diameter; 1.20 m high posts	0.15	9.08	2.59	182.42	m	**194.09**
Extra for concrete footings for handrail support posts	0.05	3.03	-	3.41	m	**6.44**
Existing guard rails						
Take from store and re-erect						
pedestrian guard railing, 3.0 m long x 1.0 m high panels	0.15	9.08	2.59	4.47	nr	**16.14**

SERIES 500: DRAINAGE AND SERVICE DUCTS

Item	Gang hours	Labour £	Plant £	Material £	Unit	Total rate £
NOTES						
General						
The re-erection cost for covers and grating complete with frames taken from store assumes that major components are in good condition; the prices below allow a sum of 10 % of the value of new materials to cover minor repairs, new fixings and touching up any coatings.						
RESOURCES - LABOUR						
Drains/sewers/culverts gang (small bore)						
1 ganger/chargehand (skill rate 4) - 50% of time		6.40				
1 skilled operative (skill rate 4)		12.05				
2 unskilled operatives (general)		22.52				
1 plant operator (skill rate 3)		14.65				
Total Gang Rate / Hour	£	**55.62**				
Drains/sewers/culverts gang (large bore)						
1 ganger/chargehand (skill rate 4) - 50% of time		6.40				
1 skilled operative (skill rate 4)		12.05				
2 unskilled operatives (general)		22.52				
1 plant operator (skill rate 3)		14.65				
1 plant operator (skill rate 3) - 30% of time		4.39				
Total Gang Rate / Hour	£	**60.02**				
Gullies gang						
1 ganger/chargehand (skill rate 4) - 50% of time		6.40				
1 skilled operative (skill rate 4)		12.05				
2 unskilled operatives (general)		22.52				
Total Gang Rate / Hour	£	**40.97**				
RESOURCES - PLANT						
Drains/sewers/culverts gang (small bore)						
0.4 m³ hydraulic excavator			16.67			
2t dumper - 30% of time			1.09			
360mm compaction plate - 30% of time			0.47			
2.80m³/min compressor, 2 tool - 30% of time			1.29			
disc saw - 30% of time			0.40			
extra 15ft/50m hose - 30% of time			0.08			
small pump - 30% of time			0.56			
Total Gang Rate / Hour		£	**20.57**			
Note: in addition to the above are the following allowances for trench struts/props/sheeting, assuming the need for close boarded earth support:						
average 1.00 m deep	-	-	1.19	-	m	**1.19**
average 1.50 m deep	-	-	1.33	-	m	**1.33**
average 2.00 m deep	-	-	1.53	-	m	**1.53**
average 2.50 m deep	-	-	1.83	-	m	**1.83**
average 3.00 m deep	-	-	2.30	-	m	**2.30**
average 3.50 m deep	-	-	2.84	-	m	**2.84**

SERIES 500: DRAINAGE AND SERVICE DUCTS

Item	Gang hours	Labour £	Plant £	Material £	Unit	Total rate £
Note: excavation in hard materials as above but with the addition of breaker attachments to the excavator as follows:						
generally : BRH91 (141kg)	-	-	2.62	-	-	**2.62**
reinforced concrete; BRH125 (310kg)	-	-	4.96	-	-	**4.96**
Drains/sewers/culverts gang (large bore)						
i.e. greater than 700 mm bore						
0.4 m³ hydraulic excavator			16.67			
2t dumper - 30% of time			1.09			
2.80 m³/min compressor, 2 tool - 30% of time			1.29			
compaction plate / roller - 30% of time			0.47			
disc saw - 30% of time			0.29			
small pump - 30% of time			0.56			
10 t crawler crane - 30% of time			6.77			
Total Gang Rate / Hour		**£**	**27.15**			
Gullies						
2t dumper - 30% of time			1.09			
Stihl disc saw - 30% of time			0.29			
Total Gang Rate / Hour		**£**	**1.38**			
RESOURCES - MATERIALS						
For the purposes of bedding widths for pipe bedding materials, trenches have been taken as exceeding 1.50 m in depth; trenches to lesser depths are generally 150 mm narrower than those given here so that the rates need to be reduced proportionately.						
DRAINS AND SERVICE DUCTS (EXCLUDING FILTER DRAINS, NARROW FILTER DRAINS AND FIN DRAINS)						
Vitrified clay pipes to BS 65, plain ends with push-fit polypropylene flexible couplings						
150 mm diameter drain or sewer in trench, depth to invert						
average 1.00 m deep	0.17	9.46	6.84	15.12	m	**31.41**
average 1.50 m deep	0.19	10.57	7.54	15.12	m	**33.22**
average 2.00 m deep	0.22	12.24	8.85	15.12	m	**36.20**
average 2.50 m deep	0.26	14.46	10.78	15.12	m	**40.36**
average 3.00 m deep	0.34	18.91	13.60	15.12	m	**47.63**
average 3.50 m deep	0.42	23.36	16.80	15.12	m	**55.28**
Extra for						
Type N sand bed 650 x 100 mm	0.04	2.22	1.61	1.04	m	**4.88**
Type T sand surround 650 wide x 100 mm	0.08	4.45	3.21	2.50	m	**10.17**
Type F granular bed 650 x 100 mm	0.05	2.78	2.02	1.51	m	**6.31**
Type S granular surround 650 wide x 100 mm	0.16	8.90	6.43	3.61	m	**18.93**
Type A concrete bed 650 x 100 mm	0.11	6.12	4.17	6.22	m	**16.51**
Type B 100 mm concrete bed and haunch	0.24	13.35	9.10	5.31	m	**27.77**
Type Z concrete surround 650 wide x 100 mm	0.22	12.24	8.35	14.92	m	**35.50**

SERIES 500: DRAINAGE AND SERVICE DUCTS

Item	Gang hours	Labour £	Plant £	Material £	Unit	Total rate £
DRAINS AND SERVICE DUCTS – cont'd						
Vitrified clay pipes to BS 65, plain ends with push-fit polypropylene flexible couplings - cont'd						
225 mm diameter drain or sewer in trench, depth to invert						
average 1.50 m deep	0.20	11.12	5.11	36.82	m	**53.05**
average 2.00 m deep	0.24	13.35	9.63	36.82	m	**59.80**
average 2.50 m deep	0.28	15.57	11.19	36.82	m	**63.58**
average 3.00 m deep	0.35	19.47	13.98	36.82	m	**70.27**
average 3.50 m deep	0.44	24.47	17.58	36.82	m	**78.88**
average 4.00 m deep	0.56	31.15	22.38	36.82	m	**90.35**
Extra for						
Type N sand bed 750 x 150 mm	0.12	6.67	4.55	1.81	m	**13.03**
Type T sand surround 750 wide x 150 mm	0.24	13.35	9.10	4.46	m	**26.91**
Type F granular bed 750 x 150 mm	0.13	7.23	4.93	2.60	m	**14.76**
Type S granular surround 750 wide x 150 mm	0.24	13.35	9.10	6.44	m	**28.89**
Type A concrete bed 750 x 150 mm	0.19	10.57	7.21	10.78	m	**28.55**
Type B 150 mm concrete bed and haunch	0.29	16.13	11.00	8.30	m	**35.43**
Type Z concrete surround 750 wide x 150 mm	0.36	20.02	13.66	26.72	m	**60.40**
300 mm diameter drain or sewer in trench, depth to invert						
average 1.50 m deep	0.21	11.68	8.45	60.38	m	**80.50**
average 2.00 m deep	0.26	14.46	10.45	60.38	m	**85.29**
average 2.50 m deep	0.30	16.69	12.00	60.38	m	**89.06**
average 3.00 m deep	0.37	20.58	14.82	60.38	m	**95.78**
average 4.00 m deep	0.61	33.93	24.39	60.38	m	**118.69**
average 5.00 m deep	1.01	56.18	40.31	60.38	m	**156.86**
Extra for						
Type N sand bed 800 x 150 mm	0.13	7.23	4.93	1.93	m	**14.09**
Type T sand surround 800 x 150 mm	0.26	14.46	9.58	6.20	m	**30.24**
Type F granular bed 800 x 150 mm	0.14	7.79	5.31	2.76	m	**15.86**
Type S granular surround 800 wide x 150 mm	0.26	14.46	9.87	8.77	m	**33.10**
Type A concrete bed 800 x 150 mm	0.20	11.12	7.59	11.50	m	**30.21**
Type B 150 mm concrete bed and haunch	0.36	20.02	13.66	10.48	m	**44.16**
Type Z concrete surround 800 wide x 150 mm	0.22	12.24	25.08	34.31	m	**71.63**
Vitrified clay pipes to BS 65, spigot and socket joints with sealing ring						
400 mm diameter drain or sewer in trench, depth to invert						
average 2.00 m deep	0.31	17.24	12.44	97.74	m	**127.42**
average 2.50 m deep	0.35	19.47	13.98	97.74	m	**131.19**
average 3.00 m deep	0.42	23.36	16.80	97.74	m	**137.90**
average 4.00 m deep	0.66	36.71	26.39	97.74	m	**160.84**
average 5.00 m deep	1.07	59.51	42.75	97.74	m	**200.01**
average 6.00 m deep	1.85	102.90	73.95	97.74	m	**274.59**
Extra for						
Type N sand bed 900 x 150 mm	0.14	7.79	5.31	2.17	m	**15.27**
Type T sand surround 900 wide x 150 mm	0.28	15.57	10.62	8.87	m	**35.06**
Type F granular bed 900 x 150 mm	0.15	8.34	5.69	3.14	m	**17.17**
Type S granular surround 900 x 150 mm	0.28	15.57	10.62	12.80	m	**39.00**
Type A concrete bed 900 x 150 mm	0.21	11.68	7.97	12.92	m	**32.57**
Type B, 150 mm concrete bed and haunch	0.43	23.92	16.31	13.80	m	**54.03**
Type Z concrete surround 900 wide x 150 mm	0.40	22.25	15.17	53.02	m	**90.45**

SERIES 500: DRAINAGE AND SERVICE DUCTS

Item	Gang hours	Labour £	Plant £	Material £	Unit	Total rate £
Concrete pipes with rebated flexible joints to BS 5911 Class L						
450 mm diameter piped culvert in trench, depth to invert						
average 2.00 m deep	0.33	18.35	13.26	20.40	m	**52.01**
average 3.00 m deep	0.44	24.47	17.41	20.40	m	**62.28**
average 4.00 m deep	0.72	40.05	28.48	20.40	m	**88.92**
average 6.00 m deep	1.65	91.77	65.76	20.40	m	**177.93**
Extra for						
Type A concrete bed 1050 x 150 mm	0.24	13.35	9.10	15.07	m	**37.53**
Type Z concrete surround 1050 wide x 150 mm	0.45	25.03	17.07	67.56	m	**109.67**
750 mm diameter piped culvert in trench, depth to invert						
average 2.00 m deep	0.38	22.81	19.21	56.02	m	**98.03**
average 3.00 m deep	0.52	31.21	26.17	56.02	m	**113.40**
average 4.00 m deep	0.92	55.22	46.33	56.02	m	**157.56**
average 6.00 m deep	2.05	123.04	103.00	56.02	m	**282.06**
Extra for						
Type A concrete bed 1250 x 150 mm	0.26	15.61	9.87	20.88	m	**46.35**
Type Z concrete surround 1250 wide x 150 mm	0.50	30.01	18.97	151.83	m	**200.81**
900 mm diameter piped culvert in trench, depth to invert						
average 2.00 m deep	0.43	25.81	21.71	76.62	m	**124.14**
average 3.00 m deep	0.62	37.21	31.29	76.62	m	**145.13**
average 4.00 m deep	1.06	63.62	53.37	76.62	m	**193.61**
average 6.00 m deep	2.25	135.04	113.02	76.62	m	**324.69**
Extra for						
Type A concrete bed 1500 x 150 mm	0.28	16.81	10.62	24.44	m	**51.87**
Type Z concrete surround 1500 wide x 150 mm	0.55	33.01	20.86	211.12	m	**265.00**
Corrugated steel pipes galvanised, hot dip bitumen coated (Armco type)						
1000 mm diameter piped culvert in trench, Type S granular surround, depth to invert						
average 2.00 m deep	0.99	59.42	49.92	110.50	m	**219.84**
1600 mm diameter piped culvert in trench, Type S granular surround, depth to invert						
average 2.00 m deep	2.13	127.84	107.44	184.36	m	**419.64**
2000 mm diameter piped culvert in trench, Type S granular surround, depth to invert						
average 3.00 m deep	2.97	178.26	149.79	347.52	m	**675.57**
2200 mm diameter piped culvert in trench, Type S granular surround, depth to invert						
average 3.00 m deep	3.29	197.47	165.88	392.58	m	**755.93**
Clay cable ducts; Hepduct						
100 mm diameter service duct in trench, Type S granular surround, depth to invert						
average 1.00 m deep	0.14	7.79	5.46	10.16	m	**23.40**
Two 100 mm diameter service ducts in trench, Type S granular surround, depth to invert						
average 1.00 m deep	0.24	13.35	9.40	20.10	m	**42.85**
Three 100 mm diameter service ducts in trench, Type S granular surround, depth to invert						
average 1.00 m deep	0.32	17.80	12.51	28.85	m	**59.16**

SERIES 500: DRAINAGE AND SERVICE DUCTS

Item	Gang hours	Labour £	Plant £	Material £	Unit	Total rate £
DRAINS AND SERVICE DUCTS – cont'd						
Four 100 mm diameter service ducts in trench, Type S granular surround, depth to invert average 1.00 m deep	0.40	22.25	15.68	39.91	m	**77.84**
Six 100 mm diameter service ducts in trench, Type S granular surround, depth to invert average 1.00 m deep	0.60	33.37	23.45	54.48	m	**111.30**
Extra for						
Type Z concrete surround on single duct	0.08	4.45	3.03	8.29	m	**15.77**
Type Z concrete surround on additional ways	0.08	4.45	3.03	8.29	m	**15.77**
150 mm diameter conduit, per way	0.01	0.56	0.38	12.07	m	**13.00**
225 mm diameter conduit, per way	0.01	0.56	0.38	37.99	m	**38.92**
FILTER DRAINS						
Vitrified clay perforated pipes to BS 65, sleeved joints						
150 mm diameter filter drain in trench with Type A bed and Type A fill filter material,						
average 1.00 m deep	0.26	14.46	10.01	22.01	m	**46.48**
average 2.00 m deep	0.30	16.69	11.51	27.77	m	**55.96**
average 3.00 m deep	0.35	19.47	13.53	33.57	m	**66.56**
150 mm pipes with Type A bed and Type B fill, depth						
average 1.00 m deep	0.26	14.46	10.01	20.99	m	**45.47**
average 2.00 m deep	0.30	16.69	11.51	25.53	m	**53.72**
average 3.00 m deep	0.35	19.47	13.53	30.13	m	**63.12**
225 mm pipes with Type A bed and Type A fill, depth						
average 1.00 m deep	0.27	15.02	10.34	33.78	m	**59.14**
average 2.00 m deep	0.31	17.24	11.82	40.97	m	**70.04**
average 3.00 m deep	0.36	20.02	13.90	47.48	m	**81.40**
225 mm pipes with Type A bed and Type B fill, depth						
average 1.00 m deep	0.27	15.02	10.34	32.63	m	**57.99**
average 2.00 m deep	0.31	17.24	11.82	38.39	m	**67.46**
average 3.00 m deep	0.36	20.02	13.90	43.55	m	**77.47**
300 mm pipes with Type A bed and Type A fill, depth						
average 1.00 m deep	0.28	15.57	10.71	55.32	m	**81.60**
average 2.00 m deep	0.32	17.80	12.28	63.60	m	**93.68**
average 3.00 m deep	0.37	20.58	14.23	71.37	m	**106.18**
300 mm pipes with Type A bed and Type B fill, depth						
average 1.00 m deep	0.28	15.57	10.71	53.98	m	**80.27**
average 2.00 m deep	0.32	17.80	12.28	60.59	m	**90.67**
average 3.00 m deep	0.37	20.58	14.23	66.83	m	**101.64**
Conc. porous pipe BS 5911 Pt 114, sleeved joints						
150 mm pipes with Type A bed and Type A fill, depth						
average 1.00 m deep	0.26	14.46	10.01	13.58	m	**38.06**
average 2.00 m deep	0.30	16.69	11.51	19.35	m	**47.54**
average 3.00 m deep	0.35	19.47	13.53	25.14	m	**58.14**

SERIES 500: DRAINAGE AND SERVICE DUCTS

Item	Gang hours	Labour £	Plant £	Material £	Unit	Total rate £
150 mm pipes with Type A bed and Type B fill, depth						
average 1.00 m deep	0.26	14.46	10.01	12.57	m	**37.04**
average 2.00 m deep	0.30	16.69	11.51	17.10	m	**45.29**
average 3.00 m deep	0.35	19.47	13.53	21.70	m	**54.70**
225 mm pipes with Type A bed and Type A fill, depth						
average 1.00 m deep	0.27	15.02	10.34	15.62	m	**40.98**
average 2.00 m deep	0.31	17.24	11.82	22.81	m	**51.88**
average 3.00 m deep	0.36	20.02	13.90	29.32	m	**63.24**
225 mm pipes with Type A bed and Type B fill, depth						
average 1.00 m deep	0.27	15.02	10.34	14.47	m	**39.83**
average 2.00 m deep	0.31	17.24	11.82	20.23	m	**49.30**
average 3.00 m deep	0.36	20.02	13.90	25.39	m	**59.31**
300 mm pipes with Type A bed and Type A fill, depth						
average 1.00 m deep	0.28	15.57	10.71	18.44	m	**44.72**
average 2.00 m deep	0.32	17.80	12.28	26.72	m	**56.80**
average 3.00 m deep	0.37	20.58	14.23	34.49	m	**69.30**
300 mm pipes with Type A bed and Type B fill, depth						
average 1.00 m deep	0.28	15.57	10.71	17.10	m	**43.39**
average 2.00 m deep	0.32	17.80	12.28	23.71	m	**53.79**
average 3.00 m deep	0.37	20.58	14.23	29.95	m	**64.76**
Filter material contiguous with filter drains, sub-base material and lightweight aggregate infill						
Type A	0.07	3.89	2.41	12.85	m³	**19.15**
Type B	0.07	3.89	2.52	10.06	m³	**16.47**
Excavate and replace filter material contiguous with filter drain						
Type A	0.47	12.93	11.25	12.85	m³	**37.04**
Type B	0.47	12.93	11.25	10.06	m³	**34.24**
FIN DRAINS AND NARROW FILTER DRAINS						
Fin Drain D.O.T. type 6 using 'Trammel' drainage fabrics and perforated clay drain; surrounding pipe with sand and granular fill						
100 mm clay perforated pipes, depth						
average 1.00 m deep	0.17	9.46	6.67	14.95	m	**31.08**
average 2.00 m deep	0.23	12.79	9.00	21.69	m	**43.48**
Fin Drain D.O.T. type 7 using 'Trammel' drainage fabrics and slotted UPVC drain; surrounding pipe with sand and granular fill, backfilling with selected suitable material						
100 mm UPVC slotted pipes, depth						
average 1.00 m deep	0.17	9.46	6.67	72.63	m	**88.76**
average 2.00 m deep	0.23	12.79	9.00	79.37	m	**101.16**

SERIES 500: DRAINAGE AND SERVICE DUCTS

Item	Gang hours	Labour £	Plant £	Material £	Unit	Total rate £
FIN DRAINS AND NARROW FILTER DRAINS – cont'd						
Narrow Filter Drain D.O.T. type 8 using 'Trammel' drainage fabrics and perforated UPVC drain; surrounding pipe with sand and granular fill, backfilling with granular material						
110 mm UPVC perforated pipes, depth						
average 1.00 m deep	0.17	9.46	6.67	20.65	m	**36.78**
average 2.00 m deep	0.22	12.24	8.62	22.40	m	**43.26**
Narrow Filter Drain D.O.T. type 9 using 'Trammel' drainage fabrics and perforated UPVC drain; surrounding pipe with sand and granular fill, backfilling with granular material						
110 mm UPVC perforated pipes, depth						
average 1.00 m deep	0.17	9.46	6.67	24.00	m	**40.13**
average 2.00 m deep	0.23	12.79	9.00	28.64	m	**50.43**
CONNECTIONS						
Note: excavation presumed covered by new trench						
Connection of pipe to existing drain, sewer or piped culvert						
150 mm	0.42	23.36	11.46	18.00	nr	**52.82**
225 mm	0.60	33.37	19.50	65.50	nr	**118.37**
300 mm	1.15	63.96	39.04	114.00	nr	**217.00**
Connection of pipes to existing chambers						
150 mm to one brick	1.20	66.74	47.01	3.68	nr	**117.43**
150 mm to precast	0.60	33.37	23.50	3.68	nr	**60.54**
300 mm to one and a half brick	2.40	133.49	93.95	5.99	nr	**233.44**
CHAMBERS AND GULLIES						
Notes						
The rates assume the most efficient items of plant (excavator) and are optimum rates, assuming continuous output with no delays caused by other operations or works. Ground conditions are assumed to be good soil with no abnormal conditions that would affect outputs and consistency of work.						
Multiplier Table for Labour and Plant for various site conditions and for working:						
out of sequence x 2.75 minimum						
in hard clay x 1.75 - 2.00						
in running sand x 2.75 minimum						
in broken rock x 2.75 - 3.50						
below water table x 2.00 minimum						

SERIES 500: DRAINAGE AND SERVICE DUCTS

Item	Gang hours	Labour £	Plant £	Material £	Unit	Total rate £
Brick construction						
Design criteria used in models: • class A engineering bricks • 215 thick walls generally; 328 thick to chambers exceeding 2.5 m deep • 225 plain concrete C20/20 base slab • 300 reinforced concrete C20/20 reducing slab • 125 reinforced concrete C20/20 top slab • maximum height of working chamber 2.0 m above benching • 750 x 750 access shaft • plain concrete C15/20 benching, 150 clay Main channel longitudinally and two 100 Branch channels • step irons at 300 mm centres, doubled if depth to invert exceeds 3000 mm • heavy duty manhole cover and frame						
750 x 700 chamber 500 depth to invert						
excavation, support, backfilling and disposal	-	-	-	-	-	37.04
concrete base	-	-	-	-	-	99.22
brickwork chamber	-	-	-	-	-	53.25
concrete cover slab	-	-	-	-	-	121.28
concrete benching, main and branch channels	-	-	-	-	-	97.24
step irons	-	-	-	-	-	11.58
access cover and frame	-	-	-	-	-	312.56
TOTAL	-	-	-	-	£	**732.17**
750 x 700 chamber 1000 depth to invert						
excavation, support, backfilling and disposal	-	-	-	-	-	60.20
concrete base	-	-	-	-	-	99.22
brickwork chamber	-	-	-	-	-	201.42
concrete cover slab	-	-	-	-	-	121.28
concrete benching; main and branch channels	-	-	-	-	-	97.24
step irons	-	-	-	-	-	17.37
access cover and frame	-	-	-	-	-	312.56
TOTAL	-	-	-	-	£	**909.29**
750 x 700 chamber 1500 depth to invert						
excavation, support, backfilling and disposal	-	-	-	-	-	85.67
concrete base	-	-	-	-	-	99.22
brickwork chamber	-	-	-	-	-	347.29
concrete cover slab	-	-	-	-	-	121.28
concrete benching; main and branch channels	-	-	-	-	-	97.24
step irons	-	-	-	-	-	28.94
access cover and frame	-	-	-	-	-	312.56
TOTAL	-	-	-	-	£	**1092.20**
900 x 700 chamber 500 depth to invert						
excavation, support, backfilling and disposal	-	-	-	-	-	41.67
concrete base	-	-	-	-	-	103.03
brickwork chamber	-	-	-	-	-	59.04
concrete cover slab	-	-	-	-	-	134.28
concrete benching; main and branch channels	-	-	-	-	-	109.98
step irons	-	-	-	-	-	11.58
access cover and frame	-	-	-	-	-	312.56
TOTAL	-	-	-	-	£	**772.14**

SERIES 500: DRAINAGE AND SERVICE DUCTS

Item	Gang hours	Labour £	Plant £	Material £	Unit	Total rate £
CHAMBERS AND GULLIES – cont'd						
Brick construction - cont'd						
900 x 700 chamber 1000 depth to invert						
excavation, support, backfilling and disposal	-	-	-	-	-	69.46
concrete base	-	-	-	-	-	103.03
brickwork chamber	-	-	-	-	-	217.63
concrete cover slab	-	-	-	-	-	134.28
concrete benching; main and branch channels	-	-	-	-	-	109.98
step irons	-	-	-	-	-	17.37
access cover and frame	-	-	-	-	-	312.56
TOTAL	-	-	-	-	£	**964.31**
900 x 700 chamber 1500 depth to invert						
excavation, support, backfilling and disposal	-	-	-	-	-	88.20
concrete base	-	-	-	-	-	103.03
brickwork chamber	-	-	-	-	-	376.23
concrete cover slab	-	-	-	-	-	134.28
concrete benching; main and branch channels	-	-	-	-	-	109.98
step irons	-	-	-	-	-	28.94
access cover and frame	-	-	-	-	-	312.56
TOTAL	-	-	-	-	£	**1153.22**
1050 x 700 chamber 1500 depth to invert						
excavation, support, backfilling and disposal	-	-	-	-	-	105.35
concrete base	-	-	-	-	-	106.50
brickwork chamber	-	-	-	-	-	404.01
concrete cover slab	-	-	-	-	-	145.87
concrete benching; main and branch channels	-	-	-	-	-	123.87
step irons	-	-	-	-	-	28.94
access cover and frame	-	-	-	-	-	312.56
TOTAL	-	-	-	-	£	**1227.10**
1050 x 700 chamber 2500 depth to invert						
excavation, support, backfilling and disposal	-	-	-	-	-	178.27
concrete base	-	-	-	-	-	106.50
brickwork chamber	-	-	-	-	-	744.36
concrete cover slab	-	-	-	-	-	145.87
concrete benching; main and branch channels	-	-	-	-	-	123.87
step irons	-	-	-	-	-	46.30
access cover and frame	-	-	-	-	-	312.56
TOTAL	-	-	-	-	£	**1657.73**
1050 x 700 chamber 3500 depth to invert						
excavation, support, backfilling and disposal	-	-	-	-	-	260.46
concrete base	-	-	-	-	-	106.50
brickwork chamber	-	-	-	-	-	740.88
concrete reducing slab	-	-	-	-	-	163.22
brickwork access shaft	-	-	-	-	-	214.16
concrete cover slab	-	-	-	-	-	98.40
concrete benching; main and branch channels	-	-	-	-	-	123.87
step irons	-	-	-	-	-	138.92
access cover and frame	-	-	-	-	-	312.56
TOTAL	-	-	-	-	£	**2158.97**

SERIES 500: DRAINAGE AND SERVICE DUCTS

Item	Gang hours	Labour £	Plant £	Material £	Unit	Total rate £
1350 x 700 chamber 2500 depth to invert						
excavation, support, backfilling and disposal	-	-	-	-	-	217.63
concrete base	-	-	-	-	-	118.07
brickwork chamber	-	-	-	-	-	865.90
concrete cover slab	-	-	-	-	-	175.96
concrete benching; main and branch channels	-	-	-	-	-	156.28
step irons	-	-	-	-	-	46.30
access cover and frame	-	-	-	-	-	312.56
TOTAL	-	-	-	-	£	**1892.70**
1350 x 700 chamber 3500 depth to invert						
excavation, support, backfilling and disposal	-	-	-	-	-	318.35
concrete base	-	-	-	-	-	118.07
brickwork chamber	-	-	-	-	-	862.43
concrete reducing slab	-	-	-	-	-	188.70
brickwork access shaft	-	-	-	-	-	214.16
concrete cover slab	-	-	-	-	-	98.40
concrete benching; main and branch channels	-	-	-	-	-	156.28
step irons	-	-	-	-	-	138.92
access cover and frame	-	-	-	-	-	312.56
TOTAL	-	-	-	-	£	**2407.87**
1350 x 700 chamber 4500 depth to invert						
excavation, support, backfilling and disposal	-	-	-	-	-	427.16
concrete base	-	-	-	-	-	118.07
brickwork chamber	-	-	-	-	-	862.43
concrete reducing slab	-	-	-	-	-	188.70
brickwork access shaft	-	-	-	-	-	515.14
concrete cover slab	-	-	-	-	-	98.40
concrete benching; main and branch channels	-	-	-	-	-	156.28
step irons	-	-	-	-	-	173.65
access cover and frame	-	-	-	-	-	312.56
TOTAL	-	-	-	-	£	**2852.39**

Precast concrete construction

Design criteria used in models:

- circular shafts
- 150 plain concrete surround
- 225 plain concrete C20/20 base slab
- precast reducing slab
- precast top slab
- maximum height of working chamber 2.0 m above benching
- 750 diameter access shaft
- plain concrete C15/20 benching, 150 clay Main channel longitudinally and two 100 branch channels
- step irons at 300 mm centres, doubled if depth to invert exceeds 3000 mm
- heavy duty manhole cover and frame
- in manholes over 6 m deep, landings at maximum intervals

SERIES 500: DRAINAGE AND SERVICE DUCTS

Item	Gang hours	Labour £	Plant £	Material £	Unit	Total rate £
CHAMBERS AND GULLIES – cont'd						
Precast concrete construction – cont'd						
675 diameter x 500 depth to invert						
excavation, support, backfilling and disposal	-	-	-	-	-	27.78
concrete base	-	-	-	-	-	41.67
main chamber rings	-	-	-	-	-	20.83
cover slab	-	-	-	-	-	82.19
concrete surround	-	-	-	-	-	42.83
concrete benching, main and branch channels	-	-	-	-	-	54.41
step irons	-	-	-	-	-	-
access cover and frame	-	-	-	-	-	312.56
TOTAL	-	-	-	-	£	**582.27**
675 diameter x 750 depth to invert						
excavation, support, backfilling and disposal	-	-	-	-	-	37.04
concrete base	-	-	-	-	-	41.67
main chamber rings	-	-	-	-	-	43.05
cover slab	-	-	-	-	-	82.19
concrete surround	-	-	-	-	-	62.52
concrete benching, main and branch channels	-	-	-	-	-	54.41
step irons	-	-	-	-	-	-
access cover and frame	-	-	-	-	-	312.56
TOTAL	-	-	-	-	£	**633.44**
675 diameter x 1000 depth to invert						
excavation, support, backfilling and disposal	-	-	-	-	-	43.00
concrete base	-	-	-	-	-	41.67
main chamber rings	-	-	-	-	-	75.24
cover slab	-	-	-	-	-	82.19
concrete surround	-	-	-	-	-	75.60
concrete benching, main and branch channels	-	-	-	-	-	54.41
step irons	-	-	-	-	-	-
access cover and frame	-	-	-	-	-	312.56
TOTAL	-	-	-	-	£	**684.67**
675 diameter x 1250 depth to invert						
excavation, support, backfilling and disposal	-	-	-	-	-	48.30
concrete base	-	-	-	-	-	41.67
main chamber rings	-	-	-	-	-	103.03
cover slab	-	-	-	-	-	82.19
concrete surround	-	-	-	-	-	98.12
concrete benching, main and branch channels	-	-	-	-	-	54.41
step irons	-	-	-	-	-	6.95
access cover and frame	-	-	-	-	-	312.56
TOTAL	-	-	-	-	£	**747.23**
900 diameter x 750 depth to invert						
excavation, support, backfilling and disposal	-	-	-	-	-	53.25
concrete base	-	-	-	-	-	57.89
main chamber rings	-	-	-	-	-	59.04
cover slab	-	-	-	-	-	99.22
concrete surround	-	-	-	-	-	78.72
concrete benching, main and branch channels	-	-	-	-	-	64.83
step irons	-	-	-	-	-	-
access cover and frame	-	-	-	-	-	312.56
TOTAL	-	-	-	-	£	**725.51**

SERIES 500: DRAINAGE AND SERVICE DUCTS

Item	Gang hours	Labour £	Plant £	Material £	Unit	**Total rate £**
900 diameter x 1000 depth to invert						
excavation, support, backfilling and disposal	-	-	-	-	-	61.74
concrete base	-	-	-	-	-	57.89
main chamber rings	-	-	-	-	-	89.30
cover slab	-	-	-	-	-	99.22
concrete surround	-	-	-	-	-	103.03
concrete benching, main and branch channels	-	-	-	-	-	64.83
step irons	-	-	-	-	-	-
access cover and frame	-	-	-	-	-	312.56
TOTAL	-	-	-	-	£	**788.57**
900 diameter x 1500 depth to invert						
excavation, support, backfilling and disposal	-	-	-	-	-	92.61
concrete base	-	-	-	-	-	57.89
main chamber rings	-	-	-	-	-	162.07
cover slab	-	-	-	-	-	99.22
concrete surround	-	-	-	-	-	152.81
concrete benching, main and branch channels	-	-	-	-	-	64.83
step irons	-	-	-	-	-	13.89
access cover and frame	-	-	-	-	-	312.56
TOTAL	-	-	-	-	£	**955.88**
1200 diameter x 1500 depth to invert						
excavation, support, backfilling and disposal	-	-	-	-	-	144.70
concrete base	-	-	-	-	-	84.50
main chamber rings	-	-	-	-	-	219.94
cover slab	-	-	-	-	-	141.22
concrete surround	-	-	-	-	-	200.27
concrete benching, main and branch channels	-	-	-	-	-	83.35
step irons	-	-	-	-	-	13.89
access cover and frame	-	-	-	-	-	312.56
TOTAL	-	-	-	-	£	**1200.43**
1200 diameter x 2000 depth to invert						
excavation, support, backfilling and disposal	-	-	-	-	-	203.74
concrete base	-	-	-	-	-	84.50
main chamber rings	-	-	-	-	-	309.09
cover slab	-	-	-	-	-	141.22
concrete surround	-	-	-	-	-	265.09
concrete benching, main and branch channels	-	-	-	-	-	83.35
step irons	-	-	-	-	-	26.63
access cover and frame	-	-	-	-	-	312.56
TOTAL	-	-	-	-	£	**1426.18**
1200 diameter x 2500 depth to invert						
excavation, support, backfilling and disposal	-	-	-	-	-	250.05
concrete base	-	-	-	-	-	84.50
main chamber rings	-	-	-	-	-	401.70
cover slab	-	-	-	-	-	141.22
concrete surround	-	-	-	-	-	329.92
concrete benching, main and branch channels	-	-	-	-	-	83.35
step irons	-	-	-	-	-	33.57
access cover and frame	-	-	-	-	-	312.56
TOTAL	-	-	-	-	£	**1636.87**

SERIES 500: DRAINAGE AND SERVICE DUCTS

Item	Gang hours	Labour £	Plant £	Material £	Unit	Total rate £
CHAMBERS AND GULLIES – cont'd						
Precast concrete construction – cont'd						
1200 diameter x 3000 depth to invert						
excavation, support, backfilling and disposal	-	-	-	-	-	321.82
concrete base	-	-	-	-	-	84.50
main chamber rings	-	-	-	-	-	495.46
cover slab	-	-	-	-	-	141.22
concrete surround	-	-	-	-	-	395.91
concrete benching, main and branch channels	-	-	-	-	-	83.35
step irons	-	-	-	-	-	94.93
access cover and frame	-	-	-	-	-	312.56
TOTAL	-	-	-	-	£	**1929.75**
1800 diameter x 1500 depth to invert						
excavation, support, backfilling and disposal	-	-	-	-	-	263.94
concrete base	-	-	-	-	-	148.18
main chamber rings	-	-	-	-	-	344.97
cover slab	-	-	-	-	-	244.26
concrete surround	-	-	-	-	-	280.15
concrete benching, main and branch channels	-	-	-	-	-	128.50
step irons	-	-	-	-	-	13.89
access cover and frame	-	-	-	-	-	312.56
TOTAL	-	-	-	-	£	**1736.45**
1800 diameter x 2000 depth to invert						
excavation, support, backfilling and disposal	-	-	-	-	-	370.44
concrete base	-	-	-	-	-	148.18
main chamber rings	-	-	-	-	-	494.31
cover slab	-	-	-	-	-	244.26
concrete surround	-	-	-	-	-	372.75
concrete benching, main and branch channels	-	-	-	-	-	128.50
step irons	-	-	-	-	-	26.63
access cover and frame	-	-	-	-	-	312.56
TOTAL	-	-	-	-	£	**2097.63**
1800 diameter x 2500 depth to invert						
excavation, support, backfilling and disposal	-	-	-	-	-	451.48
concrete base	-	-	-	-	-	148.18
main chamber rings	-	-	-	-	-	642.48
cover slab	-	-	-	-	-	244.26
concrete surround	-	-	-	-	-	465.37
concrete benching, main and branch channels	-	-	-	-	-	128.50
step irons	-	-	-	-	-	33.57
access cover and frame	-	-	-	-	-	312.56
TOTAL	-	-	-	-	£	**2426.40**
1800 diameter x 3000 depth to invert						
excavation, support, backfilling and disposal	-	-	-	-	-	581.13
concrete base	-	-	-	-	-	148.18
main chamber rings	-	-	-	-	-	791.82
cover slab	-	-	-	-	-	244.26
concrete surround	-	-	-	-	-	556.81
concrete benching, main and branch channels	-	-	-	-	-	128.50
step irons	-	-	-	-	-	94.93
access cover and frame	-	-	-	-	-	312.56
TOTAL	-	-	-	-	£	**2858.19**

SERIES 500: DRAINAGE AND SERVICE DUCTS

Item	Gang hours	Labour £	Plant £	Material £	Unit	**Total rate £**
1800 diameter x 3500 depth to invert						
excavation, support, backfilling and disposal	-	-	-	-	-	665.64
concrete base	-	-	-	-	-	148.18
main chamber rings	-	-	-	-	-	655.21
reducing slab	-	-	-	-	-	261.63
access shaft	-	-	-	-	-	93.77
cover slab	-	-	-	-	-	99.22
concrete surround	-	-	-	-	-	577.66
concrete benching, main and branch channels	-	-	-	-	-	128.50
step irons	-	-	-	-	-	121.55
access cover and frame	-	-	-	-	-	312.56
TOTAL	-	-	-	-	£	**3063.92**
1800 diameter x 4000 depth to invert						
excavation, support, backfilling and disposal	-	-	-	-	-	808.02
concrete base	-	-	-	-	-	148.18
main chamber rings	-	-	-	-	-	655.21
reducing slab	-	-	-	-	-	261.63
access shaft	-	-	-	-	-	162.07
cover slab	-	-	-	-	-	99.22
concrete surround	-	-	-	-	-	629.75
concrete benching, main and branch channels	-	-	-	-	-	128.50
step irons	-	-	-	-	-	135.44
access cover and frame	-	-	-	-	-	312.56
TOTAL	-	-	-	-	£	**3340.58**
2400 diameter x 1500 depth to invert						
excavation, support, backfilling and disposal	-	-	-	-	-	420.22
concrete base	-	-	-	-	-	229.22
main chamber rings	-	-	-	-	-	662.16
cover slab	-	-	-	-	-	714.25
concrete surround	-	-	-	-	-	362.33
concrete benching, main and branch channels	-	-	-	-	-	186.38
step irons	-	-	-	-	-	13.89
access cover and frame	-	-	-	-	-	312.56
TOTAL	-	-	-	-	£	**2901.01**
2400 diameter x 3000 depth to invert						
excavation, support, backfilling and disposal	-	-	-	-	-	912.21
concrete base	-	-	-	-	-	229.22
main chamber rings	-	-	-	-	-	1538.48
cover slab	-	-	-	-	-	714.25
concrete surround	-	-	-	-	-	717.73
concrete benching, main and branch channels	-	-	-	-	-	186.38
step irons	-	-	-	-	-	94.93
access cover and frame	-	-	-	-	-	312.56
TOTAL	-	-	-	-	£	**4705.76**
2400 diameter x 4500 depth to invert						
excavation, support, backfilling and disposal	-	-	-	-	-	1405.35
concrete base	-	-	-	-	-	229.22
main chamber rings	-	-	-	-	-	1272.23
reducing slab	-	-	-	-	-	716.57
access shaft	-	-	-	-	-	230.37
cover slab	-	-	-	-	-	99.22
concrete surround	-	-	-	-	-	821.91
concrete benching, main and branch channels	-	-	-	-	-	186.38
step irons	-	-	-	-	-	162.07
access cover and frame	-	-	-	-	-	312.56
TOTAL	-	-	-	-	£	**5435.88**

SERIES 500: DRAINAGE AND SERVICE DUCTS

Item	Gang hours	Labour £	Plant £	Material £	Unit	Total rate £
CHAMBERS AND GULLIES – cont'd						
Precast concrete construction – cont'd						
2700 diameter x 1500 depth to invert						
excavation, support, backfilling and disposal	-	-	-	-	-	512.83
concrete base	-	-	-	-	-	277.83
main chamber rings	-	-	-	-	-	759.40
cover slab	-	-	-	-	-	893.69
concrete surround	-	-	-	-	-	404.01
concrete benching, main and branch channels	-	-	-	-	-	219.94
step irons	-	-	-	-	-	13.89
access cover and frame	-	-	-	-	-	312.56
TOTAL	-	-	-	-	£	**3394.15**
2700 diameter x 3000 depth to invert						
excavation, support, backfilling and disposal	-	-	-	-	-	1109.00
concrete base	-	-	-	-	-	277.83
main chamber rings	-	-	-	-	-	1786.22
cover slab	-	-	-	-	-	893.69
concrete surround	-	-	-	-	-	799.92
concrete benching, main and branch channels	-	-	-	-	-	219.94
step irons	-	-	-	-	-	94.93
access cover and frame	-	-	-	-	-	312.56
TOTAL	-	-	-	-	£	**5494.09**
2700 diameter x 4500 depth to invert						
excavation, support, backfilling and disposal	-	-	-	-	-	1704.02
concrete base	-	-	-	-	-	277.83
main chamber rings	-	-	-	-	-	1477.13
reducing slab	-	-	-	-	-	887.90
access shaft	-	-	-	-	-	230.37
cover slab	-	-	-	-	-	99.22
concrete surround	-	-	-	-	-	893.69
concrete benching, main and branch channels	-	-	-	-	-	219.94
step irons	-	-	-	-	-	162.07
access cover and frame	-	-	-	-	-	312.56
TOTAL	-	-	-	-	£	**6264.73**
3000 diameter x 3000 depth to invert						
excavation, support, backfilling and disposal	-	-	-	-	-	1337.06
concrete base	-	-	-	-	-	332.24
main chamber rings	-	-	-	-	-	2390.49
cover slab	-	-	-	-	-	1126.37
concrete surround	-	-	-	-	-	887.90
concrete benching, main and branch channels	-	-	-	-	-	237.31
step irons	-	-	-	-	-	94.93
access cover and frame	-	-	-	-	-	312.56
TOTAL	-	-	-	-	£	**6718.86**
3000 diameter x 4500 depth to invert						
excavation, support, backfilling and disposal	-	-	-	-	-	2053.63
concrete base	-	-	-	-	-	332.24
main chamber rings	-	-	-	-	-	1977.22
reducing slab	-	-	-	-	-	1016.40
access shaft	-	-	-	-	-	230.37
cover slab	-	-	-	-	-	99.22
concrete surround	-	-	-	-	-	967.77
concrete benching, main and branch channels	-	-	-	-	-	237.31
step irons	-	-	-	-	-	162.07
access cover and frame	-	-	-	-	-	312.56
TOTAL	-	-	-	-	£	**7388.79**

SERIES 500: DRAINAGE AND SERVICE DUCTS

Item	Gang hours	Labour £	Plant £	Material £	Unit	**Total rate £**
3000 diameter x 6000 depth to invert						
excavation, support, backfilling and disposal	-	-	-	-	-	3041.08
concrete base	-	-	-	-	-	332.24
main chamber rings	-	-	-	-	-	1977.22
reducing slab	-	-	-	-	-	1016.40
access shaft	-	-	-	-	-	436.42
cover slab	-	-	-	-	-	99.22
concrete surround	-	-	-	-	-	1130.89
concrete benching, main and branch channels	-	-	-	-	-	237.31
step irons	-	-	-	-	-	230.37
access cover and frame	-	-	-	-	-	312.56
TOTAL	-	-	-	-	£	**8813.71**
Clayware vertical pipe complete with rest bend at base and tumbling bay junction to main drain complete with stopper; concrete grade C20 surround, 150 mm thick; additional excavation and disposal						
100 pipe						
1.15 m to invert	-	-	-	-	nr	**96.09**
2.15 m to invert	-	-	-	-	nr	**122.70**
3.15 m to invert	-	-	-	-	nr	**148.18**
4.15 m to invert	-	-	-	-	nr	**175.96**
150 pipe						
1.15 m to invert	-	-	-	-	nr	**148.18**
2.15 m to invert	-	-	-	-	nr	**178.27**
3.15 m to invert	-	-	-	-	nr	**213.00**
4.15 m to invert	-	-	-	-	nr	**247.74**
225 pipe						
1.15 m to invert	-	-	-	-	nr	**295.20**
2.15 m to invert	-	-	-	-	nr	**346.13**
3.15 m to invert	-	-	-	-	nr	**399.38**
4.15 m to invert	-	-	-	-	nr	**454.94**
Vitrified clay; set in concrete grade C20, 150 mm thick; additional excavation and disposal						
Road gulley						
450 mm diameter x 900 mm deep, 100 mm or 150 mm outlet; cast iron road gulley grating and frame group 4, 434 x 434 mm, on Class B engineering brick seating	0.50	20.48	0.69	271.56	nr	**292.74**
Yard gulley (mud); trapped with rodding eye; galvanised bucket; stopper						
225 mm diameter, 100 mm diameter outlet, cast iron hinged grate and frame	0.30	12.29	0.44	225.17	nr	**237.91**
Grease interceptors; internal access and bucket						
600 x 450 mm, metal tray and lid, square hopper with horizontal inlet	0.35	14.34	0.48	873.43	nr	**888.26**

SERIES 500: DRAINAGE AND SERVICE DUCTS

Item	Gang hours	Labour £	Plant £	Material £	Unit	Total rate £
CHAMBERS AND GULLIES – cont'd						
Precast concrete; set in concrete grade C20P, 150 mm thick; additional excavation and disposal						
Road gulley; trapped with rodding eye; galvanised bucket; stopper						
450 mm diameter x 900 mm deep, cast iron road gulley grating and frame group 4, 434 x 434 mm, on Class B engineering brick seating	0.54	22.12	0.75	183.48	nr	**206.35**
SOFT SPOTS AND OTHER VOIDS						
Excavation of soft spots and other voids in bottom of trenches, chambers and gullies	0.07	3.89	2.75	-	m³	**6.64**
Filling of soft spots and other voids in bottom of trenches, chambers and gullies with imported selected sand	0.09	5.01	3.32	18.80	m³	**27.12**
Filling of soft spots and other voids in bottom of trenches, chambers and gullies with imported natural gravel	0.09	5.01	3.32	25.20	m³	**33.53**
Filling of soft spots and other voids in bottom of trenches, chambers and gullies with concrete Grade C15, 40 mm aggregate	0.09	5.01	3.28	77.37	m³	**85.66**
Filling of soft spots and other voids in bottom of trenches, chambers and gullies with concrete Grade C20, 20 mm aggregate	0.09	5.01	3.28	75.30	m³	**83.58**
SUPPORTS LEFT IN EXCAVATION						
Timber close boarded supports left in						
trench	-	-	8.10	-	m²	**8.10**
pits	-	-	8.10	-	m²	**8.10**
Steel trench sheeting supports left in						
trench	-	-	33.96	-	m²	**33.96**
pits	-	-	33.96	-	m²	**33.96**
REPLACEMENT, RAISING OR LOWERING OF COVERS AND GRATINGS ON EXISTING CHAMBERS AND GULLIES						
Raising the level of 100 x 100 mm water stop tap boxes on 100 x 100 mm brick chambers						
by 150 mm or less	0.06	2.46	0.28	10.53	nr	**13.27**
Lowering the level of 100 x 100 mm water stop tap boxes on 100 x 100 mm brick chambers						
by 150 mm or less	0.04	1.64	0.18	6.32	nr	**8.14**
Raising the level of 420 x 420 mm cover and frame on 420 x 420 mm in-situ concrete chamber						
by 150 mm or less	0.10	4.10	0.46	18.25	nr	**22.81**

SERIES 500: DRAINAGE AND SERVICE DUCTS

Item	Gang hours	Labour £	Plant £	Material £	Unit	**Total rate £**
Lowering the level of 420 x 420 mm British Telecom cover and frame on 420 x 420 mm in-situ concrete chamber						
by 150 mm or less	0.08	3.07	1.78	11.22	nr	**16.07**
Raising the level of 700 x 500 mm cover and frame on 700 x 500 mm in-situ concrete chamber						
by 150 mm or less	0.17	6.96	0.77	13.03	nr	**20.76**
Raising the level of 600 x 600 mm grade A heavy duty manhole cover and frame on 600 x 600 mm brick chamber						
by 150 mm or less	0.17	6.96	0.77	26.66	nr	**34.39**
by 150 - 300 mm	0.21	8.60	0.95	33.67	nr	**43.22**
Lowering the level of 600 x 600 mm grade A heavy duty manhole cover and frame on 600 x 600 mm brick chamber						
by 150 mm or less	0.10	4.10	0.46	16.15	nr	**20.71**
REMOVE FROM STORE AND REINSTALL CHAMBER COVERS AND FRAMES AND GULLEY GRATINGS AND FRAMES						
600 x 600 mm; Group 5; super heavy duty E600 cast iron	0.25	10.24	0.42	38.28	nr	**48.95**
600 x 600 mm; Group 4; heavy duty triangular D400 cast iron	0.25	10.24	0.42	17.65	nr	**28.31**
600 x 600 x 75 mm; Group 2; light duty single seal B125 cast iron	0.25	10.24	0.42	18.39	nr	**29.06**
600 x 600 x 100mm; Group 2; medium duty single seal B125 cast iron	0.25	10.24	0.42	25.99	nr	**36.65**
GROUTING UP OF EXISTING DRAINS, SEWERS						
Concrete Grade C15						
Sealing redundant road gullies	0.02	0.82	0.08	11.47	m³	**12.37**
Filling redundant chambers with						
ne 1.0 m deep to invert	0.09	5.01	2.96	46.01	nr	**53.97**
1.0 - 2.0 m deep to invert	0.21	11.68	6.89	74.36	nr	**92.93**
2.0 - 3.0 m deep to invert	0.55	30.59	18.04	123.91	nr	**172.54**
Grouting up of existing drains and service ducts						
100 mm diameter	0.03	1.67	1.00	2.86	m	**5.53**
300 mm diameter	0.13	7.23	4.34	12.11	m	**23.68**
450 mm diameter	0.26	14.46	8.69	25.14	m	**48.29**
600 mm diameter	0.50	27.81	16.69	51.05	m	**95.56**
1200 mm diameter	1.70	94.55	56.73	142.73	m	**294.02**
EXCAVATION IN HARD MATERIAL						
Extra over excavation for excavation in Hard Material in drainage:						
existing pavement, brickwork, concrete, masonry and the like	0.15	8.34	6.22	-	m³	**14.56**
rock	0.35	19.47	14.51	-	m³	**33.98**
reinforced concrete	0.60	33.37	26.29	-	m³	**59.66**

SERIES 500: DRAINAGE AND SERVICE DUCTS

Item	Gang hours	Labour £	Plant £	Material £	Unit	Total rate £
EXCAVATION IN HARD MATERIAL – cont'd						
Reinstatement of pavement construction; extra over excavation for breaking up and subsequently reinstating 150 mm flexible surfacing and 280 mm sub-base						
100 mm diameter sewer, drain or service duct	0.09	5.01	3.68	13.33	m	**22.01**
150 mm diameter sewer, drain or service duct	0.10	5.56	4.09	14.22	m	**23.87**
225 mm diameter sewer, drain or service duct	0.10	5.56	4.09	15.56	m	**25.21**
300 mm diameter sewer, drain or service duct	0.10	5.56	4.09	16.00	m	**25.65**
375 mm diameter sewer, drain or service duct	0.10	5.56	4.09	16.44	m	**26.09**
450 mm diameter sewer, drain or service duct	0.12	6.67	4.48	17.33	m	**28.48**
2 way 100 mm diameter service ducts	0.10	5.56	4.09	13.77	m	**23.43**
3 way 100 mm diameter service ducts	0.10	5.56	4.09	14.22	m	**23.87**
4 way 100mm diameter service ducts	0.10	5.56	4.09	15.56	m	**25.21**

SERIES 600: EARTHWORKS

Item	Gang hours	Labour £	Plant £	Material £	Unit	Total rate £
GENERAL						

Notes

The cost of earth moving and other associated works is dependent on matching the overall quantities and production rates called for by the programme of works with the most appropriate plant and assessing the most suitable version of that plant that will:

- deal with the site conditions (e.g. type of ground, type of excavation, length of haul, prevailing weather, etc.);
- comply with the specification requirements (e.g. compaction, separation of materials, surface tolerances, etc.);
- complete the work economically (e.g. provide surface tolerances which will avoid undue excessive thickness of expensive imported materials).

Excavation rates

Unless stated the units for excavation in material other than topsoil rock or artificial hard materials are based on excavation in firm gravel soils.

Factors for alternative types of soil:

Multiply the rates by:-

	Scrapers Dozers And Loaders	Tractor	Backacters (minimum bucket size 0.5 m³)
Stiff clay	1.5	2.0	1.7
Chalk	2.5	3.0	2.0
Soft rock	3.5	2.5	2.0
Broken rock	3.7	2.5	1.7

Disposal rates

The other important consideration in excavation of material is bulkage of material after it is dug and loaded onto transport.

All pricing and estimating for disposal is based on the volume of solid material excavated and rates for disposal should be adjusted by the following factors for bulkage:

Sand bulkage	x1.10
Gravel bulkage	x1.20
Compacted soil bulkage	x1.30
Compacted sub-base, acceptable fill etc. bulkage	x1.30
Stiff clay bulkage	x1.20

Fill rates

The price for filling presume the material being supplied in loose state.

SERIES 600: EARTHWORKS

Item	Gang hours	Labour £	Plant £	Material £	Unit	Total rate £
RESOURCES - LABOUR						
Scraper gang						
1 plant operator (skill rate 2)		16.35				
Total Gang Rate / Hour	£	**16.35**				
Scraper and ripper bulldozer gang (hard material)						
2 plant operator (skill rate 2)		32.70				
Total Gang Rate / Hour	£	**32.70**				
General excavation gang						
1 plant operator (skill rate 3)		14.65				
1 plant operator (skill rate 3) - 25% of time		3.66				
1 banksman (skill rate 4)		12.05				
Total Gang Rate / Hour	£	**30.36**				
Shore defences (armour stones) gang						
1 plant operator (skill rate 2)		16.35				
1 banksman (skill rate 4)		12.05				
Total Gang Rate / Hour	£	**28.40**				
Filling gang						
1 plant operator (skill rate 4)		13.17				
2 unskilled operatives (general)		22.52				
Total Gang Rate / Hour	£	**35.69**				
Treatment of filled surfaces gang						
1 plant operator (skill rate 2)		16.35				
Total Gang Rate / Hour	£	**16.35**				
Geotextiles (light sheets) gang						
1 ganger/chargehand (skill rate 4) - 50% of time		6.40				
2 unskilled operatives (general)		22.52				
Total Gang Rate / Hour	£	**28.92**				
Geotextiles (medium sheets) gang						
1 ganger/chargehand (skill rate 4) - 20% of time		2.56				
3 unskilled operatives (general)		33.78				
Total Gang Rate / Hour	£	**36.34**				
Geotextiles (heavy sheets) gang						
1 ganger/chargehand (skill rate 4) - 50% of time		6.40				
2 unskilled operatives (general)		22.52				
1 plant operator (skill rate 4)		13.17				
Total Gang Rate / Hour	£	**42.09**				
Horticultural works gang						
1 skilled operative (skill rate 4)		12.05				
1 unskilled operative (general)		11.26				
Total Gang Rate / Hour	£	**23.31**				
RESOURCES - PLANT						
Scraper excavation						
motor scraper, 16.80 m³, elevating			77.50			
Total Gang Rate / Hour		£	**77.50**			
Scraper and ripper bulldozer excavation						
motor scraper, 16.80 m³, elevating			77.50			
D8 tractor dozer			79.09			
dozer attachment: triple shank ripper			5.90			
Total Gang Rate / Hour		£	**162.49**			

SERIES 600: EARTHWORKS

Item	Gang hours	Labour £	Plant £	Material £	Unit	Total rate £
General excavation						
hydraulic crawler backacter, 0.40 m³			16.67			
backacter attachments: percussion breaker (25% of time)			1.25			
tractor loader, 1.50 m³ (25% of time)			7.81			
loader attachments: ripper (25% of time)			1.62			
Total Gang Rate / Hour		£	**27.35**			
Shore defences (armour stones)						
hydraulic crawler backacter, 1.20 m³			40.76			
backacter attachments: rock bucket			2.74			
backacter attachments: clamshell grab			3.67			
Total Gang Rate / Hour		£	**47.17**			
Geotextiles (heavy sheets)						
tractor loader, 1.5 m³			31.23			
Total Gang Rate / Hour		£	**31.23**			
Filling						
tractor loader, 1.5 m³			31.23			
Total Gang Rate / Hour		£	**31.23**			
Treatment of filled surfaces						
tractor loader, 1.50 m³			31.23			
3 wheel deadweight roller, 10 tonne			13.15			
Total Gang Rate / Hour		£	**44.38**			
EXCAVATION						
Typical motorway cutting generally using motorised scrapers and/or dozers on an average haul of 2000 m (one way)						
Excavation of acceptable material Class 5A	0.01	0.13	0.62	-	m³	**0.75**
Excavation of acceptable material excluding Class 5A in cutting and other excavation	0.01	0.20	1.57	-	m³	**1.77**
General excavation using backacters						
Excavation of acceptable material excluding Class 5A in new watercourses	0.06	1.82	1.90	-	m³	**3.72**
Excavation of unacceptable material Class U1 / U2 in new watercourses	0.07	1.97	1.94	-	m³	**3.91**
General excavation using backacters						
Excavation of unacceptable material Class U1/U2 in clearing abandoned watercourses	0.06	1.67	1.65	-	m³	**3.32**
General excavation using backacters and tractor loaders						
Excavation of acceptable material Class 5A	0.04	1.21	1.15	-	m³	**2.37**
Excavation of acceptable material excluding Class 5A in structural foundations						
ne 3.0 m deep	0.08	2.28	2.23	-	m³	**4.51**
ne 6.0 m deep	0.20	6.07	5.82	-	m³	**11.89**
Excavation of acceptable material excluding Class 5A in foundations for corrugated steel buried structures and the like						
ne 3.0 m deep	0.08	2.28	2.23	-	m³	**4.51**
ne 6.0 m deep	0.20	6.07	5.82	-	m³	**11.89**
Excavation of unacceptable material Class U1 / U2 in structural foundations						
ne 3.0 m deep	0.09	2.58	2.55	-	m³	**5.13**
ne 6.0 m deep	0.21	6.38	6.15	-	m³	**12.53**

SERIES 600: EARTHWORKS

Item	Gang hours	Labour £	Plant £	Material £	Unit	Total rate £
EXCAVATION – cont'd						
General excavation using backacters and tractor loaders – cont'd						
Excavation of unacceptable material Class U1 / U2 in foundations for corrugated steel buried structures and the like						
ne 3.0 m deep	0.09	2.58	2.55	-	m³	**5.13**
ne 6.0 m deep	0.21	6.38	6.15	-	m³	**12.53**
General excavation using backacters and tractor loader with ripper						
Excavation of acceptable material Class 5A	0.03	0.91	0.86	-	m³	**1.77**
Excavation of acceptable material excluding Class 5A in cutting and other excavation	0.04	1.21	1.15	-	m³	**2.37**
Excavate unacceptable material Class U1 / U2 in cutting and other excavation	0.04	1.21	1.15	-	m³	**2.37**
EXCAVATION IN HARD MATERIAL						
Typical motorway cutting generally using motorised scrapers and/or dozers on an average haul of 2000 m (one way)						
Excavate in hard material; using scraper and ripper bulldozer						
mass concrete/medium hard rock	0.09	2.94	14.62	-	m³	**17.57**
reinforced concrete/hard rock	0.15	4.91	24.37	-	m³	**29.28**
tarmacadam	0.16	5.23	2.60	-	m³	**7.83**
General excavation using backacters						
Extra over excavation in new watercourses for excavation in hard material						
rock	0.74	22.47	21.08	-	m³	**43.55**
pavements, brickwork, concrete and masonry	0.69	20.95	19.64	-	m³	**40.58**
reinforced concrete	1.12	34.00	31.85	-	m³	**65.86**
General excavation using backacters and tractor loaders						
Extra over excavation in structural foundations for excavation in hard material						
rock	0.74	22.47	21.08	-	m³	**43.55**
pavements, brickwork, concrete and masonry	0.69	20.95	19.64	-	m³	**40.58**
reinforced concrete	1.12	34.00	31.85	-	m³	**65.86**
Extra over excavation for excavation in foundations for corrugated steel buried structures and the like for excavation in hard material						
rock	0.74	22.47	21.08	-	m³	**43.55**
pavements, brickwork, concrete and masonry	0.69	20.95	19.64	-	m³	**40.58**
reinforced concrete	1.12	34.00	31.85	-	m³	**65.86**
General excavation using backacters and tractor loader with ripper						
Extra over excavation for excavation in cutting and other excavation for excavation in hard material						
rock	0.71	21.56	20.20	-	m³	**41.75**
pavements, brickwork, concrete and masonry	0.66	20.04	18.78	-	m³	**38.82**
reinforced concrete	1.04	31.57	29.59	-	m³	**61.16**

SERIES 600: EARTHWORKS

Item	Gang hours	Labour £	Plant £	Material £	Unit	**Total rate £**
DEPOSITION OF FILL						
Deposition of acceptable material Class 1C/6B in						
embankments and other areas of fill	0.01	0.39	0.34	-	m³	**0.74**
strengthened embankments	0.01	0.39	0.34	-	m³	**0.74**
reinforced earth structures	0.01	0.39	0.34	-	m³	**0.74**
anchored earth structures	0.01	0.39	0.34	-	m³	**0.74**
landscaped areas	0.01	0.39	0.34	-	m³	**0.74**
environmental bunds	0.01	0.39	0.34	-	m³	**0.74**
fill to structures	0.01	0.39	0.34	-	m³	**0.74**
fill above structural concrete foundations	0.01	0.39	0.34	-	m³	**0.74**
DISPOSAL OF MATERIAL						
Disposal of acceptable material excluding Class 5A						
using 10 tonnes capacity tipping lorry for on-site or off-site use; haul distance to tip not exceeding 1 Km	0.06	0.94	1.16	-	m³	**2.10**
ADD per further Km haul	0.03	0.47	0.58	-	m³	**1.05**
using 10 - 15 tonnes capacity tipping lorry for on-site or off-site use; haul distance to tip not exceeding 1 Km	0.05	0.78	0.97	-	m³	**1.75**
ADD per further Km haul	0.03	0.39	0.48	-	m³	**0.87**
using 15 - 25 tonnes capacity tipping lorry for on-site or off-site use; haul distance to tip not exceeding 1 Km	0.05	0.70	0.87	-	m³	**1.57**
ADD per further Km haul	0.02	0.34	0.43	-	m³	**0.77**
Disposal of acceptable material Class 5A (excluding resale value of soil)						
using 10 tonnes capacity tipping lorry for on-site or off-site use; haul distance to tip not exceeding 1 Km	0.06	0.94	1.16	-	m³	**2.10**
ADD per further Km haul	0.03	0.47	0.58	-	m³	**1.05**
using 10 - 15 tonnes capacity tipping lorry for on-site or off-site use; haul distance to tip not exceeding 1 Km	0.05	0.78	0.97	-	m³	**1.75**
ADD per further Km haul	0.03	0.39	0.48	-	m³	**0.87**
using 15 - 25 tonnes capacity tipping lorry for on-site or off-site use; haul distance to tip not exceeding 1 Km	0.05	0.70	0.87	-	m³	**1.57**
ADD per further Km haul	0.02	0.34	0.43	-	m³	**0.77**
Disposal of unacceptable material Class U1						
using 10 tonnes capacity tipping lorry for on-site or off-site use; haul distance to tip not exceeding 1 Km	0.06	0.94	1.16	-	m³	**2.10**
ADD per further Km haul	0.03	0.47	0.58	-	m³	**1.05**
using 10 - 15 tonnes capacity tipping lorry for on-site or off-site use; haul distance to tip not exceeding 1 Km	0.05	0.78	0.97	-	m³	**1.75**
ADD per further Km haul	0.03	0.39	0.48	-	m³	**0.87**
using 15 - 25 tonnes capacity tipping lorry for on-site or off-site use; haul distance to tip not exceeding 1 Km	0.05	0.70	0.87	-	m³	**1.57**
ADD per further Km haul	0.02	0.34	0.43	-	m³	**0.77**

SERIES 600: EARTHWORKS

Item	Gang hours	Labour £	Plant £	Material £	Unit	Total rate £
DISPOSAL OF MATERIAL – cont'd						
Disposal of unacceptable material Class U2						
using 10 tonnes capacity tipping lorry for on-site or off-site use; haul distance to tip not exceeding 1 Km	0.06	0.94	1.16	-	m³	**2.10**
ADD per further Km haul	0.03	0.47	0.58	-	m³	**1.05**
using 10 - 15 tonnes capacity tipping lorry for on-site or off-site use; haul distance to tip not exceeding 1 Km	0.05	0.78	0.97	-	m³	**1.75**
ADD per further Km haul	0.03	0.39	0.48	-	m³	**0.87**
using 15 - 25 tonnes capacity tipping lorry for on-site or off-site use; haul distance to tip not exceeding 1 Km	0.05	0.70	0.87	-	m³	**1.57**
ADD per further Km haul	0.02	0.34	0.43	-	m³	**0.77**
Add to the above rates where tipping charges apply:						
non-hazardous waste	-	-	-	-	m³	**26.93**
hazardous waste	-	-	-	-	m³	**32.92**
special waste	-	-	-	-	m³	**55.38**
contaminated liquid	-	-	-	-	m³	**82.31**
contaminated sludge	-	-	-	-	m³	**134.66**
Add to the above rates where Landfill Tax applies:						
exempted material	-	-	-	-	m³	**-**
inactive or inert material	-	-	-	-	m³	**3.15**
other material	-	-	-	-	m³	**22.05**
IMPORTED FILL						
Imported graded granular fill, natural gravels DTp 1A/B/C						
Imported acceptable material in						
embankments and other areas of fill	0.02	0.64	0.56	18.50	m³	**19.71**
extra for Aggregate Tax	-	-	-	3.52	m³	**3.52**
Imported graded granular fill, crushed gravels or rock DTp 1A/B/C; using tractor loader						
Imported acceptable material in						
embankments and other areas of fill	0.02	0.64	0.56	19.98	m³	**21.19**
extra for Aggregate Tax	-	-	-	3.52	m³	**3.52**
Cohesive material DTp 2A/B/C/D; using tractor loader						
Imported acceptable material in						
embankments and other areas of fill	0.02	0.64	0.56	17.08	m³	**18.29**
landscaped areas	0.02	0.64	0.56	17.08	m³	**18.29**
environmental bunds	0.02	0.75	0.66	17.08	m³	**18.49**
fill to structures	0.02	0.86	0.75	17.08	m³	**18.69**
fill above structural concrete foundations	0.02	0.68	0.59	17.08	m³	**18.36**
Reclaimed blast furnace slag DTp 2E 2.8; using tractor loader						
Imported acceptable material in						
embankments and other areas of fill	0.02	0.64	0.56	16.21	m³	**17.42**

SERIES 600: EARTHWORKS

Item	Gang hours	Labour £	Plant £	Material £	Unit	Total rate £
Reclaimed quarry waste DTp 2E; using tractor loader						
Imported acceptable material in						
embankments and other areas of fill	0.02	0.64	0.56	9.92	m³	**11.12**
extra for Aggregate Tax (fill other than from acceptable process)	-	-	-	3.33	m³	**3.33**
Imported topsoil DTp 5B; using tractor loader						
Imported topsoil Class 5B	0.02	0.64	0.56	16.51	m³	**17.71**
Imported selected well graded granular fill DTp 6A; using tractor loader						
Imported acceptable material						
embankments and other areas of fill	-	-	0.66	24.81	m³	**25.46**
landscape areas	0.02	0.75	0.66	24.81	m³	**26.21**
environmental bunds	0.02	0.64	0.56	24.81	m³	**26.01**
fill to structures	0.02	0.86	0.75	24.81	m³	**26.41**
fill above structural concrete foundations	0.02	0.68	0.59	24.81	m³	**26.08**
extra for Aggregate Tax	-	-	-	3.52	m³	**3.52**
Imported selected granular fill, DTp 6F; using tractor loader						
Imported acceptable material in						
embankments and other areas of fill	0.10	3.39	2.97	17.59	m³	**23.95**
extra for Aggregate Tax	-	-	-	3.52	m³	**3.52**
Imported selected graded fill DTp 6I; using tractor loader						
Imported acceptable material						
reinforced earth structures	0.02	0.79	0.69	18.03	m³	**19.51**
extra for Aggregate Tax	-	-	-	3.52	m³	**3.52**
Imported rock fill; using tractor loader						
Imported acceptable material						
in embankments and other areas of fill	0.02	0.71	0.62	29.82	m³	**31.15**
extra for Aggregate Tax	-	-	-	3.52	m³	**3.52**
Imported well graded granular material; (bedding/free draining materials under shore protection) using tractor loader						
Imported acceptable material						
in embankments and other areas of fill	0.02	0.71	0.62	22.55	m³	**23.89**
Imported rock fill (as a core embankment); using tractor loader						
Imported acceptable material						
in embankments and other areas of fill	0.03	1.18	1.03	29.82	m³	**32.02**
extra for Aggregate Tax	-	-	-	3.52	m³	**3.52**
Imported Armour Stones; (shore protection of individual rocks up to 0.5 t each); using backacter						
Imported acceptable material						
in embankments and other areas of fill	0.06	1.59	2.64	40.57	m³	**44.80**
extra for Aggregate Tax	-	-	-	3.52	m³	**3.52**

SERIES 600: EARTHWORKS

Item	Gang hours	Labour £	Plant £	Material £	Unit	Total rate £
IMPORTED FILL – cont'd						
Imported Armour Stones (shore protection of individual rocks up to 1.0 t each); using backacter						
Imported acceptable material						
in embankments and other areas of fill	0.03	0.91	1.51	59.30	m³	**61.72**
extra for Aggregate Tax	-	-	-	3.52	m³	**3.52**
Imported Armour Stones (shore protection of individual rocks up to 3.0 t each); using backacter						
Imported acceptable material						
in embankments and other areas of fill	0.03	0.74	1.24	65.81	m³	**67.78**
extra for Aggregate Tax	-	-	-	3.52	m³	**3.52**
COMPACTION OF FILL						
Compaction of granular fill material						
in embankments and other areas of fill	0.01	0.05	0.22	-	m³	**0.27**
adjacent to structures	0.01	0.08	0.35	-	m³	**0.44**
above structural concrete foundations	0.01	0.12	0.53	-	m³	**0.66**
Compaction of fill material						
in sub-base or capping layers under verges, central reserves or side slopes	0.02	0.21	0.89	-	m³	**1.09**
adjacent to structures	0.04	0.36	1.55	-	m³	**1.92**
above structural concrete foundations	0.04	0.41	1.78	-	m³	**2.19**
Compaction of graded fill material						
in embankments and other areas of fill	0.01	0.06	0.27	-	m³	**0.33**
adjacent to structures	0.01	0.10	0.44	-	m³	**0.55**
above structural concrete foundations	0.02	0.16	0.67	-	m³	**0.82**
Compaction of rock fill materials						
in embankments and other areas of fill	0.01	0.09	0.40	-	m³	**0.49**
adjacent to structures	0.01	0.14	0.62	-	m³	**0.77**
above structural concrete foundations	0.02	0.18	0.75	-	m³	**0.93**
Compaction of clay fill material						
in embankments and other areas of fill	0.03	0.31	1.33	-	m³	**1.64**
adjacent to structures	0.05	0.50	2.13	-	m³	**2.63**
above structural concrete foundations	0.05	0.50	2.13	-	m³	**2.63**
GEOTEXTILES						
Notes						
The geotextile products mentioned below are not specifically confined to the individual uses stated but are examples of one of many scenarios to which they may be applied. Conversely, the scenarios are not limited to the geotextile used as an example.						
The heavier grades of sheeting will need to be manipulated into place by machine and cutting would be by hacksaw rather than knife.						
Care should be taken in assessing the wastage of the more expensive sheeting.						
The prices include for preparing surfaces, overlaps and turnups, jointing and sealing, fixing material in place if required and reasonable waste between 5% and 10%.						

SERIES 600: EARTHWORKS

Item	Gang hours	Labour £	Plant £	Material £	Unit	**Total rate £**
Stabilisation applications for reinforcement of granular sub-bases and capping layers placed over weak and variable soils.						
For use in weak soils with moderate traffic intensities e.g. light access roads: Tensar SS20 Polypropylene Geogrid						
horizontal	0.04	1.21	-	1.54	m²	**2.76**
inclined at an angle 10-45 degrees to the horizontal	0.05	1.53	-	1.54	m²	**3.08**
For use in weak soils with high traffic intensities and/or high axle loadings: Tensar SS30 Polypropylene Geogrid						
horizontal	0.05	1.64	-	2.35	m²	**3.98**
inclined at an angle 10-45 degrees to the horizontal	0.06	2.04	-	2.35	m²	**4.38**
For construction over very weak soils e.g. alluvium, marsh or peat, or firmer soil subject to exceptionally high axle loadings: Tensar SS40 Polypropylene Geogrid						
horizontal	0.05	1.89	1.41	3.84	m²	**7.14**
inclined at an angle 10-45 degrees to the horizontal	0.06	2.36	1.75	3.84	m²	**7.95**
For trafficked areas where fill comprises of aggregate exceeding 100 mm: Tensar SSLA20 Polypropylene Geogrid						
horizontal	0.04	1.21	-	2.11	m²	**3.32**
inclined at an angle 10-45 degrees to the horizontal	0.05	1.53	-	2.11	m²	**3.64**
For stabilisation and separation of granular fill from soft sub grade to prevent intermixing: Terram 1000						
horizontal	0.05	1.85	-	0.44	m²	**2.30**
inclined at an angle 10-45 degrees to the horizontal	0.06	2.33	-	0.44	m²	**2.77**
For stabilisation and separation of granular fill from soft sub grade to prevent intermixing: Terram 2000						
horizontal	0.04	1.77	1.31	1.28	m²	**4.36**
inclined at an angle 10-45 degrees to the horizontal	0.05	2.23	1.66	1.28	m²	**5.17**

SERIES 600: EARTHWORKS

Item	Gang hours	Labour £	Plant £	Material £	Unit	Total rate £
GEOTEXTILES – cont'd						
Reinforcement applications for asphalt pavements						
For roads, hardstandings and airfield pavements: Tensar AR-1 grid bonded to a geotextile						
horizontal	0.05	1.64	-	3.58	m²	**5.22**
inclined at an angle 10-45 degrees to the horizontal	0.06	2.04	-	3.58	m²	**5.62**
Slope Reinforcement and Embankment Support; For use where soils can only withstand limited shear stresses, therefore steep slopes require external support						
Paragrid 30/155; 330g/m²						
horizontal	0.04	1.21	-	1.84	m²	**3.05**
inclined at an angle 10-45 degrees to the horizontal	0.05	1.53	-	1.84	m²	**3.37**
Paragrid 100/255; 330g/m²						
horizontal	0.04	1.21	-	2.50	m²	**3.72**
inclined at an angle 10-45 degrees to the horizontal	0.05	1.53	-	2.50	m²	**4.04**
Paralink 200s; 1120g/m²						
horizontal	0.05	2.27	1.69	4.92	m²	**8.88**
inclined at an angle 10-45 degrees to the horizontal	0.07	2.86	2.12	4.92	m²	**9.91**
Paralink 600s; 2040g/m²						
horizontal	0.06	2.65	1.97	10.40	m²	**15.02**
inclined at an angle 10-45 degrees to the horizontal	0.08	3.33	2.47	10.40	m²	**16.19**
Terram grid 30/30						
horizontal	0.06	2.40	1.78	0.01	m²	**4.19**
inclined at an angle 10-45 degrees to the horizontal	0.07	2.99	2.22	0.01	m²	**5.21**
Scour and Erosion Protection						
For use where erosion protection is required to the surface of a slope once its geotechnical stability has been achieved, and to allow grass establishment: Tensar 'Mat' Polyethylene mesh, fixed with Tensar pegs						
horizontal	0.04	1.49	-	5.40	m²	**6.89**
inclined at an angle 10-45 degrees from the horizontal	0.05	1.85	-	5.70	m²	**7.55**
For use where hydraulic action exists, such as coastline protection from pressures exerted by waves, currents and tides: Typar SF56						
horizontal	0.05	2.15	1.59	0.53	m²	**4.26**
inclined at an angle 10-45 degrees from the horizontal	0.06	2.69	2.00	0.53	m²	**5.22**

SERIES 600: EARTHWORKS

Item	Gang hours	Labour £	Plant £	Material £	Unit	Total rate £
For protection against puncturing to reservoir liner: Typar SF56						
horizontal	0.05	2.15	1.59	0.53	m²	**4.26**
inclined at an angle 10-45 degrees from the horizontal	0.06	2.69	2.00	0.53	m²	**5.22**
Temporary parking areas						
For reinforcement of grassed areas subject to wear from excessive pedestrian and light motor vehicle traffic: Netlon CE131 high density polyethylene geogrid, including fixing pegs						
sheeting	0.04	1.27	-	3.63	m²	**4.91**
Landscaping applications						
For prevention of weed growth in planted areas by incorporating a geotextile over topsoil: Typar SF20, including pegs						
horizontal	0.08	2.73	-	0.61	m²	**3.34**
inclined at an angle 10-45 degrees to the horizontal	0.09	3.42	-	0.61	m²	**4.03**
For root growth control-Prevention of lateral spread of roots and mixing of road base and humus: Typar SF20						
horizontal	0.08	2.73	-	0.31	m²	**3.04**
inclined at an angle 10-45 degrees to the horizontal	0.09	3.42	-	0.31	m²	**3.73**
SOFT SPOTS AND OTHER VOIDS						
Excavation of soft spots and other voids using motorised scrapers and/or dozers						
Excavate below cuttings or under embankments	0.05	0.82	3.87	-	m³	**4.69**
Excavate in side slopes	0.06	0.98	4.65	-	m³	**5.63**
Excavation of soft spots and other voids using backacters and tractor loader						
Excavate below structural foundations and foundations for corrugated steel buried structures	0.09	1.32	1.15	-	m³	**2.47**
Imported graded granular fill; deposition using tractor loader and towed roller						
Filling of soft spots and other voids below cuttings or under embankments	0.04	0.51	0.80	18.70	m³	**20.01**
Filling of soft spots and other voids in side slopes	0.04	0.62	0.98	18.70	m³	**20.30**
Filling of soft spots and other voids below structural foundations and foundations for corrugated steel buried structures	0.04	0.51	0.80	18.70	m³	**20.01**
Imported rock fill; 1.9 t/m³; using tractor loader						
Deposition into soft areas (rock punching)	0.03	0.89	0.78	29.82	m³	**31.49**

SERIES 600: EARTHWORKS

Item	Gang hours	Labour £	Plant £	Material £	Unit	Total rate £
DISUSED SEWERS ETC.						
Removal of disused sewer or drain						
100 mm internal diameter; with less than 1 m of cover to formation level	0.16	2.14	2.22	-	m	**4.35**
150 mm internal diameter; with 1 to 2 m of cover to formation level	0.20	2.62	2.72	-	m	**5.33**
Backfilling with acceptable material of disused sewer or drain						
100 mm internal diameter; with less than 1 m of cover to formation level	0.23	3.07	3.19	-	m³	**6.26**
150 mm internal diameter; with 1 to 2 m of cover to formation level	0.23	3.07	3.19	-	m³	**6.26**
Backfilling of disused basements, cellars and the like						
with acceptable material	0.26	3.47	3.60	-	m³	**7.08**
Backfilling of disused gullies						
with concrete Grade C15	0.10	1.33	1.39	8.88	nr	**11.60**
SUPPORTS LEFT IN EXCAVATION						
Timber close boarded supports left in excavation	-	-	8.10	-	m²	**8.10**
Steel trench sheeting supports left in excavation	-	-	33.96	-	m²	**33.96**
TOPSOILING AND STORAGE OF TOPSOIL						
Topsoiling 150 mm thick to surfaces						
at 10 degrees or less to horizontal	0.04	0.57	2.71	-	m²	**3.28**
more than 10 degrees to horizontal	0.04	0.65	3.10	-	m²	**3.75**
Topsoiling 350 mm thick to surfaces						
at 10 degrees or less to horizontal	0.05	0.74	3.49	-	m²	**4.22**
more than 10 degrees to horizontal	0.05	0.82	3.87	-	m²	**4.69**
Topsoiling 450 mm thick to surfaces						
at 10 degrees or less to horizontal	0.05	0.82	3.87	-	m²	**4.69**
more than 10 degrees to horizontal	0.06	0.90	4.26	-	m²	**5.16**
Topsoiling 600 mm thick to surfaces						
at 10 degrees or less to horizontal	0.05	0.82	3.87	-	m²	**4.69**
more than 10 degrees to horizontal	0.06	0.90	4.26	-	m²	**5.16**
Permanent storage of topsoil	0.06	0.98	4.65	-	m³	**5.63**
COMPLETION OF FORMATION AND SUB-FORMATION						
Completion of sub-formation						
on material other than Class 1C, 6B or rock in cuttings	0.01	0.18	0.47	-	m²	**0.65**
Completion of formation						
on material other than Class 1C, 6B or rock in cuttings	0.01	0.19	0.39	-	m²	**0.58**

SERIES 600: EARTHWORKS

Item	Gang hours	Labour £	Plant £	Material £	Unit	Total rate £
LINING OF WATERCOURSES						
Lining new watercourse invert with precast concrete units 63 mm thick	0.21	4.90	0.25	8.92	m^2	**14.07**
Lining new watercourse side slopes with precast concrete units 63 mm thick	0.25	5.83	0.28	8.00	m^2	**14.11**
Lining enlarged watercourse invert with precast concrete units 63 mm thick	0.24	5.59	0.29	8.92	m^2	**14.80**
Lining enlarged watercourse side slopes with precast concrete units 63 mm thick	0.29	6.76	0.35	8.00	m^2	**15.11**
GROUND IMPROVEMENT - DYNAMIC COMPACTION						
Ground consolidation by dynamic compaction is a technique which involves the dropping of a steel or concrete pounder several times in each location in a grid pattern that covers the whole site. For a ground compaction of up to 10 m, a 15 t pounder with a free fall of 20 m would be typical. Several passes over the site are normally required to achieve full compaction. The process is recommended for naturally cohesive soils and is usually uneconomic for areas of less than 4000 m^2, for sites with granular or mixed granular cohesive soils and 6000 m^2 for a site with weak cohesive soils. The main considerations to be taken into account when using this method of consolidation are: • sufficient area to be viable • proximity and condition of adjacent property and services • need for blanket layer of granular material for a working surface and as backfill to offset induced settlement • water table level The granular blanket layer performs the dual functions of working surface and backfill material. Generally 300 mm thickness is required. The final bearing capacity and settlement criteria that can be achieved depends on the nature of the material being compacted. Allowable bearing capacity may be increased by up to twice the pre-treated value for the same settlement. Control testing can be by crater volume measurements, site levelling between passes, penetration tests or plate loading tests. The following range of costs are averages based on treating an area of about 10,000 m^2 at 5 - 6 m compaction depth. Typical progress would be 1,500 - 2,000 m^2 per week.						
GROUND IMPROVEMENT - ESTABLISHMENT OF PLANT						
Establishment of dynamic compaction plant	-	-	-	-	sum	**33600.00**

SERIES 600: EARTHWORKS

Item	Gang hours	Labour £	Plant £	Material £	Unit	Total rate £
GROUND IMPROVEMENT - DYNAMIC COMPACTION						
Dynamic compaction in main compaction with a 15 t pounder	-	-	-	-	m²	**8.35**
Dynamic compaction plant standing time	-	-	-	-	hr	**13.25**
Free-draining granular material in granular blanket	-	-	-	-	t	**10.31**
Control testing including levelling, piezometers and penetrameter testing	-	-	-	-	m²	**3.43**
Kentledge load test	-	-	-	-	nr	**11025.00**
GABION WALLING AND MATTRESSES						
Gabion walling with plastic coated galvanised wire mesh, wire laced; filled with 50 mm Class 6G material						
2.0 x 1.0 x 1.0 m module sizes	3.98	41.34	7.08	40.69	m³	**89.10**
2.0 x 1.0 x 0.5 m module sizes	5.20	52.13	17.42	49.84	m³	**119.39**
Gabion walling with heavily galvanised woven wire mesh, wire laced; filled with 50 mm Class 6G material						
2.0 x 1.0 x 1.0 m module sizes	4.22	42.36	14.15	32.45	m³	**88.95**
2.0 x 1.0 x 0.5 m module sizes	5.20	52.13	17.42	40.31	m³	**109.86**
Mattress with plastic coated galvanised wire mesh; filled with 50 mm Class 6G material installed at 10 degrees or less to the horizontal						
6.0 x 2.0 x 0.23 m module sizes	4.22	42.36	14.15	41.59	m³	**98.10**
CRIB WALLING						
Notes						
There are a number of products specially designated for large scale earth control. Crib walling consists of a rigid unit built of rectangular interlocking timber or precast concrete members forming a skeleton of cells laid on top of each other and filled with earth or rock. Prices for these items depend on quantity, difficulty of access to the site and availability of suitable filling material; estimates should be obtained from the manufacturer when the site conditions have been determined.						
Crib walling of timber components laid with a battering face; stone infill						
ne 1.5 m high	5.28	55.82	9.68	83.56	m²	**149.06**
ne 3.7 m high	7.11	75.16	11.41	87.47	m²	**174.05**
ne 4.2 m high	8.13	85.95	12.38	91.48	m²	**189.81**
ne 5.9 m high	10.71	113.22	14.82	102.33	m²	**230.37**
ne 7.4 m high	17.16	181.41	20.93	132.01	m²	**334.34**
Crib walling of precast concrete crib units, laid dry jointed with a battered face; stone infill						
1.0 m high; no dowels	6.75	71.36	23.60	92.75	m	**187.71**
1.5 m high; no dowels	9.36	98.95	32.72	139.13	m	**270.80**
2.5 m high; no dowels	22.11	233.74	78.47	275.80	m	**588.01**
4.0 m high; no dowels	34.11	360.60	121.06	441.28	m	**922.94**

SERIES 600: EARTHWORKS

Item	Gang hours	Labour £	Plant £	Material £	Unit	**Total rate £**
GROUND ANCHORAGES - GENERALLY						
Ground anchorages consist of the installation of a cable or solid bar tendon fixed in the ground by grouting and tensioned to exceed the working load to be carried. Ground anchors may be of a permanent or temporary nature and can be used in conjunction with diaphragm walling or sheet piling to eliminate the use of strutting etc.						
The following costs are based on the installation of 50 nr ground anchors						
GROUND ANCHORAGES - GROUND ANCHORAGE PLANT						
Establishment of ground anchorage plant	-	-	-	-	sum	**9765.00**
Ground anchorage plant standing time	-	-	-	-	hr	**133.48**
GROUND ANCHORAGES						
Ground anchorages; temporary or permanent						
15.0 m maximum depth; in rock, alluvial or clay; 0 - 50 t load	-	-	-	-	nr	**78.24**
15.0 m maximum depth; in rock or alluvial; 50 – 90 t load	-	-	-	-	nr	**93.61**
15.0 m maximum depth; in rock only; 90 - 150 t load	-	-	-	-	nr	**125.69**
Temporary tendons						
in rock, alluvial or clay; 0 - 50 t load	-	-	-	-	nr	**61.36**
in rock or alluvial; 50 - 90 t load	-	-	-	-	nr	**93.61**
in rock only; 90 - 150 t load	-	-	-	-	nr	**125.69**
Permanent tendons						
in rock, alluvial or clay; 0 - 50 t load	-	-	-	-	nr	**91.99**
in rock or alluvial; 50 - 90 t load	-	-	-	-	nr	**12.59**
in rock only; 90 - 150 t load	-	-	-	-	nr	**157.24**
GROUND WATER LOWERING						
The following unit costs are for dewatering pervious ground only and are for sets of equipment comprising:						
• hire of 1 nr diesel driven pump (WP 150/60 or similar) complete with allowance of £30 for fuel	-	-	-	-	day	**71.14**
• hire of 50 m of 150 mm diameter header pipe	-	-	-	-	day	**17.34**
• purchase of 35 nr of disposable well points	-	-	-	-	sum	**8873.20**
• hire of 18 m of delivery pipe	-	-	-	-	day	**6.88**
• hire of 1 nr diesel driven standby pump	-	-	-	-	day	**32.27**
• hire of 1 nr jetting pump with hoses (for installation of wellpoints only)	-	-	-	-	day	**53.79**
• cost of attendant labour and plant (2 hrs per day) inclusive of small dumper and bowser	-	-	-	-	day	**89.05**
Costs are based on 24 hr operation in 12 hr shifts with attendant operators (specialist advice)						

SERIES 600: EARTHWORKS

Item	Gang hours	Labour £	Plant £	Material £	Unit	Total rate £
GROUND WATER LOWERING – cont'd						
Guide price for single set of equipment comprising pump, 150 mm diameter header pipe, 35 nr well points, delivery pipe and attendant labour and plant						
Bring to site equipment and remove upon completion	-	-	-	-	sum	**2147.40**
Installation: 3 day hire of jetting pump	-	-	-	-	sum	**161.37**
Operating costs						
purchase of well points	-	-	-	-	nr	**253.52**
hire of pump, header pipe, delivery pipe and standby pump complete with fuel etc. and attendant labour and plant	-	-	-	-	day	**216.68**
TRIAL PITS						
The following costs assume the use of mechanical plant and excavating and backfilling on the same day						
Trial pit						
ne 1.0 m deep	1.44	15.22	7.18	-	nr	**22.40**
1.0 - 2.0 m deep	2.62	27.56	14.06	-	nr	**41.62**
over 2.0 m deep	3.18	33.62	15.86	-	nr	**49.47**
BREAKING UP AND PERFORATION OF REDUNDANT PAVEMENTS						
Using scraper and ripper bulldozer						
Breaking up of redundant concrete slab						
ne 100 mm deep	0.05	0.82	8.12	-	m²	**8.94**
100 to 200 mm deep	0.10	1.64	16.25	-	m²	**17.88**
Breaking up of redundant flexible pavement						
ne 100 mm deep	0.02	0.33	3.25	-	m²	**3.58**
100 to 200 mm deep	0.03	0.49	4.87	-	m²	**5.37**
Using backacters and tractor loader with ripper						
Breaking up of redundant reinforced concrete pavement						
ne 100 mm deep	0.28	3.79	3.29	-	m²	**7.08**
100 to 200 mm deep	0.49	6.63	5.76	-	m²	**12.39**
200 to 300 mm deep	0.70	9.48	8.24	-	m²	**17.72**
Using scraper and ripper bulldozer						
Breaking up of redundant flexible pavement using scraper and ripper bulldozer						
ne 100 mm deep	0.02	0.33	3.25	-	m²	**3.58**
100 to 200 mm deep	0.03	0.49	4.87	-	m²	**5.37**
Using backacters and breakers						
Perforation of redundant reinforced concrete pavement						
ne 100 mm deep	0.02	0.29	0.35	-	m²	**0.65**
100 to 200 mm deep	0.03	0.44	0.54	-	m²	**0.98**
200 to 300 mm deep	0.05	0.73	0.89	-	m²	**1.63**

SERIES 600: EARTHWORKS

Item	Gang hours	Labour £	Plant £	Material £	Unit	**Total rate £**
Perforation of redundant flexible pavement						
ne 100 mm deep	0.01	0.15	0.19	-	m²	**0.33**
100 to 200 mm deep	0.02	0.22	0.27	-	m²	**0.49**
PERFORATION OF REDUNDANT SLABS, BASEMENTS AND THE LIKE						
Using backacters and breakers						
Perforation of redundant reinforced concrete slab						
ne 100 mm deep	0.02	0.29	0.35	-	m²	**0.65**
100 to 200 mm deep	0.03	0.44	0.54	-	m²	**0.98**
200 to 300 mm deep	0.05	0.73	0.89	-	m²	**1.63**
Perforation of redundant reinforced concrete basement						
ne 100 mm deep	0.02	0.31	0.37	-	m²	**0.68**
100 to 200 mm deep	0.03	0.46	0.56	-	m²	**1.03**
200 to 300 mm deep	0.05	0.77	0.93	-	m²	**1.70**
REINFORCED AND ANCHORED EARTH STRUCTURES						
SPECIALIST ADVICE						
As each structure is different, it is virtually impossible to give accurate unit cost prices, as they will vary with the following parameters: • Type of structure • Where located (in water, dry condition) • Where geographically in the country • Type of fill • Duration of structure • Size of structure, etc.						
To arrive at the unit costs below assumptions have been made for a structure with the following characteristics : • Structure - retaining wall 6 m high x 150 m in length • Construction - as DTp Specification BE 3/78 • Site conditions - good foundations • Fill - 5 m³ per m² of wall face • Fill costs - DTp Specification average £5.00 / tonne						
Therefore specialist advice should be sought in order to give accurate budget costings for individual projects						
Retaining wall (per m² of face)						
concrete faced using ribbed strip	1.00	8.14	4.08	169.05	m²	**181.27**
concrete faced using flat strip	1.00	8.14	4.08	215.15	m²	**227.38**
concrete faced using polyester strip	1.00	21.60	8.41	184.41	m²	**214.42**
concrete faced using geogrid reinforcement	1.00	18.06	9.25	192.12	m²	**219.43**
preformed mesh using ribbed strip	1.00	9.58	6.35	138.31	m²	**154.24**

SERIES 700: PAVEMENTS

Item	Gang hours	Labour £	Plant £	Material £	Unit	Total rate £
RESOURCES - LABOUR						
Sub-base laying gang						
1 ganger/chargehand (skill rate 4)		12.80				
1 skilled operative (skill rate 4)		12.05				
2 unskilled operatives (general)		22.52				
1 plant operator (skill rate 2)		16.35				
1 plant operator (skill rate 3)		14.65				
Total Gang Rate / Hour	£	**78.37**				
Flexible paving gang						
1 ganger/chargehand (skill rate 4)		12.80				
2 skilled operatives (skill rate 4)		24.10				
4 unskilled operatives (general)		45.04				
4 plant operators (skill rate 3)		58.60				
Total Gang Rate / Hour	£	**140.54**				
Concrete paving gang						
1 ganger/chargehand (skill rate 4)		12.80				
2 skilled operatives (skill rate 4)		24.10				
4 unskilled operatives (general)		45.04				
1 plant operator (skill rate 2)		16.35				
1 plant operator (skill rate 3)		14.65				
Total Rate / Hour	£	**112.94**				
Road surface spraying gang						
1 plant operator (skill rate 3)		14.65				
Total Gang Cost / Hour	£	**14.65**				
Road chippings gang						
1 ganger/chargehand (skill rate 4) - 50% of time		6.40				
1 skilled operative (skill rate 4)		12.05				
2 unskilled operatives (general)		22.52				
3 plant operators (skill rate 3)		43.95				
Total Gang Rate / Hour	£	**84.92**				
Cutting slabs gang						
1 unskilled operative (general)		11.26				
Total Gang Rate / Hour	£	**11.26**				
Concrete filled joints gang						
1 ganger/chargehand - 50% of time		6.40				
1 skilled operative (skill rate 4)		12.05				
2 unskilled operatives (general)		22.52				
Total Gang Rate / Hour	£	**40.97**				
Milling gang						
1 ganger/chargehand (skill rate 4)		12.80				
2 skilled operatives (skill rate 4)		24.10				
4 unskilled operatives (general)		45.04				
1 plant operator (skill rate 2)		16.35				
1 plant operator (skill rate 3)		14.65				
Total Gang Rate / Hour	£	**112.94**				
Rake and compact planed material gang						
1 ganger/chargehand (skill rate 4)		12.80				
1 skilled operative (skill rate 4)		12.05				
3 unskilled operatives (general)		33.78				
1 plant operator (skill rate 4)		13.17				
1 plant operator (skill rate 3)		14.65				
Total Gang Rate / Hour	£	**86.45**				

SERIES 700: PAVEMENTS

Item	Gang hours	Labour £	Plant £	Material £	Unit	Total rate £
RESOURCES - PLANT						
Sub-base laying						
93 kW motor grader			22.12			
0.8 m³ tractor loader			18.28			
6 t towed vibratory roller			8.45			
Total Gang Rate / Hour		£	**48.85**			
Flexible paving						
2 asphalt pavers, 35 kW, 4.0 m			85.83			
2 deadweight rollers, 3 point, 10 t			26.29			
tractor with front bucket and integral 2 tool compressor			16.31			
compressor tools : rammer			0.41			
compressor tools : poker vibrator			1.44			
compressor tools : extra 15 m hose			0.26			
tar sprayer, 100 litre			4.69			
self propelled chip spreader			7.55			
channel (heat) iron			1.12			
Total Gang Rate / Hour		£	**143.91**			
Concrete paving						
wheeled loader, 2.60 m³			51.00			
concrete paver, 6.0 m			69.58			
concrete slipform finisher			15.86			
Total Gang Rate / Hour		£	**136.45**			
Road surface spraying						
tar sprayer; 100 litre			4.69			
Total Gang Rate / Hour		£	**4.69**			
Road chippings						
deadweight roller, 3 point, 10 t			13.15			
tar sprayer, 100 litre			4.69			
self propelled chip spreader			7.55			
channel (heat) iron			1.12			
Total Gang Rate / Hour		£	**26.50**			
Cutting slabs						
compressor, 65 cfm			3.44			
12" disc cutter			1.33			
Total Gang Rate / Hour		£	**4.77**			
Cold milling						
cold planer, 2.10 m			49.72			
wheeled loader, 2.60 m³			51.00			
Total Gang Rate / Hour		£	**100.73**			
Heat planing						
heat planer, 4.5 m			72.82			
wheeled loader, 2.60 m³			51.00			
Total Gang Rate / Hour		£	**123.82**			
Rake and compact planed material						
deadweight roller, 3 point, 10 t			13.15			
tractor with front bucket and integral 2 tool compressor			16.31			
channel (heat) iron			1.12			
Total Gang Rate / Hour		£	**30.58**			

SERIES 700: PAVEMENTS

Item	Gang hours	Labour £	Plant £	Material £	Unit	Total rate £
SUB-BASE						
The following unit costs are generally based on the Department of Transport (DTp) Specification for Highway Works and reference is made throughout this Section to clauses within that specification.						
Granular material DTp specified type 1; Sub-base in carriageway, hardshoulder and hardstrip						
75 mm deep	0.04	2.74	1.71	21.47	m³	**25.92**
100 mm deep	0.04	3.13	1.95	21.47	m³	**26.56**
150 mm deep	0.05	3.53	2.20	21.47	m³	**27.19**
200 mm deep	0.05	3.92	2.44	21.47	m³	**27.83**
Granular material DTp specified type 2; Sub-base; spread and graded						
75 mm deep	0.04	2.74	1.71	22.91	m³	**27.36**
100 mm deep	0.04	3.13	1.95	22.91	m³	**28.00**
150 mm deep	0.05	3.53	2.20	22.91	m³	**28.64**
200 mm deep	0.05	3.92	2.44	22.91	m³	**29.27**
Wet lean concrete DTp specified strength mix C20, 20 mm aggregate; Sub-base; spread and graded						
100 mm deep	0.05	3.53	2.20	84.44	m³	**90.16**
200 mm deep	0.05	3.92	2.44	84.44	m³	**90.80**
Hardcore; Sub-base; spread and graded						
100 mm deep	0.04	3.13	1.95	33.83	m³	**38.92**
150 mm deep	0.05	3.53	2.20	33.83	m³	**39.56**
200 mm deep	0.05	3.92	2.44	33.83	m³	**40.19**
Wet mix macadam; DTp clause 808; Sub-base; spread and graded						
75 mm deep	0.04	2.74	2.59	22.60	m³	**27.94**
100 mm deep	0.04	3.13	2.69	22.60	m³	**28.43**
200 mm deep	0.05	3.92	2.83	22.60	m³	**29.36**
PAVEMENT (FLEXIBLE)						
Notes - Labour and Plant All outputs are based on clear runs without undue delay to two pavers with a 75% utilisation The outputs can be adjusted as follows to take account of space or time influences on the utilisation						
Factors for varying utilisation of Labour and Plant:						
1 paver @ 75% utilisation = x 2.00						
1 paver @ 100% utilisation = x 1.50						
2 pavers @ 100% utilisation = x 0.75						

SERIES 700: PAVEMENTS

Item	Gang hours	Labour £	Plant £	Material £	Unit	**Total rate £**
Dense Bitumen Macadam						
Road Base to DTp clause 903						
100 mm deep	0.02	2.81	2.88	6.05	m^2	**11.74**
150 mm deep	0.03	3.51	3.60	9.07	m^2	**16.18**
200 mm deep	0.03	4.22	4.32	12.09	m^2	**20.63**
Road Base in emergency crossing						
100 mm deep	0.02	2.81	2.88	6.05	m^2	**11.74**
150 mm deep	0.03	3.51	3.60	9.07	m^2	**16.18**
200 mm deep	0.03	4.22	4.32	12.09	m^2	**20.63**
Road Base in lay-by and bus bay						
100 mm deep	0.02	2.81	2.88	6.05	m^2	**11.74**
150 mm deep	0.03	3.51	3.60	9.07	m^2	**16.18**
200 mm deep	0.03	4.22	4.32	12.09	m^2	**20.63**
Base Course to DTp clause 906						
50 mm deep	0.02	2.11	2.16	2.55	m^2	**6.82**
100 mm deep	0.02	2.81	2.88	5.10	m^2	**10.79**
Base Course in emergency crossing						
50 mm deep	0.02	2.11	2.16	2.55	m^2	**6.82**
100 mm deep	0.02	2.81	2.88	5.10	m^2	**10.79**
Base Course in lay-by and bus bay						
50 mm deep	0.02	2.11	2.16	2.55	m^2	**6.82**
100 mm deep	0.02	2.81	2.88	5.10	m^2	**10.79**
Wearing Course to DTp clause 912						
30 mm deep	0.01	1.41	1.44	2.17	m^2	**5.02**
50 mm deep	0.02	2.11	2.16	3.62	m^2	**7.89**
Wearing Course in emergency crossing						
30 mm deep	0.01	1.41	1.44	2.17	m^2	**5.02**
50 mm deep	0.02	2.11	2.16	3.62	m^2	**7.89**
Wearing Course in lay-by and bus bay						
30 mm deep	0.01	1.41	1.44	2.17	m^2	**5.02**
50 mm deep	0.02	2.11	2.16	3.62	m^2	**7.89**
Bitumen Macadam						
Base Course to DTp clause 908						
35 mm deep	0.01	1.41	1.44	2.31	m^2	**5.16**
70 mm deep	0.02	2.11	2.16	4.62	m^2	**8.89**
Base Course in emergency crossing						
35 mm deep	0.01	1.41	1.44	2.31	m^2	**5.16**
70 mm deep	0.02	2.11	2.16	4.62	m^2	**8.89**
Base Course in lay-by and bus bay						
35 mm deep	0.01	1.41	1.44	2.31	m^2	**5.16**
70 mm deep	0.02	2.11	2.16	4.62	m^2	**8.89**

SERIES 700: PAVEMENTS

Item	Gang hours	Labour £	Plant £	Material £	Unit	Total rate £
PAVEMENT (FLEXIBLE) – cont'd						
Dense Tarmacadam						
Road Base to DTp clause 902						
50 mm deep	0.02	2.11	2.16	2.99	m²	**7.26**
100 mm deep	0.02	2.11	2.16	5.97	m²	**10.24**
Road Base in emergency crossing						
50 mm deep	0.02	2.11	2.16	2.99	m²	**7.26**
100 mm deep	0.02	2.11	2.16	5.97	m²	**10.24**
Road Base in lay-by and bus bay						
50 mm deep	0.02	2.11	2.16	2.99	m²	**7.26**
100 mm deep	0.02	2.11	2.16	5.97	m²	**10.24**
Base Course to DTp clause 907						
60 mm deep	0.02	2.11	2.16	3.89	m²	**8.16**
80 mm deep	0.02	2.11	2.16	5.19	m²	**9.46**
Base Course in emergency crossing						
60 mm deep	0.02	2.11	2.16	3.89	m²	**8.16**
80 mm deep	0.02	2.11	2.16	5.19	m²	**9.46**
Base Course in lay-by and bus bay						
60 mm deep	0.02	2.11	2.16	3.89	m²	**8.16**
80 mm deep	0.02	2.11	2.16	5.19	m²	**9.46**
Dense Tar Surfacing						
Wearing Course to DTp clause 913						
30 mm deep	0.01	1.41	1.44	2.21	m²	**5.06**
50 mm deep	0.02	2.11	2.16	3.69	m²	**7.96**
Base Course in emergency crossing						
30 mm deep	0.01	1.41	1.44	2.21	m²	**5.06**
50 mm deep	0.02	2.11	2.16	3.69	m²	**7.96**
Base Course in lay-by and bus bay						
30 mm deep	0.01	1.41	1.44	2.21	m²	**5.06**
50 mm deep	0.02	2.11	2.16	3.69	m²	**7.96**
Cold Asphalt						
Wearing Course to DTp clause 914						
15 mm deep	0.01	1.41	1.44	1.13	m²	**3.98**
30 mm deep	0.01	1.41	1.44	2.24	m²	**5.09**
Base Course in emergency crossing						
15 mm deep	0.01	1.41	1.44	1.13	m²	**3.98**
30 mm deep	0.01	1.41	1.44	2.24	m²	**5.09**
Base Course in lay-by and bus bay						
15 mm deep	0.01	1.41	1.44	1.13	m²	**3.98**
30 mm deep	0.01	1.41	1.44	2.24	m²	**5.09**
Rolled Asphalt						
Base Course to DTp clause 905						
60 mm deep	0.02	2.11	2.16	3.89	m²	**8.16**
80 mm deep	0.02	2.11	2.16	5.19	m²	**9.46**
Base Course in emergency crossing						
60 mm deep	0.02	2.11	2.16	3.89	m²	**8.16**
80 mm deep	0.02	2.11	2.16	5.19	m²	**9.46**
Base Course in lay-by and bus bay						
60 mm deep	0.02	2.11	2.16	3.89	m²	**8.16**
80 mm deep	0.02	2.11	2.16	5.19	m²	**9.46**

SERIES 700: PAVEMENTS

Item	Gang hours	Labour £	Plant £	Material £	Unit	**Total rate £**
Wearing Course to DTp clause 911						
40 mm deep	0.02	2.11	2.16	3.52	m²	**7.79**
60 mm deep	0.02	2.11	2.16	5.29	m²	**9.55**
Wearing Course in emergency crossing						
40 mm deep	0.02	2.11	2.16	3.52	m²	**7.79**
60 mm deep	0.02	2.11	2.16	5.29	m²	**9.55**
Wearing Course in lay-by and bus bay						
40 mm deep	0.02	2.11	2.16	3.52	m²	**7.79**
60 mm deep	0.02	2.11	2.16	5.29	m²	**9.55**
PAVEMENT (CONCRETE)						
The following unit costs are for jointed reinforced concrete slabs, laid in reasonable areas (over 200 m²) by paver train/slipformer						
Designed mix; cement to BS 12; grade C30, 20 mm aggregate						
Slab, runway, access roads or similar						
180 mm deep	0.02	1.69	2.05	13.79	m²	**17.53**
220 mm deep	0.02	2.03	2.46	16.86	m²	**21.35**
260 mm deep	0.02	2.48	3.00	19.92	m²	**25.41**
300 mm deep	0.03	2.82	3.41	22.99	m²	**29.22**
Fabric reinforcement						
Steel fabric reinforcement to BS4483						
Ref A142 nominal mass 2.22 kg	0.03	3.39	-	1.30	m²	**4.69**
Ref A252 nominal mass 3.95 kg	0.04	4.52	-	2.30	m²	**6.82**
Ref B385 nominal mass 4.53 kg	0.04	4.52	-	2.65	m²	**7.16**
Ref C636 nominal mass 5.55 kg	0.05	5.65	-	3.25	m²	**8.90**
Ref B503 nominal mass 5.93 kg	0.05	5.65	-	3.47	m²	**9.12**
Mild Steel bar reinforcement BS 4449						
Bars; supplied in bent and cut lengths						
6 mm nominal size	8.00	903.52	-	467.63	tonne	**1371.15**
8 mm nominal size	6.74	761.22	-	456.52	tonne	**1217.74**
10 mm nominal size	6.74	761.22	-	446.42	tonne	**1207.64**
12 mm nominal size	6.74	761.22	-	444.40	tonne	**1205.62**
16 mm nominal size	6.15	694.58	-	440.93	tonne	**1135.51**
High yield steel bar reinforcement BS 4449 or 4461						
Bars; supplied in bent and cut lengths						
6 mm nominal size	8.00	903.52	-	438.34	tonne	**1341.86**
8 mm nominal size	6.74	761.22	-	428.24	tonne	**1189.46**
10 mm nominal size	6.74	761.22	-	417.13	tonne	**1178.35**
12 mm nominal size	6.74	761.22	-	419.15	tonne	**1180.37**
16 mm nominal size	6.15	694.58	-	414.10	tonne	**1108.68**
Sheeting to prevent moisture loss						
Polyethelene sheeting; lapped joints; horizontal below concrete pavements						
1000 gauge	0.01	1.13	-	0.81	m²	**1.94**
2000 gauge	0.01	1.13	-	1.47	m²	**2.60**

SERIES 700: PAVEMENTS

Item	Gang hours	Labour £	Plant £	Material £	Unit	Total rate £
PAVEMENT (CONCRETE) – cont'd						
Joints in concrete slabs						
Longitudinal joints						
180 mm deep concrete	0.01	1.36	1.64	18.77	m	**21.77**
220 mm deep concrete	0.01	1.36	1.64	19.79	m	**22.79**
260 mm deep concrete	0.01	1.36	1.64	23.64	m	**26.63**
300 mm deep concrete	0.01	1.36	1.64	25.01	m	**28.00**
Expansion joints						
180 mm deep concrete	0.01	1.36	1.64	28.83	m	**31.83**
220 mm deep concrete	0.01	1.36	1.64	34.05	m	**37.04**
260 mm deep concrete	0.01	1.36	1.64	39.26	m	**42.25**
300 mm deep concrete	0.01	1.36	1.64	40.29	m	**43.28**
Contraction joints						
180 mm deep concrete	0.01	1.36	1.64	15.61	m	**18.61**
220 mm deep concrete	0.01	1.36	1.64	16.67	m	**19.67**
260 mm deep concrete	0.01	1.36	1.64	19.59	m	**22.59**
300 mm deep concrete	0.01	1.36	1.64	21.88	m	**24.87**
Construction joints						
180 mm deep concrete	0.01	1.36	1.64	9.32	m	**12.32**
220 mm deep concrete	0.01	1.36	1.64	10.43	m	**13.42**
260 mm deep concrete	0.01	1.36	1.64	11.47	m	**14.46**
300 mm deep concrete	0.01	1.36	1.64	12.49	m	**15.49**
Open joints with filler						
ne 0.5 m; 10 mm flexcell joint filler	0.11	4.51	-	3.22	m	**7.73**
0.5 - 1 m; 10 mm flexcell joint filler	0.11	4.51	-	4.63	m	**9.14**
Joint sealants						
10 x 20 mm hot bitumen sealant	0.14	5.74	-	2.90	m	**8.63**
20 x 20 mm cold polysulphide sealant	0.18	7.37	-	5.76	m	**13.14**
Trimming edges only of existing slabs, floors or similar surfaces (wet or dry); 6 mm cutting width						
50 mm deep	0.02	0.23	0.10	1.33	m	**1.65**
100 mm deep	0.03	0.34	0.15	3.01	m	**3.50**
Cutting existing slabs, floors or similar surfaces (wet or dry); 8 mm cutting width						
50 mm deep	0.03	0.28	0.12	3.18	m	**3.58**
100 mm deep	0.06	0.68	0.29	3.35	m	**4.31**
150 mm deep	0.08	0.90	0.38	3.71	m	**4.99**
F SURFACE TREATMENT						
Slurry sealing; BS 434 class K3						
Slurry sealing to DTp clause 918						
3 mm deep	0.02	0.22	0.07	1.19	m²	**1.48**
4 mm deep	0.02	0.22	0.07	1.65	m²	**1.94**
Coated chippings, 9 - 11 kg/m²						
Surface dressing to DTp clause 915						
6 mm nominal size	0.01	0.85	0.27	0.73	m²	**1.84**
8 mm nominal size	0.01	0.85	0.27	0.75	m²	**1.86**
10 mm nominal size	0.01	0.85	0.27	0.76	m²	**1.87**
12 mm nominal size	0.01	0.85	0.27	0.83	m²	**1.94**

SERIES 700: PAVEMENTS

Item	Gang hours	Labour £	Plant £	Material £	Unit	**Total rate £**
TACK COAT						
Bituminous spray; BS 434 K1 - 40						
Tack coat to DTp clause 920						
large areas; over 20 m²	0.02	0.22	0.07	0.25	m²	**0.54**
COLD MILLING (PLANING)						
Milling pavement (assumes disposal on site or re-use as fill but excludes transport if required)						
75 mm deep	0.03	3.05	2.72	-	m²	**5.77**
100 mm deep	0.04	4.07	3.63	-	m²	**7.69**
50 mm deep; scarifying surface	0.02	2.48	2.22	-	m²	**4.70**
75 mm deep; scarifying surface	0.04	4.18	3.73	-	m²	**7.91**
25 mm deep; heat planing for re-use	0.03	3.61	3.96	-	m²	**7.58**
50 mm deep; heat planing for re-use	0.06	6.32	6.93	-	m²	**13.26**
INSITU RECYCLING						
Raking over scarified or heat planed material; compacting with 10 t roller						
50 mm deep	0.01	0.87	0.31	-	m²	**1.17**

SERIES 1100: KERBS, FOOTWAYS & PAVED AREAS

Item	Gang hours	Labour £	Plant £	Material £	Unit	Total rate £
NOTES						
Measurement Note: sub-bases are shown separate from their associated paving to simplify the presentation of cost alternatives.						
Measurement Note: bases are shown separate from their associated kerb etc. to simplify the presentation of cost alternatives.						
Kerb quadrants, droppers are shown separately						
The re-erection cost for kerbs, channels and edgings etc. taken from store assumes that major components are in good condition; the prices below allow a sum of 20 % of the value of new materials to cover minor repairs together with an allowance for replacing a proportion of units.						
RESOURCES - LABOUR						
Kerb laying gang						
3 skilled operatives (skill rate 4)		36.15				
1 unskilled operative (general)		11.26				
1 plant operator (skill rate 3) - 25% of time		3.66				
Total Gang Rate / Hour	£	**51.07**				
Path sub-base, bitmac and gravel laying gang						
1 skilled operative (skill rate 4)		12.05				
2 unskilled operatives (general)		22.52				
1 plant operator (skill rate 3)		14.65				
Total Gang Rate / Hour	£	**49.22**				
Paviors and flagging gang						
1 skilled operative (skill rate 4)		12.05				
1 unskilled operative (general)		11.26				
Total Gang Rate / Hour	£	**23.31**				
RESOURCES - PLANT						
Kerb laying						
backhoe JCB 3CX (25% of time)			3.90			
12" Stihl saw			0.96			
road forms			1.62			
Total Gang Rate / Hour		£	**6.49**			
Path sub-base, bitmac and gravel laying						
backhoe JCB3CX			15.62			
2 t dumper			3.64			
pedestrian roller `Bomag BW90S			4.16			
Total Gang Rate / Hour		£	**23.41**			
Paviors and flagging						
2 t dumper (33% of time)			1.21			
Total Gang Rate / Hour		£	**1.21**			

SERIES 1100: KERBS, FOOTWAYS & PAVED AREAS

Item	Gang hours	Labour £	Plant £	Material £	Unit	**Total rate £**
KERBS, CHANNELS, EDGINGS, COMBINED DRAINAGE AND KERB BLOCKS AND LINEAR DRAINAGE CHANNEL SYSTEMS						
Foundations to kerbs etc.						
Mass concrete						
200 x 100 mm	0.01	0.51	0.06	1.51	m	**2.08**
300 x 150 mm	0.02	0.77	0.10	3.46	m	**4.33**
450 x 150 mm	0.02	1.02	0.13	5.12	m	**6.27**
100 x 100 mm haunching, per side	0.01	0.26	0.03	0.37	m	**0.65**
Precast concrete units; BS 7263; bedded jointed and pointed in cement mortar						
Kerbs; bullnosed, splayed or half battered; laid straight or curved exceeding 12 m radius						
125 x 150 mm	0.06	3.06	0.39	3.51	m	**6.96**
125 x 255 mm	0.07	3.57	0.46	4.33	m	**8.36**
150 x 305 mm	0.07	3.57	0.46	7.08	m	**11.11**
Kerbs; bullnosed, splayed or half battered; laid to curves not exceeding 12 m radius						
125 x 150 mm	0.07	3.32	0.42	3.51	m	**7.25**
125 x 255 mm	0.08	3.83	0.49	4.33	m	**8.64**
150 x 305 mm	0.08	3.83	0.49	7.08	m	**11.40**
Quadrants (normally included in general rate for kerbs; shown separately for estimating purposes)						
305 x 305 x 150 mm	0.08	4.09	0.52	7.89	nr	**12.50**
455 x 455 x 255 mm	0.10	5.11	0.65	9.39	nr	**15.15**
Drop kerbs (normally included in general rate for kerbs; shown separately for estimating purposes)						
125 x 255 mm	0.07	3.57	0.46	5.87	nr	**9.90**
150 x 305 mm	0.07	3.57	0.46	12.29	nr	**16.32**
Channels; laid straight or curved exceeding 12 m radius						
125 x 255 mm	0.07	3.57	0.46	5.74	m	**9.77**
Channels; laid to curves not exceeding 12 m radius						
255 x 125 mm	0.07	3.57	0.46	5.74	m	**9.77**
Edgings; laid straight or curved exceeding 12 m radius						
150 x 50 mm	0.04	2.04	0.26	2.31	m	**4.61**
Edgings; laid to curves not exceeding 12 m radius						
150 x 50 mm	0.05	2.30	0.29	3.09	m	**5.68**
Precast concrete drainage channels; Charcon 'Safeticurb'; channels jointed with plastic rings and bedded, jointed and pointed in cement mortar						
Channel unit; Type DBA/3; laid straight or curved exceeding 12 m radius						
250 x 250 mm; medium duty	0.08	3.83	0.49	27.96	m	**32.27**
305 x 305 mm; heavy duty	0.10	4.85	0.62	62.22	m	**67.69**

SERIES 1100: KERBS, FOOTWAYS & PAVED AREAS

Item	Gang hours	Labour £	Plant £	Material £	Unit	Total rate £
KERBS, CHANNELS, EDGINGS, COMBINED DRAINAGE AND KERB BLOCKS AND LINEAR DRAINAGE CHANNEL SYSTEMS – cont'd						
Precast concrete 'Ellis Trief' safety kerb; bedded jointed and pointed in cement mortar						
Kerbs; laid straight or curved exceeding 12 m radius						
415 x 380 mm	0.23	11.49	1.46	57.65	m	**70.60**
Kerbs; laid to curves not exceeding 12 m radius						
415 x 380 mm	0.25	12.77	1.62	57.65	m	**72.04**
Precast concrete combined kerb and drainage block 'Beany Block System'; bedded jointed and pointed in cement mortar						
Kerb; top block, shallow base unit, standard cover plate and frame						
laid straight or curved exceeding 12 m radius	0.15	7.66	0.97	105.55	m	**114.18**
laid to curves not exceeding 12 m radius	0.20	10.21	1.30	147.51	m	**159.02**
Kerb; top block, standard base unit, standard cover plate and frame						
laid straight or curved exceeding 12 m radius	0.15	7.66	0.97	111.32	m	**119.96**
laid to curves not exceeding 12 m radius	0.20	10.21	1.30	155.56	m	**167.07**
Kerb; top block, deep base unit, standard cover plate and frame						
Straight or curved over 12 m radius	0.15	7.66	0.97	138.97	m	**147.61**
laid to curves not exceeding 12 m radius	0.20	10.21	1.30	194.27	m	**205.78**
Base block depth tapers	0.10	5.11	0.65	58.32	m	**64.08**
Extruded asphalt edgings to pavings; slip formed BS 5931						
Kerb; laid straight or curved exceeding 12 m radius						
75 mm kerb height	-	-	-	5.74	m	**5.74**
100 mm kerb height	-	-	-	10.02	m	**10.02**
125 mm kerb height	-	-	-	11.86	m	**11.86**
Channel; laid straight or curved exceeding 12 m radius						
300 mm channel width	-	-	-	12.35	m	**12.35**
250 mm channel width	-	-	-	12.35	m	**12.35**
Kerb; laid to curves not exceeding 12 m radius						
75 mm kerb height	-	-	-	11.48	m	**11.48**
100 mm kerb height	-	-	-	10.02	m	**10.02**
125 mm kerb height	-	-	-	10.48	m	**10.48**
Channel; laid to curves not exceeding 12 m radius						
300 mm channel width	-	-	-	16.77	m	**16.77**
250 mm channel width	-	-	-	12.78	m	**12.78**
Extruded concrete; slip formed						
Kerb; laid straight or curved exceeding 12 m radius						
100 mm kerb height	-	-	-	10.26	m	**10.26**
125 mm kerb height	-	-	-	11.86	m	**11.86**
Kerb; laid to curves not exceeding 12 m radius						
100 mm kerb height	-	-	-	10.02	m	**10.02**
125 mm kerb height	-	-	-	10.48	m	**10.48**

SERIES 1100: KERBS, FOOTWAYS & PAVED AREAS

Item	Gang hours	Labour £	Plant £	Material £	Unit	**Total rate £**
ADDITIONAL CONCRETE FOR KERBS, CHANNELS, EDGINGS, COMBINED DRAINAGE AND KERB BLOCKS AND LINEAR DRAINAGE CHANNEL SYSTEMS						
Additional in situ concrete concrete						
kerbs	0.50	25.54	2.92	75.30	m³	**103.75**
channels	0.50	25.54	2.92	75.30	m³	**103.75**
edgings	0.50	25.54	2.92	75.30	m³	**103.75**
REMOVE FROM STORE AND RELAY KERBS, CHANNELS, EDGINGS, COMBINED DRAINAGE AND KERB BLOCKS AND LINEAR DRAINAGE CHANNEL SYSTEMS						
Remove from store and relay precast concrete units; bedded jointed and pointed in cement mortar						
Kerbs; laid straight or curved exceeding 12 m radius						
125 x 150 mm	0.06	3.06	0.39	1.28	m	**4.73**
125 x 255 mm	0.07	3.57	0.46	1.44	m	**5.47**
150 x 305 mm	0.07	3.57	0.46	1.99	m	**6.02**
Kerbs; laid to curves not exceeding 12 m radius						
125 x 150 mm	0.07	3.32	0.42	1.28	m	**5.02**
125 x 255 mm	0.08	3.83	0.49	1.44	m	**5.76**
150 x 305 mm	0.08	3.83	0.49	1.99	m	**6.31**
Quadrants (normally included in general rate for kerbs; shown separately for estimating purposes)						
305 x 305 x 150 mm	0.08	4.09	0.52	2.15	nr	**6.76**
455 x 455 x 255 mm	0.10	5.11	0.65	2.45	nr	**8.21**
Drop kerbs (normally included in general rate for kerbs; shown separately for estimating purposes)						
125 x 255 mm	0.07	3.57	0.46	1.75	nr	**5.78**
150 x 305 mm	0.07	3.57	0.46	3.03	nr	**7.06**
Channels; laid straight or curved exceeding 12 m radius						
125 x 255 mm	0.07	3.57	0.46	1.72	m	**5.75**
Channels; laid to curves not exceeding 12 m radius						
255 x 125 mm	0.07	3.57	0.46	1.72	m	**5.75**
Edgings; laid straight or curved exceeding 12 m radius						
150 x 50 mm	0.04	2.04	0.26	1.04	m	**3.34**
Edgings; laid to curves not exceeding 12 m radius						
150 x 50 mm	0.05	2.30	0.29	1.04	m	**3.62**
Remove from store and relay precast concrete drainage channels;Charcon 'Safeticurb'; channels jointed with plastic rings and bedded, jointed and pointed in cement mortar						
Channel unit; Type DBA/3; laid straight or curved exceeding 12 m radius						
250 x 254 mm; medium duty	0.08	3.83	0.49	6.17	m	**10.48**
305 x 305 mm; heavy duty	0.10	4.85	0.62	13.02	m	**18.49**

SERIES 1100: KERBS, FOOTWAYS & PAVED AREAS

Item	Gang hours	Labour £	Plant £	Material £	Unit	Total rate £
REMOVE FROM STORE AND RELAY KERBS, CHANNELS, EDGINGS, COMBINED DRAINAGE AND KERB BLOCKS AND LINEAR DRAINAGE CHANNEL SYSTEMS – cont'd						
Remove from store and relay precast concrete 'Ellis Trief' safety kerb; bedded jointed and pointed in cement mortar						
Kerbs; laid straight or curved exceeding 12 m radius						
415 x 380 mm	0.23	11.49	1.46	12.11	m	**25.06**
Kerbs; laid to curves not exceeding 12 m radius						
415 x 380 mm	0.25	12.77	1.62	12.11	m	**26.50**
Remove from store and relay precast concrete combined kerb and drainage block 'Beany Block System'; bedded jointed and pointed in cement mortar						
Kerb; top block, shallow base unit, standard cover plate and frame						
laid straight or curved exceeding 12 m radius	0.15	7.66	0.97	21.68	m	**30.32**
laid to curves not exceeding 12 m radius	0.20	10.21	1.30	30.08	m	**41.59**
Kerb; top block, standard base unit, standard cover plate and frame						
laid straight or curved exceeding 12 m radius	0.15	7.66	0.97	22.84	m	**31.47**
laid to curves not exceeding 12 m radius	0.20	10.21	1.30	31.69	m	**43.20**
Kerb; top block, deep base unit, standard cover plate and frame						
Straight or curved over 12 m radius	0.15	7.66	0.97	28.37	m	**37.00**
laid to curves not exceeding 12 m radius	0.20	10.21	1.30	39.43	m	**50.94**
Base block depth tapers	0.10	5.11	0.65	12.24	m	**17.99**
FOOTWAYS AND PAVED AREAS						
Sub-bases						
To paved area; sloping not exceeding 10 degrees to the horizontal						
100 mm thick sand	0.01	0.44	0.21	3.41	m²	**4.07**
150 mm thick sand	0.01	0.59	0.28	5.12	m²	**5.99**
100 mm thick gravel	0.01	0.44	0.21	2.58	m²	**3.23**
150 mm thick gravel	0.01	0.59	0.28	3.86	m²	**4.74**
100 mm thick hardcore	0.01	0.44	0.21	0.91	m²	**1.57**
150 mm thick hardcore	0.01	0.59	0.28	1.37	m²	**2.24**
100 mm thick concrete grade 20/20	0.02	1.03	0.49	7.54	m²	**9.06**
150 mm thick concrete grade 20/20	0.03	1.57	0.75	11.30	m²	**13.63**
Bitumen macadam surfacing; BS 4987; base course of 20 mm open graded aggregate to clause 2.6.1 tables 5 - 7; wearing course of 6 mm medium graded aggregate to clause 2.7.6 tables 32 - 33; excluding sub-base						
Paved area 60 mm thick; comprising base course 40 mm thick wearing course 20 mm thick						
sloping at 10 degrees or less to the horizontal	0.09	4.18	1.99	9.96	m²	**16.13**
sloping at more than 10 degrees to the horizontal	0.10	4.68	2.22	9.96	m²	**16.86**

SERIES 1100: KERBS, FOOTWAYS & PAVED AREAS

Item	Gang hours	Labour £	Plant £	Material £	Unit	Total rate £
Bitumen macadam surfacing; red additives; BS 4987; base course of 20 mm open graded aggregate to clause 2.6.1 tables 5 - 7; wearing course of 6 mm medium graded aggregate to clause 2.7.6 tables 32 - 33; excluding sub-base						
Paved area 60 mm thick; comprising base course 40 mm thick wearing course 20 mm thick						
sloping at 10 degrees or less to the horizontal	0.09	4.18	1.99	11.85	m²	**18.02**
sloping at more than 10 degrees to the horizontal	0.10	4.68	2.22	11.85	m²	**18.75**
Bitumen macadam surfacing; green additives; BS 4987; base course of 20 mm open graded aggregate to clause 2.6.1 tables 5 - 7; wearing course of 6 mm medium graded aggregate to clause 2.7.6 tables 32 - 33; excluding sub-base						
Paved area 60 mm thick; comprising base course 40 mm thick wearing course 20 mm thick						
sloping at 10 degrees or less to the horizontal	0.09	4.18	1.99	13.14	m²	**19.31**
sloping at more than 10 degrees to the horizontal	0.10	4.68	2.22	13.14	m²	**20.04**
Granular base surfacing; Central Reserve Treatments Limestone, graded 10 mm down laid and compacted; excluding sub-base						
Paved area 100 mm thick; surface sprayed with two coats of cold bituminous emulsion; blinded with 6 mm quarzite fine gravel						
sloping not exceeding 10 degrees to the horizontal	0.02	0.98	0.47	5.47	m²	**6.92**
Breedon plc Golden gravel; graded 13mm to fines; rolled wet						
Paved area 50 mm thick; single layer						
sloping not exceeding 10 degrees to the horizontal	0.03	1.48	0.70	7.28	m²	**9.45**
Precast concrete slabs; BS 7263; grey; 5 point bedding and pointing joints in cement mortar; excluding sub-base						
Paved area 50 mm thick; comprising 600 x 450 x 50 mm units						
sloping at 10 degrees or less to the horizontal	0.28	6.53	0.34	11.36	m²	**18.23**
Paved area 50 mm thick; comprising 600 x 600 x 50 mm units						
sloping at 10 degrees or less to the horizontal	0.24	5.59	0.29	7.82	m²	**13.70**
Paved area 50 mm thick; comprising 900 x 600 x 50 mm units						
sloping at 10 degrees or less to the horizontal	0.20	4.66	0.24	6.73	m²	**11.64**
Extra for coloured, 50 mm thick	-	-	-	3.31	m²	**3.31**
Paved area 63 mm thick; comprising 600 x 600 x 63 mm units						
sloping at 10 degrees or less to the horizontal	0.25	5.83	0.30	9.31	m²	**15.44**
Paved area 63 mm thick; comprising 900 x 600 x 63 mm units						
sloping at 10 degrees or less to the horizontal	0.21	4.90	0.25	7.62	m²	**12.77**

SERIES 1100: KERBS, FOOTWAYS & PAVED AREAS

Item	Gang hours	Labour £	Plant £	Material £	Unit	Total rate £
FOOTWAYS AND PAVED AREAS – cont'd						
Precast concrete rectangular paving blocks; BS 6717; grey; bedding on 50 mm thick dry sharp sand; filling joints; excluding sub-base						
Paved area 80 mm thick; comprising 200 x 100 x 80 mm units						
sloping at 10 degrees or less to the horizontal	0.30	6.99	0.36	12.78	m²	**20.13**
Precast concrete rectangular paving blocks; BS 6717; coloured; bedding on 50 mm thick dry sharp sand; filling joints; excluding sub-base						
Paved area 80 mm thick; comprising 200 x 100 x 80 mm units						
sloping at 10 degrees or less to the horizontal	0.30	6.99	0.36	13.50	m²	**20.85**
Brick paviors delivered to site; bedding on 20 mm thick mortar; excluding sub-base						
Paved area 85 mm thick; comprising 215 x 103 x 65 mm units						
sloping at 10 degrees or less to the horizontal	0.30	6.99	0.36	21.40	m²	**28.76**
Granite setts (2.88 kg/mm thickness/m²); bedding on 25 mm cement mortar; excluding sub-base						
Paved area 100 mm thick; comprising 100 x 100 x 100 mm units; laid to random pattern						
sloping at 10 degrees or less to the horizontal	0.90	20.98	1.09	45.19	m²	**67.26**
Paved area 100 mm thick; comprising 100 x 100 x 100 mm units; laid to specific pattern						
sloping at 10 degrees or less to the horizontal	1.20	27.97	1.46	45.19	m²	**74.62**
Cobble paving; 50 - 75 mm stones; bedding on 25 mm cement mortar; filling joints; excluding sub-base						
Paved area; comprising 50 - 75 mm stones; laid to random pattern						
sloping at 10 degrees or less to the horizontal	1.00	23.31	1.20	13.29	m²	**37.80**

SERIES 1100: KERBS, FOOTWAYS & PAVED AREAS

Item	Gang hours	Labour £	Plant £	Material £	Unit	Total rate £
REMOVE FROM STORE AND RELAY PAVING FLAGS, SLABS AND BLOCKS						
Note						
An allowance of 20 % of the cost of providing new pavings has been included in the rates below to allow for units which have to be replaced through unacceptable damage.						
Remove from store and relay precast concrete units; bedded jointed and pointed in cement mortar. Remove from store and relay precast concrete slabs; 5 point bedding and pointing joints in cement mortar; excluding sub-base						
Paved area 50 mm thick; comprising 600 x 450 x 50 mm units						
sloping at 10 degrees or less to the horizontal	0.28	6.53	0.34	2.60	m²	**9.47**
Paved area 50 mm thick; comprising 600 x 600 x 50 mm units						
sloping at 10 degrees or less to the horizontal	0.24	5.59	0.29	1.89	m²	**7.78**
Paved area 50 mm thick; comprising 900 x 600 x 50 mm units						
sloping at 10 degrees or less to the horizontal	0.20	4.66	0.24	1.67	m²	**6.58**
Paved area 63 mm thick; comprising 600 x 600 x 63 mm units						
sloping at 10 degrees or less to the horizontal	0.25	5.83	0.30	2.19	m²	**8.32**
Paved area 63 mm thick; comprising 900 x 600 x 63 mm units						
sloping at 10 degrees or less to the horizontal	0.21	4.90	0.25	1.85	m²	**7.00**
Remove from store and relay precast concrete rectangular paving blocks; bedding on 50 mm thick dry sharp sand; filling joints; excluding sub-base						
Paved area 80 mm thick; comprising 200 x 100 x 80 mm units						
sloping at 10 degrees or less to the horizontal	0.30	6.99	0.36	3.36	m²	**10.72**
Remove from store and relay brick paviors; bedding on 20 mm thick mortar; excluding sub-base						
Paved area 85 mm thick; comprising 215 x 103 x 65 mm units						
sloping at 10 degrees or less to the horizontal	0.30	6.99	0.36	5.22	m²	**12.58**

SERIES 1200: TRAFFIC SIGNS AND MARKINGS

Item	Gang hours	Labour £	Plant £	Material £	Unit	Total rate £
NOTES						
The re-erection cost for traffic signs taken from store assumes that major components are in good condition; the prices below allow a sum of 20 % of the value of new materials to cover minor repairs, new fixings and touching up any coatings.						
RESOURCES - LABOUR						
Traffic signs gang						
1 ganger/chargehand (skill rate 3)		14.21				
1 skilled operative (skill rate 3)		13.46				
2 unskilled operatives (general)		22.52				
1 plant operator (skill rate 3) - 25% of time		3.66				
Total Gang Rate / Hour	£	**53.85**				
Bollards, furniture gang						
1 ganger/chargehand (skill rate 4)		12.80				
1 skilled operative (skill rate 4)		12.05				
2 unskilled operatives (general)		22.52				
Total Gang Rate / Hour	£	**47.37**				
RESOURCES - PLANT						
Traffic signs						
JCB 3CX backhoe - 50% of time			7.81			
125 cfm compressor - 50% of time			2.42			
compressor tools: hand held hammer drill - 50% of time			0.44			
compressor tools: clay spade - 50% of time			0.20			
compressor tools: extra 15 m hose - 50% of time			0.13			
8 t lorry with hiab lift - 50% of time			10.09			
Total Rate / Hour		£	**21.09**			
Bollards, furniture						
125 cfm compressor - 50% of time			2.42			
compressor tools: hand held hammer drill - 50% of time			0.44			
compressor tools: clay spade - 50% of time			0.20			
compressor tools: extra 15 m hose - 50% of time			0.13			
8 t lorry with hiab lift - 25% of time			5.05			
Total Rate / Hour		£	**8.24**			
TRAFFIC SIGNS						
In this section prices will vary depending upon the diagram configurations. The following are average costs of signs and Bollards. Diagram numbers refer to the Traffic Signs Regulations and General Directions 1989 and the figure numbers refer to the Traffic Signs Manual.						

SERIES 1200: TRAFFIC SIGNS AND MARKINGS

Item	Gang hours	Labour £	Plant £	Material £	Unit	Total rate £
Examples of Prime Costs for Class 1 (High Intensity) traffic and road signs (ex works) for orders exceeding £1,000.						
Permanent traffic sign as non-Lit unit on						
600 x 450 mm	-	-	-	78.53	nr	**78.53**
600 mm diameter	-	-	-	98.62	nr	**98.62**
600 mm triangular	-	-	-	82.27	nr	**82.27**
500 x 500 mm	-	-	-	69.34	nr	**69.34**
450 x 450 mm	-	-	-	59.21	nr	**59.21**
450 x 300 mm	-	-	-	49.00	nr	**49.00**
1200 x 400 mm (CHEVRONS)	-	-	-	125.95	nr	**125.95**
Examples of Prime Costs for Class 2 (Engineering Grade) traffic and road signs (ex works) for orders exceeding £600.						
600 x 450 mm	-	-	-	63.61	nr	**63.61**
600 mm diameter	-	-	-	107.39	nr	**107.39**
600 mm triangular	-	-	-	89.59	nr	**89.59**
500 x 500 mm	-	-	-	56.25	nr	**56.25**
450 x 450 mm	-	-	-	47.73	nr	**47.73**
450 x 300 mm	-	-	-	41.12	nr	**41.12**
1200 x 400 mm (CHEVRONS)	-	-	-	137.16	nr	**137.16**
Standard reflectorised traffic signs						
Note: Unit costs do not include concrete foundations (see Series 1700)						
Standard one post signs; 600 x 450 mm type C1 signs						
fixed back to back to another sign (measured separately) with aluminium clips to existing post (measured separately)	0.04	2.15	0.84	82.46	nr	**85.45**
Extra for fixing singly with aluminium clips	0.01	0.54	0.15	1.53	nr	**2.22**
Extra for fixing singly with stainless steel clips	0.01	0.54	0.27	10.75	nr	**11.56**
fixed back to back to another sign (measured separately) with stainless steel clips to one new 76 mm diameter plastic coated steel posts 1.75 m long	0.27	14.54	5.70	121.17	nr	**141.41**
Extra for fixing singly to one face only	0.01	0.54	0.15	-	nr	**0.69**
Extra for 76 mm diameter 1.75 m long aluminium post	0.02	1.08	0.33	11.44	nr	**12.85**
Extra for 76 mm diameter 3.5 m long plastic coated steel post	0.02	1.08	0.33	25.85	nr	**27.25**
Extra for 76 mm diameter 3.5 m long aluminium post	0.02	1.08	0.33	48.99	nr	**50.39**
Extra for excavation for post, in hard material	1.10	59.23	17.40	-	nr	**76.64**
Extra for single external illumination unit with fitted photo cell (excluding trenching and cabling - see Series 1400); unit cost per face illuminated	0.33	17.77	5.21	62.26	nr	**85.25**
Standard two post signs; 1200 x 400 mm, signs						
fixed back to back to another sign (measured separately) with stainless steel clips to two new 76 mm diameter plastic coated steel posts 1.75 m long	0.51	27.46	10.76	206.07	nr	**244.29**
Extra for fixing singly to one face only	0.02	1.08	0.33	-	nr	**1.40**
Extra for two 76 mm diameter 1.75 m long aluminium posts	0.04	2.15	0.63	22.89	nr	**25.68**

SERIES 1200: TRAFFIC SIGNS AND MARKINGS

Item	Gang hours	Labour £	Plant £	Material £	Unit	Total rate £
TRAFFIC SIGNS – cont'd						
Standard reflectorised traffic signs – cont'd						
Extra for two 76 mm diameter 3.5 m long plastic coated steel posts	0.04	2.15	0.63	51.70	nr	**54.49**
Extra for two 76 mm diameter 3.5 m long aluminium post	0.04	2.15	0.63	97.97	nr	**100.76**
Extra for excavation for post, in hard material	1.10	59.23	17.40	-	nr	**76.64**
Extra for single external illumination unit with fitted photo cell (excluding trenching and cabling - see Series 1400); unit cost per face illuminated	0.58	31.23	9.18	87.21	nr	**127.62**
Standard internally illuminated traffic signs						
Bollard with integral mould-in translucent graphics (excluding trenching and cabling)						
fixing to concrete base	0.48	25.85	10.12	146.00	nr	**181.97**
Special traffic signs						
Note: Unit costs do not include concrete foundations (see Series 1700) or trenching and cabling (see Series 1400)						
Externally illuminated reflectorised traffic signs manufactured to order						
special signs, surface area 1.50 m² on two 100 mm diameter steel posts	-	-	-	-	nr	**555.50**
special signs, surface area 4.00 m² on three 100 mm diameter steel posts	-	-	-	-	nr	**914.51**
Internally illuminated traffic signs manufactured to order						
special signs, surface area 0.25 m² on one new 76 mm diameter steel post	-	-	-	-	nr	**199.94**
special signs, surface area 0.75 m² on one new 100 mm diameter steel post	-	-	-	-	nr	**280.19**
special signs, surface area 4.00 m² on four new 120 mm diameter steel posts	-	-	-	-	nr	**713.18**
Signs on gantries						
Externally illuminated reflectorised signs						
1.50 m²	1.78	96.12	48.17	186.88	nr	**331.18**
2.50 m²	2.15	115.78	58.03	193.70	nr	**367.51**
3.00 m²	3.07	165.32	82.86	204.62	nr	**452.79**
Internally illuminated sign with translucent optical reflective sheeting and remote light source						
0.75 m²	1.56	84.01	42.10	1282.60	nr	**1408.70**
1.00 m²	1.70	91.55	45.88	1710.12	nr	**1847.54**
1.50 m²	2.41	129.78	65.04	2565.19	nr	**2760.01**
Existing signs						
Take from store and re-erect						
3.0 m high road sign	0.28	15.08	7.56	60.10	nr	**82.73**
road sign on two posts	0.50	26.93	13.49	120.20	nr	**160.62**

SERIES 1200: TRAFFIC SIGNS AND MARKINGS

Item	Gang hours	Labour £	Plant £	Material £	Unit	Total rate £
ROAD MARKINGS						
Thermoplastic screed or spray						
Note: Unit costs based upon new road with clean surface closed to traffic)						
Continuous line in reflectorised white						
150 mm wide	-	-	-	-	m	**0.86**
200 mm wide	-	-	-	-	m	**1.16**
Continuous line in reflectorised yellow						
100 mm wide	-	-	-	-	m	**0.58**
150 mm wide	-	-	-	-	m	**0.58**
Intermittent line in reflectorised white						
60 mm wide with 0.60 m line and 0.60 m gap	-	-	-	-	m	**0.69**
100 mm wide with 1.0 m line and 5.0 m gap	-	-	-	-	m	**0.69**
100 mm wide with 2.0 m line and 7.0 m gap	-	-	-	-	m	**0.69**
100 mm wide with 4.0 m line and 2.0 m gap	-	-	-	-	m	**0.69**
100 mm wide with 6.0 m line and 3.0 m gap	-	-	-	-	m	**0.69**
150 mm wide with 1.0 m line and 5.0 m gap	-	-	-	-	m	**1.04**
150 mm wide with 6.0 m line and 3.0 m gap	-	-	-	-	m	**1.04**
150 mm wide with 0.60 m line and 0.30 m gap	-	-	-	-	m	**1.04**
200 mm wide with 0.60 m line and 0.30 m gap	-	-	-	-	m	**1.39**
200 mm wide with 1.0 m line and 1.0 m gap	-	-	-	-	m	**1.39**
Ancillary line in reflectorised white						
150 mm wide in hatched areas	-	-	-	-	m	**0.86**
200 mm wide in hatched areas	-	-	-	-	m	**1.39**
Ancillary line in yellow						
150 mm wide in hatched areas	-	-	-	-	m	**0.86**
Triangles in reflectorised white						
1.6 m high	-	-	-	-	nr	**8.06**
2.0 m high	-	-	-	-	nr	**10.95**
3.75 m high	-	-	-	-	nr	**14.41**
Circles with enclosing arrows in reflectorised						
1.6 m diameter	-	-	-	-	nr	**57.59**
Arrows in reflectorised white						
4.0 m long straight or turning	-	-	-	-	nr	**23.04**
6.0 m long straight or turning	-	-	-	-	nr	**28.79**
6.0 m long curved	-	-	-	-	nr	**28.79**
6.0 m long double headed	-	-	-	-	nr	**40.31**
8.0 m long double headed	-	-	-	-	nr	**57.59**
16.0 m long double headed	-	-	-	-	nr	**86.38**
32.0 m long double headed	-	-	-	-	nr	**115.17**
Kerb markings in yellow						
250 mm long	-	-	-	-	nr	**0.58**
Letters or numerals in reflectorised white						
1.6 m high	-	-	-	-	nr	**7.50**
2.0 m high	-	-	-	-	nr	**10.95**
3.75 m high	-	-	-	-	nr	**19.02**
Verynyl strip markings						
Note: Unit costs based upon new road with clean surface closed to traffic)						
'Verynyl' strip markings (pedestrian crossings and similar locations)						
200 mm wide line	-	-	-	-	m	**7.55**
600 x 300 mm single stud tile	-	-	-	-	nr	**12.89**

SERIES 1200: TRAFFIC SIGNS AND MARKINGS

Item	Gang hours	Labour £	Plant £	Material £	Unit	Total rate £
ROAD STUDS						
Reflecting Road Studs						
100 x 100 mm square bi-directional reflecting road stud with amber corner cube reflectors	-	-	-	-	nr	**5.60**
140 x 254 mm rectangular one way reflecting road stud with red catseye reflectors	-	-	-	-	nr	**12.86**
140 x 254 mm rectangular one way reflecting road stud with green catseye reflectors	-	-	-	-	nr	**12.86**
140 x 254 mm rectangular bi-directional reflecting road stud with white catseye reflectors	-	-	-	-	nr	**12.86**
140 x 254 mm rectangular bi-directional reflecting road stud with amber catseye reflectors	-	-	-	-	nr	**12.86**
140 x 254 mm rectangular bi-directional reflecting road stud without catseye reflectors	-	-	-	-	nr	**8.39**
REMOVE FROM STORE AND RE-INSTALL ROAD STUDS						
Remove from store and re-install 100 x 100 mm square bi-directional reflecting road stud with corner cube reflectors	-	-	-	-	nr	**2.79**
Remove from store and re-install 140 x 254 mm rectangular one way reflecting road stud with catseye reflectors	-	-	-	-	nr	**6.71**
TRAFFIC SIGNAL INSTALLATIONS						
Traffic signal installation is carried out exclusively by specialist contractors, although certain items are dealt with by the main contractor or a sub-contractor.						
The following detailed prices are given to assist in the calculation of the total installation cost.						
Installation of signal pedestals, loop detector unit pedestals, controller unit boxes and cable connection pillars						
signal pedestal	-	-	-	-	nr	**30.18**
loop detector unit pedestal	-	-	-	-	nr	**18.69**
controller unit box	-	-	-	-	nr	**46.00**
Excavate trench for traffic signal cable, depth ne 1.50 m; supports, backfilling						
450 mm wide	-	-	-	-	m	**5.75**
Extra for excavating in hard material	-	-	-	-	m³	**31.63**
Saw cutting grooves in pavement for detector loops and feeder cables; seal with hot bitumen sealant after installation						
25 mm deep	-	-	-	-	m	**5.02**

SERIES 1200: TRAFFIC SIGNS AND MARKINGS

Item	Gang hours	Labour £	Plant £	Material £	Unit	**Total rate £**
MARKER POSTS						
Glass reinforced plastic marker posts						
types 1,2,3 or 4	-	-	-	-	nr	**12.29**
types 5,6,7 or 8	-	-	-	-	nr	**11.65**
Line posts for emergency crossing	-	-	-	-	nr	**6.78**
Standard reflectorised traffic cylinder 1000 mm high 125 mm diameter; mounted in cats eye base (deliniator)	-	-	-	-	nr	**21.15**
PERMANENT BOLLARDS						
Permanent bollard; non-illuminated; precast concrete						
150 mm minimum diameter 750 mm high	0.80	37.90	6.54	41.04	nr	**85.47**
300 mm minimum diameter 750 mm high	0.80	37.90	6.54	71.92	nr	**116.36**
Extra for exposed aggregate finish	-	-	-	11.03	nr	**11.03**
Permanent bollard; non-illuminated; galvanised steel						
removable and lockable pattern	0.80	37.90	9.66	85.84	nr	**133.40**
MISCELLANEOUS FURNITURE						
Galvanised steel lifting traffic barrier						
4.0 m wide	2.40	113.69	19.61	946.45	nr	**1079.75**
Precast concrete seats						
bench seat 2.0 m long	0.75	35.53	6.13	167.08	nr	**208.74**
bench seat with concrete ends and timber slats 2.0 m long	0.75	35.53	6.13	196.78	nr	**238.44**
Timber seat fixed to concrete base						
bench seat 2.0 m long	0.45	21.32	3.45	263.52	nr	**288.29**
Metal seat						
bench seat 2.0 m long	0.75	35.53	6.13	281.06	nr	**322.73**

SERIES 1300: ROAD LIGHTING COLUMNS, BRACKETS AND CCTV MASTS

Item	Gang hours	Labour £	Plant £	Material £	Unit	Total rate £
NOTES						
For convenience in pricing this section departs from Series 1300 requirements and shows lighting column costs broken down into main components of columns, brackets, and lamps and cabling.						
The outputs assume operations are continuous and are based on at least 10 complete units and do not include any allowance for on site remedial works after erection of the columns.						
Painting and protection of the columns apart from galvanising is not included in the following prices or outputs.						
The re-erection cost for lighting columns taken from store assumes that major components are in good condition; the prices below allow a sum of 20 % of the value of new materials to cover minor repairs, new fixings and touching up any coatings.						
RESOURCES - LABOUR						
Column erection gang						
1 ganger/chargehand (skill rate 4)		12.80				
1 skilled operative (skill rate 4)		12.05				
1 unskilled operative (general)		11.26				
1 plant operator (craftsman) - 50% of time		9.37				
Total Gang Rate / Hour	£	**45.48**				
Bracket erection gang						
1 ganger/chargehand (skill rate 4)		12.80				
1 skilled operative (skill rate 4)		12.05				
1 plant operator (skill rate 4)		13.17				
1 plant operator (craftsman)		18.73				
Total Gang Rate / Hour	£	**56.75**				
Lanterns gang						
1 skilled operative (skill rate 3)		13.46				
1 skilled operative (skill rate 4)		12.05				
1 plant operator (skill rate 4)		13.17				
1 plant operator (craftsman)		18.73				
Total Gang Rate / Hour	£	**57.41**				
RESOURCES - PLANT						
Columns and bracket arms						
15 t mobile crane - 50% of time			14.10			
125 cfm compressor - 50% of time			2.42			
compressor tools: 2 single head scabbler - 50% of time			1.78			
2 t dumper - 50% of time			1.82			
Total Rate / Hour	£		**20.12**			
Bracket arms						
15 t mobile crane			28.20			
access platform, Simon hoist (50 ft)			30.20			
Total rate / Hour	£		**58.40**			
Lanterns						
15 t mobile crane			28.20			
access platform, Simon hoist (50 ft)			30.20			
Total Rate / Hour		£	**58.40**			

SERIES 1300: ROAD LIGHTING COLUMNS, BRACKETS AND CCTV MASTS

Item	Gang hours	Labour £	Plant £	Material £	Unit	Total rate £
ROAD LIGHTING COLUMNS, BRACKETS, WALL MOUNTINGS AND CCTV MASTS						
Galvanised steel road lighting columns to BS 5649 with flange plate base (including all control gear, switching, fuses and internal wiring)						
4.0 m nominal height	0.75	34.11	15.09	168.23	nr	**217.43**
6.0 m nominal height	0.80	36.38	22.34	306.69	nr	**365.41**
8.0 m nominal height	0.96	43.66	26.80	384.14	nr	**454.60**
10.0 m nominal height	1.28	58.21	35.74	455.58	nr	**549.53**
12.0 m nominal height	1.44	65.49	40.20	641.69	nr	**747.39**
15.0 m nominal height	1.76	80.04	49.14	943.96	nr	**1073.14**
3.0 m cast iron column (pedestrian / landscape area)	0.75	34.11	20.95	266.33	nr	**321.39**
Precast concrete lighting columns to BS 1308 with flange plate base (including all control gear, switching, fuses and internal wiring)						
5.0 m nominal height	0.75	34.11	20.95	207.39	nr	**262.45**
10.0 m nominal height	1.28	58.21	35.77	500.80	nr	**594.79**
Galvanised steel bracket arm to BS 1840 and 5649; with 5 degrees uplift						
0.5 m projection, single arm	0.16	9.08	17.61	34.31	nr	**61.00**
1.0 m projection, single arm	0.19	10.78	11.10	73.35	nr	**95.23**
1.5 m projection, single arm	0.21	9.55	12.19	80.41	nr	**102.14**
2.0 m projection, single arm	0.27	15.32	15.77	97.31	nr	**128.41**
1.0 m projection, double arm	0.29	16.46	16.94	121.69	nr	**155.09**
2.0 m projection, double arm	0.32	18.16	18.69	146.75	nr	**183.60**
Precast concrete bracket arm to BS 1840 and 5649; with 5 degrees uplift						
1.0 m projection, single arm	0.24	13.62	14.02	141.61	nr	**169.25**
2.0 m projection, single arm	0.32	18.16	18.69	169.91	nr	**206.76**
1.0 m projection, double arm	0.35	19.86	20.44	164.42	nr	**204.72**
2.0 m projection, doube arm	0.37	21.00	21.61	199.81	nr	**242.42**
Lantern unit with photo-electric control set to switch on at 100 lux; lamps						
55W SON (P226); to suit 4 m and 5 m columns	0.40	22.96	17.72	264.34	nr	**305.02**
70W SON (P236); to suit 5 m and 6 m columns	0.40	22.96	17.72	267.12	nr	**307.80**
250W SON (P426); to suit 8 m, 10 m and 12 m columns	0.50	28.70	22.15	400.67	nr	**451.53**
Sphere 70W SON; to suit 3 m columns (P456)	0.50	28.70	22.15	479.98	nr	**530.83**
400W SON High pressure sodium; to suit 12 m and 15 m columns	0.50	28.70	22.15	556.47	nr	**607.33**

SERIES 1300: ROAD LIGHTING COLUMNS, BRACKETS AND CCTV MASTS

Item	Gang hours	Labour £	Plant £	Material £	Unit	Total rate £
REMOVE FROM STORE AND RE-ERECT ROAD LIGHTING COLUMNS, BRACKETS AND WALL MOUNTINGS						
Re-erection of galvanised steel road lighting columns with flange plate base; including all control gear, switching, fuses and internal wiring						
4.0 m nominal height	0.75	34.11	15.11	17.14	nr	**66.35**
6.0 m nominal height	0.80	36.38	16.12	22.43	nr	**74.93**
8.0 m nominal height	0.96	43.66	19.31	27.70	nr	**90.67**
10.0 m nominal height	1.28	58.21	25.78	32.97	nr	**116.96**
12.0 m nominal height	1.44	65.49	29.00	38.26	nr	**132.74**
15.0 m nominal height	1.76	80.04	35.43	46.16	nr	**161.64**
3.0 m cast iron column (pedestrian / landscape area)	0.75	34.11	15.13	14.50	nr	**63.74**
Re-erection of precast concrete lighting columns with flange plate base; including all control gear, switching, fuses and internal wiring						
5.0 m nominal height	0.75	34.11	15.11	19.79	nr	**69.00**
10.0 m nominal height	1.28	58.21	25.78	32.97	nr	**116.96**
Re-erection of galvanised steel bracket arms						
0.5 m projection, single arm	0.16	9.08	9.34	19.84	nr	**38.27**
1.0 m projection, single arm	0.19	10.78	11.10	19.84	nr	**41.72**
1.5 m projection, single arm	0.21	11.92	12.26	19.84	nr	**44.03**
2.0 m projection, single arm	0.27	15.32	15.77	19.84	nr	**50.94**
1.0 m projection, double arm	0.29	16.46	16.94	19.84	nr	**53.24**
2.0 m projection, double arm	0.32	18.16	18.69	19.84	nr	**56.69**
Re-erection of precast concrete bracket arms						
1.0 m projection, single arm	0.24	13.62	14.02	19.84	nr	**47.48**
2.0 m projection, single arm	0.32	18.16	18.69	19.84	nr	**56.69**
1.0 m projection, double arm	0.35	19.86	20.70	19.84	nr	**60.41**
2.0 m projection, double arm	0.37	21.00	20.44	19.84	nr	**61.28**
Re-installing lantern unit with photo-electric control set to switch on at 100 lux; lamps						
55W SON (P226); to suit 4 m and 5 m columns	0.40	22.96	17.72	-	nr	**40.68**
70W SON (P236); to suit 5 m and 6 m columns	0.22	12.63	17.72	-	nr	**30.35**
250W SON (P426); to suit 8 m, 10 m and 12 m columns	0.50	28.70	22.15	-	nr	**50.86**
Sphere 70W SON; to suit 3 m (P456)	0.50	28.70	22.15	-	nr	**50.86**
400W SON High pressure sodium; to suit 12 m and 15 m columns	0.50	28.70	22.15	-	nr	**50.86**

SERIES 1400: ELECTRICAL WORK FOR ROAD LIGHTING AND TRAFFIC SIGNS

Item	Gang hours	Labour £	Plant £	Material £	Unit	Total rate £
RESOURCES - LABOUR						
Trenching gang						
1 ganger/chargehand (skill rate 4)		12.05				
1 skilled operative (skill rate 4)		12.05				
2 unskilled operatives (general)		22.52				
1 plant operator (skill rate 3) - 75% of time		10.99				
Total Gang Rate / Hour	£	**57.61**				
Cable laying gang						
1 ganger/chargehand (skill rate 4)		12.05				
2 skilled operatives (skill rate 3)		26.92				
1 skilled operative (skill rate 4)		12.05				
Total Gang Rate / Hour	£	**51.02**				
RESOURCES - PLANT						
Service trenching						
JCB 3CX backhoe - 50% of time			7.81			
125 cfm compressor - 50% of time			2.42			
compressor tools: 2 single head scabbler - 50% of time			1.78			
2 t dumper - 50% of time			1.82			
trench excavator - 25% of time			14.44			
Total Rate / Hour		£	**28.27**			
Cable laying						
8 t Leyland Road Runner chassis - 50% of time			4.81			
Total Rate / Hour		£	**4.81**			
LOCATING BURIED ROAD LIGHTING AND TRAFFIC SIGNS CABLE						
Locating buried road lighting and traffic signs cable						
in carriageways, footways, bridge decks and paved areas	0.25	14.40	7.06	-	m	**21.47**
in verges and central reserves	0.20	11.52	5.65	-	m	**17.17**
in side slopes of cuttings or side slopes of embankments	0.15	8.64	4.25	-	m	**12.89**
TRENCH FOR CABLE OR DUCT						
Trench for cable						
300 to 450 mm wide; depth not exceeding 1.5 m	0.15	8.64	4.25	-	m	**12.89**
450 to 600 mm wide; depth not exceeding 1.5 m	0.20	11.52	5.65	-	m	**17.17**
Extra for excavating rock or reinforced concrete in trench	0.50	28.80	14.14	-	m3	**42.94**
Extra for excavating brickwork or mass concrete in trench	0.40	23.04	11.30	-	m3	**34.35**
Extra for backfilling with pea gravel	0.02	1.15	0.57	18.31	m3	**20.03**
Extra for 450 x 100 mm sand cable bedding and covering	0.02	1.15	0.57	1.21	m	**2.93**
Extra for PVC marker tape	0.01	0.29	0.15	0.17	m	**0.61**
Extra for 150 x 300 clay cable tiles	0.05	2.88	1.41	4.39	m	**8.68**
Extra for 150 x 900 concrete cable tiles	0.03	1.73	0.85	7.46	m	**10.05**

SERIES 1400: ELECTRICAL WORK FOR ROAD LIGHTING AND TRAFFIC SIGNS

Item	Gang hours	Labour £	Plant £	Material £	Unit	Total rate £
CABLE AND DUCT						
600/1000V 2 core, PVC/SWA/PVC cable with copper conductors						
Cable; in trench not exceeding 1.5 m deep						
2.5 mm²	0.01	0.51	0.05	2.66	m	**3.22**
4 mm²	0.02	1.02	0.10	3.53	m	**4.65**
6 mm²	0.02	1.02	0.10	4.13	m	**5.24**
10 mm²	0.02	1.02	0.10	5.67	m	**6.79**
16 mm²	0.02	1.02	0.10	7.10	m	**8.22**
25 mm²	0.03	1.53	0.14	9.00	m	**10.67**
600/1000V 4 core, PVC/SWA/PVC cable with copper conductors						
Cable; in trench not exceeding 1.5 m deep						
16 mm²	0.03	1.53	0.14	10.22	m	**11.89**
35 mm²	0.13	6.63	0.63	12.28	m	**19.54**
70 mm²	0.15	7.65	0.72	16.63	m	**25.00**
600/1000V 2 core, PVC/SWA/PVC cable drawn into ducts, pipe bays or troughs						
Cable; in trench not exceeding 1.5 m deep						
2.5 mm²	0.01	0.51	0.05	2.54	m	**3.10**
4 mm²	0.02	1.02	0.10	3.39	m	**4.51**
6 mm²	0.02	1.02	0.10	4.01	m	**5.13**
10 mm²	0.02	1.02	0.10	5.54	m	**6.65**
16 mm²	0.03	1.53	0.14	6.97	m	**8.64**
35 mm²	0.12	6.12	0.58	8.88	m	**15.58**
70 mm²	0.14	7.14	0.67	13.46	m	**21.28**
CABLE JOINTS AND CABLE TERMINATIONS						
Straight joint in 2 core PVC/SWA/PVC cable						
2.5 mm²	0.30	15.31	1.44	48.48	nr	**65.23**
4 mm²	0.30	15.31	1.44	48.48	nr	**65.23**
6 mm²	0.32	16.33	1.54	48.48	nr	**66.35**
10 mm²	0.42	21.43	2.02	49.49	nr	**72.94**
16 mm²	0.50	25.51	2.40	51.91	nr	**79.82**
35 mm²	0.80	40.82	3.85	66.02	nr	**110.68**
70 mm²	1.15	58.67	5.53	73.05	nr	**137.25**
Tee joint in 2 core PVC/SWA/PVC cable						
2.5 mm²	0.46	23.47	2.21	72.36	nr	**98.04**
4 mm²	0.46	23.47	2.21	72.36	nr	**98.04**
6 mm²	0.48	24.49	2.31	72.36	nr	**99.15**
10 mm²	0.62	31.63	2.98	74.22	nr	**108.83**
16 mm²	0.74	37.75	3.56	77.48	nr	**118.79**
25 mm²	0.91	46.43	4.38	82.99	nr	**133.79**
Tee joint in 4 core PVC/SWA/PVC cable						
16 mm²	1.10	56.12	5.29	85.22	nr	**146.63**
35 mm²	1.30	66.33	6.25	95.42	nr	**168.00**
70 mm²	1.60	81.63	7.69	137.03	nr	**226.35**

SERIES 1400: ELECTRICAL WORK FOR ROAD LIGHTING AND TRAFFIC SIGNS

Item	Gang hours	Labour £	Plant £	Material £	Unit	**Total rate £**
Looped terminations of 2 core PVC/SWA/PVC cable in lit sign units, traffic signals installation control unit, pedestrian crossing control unit, road lighting column, wall mounting, subway distribution box, gantry distribution box or feeder pillar.						
2.5 mm²	0.15	7.65	0.72	7.10	nr	**15.48**
6 mm²	0.15	7.65	0.72	9.61	nr	**17.99**
10 mm²	0.16	8.16	0.77	12.50	nr	**21.43**
16 mm²	0.25	12.76	1.20	12.75	nr	**26.71**
25 mm²	0.30	15.31	1.44	17.79	nr	**34.54**
Terminations of 4 core PVC/SWA/PVC cable in lit sign units, traffic signals installation control unit, pedestrian crossing control unit, road lighting column, wall mounting, subway distribution box, gantry distribution box or feeder pillar.						
16 mm²	0.35	17.86	1.68	13.79	nr	**33.33**
35 mm²	0.55	28.06	2.64	25.14	nr	**55.85**
70 mm²	0.68	34.69	3.27	47.70	nr	**85.66**
FEEDER PILLARS						
Galvanised steel feeder pillars						
411 x 610 mm	4.64	236.73	22.31	203.30	nr	**462.35**
611 x 810 mm	4.24	216.32	20.39	309.25	nr	**545.96**
811 x 1110	4.96	253.06	23.85	399.49	nr	**676.40**
1111 x 1203 mm	4.19	213.77	20.20	479.29	nr	**713.26**
EARTH ELECTRODES						
Earth electrodes providing minimal protection using earth rods, plates or stops and protective tape and joint						
to suit columns ne 12.0 m	0.40	20.41	-	269.72	nr	**290.13**
to suit columns ne 15.0 m	0.40	20.41	-	357.71	nr	**378.12**
to suit Superstructure or Buildings using copper lead conductor (per 23 m height)	1.00	51.02	-	608.37	nr	**659.39**
CHAMBERS						
Brick chamber with galvanised steel cover and frame; depth to uppermost surface of base slab						
ne 1.0 m deep	-	-	-	-	nr	**586.31**

SERIES 1500: MOTORWAY COMMUNICATIONS

Item	Gang hours	Labour £	Plant £	Material £	Unit	**Total rate £**
RESOURCES - LABOUR						
Service trenching gang						
1 ganger/chargehand (skill rate 4)		12.80				
1 skilled operative (skill rate 4)		12.05				
2 unskilled operatives (general)		22.52				
1 plant operator (skill rate 3) - 75% of time		10.99				
Total Gang Rate / Hour	£	**58.36**				
RESOURCES - PLANT						
Trenching						
JCB 3CX backhoe (50% of time)			7.81			
125 cfm compressor (50% of time)			2.42			
compressor tools: 2 single head scabbler (50% of time)			1.78			
2 t dumper (50% of time)			1.82			
trench excavator (25% of time)			14.44			
Total Gang Rate / Hour	£		**28.27**			
LOCATING BURIED COMMUNICATIONS CABLE						
Locating buried road lighting and traffic signs cable						
in carriageways, footways, bridge decks and paved areas	0.25	14.40	7.06	-	m	**21.47**
in verges and central reserves	0.20	11.52	5.65	-	m	**17.17**
in side slopes of cuttings or side slopes of embankments	0.15	8.64	4.25	-	m	**12.89**
TRENCH FOR COMMUNICATIONS CABLE OR DUCT						
Trench for cable						
300 to 450 mm wide; depth not exceeding 1.5 m	0.15	8.64	4.25	-	m	**12.89**
450 to 600 mm wide; depth not exceeding 1.5 m	0.20	11.52	5.65	-	m	**17.17**
Extra for excavating rock or reinforced concrete in trench	0.50	28.80	14.14	-	m3	**42.94**
Extra for excavating brickwork or mass concrete in trench	0.40	23.04	11.30	-	m3	**34.35**
Extra for backfilling with pea gravel	0.02	1.15	0.57	18.31	m3	**20.03**
Extra for 450 x 100 mm sand cable bedding and covering	0.02	1.15	0.57	1.21	m	**2.93**
Extra for PVC marker tape	0.01	0.29	0.15	0.17	m	**0.61**
Extra for 150 x 300 clay cable tiles	0.05	2.88	1.41	4.39	m	**8.68**
Extra for 150 x 900 concrete cable tiles	0.03	1.73	0.85	7.46	m	**10.05**
COMMUNICATIONS CABLING AND COMMUNICATIONS DUCT						
Communication cables laid in trench						
type A1 2 pair 0.9 mm² armoured multi-pair	-	-	-	-	m	**4.05**
type A2 20 pair 0.9 mm² armoured multi-pair	-	-	-	-	m	**7.29**
type A3 30 pair 0.9 mm² armoured multi-pair	-	-	-	-	m	**8.90**
Power cables laid in trench						
type A4 10.0 mm² armoured split concentric	-	-	-	-	m	**6.38**

SERIES 1500: MOTORWAY COMMUNICATIONS

Item	Gang hours	Labour £	Plant £	Material £	Unit	Total rate £
Detector feeder cables laid in trench						
type A6 50/0.25m² single core detector feeder cable	-	-	-	-	m	**4.87**
type A7 50/0.25m² single core detector feeder cable	-	-	-	-	m	**4.87**
COMMUNICATIONS CABLE JOINTS AND TERMINATIONS						
Cable terminations						
of type 1 cable	-	-	-	-	nr	**57.92**
of type 2 cable	-	-	-	-	nr	**258.64**
of type 3 cable	-	-	-	-	nr	**370.98**
of type 4 cable	-	-	-	-	nr	**62.06**
of type 6 cable	-	-	-	-	nr	**36.96**
of type 7 cable	-	-	-	-	nr	**36.96**
COMMUNICATIONS EQUIPMENT						
Cabinet bases						
foundation plinth	-	-	-	-	nr	**118.25**
Matrix signal post bases						
foundation plinth	-	-	-	-	nr	**122.44**
CCTV camera bases						
foundation plinth	-	-	-	-	nr	**118.25**
Wall mounted brackets						
at maximum 15.0 m height	-	-	-	-	nr	**39.44**
Fix only the following equipment						
communication equipment cabinet, 600 type series	-	-	-	-	nr	**127.39**
terminator type II	-	-	-	-	nr	**135.03**
emergency telephone post	-	-	-	-	nr	**67.69**
telephone housing	-	-	-	-	nr	**21.16**
matrix signal post	-	-	-	-	nr	**67.69**
Motorwarn / fogwarn	-	-	-	-	nr	**135.23**
distributor on gantry	-	-	-	-	nr	**172.16**
isolator switch for gantry	-	-	-	-	nr	**374.90**
heater unit mounted on gantry; Henleys' 65 W type 22501	-	-	-	-	nr	**25.69**
Terminal blocks						
Klippon BK6	-	-	-	-	nr	**6.82**
Klippon BK12	-	-	-	-	nr	**7.61**
Work to pavement for loop detection circuits						
cut or form grooves in pavement for detector loops and feeders	-	-	-	-	m	**4.53**
additional cost for sealing with hot bitumen sealant	-	-	-	-	m	**0.47**
CHAMBERS						
Brick chamber with galvanised steel cover and frame; depth to uppermost surface of base slab						
ne 1.0 m deep	-	-	-	-	nr	**586.31**

SERIES 1600: PILING & EMBEDDED RETAINING WALLS

Item	Gang hours	Labour £	Plant £	Material £	Unit	Total rate £
GENERAL						
Notes						
There are a number of different types of piling which are available for use in differing situations. Selection of the most suitable type of piling for a particular site will depend on a number of factors including the physical conditions likely to be encountered during driving, the loads to be carried, the design of superstructure, etc.						
The most commonly used systems are included in this section.						
It is essential that a thorough and adequate site investigation is carried out to ascertain details of the ground strata and bearing capacities to enable a proper assessment to be made of the most suitable and economical type of piling to be adopted.						
There are so many factors, apart from design considerations, which influence the cost of piling that it is not possible to give more than an approximate indication of costs. To obtain reliable costs for a particular contract advice should be sought from a company specialising in the particular type of piling proposed. Some Specialist Contractors will also provide a design service if required.						
PILING PLANT						
Driven precast concrete reinforced piles						
Establishment of piling plant for						
235 x 235 mm precast reinforced and prestressed concrete piles in main piling	-	-	-	-	item	**3093.30**
275 x 275 mm precast reinforced and prestressed concrete piles in main piling	-	-	-	-	item	**3093.30**
350 x 350 mm precast reinforced and prestressed concrete piles in main piling	-	-	-	-	item	**3748.50**
Moving piling plant for						
235 x 235 mm precast reinforced and prestressed concrete piles in main piling	-	-	-	-	nr	**34.82**
275 x 275 mm precast reinforced and prestressed concrete piles in main piling	-	-	-	-	nr	**34.82**
350 x 350 mm precast reinforced and prestressed concrete piles in main piling	-	-	-	-	nr	**55.71**
Bored in-situ reinforced concrete piling (tripod rig)						
Establishment of piling plant for 500 mm diameter cast-in-place concrete piles (tripod rig) in main piling	-	-	-	-	item	**6063.75**
Moving piling plant for 500 mm diameter cast-in-place concrete piles (tripod rig) in main piling	-	-	-	-	nr	**38.91**

SERIES 1600: PILING & EMBEDDED RETAINING WALLS

Item	Gang hours	Labour £	Plant £	Material £	Unit	Total rate £
Bored in-situ reinforced concrete piling (mobile rig)						
Establishment of piling plant for 500 mm diameter cast-in-place concrete piles (mobile rig) in main piling	-	-	-	-	item	**9635.85**
Moving piling plant for 500 mm diameter cast-in-place concrete piles in (mobile rig) main piling	-	-	-	-	nr	**55.57**
Concrete injected piles (continuous flight augered)						
Establishment of piling plant for cast-in-place concrete piles (CFA) in main piling						
450 mm diameter; 650kN	-	-	-	-	item	**4380.24**
600 mm diameter; 1400kN	-	-	-	-	item	**4380.38**
750 mm diameter; 2200kN	-	-	-	-	item	**5006.15**
Moving piling plant for cast-in-place concrete piles (CFA) in main piling						
450 mm diameter; 650kN	-	-	-	-	nr	**49.07**
600 mm diameter 1400kN	-	-	-	-	nr	**49.07**
750 mm diameter 2200kN	-	-	-	-	nr	**49.07**
Driven cast in place piles; segmental casing method						
Establishment of piling plant for cast-in-place concrete piles in main piling	-	-	-	-	item	**13230.00**
Moving piling plant for cast-in-place concrete piles in main piling	-	-	-	-	nr	**126.00**
Establishment of piling plant for cast-in-place concrete piles in main piling						
bottom driven	-	-	-	-	item	**6000.00**
top driven	-	-	-	-	item	**6000.00**
Moving piling plant for 430 mm diameter cast-in-place concrete piles in main piling						
bottom driven	-	-	-	-	nr	**86.44**
top driven	-	-	-	-	nr	**59.80**
Steel bearing piles						
Establishment of piling plant for steel bearing piles in main piling						
maximum 100 miles radius from base	-	-	-	-	item	**7188.30**
maximum 250 miles radius from base	-	-	-	-	item	**17970.75**
Moving piling plant for steel bearing piles in main piling	-	-	-	-	nr	**234.37**
Frodingham steel piles						
Provision of all plant, equipment and labour including transport to and from the site and establishing and dismantling for						
driving of sheet piling	-	-	-	-	item	**5896.25**
extraction of sheet piling	-	-	-	-	item	**5240.00**

SERIES 1600: PILING & EMBEDDED RETAINING WALLS

Item	Gang hours	Labour £	Plant £	Material £	Unit	Total rate £
PILING PLANT – cont'd						
Larssen steel piles						
Provision of plant, equipment and labour including transport to and from the site and establishing and dismantling for						
driving of sheet piling	-	-	-	-	item	**5918.75**
extraction of sheet piling	-	-	-	-	item	**5241.25**
Establishment of piling plant for steel tubular piles in main piling						
maximum 100 miles radius from base	-	-	-	-	item	**7083.51**
maximum 250 miles radius from base	-	-	-	-	item	**17708.78**
Moving piling plant for steel tubular piles in main piling	-	-	-	-	nr	**257.36**
PRECAST CONCRETE PILES						
Driven precast reinforced concrete piles						
The following unit costs cover the installation of driven precast concrete piles by using a hammer acting on a shoe fitted onto or cast into the pile unit.						
The costs are based installing 100 piles of nominal sizes stated, and a concrete strength of 50N/mm² suitably reinforced for a working load not exceeding 600kN, with piles average 15m long, on a clear site with reasonable access.						
Single pile lengths are normally a maximum of 13m long, at which point, a mechanical interlocking joint is required to extend the pile. These joints are most economically and practically formed at works.						
Lengths, sizes of sections, reinforcement details and concrete mixes vary for differing contractors, whose specialist advice should be sought for specific designs.						
Precast concrete piles; concrete 50N/mm²						
235 x 235 mm; 5 - 10 m in length; main piling	-	-	-	-	m	**18.47**
275 x 275 mm; 5 - 10 m in length; main piling	-	-	-	-	m	**19.41**
350 x 350 mm; 5 - 10 m in length; main piling	-	-	-	-	m	**37.20**
Mechanical Interlocking joint						
235 x 235 mm	-	-	-	-	nr	**45.32**
275 x 275 mm	-	-	-	-	nr	**51.79**
350 x 350 mm	-	-	-	-	nr	**64.74**
Driving vertical precast piles						
235 x 235 mm; 5 - 10 m in length; in main piling	-	-	-	-	m	**2.77**
275 x 275 mm; 5 - 10 m in length; in main piling	-	-	-	-	m	**3.23**
350 x 350 mm; 5 - 10 m in length; in main piling	-	-	-	-	m	**3.89**
Stripping vertical precast concrete pile heads						
235 x 235 mm piles in main piling	-	-	-	-	nr	**28.11**
275 x 275 mm piles in main piling	-	-	-	-	nr	**36.57**
350 x 350 mm piles in main piling	-	-	-	-	nr	**56.24**
Standing time						
275 x 275 mm	-	-	-	-	hr	**174.12**
350 x 350 mm	-	-	-	-	hr	**208.94**

SERIES 1600: PILING & EMBEDDED RETAINING WALLS

Item	Gang hours	Labour £	Plant £	Material £	Unit	**Total rate £**
CAST IN PLACE PILES						
Bored in-situ reinforced concrete piling (tripod rig)						
The following unit costs cover the construction of small diameter bored piling using light and compact tripod rigs requiring no expensive site levelling or access ways. Piling can be constructed in very restricted headroom or on confined and difficult sites. Standard diameters are between 400 and 600 mm with a normal maximum depth of 30 m.						
The costs are based on installing 100 piles of 500 mm nominal diameter, a concrete strength of 20N/mm² with nominal reinforcement, on a clear site with reasonable access. Disposal of excavated material is included separately.						
Vertical 500 mm diameter cast in place piles; 20N/mm² concrete; nominal reinforcement; in main piling	-	-	-	-	m	**93.71**
Vertical 500 mm diameter empty bores in main piling	-	-	-	-	m	**39.13**
Add for boring through obstructions	-	-	-	-	hr	**104.47**
Standing time	-	-	-	-	hr	**104.47**
Bored in-situ reinforced concrete piling (mobile rig)						
The following unit costs cover the construction of small diameter bored piles using lorry or crawler mounted rotary boring rigs. This type of plant is more mobile and faster in operation than the tripod rigs and is ideal for large contracts in cohesive ground. Construction of piles under bentonite suspension can be carried out to obviate the use of liners. Standard diameters of 450 to 900 mm diameter can be constructed to depths of 30 m.						
The costs are based on installing 100 piles of 500 mm nominal diameter, a concrete strength of 20N/mm² with nominal reinforcement, on a clear site with reasonable access. Disposal of excavated material is included separately.						
Vertical 500 mm diameter cast in place piles; 20N/mm² concrete; nominal reinforcement	-	-	-	-	m	**48.46**
Vertical 500 mm diameter empty bores	-	-	-	-	m	**29.94**
Add for boring through obstructions	-	-	-	-	hr	**212.97**
Standing time	-	-	-	-	hr	**212.97**

SERIES 1600: PILING & EMBEDDED RETAINING WALLS

Item	Gang hours	Labour £	Plant £	Material £	Unit	Total rate £
CAST IN PLACE PILES – cont'd						
Concrete injected piles (continuous flight augered)						
The following unit costs cover the construction of piles by screwing a continuous flight auger into the ground to a design depth (Determined prior to commencement of piling operations and upon which the rates are based and subsequently varied to actual depths). Concrete is then pumped through the hollow stem of the auger to the bottom and the pile formed as the auger is withdrawn. Spoil is removed by the auger as it is withdrawn. This is a fast method of construction without causing disturbance or vibration to adjacent ground. No casing is required even in unsuitable soils. Reinforcement can be placed after grouting is complete.						
The costs are based on installing 100 piles on a clear site with reasonable access. Disposal of excavated material is included separately.						
Vertical cast in place piles; 20N/mm² concrete						
450 mm diameter; 650kN; 10 - 15m in length; main piling	-	-	-	-	m	**29.97**
600 mm diameter; 1400kN; 10 - 15m in length; main piling	-	-	-	-	m	**40.47**
750 mm diameter; 2200kN; 10 - 15m in length; main piling	-	-	-	-	m	**69.56**
Vertical empty bores						
450 mm diameter	-	-	-	-	m	**23.85**
600 mm diameter	-	-	-	-	m	**31.56**
750 mm diameter	-	-	-	-	m	**50.90**
Standing time / Boring through obstructions time						
450 mm diameter	-	-	-	-	hr	**140.18**
600 mm diameter	-	-	-	-	hr	**140.18**
750 mm diameter	-	-	-	-	hr	**161.19**
Driven cast in place piles; segmental casing method						
The following unit costs cover the construction of piles by driving into hard material using a serrated thick wall tube. It is oscillated and pressed into the hard material using a hydraulic attachment to the piling rig. The hard material is broken up using chiselling methods and is then removed by mechanical grab.						
Vertical cast in place piles; 20N/mm² concrete						
620 mm diameter; 10 - 15m in length; main piling	-	-	-	-	m	**98.44**
1180 mm diameter; 10 - 15m in length; main piling	-	-	-	-	m	**157.50**
1500 mm diameter; 10 - 15m in length; main piling	-	-	-	-	m	**210.00**
Standing time	-	-	-	-	hr	**276.99**
Add for driving through obstructions	-	-	-	-	hr	**295.47**

SERIES 1600: PILING & EMBEDDED RETAINING WALLS

Item	Gang hours	Labour £	Plant £	Material £	Unit	Total rate £
Driven in-situ reinforced concrete piling						
The following unit costs cover the construction of piles by driving a tube into the ground either by using an internal hammer acting on a gravel or concrete plug or, as is more usual, by using an external hammer on a driving helmet at the top of the tube. After driving to the required depth an enlarged base is formed by hammering out sucessive charges of concrete down the tube. The tube is then filled with concrete which is compacted as the tube is vibrated and withdrawn. Piles of 350 to 500 mm diameter can be constructed with rakes up to 1 in 4 to carry working loads up to 120t per pile.						
The costs are based on installing 100 piles of 430 mm nominal diameter, a concrete strength of 20N/mm² suitably reinforced for a working load not exceeding 750kN, on a clear site with reasonable access.						
SEE SECTION PILING PLANT Establishment of piling plant for cast in place concrete piles in main piling						
bottom driven	-	-	-	-	item	**5694.19**
top driven	-	-	-	-	item	**5694.19**
SEE SECTION PILING PLANT Moving piling plant for 430 mm diameter cast in place concrete piles in main piling						
bottom driven	-	-	-	-	nr	**81.80**
top driven	-	-	-	-	nr	**56.59**
Vertical 430 mm diameter cast in place piles 20N/mm² concrete; reinforcement for 750kN maximum load						
bottom driven	-	-	-	-	m	**28.96**
top driven	-	-	-	-	m	**28.62**
Standing time	-	-	-	-	hr	**167.41**
Add for driving through obstructions where within the capabilities of the normal plant	-	-	-	-	hr	**174.39**
Stripping vertical precast concrete pile heads						
430 mm diameter heads	-	-	-	-	nr	**52.15**
REINFORCEMENT FOR CAST IN PLACE PILES						
Mild steel						
Steel bar reinforcement						
6 mm nominal size; not exceeding 12 in length	-	-	-	-	tonne	**542.43**
8 mm nominal size; not exceeding 12 in length	-	-	-	-	tonne	**529.20**
10 mm nominal size; not exceeding 12 in length	-	-	-	-	tonne	**516.80**
12 mm nominal size; not exceeding 12 in length	-	-	-	-	tonne	**514.04**
16 mm nominal size; not exceeding 12 in length	-	-	-	-	tonne	**509.91**
25 mm nominal size; not exceeding 12 in length	-	-	-	-	tonne	**509.91**
32 mm nominal size; not exceeding 12 in length	-	-	-	-	tonne	**514.04**
12 mm nominal size; not exceeding 12 in length	-	-	-	-	tonne	**531.96**
Steel helical reinforcement						
6 mm nominal size; not exceeding 12 in length	-	-	-	-	tonne	**556.76**
8 mm nominal size; not exceeding 12 in length	-	-	-	-	tonne	**542.99**

SERIES 1600: PILING & EMBEDDED RETAINING WALLS

Item	Gang hours	Labour £	Plant £	Material £	Unit	Total rate £
REINFORCEMENT FOR CAST IN PLACE PILES - cont'd						
Mild Steel – cont'd						
10 mm nominal size; not exceeding 12 in length	-	-	-	-	tonne	**516.80**
12 mm nominal size; not exceeding 12 in length	-	-	-	-	tonne	**527.82**
High tensile steel						
Steel bar reinforcement						
6 mm nominal size; not exceeding 12 in length	-	-	-	-	tonne	**560.90**
8 mm nominal size; not exceeding 12 in length	-	-	-	-	tonne	**547.11**
10 mm nominal size; not exceeding 12 in length	-	-	-	-	tonne	**536.09**
12 mm nominal size; not exceeding 12 in length	-	-	-	-	tonne	**531.96**
16 mm nominal size; not exceeding 12 in length	-	-	-	-	tonne	**509.91**
25 mm nominal size; not exceeding 12 in length	-	-	-	-	tonne	**509.91**
32 mm nominal size; not exceeding 12 in length	-	-	-	-	tonne	**531.96**
12 mm nominal size; not exceeding 12 in length	-	-	-	-	tonne	**551.25**
Steel helical reinforcement						
6 mm nominal size; not exceeding 12 in length	-	-	-	-	tonne	**574.68**
8 mm nominal size; not exceeding 12 in length	-	-	-	-	tonne	**560.90**
10 mm nominal size; not exceeding 12 in length	-	-	-	-	tonne	**549.87**
12 mm nominal size; not exceeding 12 in length	-	-	-	-	tonne	**545.74**
STEEL BEARING PILES						
Steel bearing piles are commonly carried out by a Specialist Contractor and whose advice should be sought to arrive at accurate costing. However the following items can be used to assess a budget cost for such work. The following unit costs are based upon driving 100nr steel bearing piles on a clear site with reasonable access. Supply is based on delivery 75 miles from works, in loads over 20t.						
Steel bearing piles						
Standing time	-	-	-	-	hr	**345.08**
203 x 203 x 45 kg/m steel bearing piles; Grade S275						
not exceeding 5m in length in main piling	-	-	-	28.73	m	**28.73**
5 - 10m in length in main piling	-	-	-	27.98	m	**27.98**
10 - 15m in length in main piling	-	-	-	27.98	m	**27.98**
15 - 20m in length in main piling	-	-	-	28.23	m	**28.23**
203 x 203 x 54 kg/m steel bearing piles; Grade S275						
not exceeding 5m in length in main piling	-	-	-	34.64	m	**34.64**
5 - 10m in length in main piling	-	-	-	33.75	m	**33.75**
10 - 15m in length in main piling	-	-	-	33.75	m	**33.75**
15 - 20m in length in main piling	-	-	-	34.04	m	**34.04**

SERIES 1600: PILING & EMBEDDED RETAINING WALLS

Item	Gang hours	Labour £	Plant £	Material £	Unit	**Total rate £**
254 x 254 x 63 kg/m steel bearing piles; Grade S275						
ne 5m in length	-	-	-	40.70	m	**40.70**
5 - 10m in length	-	-	-	39.66	m	**39.66**
10 - 15m in length	-	-	-	39.66	m	**39.66**
15 - 20m in length	-	-	-	40.01	m	**40.01**
254 x 254 x 71 kg/m steel bearing piles; Grade S275						
ne 5m in length	-	-	-	45.87	m	**45.87**
5 - 10m in length	-	-	-	44.70	m	**44.70**
10 - 15m in length	-	-	-	44.70	m	**44.70**
15 - 20m in length	-	-	-	45.09	m	**45.09**
254 x 254 x 85 kg/m steel bearing piles; Grade S275						
ne 5m in length	-	-	-	54.92	m	**54.92**
5 - 10m in length	-	-	-	53.51	m	**53.51**
10 - 15m in length	-	-	-	53.51	m	**53.51**
15 - 20m in length	-	-	-	53.98	m	**53.98**
305 x 305 x 79 kg/m steel bearing piles; Grade S275						
ne 5m in length	-	-	-	52.26	m	**52.26**
5 - 10m in length	-	-	-	50.95	m	**50.95**
10 - 15m in length	-	-	-	50.95	m	**50.95**
15 - 20m in length	-	-	-	51.39	m	**51.39**
305 x 305 x 95kg/m steel bearing piles; Grade S275						
ne 5m in length	-	-	-	62.84	m	**62.84**
5 - 10m in length	-	-	-	61.27	m	**61.27**
10 - 15m in length	-	-	-	61.27	m	**61.27**
15 - 20m in length	-	-	-	61.79	m	**61.79**
305 x 305 x 110 kg/m steel bearing piles; Grade S275						
ne 5m in length	-	-	-	72.76	m	**72.76**
5 - 10m in length	-	-	-	70.95	m	**70.95**
10 - 15m in length	-	-	-	70.95	m	**70.95**
15 - 20m in length	-	-	-	71.55	m	**71.55**
305 x 305 x 126 kg/m steel bearing piles; Grade S275						
ne 5m in length	-	-	-	83.34	m	**83.34**
5 - 10m in length	-	-	-	81.27	m	**81.27**
10 - 15m in length	-	-	-	81.27	m	**81.27**
15 - 20m in length	-	-	-	81.96	m	**81.96**
305 x 305 x 149 kg/m steel bearing piles; Grade S275						
ne 5m in length	-	-	-	98.56	m	**98.56**
5 - 10m in length	-	-	-	96.10	m	**96.10**
10 - 15m in length	-	-	-	96.10	m	**96.10**
15 - 20m in length	-	-	-	96.92	m	**96.92**

SERIES 1600: PILING & EMBEDDED RETAINING WALLS

Item	Gang hours	Labour £	Plant £	Material £	Unit	**Total rate £**
STEEL BEARING PILES – cont'd						
Steel bearing piles – cont'd						
305 x 305 x 186 kg/m steel bearing piles; Grade S275						
ne 5m in length	-	-	-	123.03	m	**123.03**
5 - 10m in length	-	-	-	119.96	m	**119.96**
10 - 15m in length	-	-	-	119.96	m	**119.96**
15 - 20m in length	-	-	-	120.99	m	**120.99**
305 x 305 x 233 kg/m steel bearing piles; Grade S275						
ne 5m in length	-	-	-	147.51	m	**147.51**
5 - 10m in length	-	-	-	143.83	m	**143.83**
10 - 15m in length	-	-	-	143.83	m	**143.83**
15 - 20m in length	-	-	-	145.05	m	**145.05**
356 x 368 x 109 kg/m steel bearing piles; Grade S275						
ne 5m in length	-	-	-	73.78	m	**73.78**
5 - 10m in length	-	-	-	71.98	m	**71.98**
10 - 15m in length	-	-	-	71.98	m	**71.98**
15 - 20m in length	-	-	-	72.58	m	**72.58**
356 x 368 x 133 kg/m steel bearing piles; Grade S275						
ne 5m in length	-	-	-	90.02	m	**90.02**
5 - 10m in length	-	-	-	87.83	m	**87.83**
10 - 15m in length	-	-	-	87.83	m	**87.83**
15 - 20m in length	-	-	-	88.56	m	**88.56**
356 x 368 x 152kg/m steel bearing piles; Grade S275						
ne 5m in length	-	-	-	102.88	m	**102.88**
5 - 10m in length	-	-	-	100.38	m	**100.38**
10 - 15m in length	-	-	-	100.38	m	**100.38**
15 - 20m in length	-	-	-	101.21	m	**101.21**
356 x 368 x 174 kg/m steel bearing piles; Grade S275						
ne 5m in length	-	-	-	117.77	m	**117.77**
5 - 10m in length	-	-	-	114.90	m	**114.90**
10 - 15m in length	-	-	-	114.90	m	**114.90**
15 - 20m in length	-	-	-	115.86	m	**115.86**
Driving vertical steel bearing piles						
section weight not exceeding 70 kg/m	-	-	-	-	m	**8.81**
section weight 70 - 90 kg/m	-	-	-	-	m	**9.44**
section weight 90 - 110 kg/m	-	-	-	-	m	**10.06**
section weight 90 - 110 kg/m	-	-	-	-	m	**10.69**
section weight 110 - 130 kg/m	-	-	-	-	m	**11.33**
section weight 150 - 170 kg/m	-	-	-	-	m	**11.95**
Driving raking steel bearing piles						
section weight not exceeding 70 kg/m	-	-	-	-	m	**10.69**
section weight 70 - 90 kg/m	-	-	-	-	m	**11.33**
section weight 90 - 110 kg/m	-	-	-	-	m	**12.58**
section weight 110 - 130 kg/m	-	-	-	-	m	**13.21**
section weight 130 - 150 kg/m	-	-	-	-	m	**14.48**
section weight 150 - 170 kg/m	-	-	-	-	m	**14.48**
section weight 170 - 190 kg/m	-	-	-	-	m	**14.48**
section weight 190 - 210 kg/m	-	-	-	-	m	**15.09**

SERIES 1600: PILING & EMBEDDED RETAINING WALLS

Item	Gang hours	Labour £	Plant £	Material £	Unit	**Total rate £**
allow 30% of the respective item above for the lengthened section only						
Welding on lengthening pieces to vertical steel bearing piles						
203 x 203 x any kg/m	-	-	-	-	nr	**100.64**
254 x 254 x any kg/m	-	-	-	-	nr	**125.81**
305 x 305 x any kg/m	-	-	-	-	nr	**150.97**
356 x 368 x any kg/m	-	-	-	-	nr	**176.96**
Cutting or burning off surplus length of vertical steel bearing piles						
203 x 203 x any kg/m	-	-	-	-	nr	**5.04**
254 x 254 x any kg/m	-	-	-	-	nr	**7.55**
305 x 305 x any kg/m	-	-	-	-	nr	**8.81**
356 x 368 x any kg/m	-	-	-	-	nr	**10.06**
STEEL TUBULAR PILES						
Steel tubular piles are commonly carried out by a Specialist Contractor and whose advice should be sought to arrive at accurate costings. However the following items can be used to assess a budget cost for each work. The following unit costs are based upon driving 100nr steel tubular piles on a clear site with reasonable access.						
Standing time	-	-	-	-	hr	**321.70**
Steel Grade 43A; delivered in 10 - 20 t loads; mass 60 - 120 kg/m						
section 508 mm x 8 mm x 98.6 kg/m	-	-	-	79.71	m	**79.71**
section 559 mm x 8 mm x 109 kg/m	-	-	-	88.11	m	**88.11**
Steel Grade 43A; delivered in 10 - 20 t loads; mass 120 - 250 kg/m						
section 508 mm x 10 mm x 123 kg/m	-	-	-	99.43	m	**99.43**
section 508 mm x 12.5 mm x 153 kg/m	-	-	-	123.68	m	**123.68**
section 508 mm x 16 mm x 194 kg/m	-	-	-	156.83	m	**156.83**
section 508 mm x 20 mm x 241 kg/m	-	-	-	194.82	m	**194.82**
section 559 mm x 10 mm x 135 kg/m	-	-	-	109.13	m	**109.13**
section 559 mm x 12.5 mm x 168 kg/m	-	-	-	135.81	m	**135.81**
section 559 mm x 16 mm x 214 kg/m	-	-	-	172.99	m	**172.99**
section 610 mm x 8 mm x 119 kg/m	-	-	-	96.20	m	**96.20**
section 610 mm x 10 mm x 148 kg/m	-	-	-	119.64	m	**119.64**
section 610 mm x 12.5 mm x 184 kg/m	-	-	-	148.74	m	**148.74**
section 610 mm x 16 mm x 234 kg/m	-	-	-	189.16	m	**189.16**
section 660 mm x 8 mm x 129 kg/m	-	-	-	104.28	m	**104.28**
section 660 mm x 10 mm x 160 kg/m	-	-	-	129.34	m	**129.34**
section 660 mm x 12.5 mm x 200 kg/m	-	-	-	161.68	m	**161.68**
section 711 mm x 8 mm x 134 kg/m	-	-	-	108.32	m	**108.32**
section 711 mm x 10 mm x 173 kg/m	-	-	-	139.85	m	**139.85**
section 711 mm x 12 mm x 215 kg/m	-	-	-	173.80	m	**173.80**
section 762 mm x 8 mm x 149 kg/m	-	-	-	120.45	m	**120.45**
section 762 mm x 10 mm x 185 kg/m	-	-	-	149.55	m	**149.55**
section 762 mm x 12.5 mm x 231 kg/m	-	-	-	186.74	m	**186.74**
Steel Grade 43A; delivered in 10 - 20 t loads; mass 250 - 500 kg/m						
section 559 mm x 20 mm x 266 kg/m	-	-	-	215.03	m	**215.03**
section 610 mm x 20 mm x 291 kg/m	-	-	-	235.24	m	**235.24**
section 660 mm x 16 mm x 254 kg/m	-	-	-	205.33	m	**205.33**
section 660 mm x 20 mm x 316 kg/m	-	-	-	255.45	m	**255.45**

SERIES 1600: PILING & EMBEDDED RETAINING WALLS

Item	Gang hours	Labour £	Plant £	Material £	Unit	Total rate £
STEEL TUBULAR PILES – cont'd						
section 660 mm x 25 mm x 392 kg/m	-	-	-	316.88	m	**316.88**
section 711 mm x 16 mm x 274 kg/m	-	-	-	221.50	m	**221.50**
section 711 mm x 20 mm x 341 kg/m	-	-	-	275.66	m	**275.66**
section 711 mm x 25 mm x 423 kg/m	-	-	-	341.94	m	**341.94**
section 762 mm x 16 mm x 294 kg/m	-	-	-	237.66	m	**237.66**
section 762 mm x 20 mm x 366 kg/m	-	-	-	295.87	m	**295.87**
section 762 mm x 25 mm x 454 kg/m	-	-	-	367.00	m	**367.00**
Driving vertical steel tubular piles						
mass 60 - 120 kg/m	-	-	-	-	m	**7.80**
mass 120 - 150 kg/m	-	-	-	-	m	**7.94**
mass 150 - 160 kg/m	-	-	-	-	m	**8.06**
mass 160 - 190 kg/m	-	-	-	-	m	**8.18**
mass 190 - 220 kg/m	-	-	-	-	m	**8.09**
mass 220 - 250 kg/m	-	-	-	-	m	**8.42**
mass 250 - 280 kg/m	-	-	-	-	m	**8.55**
mass 280 - 310 kg/m	-	-	-	-	m	**8.68**
mass 310 - 340 kg/m	-	-	-	-	m	**8.80**
mass 340 -370 kg/m	-	-	-	-	m	**9.40**
mass 370 - 400 kg/m	-	-	-	-	m	**9.64**
mass 400 - 430 kg/m	-	-	-	-	m	**9.31**
mass 430 - 460 kg/m	-	-	-	-	m	**10.03**
Driving raking steel tubular piles						
mass 60 - 120 kg/m	-	-	-	-	m	**10.14**
mass 120 - 150 kg/m	-	-	-	-	m	**10.31**
mass 150 - 160 kg/m	-	-	-	-	m	**10.47**
mass 160 - 190 kg/m	-	-	-	-	m	**10.64**
mass 190 - 220 kg/m	-	-	-	-	m	**10.80**
mass 220 - 250 kg/m	-	-	-	-	m	**10.95**
mass 250 - 280 kg/m	-	-	-	-	m	**11.11**
mass 280 - 310 kg/m	-	-	-	-	m	**11.29**
mass 310 - 340 kg/m	-	-	-	-	m	**11.43**
mass 340 -370 kg/m	-	-	-	-	m	**11.60**
mass 370 - 400 kg/m	-	-	-	-	m	**11.92**
mass 400 - 430 kg/m	-	-	-	-	m	**12.09**
mass 430 - 460 kg/m	-	-	-	-	m	**12.41**
Driving lengthened vertical steel tubular piles						
mass 60 - 120 kg/m	-	-	-	-	m	**10.68**
mass 120 - 150 kg/m	-	-	-	-	m	**10.85**
mass 150 - 160 kg/m	-	-	-	-	m	**11.03**
mass 160 - 190 kg/m	-	-	-	-	m	**11.18**
mass 190 - 220 kg/m	-	-	-	-	m	**11.36**
mass 220 - 250 kg/m	-	-	-	-	m	**11.53**
mass 250 - 280 kg/m	-	-	-	-	m	**11.70**
mass 280 - 310 kg/m	-	-	-	-	m	**11.87**
mass 310 - 340 kg/m	-	-	-	-	m	**12.04**
mass 340 -370 kg/m	-	-	-	-	m	**12.20**
mass 370 - 400 kg/m	-	-	-	-	m	**12.54**
mass 400 - 430 kg/m	-	-	-	-	m	**12.72**
mass 430 - 460 kg/m	-	-	-	-	m	**13.04**
Driving lengthened raking steel tubular piles						
mass 60 - 120 kg/m	-	-	-	-	m	**9.65**
mass 120 - 150 kg/m	-	-	-	-	m	**9.79**
mass 150 - 160 kg/m	-	-	-	-	m	**9.91**
mass 160 - 190 kg/m	-	-	-	-	m	**10.10**
mass 190 - 220 kg/m	-	-	-	-	m	**10.26**
mass 220 - 250 kg/m	-	-	-	-	m	**10.41**

SERIES 1600: PILING & EMBEDDED RETAINING WALLS

Item	Gang hours	Labour £	Plant £	Material £	Unit	Total rate £
mass 250 - 280 kg/m	-	-	-	-	m	10.56
mass 280 - 310 kg/m	-	-	-	-	m	10.72
mass 310 - 340 kg/m	-	-	-	-	m	10.87
mass 340 -370 kg/m	-	-	-	-	m	11.01
mass 370 - 400 kg/m	-	-	-	-	m	11.32
mass 400 - 430 kg/m	-	-	-	-	m	11.48
mass 430 - 460 kg/m	-	-	-	-	m	11.78
Welding on lengthening piece to steel tubular piles						
section diameter 508 x any thickness	-	-	-	-	nr	173.54
section diameter 559 x any thickness	-	-	-	-	nr	183.46
section diameter 610 x any thickness	-	-	-	-	nr	193.38
section diameter 660 x any thickness	-	-	-	-	nr	203.29
section diameter 711 x any thickness	-	-	-	-	nr	213.21
section diameter 762 x any thickness	-	-	-	-	nr	223.13
Cutting or burning off surplus length of steel tubular piles						
section diameter 508 x any thickness	-	-	-	-	nr	11.88
section diameter 559 x any thickness	-	-	-	-	nr	11.95
section diameter 610 x any thickness	-	-	-	-	nr	12.00
section diameter 660 x any thickness	-	-	-	-	nr	12.07
section diameter 711 x any thickness	-	-	-	-	nr	12.16
section diameter 762 x any thickness	-	-	-	-	nr	12.23
PROOF LOADING OF PILES						
Driven precast concrete piles						
Establishment of proof loading equipment; proof loading of vertical precast concrete piles with maintained load to 900 kN	-	-	-	-	item	2415.00
Establishment of proof loading equipment; proof loading of vertical precast concrete piles by dynamic testing with piling hammer	-	-	-	-	nr	569.52
Bored in-situ reinforced concrete piling (tripod rig)						
Establishment of proof loading equipment for bored cast-in-place piles	-	-	-	-	item	1732.50
Proof loading of vertical cast-in-place piles with maximum test load of 600kN on a working pile 500mm diameter using tension piles as reaction	-	-	-	-	nr	3360.00
Bored in-situ reinforced concrete piling (mobile rig)						
Establishment of proof loading equipment for bored cast-in-place piles	-	-	-	-	item	1732.50
Proof loading of vertical cast-in-place piles with maximum test load of 600kN on a working pile 500mm diameter using tension piles as reaction	-	-	-	-	nr	3360.00

SERIES 1600: PILING & EMBEDDED RETAINING WALLS

Item	Gang hours	Labour £	Plant £	Material £	Unit	Total rate £
PROOF LOADING OF PILES – cont'd						
Concrete injected piles (continuous flight augered)						
Establishment of proof loading equipment for bored cast-in-place piles						
450 mm diameter; 650kN	-	-	-	-	item	**2520.00**
600 mm diameter; 1400kN	-	-	-	-	item	**2520.00**
750 mm diameter; 2200kN	-	-	-	-	item	**2520.00**
Proof loading of vertical cast-in-place piles to 1.5 times working load						
450 mm diameter; 650kN	-	-	-	-	nr	**1260.00**
600 mm diameter; 1400kN	-	-	-	-	nr	**1879.50**
750 mm diameter; 2200kN	-	-	-	-	nr	**2415.00**
Electronic integrity testing						
cost per pile (minimum 40 piles per visit)	-	-	-	-	nr	**15.62**
Segmental casing method piles						
Establishment of proof loading equipment for driven cast-in-place piles	-	-	-	-	item	**1260.00**
Proof loading of vertical cast-in-place piles with maximum test load of 600kN on a working pile 500 mm diameter using non-working tension piles as reaction	-	-	-	-	nr	**3675.00**
Driven in-situ reinforced concrete piling						
Establishment of proof loading equipment for driven cast-in-place piles						
bottom driven	-	-	-	-	item	**3570.00**
top driven	-	-	-	-	item	**2730.00**
Proof loading of vertical cast-in-place piles with maximum test load of 1125kN on a working pile 430 mm diameter using non-working tension piles as reaction						
bottom driven	-	-	-	-	nr	**1139.04**
top driven	-	-	-	-	nr	**1139.04**
Electronic integrity testing						
Cost per pile (minimum 40 piles per visit)	-	-	-	-	nr	**18.38**
Steel bearing piles						
Establishment of proof loading equipment for steel bearing piles in main piling	-	-	-	-	item	**18826.50**
Proof loading of vertical steel bearing piles with maximum test load of 108 t load on a working pile using non-working tension piles as reaction	-	-	-	-	nr	**6846.00**
Steel tubular piles						
Establishment of proof loading equipment for steel tubular piles	-	-	-	-	item	**7673.40**
Proof loading of steel tubular piles with maximum test load of 108 t load on a working pile using non-working tension piles as reaction	-	-	-	-	nr	**10391.01**

SERIES 1600: PILING & EMBEDDED RETAINING WALLS

Item	Gang hours	Labour £	Plant £	Material £	Unit	**Total rate £**
STEEL SHEET PILES						
Sheet steel piling is commonly carried out by a Specialist Contractor, whose advice should be sought to arrive at accurate costings. However, the following items can be used to assess a budget for such work.						
The following unit costs are based on driving/extracting 1,500m² of sheet piling on a clear site with reasonable access.						
Note: area of driven piles will vary from area supplied dependant upon pitch line of piling and provision for such allowance has been made in PC for supply.						
The materials cost below includes the manufacturers tariffs for a 200 mile delivery radius from works, delivery in 5-10t loads and with an allowance of 10% to cover waste / projecting piles etc.						
ProfilARBED Z section steel piles; EN 10248 grade S270GP steel						
Provision of all plant, equipment and labour including transport to and from the site and establishing and dismantling for						
driving of sheet piling	-	-	-	-	sum	**5896.25**
extraction of sheet piling	-	-	-	-	sum	**5240.00**
Standing time	-	-	-	-	hr	**283.90**
Section modulus 800 - 1200 cm³/m; section reference AZ 12; mass 98.7 kg/m², sectional modulus 1200 cm³/m; EN 10248 grade S270GP steel						
length of welded corner piles	-	-	-	-	m	**68.58**
length of welded junction piles	-	-	-	-	m	**96.00**
driven area	-	-	-	-	m²	**33.44**
area of piles of length not exceeding 14 m	-	-	-	-	m²	**69.95**
length 14- 24 m	-	-	-	-	m²	**75.19**
area of piles of length exceeding 24 m	-	-	-	-	m²	**77.70**
Section modulus 1200 - 2000 cm³/m; section reference AZ 17; mass 108.6 kg/m²; sectional modulus 1665 cm³/m; EN 10248 grade S270GP steel						
length of welded corner piles	-	-	-	-	m	**68.58**
length of welded junction piles	-	-	-	-	m	**96.00**
driven area	-	-	-	-	m²	**30.38**
area of piles of length not exceeding 14 m	-	-	-	-	m²	**73.94**
length 14- 24 m	-	-	-	-	m²	**75.19**
area of piles of length exceeding 24 m	-	-	-	-	m²	**77.70**
Section modulus 2000 - 3000 cm³/m; section reference AZ 26; mass 155.2 kg/m²; sectional modulus 2600 cm³/m; EN 10248 grade S270GP steel						
driven area	-	-	-	-	m²	**26.95**
area of piles of length 6 - 18 m	-	-	-	-	m²	**83.78**
area of piles of length 18 - 24 m	-	-	-	-	m²	**85.20**

SERIES 1600: PILING & EMBEDDED RETAINING WALLS

Item	Gang hours	Labour £	Plant £	Material £	Unit	Total rate £
STEEL SHEET PILES – cont'd						
Section modulus 3000 - 4000 cm³/m; section reference AZ 36; mass 194.0 kg/m²; sectional modulus 3600 cm³/m; EN 10248 grade S270GP steel						
driven area	-	-	-	-	m²	**29.68**
area of piles of length 6 - 18 m	-	-	-	-	m²	**102.29**
area of piles of length 18 - 24 m	-	-	-	-	m²	**104.03**
Straight section modulus ne 500 cm³/m; section reference AS 500-12 mass 149 kg/m²; sectional modulus 51 cm³/m; EN 10248 grade S270GP steel						
driven area	-	-	-	-	m²	**26.95**
area of piles of length 6 - 18 m	-	-	-	-	m²	**138.58**
area of piles of length 18 - 24 m	-	-	-	-	m²	**141.82**
For associated black tar vinyl (PC1) protective treatment applications see Civil Engineering Works, Section P, page 267						
For associated black high build isocyanate cured epoxy pitch (PC2) protective treatment applications see Civil Engineering Works, Section P, page 268						
ProfilARBED U section steel piles; EN 10248 grade S270GP steel						
The following unit costs are based on driving/extracting 1,500 m² of sheet piling on a clear site with reasonable access.						
Provision of plant, equipment and labour including transport to and from the site and establishing and dismantling						
driving of sheet piling	-	-	-	-	sum	**5918.75**
extraction of sheet piling	-	-	-	-	sum	**5241.25**
Standing time	-	-	-	-	hr	**283.90**
Section modulus 500 - 800 cm³/m; section reference PU 6; mass 76.0 kg/m²; sectional modulus 600 cm³/m						
driven area	-	-	-	-	m²	**36.85**
area of piles of length 6 - 18 m	-	-	-	-	m²	**58.51**
area of piles of length 18 - 24 m	-	-	-	-	m²	**59.50**
Section modulus 800 - 1200 cm³/m; section reference PU 8; mass 90.9 kg/m²; sectional modulus 830 cm³/m						
driven area	-	-	-	-	m²	**31.40**
area of piles of length 6 - 18 m	-	-	-	-	m²	**58.51**
area of piles of length 18 - 24 m	-	-	-	-	m²	**59.56**
Section modulus 1200 - 2000 cm³/m; section reference PU 12; mass 110.1 kg/m²; sectional modulus 1200 cm³/m						
driven area	-	-	-	-	m²	**27.99**
area of piles of length 6 - 18 m	-	-	-	-	m²	**71.09**
area of piles of length 18 - 24 m	-	-	-	-	m²	**72.32**

SERIES 1600: PILING & EMBEDDED RETAINING WALLS

Item	Gang hours	Labour £	Plant £	Material £	Unit	**Total rate £**
Section modulus 1200 - 2000 cm³/m; section reference PU 18; mass 128.2 kg/m²; sectional modulus 1800 cm³/m						
driven area	-	-	-	-	m²	**24.90**
area of piles of length 6 - 18 m	-	-	-	-	m²	**82.57**
area of piles of length 18 - 24 m	-	-	-	-	m²	**84.00**
Section modulus 2000 - 3000 cm³/m; section reference PU 22; mass 143.6 kg/m²; sectional modulus 2200 cm³/m						
driven area	-	-	-	-	m²	**22.86**
area of piles of length 6 - 18 m	-	-	-	-	m²	**99.94**
area of piles of length 18 - 24 m	-	-	-	-	m²	**102.35**
Section modulus 3000 - 4000 cm³/m; section reference PU 32; mass 190.2 kg/m²; sectional modulus 3200 cm³/m						
driven area	-	-	-	-	m²	**19.79**
area of piles of length 6 - 18 m	-	-	-	-	m²	**112.98**
area of piles of length 18 - 24 m	-	-	-	-	m²	**115.70**
For associated black tar vinyl (PC1) protective treatment applications see Civil Engineering Works, Section P, page 267						
For associated black high build isocyanate cured epoxy pitch (PC2) protective treatment applications see Civil Engineering Works, Section P, page 268						
EMBEDDED RETAINING WALL PLANT						
Diaphgram walls are the construction of vertical walls, cast in place in a trench excavation. They can be formed in reinforced concrete to provide structural elements for temporary or permanent retaining walls. Wall thicknesses of 500 to 1,500mm up to 40m deep may be constructed. Special equipment such as the Hydrofraise can construct walls up to 100 m deep. Restricted urban sites will significantly increase the costs. The following costs are based on constructing a diaphgram wall with an excavated volume of 4000 m³ using a grab. Typical progress would be up to 500 m per week. Disposal of excavated material is not included in the unit costs.						
Establishment of standard diaphragm walling plant, including bentonite storage tanks.	-	-	-	-	item	**110000.00**
Standing time	-	-	-	-	hr	**830.00**
Guide walls (twin)	-	-	-	-	m	**385.10**
Waterproofed joints	-	-	-	-	m	**5.00**
J DIAPHRAGM WALLS						
Excavation, disposal of soil and placing concrete	-	-	-	-	m³	**400.00**
Provide and place reinforcement cages	-	-	-	-	t	**650.00**
Excavate/chisel in hard materials/rock	-	-	-	-	hr	**950.00**

SERIES 1700: STRUCTURAL CONCRETE

Item	Gang hours	Labour £	Plant £	Material £	Unit	Total rate £
GENERAL						
Notes						
Refer also to Civil Engineering - Concrete, Formwork, Reinforcement and Precast Concrete, although this section is fundamentally different in that the provision of concrete of different classes and its placement is combined in the unit costs.						
RESOURCES - LABOUR						
Concreting gang						
1 ganger/chargehand (skill rate 4)		12.80				
2 skilled operatives (skill rate 4)		24.10				
4 unskilled operatives (general)		45.04				
1 plant operator (skill rate 3) - 25% of time		3.66				
Total Gang Rate / Hour	£	**85.60**				
Formwork gang						
1 foreman (craftsman)		18.47				
2 joiners (craftsman)		34.70				
1 unskilled operative (general)		11.26				
1 plant operator (craftsman) - 25% of time		4.68				
Total Gang Rate / Hour	£	**69.11**				
Reinforcement gang						
1 foreman (craftsman)		18.47				
4 steel fixers (craftsman)		69.40				
1 unskilled operative (general)		11.26				
1 plant operator (craftsman) - 25% of time		4.68				
Total Gang Rate / Hour	£	**103.81**				
RESOURCES - PLANT						
Concreting						
22RB crane- 25% of time			5.54			
gas oil for ditto			0.10			
0.76 m³ concrete skip - 25% of time			0.23			
11.3 m³/min compressor, 4 tool			10.11			
gas oil for ditto			2.55			
4 poker vibrators P5475 mm or less in thickness			4.96			
Total Gang Rate / Hour		£	**23.49**			
Formwork						
20 t crawler crane - 25% of time			5.07			
gas oil for ditto			0.11			
22" saw bench			1.16			
gas oil for ditto			0.15			
small power tools (formwork)			1.17			
Total Gang Rate / Hour		£	**7.65**			
Reinforcement						
30 t crawler crane (25% of time)			6.60			
gas oil for ditto			0.11			
bar cropper			1.86			
small power tools (reinforcement)			0.63			
tirfors, kentledge etc.			0.81			
Total Gang Rate / Hour		£	**10.01**			

SERIES 1700: STRUCTURAL CONCRETE

Item	Gang hours	Labour £	Plant £	Material £	Unit	**Total rate £**
RESOURCES - MATERIALS						
Wastage allowances 5% onto delivered ready mixed concrete						
IN-SITU CONCRETE						
In-situ concrete Grade C10						
Blinding						
75 mm or less in thickness	0.18	15.31	4.23	78.48	m^3	**98.01**
Blinding; in narrow widths up to 1.0 m wide or in bottoms of trenches up to 2.5 m wide; excluding formwork						
75 mm or less in thickness	0.20	17.01	4.70	78.48	m^3	**100.19**
In-situ concrete Grade C15						
Blinding; excluding formwork						
75 mm or less in thickness	0.16	13.61	3.76	80.11	m^3	**97.48**
Blinding; in narrow widths up to 1.0 m wide or in bottoms of trenches up to 2.5 m wide; excluding formwork						
75 mm or less in thickness	0.18	15.31	4.23	80.11	m^3	**99.65**
In-situ concrete Grade C20/20						
Bases, footings, pile caps and ground beams; thickness						
ne 150 mm	0.20	17.01	4.70	77.07	m^3	**98.78**
150 - 300 mm	0.17	14.46	4.01	77.07	m^3	**95.54**
300 - 500 mm	0.15	12.76	3.54	77.07	m^3	**93.37**
exceeding 500 mm	0.14	11.91	3.29	77.07	m^3	**92.27**
Walls; thickness						
ne 150 mm	0.21	17.86	4.95	77.07	m^3	**99.88**
150 - 300 mm	0.15	12.76	3.54	77.07	m^3	**93.37**
300 - 500 mm	0.13	11.06	3.07	77.07	m^3	**91.20**
exceeding 500 mm	0.12	10.21	2.82	77.07	m^3	**90.10**
Suspended slabs; thickness						
ne 150 mm	0.27	22.97	6.36	77.07	m^3	**106.40**
150 - 300 mm	0.21	17.86	4.95	77.07	m^3	**99.88**
300 - 500 mm	0.19	16.16	4.48	77.07	m^3	**97.71**
exceeding 500 mm	0.19	16.16	4.48	77.07	m^3	**97.71**
Columns, piers and beams; cross-sectional area						
ne 0.03 m^2	0.50	42.53	11.74	77.07	m^3	**131.34**
0.03 - 0.10 m^2	0.40	34.02	9.39	77.07	m^3	**120.49**
0.10 - 0.25 m^2	0.35	29.77	8.24	77.07	m^3	**115.08**
0.12 - 1.00 m^2	0.35	29.77	8.24	77.07	m^3	**115.08**
exceeding 1 m^2	0.28	23.82	6.58	77.07	m^3	**107.46**
ADD to the above prices for						
sulphate resisting cement	-	-	-	6.44	m^3	**6.44**
air entrained concrete	-	-	-	4.89	m^3	**4.89**
water repellant concrete	-	-	-	4.93	m^3	**4.93**
In-situ concrete Grade C30/20						
Bases, footings, pile caps and ground beams; thickness						
ne 150 mm	0.21	17.86	4.95	79.64	m^3	**102.46**
150 - 300 mm	0.18	15.31	4.23	79.64	m^3	**99.18**
300 - 500 mm	0.15	12.76	3.54	79.64	m^3	**95.94**
exceeding 500 mm	0.14	11.91	3.29	79.64	m^3	**94.84**

SERIES 1700: STRUCTURAL CONCRETE

Item	Gang hours	Labour £	Plant £	Material £	Unit	Total rate £
IN-SITU CONCRETE – cont'd						
In-situ concrete Grade C30/20 – cont'd						
Walls; thickness						
ne 150 mm	0.22	18.71	5.17	79.64	m³	**103.52**
150 - 300 mm	0.16	13.61	4.70	79.64	m³	**97.95**
300 - 500 mm	0.13	11.06	3.83	79.64	m³	**94.53**
exceeding 500 mm	0.12	10.21	3.52	79.64	m³	**93.37**
Suspended slabs; thickness						
ne 150 mm	0.28	23.82	8.22	79.64	m³	**111.68**
150 - 300 mm	0.22	18.71	6.46	79.64	m³	**104.81**
300 - 500 mm	0.19	16.16	5.60	79.64	m³	**101.40**
exceeding 500 mm	0.18	15.31	5.28	79.64	m³	**100.24**
Columns, piers and beams; cross-sectional area						
ne 0.03 m²	0.53	45.08	15.57	79.64	m³	**140.30**
0.03 - 0.10 m²	0.42	35.73	12.33	79.64	m³	**127.69**
0.10 - 0.25 m²	0.36	30.62	10.56	79.64	m³	**120.83**
0.25 - 1.00 m²	0.35	29.77	10.29	79.64	m³	**119.71**
exceeding 1 m²	0.28	23.82	8.22	79.64	m³	**111.68**
ADD to the above prices for						
sulphate resisting cement	-	-	-	6.44	m³	**6.44**
air entrained concrete	-	-	-	4.89	m³	**4.89**
water repellant concrete	-	-	-	4.93	m³	**4.93**
PRECAST CONCRETE						
The cost of precast concrete item is very much dependant on the complexity of the moulds, the number of units to be cast from each mould and the size and the weight of the unit to be handled. The unit rates below are for standard precast items that are often to be found on a Civil Engineering project. It would be misleading to quote for indicative costs for tailor-made precast concrete units and it is advisable to contact specialist maunfacturers for guide prices.						
Pretensioned prestressed beams; concrete Grade C20						
Beams						
100 x 65 x 1050 mm long	1.00	3.01	-	6.74	nr	**9.75**
265 x 65 x 1800 mm long	1.00	3.76	2.53	25.46	nr	**31.75**
Inverted 'T' beams, flange width 495 mm						
section T1; 8 m long, 380 mm deep; mass 1.88t	-	-	-	-	nr	**661.04**
section T2; 9 m long, 420 mm deep; mass 2.29t	-	-	-	-	nr	**791.87**
section T3; 11 m long, 535 mm deep; mass 3.02t	-	-	-	-	nr	**929.59**
section T4; 12 m long, 575 mm deep; mass 3.54t	-	-	-	-	nr	**1032.87**
section T5; 13 m long, 615 mm deep; mass 4.08t	-	-	-	-	nr	**1067.30**
section T6; 13 m long, 655 mm deep; mass 4.33t	-	-	-	-	nr	**1067.30**
section T7; 12 m long, 695 mm deep; mass 4.95t	-	-	-	-	nr	**1239.44**
section T8; 15 m long, 735 mm deep; mass 5.60t	-	-	-	-	nr	**1342.75**
section T9; 16 m long, 775 mm deep; mass 6.28t	-	-	-	-	nr	**1446.04**
section T10; 18 m long, 815 mm deep; mass 7.43t	-	-	-	-	nr	**1618.18**

SERIES 1700: STRUCTURAL CONCRETE

Item	Gang hours	Labour £	Plant £	Material £	Unit	**Total rate £**
'M' beams, flange width 970 mm						
section M2 ; 17 m long, 720 mm deep; mass 12.95t	-	-	-	-	nr	**3511.77**
section M3 ; 18 m long, 800 mm deep; mass 15.11t	-	-	-	-	nr	**3305.21**
section M6 ; 22 m long, 1040 mm deep; mass 20.48t	-	-	-	-	nr	**5302.10**
section M8 ; 25 m long, 1200 mm deep; mass 23.68t	-	-	-	-	nr	**6885.84**
'U' beams, base width 970 mm						
section U3 ; 16 m long, 900 mm deep; mass 19.24t	-	-	-	-	nr	**5921.81**
section U5 ; 20 m long, 1000 mm deep; mass 25.64t	-	-	-	-	nr	**7574.42**
section U8 ; 24 m long, 1200 mm deep; mass 34.56t	-	-	-	-	nr	**10259.92**
section U12 ; 30 m long, 1600 mm deep; mass 52.74t	-	-	-	-	nr	**14047.12**
Precast concrete culverts, cattle creeps and subway units; rebated joints						
Rectangular cross section						
500 mm high x 1000 mm wide	1.00	4.08	4.42	199.36	m	**207.86**
1000 mm high x 1500 mm wide	1.00	9.08	10.14	367.89	m	**387.11**
1500 mm high x 1500 mm wide	1.00	16.46	18.20	317.68	m	**352.35**
2000 mm high x 2750 mm wide	1.00	30.89	28.96	1333.04	m	**1392.89**
2750 mm high x 3000 mm wide	1.00	40.88	58.14	1486.39	m	**1585.41**
Extra for units curved on plan to less than 20 m radius	-	-	-	241.64	m	**241.64**

SERIES 1700: STRUCTURAL CONCRETE

Item	Gang hours	Labour £	Plant £	Material £	Unit	**Total rate £**
SURFACE FINISH OF CONCRETE – FORMWORK						
Materials						
Formwork materials include for shutter, bracing, ties, support, kentledge and all consumables. These unit costs are based upon those outputs and prices detailed in Civil Engineering - Concrete Formwork but are referenced to The Specification for Highway Works, clause 1708. The following unit rates do not include for formwork outside the payline and are based on an optimum of a minimum 8 uses with 10% per use towards the cost of repairs / replacement of components damaged during disassembly.						
ADJUST formwork material costs generally depending on the number of uses:						
Nr of Uses / % Adjustment / Inclusion for waste						
1 / Add 90 - 170% / 7%						
2 / Add 50 - 180% / 7%						
3 / Add 15 - 30% / 6%						
6 / Add 5 - 10% / 5%						
8 / No change / 5%						
10 / Deduct 5 - 7% / 5%						
Definitions						
'Class F1' formwork is rough finish						
'Class F2' formwork is fair finish						
'Class F3' formwork is extra smooth finish						
Formwork Class F1						
Horizontal more than 300 mm wide	0.52	35.94	3.98	5.94	m²	**45.85**
Inclined more than 300 mm wide	0.55	38.01	4.21	8.10	m²	**50.33**
Vertical more than 300 mm wide	0.61	42.16	4.67	8.70	m²	**55.53**
300 mm wide or less at any inclination	0.72	49.76	5.51	8.08	m²	**63.35**
Curved of both girth and width more than 300 mm at any inclination	0.95	65.65	7.27	9.77	m²	**82.70**
Curved of girth or width of 300 mm or less at any inclination	0.72	49.76	5.51	9.77	m²	**65.04**
Domed	1.20	82.93	9.18	12.29	m²	**104.41**
Void former cross-section 100 x 100 mm	0.07	4.84	0.18	2.85	m	**7.86**
Void former cross-section 250 x 250 mm	0.12	8.29	0.30	7.26	m	**15.85**
Void former cross-section 500 x 500 mm	0.30	20.73	0.74	14.46	m	**35.94**
Formwork Class F2						
Horizontal more than 300 mm wide	0.54	37.32	4.13	10.92	m²	**52.38**
Inclined more than 300 mm wide	0.57	39.39	4.37	16.93	m²	**60.69**
Vertical more than 300 mm wide	0.63	43.54	4.82	17.55	m²	**65.92**
300 mm wide or less at any inclination	0.74	51.14	5.66	16.93	m²	**73.74**
Curved of both girth and width more than 300 mm at any inclination	0.98	67.73	7.50	20.56	m²	**95.79**
Curved of girth or width of 300 mm or less at any inclination	0.75	51.83	5.74	20.56	m²	**78.13**
Domed	1.40	96.75	10.72	26.62	m²	**134.09**

SERIES 1700: STRUCTURAL CONCRETE

Item	Gang hours	Labour £	Plant £	Material £	Unit	**Total rate £**
Formwork Class F3						
Horizontal more than 300 mm wide	0.56	38.70	4.29	12.94	m²	**55.92**
Inclined more than 300 mm wide	0.59	40.77	4.52	18.94	m²	**64.23**
Vertical more than 300 mm wide	0.65	44.92	4.98	19.56	m²	**69.46**
300 mm wide or less at any inclination	0.76	52.52	5.82	18.94	m²	**77.28**
Curved of both girth and width more than 300 mm at any inclination	0.99	68.42	7.58	22.57	m²	**98.57**
Curved of girth or width of 300 mm or less at any inclination	0.77	53.21	5.90	22.57	m²	**81.68**
Domed	1.45	100.21	11.10	28.63	m²	**139.94**
Formwork ancillaries						
Allowance for additional craneage and rub up where required	0.13	8.98	1.00	0.17	m²	**10.15**
STEEL REINFORCEMENT FOR STRUCTURES						
Stainless steel bars						
Bar reinforcement nominal size 16 mm and under not exceeding 12 m in length	6.74	699.68	67.46	2500.00	tonne	**3267.14**
Bar reinforcement nominal size 20 mm and over not exceeding 12 m in length						
20 mm nominal size	4.44	460.92	44.44	2360.00	tonne	**2865.36**
25 mm nominal size	4.44	460.92	44.44	2300.00	tonne	**2805.36**
32 mm nominal size	4.44	460.92	44.44	2300.00	tonne	**2805.36**
ADD to the above for bars						
12 - 13.5 m long	-	-	-	15.25	tonne	**15.25**
13.5 - 15 m long	-	-	-	15.25	tonne	**15.25**
over 15 m long; per 500 mm increment	-	-	-	3.40	tonne	**3.40**
High yield steel bars BS 4449; deformed, Grade 460, type 2						
Bar reinforcement nominal size 16 mm and under not exceeding 12 m in length	6.74	699.68	67.46	370.00	tonne	**1137.14**
Bar reinforcement nominal size 20 mm and over not exceeding 12 m in length						
20 mm nominal size	4.44	460.92	44.44	410.00	tonne	**915.36**
25 mm nominal size	4.44	460.92	44.44	410.00	tonne	**915.36**
32 mm nominal size	4.44	460.92	44.44	410.00	tonne	**915.36**
40 mm nominal size	4.44	460.92	44.44	418.00	tonne	**923.36**
ADD to the above for bars						
12 - 13.5 m long	-	-	-	15.25	-	**15.25**
13.5 - 15 m long	-	-	-	15.25	-	**15.25**
over 15 m long; per 500 mm increment	-	-	-	3.40	-	**3.40**
Helical reinforcement nominal size 16 mm and under ne 12 m in length	6.74	699.68	67.46	426.97	tonne	**1194.12**
Helical reinforcement nominal size 20 mm and over						
20 mm nominal size	4.44	460.92	44.44	414.10	tonne	**919.46**
25 mm nominal size	4.44	460.92	44.44	414.10	tonne	**919.46**
32 mm nominal size	4.44	460.92	44.44	414.10	tonne	**919.46**
40 mm nominal size	4.44	460.92	44.44	422.18	tonne	**927.54**

SERIES 1700: STRUCTURAL CONCRETE

Item	Gang hours	Labour £	Plant £	Material £	Unit	Total rate £
STEEL REINFORCEMENT FOR STRUCTURES – cont'd						
Dowels						
16 mm diameter x 600 mm long	0.10	10.38	1.00	1.93	nr	**13.31**
20 mm diameter x 600 mm long	0.10	10.38	1.00	2.59	nr	**13.98**
25 mm diameter x 600 mm long	0.10	10.38	1.00	3.70	nr	**15.08**
32 mm diameter x 600 mm long	0.10	10.38	1.00	5.69	nr	**17.07**
Mild steel bars BS 4449; Grade 250						
Bar reinforcement nominal size 16 mm and under ne 12 m in length	6.74	699.68	67.46	456.04	tonne	**1223.18**
Bar reinforcement nominal size 20 mm and over; not exceeding 12 m in length						
20 mm nominal size	4.44	460.92	44.44	440.93	tonne	**946.29**
25 mm nominal size	4.44	460.92	44.44	440.93	tonne	**946.29**
32 mm nominal size	4.44	460.92	44.44	434.30	tonne	**939.66**
40 mm nominal size	4.44	460.92	44.44	462.64	tonne	**968.00**
ADD to the above for bars						
12 - 13.5 m long	-	-	-	15.25	tonne	**15.25**
13.5 - 15 m long	-	-	-	15.25	tonne	**15.25**
over 15 m long, per 500 mm increment	-	-	-	3.40	tonne	**3.40**
ADD for cutting, bending, tagging and baling reinforcement on site						
6 mm nominal size	4.87	162.12	48.75	1.20	tonne	**212.07**
8 mm nominal size	4.58	152.47	45.84	1.20	tonne	**199.51**
10 mm nominal size	3.42	113.85	34.23	1.20	tonne	**149.28**
12 mm nominal size	2.55	84.89	25.53	1.20	tonne	**111.62**
16 mm nominal size	2.03	67.58	20.32	1.20	tonne	**89.10**
20 mm nominal size	1.68	55.93	16.82	1.20	tonne	**73.94**
25 mm nominal size	1.68	55.93	16.82	1.20	tonne	**73.94**
32 mm nominal size	1.39	46.27	13.92	1.20	tonne	**61.39**
40 mm nominal size	1.39	46.27	13.92	1.20	tonne	**61.39**
Fabric reinforcement; high yield steel BS 4483						
Fabric reinforcement						
BS ref A98; nominal mass 1.54 kg/m²	0.03	3.11	0.30	1.25	m²	**4.67**
BS ref A142; nominal mass 2.22 kg/m²	0.03	3.11	0.30	1.30	m²	**4.72**
BS ref A193; nominal mass 3.02 kg/m²	0.04	4.15	0.40	1.77	m²	**6.32**
BS ref A252; nominal mass 3.95 kg/m²	0.04	4.15	0.40	2.30	m²	**6.85**
BS ref A393; nominal mass 6.16 kg/m²	0.07	7.27	0.70	3.60	m²	**11.57**
BS ref B196; nominal mass 3.05 kg/m²	0.04	4.15	0.40	1.78	m²	**6.34**
BS ref B283; nominal mass 3.73 kg/m²	0.04	4.15	0.40	2.20	m²	**6.75**
BS ref B385; nominal mass 4.53 kg/m²	0.05	5.19	0.50	2.65	m²	**8.34**
BS ref B503; nominal mass 5.93 kg/m²	0.05	5.19	0.50	3.47	m²	**9.17**
BS ref B785; nominal mass 8.14 kg/m²	0.08	8.30	0.80	4.76	m²	**13.87**
BS ref B1131; nominal mass 10.90 kg/m²	0.09	9.34	0.90	6.37	m²	**16.62**
BS ref C282; nominal mass 2.61 kg/m²	0.03	3.11	0.30	1.53	m²	**4.95**
BS ref C385; nominal mass 3.41 kg/m²	0.04	4.15	0.40	2.00	m²	**6.55**
BS ref C503; nominal mass 4.34 kg/m²	0.05	5.19	0.50	2.54	m²	**8.24**
BS ref C636; nominal mass 5.55 kg/m²	0.05	5.19	0.50	3.25	m²	**8.95**
BS ref C785; nominal mass 6.72 kg/m²	0.07	7.27	0.70	3.91	m²	**11.88**
BS ref D49; nominal mass 0.77 kg/m²	0.02	2.08	0.20	1.37	m²	**3.64**
BS ref D98; nominal mass 1.54 kg/m²	0.02	2.08	0.20	1.37	m²	**3.64**

SERIES 1800: STEELWORK FOR STRUCTURES

Item	Gang hours	Labour £	Plant £	Material £	Unit	**Total rate £**
FABRICATION OF STEELWORK						
Steelwork to BS 7613; Grade S275						
Fabrication of main members						
rolled sections	-	-	-	-	tonne	**1163.69**
plated rolled sections	-	-	-	-	tonne	**2999.90**
plated girders	-	-	-	-	tonne	**2501.18**
box girders	-	-	-	-	tonne	**1186.50**
Fabrication of deck panels						
rolled sections	-	-	-	-	tonne	**2002.46**
plated rolled sections	-	-	-	-	tonne	**1828.66**
plated girders	-	-	-	-	tonne	**1828.66**
Fabrication of subsiduary steelwork						
rolled sections	-	-	-	-	tonne	**1828.66**
plated rolled sections	-	-	-	-	tonne	**1828.66**
plated girders	-	-	-	-	tonne	**1911.78**
ERECTION OF STEELWORK						
Trial erection at the place of fabrication	-	-	-	-	tonne	**290.71**
Permanent erection of steelwork; substructure	-	-	-	-	tonne	**221.48**
Permanent erection of steelwork; superstructure	-	-	-	-	tonne	**221.48**
MISCELLANEOUS METALWORK						
Mild steel						
Ladders						
Cat ladder; 64 x 13 mm bar strings; 19mm rungs at 250mm centres; 450 mm wide with safety hoops	-	-	-	-	m	**318.38**
Handrails						
Galvanised tubular metal; 76 mm diameter handrail, 48 mm diameter standards at 750 mm centres, 48 mm diameter rail; 1070 mm high overall	-	-	-	-	m	**119.76**
Metal access cover and frame						
Group 4, ductile iron, single seal 610 x 610 x 100 mm depth; D400	-	-	-	-	nr	**161.30**
Group 2, ductile iron, double seal single piece cover 600 x 450 mm; B125	-	-	-	-	nr	**130.77**

SERIES 1900: PROTECTION OF STEELWORK AGAINST CORROSION

Item	Gang hours	Labour £	Plant £	Material £	Unit	Total rate £
RESOURCES - LABOUR						
Protective painting gang						
1 ganger/chargehand (skill rate 4)		12.80				
2 skilled operatives (skill rate 4)		24.10				
2 unskilled operatives (general)		22.52				
Total Gang rate / Hour	£	**59.42**				
RESOURCES - PLANT						
Protective painting						
power tools (protection of steelwork)			2.92			
access scaffolding, trestles and ladders			1.71			
5 t transit van (50% of time)			3.40			
gas oil for ditto			0.56			
Total Gang Rate / Hour		£	**8.59**			
RESOURCES - MATERIALS						
Resources - materials						
All coats applied off site except as noted						
The external environment has been taken as Inland 'B' Exposed.						
PROTECTIVE SYSTEM						
Galvanising to BS 729; apply protective coatings comprising: 1st coat: Mordant T wash; 2nd coat: Zinc rich epoxy primer; 3rd coat: Zinc phosphate, CR/Alkyd Undercoat; 4th coat; MIO CR Undercoat-on site externally; 5th coat: CR coloured finish-on site externally						
To metal parapets and fencing, lighting columns, brackets						
by brush or airless spray to dry film thickness 200 microns	0.20	11.88	1.72	15.41	m²	**29.01**
Blast clean to BS 7079 (surface preparation); apply protective coatings comprising: 1st coat: Zinc Chromate,Red Oxide Blast Primer; 2nd coat: Zinc Phosphate, Epoxy Ester Undercoat; 3rd coat: MIO Undercoat; 4th coat: MIO coloured finish-on site externally						
To subsidiary steelwork, interior finishes						
By brush or airless spray to dry film thickness 175 microns	0.15	8.91	1.29	11.12	m²	**21.33**

SERIES 1900: PROTECTION OF STEELWORK AGAINST CORROSION

Item	Gang hours	Labour £	Plant £	Material £	Unit	**Total rate £**
Blast clean to BS 7079 (surface preparation); apply metal coating of aluminium spray at works; apply protective coatings comprising:						
1st coat: Zinc Chromate Etch Primer (2 pack)						
2nd coat: Zinc Phosphate, CR/Aalkyd Undercoat						
3rd coat: Zinc Phosphate, CR/Aalkyd Undercoat						
4th coat: MIO CR Undercoat						
5th coat: CR finish-on site externally						
To main steel members						
By brush or airless spray to dry film thickness 250 microns	0.18	10.70	1.55	12.60	m²	**24.84**
Blast clean to BS 7079 (second quality surface preparation); remove all surface defects to BS 7613;apply protective coatings comprising:						
1st coat: Zinc rich primer (2 pack)						
2nd coat: Epoxy High Build M10 (2 pack)						
3rd coat: Polyurethane Undercoat (2 pack) -on site internally						
4th coat: Finish coat polyurethane (2 pack) -on site externally						
To internal steel members						
By brush or airless spray	0.15	8.91	1.29	17.42	m²	**27.63**
ALTERNATIVE SURFACE TREATMENTS						
Galvanising (Hot dip) to BS 729, assuming average depth 20 m² per tonne of steel	-	-	-	-	m²	**12.85**
Shot blasting (at works)	-	-	-	-	m²	**2.59**
Grit blasting (at works)	-	-	-	-	m²	**3.69**
Sand blasting (at works)	-	-	-	-	m²	**5.54**
Shot blasting (on site)	-	-	-	-	m²	**3.62**
Grit blasting (on site)	-	-	-	-	m²	**5.16**
Sand blasting (on site)	-	-	-	-	m²	**7.76**

SERIES 2000:WATERPROOFING FOR STRUCTURES

Item	Gang hours	Labour £	Plant £	Material £	Unit	Total rate £
GENERAL						
Notes						
This section is based around the installation of proprietary systems to new/recently completed works as part of major scheme, for minor works/repairs outputs will be many times more. Outputs are also based on use of skilled labour, therefore effieciency is high and wastage low (5-7% only allowed) excepting laps where required.						
RESOURCES - LABOUR						
Asphalting gang						
1 ganger/chargehand (skill rate 4)		12.80				
2 unskilled operative (general)		22.52				
Total Gang rate / Hour	£	**35.32**				
Damp proofing gang						
1 ganger/chargehand (skill rate 4)		12.80				
1 skilled operative (skill rate 4)		12.05				
1 unskilled operative (general)		11.26				
Total Gang rate / Hour	£	**36.11**				
Sprayed/brushed waterproofing gang						
1 ganger/chargehand (skill rate 4) - 30% of time		3.84				
1 skilled operative (skill rate 4)		12.05				
Total Gang rate / Hour	£	**15.89**				
Protective layers - screed gang						
1 ganger/chargehand (skill rate 4)		12.80				
1 skilled operative (skill rate 4)		12.05				
1 unskilled operative (general)		11.26				
Total Gang rate / Hour	£	**36.11**				
RESOURCES - PLANT						
Asphalting						
45 litre portable tar boiler including sprayer (50 % of time)			0.12			
2 t dumper (50% of time)			1.58			
gas oil for ditto			0.24			
Total Gang Rate / Hour		£	**1.94**			
Damp proofing						
2 t dumper (50% of time)			1.58			
gas oil for ditto			0.24			
Total Gang Rate / Hour		£	**1.82**			
WATERPROOFING						
Mastic asphalt; BS 6925 Type T 1097; 20 mm thick; two coats						
over 300 mm wide; ne 30 degrees to horizontal	0.33	11.66	0.97	9.78	m²	**22.40**
over 300 mm wide; 30 - 90 degrees to horizontal	0.50	17.66	0.64	9.78	m²	**28.08**
ne 300 mm wide; at any inclination	0.60	21.19	0.64	9.78	m²	**31.61**
to domed surfaces	0.75	26.49	0.64	9.78	m²	**36.91**

SERIES 2000:WATERPROOFING FOR STRUCTURES

Item	Gang hours	Labour £	Plant £	Material £	Unit	**Total rate £**
'Bitu-thene' 1000 X; lapped joints						
over 300 mm wide; ne 30 degrees to horizontal	0.05	1.81	0.09	7.08	m²	**8.98**
over 300 mm wide; 30 - 90 degrees to horizontal	0.06	2.17	0.11	9.03	m²	**11.31**
ne 300 mm wide; at any inclination	0.08	2.89	0.15	9.04	m²	**12.07**
Extra; one coat primer on vertical surfaces	0.03	1.08	0.05	0.01	m²	**1.15**
'Bitu-thene' 4100 X; lapped joints						
over 300 mm wide; ne 30 degrees to horizontal	0.05	1.81	0.09	8.64	m²	**10.54**
over 300 mm wide; 30 - 90 degrees to horizontal	0.06	2.17	0.60	11.52	m²	**14.29**
ne 300 mm wide; at any inclination	0.08	2.89	0.69	11.52	m²	**15.10**
'Famguard' (hot applied) with Fam-primer						
over 300 mm wide; ne 30 degrees to horizontal	0.32	11.56	0.58	15.87	m²	**28.01**
over 300 mm wide; 30 - 90 degrees to horizontal	0.35	12.64	0.64	16.18	m²	**29.46**
ne 300 mm wide; at any inclination	0.40	14.44	0.73	18.47	m²	**33.64**
'Famflex' (hot applied) with Fam-primer						
over 300 mm wide; ne 30 degrees to horizontal	0.32	11.56	0.58	12.23	m²	**24.36**
over 300 mm wide; 30 - 90 degrees to horizontal	0.34	12.28	0.62	12.46	m²	**25.35**
ne 300 mm wide; at any inclination	0.38	13.72	0.69	14.14	m²	**28.55**
One coat of Ventrot primer; one coat Ventrot hot applied damp proof membrane sprayed or brushed on						
over 300 mm wide; ne 30 degrees to horizontal	0.03	0.48	0.05	5.82	m²	**6.35**
over 300 mm wide; 30 - 90 degrees to horizontal	0.04	0.64	0.07	5.82	m²	**6.53**
ne 300 mm wide; at any inclination	0.05	0.79	0.09	5.82	m²	**6.71**
Two coats of RIW liquid asphaltic composition sprayed or brushed on						
over 300 mm wide; ne 30 degrees to horizontal	0.03	0.48	0.05	3.49	m²	**4.02**
over 300 mm wide; 30 - 90 degrees to horizontal	0.03	0.48	0.05	3.49	m²	**4.02**
ne 300 mm wide; at any inclination	0.04	0.64	0.07	3.49	m²	**4.19**
Two coats of 'Mulseal' sprayed or brushed on						
any inclination	0.07	1.11	0.13	0.01	m²	**1.25**
20 mm thick red tinted sand asphalt layer						
onto bridge deck	0.02	0.72	0.22	9.98	m²	**10.92**
SURFACE IMPREGNATION OF CONCRETE						
Silane waterproofing						
Surface impregnation to plain surfaces	-	-	-	-	m²	**3.29**
REMOVALOF EXISTING WATERPROOFING						
Removal of existing asphalt waterproofing						
over 300 mm wide; ne 30 degrees to horizontal	0.13	4.59	-	-	m²	**4.59**
over 300 mm wide; 30-90 degrees to horizontal	0.18	6.36	-	-	m²	**6.36**
over 300 mm wide; at any inclination	0.20	7.06	-	-	m²	**7.06**
to domed surfaces	0.25	8.83	-	-	m²	**8.83**

SERIES 2100: BRIDGE BEARINGS

Item	Gang hours	Labour £	Plant £	Material £	Unit	Total rate £
GENERAL						
Notes						
Bridge bearings are manufactured and installed to individual specifications. The following guide prices are for different sizes of simple bridge bearings. If requirements are known, then advice ought to be obtained from specialist manufacturers such as CCL.						
RESOURCES - LABOUR						
Bridge bearing gang						
1 ganger/chargehand (skill rate 4)		12.80				
2 unskilled operatives (general)		22.52				
Total Gang Rate / Hour	£	**35.32**				
BEARINGS						
Supply plain rubber bearings (3 m and 5 m lengths)						
150 x 20 mm	0.35	12.36	-	33.30	m	**45.66**
150 x 25 mm	0.35	12.36	-	42.89	m	**55.25**
Supply and place in position laminated elastomeric rubber bearing						
250 x 150 x 19 mm	0.25	8.83	-	11.85	nr	**20.68**
300 x 200 x 19 mm	0.25	8.83	-	17.78	nr	**26.61**
300 x 200 x 30 mm	0.27	9.54	-	28.44	nr	**37.98**
300 x 200 x 41 mm	0.27	9.54	-	37.93	nr	**47.46**
300 x 250 x 41 mm	0.30	10.60	-	47.41	nr	**58.00**
300 x 250 x 63 mm	0.30	10.60	-	73.48	nr	**84.08**
400 x 250 x 19 mm	0.32	11.30	-	29.63	nr	**40.93**
400 x 250 x 52 mm	0.32	11.30	-	80.59	nr	**91.90**
400 x 300 x 19 mm	0.32	11.30	-	35.56	nr	**46.86**
600 x 450 x 24 mm	0.35	12.36	-	99.56	nr	**111.92**
Adhesive fixings to laminated elastomeric rubber bearings						
2 mm thick epoxy adhesive	1.00	35.32	-	46.50	m²	**81.82**
15 mm thick epoxy mortar	1.50	52.98	-	259.86	m²	**312.84**
15 mm thick epoxy pourable grout	2.00	70.64	-	252.29	m²	**322.94**
Supply and install mechanical guides for laminated elastomeric rubber bearings						
500kN SLS design load; FP50 fixed pin Type 1	2.00	70.64	-	441.00	nr	**511.64**
500kN SLS design load; FP50 fixed pin Type 2	2.00	70.64	-	551.25	nr	**621.89**
750kN SLS design load; FP75 fixed pin Type 1	2.10	74.17	-	716.63	nr	**790.80**
750kN SLS design load; FP75 fixed pin Type 2	2.10	74.17	-	826.88	nr	**901.05**
300kN SLS design load; UG300 Uniguide Type 1	2.00	70.64	-	441.00	nr	**511.64**
300kN SLS design load; UG300 Uniguide Type 2	2.00	70.64	-	551.25	nr	**621.89**
Supply and install fixed pot bearings						
355 x 355; PF200	2.00	70.64	-	551.25	nr	**621.89**
425 x 425; PF300	2.10	74.17	-	606.38	nr	**680.55**

SERIES 2100: BRIDGE BEARINGS

Item	Gang hours	Labour £	Plant £	Material £	Unit	**Total rate £**
Supply and install free sliding pot bearings						
445 x 345; PS200	2.10	74.17	-	441.00	nr	**515.17**
520 x 415; PS300	2.20	77.70	-	661.50	nr	**739.20**
Supply and install guided sliding pot bearings						
455 x 375; PG200	2.20	77.70	-	716.63	nr	**794.33**
545 x 435; PG300	2.30	81.24	-	826.88	nr	**908.11**
TESTING BEARINGS						
If there is a requirement for testing bridge bearings prior to their being installed then the tests should be enumerated separately. Specialist advice should be sought once details are known.						
Compression test for laminated elastomeric bearings						
generally	-	-	-	-	nr	**52.50**
Shear test for laminated elastomeric bearings						
generally	-	-	-	-	nr	**68.25**
Bond test for elastomeric bearings (Exclusive of cost of bearings as this is a destructive test)						
generally	-	-	-	-	nr	**283.50**

SERIES 2200: PARAPETS

Item	Gang hours	Labour £	Plant £	Material £	Unit	Total rate £
GENERAL						
Notes						
The heights of the following parapets are in accordance with the Standard Designs and DTp requirements. The rates include for all anchorages and fixings and in the case of steel, galvanising at works and painting four coat paint system on site together with etching the galvanised surface, as necessary.						
RESOURCES - LABOUR						
Parapet gang						
1 ganger/chargehand (skill rate 4)		12.80				
1 skilled operative (skill rate 4)		12.05				
2 unskilled operatives (general)		22.52				
1 plant operator (skill rate 4)		13.17				
Total Gang Rate / Hour	£	**60.54**				
RESOURCES - PLANT						
Parapets						
agricultural type tractor; front bucket - 50% of time			6.50			
gas oil for ditto			0.70			
2.80 m3/min (100 cfm) compressor; two tool			3.28			
gas oil for ditto			1.02			
compressor tools: heavy rock drill 33, 84 cfm			1.94			
compressor tools: rotary drill, 10 cfm			0.98			
8 t lorry with 1 t hiab - 50% of time			9.53			
gas oil for ditto			0.57			
Total Gang Rate / Hour		£	**24.52**			
STEEL PARAPETS						
Metal parapet Group P1; 1.0 m high; comprising steel yielding posts and steel horizontal rails						
straight or curved exceeding 50 m radius	-	-	-	155.95	m	**155.95**
curved not exceeding 50 m radius	-	-	-	168.94	m	**168.94**
Metal parapet Group P2 (48 Kph); 1.0 m high; comprising steel yielding posts and steel horizontal rails with vertical infill bars						
straight or curved exceeding 50 m radius	-	-	-	220.93	m	**220.93**
curved not exceeding 50 m radius	-	-	-	233.92	m	**233.92**
Metal parapet Group P2 (80 Kph); 1.0 m high; comprising steel yielding posts and steel horizontal rails						
straight or curved exceeding 50 m radius	-	-	-	181.94	m	**181.94**
curved not exceeding 50 m radius	-	-	-	194.94	m	**194.94**
Metal parapet Group P2 (113 Kph); 1.0 m high; comprising steel yielding posts and steel horizontal rails						
straight or curved exceeding 50 m radius	-	-	-	181.94	m	**181.94**
curved not exceeding 50 m radius	-	-	-	194.94	m	**194.94**

SERIES 2200: PARAPETS

Item	Gang hours	Labour £	Plant £	Material £	Unit	Total rate £
Metal parapet Group P4; 1.15 m high; comprising steel yielding posts and steel horizontal rails						
straight or curved exceeding 50 m radius	-	-	-	155.95	m	**155.95**
curved not exceeding 50 m radius	-	-	-	168.94	m	**168.94**
Metal parapet Group P5; 1.25 m high; comprising steel yielding posts and steel horizontal rails						
straight or curved exceeding 50 m radius	-	-	-	197.91	m	**197.91**
curved not exceeding 50 m radius	-	-	-	220.93	m	**220.93**
Metal parapet Group P5; 1.50 m high; comprising steel yielding posts and steel horizontal rails						
straight or curved exceeding 50 m radius	-	-	-	220.93	m	**220.93**
curved not exceeding 50 m radius	-	-	-	233.92	m	**233.92**
Metal parapet Group P6; 1.50 m high; comprising steel yielding posts and steel horizontal rails						
straight or curved exceeding 50 m radius	-	-	-	1033.16	m	**1033.16**
curved not exceeding 50 m radius	-	-	-	1117.63	m	**1117.63**
ALUMINIUM PARAPETS						
Metal parapet Group P1; 1.0 m high; comprising aluminium yielding / frangible posts and aluminium horizontal rails						
straight or curved exceeding 50 m radius	-	-	-	155.95	m	**155.95**
curved not exceeding 50 m radius	-	-	-	168.94	m	**168.94**
Metal parapet Group P2 (80 Kph); 1.0 m high; comprising aluminium yielding / frangible posts and aluminium horizontal rails with mesh infill						
straight or curved exceeding 50 m radius	-	-	-	168.94	m	**168.94**
curved not exceeding 50 m radius	-	-	-	181.94	m	**181.94**
Metal parapet Group P2 (113 Kph); 1.0 m high; comprising aluminium yielding / frangible posts and aluminium horizontal rails with mesh infill						
straight or curved exceeding 50 m radius	-	-	-	168.94	m	**168.94**
curved not exceeding 50 m radius	-	-	-	177.50	m	**177.50**
Metal parapet Group P4; 1.15 m high; comprising aluminium yielding / frangible posts and aluminium horizontal rails						
straight or curved exceeding 50 m radius	-	-	-	162.44	m	**162.44**
curved not exceeding 50 m radius	-	-	-	175.45	m	**175.45**
Metal parapet Group P5; 1.25 m high; comprising aluminium yielding / frangible posts and aluminium horizontal rails with solid sheet infill, anti-access panels						
straight or curved exceeding 50 m radius	-	-	-	194.94	m	**194.94**
curved not exceeding 50 m radius	-	-	-	207.93	m	**207.93**
Metal parapet Group P5; 1.50 m high; comprising aluminium yielding / frangible posts and aluminium horizontal rails with solid sheet infill, anti-access panels						
straight or curved exceeding 50 m radius	-	-	-	207.93	m	**207.93**
curved not exceeding 50 m radius	-	-	-	220.93	m	**220.93**

SERIES 2300: BRIDGE EXPANSION JOINTS AND SEALING OF GAPS

Item	Gang hours	Labour £	Plant £	Material £	Unit	Total rate £
GENERAL						
Notes						
Major movement joints to bridge and viaduct decks are manufactured and installed to individual specifications determined by the type of structure location in the deck, amount of movement to be expected, and many other variables. The following unit rates for other types of movement joints found in structures.						
RESOURCES - LABOUR						
Jointing gang						
1 ganger/chargehand (skill rate 3)		14.21				
1 skilled operative (skill rate 3)		13.46				
1 unskilled operative		11.26				
TOTAL	£	**38.93**				
Bridge jointing gang						
1 ganger/chargehand (skill rate 4)		12.80				
1 skilled operative (skill rate 4)		12.05				
1 unskilled operative		11.26				
TOTAL	£	**36.11**				
SEALING OF GAPS						
Flexcell joint filler board						
10 mm thick	0.10	3.61	-	3.30	m²	**6.92**
19 mm thick	0.16	5.78	-	5.56	m²	**11.34**
25 mm thick	0.16	5.78	-	6.96	m²	**12.74**
Building paper slip joint to abutment toe	0.01	0.40	-	1.02	m²	**1.42**
Bond breaking agent	0.03	0.90	-	3.71	m²	**4.61**
Hot poured rubber bitumen joint sealant						
10 x 20 mm	0.03	1.19	-	1.01	m	**2.20**
20 x 20 mm	0.04	1.44	-	2.15	m	**3.59**
25 x 15 mm	0.07	2.35	-	2.06	m	**4.41**
25 x 25 mm	0.07	2.64	-	3.43	m	**6.07**
Cold applied polysulphide joint sealant						
20 x 20 mm	0.07	2.35	-	3.50	m	**5.85**
Gun grade cold applied elastomeric joint sealant						
25 x 25 mm on 3 mm foam strip	0.07	2.35	-	9.25	m	**11.60**
50 x 25 mm on 3 mm foam strip	0.09	3.25	-	17.78	m	**21.03**
PVC centre bulb waterstop						
160 mm wide	0.08	2.89	-	3.44	m	**6.33**
210 mm wide	0.09	3.25	-	4.92	m	**8.17**
260 mm wide	0.11	3.97	-	5.76	m	**9.73**
PVC flat dumbell waterstop						
170 mm wide	0.08	2.89	-	20.74	m	**23.63**
210 mm wide	0.10	3.61	-	31.00	m	**34.61**
250 mm wide	0.12	4.33	-	50.80	m	**55.13**
Dowels, plain or greased						
12 mm mild steel 450 mm long	0.04	1.44	-	1.63	nr	**3.08**
16 mm mild steel 750 mm long	0.05	1.62	-	3.12	nr	**4.75**
16 mm mild steel 750 mm long with debonding agent for 375 mm	0.05	1.91	-	3.79	nr	**5.70**
Dowels, sleeved or capped						
12 mm mild steel 450 mm long with debonding agent for 225 mm and PVC dowel cap	0.05	1.62	-	1.60	nr	**3.22**

SERIES 2400: BRICKWORK, BLOCKWORK AND STONEWORK

Item	Gang hours	Labour £	Plant £	Material £	Unit	Total rate £
RESOURCES - LABOUR						
Masonry gang						
1 foreman bricklayer (craftsman)		18.47				
4 bricklayers (craftsman)		69.40				
1 unskilled operative (general)		11.26				
Total Gang Rate / Hour	£	**99.13**				
RESOURCES - PLANT						
Masonry						
dumper 2 t (50% of time)			1.58			
gas oil for ditto			0.24			
cement mixer 4/3 (50% of time)			0.47			
petrol for ditto			0.42			
small power tools (masonry)			1.74			
minor scaffolding (masonry)			1.28			
Total Rate / Hour		£	**5.73**			
RESOURCES - MATERIALS						
Half brick thick walls are in stretcher bond, thicker than this in English bond (3 stretchers : 1 header) unless otherwise stated.						
DTp Table 24/1: Mortar Proportions by Volume :-						
Mortar Type / Cement: Lime:sand / Masonry Cement:sand / Cement: sand						
(i) 1.0 to 1/4.3						
(ii) 1:½:4 to 4½ / 1:2½ to 3½ / 1:3 to 4						
(iii) 1:1:5 to 6 / 1:4½ / 1:5 to 6						
BRICKWORK						
Common bricks; PC £130.00/1000; in stretcher bond; in cement mortar designation (ii)						
Walls						
half brick thick	0.23	22.60	1.31	10.71	m²	**34.62**
one brick thick	0.44	43.82	2.53	21.90	m²	**68.25**
one and a half bricks thick	0.64	63.84	3.69	33.45	m²	**100.99**
two bricks thick	0.83	82.62	4.78	44.86	m²	**132.26**
Walls, curved on plan						
half brick thick	0.30	29.34	1.68	10.71	m²	**41.74**
one brick thick	0.57	56.50	3.27	21.90	m²	**81.67**
one and a half bricks thick	0.82	81.29	4.72	33.45	m²	**119.46**
two bricks thick	1.06	104.88	6.06	44.86	m²	**155.80**
Walls, with a battered face						
half brick thick	0.33	32.61	1.89	10.71	m²	**45.21**
one brick thick	0.63	62.45	3.61	21.90	m²	**87.97**
one and a half bricks thick	0.91	89.91	5.20	33.45	m²	**128.56**
two bricks thick	1.16	115.19	6.66	44.86	m²	**166.71**
Facework to concrete						
half brick thick	0.25	24.49	1.42	18.18	m²	**44.08**
one brick thick	0.48	47.38	2.74	29.37	m²	**79.50**
one and a half bricks thick	0.69	68.80	3.98	40.92	m²	**113.70**
two bricks thick	0.90	89.02	5.15	52.24	m²	**146.41**

SERIES 2400: BRICKWORK, BLOCKWORK AND STONEWORK

Item	Gang hours	Labour £	Plant £	Material £	Unit	Total rate £
BRICKWORK – cont'd						
Common bricks; PC £130.00/1000; in stretcher bond; in cement mortar designation (ii) – cont'd						
In alteration work						
half brick thick	0.30	29.34	1.70	18.18	m²	**49.22**
one brick thick	0.57	56.50	3.27	29.37	m²	**89.14**
one and a half bricks thick	0.82	81.58	4.72	40.92	m²	**127.23**
two bricks thick	1.06	104.88	6.06	52.24	m²	**163.19**
ADD or DEDUCT to materials costs for variation of £10.00/1000 in PC of common bricks						
half brick thick	-	-	-	0.63	m²	**0.63**
one brick thick	-	-	-	1.26	m²	**1.26**
one and a half bricks thick	-	-	-	1.89	m²	**1.89**
two bricks thick	-	-	-	2.52	m²	**2.52**
Copings; standard header-on-edge; PC £13.00/100						
215 mm wide x 103 mm high	0.11	10.41	0.60	2.40	m	**13.41**
ADD or DEDUCT to copings for variation of £1.00/100 in PC of Common bricks	-	-	-	0.13	m	**0.13**
Class A engineering bricks, perforated; PC £275.00/1000; in cement mortar designation (ii)						
Walls						
half brick thick	0.27	26.67	1.54	20.51	m²	**48.72**
one brick thick	0.52	51.55	2.98	41.50	m²	**96.03**
one and a half bricks thick	0.75	74.64	4.32	62.85	m²	**141.81**
two bricks thick	0.97	96.26	5.57	84.06	m²	**185.88**
Walls, curved on plan						
half brick thick	0.37	36.68	2.12	20.51	m²	**59.31**
one brick thick	0.70	69.89	4.04	41.50	m²	**115.43**
one and a half bricks thick	1.01	100.12	5.79	62.85	m²	**168.76**
two bricks thick	1.29	127.68	7.38	84.06	m²	**219.12**
Walls, with a battered face						
half brick thick	0.37	36.68	2.12	20.51	m²	**59.31**
one brick thick	0.70	69.89	4.04	41.50	m²	**115.43**
one and a half bricks thick	1.01	100.12	5.79	62.85	m²	**168.76**
two bricks thick	1.29	127.68	7.38	84.06	m²	**219.12**
Facework to concrete						
half brick thick	0.32	31.23	1.80	27.98	m²	**61.01**
one brick thick	0.60	59.97	3.47	48.97	m²	**112.42**
ADD or DEDUCT to materials costs for variation of £10.00/1000 in PC of engineering bricks						
half brick thick	-	-	-	0.63	m²	**0.63**
one brick thick	-	-	-	1.26	m²	**1.26**
one and a half bricks thick	-	-	-	1.89	m²	**1.89**
two bricks thick	-	-	-	2.52	m²	**2.52**
Brick coping in standard bricks in headers on edge; PC £27.50/100						
215 mm wide x 103 mm high	0.29	28.75	1.66	6.23	m	**36.64**
ADD or DEDUCT to copings for variation of £1.00/100 in PC of Class A engineering bricks	-	-	-	0.13	m	**0.13**

SERIES 2400: BRICKWORK, BLOCKWORK AND STONEWORK

Item	Gang hours	Labour £	Plant £	Material £	Unit	Total rate £
Class B engineering bricks, perforated;PC £200.00/1000; in cement mortar designation (ii)						
Walls						
half brick thick	0.29	28.75	1.66	14.49	m²	**44.90**
one brick thick	0.48	48.08	2.78	29.94	m²	**80.80**
one and a half bricks thick	0.69	68.90	3.98	43.81	m²	**116.69**
two bricks thick	0.91	89.71	5.19	56.66	m²	**151.56**
Walls, curved on plan						
half brick thick	0.42	41.24	2.39	14.49	m²	**58.11**
one brick thick	0.78	77.32	4.47	29.94	m²	**111.74**
one and a half bricks thick	1.04	103.10	5.96	43.81	m²	**152.87**
two bricks thick	1.30	128.87	7.45	56.66	m²	**192.99**
Walls, with battered face						
half brick thick	0.42	41.24	2.39	14.49	m²	**58.11**
one brick thick	0.78	77.32	4.47	29.94	m²	**111.74**
one and a half bricks thick	1.04	103.10	5.96	43.81	m²	**152.87**
two bricks thick	1.30	128.87	7.45	56.66	m²	**192.99**
Facework to concrete						
half brick thick	0.32	31.52	1.82	21.93	m²	**55.27**
one brick thick	0.56	55.51	3.21	37.38	m²	**96.10**
ADD or DEDUCT to materials costs for variation of £10.00/1000 in PC of engineering bricks						
half brick thick	-	-	-	0.63	m²	**0.63**
one brick thick	-	-	-	1.26	m²	**1.26**
one and a half bricks thick	-	-	-	1.89	m²	**1.89**
two bricks thick	-	-	-	2.52	m²	**2.52**
Brick coping in standard bricks in headers on edge; PC £20.00/100						
215 mm wide x 103 mm high	0.13	13.28	0.77	2.80	m	**16.85**
ADD or DEDUCT to copings for variation of £1.00/100 in PC of Class B engineering bricks	-	-	-	0.13	m	**0.13**
Facing bricks; PC £390.00/1000; in lime mortar designation (ii)						
Walls						
half brick thick	0.34	33.70	1.95	28.97	m²	**64.63**
one brick thick	0.57	56.90	3.47	58.43	m²	**118.80**
one and a half bricks thick	0.83	81.98	4.74	88.24	m²	**174.96**
two bricks thick	1.08	107.06	6.19	117.91	m²	**231.16**
Walls, curved on plan						
half brick thick	0.45	44.61	2.58	28.97	m²	**76.16**
one brick thick	0.84	83.27	4.82	58.43	m²	**146.51**
one and a half bricks thick	1.11	110.53	6.39	88.24	m²	**205.16**
two bricks thick	1.39	137.79	7.97	117.91	m²	**263.67**
Walls, with a battered face						
half brick thick	0.45	44.61	2.58	28.97	m²	**76.16**
one brick thick	0.84	83.27	4.82	58.43	m²	**146.51**
one and a half bricks thick	1.11	110.53	6.39	88.24	m²	**205.16**
two bricks thick	1.39	137.79	7.97	117.91	m²	**263.67**
Facework to concrete						
half brick thick	0.37	36.48	2.05	36.44	m²	**74.97**
one brick thick	0.66	65.43	3.79	65.90	m²	**135.11**
Extra over common brickwork in English bond for facing with facing bricks in lime mortar designation (ii)	0.11	11.10	0.69	19.05	m²	**30.85**

SERIES 2400: BRICKWORK, BLOCKWORK AND STONEWORK

Item	Gang hours	Labour £	Plant £	Material £	Unit	**Total rate £**
BRICKWORK – cont'd						
Facing bricks; PC £390.00/1000; in lime mortar designation (ii) – cont'd						
ADD or DEDUCT to materials costs for variation of £10.00/1000 in PC of facing bricks						
half brick thick	-	-	-	0.63	m²	**0.63**
one brick thick	-	-	-	1.26	m²	**1.26**
one and a half bricks thick	-	-	-	1.89	m²	**1.89**
two bricks thick	-	-	-	2.52	m²	**2.52**
Brick coping in standard bricks in headers on edge; PC £39.00/100						
215 mm wide x 103 mm high	0.13	13.28	0.77	6.19	m	**20.24**
Flat arches in standard stretchers on end; PC £39.00/100						
103 mm wide x 215 mm high	0.21	20.82	1.20	5.98	m	**28.00**
Flat arches in bullnose stretchers on end; PC £120.00/100						
103 mm x 215 mm high	0.22	21.81	1.26	39.64	m	**62.71**
Segmental arches in single ring stretchers on end; PC £39.00/100						
103 mm wide x 215 mm high	0.37	36.68	2.12	5.98	m	**44.77**
Segmental arches in double ring stretchers on end; PC £39.00/100						
103 mm wide x 440 mm high	0.49	48.57	2.81	24.14	m	**75.52**
Segmental arches; cut voussoirs PC £250.00/100						
103 mm wide x 215 mm high	0.39	38.66	2.24	176.78	m	**217.67**
ADD or DEDUCT to copings and arches for variation of £1.00/100 in PC of facing bricks						
header-on-edge	-	-	-	0.13	m	**0.13**
stretcher-on-end	-	-	-	0.13	m	**0.13**
stretcher-on-end bullnose specials	-	-	-	0.13	m	**0.13**
single ring	-	-	-	0.13	m	**0.13**
two ring	-	-	-	0.27	m	**0.27**
BLOCKWORK AND STONEWORK						
Lightweight concrete blocks; solid; 3.5 N/mm²; in cement-lime mortar						
Walls						
100 mm thick; PC £4.94/m²	0.17	17.25	1.00	5.61	m²	**23.86**
140 mm thick; PC £6.98/m²	0.23	22.30	1.29	7.46	m²	**31.05**
215 mm thick; PC £12.50/m²	0.28	27.26	1.58	11.91	m²	**40.74**
Walls, curved on plan						
100 mm thick; PC £4.94/m²	0.23	22.90	1.32	5.61	m²	**29.83**
140 mm thick; PC £6.98/m²	0.30	29.64	1.71	7.46	m²	**38.81**
215 mm thick; PC £12.50/m²	0.37	36.28	2.10	11.91	m²	**50.29**
Facework to concrete						
100 mm thick; PC £4.94/m²	0.18	17.74	1.02	13.08	m²	**31.85**
140 mm thick; PC £6.98/m²	0.23	23.00	1.33	14.93	m²	**39.25**
215 mm thick; PC £12.50/m²	0.28	28.05	1.62	19.38	m²	**49.05**
In alteration work						
100 mm thick; PC £4.94/m²	0.17	17.25	1.00	5.61	m²	**23.86**
140 mm thick; PC £6.98/m²	0.23	22.30	1.29	7.46	m²	**31.05**
215 mm thick; PC £12.50/m²	0.28	27.26	1.58	11.91	m²	**40.74**

SERIES 2400: BRICKWORK, BLOCKWORK AND STONEWORK

Item	Gang hours	Labour £	Plant £	Material £	Unit	Total rate £
Dense concrete blocks; solid; 3.5 or 7 N/mm²; in cement-lime mortar						
Walls						
100 mm thick; PC £4.79/m²	0.17	16.85	0.97	4.73	m²	**22.56**
140 mm thick; PC £6.93/m²	0.20	19.83	1.15	6.78	m²	**27.75**
215 mm thick; PC £13.90/m²	0.24	23.79	1.38	13.71	m²	**38.88**
Walls, curved on plan						
100 mm thick; PC £4.79/m²	0.23	22.40	1.29	4.73	m²	**28.43**
140 mm thick; PC £6.93/m²	0.23	22.40	1.52	6.78	m²	**30.71**
215 mm thick; PC £13.90/m²	0.32	31.62	1.83	13.43	m²	**46.88**
Facework to concrete						
100 mm thick; PC £4.79/m²	0.17	17.35	1.00	12.20	m²	**30.55**
140 mm thick; PC £6.93/m²	0.21	20.42	1.18	14.25	m²	**35.85**
215 mm thick; PC £13.90/m²	0.35	34.70	1.73	20.90	m²	**57.32**
In alteration work						
100 mm thick; PC £4.79/m²	0.27	26.37	1.29	4.73	m²	**32.40**
140 mm thick; PC £6.93/m²	0.26	25.77	1.52	6.78	m²	**34.08**
215 mm thick; PC £13.90/m²	0.32	31.62	1.83	13.43	m²	**46.88**
Reconstituted stone; Bradstone 100 bed weathered Cotswold or North Cerney masonary blocks; rough hewn rockfaced blocks; in coloured cement-lime mortar designation (1:2:9) (iii)						
Walls, thickness 100mm						
vertical and straight	0.30	29.74	1.72	56.63	m²	**88.09**
curved on plan	0.39	38.66	2.24	56.63	m²	**97.52**
with a battered face	0.34	34.20	1.98	56.63	m²	**92.80**
in arches	0.57	57.00	3.30	56.63	m²	**116.92**
Facing to concrete; wall ties						
vertical and straight	0.24	23.49	1.36	94.39	m²	**119.25**
curved on plan	0.32	31.23	1.80	94.39	m²	**127.42**
with a battered face	0.36	35.29	2.04	94.39	m²	**131.72**
Reconstituted stone; Bradstone Architectural dressings in weathered Cotswold or North Cerney shades; in coloured cement-lime mortar designation (1:2:9) (iii)						
Copings; twice weathered and throated						
152 x 76 mm ; PC £12.38/m	0.08	7.93	0.46	16.28	m	**24.67**
152 x 76 mm; curved on plan; PC £45.92/m	0.11	10.51	0.61	60.38	m	**71.50**
305 x 76 mm; PC £22.61/m	0.10	9.91	0.58	31.52	m	**42.01**
305 x 76 mm; curved on plan; PC £58.06/m	0.13	13.18	0.76	77.35	m	**91.30**
Corbels						
479 x 100 x 215 mm, splayed	0.49	48.57	2.81	89.20	nr	**140.58**
665 x 100 x 215 mm, splayed	0.55	54.52	3.15	123.06	nr	**180.74**
Pier caps						
305 x 305 mm	0.09	8.92	0.52	12.98	nr	**22.42**
381 x 381 mm	0.11	10.90	0.63	18.34	nr	**29.88**
457 x 457 mm	0.13	12.89	0.74	25.10	nr	**38.73**
533 x 533 mm	0.15	14.87	0.86	34.95	nr	**50.68**
Lintels						
100 x 140 mm	0.11	10.90	0.63	27.65	m	**39.19**
100 x 215 mm	0.16	15.86	0.92	36.47	m	**53.25**

SERIES 2400: BRICKWORK, BLOCKWORK AND STONEWORK

Item	Gang hours	Labour £	Plant £	Material £	Unit	Total rate £
BLOCKWORK AND STONEWORK – cont'd						
Natural stone ashlar; Portland Whitbed limestone; in cement-lime mortar designation (iii)						
Walls						
vertical and straight	12.70	1258.95	61.74	2680.85	m³	**4001.54**
curved on plan	19.30	1913.21	110.64	4012.71	m³	**6036.56**
with a battered face	19.30	1913.21	110.64	4012.71	m³	**6036.56**
Facing to concrete; wall ties						
vertical and straight	17.00	1685.21	97.46	2720.69	m³	**4503.36**
curved on plan	25.80	2557.55	147.91	4052.55	m³	**6758.01**
with a battered face	25.80	2557.55	147.91	4052.55	m³	**6758.01**
Copings; twice weathered and throated						
250 x 150 mm	0.45	44.61	2.58	142.63	m	**189.82**
250 x 150 mm; curved on plan	0.45	44.61	2.58	171.12	m	**218.31**
400 x 150 mm	0.49	48.57	2.81	209.66	m	**261.04**
400 x 150 mm; curved on plan	0.49	48.57	2.81	251.57	m	**302.95**
Shaped and dressed string courses						
75 mm projection x 150 mm high	0.45	44.61	2.58	112.68	m	**159.87**
Corbel						
500 x 450 x 300 mm	0.55	54.52	3.15	161.42	nr	**219.10**
Keystone						
750 x 900 x 300 mm (extreme)	1.30	128.87	7.45	630.45	nr	**766.78**
Random rubble uncoursed , weighing 2.0 t/m³ of wall; in cement-lime mortar designation (iii)						
Walls						
vertical and straight	4.17	413.37	23.90	277.24	m³	**714.52**
curved on plan	4.67	462.94	26.77	308.65	m³	**798.36**
with a battered face	4.67	462.94	26.77	308.65	m³	**798.36**
in arches	8.53	845.58	48.90	308.65	m³	**1203.13**
Facework to concrete						
vertical and straight	4.17	413.37	23.90	316.12	m³	**753.40**
curved on plan	4.67	462.94	26.77	316.12	m³	**805.83**
with a battered face	4.67	462.94	26.77	316.12	m³	**805.83**
in arches	8.53	845.58	48.90	316.12	m³	**1210.60**
Copings						
500 x 125 mm	0.49	48.57	2.81	236.88	m	**288.26**
Squared random rubble uncoursed , weighing 2.0 t/m³ of wall; in cement-lime mortar designation (iii)						
Walls						
vertical and straight	4.17	413.37	23.90	441.58	m³	**878.86**
curved on plan	4.67	462.94	26.77	441.58	m³	**931.29**
with a battered face	4.67	462.94	26.77	441.58	m³	**931.29**
in arches	8.53	845.58	48.90	441.58	m³	**1336.06**
Facework to concrete						
vertical and straight	4.17	413.37	23.90	449.05	m³	**886.33**
curved on plan	4.67	462.94	26.77	449.05	m³	**938.76**
with a battered face	4.67	462.94	26.77	449.05	m³	**938.76**
in arches	8.53	845.58	48.90	449.05	m³	**1343.53**
Copings						
500 x 125 mm	0.49	48.57	2.81	236.88	m	**288.26**

SERIES 2400: BRICKWORK, BLOCKWORK AND STONEWORK

Item	Gang hours	Labour £	Plant £	Material £	Unit	Total rate £
Dry rubble, weighing 2.0 t/m³ of wall						
Walls						
vertical and straight	3.83	379.67	21.96	321.62	m³	**723.24**
curved on plan	4.33	429.23	24.82	321.62	m³	**775.67**
with a battered face	4.33	429.23	24.82	321.62	m³	**775.67**
Copings formed of rough stones						
275 x 200 mm (average) high	0.45	44.61	2.58	26.38	m	**73.57**
500 x 200 mm	0.55	54.52	3.15	49.48	m	**107.15**

SERIES 2500: SPECIAL STRUCTURES

Item	Gang hours	Labour £	Plant £	Material £	Unit	Total rate £
SPECIAL STRUCTURES DESIGNED BY THE CONTRACTOR						
Notes This section envisages the following types of structure which may be required to be designed by the Contractor based on stipulated performance criteria: • Buried structures • Earth retaining structures • Environmental barriers • Underbridges up to 8 m span • Footbridges • Piped culverts • Box culverts • Drainage exceeding 900 mm diameter • Other structures						
Naturally, this work cannot be catered for directly in this section and will require the preparation of a sketch solution and approximate quantities to allow pricing using the various other Unit Costs sections as well as the Approximate Estimates section.						
An allowance must be added to such an estimate to cover the Contractor's design fee(s) and expenses.						

SERIES 3000: LANDSCAPING AND ECOLOGY

Item	Gang hours	Labour £	Plant £	Material £	Unit	**Total rate £**
GROUND PREPARATION AND CULTIVATION						
Supply and apply granular cultivation treatments (PC £1.20/kg) by hand						
35 grammes / m²	0.50	11.65	-	31.53	100m²	**43.19**
50 grammes / m²	0.65	15.15	-	45.05	100m²	**60.20**
75 grammes / m²	0.85	19.81	-	67.57	100m²	**87.38**
100 grammes / m²	1.00	23.31	-	90.09	100m²	**113.40**
150 grammes / m²	1.20	27.97	-	135.13	100m²	**163.11**
Supply and apply granular cultivation treatments PC £1.20/kg by machine in suitable economically large areas						
100 grammes / m²	-	-	3.68	90.09	100m²	**93.77**
ADD to above for:						
granular treatments per £0.10/kg PC variation +10%						
selective weedkiller PC £3.25/kg +171%						
herbicide PC £8.00/kg +567%						
fertilizer +100%						
Supply and incorporate cultivation additives into top 150 mm topsoil by hand						
1 m³ / 10m²	20.00	466.20	-	11.55	100m²	**477.75**
1 m³ / 13m²	20.00	466.20	-	8.88	100m²	**475.08**
1 m³ / 20m²	19.00	442.89	-	5.78	100m²	**448.66**
1 m³ / 40m²	17.00	396.27	-	2.89	100m²	**399.16**
Supply and incorporate cultivation additives into top 150 mm topsoil by machine in suitable economically large areas						
1 m³ / 10m²	-	-	126.36	11.55	100m²	**137.91**
1 m³ / 13m²	-	-	116.64	8.88	100m²	**125.52**
1 m³ / 20m²	-	-	103.68	5.78	100m²	**109.45**
1 m³ / 40m²	-	-	95.58	2.89	100m²	**98.46**
ADD to above for						
cultavation additives per £0.10/m² PCvariation +10%						
Compost +300%						
Manure +166%						
Peat +1800%						
SEEDING AND TURFING						
Selected grass seed; at the rate of 0.050 kg/m² in two operations						
Grass seeding; by conventional sowing						
at 10 degrees or less to the horizontal	0.01	0.23	-	0.37	m²	**0.60**
more than 10 degrees to horizontal	0.02	0.35	-	0.37	m²	**0.72**
Wild Flora mixture BSH ref WF1 combined with Low maintenance conservation grass BSH ref A4 (20%:80%); at the rate of 80 g/m²						
Wildflower seeding; by conventional sowing						
at 10 degrees or less to the horizontal	0.50	11.65	-	72.15	100 m²	**83.80**
more than 10 degrees to horizontal	0.02	0.35	-	0.37	m²	**0.72**
Hydraulic mulch grass seed						
Grass seeding by hydraulic seeding						
at 10 degrees or less to the horizontal	0.01	0.12	0.07	0.13	m²	**0.32**
at more than 10 degrees to the horizontal	0.01	0.16	0.07	0.13	m²	**0.37**

SERIES 3000: LANDSCAPING AND ECOLOGY

Item	Gang hours	Labour £	Plant £	Material £	Unit	Total rate £
SEEDING AND TURFING – cont'd						
Imported turf						
Turfing to surfaces						
at 10 degrees or less to the horizontal	0.12	2.80	-	1.96	m²	**4.76**
more than 10 to the horizontal; pegging down	0.17	3.96	-	1.96	m²	**5.92**
PLANTING						
Trees						
The cost of planting semi-mature trees will depend on the size and species, and on the access to the site for tree handling machines.						
Prices should be obtained for individual trees and planting.						
Break up subsoil to a depth of 200mm						
in treepit	0.05	1.17	-	-	nr	**1.17**
Supply and plant tree in prepared pit; backfill with excavated topsoil minimum 600 mm deep						
light standard; in pits	0.25	5.83	-	10.08	nr	**15.91**
standard tree	0.45	10.49	-	15.12	nr	**25.61**
selected standard tree	0.75	17.48	-	20.16	nr	**37.64**
heavy standard tree	0.85	19.81	-	35.28	nr	**55.09**
extra heavy standard tree	1.50	34.97	-	63.00	nr	**97.97**
extra for filling with topsoil from spoil heap ne 100m distant	0.15	3.50	0.55	-	m³	**4.04**
extra for filling with imported topsoil	0.08	1.86	0.29	18.17	m³	**20.33**
extra for incorporating manure or compost into top soil at the rate of 1 m³ per 5 m³ +60%						
Supply tree stake and drive 500 mm into firm ground and trim to approved height, including two tree ties to approved pattern						
one stake; 2.4 m long, 100 mm diameter	0.16	3.73	-	6.49	nr	**10.22**
one stake; 3.0 m long, 100 mm diameter	0.20	4.66	-	8.18	nr	**12.84**
two stakes; 2.4 m long, 100 mm diameter	0.24	5.59	-	12.98	nr	**18.57**
two stakes; 3.0 m long, 100 mm diameter	0.30	6.99	-	16.35	nr	**23.35**
Supply and fit tree support comprising three collars and wire guys; including pickets						
galvanised steel 50 x 600 mm	1.50	34.97	-	27.19	nr	**62.16**
hardwood 75 x 600 mm	1.50	34.97	-	27.82	nr	**62.78**
Supply and fix standard steel tree guard	0.30	6.99	-	22.72	nr	**29.71**
Hedge plants						
Excavate trench by hand for hedge and deposit soil alongside trench						
300 wide x 300 mm deep	0.10	2.33	-	-	m	**2.33**
450 wide x 300 mm deep	0.13	3.03	-	-	m	**3.03**
Excavate trench by machine for hedge and deposit soil alongside trench						
300 wide x 300 mm deep	0.02	0.26	0.31	-	m	**0.58**
450 wide x 300 mm deep	0.02	0.26	0.31	-	m	**0.58**
Set out, nick out and excavate trench and break up subsoil to minimum depth of 300 mm						
400 mm minimum deep	0.15	3.50	-	-	m	**3.50**
Supply and plant hedging plants; backfill with excavated topsoil						
single row plants at 200 mm centres	0.25	5.83	-	5.01	m	**10.84**
single row plants at 300 mm centres	0.17	3.96	-	3.34	m	**7.30**

SERIES 3000: LANDSCAPING AND ECOLOGY

Item	Gang hours	Labour £	Plant £	Material £	Unit	**Total rate £**
single row plants at 400 mm centres	0.13	2.91	-	2.51	m	**5.42**
single row plants at 500 mm centres	0.10	2.33	-	2.00	m	**4.34**
single row plants at 600 mm centres	0.08	1.86	-	1.66	m	**3.53**
double row plants at 200 mm centres	0.50	11.65	-	10.02	m	**21.68**
double row plants at 300 mm centres	0.34	7.93	-	6.68	m	**14.60**
double row plants at 400 mm centres	0.25	5.83	-	5.01	m	**10.84**
double row plants at 500 mm centres	0.20	4.66	-	4.01	m	**8.67**
double row plants at 600 mm centres	0.16	3.73	-	3.33	m	**7.06**
Extra for incorporating manure at 1 m³ / 30m³	0.60	6.76	-	0.22	-	**6.97**
Shrubs						
Form planting hole in previously cultivated area, supply and plant specified shrub and backfill with excavated material						
shrub 300 mm high	0.10	2.33	-	2.39	each	**4.72**
shrub 600 mm high	0.10	2.33	-	3.44	each	**5.77**
shrub 900 mm high	0.10	2.33	-	4.11	each	**6.44**
shrub 1.0 m high and over	0.10	2.33	-	5.49	each	**7.82**
Supply and fix shrub stake including two ties						
one stake; 1.5 m long, 75 mm diameter	0.12	2.80	-	5.08	each	**7.88**
Extra for the above items for planting in prefabricated or in-situ planters	+20%					
Herbaceous plants						
Form planting hole in previously cultivated area, supply and plant specified herbaceous plants and backfill with excavated material						
5 plants / m²	0.05	1.17	-	9.06	m²	**10.22**
10 plants / m²	0.16	3.73	-	18.11	m²	**21.84**
25 plants / m²	0.42	9.79	-	45.28	m²	**55.07**
35 plants / m²	0.58	13.52	-	63.39	m²	**76.91**
50 plants / m²	0.83	19.35	-	90.56	m²	**109.91**
Supply and fix plant support netting on 50 mm diameter stakes 750 mm long driven into the ground at 1.5 m centres						
1.15 m high green extruded plastic mesh, 125 mm square mesh, PC £45.00 / 300 m roll	0.06	1.40	-	1.94	m²	**3.33**
Extra to the above items for planting in prefabricated or in-situ planters	+20%					
Form planting hole in previously cultivated area; supply and plant bulbs and backfill with excavated material						
small	0.01	0.23	-	0.16	each	**0.39**
medium	0.01	0.23	-	0.25	each	**0.49**
large	0.01	0.23	-	0.30	each	**0.54**
Supply and plant bulb in grassed area using bulb planter and backfill with screened topsoil or peat and cut turf plug						
small	0.01	0.23	-	0.16	each	**0.39**
medium	0.01	0.23	-	0.25	each	**0.49**
large	0.01	0.23	-	0.30	each	**0.54**
Extra to the above items for planting in prefabricated or in-situ planters	+15%					

SERIES 3000: LANDSCAPING AND ECOLOGY

Item	Gang hours	Labour £	Plant £	Material £	Unit	Total rate £
MULCHING						
Organic mulching of medium bark mulch to a depth of 50 mm in planting areas to surfaces						
at 10 degrees or less to the horizontal	0.01	0.23	-	1.74	m²	**1.97**
more than 10 degrees to horizontal	0.02	0.35	-	1.74	m²	**2.09**
Organic mulching of timber mulch to a depth of 75 mm in planting areas to surfaces						
at 10 degrees or less to the horizontal	0.02	0.35	-	3.47	m²	**3.82**
more than 10 degrees to horizontal	0.02	0.47	-	3.47	m²	**3.94**
WEED CONTROL						
Weed and handfork planted areas including removing and dumping weed and debris on site	0.07	1.63	-	-	m²	**1.63**
Supply and apply selective weed killer						
35 grammes / m²	0.01	0.12	-	0.13	m²	**0.24**
70 grammes / m²	0.01	0.16	-	0.25	m²	**0.41**
100 grammes / m²	0.01	0.23	-	0.36	m²	**0.59**
MAINTENANCE OF ESTABLISHED GRASSED AREAS						
Grass cutting at medium frequency						
on central reserves	-	-	-	-	100m²	**5.17**
MAINTENANCE OF ESTABLISHED TREES+SHRUBS						
Initial cut back to shrubs and hedge plants, clear away all cuttings	0.04	0.93	-	-	m	**0.93**
Protect planted areas with windbreak fencing fixed to stakes 1.5 m x 50 mm diameter at 2.0 m centres and clear away on completion						
1.0 m high	0.06	1.40	-	6.19	m	**7.58**
Cut out dead and diseased wood, prune, trim and cut to shape; treat wounds with sealant; clear away cuttings						
shrubs ne 1.0 m high	0.20	4.66	-	-	each	**4.66**
shrubs 1.0 - 2.0 m high	0.30	6.99	-	-	each	**6.99**
shrubs 2.0 - 3.0 m high	0.40	9.32	-	-	each	**9.32**
Cut and trim ornamental hedge to specified profile; clear away cuttings						
ornamental hedge 2.0 m high	0.60	13.99	-	-	m	**13.99**
ornamental hedge 4.0 m high	0.90	20.98	-	-	m	**20.98**
Trim field hedge to specified height and shape; clear away cuttings						
using flail	0.05	1.17	-	-	m	**1.17**
using cutting bar	0.07	1.63	-	-	m	**1.63**

PART 6

Land Remediation

The purpose of this part is to review the general background of ground contamination, the cost implications of current legislation and to consider the various remedial measures and to present helpful guidance on the cost of Land Remediation.

The introduction of the Landfill Directive in July 2004 has had a considerable impact on the cost of Remediation works in general and particularly on the practice of Dig and Dump. The number of Landfill sites licensed to accept Hazardous Waste has drastically reduced and inevitably this has led to increased costs.

Market forces will determine future increases in cost resulting from the introduction of the Landfill Directive and the cost guidance given within this section will require review in light of these factors.

It must be emphasised that the cost advice given is an average and that costs can vary considerably from contract to contract depending on individual Contractors, site conditions, type and extent of contamination, methods of working and various other factors as diverse as difficulty of site access and distance from approved tips.

We have structured this Unit Cost section to cover as many aspects of Land Remediation works as possible.

GUIDANCE NOTES

Generally

Adjustments should be made to the rates shown for time, location, local conditions, site constraints and any other factors likely to affect the costs of a specific scheme.

Method of Measurement

Although this part of the book is primarily based on CESMM3, the specific rules have been varied from in cases where it has been felt that an alternative presentation would be of value to the book's main purpose of providing guidance on prices. This is especially so with a number of specialist contractors.

DEFINITION

Generally

Contaminated land refers generally to land which contains contaminants in sufficient quantities to harm people, animals, the environment or structures. There is now a statutory definition of 'contaminated land' contained within Part IIA of the Environmental Protection Act.

Contaminants comprise hazardous substances (solids, liquids or gases) that are not naturally occurring in the site. They arise from previous site usage, although sites can be affected by pollutants arising from adjoining sites through the movement of water and air. A contaminated site can similarly pose a risk to surrounding land by offsite migration of contaminants.

Contaminated sites can be sold on, although the new owner would take on the responsibility for the contamination and would obviously take this into account in the offered price.

The extent of remediation works required to address contamination varies dependent on the intended future use of the site – with industrial uses calling for a lower level of work than if the site were intended to be used for residential or agricultural purposes. In a commercial world, expenditure on decontaminating the land would usually need to be balanced against the release of the latent site value – unless of course the contamination contravened statutory limits.

Hazardous contaminants fall into three broad categories:

- Chemical contamination of land or water
- Biological contamination of land or water (e.g. samples containing pathological bacteria potentially harmful to health).
- Contamination of a physical nature (e.g. radioactive material, unsuitable fill materials, flammable gas or combustible material e.g. wood dust)

These can also be listed in the following sub-groups:

Gases	Toxic, flammable and explosive gases
	e.g. hydrogen cyanide
	hydrogen sulphide
	Flammable and explosive gases
	e.g. methane
Liquids	Flammable liquids and solids
	Fuels, oils and other hydrocarbons
	Solvents
Combustible materials	Timber
	Ash
	Coal residue
Heavy metals	Arsenic, lead, mercury, cadmium, chromium
Corrosive substances	Acids
	Alkalis
Toxic substances	Hydrocarbons
	Inorganic salts
Asbestos	

DEFINITION - continued

Generally - continued

☐	Substituted aromatic compounds	PCBs
		Dioxins
		Furans
☐	Biological agents	Anthrax
		Tetanus
		BSE
		Genetically modified organisms

The following is a brief list of some of the main industrial sectors and their potential contaminants.

	Sector	Contaminant type	Example
☐	**Gasworks**	Coal tar	polyaromatic hydrocarbons (PAH's)
			phenol
		Cyanide	free / complex
		Sulphur	sulphide / sulphate
		Metals	lead, cadmium, mercury
		Aromatic hydrocarbons	benzene
☐	**Iron + Steel works**	Metals	copper, nickel, lead
		Acids	sulphuric, hydrochloric
		Mineral oils	-
		Coking works residues	(as for gasworks)
☐	**Metal finishing**	Metals	cadmium, chromium, copper, nickel, zinc
		Acids	sulphuric, hydrochloric
		Plating salts	cyanide
		Aromatic hydrocarbons	benzene
		Chlorinated hydrocarbons	trichloroethane
☐	**Non ferrous metal processing**	Metals	Copper, cadmium, lead, zinc
		Impurity metals	antimony, arsenic
		Other wastes	battery cases, acids
☐	**Oil refineries**	Hydrocarbons	various fractions
		Acids, alkalis	sulphuric, caustic soda
		Lagging, insulation	asbestos
		Spent catalysts	lead, nickel, chromium
☐	**Paints**	Metals	lead, cadmium, barium
		Alcohols	toluol, xylol
		Chlorinated hydrocarbons	methylene chloride
		Fillers, extenders	silica, titanium dioxide, talc
☐	**Petrochemical plants**	Acids, alkalis	sulphuric, caustic soda

	Metals	copper, cadmium, mercury
	Reactive monomers	styrene, acrylate, VCM
	Cyanide	toluene di-isocyanate
	Amines	analine
	Aromatic hydrocarbons	benzene, toluene
☐ **Petrol stations**	Metals	copper, cadmium, lead, nickel, zinc
	Aromatic hydrocarbons	benzene
	Octane boosters	lead, MTBE
	Mineral oil	-
	Paint, plastic residues	barium, cadmium, lead
☐ **Rubber processing**	Metals	zinc, lead
	Sulphur compounds	sulphur, thiocarbonate
	Reactive monomers	isoprene, isobutylene
	Acids	sulphuric. hydrochloric
	Aromatic hydrocarbons	xylene, Toluene
☐ **Semi-conductors**	Metals	copper, nickel, cadmium
	Metalloids	arsenic, antimony, zinc
	Acids	nitric, hydrofluoric
	Chlorinated hydrocarbons	trichloroethylene
	Alcohols	methanol
	Aromatic hydrocarbons	xylene, toluene
☐ **Tanneries**	Acids	hydrochloric
	Metals	trivalent chromium
	Salts	chlorides, sulphides
	Solvents	kerosene, white spirit
	Cyanide	methyl isocyanate
	Degreasers	trichloroethylene
	Dyestuff residues	cadmium, benzidine
☐ **Textiles**	Metals	aluminium, tin, titanium, zinc
	Acids, alkalines	sulphuric, caustic soda
	Salts	sodium hypochlorite
	Chlorinated hydrocarbons	perchloroethylene
	Aromatic hydrocarbons	phenol
	Pesticides	dieldrin, aldrin, endrine
	Dyestuff residues	cadmium, Benzedrine
☐ **Wood processing**	Coal tar based preservatives	creosote
	Chlorinated hydrocarbons	pentachlorophenol
	Metalloids / metals	arsenic, copper, chromium
☐ **Hat making**	Mercury	

DEFINITION - continued

Generally - continued

	Sector	Contaminant type	Example
☐	**Tin smelting**	Radioactivity	
☐	**Vehicle parking areas**	Metals	copper, cadmium, lead, nickel, zinc
		Aromatic hydrocarbons	benzene
		Octane boosters	lead, MTBE
		Mineral oil	-

BACKGROUND

Statutory framework

April 2000 saw the introduction of new contaminated land provisions, contained in Part IIA of the Environmental Protection Act 1990. A primary objective of the legislation is to identify and remediate contaminated land.

Under the Act action to remediate land is required only where there are unacceptable actual or potential risks to health or the environment. Sites that have been polluted from previous land use may not need remediating until the land use is changed to a more sensitive end-use. In addition, it may be necessary to take action only where there are appropriate, cost-effective remediation processes that take the use of the site into account.

The Environmental Act 1995 amended the Environment Protection Act 1990 by introducing a new regime for dealing with 'contaminated land' as defined. The regime became operational on 1 April 2000. Local authorities are the main regulators of the new regime although the Environment Agency regulates seriously contaminated sites which are known as "special sites".

The contaminated land regime incorporates statutory guidance on the inspection, definition, remediation, apportionment of liabilities and recovery of costs of remediation. The measures are to be applied in accordance with the following criteria:

- ☐ the planning system
- ☐ the standard of remediation should relate to the present use
- ☐ the costs of remediation should be reasonable in relation to the seriousness of the potential harm
- ☐ the proposals should be practical in relation to the availability of remediation technology, impact of site constraints and the effectiveness of the proposed clean-up method.

The contaminated land provisions of the Environmental Protection Act 1990 are only one element of a series of statutory measures dealing with pollution and land remediation that have been introduced. Others include:

- ☐ groundwater regulations, including pollution prevention measures
- ☐ an integrated prevention and control regime for pollution
- ☐ sections of the Water Resources Act 1991, which deals with works notices for site controls, restoration and clean up.

The risks involved in the purchase of potentially contaminated sites are high, particularly considering that a transaction can result in the transfer of liability for historic contamination from the vendor to the purchaser.

The ability to forecast the extent and cost of remedial measures is essential for both parties, so that they can be accurately reflected in the price of the land.

The EU Landfill Directive

The Landfill (England and Wales) Regulations 2002 came into force on 15 June 2002 followed by Amendments in 2004 and 2005. These new regulations implement the Landfill Directive (Council Directive 1999/31/EC), which aims to prevent, or to reduce as far as possible, the negative environmental effects of landfill. The regulations will have a major impact on waste regulation and the waste management industry in the UK.

The Scottish Executive and the Northern Ireland Assembly will be bringing forward separate legislation to implement the Directive within their regions.

In summary, the Directive requires that:

- Sites are to be classified into one of three categories: hazardous, non-hazardous or inert, according to the type of waste they will receive.
- Higher engineering and operating standards will be followed.
- Biodegradable waste will be progressively diverted away from landfills.
- Certain hazardous and other wastes, including liquids, explosive waste and tyres will be prohibited from landfills.
- Pre-treatment of wastes prior to landfilling will become a requirement.

On 15 July 2004 the co-disposal of hazardous and non-hazardous waste in the same landfill site ended and in July 2005 new waste acceptance criteria (WAC) were introduced which also prevents the disposal of materials contaminated by coal tar.

The effect of this Directive has been to dramatically reduce the hazardous disposal capacity post July 2004, resulting in a **SIGNIFICANT** increase in remediating costs. There are now less than 20 commercial landfills licensed to accept hazardous waste as a direct result of the implementation of the Directive. There are no sites in Wales and only limited capacity in the South of England. This has significantly increased travelling distance and cost for disposal to landfill. The increase in operating expenses incurred by the landfill operators has also resulted in higher tipping costs.

All hazardous materials designated for disposal off-site are subject to WAC tests. Samples of these materials are taken from site to laboratories in order to classify the nature of the contaminants. These tests, which cost approximately £200 each, have resulted in increased costs for site investigations and as the results may take up to 3 weeks this can have a detrimental effect on programme.

There has been a marked slowdown in brownfield development in the UK with higher remediation costs, longer clean-up programmes and a lack of viable treatment options for some wastes.

The UK Government established the Hazardous Waste Forum in December 2002 to bring together key stakeholders to advise on the way forward on the management of hazardous waste.

Effect on Disposal Costs

Although most landfills are reluctant to commit to tipping prices tipping costs during the first half of 2006 have generally stabilised. However, there are significant geographical variances, with landfill tip costs in the North of England being substantially less than their counterparts in the Southern regions.

For most projects to remain viable there will be an increasing need to treat soil in-situ by bioremediation, soil washing or other alternative long-term remediation measures. Waste untreatable on-site such as coal tar will remain a problem. Development costs and programmes will need to reflect this change in methodology.

Types of hazardous waste

- Sludges, acids and contaminated wastes from the oil and gas industry
- Acids and toxic chemicals from chemical and electronics industries
- Pesticides from the agrochemical industry
- Solvents, dyes and sludges from leather and textile industries
- Hazardous compounds from metal industries
- Oil, oil filters and brake fluids from vehicles and machines
- Mercury-contaminated waste from crematoria

BACKGROUND - continued

Types of hazardous waste - continued

- Explosives from old ammunition, fireworks and airbags
- Lead, nickel, cadmium and mercury from batteries
- Asbestos from the building industry
- Amalgam from dentists
- Veterinary medicines

Source: Sepa

INITIAL STUDY

Approach

Part IIA of the Environmental Protection Act (1990) [EPA], which was introduced by Section 57 of the Environment Act 1995, requires an overall risk-based approach to dealing with contaminated sites, which is consistent with the general good practice approach to managing land subject to contamination.

The regulatory regime set out in Part IIA is based on the following activities:

- ☐ identify the problem
- ☐ assess the risks
- ☐ determine the appropriate remediation requirements
- ☐ consider the costs
- ☐ establish who should pay
- ☐ implementation and remediation

These are examined more fully below:

Identify the problem

Site investigations comprising desk study research and intrusive investigations are necessary to provide information on the soil conditions and possible contaminants located on the site in order that an assessment of the risks may be carried out. Based on these, it should then be possible to ensure that the most appropriate remedial measures are used. From a purely cost angle, the more complete the study is at this stage, the greater the reliability of the cost estimate.

Initially:

- ☐ research previous use(s) of the site by reference to historical maps, local records, interviewing local inhabitants, previous employees etc.
- ☐ study geological maps and local records to determine ground strata, water table, underground aquifers, direction of movement of ground water, presence of water extraction wells locally which may be at risk from contamination.
- ☐ examine local records to try and determine underground and above ground service routes and whether still live

Ideally such studies should also take in surrounding land as the site may be at risk of contamination from an adjoining site problem.

The aim would be to establish the previous pattern of development of the site. Based on the sites various uses, areas of previous development and likely contaminants may be identifiable. It may well be possible to categorize the site into areas each with possibly differing problems and occurring at differing depths. In all probability, a proportion of even a contaminated site could prove to be trouble free.

Even though a site may have no history of potentially contaminative use, previous owners may have inadvertently created a problem by importing fill materials from a contaminated source.

A site 'conceptual model' should be developed from the desk study research, showing the 'receptors' potentially at risk and the 'pathway' between the contamination source and each receptor.

The next step should be soil sampling of the various areas and laboratory analysis of the materials to determine if the soil contains a contaminant (or contaminants) as well as obtaining an idea of the distribution of the contaminants over the site (in areas as well as depth) and the associated concentrations.

The number of sampling points taken on any given site will largely depend on engineering judgement, economics and time. The use of an accurate and regularly spaced sampling grid allows the site to be categorized into areas of high and low risk (i.e. areas exceeding or falling below the trigger values). It may then be possible to use different remedial measures for the differing areas of concentration.

Consideration should be given to the proposed development. There is little point in carrying out expensive remedial works to a site merely for it to be then subjected to the same use. On the other hand, work is very much called for if the proposed use is for agricultural or recreational purpose or else the contaminants threaten the water regime. If the contaminants are static and not putting development at risk, there may be a strong economic argument for leaving them alone.

Bearing in mind the history of the site, the nature and concentration of the contaminant(s) and the most likely path, it should then be possible to choose an appropriate remedial technique.

Assess the risks

With an understanding of the site geology and of the nature, distribution and magnitude of contaminants within the site, an assessment can then be made of the risk that they pose, which will depend largely on the sensitivity of the site and the future land use.

A qualitative risk assessment used to involve a comparison of the observed contaminant concentrations with screening values, such as the ICRCL trigger concentrations. A quantitative risk assessment looks at the particular site, the potential of the hazards to migrate and then a calculation of the likely dose of contaminant at the receptor to compare with EEC and other Regulator limits.

It can be difficult to arrive at an adequate definition of the acceptable levels of risk for the particular site, in view of the complications caused by differences in geology on a regional scale and local site variations. Individual site variations can affect chemical and physical properties and substances themselves can interact to increase or even reduce the risk.

Threats to aquifers would force a vigorous approach to treatment.

Determine the appropriate remediation requirements

Consideration should be given to the most appropriate treatment suited to the location and type of contaminant, the end use of the site and the risks.

Consider the costs

The estimate should review the chosen treatment and take into account:

- a careful measurement/estimate of the quantities involved, e.g. the volume(s) of contaminants which may need removal, length and depth of containment walls, size distribution and depth of boreholes for the extraction of contaminated liquids
- site location - a study of the surrounding roads, built-up areas etc. which may restrict access to and from the site for construction vehicles; this could well influence the size of the vehicles used

INITIAL STUDY – continued

Consider the costs – continued

- ☐ appropriate landfill sites / haulage distance - a review of suitable tips which can take the contaminants, checking their rate of accepting such material, which may well be limited by their licence; bear in mind that tipping charges can vary significantly over a period of time but preferential rates can be obtained for a programme of tipping. Large quantities of material may force a proportion being disposed of not just at the closest, but also at the more distant tips as well – if time is not at a premium, it may well prove cheaper to extend the length of the programme.
- ☐ the location / haulage distance of sites of suitable fill materials
- ☐ a review of any set time scales and a calculation of a practical contract period, perhaps taking into account any requirements for phased hand over of parts of the site.
- ☐ an assessment of the implications of the contaminants on the site establishment – protective clothing and footwear and shower and messing facilities for personnel, wheel wash facilities for vehicles and the careful disposal of contaminated wash water etc.

It is essential that the estimate is a practical all-embracing exercise to help ensure that a realistic budget is set for the project. Once the work commences on site, it cannot be halted even if costs overrun the anticipated budget.

If it does not prove possible to survey the site fully, part or all of the estimate could be at risk and the client should be made aware of the potential risk.

Cost minimisation

Apart from careful measurement and rate evaluation, there are a number of methods of helping to minimise costs, such as:

- ☐ On-site testing to reduce off site volumes
  To help ensure that low risk materials are not removed unnecessarily
- ☐ Correct classification of waste
  To help ensure that the contaminant incurs the lowest tipping charge
- ☐ Use of clean site rubble as far as possible for fill
- ☐ Use of existing contaminated solids for the stabilisation of contaminated liquids/sludges.
  The removal of such a mixture by lorry may be much cheaper than removing liquid by tanker.
- ☐ Ensure the fill and rate of compaction are suitable for the end use
  To avoid future cost.
- ☐ Encapsulation of non-mobile contaminants
  Cheaper than removing from site or treating.
- ☐ Back-hauling of fill materials
  Where possible, organising the truck taking the material to the tip to return with the fill materials, avoiding the cost of travelling empty.

Establish who should pay

Liability for the costs of remediation rests with either the party that "caused or knowingly permitted" contamination, or with the current owners or occupiers of the land.

Apportionment of liability, where shared, is determined by the local authority. Although owners or occupiers become liable only if the polluter cannot be identified, the liability for contamination is commonly passed on when land is sold.

If neither the polluter nor owner can be found, the clean-up is funded from public resources.

Implementation and remediation

Tenders should be sought on clear documentation with the ability for the subsequent work to be remeasured and revalued.

There should be a stringent site monitoring system implemented. This should include monitoring the effectiveness of operations (especially in the case of remedial operations other than dig and dump) as well as checking the chemical profile of all imported fill materials to avoid the obvious.

Approvals involve both the Environment Agency and the local authority Environmental Health Officer / Contaminated Land Officer.

A Health and Safety file handed to the site owner at completion should cover all the checking procedures in detail. Arrange for suitable warranties, which would be required by any future purchaser of the site.

LAND REMEDIATION TECHNIQUES

There are two principal approaches to remediation - dealing with the contamination in situ or off site. The selection of the approach will be influenced by factors such as: initial and long term cost, timeframe for remediation, types of contamination present, depth and distribution of contamination, the existing and planned topography, adjacent land uses, patterns of surface drainage, the location of existing on-site services, depth of excavation necessary for foundations and below-ground services, environmental impact and safety, prospects for future changes in land use and long-term monitoring and maintenance of in situ treatment.

On most sites, contamination can be restricted to the top couple of metres, although gasholder foundations for example can go down 8 or 10 metres. Underground structures can interfere with the normal water regime and trap water pockets.

There could be a problem if contaminants get into fissures in bedrock.

In situ techniques

A range of in situ techniques is available for dealing with contaminants, including:

- ☐ Dilution - the reduction of the concentrations of contaminants to below trigger levels by on-site mixing with cleaner material – this is unlikely to be acceptable to the regulators.

- ☐ Clean cover - a layer of clean soil is used to segregate contamination from receptor. This technique is best suited to sites with widely dispersed contamination. Costs will vary according to the need for barrier layers to prevent migration of the contaminant.

- ☐ On-site encapsulation - the physical containment of contaminants using barriers such as slurry trench cut-off walls. The cost of on-site encapsulation varies in relation to the type and extent of barriers required, the costs of which range from £50/m^2 to more than £175/m^2.

There are also in situ techniques for treating more specific contaminants, including:

- ☐ Bio-remediation - for removal of oily, organic contaminants through natural digestion by micro-organisms. Most bio-remediation is ex-situ, i.e. it is dug out and then treated on site in bio-piles. The process can be slow, perhaps taking as much as one to three years depending upon the scale of the problem, but is particularly effective for the long-term improvement of a site, prior to a change of use.

- ☐ Soil washing - involving the separation of a contaminated soil fraction or oily residue through a washing process. This also involves the excavation of the material for washing ex-situ. The de-watered contaminant still requires disposal to landfill. In order to be cost effective, 70 - 90% of soil mass needs to be recovered. It will involve constructing a hard area for the washing, intercepting the now-contaminated water and taking it away in tankers.

LAND REMEDIATION TECHNIQUES - continued

In situ techniques – continued

- Vacuum extraction - involving the extraction of volatile organic compounds (e.g. benzene) from soil and groundwater by vacuum.
- Stabilisation - cement or lime, is used to physically or chemically bound oily or metal contaminants to prevent leaching or migration. Stabilisation can be used in both in situ and off-site locations.
- Aeration – if the ground contamination is highly volatile, e.g. fuel oils, then the ground can be ploughed and rotovated to allow the substance to vaporize.
- Pumping – to remove liquid contaminants from boreholes or excavations. Contaminated water can be pumped into holding tanks and allowed to settle; testing may well prove it to be suitable for discharging into the foul sewer subject to payment of a discharge fee to the local authority of 35p to 65p per m³. It may be necessary to process the water through an approved water treatment system to render it suitable for discharge.

Off-site techniques

Removal for landfill disposal has, historically, been the most common and cost-effective approach to remediation in the UK, providing a broad spectrum solution by dealing with all contaminants. As discussed above, the implementation of the Landfill Directive will result in other techniques becoming more competitive and enjoy a wider usage. Removal to Landfill is suited to sites where sources of contamination can be easily identified and it is local to an approved Landfill site.

If used in combination with material-handling techniques such as soil washing, the volume of material disposed at landfill sites can be significantly reduced. The disadvantages of the technique include the fact that the contamination is not destroyed, there are risks of pollution during excavation and transfer; road haulage may also cause a local nuisance.

COST CONSIDERATIONS

Cost drivers

Cost drivers relate to the selected remediation technique, site conditions and the size and location of a project.

The wide variation of indicative costs of land remediation techniques shown below is largely because of differing site conditions.

Indicative costs of land remediation techniques (excluding landfill tax)		
Remediation technique	**Unit**	**Rate (£/unit)**
Removal – non-hazardous Removal – hazardous Note: excluding any pre-treatment of material	disposed material (m³) disposed material (m³)	60 - 120 100 – 225
Clean cover	surface area of site (m²)	25 - 60
On-site encapsulation	encapsulated material (m³)	30 - 95
Bio-remediation	treated material (tonne)	35 - 100
Soil washing	treated material (tonne)	50 - 100
Soil flushing	treated material (tonne)	70 - 130
Vacuum extraction	treated material (tonne)	60 - 130
Thermal treatment	treated material (tonne)	200 - 900

Factors that need to be considered include:

- waste classification of the material
- underground obstructions, pockets of contamination and live services
- ground water flows and the requirement for barriers to prevent the migration of contaminants
- health and safety requirements and environmental protection measures
- location, ownership and land use of adjoining sites
- distance from landfill tips, capacity of the tip to accept contaminated materials, and transport restrictions
- the escalating cost of diesel fuel, currently nearing £1 per litre

Other project related variables include size, access to disposal sites and tipping charges; the interaction of these factors can have a substantial impact on overall unit rates.

The tables below set out the costs of remediation using *dig-and-dump* methods for different sizes of project, differentiated by the disposal of non-hazardous and hazardous material. Variation in site establishment and disposal cost accounts for 60 - 70% of the range in cost.

Variation in the costs of land remediation by removal: Non-hazardous Waste			
Item	**Disposal Volume (less than 3000 m³) (£/m³)**	**Disposal Volume (3000 - 10 000 m³) (£/m³)**	**Disposal Volume (more than 10 000 m³) (£/m³)**
General items and site organisation costs	55 - 90	25 - 40	7 - 20
Site investigation and testing	5 - 10	2 - 6	2 – 5
Excavation and backfill	18 - 35	12 - 25	10 – 20
Disposal costs (including tipping charges but not landfill tax)	20 - 35	20 - 35	20 - 35
Haulage	15 - 30	15 - 30	15 – 30
Total (£/m³)	**113 - 200**	**76 - 136**	**54 – 110**
Allowance for site abnormals	*0 - 10 +*	*0 - 15 +*	*0 - 10 +*

Variation in the costs of land remediation by removal: Hazardous Waste			
Item	**Disposal Volume (less than 3000 m³) (£/m³)**	**Disposal Volume (3000 - 10 000 m³) (£/m³)**	**Disposal Volume (more than 10 000 m³) (£/m³)**
General items and site organisation costs	55 - 90	25 - 40	7 - 20
Site investigation and testing	10 - 15	5 - 10	5 - 10
Excavation and backfill	18 - 35	12 - 25	10 - 20
Disposal costs (including tipping charges but not landfill tax)	50 - 170	50 - 170	50 - 170
Haulage	20 - 100	20 - 100	20 - 100
Total (£/m³)	**153 - 410**	**112 - 345**	**92 - 340**
Allowance for site abnormals	*0 - 10 +*	*0 - 15 +*	*0 - 10 +*

COST CONSIDERATIONS – continued

Cost drivers – continued

The strict Health and Safety requirements of remediation works can produce quite high site organisation costs as a % of the overall project cost (see the table above). A high proportion of these costs are fixed and, as a result, the unit costs of site organisation increase disproportionately on smaller projects.

Haulage costs are largely determined by the distances to a licensed tip. Current average haulage rates, based on a return journey range from £1.55 to £2.75 per mile. Short journeys to tips, which involve proportionally longer standing times, typically incur higher mileage rates, up to £6.00 per mile.

A further source of cost variation relates to tipping charges. The table below summarises indicative tipping charges for 2006, exclusive of landfill tax:

Typical 2006 tipping charges (excluding landfill tax)	
Waste classification	**Charges (£/tonne)**
Non-hazardous wastes	10 - 15
Hazardous wastes	20 - 85
Contaminated liquid	20 - 60
Contaminated sludge	45 - 75

Tipping charges fluctuate in relation to the grades of material a tip can accept at any point in time. This fluctuation is a further source of cost risk. Furthermore, tipping charges in the North of England are generally less than in the rest of the country.

Prices at licensed tips can vary by as much as 50%. In addition, landfill tips generally charge a tip administration fee of approximately £20 per load, equivalent to £1 per tonne. This charge does not apply to non-hazardous wastes.

Cost Studies

Site study 1

A recently completed project involving site remedial work to a former gas works site by the dig and dump approach (1,000m^3 sent to hazardous landfill) analyses as follows (the Class references are from CESMM3):

Class A	General Items	23	%
Class B	Ground Investigation	8	%
Class C	Geotechnical Services	0	%
Class D	Demolition and Site Clearance	1	%
Class E	Earthworks		
	- Excavation	4	%
	- Haulage	18	%
	- Disposal	34	%
	- Backfilling	9	%
Class F-X	(A number of minor work classes)	3	%
-	Provisional Sums	0	%
-	Abnormal Costs	0	%
		100.00	%

Site study 2

A recently completed project involving site remedial work to a former gas works site by the dig and dump and soil washing approach (1,500m^3 sent to hazardous landfill) analyses as follows (the Class references are from CESMM3):

Class A	General Items	19	%
Class B	Ground Investigation	6	%
Class C	Geotechnical Services (alternative technique)	40	%
Class D	Demolition and Site Clearance	3	%
Class E	Earthworks		
	- Excavation	5	%
	- Haulage	5	%
	- Disposal	12	%
	- Backfilling	4	%
Class F-X	(A number of minor work classes)	0	%
-	Provisional Sums	5	%
-	Abnormal Costs	1	%
		100.00	%

Site study 3

A recently completed project involving site remedial work to a former gas works site by the dig and dump and soil washing/bioremediation approach (5,000m^3 sent to hazardous landfill) analyses as follows (the Class references are from CESMM3):

Class A	General Items	14	%
Class B	Ground Investigation	5	%
Class C	Geotechnical Services (alternative technique)	34	%
Class D	Demolition and Site Clearance	5	%
Class E	Earthworks		
	- Excavation	6	%
	- Haulage	7	%
	- Disposal	15	%
	- Backfilling	5	%
Class F-X	(A number of minor work classes)	2	%
-	Provisional Sums	6	%
-	Abnormal Costs	1	%
		100.00	%

COST CONSIDERATIONS – continued

Class A General Items

Remedial works contracts generally show a high level of preliminaries, perhaps around the 30-40% mark for relatively small projects, mainly due to the high costs of Health and Safety and temporary works when dealing with contamination. The tables below demonstrate the spread of costs on three site studies included as examples. Over and above the normal site establishment costs included within Part 4 Class A; a number of the following items may need to be included.

- Protective clothing and footwear / site safety inductions
- Hygiene / Decontamination Unit
- Occupational health checks (office staff as well as site labour)
- Bath wheel wash facility
- Weighbridge facility with auto ticketing
- Scaffold gantry for safe covering of wagons
- Steel storage tanks for contaminated liquids
- Temporary fencing between clean and dirty areas and around excavations
- Administration connected with special waste taxes, licenses, etc.
- Portable on-site laboratory
- Dust suppression
- Odour suppression
- Vibration monitoring

Class B Ground Investigation

A further 3 - 9% can be spent on carrying out the site testing and validation sampling. There is a direct connection between a client spending more money on adequate site investigations and reducing the risk of there being something unforeseen and untoward on the site. This section covers items such as trial pits and trenches, light cable percussion boreholes, laboratory testing which are carried out as part of the remediation contract to prove the ground has been cleaned to an acceptable standard. Indicative costs for testing can be found in Part 4 Class B. Long term ground water monitoring may also be required.

Class C Geotechnical Services

As part of the remediation strategy it may be deemed necessary by the site engineer to use diaphragm walls, ground anchorages, ground consolidation techniques, grout holes, etc. Indicative costs are included within Part 4 Class C.

Class D Demolition and Site Clearance

The costs associated with this section are site dependant and can found in Part 4 Class D.

Class E Excavation

Excavation and backfilling costs can be found within Part 4 Class E. As part of the remediation strategy contaminated material may be deemed acceptable as backfill by mixing with inert material. Costs for the rotovation of material within stockpiles are approximately £1.00/m^3. Use of clean material to stabilise contaminated sludges/liquids to allow transportation off site will vary depending on the ratio. As a guide, mixing on a 1:1 ratio will cost approximately £4/m^3. Costs for crushing and screening excavated material are usually expected to be around £6/m^3 and £4/m^3 respectively.

Haulage costs are largely determined by the distances to a licensed tip. A frequently used haulage vehicle will be a 19 tonne payload articulated vehicle, costing say £560 per day including driver and fuel. The cost per load naturally reflects the difficulty of the route. An average 75 miles round trip, with 5 trips being carried out per day would cost £112 per load, or £1.50 per mile. For short distances, the cost per mile could rise to £6.00, reflecting the greater number of trips and hence a greater amount of time in loading and dumping. The haulage cost per m^3 of the disposed material naturally depends on its density. Disposal costs need to be established with the nearest licensed landfill site as they vary depending on the locality and waste classification. On top of this must be included the Special Waste Regulation Charge and consideration as to whether landfill tax is applicable.

THE LANDFILL TAX

The Tax

The Landfill Tax came into operation on 1 October 1996. It is currently levied on operators of licensed landfill sites at the following rates:

£2 per tonne - Inactive or inert wastes.
Included are soil, stones, brick, plain and reinforced concrete, plaster and glass (refer to the table below for a fuller list),

£21 per tonne - All other taxable wastes.
Included are timber, paint and other organic wastes generally found in demolition work, builders skips etc..

From 1st April 2006 the rate of £18 per tonne for "all other taxable wastes" was increased to £21 per tonne, whilst the rate for "inactive or inert wastes" remained at £2 per tonne.

The government has stated that the standard rate of tax will increase by at least £3 per tonne in subsequent years to a rate of £35 per tonne by 2010 in the medium to long term.

Mixtures containing wastes not classified as inactive or inert will not qualify for the lower rate of tax unless the amount of non-qualifying material is small and there is no potential for pollution. Water can be ignored and the weight discounted.

Following the introduction of new Waste Acceptance Criteria from July 2005, certain wastes are not be accepted by licensed landfills. These may include high concentration coal tars. The full definition of acceptable waste has yet to be defined.

Waste liable at the lower rate

Group	Description of material	Conditions	
1	Rocks and soils	Naturally occurring	Rocks and soils includes clay, sand, gravel, sandstone, limestone, crushed stone, china clay, construction stone, stone from the demolition of buildings or structures, slate, topsoil, peat, silt and dredgings Glass includes fritted enamel, but excludes glass fibre and glass reinforced plastics
2	Ceramic or concrete materials		Ceramics includes bricks, bricks and mortar, tiles, clay ware, pottery, china and refractories Concrete includes reinforced concrete, concrete blocks, breeze blocks and aircrete blocks, but excludes concrete plant washings

THE LANDFILL TAX – continued

Waste liable at the lower rate - continued

Group	Description of material	Conditions	
3	Minerals	Processed or prepared, not used	Moulding sands excludes sands containing organic binders Clays includes moulding clays and clay absorbents, including Fuller's earth and bentonite Man-made mineral fibres includes glass fibres, but excludes glass-reinforced plastic and asbestos Silica, mica and mineral abrasives
4	Furnace slags		Vitrified wastes and residues from thermal processing of minerals where, in either case, the residue is both fused and insoluble slag from waste incineration
5	Ash		Comprises only bottom ash and fly ash from wood, coal or Waste combustion Excludes fly ash from municipal, clinical, and hazardous waste incinerators and sewage sludge incinerators
6	Low activity inorganic compound		Comprises only titanium dioxide, calcium carbonate, magnesium carbonate, magnesium oxide, magnesium hydroxide, iron oxide, ferric hydroxide, aluminium oxide, aluminium hydroxide & zirconium dioxide
7	Calcium sulphate	Disposed of either at a site not licensed to take putrescible waste or in a containment cell which takes only calcium sulphate	Includes gypsum and calcium sulphate based plasters, but excludes plasterboard
8	Calcium hydroxide and brine	Deposited in brine cavity	
9	Water	Containing other qualifying material in suspension	

Volume to weight conversion factors (for estimating purposes)

To convert inactive or inert waste (i.e. largely water insoluble and non or very slowly biodegradable: e.g. sand, subsoil, concrete, bricks, mineral fibres, fibreglass etc.), multiply the measured volume in cubic metres by 1.9 to calculate the weight in tonnes.

Calculating the weight of waste

There are two options:

- If licensed sites have a weighbridge, tax will be levied on the actual weight of waste.
- If licensed sites do not have a weighbridge, tax will be levied on the permitted weight of the lorry based on an alternative method of calculation based on volume to weight factors for various categories of waste.

Effect on prices

The tax is paid by Landfill site operators only. Tipping charges reflect this additional cost.

Apart from the possible incidence of Landfill Tax, the cost of disposal will generally comprise the haulage cost plus a tipping charge which will vary according to the toxicity of the material.

Exemptions

The following disposals are exempt from Landfill Tax:

- dredgings which arise from the maintenance of inland waterways and harbours,
- naturally occurring materials arising from mining or quarrying operations,
- waste resulting from the cleaning up of historically contaminated land (although to obtain an exemption it is first necessary to obtain a contaminated land certificate from Customs and Excise),
- waste removed from one site to be used on another or to be recycled or incinerated.

An additional exemption was introduced from 1st October 1999 for inert waste used to restore landfill sites and to fill working and old quarries with a planning condition or obligation in existence to fill the void.

PART 7

Unit Costs (Ancillary Building Works)

INTRODUCTORY NOTES

This part enables the user to include within the estimate for Ancillary Building Works which may be associated with a Civil / Engineering project but, because of the diversity which can occur on the Specification for these Works, cannot be priced as accurately as Unit Cost Items. Additionally, as such works form only a minor percentage of an overall Civil Engineering Budget, then the need for such accuracy in the budget is not as critical. Therefore, the rates given within this Part are based upon an average range for each item to allow the user discretionary use based upon more detailed knowledge of the specific project. Should however more detailed pricing information be required then reference should be made to the latest edition of:

SPON'S ARCHITECTS' AND BUILDERS' PRICE BOOK

SPON'S LANDSCAPE AND EXTERNAL WORKS PRICE BOOK

SPON'S MECHANICAL AND ELECTRICAL SERVICES PRICE BOOK

Also included in this part are items which allow the user to prepare order of cost estimates for various areas of Civil Engineering Works more accurately then by using Part 12: Approximate Estimates but without the necessity to complete the full unit cost estimate.

Adjustments should be made to the rates shown to allow for time, location, local conditions, site constraints and any other factors likely to affect the cost of the specific scheme.

SUBSTRUCTURE

Trench fill foundations

Description	Unit		WIDTH & DEPTH OF CONCRETE 600x1200	750x1200
Machine excavation, disposal, plain insitu concrete 21 N/mm² - 20 mm aggregate (1:2:4) trench fill, 300 mm high brickwork in cement mortar (1:3), pitch polymer damp proof course with common bricks PC £120.00/1000 in				
half brick wall	m	£	83.45	103.87
one brick wall	m	£	95.19	113.56
cavity wall	m	£	97.10	113.56

Strip foundations

Description	Unit		WALL THICKNESS / FOOTING WIDTH / BRICK BOND: One brick 600 mm English	Cavity 750 mm Stretcher
Excavate trench, partial backfill, partial disposal, earthwork support (risk item), compact base of trench, plain insitu concrete 21 N/mm² - 20 mm aggregate (1:2:4) 250 mm thick, formwork, common brickwork (PC £120.00/1000) in cement mortar (1:3), pitch polymer damp proof course				
hand excavation, depth of wall				
600 mm	m	£	84.47	83.20
900 mm	m	£	103.87	124.28
1200 mm	m	£	133.97	158.22
1500 mm	m	£	158.22	188.34
machine excavation, depth of wall				
600 mm	m	£	73.75	84.47
900 mm	m	£	99.02	108.72
1200 mm	m	£	118.41	138.83
1500 mm	m	£	138.83	163.07
Extra over for three courses of facing bricks PC £ 275.00/1000	m	£	2.91	5.92

Column bases

Description	Unit		DEPTH OF PIT 1200mm	1800mm
Excavate pit in firm ground, partial backfill, partial disposal, earth work support, compact base of pit, plain insitu concrete 21 N/m² - 20 mm aggregate (1:2:4), formwork				
hand excavation, base size				
600 x 600 x 300 mm	nr	£	64.06	81.54
900 x 900 x 450 mm	nr	£	131.05	152.35
1200 x 1200 x 450 mm	nr	£	201.86	253.41
1500 x 1500 x 600 mm	nr	£	364.04	424.27
machine excavation, base size				
600 x 600 x 300 mm	nr	£	50.53	63.16
900 x 900 x 450 mm	nr	£	110.63	120.33
1200 x 1200 x 450 mm	nr	£	162.18	182.47
1500 x 1500 x 600 mm	nr	£	302.92	330.10

SUBSTRUCTURE - continued					
Column bases - continued					
Excavate pit in firm ground by machine, partial backfill, partial disposal, earthwork support, compact base of pit, reinforced insitu concrete 21N/mm² - 20 mm aggregate (1:2:4), formwork			DEPTH OF PIT		
			1200mm		1800mm
Reinforcement at 50 kg/m³ concrete, base size					
1750 x 1750 x 500 mm	nr	£	343.24		384.08
2000 x 2000 x 500 mm	nr	£	423.63		475.95
2200 x 2200 x 600 mm	nr	£	626.52		687.76
2400 x 2400 x 600 mm	nr	£	727.32		788.57
Reinforcement at 75 kg/m³ concrete, base size					
1750 x 1750 x 500 mm	nr	£	363.66		403.22
2000 x 2000 x 500 mm	nr	£	455.53		505.30
2200 x 2200 x 600 mm	nr	£	657.14		706.90
2400 x 2400 x 600 mm	nr	£	769.43		828.12
Pile caps					
Excavate pit in firm ground by machine, partial backfill, partial disposal, earthwork support, compact base of pit, cut off top of pile and prepare reinforcement, reinforced insitu concrete 26 N/mM² - 20 mm aggregate(1:2:4), formwork			DEPTH OF PIT		
			1500mm		2100mm
Reinforcement at 50 kg/m³ concrete, cap size					
900 x 900 x 1400 mm, one pile	nr	£	322.83		343.24
2700 x 900 x 1400 mm, two piles	nr	£	807.71		828.12
*2700 x 2475 x 1400 mm, three piles	nr	£	1466.12		1515.89
2700 x 2700 x 1400 mm, four piles	nr	£	1869.34		1933.14
3700 x 2700 x 1400 mm, six piles	nr	£	2526.48		2673.22
Reinforcement at 75 kg/m³ concrete, cap size					
900 x 900 x 1400 mm, one pile	nr	£	322.83		341.97
2700 x 900 x 1400 mm, two piles	nr	£	807.71		857.47
*2700 x 2475 x 1400 mm, three piles	nr	£	1485.26		1614.14
2700 x 2700 x 1400 mm, four piles	nr	£	2018.63		2042.88
3700 x 2700 x 1400 mm, six piles	nr	£	2673.22		2831.44
* = triangular on plan, overall size given					
			30 N/m³		40 N/m³
Additional cost of alternative strength concrete	m³	£	1.40		3.45
Strip or base foundations					
Foundations in good ground; reinforced concrete bed; for one storey development					
shallow foundations per m² ground floor plan area	m²	£	60.23	to £	81.03
deep foundations per m² ground floor plan area	m²	£	100.93	to £	120.84

Description	Unit				
Foundations in good ground; reinforced concrete bed; for two storey development					
shallow foundations per m² ground floor plan area	m²	£	70.44	to £	110.63
deep foundations per m² ground floor plan area	m²	£	110.63	to £	172.26
Extra for each additional storey	m²	£	24.24	to £	28.20
Raft foundations					
Raft on poor ground for development					
one storey per m² ground floor plan area	m²	£	81.03	to £	140.74
two storey per m² ground floor plan area	m²	£	130.53	to £	192.17
Extra for each additional storey	m²	£	24.24	to £	28.20
Piled foundations					
Foundations in poor ground; reinforced concrete slab; for one storey commercial development per m² ground floor plan area					
short bore piles to columns only	m²	£	100.93	to £	152.35
short bore piles	m²	£	130.53	to £	172.26
fully piled	m²	£	172.26	to £	241.67
Ground slabs					
Mechanical excavation to reduce levels, disposal, level and compact, hardcore bed blinded with sand, 1200 gauge polythene damp proof membrane, concrete 21.00N/mm² - 20 mm aggregate (1:2:4) ground slab, tamped finish			THICKNESS OF SLAB		
			200mm	300mm	450mm
thickness of hardcore bed per m² ground floor plan area :					
150 mm	m²	£	40.83	48.49	63.16
175 mm	m²	£	42.75	50.53	65.08
200 mm	m²	£	42.75	50.53	65.08
Add to the foregoing prices for fabric reinforcement BS 4483, lapped; per m² ground floor plan area					
A142 (2.22 kg/m²); 1 layer	m²	£	4.08	40.83	4.08
A142 (2.22 kg/m²); 2 layers	m²	£	6.00	7.02	7.02
A193 (3.02 kg/m²); 1 layer	m²	£	4.08	4.08	5.10
A193 (3.02 kg/m²); 2 layers	m²	£	7.02	8.17	8.17
A252 (3.95 kg/m²); 1 layer	m²	£	5.10	5.10	6.00
A252 (3.95 kg/m²); 2 layers	m²	£	8.17	9.19	10.08
A393 (6.16 kg/m²); 1 layer	m²	£	6.00	6.00	7.02
A393 (6.16 kg/m²); 2 layers	m²	£	11.99	11.99	11.99
High yield bent bar reinforcement BS 4449 ; per m² ground floor plan area at a rate of			THICKNESS OF SLAB		
			200mm	300mm	450mm
25 kg/m³	m²	£	5.10	8.17	11.10
50 kg/m³	m²	£	9.19	14.16	21.18
75 kg/m³	m²	£	14.16	20.42	30.24
100 kg/m³	m²	£	18.25	26.16	40.58

SUBSTRUCTURE - continued					
Alternative concrete mixes in lieu of 21.00N/m² - 20 mm aggregate (1:2:4); per m² ground floor plan area					
25 N/mm²	m²	£	0.51	0.75	1.15
30 N/mm²	m²	£	0.71	1.07	1.60
40 N/mm²	m²	£	1.26	1.86	2.79
Other foundations/alternative slabs/extras					
Reinforced concrete bed including excavation and hardcore under; per m² ground floor plan area					
150 mm thick	m²	£	36.37	to £	46.57
200 mm thick	m²	£	48.49	to £	62.65
300 mm thick	m²	£	60.23	to £	84.98
Extra per m² ground floor plan area for					
sound reducing quilt in screed	m²	£	3.83	to £	6.76
50 mm insulation under slab and at edges	m²	£	6.25	to £	9.19
75 mm insulation under slab and at edges	m²	£	7.78	to £	11.61
suspended precast concrete slabs in lieu of insitu slab	m²	£	16.46	to £	20.93
SUPERSTRUCTURE					
FRAME TO ROOF					
Reinforced insitu concrete columns, bar reinforcement at 200 kg/m³; basic formwork (assumed four uses) ; column size			Strength N/mm²		
			21	30	40
225 x 225 mm	m	£	47.21	47.21	48.49
300 x 300 mm	m	£	65.08	66.35	70.18
300 x 450 mm	m	£	91.87	91.87	95.70
300 x 600 mm	m	£	112.29	112.29	118.67
450 x 450 mm	m	£	116.12	118.67	119.94
			Strength N/mm²		
			21	30	40
Insitu concrete casing to steel column, basic formwork (assumed four uses), column size					
225 x 225 mm	m	£	44.66	44.66	47.21
300 x 300 mm	m	£	61.25	62.52	65.08
300 x 450 mm	m	£	81.66	82.94	85.49
300 x 600 mm	m	£	96.98	99.53	103.36
450 x 450 mm	m	£	100.80	100.80	108.46
450 x 600 mm	m	£	122.50	125.05	131.43
450 x 900 mm	m	£	169.71	169.71	181.19

			Strength N/mm²		
			21	30	40
Reinforced insitu concrete isolated beams; bar reinforcement at 200 kg/m³ basic formwork (assumed four uses); beam size					
225 x 450 mm	m	£	81.66	82.94	85.49
225 x 600 mm	m	£	103.36	103.36	108.46
300 x 600 mm	m	£	118.67	118.67	122.50
300 x 900 mm	m	£	164.60	168.43	172.26
300 x 1200 mm	m	£	211.82	213.09	220.75
450 x 600 mm	m	£	151.84	151.84	160.78
450 x 900 mm	m	£	209.26	211.82	220.75
450 x 1200 mm	m	£	265.41	269.24	279.44
600 x 600 mm	m	£	186.30	188.85	197.78
600 x 900 mm	m	£	251.37	253.92	269.24
600 x 1200 mm	m	£	320.28	325.38	340.69
Insitu concrete casing to steel attached beams; basic formwork (assumed four uses) ; beam size					
225 x 450 mm	m	£	75.28	75.28	79.11
225 x 600 mm	m	£	95.70	95.70	99.53
300 x 600 mm	m	£	108.46	108.46	112.29
300 x 900 mm	m	£	148.02	149.29	155.67
300 x 1200 mm	m	£	186.30	188.85	197.78
450 x 600 mm	m	£	131.43	131.43	141.64
450 x 900 mm	m	£	181.19	182.47	192.68
450 x 1200 mm	m	£	225.85	229.68	241.16
600 x 600 mm	m	£	160.78	163.33	172.26
600 x 900 mm	m	£	211.82	215.64	229.68
600 x 1200 mm	m	£	265.41	269.24	285.82
Extra for smooth finish formwork; all categories	m	£	5.10	to £	14.67
Softwood joisted floor; no frame					
Joisted floor; no frame; 22 mm chipboard t & g flooring; herring bone strutting; no coverings or finishes ; per m² of upper floor area					
150 x 50 mm joists	m²	£	28.07	to £	33.18
175 x 50 mm joists	m²	£	31.90	to £	38.28
200 x 50 mm joists	m²	£	33.18	to £	38.28
225 x 50 mm joists	m²	£	35.73	to £	39.56
250 x 50 mm joists	m²	£	38.28	to £	42.11
275 x 50 mm joists	m²	£	42.11	to £	47.21

Item	Unit		Rate		Rate
SUPERSTRUCTURE - continued					
FRAME TO ROOF - continued					
Softwood joisted floor; average depth; plasterboard; skim; emulsion; vinyl flooring and painted softwood skirtings ; per m² of upper floor area	m²	£	66.35	to £	85.49
Joisted floor; no frame; 22 mm chipboard t & g flooring; herring bone strutting; no coverings or finishes ; per m² of upper floor area					
150 x 50 mm joists	m²	£	28.07	to £	33.18
175 x 50 mm joists	m²	£	31.90	to £	38.28
200 x 50 mm joists	m²	£	33.18	to £	38.28
225 x 50 mm joists	m²	£	35.73	to £	39.56
250 x 50 mm joists	m²	£	38.28	to £	42.11
275 x 50 mm joists	m²	£	42.11	to £	47.21
Softwood joisted floor; average depth; plasterboard; skim; emulsion; vinyl flooring and painted softwood skirtings ; per m² of upper floor area	m²	£	66.35	to £	85.49
Reinforced concrete floors; no frame					
Suspended slab; no coverings or finishes; per m² of upper floor area					
2.75 m span; 8.00 KN/m² loading	m²	£	54.87	to £	65.08
3.35 m span; 8.00 KN/m² loading	m²	£	62.52	to £	74.01
4.25 m span; 8.00 KN/m² loading	m²	£	77.84	to £	91.87
Suspended slab; no coverings or finishes; per m² of upper floor area					
150 mm thick	m²	£	74.01	to £	107.18
225 mm thick	m²	£	116.12	to £	131.43
Reinforced concrete floors and frame					
Suspended slab; average depth; no coverings or finishes ; per m² of upper floor area					
up to six storeys	m²	£	141.64	to £	190.12
Wide span suspended slab with frame; per m² of upper floor area					
up to six storeys	m²	£	159.50	to £	196.50
Reinforced concrete floors; steel frame					
Suspended slab; average depth; 'Holorib' permanent steel shuttering; protected steel frame; no coverings or finishes; per m² of upper floor area					
up to six storeys	m²	£	190.12	to £	233.51
Extra for spans 7.5 to 15 m	m²	£	21.69	to £	58.70
Suspended slab; average depth; protected steel frame; no coverings or finishes; per m² of upper floor area					
up to six storeys	m²	£	181.19	to £	242.44

Suspended slab; 75 mm screed; no coverings or finishes ; per m² of upper floor area					
3 m span; 8.50 KN/m² loading	m²	£	58.70	to £	67.63
6 m span; 8.50 KN/m² loading	m²	£	62.52	to £	71.46
7.5 m span; 8.50 KN/m² loading	m²	£	65.08	to £	74.01
3 m span; 12.50 KN/m² loading	m²	£	70.18	to £	77.84
6 m span; 12.50 KN/m² loading	m²	£	57.42	to £	82.94
Precast concrete floors; reinforced concrete frame					
Suspended slab; average depth; no coverings or finishes ; per m² of upper floor area	m²	£	93.15	to £	122.50
Precast concrete floors and frame					
Suspended slab; average depth; no coverings or finishes ; per m² of upper floor area	m²	£	93.15	to £	176.09
Precast concrete floors; steel frame					
Suspended slabs; average depth; protected steel frame; no coverings or finishes; per m² of upper floor area					
up to six storeys	m²	£	168.43	to £	215.64
Extra per m² of upper floor area for					
wrought formwork	m²	£	3.96	to £	4.72
sound reducing quilt in screed	m²	£	4.21	to £	7.53
insulation to avoid cold bridging	m²	£	7.53	to £	10.34
ROOF					
Softwood flat roofs					
Roof joists; average depth; 25 mm softwood boarding; PVC rainwater goods; plasterboard; skim and emulsion ; per m² of roof plan area					
three layer felt and chippings	m²	£	89.32	to £	112.29
two coat asphalt and chippings	m²	£	85.49	to £	123.77
Softwood trussed pitched roofs					
Structure only comprising 100 x 38 mm Fink roof trusses @ 600 mm centres (measured on plan) ; per m² of roof plan area					
30 degrees pitch	m²	£	21.69	to £	25.52
35 degrees pitch	m²	£	22.97	to £	25.52
40 degrees pitch	m²	£	24.24	to £	29.35

Description	Unit		Rate		
SUPERSTRUCTURE - continued					
ROOF - continued					
Softwood trussed pitched roofs - continued					
Fink roof trusses; narrow span; 100 mm insulation; PVC rainwater goods; plasterboard; skim and emulsion per m² of roof plan area					
concrete interlocking tile coverings	m²	£	91.87	to £	123.77
clay pantile coverings	m²	£	99.53	to £	131.43
composition slate coverings	m²	£	104.63	to £	136.53
plain clay tile coverings	m²	£	123.77	to £	156.95
natural slate coverings	m²	£	131.43	to £	164.60
reconstructed stone coverings	m²	£	107.18	to £	170.98
Monopitch roof trusses; 100 mm insulation; PVC rainwater goods; plasterboard; skim and emulsion ; per m² of roof plan area					
concrete interlocking tile coverings	m²	£	111.01	to £	131.43
clay pantile coverings	m²	£	108.46	to £	136.53
composition slate coverings	m²	£	114.84	to £	145.46
plain clay tile coverings	m²	£	131.43	to £	167.16
natural slate coverings	m²	£	136.53	to £	170.98
reconstructed stone coverings	m²	£	112.29	to £	178.64
Steel trussed pitched roofs					
Steel roof trusses and beams; thermal and accoustic insulation; per m² of roof plan area					
aluminium profiled composite cladding	m²	£	237.34	to £	282.00
Steel roof and glulam beams; thermal and accoustic insulation; per m² of roof plan area					
aluminium profiled composite cladding	m²	£	237.34	to £	313.90
Concrete flat roofs					
Structure only comprising reinforced concrete suspended slab; no coverings or finishes ; per m² of roof plan area					
3.65 m span; 8.00 KN/m² loading	m²	£	58.70	to £	61.25
4.25 m span; 8.00 KN/m² loading	m²	£	70.18	to £	74.01
Precast concrete suspended slab; average depth; 100 mm insulation; PVC rainwater goods; per m² of roof plan area					
two coat asphalt coverings and chippings	m²	£	107.18	to £	149.29
Reinforced concrete or waffle suspended slabs; average depth; 100 mm insulation; PVC rainwater goods; per m² of roof plan area					
two coat asphalt coverings and chippings	m²	£	116.12	to £	153.12
Reinforced concrete suspended slabs; on `Holorib' permanent steel shuttering; average depth; 100 mm insulation; PVC rainwater goods ;per m² of roof plan area					
two coat asphalt coverings and chippings	m²	£	108.46	to £	130.15

Item	Unit		Rate		
Flat roof decking and finishes					
Woodwool roof decking; per m² of roof plan area					
50 mm thick; two coat asphalt coverings to BS 6925 and chippings	m²	£	48.49	to £	61.25
Galvanised steel roof decking; 100 mm insulation; three layer felt roofing and chippings ; per m² of roof plan area					
0.7 mm thick; 3.74 m span	m²	£	53.59	to £	66.35
0.7 mm thick; 5.13 m span	m²	£	54.87	to £	68.90
Aluminium roof decking; 100 mm insulation; three layer felt roofing and chippings ; per m² of roof plan area					
0.9 mm thick; 2.34 m span	m²	£	62.52	to £	76.56
Metal decking; 100 mm insulation; on wood/steel open lattice beams; per m² of roof plan area					
three layer felt roofing and chippings	m²	£	93.15	to £	114.84
two layer high performance felt roofing and chippings	m²	£	96.98	to £	116.12
Roof claddings					
Non-asbestos profiled cladding; per m² of roof plan area					
'profile 3'; natural	m²	£	16.08	to £	19.91
'profile 6'; natural	m²	£	17.86	to £	21.82
'profile 6'; natural; insulated; inner lining panel	m²	£	19.91	to £	21.82
Extra for colours	m²	£	1.79	to £	2.30
Non-asbestos profiled cladding on steel purlins; per m² of roof plan area					
insulated	m²	£	31.52	to £	39.81
insulated; with 10% translucent sheets	m²	£	35.47	to £	41.73
insulated; plasterboard inner lining on metal tees	m²	£	51.42	to £	60.23
PVF2 coated galvanised steel profiled cladding on steel purlins; per m² of roof plan area					
cladding only; 0.72mm thick	m²	£	21.82	to £	31.52
insulated	m²	£	38.41	to £	57.29
insulated; plasterboard inner lining on metal tees	m²	£	57.29	to £	81.54
insulated; plasterboard inner lining on metal tees; with 1% fire vents	m²	£	66.99	to £	100.93
insulated; plasterboard inner lining on metal tees; with 2.5% fire vents	m²	£	83.45	to £	116.50
insulated; coloured inner lining panel	m²	£	59.21	to £	83.45
insulated; coloured inner lining panel; with 1% fire vents	m²	£	66.99	to £	93.15
insulated; coloured inner lining panel; with 2.5% fire vents	m²	£	83.45	to £	108.72
insulated; sandwich panel	m²	£	124.28	to £	217.43

SUPERSTRUCTURE - continued					
ROOF - continued					
Rooflights/patent glazing and glazed roofs					
Rooflights					
standard pvc	m²	£	139.08	to £	248.82
feature/ventilating	m²	£	248.82	to £	456.81
Patent glazing; including flashings					
standard aluminium georgian wired; single glazed	m²	£	178.64	to £	248.82
standard aluminium georgian wired; double glazed	m²	£	207.99	to £	287.10
Comparative over/underlays					
Roofing felt; unreinforced					
sloping (measured on face)	m²	£	1.60	to £	2.08
Roofing felt; reinforced					
sloping (measured on face)	m²	£	1.98	to £	2.27
sloping (measured on plan); 20 degrees pitch	m²	£	2.27	to £	2.87
sloping (measured on plan); 30 degrees pitch	m²	£	2.49	to £	3.20
sloping (measured on plan); 40 degrees pitch	m²	£	3.20	to £	3.59
Building paper; per m² of roof plan area	m²	£	1.38	to £	2.87
Vapour barrier; per m² of roof plan area	m²	£	2.08	to £	6.21
Insulation quilt; laid over ceiling joists ; per m² of roof plan area					
100 mm thick	m²	£	4.66	to £	4.95
150 mm thick	m²	£	6.41	to £	7.18
200 mm thick	m²	£	8.54	to £	9.31
Wood fibre insulation boards; impregnated; density 220 - 350 kg/m³; per m² of roof plan area					
12.7 mm thick	m²	£	5.73	to £	7.96
Limestone ballast; per m² of roof plan area	m²	£	6.25	to £	10.72
Polyurethane insulation boards; density 32 kg/m³ ; per m² of roof plan area					
30 mm thick	m²	£	8.93	to £	10.72
50 mm thick	m²	£	10.72	to £	11.10
Glass fibre insulation boards; density 120 - 130 kg/m²; per m² of roof plan area					
60 mm thick	m²	£	16.46	to £	19.91
Extruded polystyrene foam boards; per m² of roof plan area					
50 mm thick	m²	£	15.57	to £	17.86
50 mm thick; with cement topping	m²	£	25.26	to £	28.58
75 mm thick	m²	£	20.42	to £	22.84
Screeds to receive roof coverings; per m² of roof plan area					
50 mm cement and sand screed	m²	£	10.34	to £	11.61
60 mm (av.) 'Isocrete K' screed; density 500 kg/m³	m²	£	11.10	to £	11.99
75 mm lightweight bituminous screed and vapour barrier	m²	£	17.35	to £	19.91
100 mm lightweight bituminous screed and vapour barrier	m²	£	21.31	to £	23.73

Description	Unit		Rate			Rate
50 mm Woodwool slabs; unreinforced						
sloping (measured on face)	m²	£	0.00	to	£	0.00
sloping (measured on plan); 20 degrees pitch	m²	£	0.00	to	£	0.00
sloping (measured on plan); 30 degrees pitch	m²	£	0.00	to	£	0.00
sloping (measured on plan); 40 degrees pitch	m²	£	0.00	to	£	0.00
50 mm Woodwool slabs; unreinforced; on and including steel purlins at 600 mm centres	m²	£	11.10	to	£	14.29
25 mm `Tanalised' softwood boarding						
sloping (measured on face)	m²	£	15.18	to	£	17.35
sloping (measured on plan); 20 degrees pitch	m²	£	16.46	to	£	18.88
sloping (measured on plan); 30 degrees pitch	m²	£	19.91	to	£	22.84
sloping (measured on plan); 40 degrees pitch	m²	£	23.73	to	£	25.78
18 mm External quality plywood boarding						
sloping (measured on face)	m²	£	19.27	to	£	23.35
sloping (measured on plan); 20 degrees pitch	m²	£	21.31	to	£	24.75
sloping (measured on plan); 30 degrees pitch	m²	£	25.78	to	£	30.62
sloping (measured on plan); 40 degrees pitch	m²	£	30.62	to	£	34.45
Comparative tiling and slating finishes/perimeter treatments (including underfelt, battening, eaves courses and ridges)						
Concrete troughed interlocking tiles; 413 x 300 mm; 75 mm lap						
sloping (measured on face)	m²	£	18.37	to	£	22.84
sloping (measured on plan); 30 degrees pitch	m²	£	23.73	to	£	28.58
sloping (measured on plan); 40 degrees pitch	m²	£	28.58	to	£	31.52
Concrete interlocking slates; 430 x 330 mm; 75 mm lap						
sloping (measured on face)	m²	£	19.27	to	£	23.35
sloping (measured on plan); 30 degrees pitch	m²	£	20.93	to	£	29.09
sloping (measured on plan); 40 degrees pitch	m²	£	28.58	to	£	32.03
Concrete bold roll interlocking tiles; 418 x 332 mm; 75 mm lap						
sloping (measured on face)	m²	£	18.37	to	£	22.84
sloping (measured on plan); 30 degrees pitch	m²	£	24.24	to	£	28.58
sloping (measured on plan); 40 degrees pitch	m²	£	27.18	to	£	31.01
Natural red pantiles; 337 x 241 mm; 76 mm head and 38 mm side laps						
sloping (measured on face)	m²	£	30.62	to	£	37.39
sloping (measured on plan); 30 degrees pitch	m²	£	39.81	to	£	46.57
sloping (measured on plan); 40 degrees pitch	m²	£	45.17	to	£	51.42
Blue composition (non-asbestos) slates; 600 x 300 mm; 75 mm lap						
sloping (measured on face)	m²	£	31.52	to	£	38.79
sloping to mansard (measured on face)	m²	£	44.66	to	£	51.42
sloping (measured on plan); 30 degrees pitch	m²	£	41.73	to	£	52.44
sloping (measured on plan); 40 degrees pitch	m²	£	47.08	to	£	76.69
Concrete plain tiles; 267 x 165 mm; 64 mm lap						
sloping (measured on face)	m²	£	36.75	to	£	44.28
sloping (measured on plan); 30 degrees pitch	m²	£	45.55	to	£	55.44
sloping (measured on plan); 40 degrees pitch	m²	£	55.00	to	£	68.78

SUPERSTRUCTURE - continued						
ROOF - continued						
Comparative tiling and slating finishes/perimeter treatments (including underfelt, battening, eaves courses and ridges) - continued						
Machine made clay plain tiles; 267 x 165 mm; 64 mm lap						
sloping (measured on face)	m²	£	48.11	to £		56.78
sloping (measured on plan); 30 degrees pitch	m²	£	59.72	to £		71.33
sloping (measured on plan); 40 degrees pitch	m²	£	72.35	to £		85.36
Welsh natural slates; 510 x 255 mm; 76 mm lap						
sloping (measured on face)	m²	£	53.34	to £		69.41
sloping (measured on plan); 30 degrees pitch	m²	£	69.41	to £		83.45
sloping (measured on plan); 40 degrees pitch	m²	£	81.54	to £		89.32
Reconstructed stone slates; random lengths; 80 mm lap						
sloping (measured on face)	m²	£	33.43	to £		63.16
sloping (measured on plan); 30 degrees pitch	m²	£	41.73	to £		81.54
sloping (measured on plan); 40 degrees pitch	m²	£	49.51	to £		93.15
Verges to sloping roofs; 250 x 25 mm painted softwood bargeboard; measured perimeter length						
6 mm 'Masterboard' soffit lining 150 mm wide	m	£	17.35	to £		20.93
19 x 150 mm painted softwood soffit	m	£	20.93	to £		23.35
Eaves to sloping roofs; 200 x 25 mm painted softwood fascia; 6 mm 'Masterboard' soffit lining 225 mm wide; measured perimeter length						
100 mm PVC gutter	m	£	23.73	to £		32.03
150 mm PVC gutter	m	£	30.11	to £		38.79
100 mm cast iron gutter; decorated	m	£	37.90	to £		45.17
150 mm cast iron gutter; decorated	m	£	46.06	to £		53.34
Eaves to sloping roofs; 200 x 25 mm painted softwood fascia; 19 x 225 mm painted softwood soffit; measured perimeter length						
100 mm PVC gutter	m	£	28.58	to £		35.86
150 mm PVC gutter	m	£	35.47	to £		42.75
100 mm cast iron gutter; decorated	m	£	42.75	to £		48.49
150 mm cast iron gutter; decorated	m	£	50.02	to £		57.80

Item	Unit		Rate		
Rainwater pipes; fixed to backgrounds; including offsets and shoe measured length of pipes					
68 mm PVC	m	£	8.29	to £	11.87
110 mm PVC	m	£	12.38	to £	27.56
75 mm cast iron; decorated	m	£	30.11	to £	35.86
100 mm cast iron; decorated	m	£	35.47	to £	42.75
Ridges measured length of ridge					
concrete half round tiles	m	£	16.84	to £	20.93
machine-made clay half round tiles	m	£	20.93	to £	24.75
Hips; including mitring roof tiles measured length of hip					
concrete half round tiles	m	£	21.82	to £	28.58
machine-made clay half round tiles	m	£	31.01	to £	34.96
Comparative cladding finishes (including underfelt, labours etc.)					
0.91 mm Aluminium roofing; commercial grade					
flat	m²	£	48.49	to £	54.87
0.91 mm Aluminium roofing; commercial grade ; fixed to boarding (included)					
sloping (measured on face)	m²	£	51.04	to £	59.97
sloping (measured on plan); 20 degrees pitch	m²	£	57.42	to £	66.35
sloping (measured on plan); 30 degrees pitch	m²	£	68.90	to £	79.11
sloping (measured on plan); 40 degrees pitch	m²	£	81.66	to £	89.32
Comparative waterproof finishes/perimeter treatments					
Liquid applied coatings					
solar reflective paint	m²	£	1.79	to £	3.19
spray applied bitumen	m²	£	7.15	to £	11.61
spray applied co-polymer	m²	£	8.42	to £	12.89
spray applied polyurethane	m²	£	13.91	to £	18.88
20 mm Two coat asphalt roofing; laid flat; on felt underlay					
to BS 6925	m²	£	14.80	to £	19.27
to BS 6577	m²	£	20.93	to £	25.78
Extra for					
solar reflective paint	m²	£	2.42	to £	3.45
limestone chipping finish	m²	£	3.06	to £	7.91
grp tiles in hot bitumen	m²	£	33.43	to £	41.73
20 mm Two coat reinforced asphaltic compound; laid flat; on felt underlay to BS 6577	m²	£	22.84	to £	28.58
Built-up bitumen felt roofing; laid flat					
three layer glass fibre roofing	m²	£	19.91	to £	24.75
three layer asbestos based roofing	m²	£	24.75	to £	28.58
Extra for granite chipping finish	m²	£	3.06	to £	7.78

SUPERSTRUCTURE - continued						
ROOF - continued						
Comparative waterproof finishes/perimeter treatments - continued						
Built-up self-finished asbestos based bitumen felt roofing; laid sloping						
two layer roofing (measured on face)	m²	£	26.67	to	£	32.03
two layer roofing (measured on plan); 40 degree pitch	m²	£	40.83	to	£	44.66
three layer roofing (measured on face)	m²	£	35.47	to	£	42.75
three layer roofing (measured on plan); 20 degree pitch	m²	£	53.34	to	£	59.21
three layer roofing (measured on plan); 30 degree pitch	m²	£	55.38	to	£	61.12
Elastomeric single ply roofing; laid flat						
EPDM membrane; laid loose	m²	£	23.35	to	£	26.67
Butyl rubber membrane; laid loose	m²	£	23.35	to	£	26.67
Extra for ballast	m²	£	7.15	to	£	11.61
Thermoplastic single ply roofing; laid flat						
laid loose	m²	£	22.33	to	£	26.67
mechanically fixed	m²	£	28.58	to	£	33.43
fully adhered	m²	£	31.52	to	£	35.86
CPE membrane; laid loose	m²	£	25.78	to	£	31.52
CSPG membrane; fully adhered	m²	£	25.78	to	£	31.52
PIB membrane; laid loose	m²	£	29.60	to	£	35.86
Extra for ballast	m²	£	7.15	to	£	11.61
High performance built-up felt roofing; laid flat						
three layer 'Ruberglas 120 GP' felt roofing; granite chipping finish	m²	£	31.52	to	£	34.45
'Andersons' three layer self-finish polyester based bitumen felt roofing	m²	£	31.52	to	£	35.86
High performance built-up felt roofing; laid flat						
three layer polyester based modified bitumen felt roofing	m²	£	32.54	to	£	35.86
three layer 'Ruberfort HP 350' felt roofing; granite chipping finish	m²	£	39.81	to	£	44.66
three layer 'Hyload 150E' elastomeric roofing; granite chipping finish	m²	£	39.81	to	£	44.66
three layer 'Polybit 350' elastomeric roofing; granite chipping finish	m²	£	41.73	to	£	46.57
Torch on roofing; laid flat						
three layer polyester-based modified bitumen roofing	m²	£	28.58	to	£	33.94
two layer polymeric isotropic roofing	m²	£	28.58	to	£	33.94
Extra for granite chipping finish	m²	£	2.68	to	£	7.15
Edges to flat felt roofs; softwood splayed fillet; 280 x 25 mm painted softwood fascia; no gutter						
aluminium edge trim	m	£	33.43	to	£	35.86

Description	Unit	Rate
Edges to flat roofs; code 4 lead drip dressed into gutter; 230 x 25 mm painted softwood fascia		
100 mm PVC gutter	m	£ 31.52 to £ 41.73
150 mm PVC gutter	m	£ 39.81 to £ 47.59
100 mm cast iron gutter; decorated	m	£ 47.59 to £ 59.21
150 mm cast iron gutter; decorated	m	£ 59.21 to £ 71.33
STAIRS		
Timber construction		
Softwood staircase; softwood balustrades and hardwood handrail; plasterboard; skim and emulsion to soffit		
2.6 m rise; standard; straight flight	nr	£ 712.01 to £ 1079.50
2.6 m rise; standard; top three treads winding	nr	£ 871.51 to £ 1198.16
2.6 m rise; standard; dogleg	nr	£ 1000.38 to £ 1268.34
Reinforced concrete construction		
Escape staircase; granolithic finish; mild steel balustrades and handrails		
3 m rise; dogleg	nr	£ 3718.26 to £ 4756.93
Plus or minus for each 300 mm variation in storey height	nr	£ 361.11 to £ 465.74
Staircase; terrazzo finish; mild steel balustrades and handrails; plastered soffit; balustrades and staircase soffit decorated		
3 m rise; dogleg	nr	£ 7426.32 to £ 9309.70
Plus or minus for each 300 mm variation in storey height	nr	£ 747.74 to £ 931.48
Metal construction		
Steel access/fire ladder		
3 m high	nr	£ 562.72 to £ 786.02
4 m high; epoxide finished	nr	£ 786.02 to £ 1348.73
Light duty metal staircase; galvanised finish; perforated treads; no risers; balustrades and handrails; decorated		
3 m rise; straight; 760 mm wide	nr	£ 2766.37 to £ 3368.64
Plus or minus for each 300 mm variation in storey height	nr	£ 278.17 to £ 339.42
Heavy duty cast iron staircase; perforated treads; no risers; balustrades and handrails; decorated		
3 m rise; straight	nr	£ 4358.82 to £ 5747.10
Plus or minus for each 300 mm variation in storey height	nr	£ 435.12 to £ 572.92
Galvanised steel catwalk; nylon coated balustrading		
450 mm wide	m	£ 306.24 to £ 386.63

SUPERSTRUCTURE - continued						
STAIRS - continued						
Metal construction - continued						
Finishes to treads and risers						
PVC floor tiles including screeds	store	£	747.74	to	£	1048.87
granolithic	store	£	1135.64	to	£	1291.31
heavy duty carpet	store	£	1582.24	to	£	1980.35
terrazzo	store	£	3174.69	to	£	4154.66
Comparative finishes/balustrading						
Wall handrails						
PVC covered mild steel rail on brackets	store	£	297.31	to	£	475.95
hardwood handrail on brackets	store	£	811.54	to	£	1348.73
stainless steel handrail on brackets	store	£	2600.49	to	£	5144.83
Balustrading and handrails						
mild steel balustrades and PVC covered handrails	store	£	931.48	to	£	1194.34
mild steel balustrades and hardwood handrails	store	£	1679.22	to	£	2378.46
stainless steel balustrades and handrails	store	£	7135.39	to	£	8639.80
stainless steel and glass balustrades	store	£	6252.40	to	£	15823.68
EXTERNAL WALLS						
Brick/block walling						
Dense aggregate block walls						
100 mm thick	m²	£	19.91	to	£	22.84
140 mm thick	m²	£	28.58	to	£	34.45
Common brick solid walls; bricks PC £ 120.00/1000						
half brick thick	m²	£	30.62	to	£	36.88
one brick thick	m²	£	55.38	to	£	63.16
one and a half brick thick	m²	£	79.62	to	£	91.23
Add or deduct for each variation of £ 10.00/1000 in PC value						
half brick thick	m²	£	0.89	to	£	1.40
one brick thick	m²	£	1.79	to	£	2.17
one and a half brick thick	m²	£	2.68	to	£	3.32
Extra for fair face one side	m²	£	1.79	to	£	2.42
Engineering brick walls; class B; bricks PC £ 210.00/1000						
half brick thick	m²	£	37.39	to	£	44.66
one brick thick	m²	£	71.33	to	£	81.54
Facing brick walls; machine-made facings; bricks PC £ 275.00/1000						
half brick thick; pointed one side	m²	£	55.38	to	£	63.16
half brick thick; built against concrete	m²	£	59.21	to	£	66.99
one brick thick; pointed both sides	m²	£	108.72	to	£	126.20

Description	Unit		Rate		
Add or deduct for each variation of £ 10.00/1000 in PC value					
half brick thick	m²	£	0.89	to £	1.40
one brick thick	m²	£	1.79	to £	2.04
Composite solid walls; facing brick on outside; bricks PC £120/1000 and common brick on inside; bricks PC £ 120.00/1000					
one brick thick; pointed one side	m²	£	86.77	to £	99.53
Extra for weather pointing as a separate operation	m²	£	4.72	to £	7.78
one and a half brick thick; pointed one side	m²	£	108.46	to £	127.60
Composite cavity wall; block outer skin; 50 mm insulation; lightweight block inner skin					
outer block rendered	m²	£	61.25	to £	81.66
Extra for					
heavyweight block inner skin	m²	£	2.42	to £	3.83
fair face one side	m²	£	11.48	to £	16.33
75 mm cavity insulation	m²	£	61.25	to £	85.49
100 mm cavity insulation	m²	£	66.35	to £	91.87
plaster and emulsion	m²	£	75.28	to £	95.70
outer block rendered; no insulation; inner skin insulating	m²	£	1.28	to £	2.42
outer block roughcast	m²	£	0.64	to £	2.42
coloured masonry outer block	m²	£	1.28	to £	2.42
Composite cavity wall; facing brick outer skin; 50 mm insulation; plasterboard on stud inner skin; emulsion					
machine-made facings; PC £ 275.00/1000	m²	£	93.15	to £	104.63
Composite cavity wall; facing brick outer skin; lightweight block inner skin; plaster and emulsion					
machine-made facings; PC £ 275.00/1000	m²	£	89.32	to £	103.36
Add or deduct for each variation of £ 10.00/1000 in PC value	m²	£	0.89	to £	1.28
Extra for					
heavyweight block inner skin	m²	£	1.28	to £	2.42
insulating block inner skin	m²	£	2.42	to £	6.25
30 mm cavity wall slab	m²	£	2.93	to £	7.15
50 mm cavity insulation	m²	£	3.45	to £	4.21
75 mm cavity insulation	m²	£	4.98	to £	5.74
100 mm cavity insulation	m²	£	6.25	to £	7.15
weather-pointing as a separate operation	m²	£	4.98	to £	7.91
purpose made feature course to windows	m²	£	6.25	to £	12.76
Composite cavity wall; facing brick outer skin; 50 mm insulation; common brick inner skin; fair face on inside					
machine-made facings; PC £ 275.00/1000	m²	£	93.15	to £	108.46
Composite cavity wall; facing brick outer skin; 50 mm insulation; common brick inner skin; plaster and emulsion					
machine-made facings; PC £ 275.00/1000	m²	£	100.80	to £	118.67

SUPERSTRUCTURE - continued						
EXTERNAL WALLS - continued						
Brick/block walling - continued						
Composite cavity wall; coloured masonry block; outer and inner skins; fair faced both sides	m²	£	111.01	to	£	156.95
Reinforced concrete walling						
Insitu reinforced concrete 25.5 N/m²; 13 kg/m² reinforcement; formwork both sides						
150 mm thick	m²	£	96.98	to	£	118.67
225 mm thick	m²	£	118.67	to	£	131.43
Wall claddings						
Non-asbestos profiled cladding						
'profile 3'; natural	m²	£	20.93	to	£	23.73
'profile 3'; coloured	m²	£	22.84	to	£	25.78
'profile 6'; natural	m²	£	22.84	to	£	25.78
'profile 6'; coloured	m²	£	23.73	to	£	26.67
insulated; inner lining of plasterboard	m²	£	40.83	to	£	51.42
'profile 6'; natural; insulated; inner lining panel	m²	£	40.83	to	£	51.42
insulated; with 2.8 m high block inner skin; emulsion	m²	£	35.47	to	£	40.83
insulated; with 2.8 m high block inner skin; plasterboard lining on metal tees; emulsion	m²	£	48.49	to	£	57.29
PVF2 coated galvanised steel profiled cladding						
0.60 mm thick; 'profile 20B'; corrugations vertical	m²	£	28.58	to	£	35.47
0.60 mm thick; 'profile 30'; corrugations vertical	m²	£	28.58	to	£	35.47
0.60 mm thick; 'profile TOP 40'; corrugations vertical	m²	£	26.67	to	£	33.43
0.60 mm thick; 'profile 60B'; corrugations vertical	m²	£	33.43	to	£	41.73
0.60 mm thick; 'profile 30'; corrugations horizontal	m²	£	29.60	to	£	37.90
0.60 mm thick; 'profile 60B; corrugations horizontal	m²	£	33.43	to	£	42.75
Extra for						
80 mm insulation and 0.4 mm thick coated inner lining sheet	m²	£	16.08	to	£	16.84
PVF2 coated galvanised steel profiled cladding on steel rails						
2.8 m high insulating block inner skin; emulsion	m²	£	58.70	to	£	71.46
2.8 m high insulated block inner skin; plasterboard lining on metal tees; emulsion	m²	£	66.35	to	£	85.49
insulated; coloured inner lining panel	m²	£	66.35	to	£	85.49
insulated; full-height insulating block inner skin; plaster and emulsion	m²	£	82.94	to	£	112.29
insulated; metal sandwich panel system	m²	£	168.43	to	£	257.75

Description	Unit		Rate from		Rate to
Other cladding systems					
vitreous enamelled insulated steel sandwich panel system; with non-asbestos fibre insulating board on inner face	m²	£	160.78	to £	197.78
Formalux sandwich panel system; with coloured lining tray; on steel cladding rails	m²	£	188.85	to £	227.13
aluminium over-cladding system rain screen	m²	£	237.34	to £	267.96
natural stone cladding on full-height insulating block inner skin; plaster and emulsion	m²	£	452.98	to £	611.20
Curtain/glazed walling					
Single glazed polyester powder coated aluminium curtain walling					
economical; including part-height block back-up wall; plaster and emulsion	m²	£	357.28	to £	510.40
Extra over single 6 mm float glass for					
double glazing unit with two 6 mm float glass skins	m²	£	38.28	to £	48.49
double glazing unit with one 6 mm 'Antisun' skin and one 6 mm float glass skin	m²	£	74.01	to £	95.70
'look-a-like' non-vision spandrel panels	m²	£	48.49	to £	71.46
10% opening lights	m²	£	11.48	to £	25.52
economical; including infill panels	m²	£	341.97	to £	408.32
Extra for					
50 mm insulation	m²	£	20.42	to £	22.97
anodised finish in lieu of polyester powder coating	m²	£	24.24	to £	84.22
bronze anodising in lieu of polyester powder coating	m²	£	48.49	to £	119.94
good quality	m²	£	558.89	to £	650.76
good quality; 35% opening lights	m²	£	701.80	to £	816.64
Extra over single 6 mm float glass for double glazing unit with Low 'E' and tinted glass	m²	£	72.73	to £	95.70
High quality structural glazing to entrance elevation	m²	£	766.88	to £	1205.82
Patent glazing systems; excluding opening lights and flashings					
7 mm georgian wired cast glass, aluminium alloy bars spanning up to 3 mat 600 - 625 mm spacing	m²	£	114.84	to £	140.36
6.4 mm laminated safety glass polyester powder coated aluminium capped glazing bars spanning up to 3 m at 600 - 625 mm spacing.	m²	£	357.28	to £	421.08
Comparative external finishes					
Comparative concrete wall finishes					
wrought formwork one side including rubbing down	m²	£	3.32	to £	6.64
shotblasting to expose aggregate	m²	£	4.21	to £	8.42
bush hammering to expose aggregate	m²	£	13.53	to £	18.63
two coats 'Sandex Matt' cement paint	m²	£	7.91	to £	11.10
cement and sand plain face rendering	m²	£	13.91	to £	20.93
three-coat Tyrolean rendering; including backing	m²	£	33.81	to £	39.81
'Mineralite' decorative rendering; including backing	m²	£	66.99	to £	77.71

SUPERSTRUCTURE - continued					
WINDOWS AND EXTERNAL DOORS					
Softwood windows and external doors					
Standard windows; painted					
single glazed	m²	£	207.99	to £	297.31
double glazed	m²	£	267.96	to £	378.97
Standard external softwood doors and hardwood frames; doors painted; including ironmongery					
solid flush door	nr	£	436.39	to £	1009.32
heavy duty solid flush door; single leaf	nr	£	659.69	to £	1009.32
heavy duty solid flush door; double leaf	nr	£	1164.99	to £	1660.08
Extra for					
emergency fire exit door	nr	£	248.82	to £	387.90
Steel windows and external doors					
Standard windows					
single glazed; galvanised; painted	m²	£	216.92	to £	306.24
single glazed; powder-coated	m²	£	227.13	to £	316.45
double glazed; galvanised; painted	m²	£	287.10	to £	367.49
double glazed; powder coated	m²	£	287.10	to £	387.90
Steel roller shutters					
Shutters; galvanised					
manual	m²	£	216.92	to £	306.24
electric	m²	£	267.96	to £	408.32
manual; insulated	m²	£	367.49	to £	465.74
electric; insulated	m²	£	427.46	to £	562.72
electric; insulated; fire-resistant	m²	£	1038.66	to £	1291.31
Hardwood windows and external doors					
Standard windows; stained or UPVC coated					
single glazed	m²	£	316.45	to £	495.09
double glazed	m²	£	408.32	to £	621.41
Pvc-U windows and external doors					
Purpose-made windows					
double glazed	m²	£	592.06	to £	796.22

Aluminium windows and external doors					
Standard windows; anodised finish					
single glazed; horizontal sliding sash	m²	£	287.10	to £	347.07
single glazed; vertical sliding sash	m²	£	446.60	to £	533.37
single glazed; casement; in hardwood sub-frame	m²	£	335.59	to £	446.60
double glazed; vertical sliding sash	m²	£	495.09	to £	572.92
double glazed; casement; in hardwood sub-frame	m²	£	408.32	to £	543.58
Purpose-made entrance screens and doors					
double glazed	m²	£	796.22	to £	1291.31
Stainless steel entrance screens and doors					
Purpose-made screen; double glazed					
with manual doors	m²	£	1291.31	to £	1942.07
with automatic doors	m²	£	1601.38	to £	2281.49
Shop fronts, shutters and grilles					
Shutters and grilles per metre of plan length					
Grilles or shutters	m	£	640.55	to £	1291.31
Fire shutters; power-operated	m	£	1135.64	to £	1621.80
INTERNAL WALLS, PARTITIONS AND DOORS					
Timber or metal stud partitions and doors					
Softwood stud and plasterboard partitions					
100 mm partition; skim and emulsioned both sides	m²	£	50.53	to £	57.29
150 mm partition as party wall; skim and emulsioned both sides	m²	£	60.23	to £	71.33
Metal stud and plasterboard partitions					
170 mm partition; one hour; taped joints; emulsioned both sides	m²	£	51.04	to £	62.52
200 mm partition; two hour; taped joints; emulsioned both sides	m²	£	71.46	to £	79.11
Metal stud and plasterboard partitions; emulsioned both sides; softwood doors and frames; painted					
170 mm partition	m²	£	66.35	to £	91.87
200 mm partition; insulated	m²	£	91.87	to £	111.01
Stud or plasterboard partitions; softwood doors and frames; painted					
partition; plastered and emulsioned both sides	m²	£	82.94	to £	108.46
Stud or plasterboard partitions; hardwood doors and frames; painted					
partition; plastered and emulsioned both sides	m²	£	108.46	to £	136.53

SUPERSTRUCTURE - continued					
INTERNAL WALLS, PARTITIONS AND DOORS - continued					
Brick/block partitions and doors					
Autoclaved aerated/lightweight block partitions					
75 mm thick	m^2	£	16.33	to £	19.40
100 mm thick	m^2	£	21.82	to £	26.16
130 mm thick; insulating	m^2	£	26.67	to £	29.60
150 mm thick	m^2	£	29.60	to £	31.52
190 mm thick	m^2	£	35.47	to £	40.83
Extra for					
fair face both sides	m^2	£	2.93	to £	5.74
curved work			+10% to +20%		
average thickness; fair face both sides	m^2	£	28.58	to £	35.47
average thickness; fair face and emulsioned both sides	m^2	£	33.43	to £	42.75
average thickness; plastered and emulsioned both sides	m^2	£	51.42	to £	61.12
Dense aggregate block partitions					
average thickness; fair face both sides	m^2	£	33.43	to £	40.83
average thickness; fair face and emulsioned both sides	m^2	£	38.79	to £	46.57
average thickness; plastered and emulsioned both sides	m^2	£	57.29	to £	65.08
Common brick partitions; bricks PC £ 120.00/1000					
half brick thick	m^2	£	30.62	to £	33.43
half brick thick; fair face both sides	m^2	£	33.43	to £	40.83
half brick thick; fair face and emulsioned both sides	m^2	£	38.79	to £	46.57
half brick thick; plastered and emulsioned both sides	m^2	£	55.38	to £	71.33
one brick thick	m^2	£	57.29	to £	65.08
one brick thick; fair face both sides	m^2	£	61.12	to £	71.33
one brick thick; fair face and emulsioned both sides	m^2	£	66.99	to £	77.20
one brick thick; plastered and emulsioned both sides	m^2	£	83.45	to £	99.02
Block partitions; softwood doors and frames; painted					
partition	m^2	£	47.59	to £	63.16
partition; fair face both sides	m^2	£	49.51	to £	66.99
partition; fair face and emulsioned both sides	m^2	£	57.29	to £	73.24
partition; plastered and emulsioned both sides	m^2	£	75.28	to £	95.19
Block partitions; hardwood doors and frames					
partition	m^2	£	71.46	to £	91.87
partition; plastered and emulsioned both sides	m^2	£	96.98	to £	122.50
Reinforced concrete walls					
Walls					
150 mm thick	m^2	£	99.53	to £	118.67
150 mm thick; plastered and emulsioned both sides	m^2	£	132.07	to £	159.24

Item	Unit		Rate			Rate
Special partitioning and doors						
Demountable partitioning; hardwood doors						
medium quality; vinyl-faced	m^2	£	122.50	to	£	159.50
high quality; vinyl-faced	m^2	£	160.78	to	£	225.85
Aluminium internal patent glazing						
single glazed	m^2	£	108.46	to	£	149.29
double glazed	m^2	£	185.02	to	£	229.68
Demountable steel partitioning and doors						
medium quality	m^2	£	197.78	to	£	248.82
high quality	m^2	£	248.82	to	£	306.24
Demountable fire partitions						
enamelled steel; half hour	m^2	£	317.72	to	£	722.22
stainless steel; half hour	m^2	£	900.86	to	£	1139.47
Soundproof partitions; hardwood doors						
luxury veneered	m^2	£	227.13	to	£	376.42
WC/Changing cubicles						
WC cubicles cost per cubicle	nr	£	316.45	to	£	703.08
INTERNAL DOORS						
Comparative doors/door linings/frames						
Standard softwood doors and frames; including ironmongery and painting						
flush; hollow core	nr	£	237.34	to	£	297.31
flush; hollow core; hardwood faced	nr	£	237.34	to	£	316.45
flush; solid core						
single leaf	nr	£	278.17	to	£	357.28
double leaf	nr	£	405.77	to	£	546.13
flush; solid core; hardwood faced	nr	£	325.38	to	£	376.42
four panel door	nr	£	405.77	to	£	495.09
Purpose-made softwood doors and hardwood frames; including ironmongery; painting and polishing						
flush; solid core; heavy duty						
single leaf	nr	£	683.94	to	£	801.33
double leaf	nr	£	930.20	to	£	1187.96
flush; solid core; heavy duty; plastic laminate faced						
single leaf	nr	£	830.68	to	£	941.69
double leaf	nr	£	1149.68	to	£	1287.48
Purpose-made softwood fire doors and hardwood frames; including ironmongery; painting and polishing						
flush; one hour fire resisting						
single leaf	nr	£	900.86	to	£	1009.32
double leaf	nr	£	1149.68	to	£	1278.55

SUPERSTRUCTURE - continued					
INTERNAL DOORS - continued					
Purpose-made softwood fire doors and hardwood frames; including ironmongery; painting and polishing - continued					
flush; one hour fire resisting; plastic laminate faced					
single leaf	nr	£	1117.78	to £	1227.51
double leaf	nr	£	1436.78	to £	1552.89
Purpose-made softwood doors and pressed steel frames;					
flush; half hour fire check; plastic laminate faced	nr	£	1038.66	to £	1247.93
Perimeter treatments					
Precast concrete lintels; in block walls					
75 mm wide	m	£	9.57	to £	20.80
100 mm wide	m	£	11.87	to £	25.78
Precast concrete lintels; in brick walls					
half brick thick	m	£	11.87	to £	25.78
one brick thick	m	£	18.25	to £	41.73
Purpose-made softwood architraves; painted; including grounds					
25 x 50 mm; to both sides of openings					
726 x 2040 mm opening	nr	£	93.15	to £	100.80
826 x 2040 mm opening	nr	£	95.70	to £	103.36
WALL FINISHES					
Dry plasterboard lining; taped joints; for direct decoration					
9.5 mm Gyproc Wallboard	m²	£	8.29	to £	12.89
Extra for insulating grade	m²	£	0.64	to £	0.77
Extra for insulating grade; plastic faced	m²	£	2.04	to £	2.42
12.5 mm Gyproc Wallboard (half-hour fire-resisting)	m²	£	9.95	to £	14.42
Extra for insulating grade	m²	£	0.64	to £	0.89
Extra for insulating grade; plastic faced	m²	£	1.79	to £	2.04
two layers of 12.5 mm Gyproc Wallboard (one hour fire-resisting)	m²	£	17.86	to £	22.84
9 mm Supalux (half-hour fire-resisting)	m²	£	17.86	to £	22.84
Dry plasterboard lining; taped joints; for direct decoration; fixed to wall on dabs					
9.5 mm Gyproc Wallboard	m²	£	9.95	to £	14.42
Dry plasterboard lining; taped joints; for direct decoration; including metal tees					
9.5 mm Gyproc Wallboard	m²	£	21.82	to £	25.26
12.5 mm Gyproc Wallboard	m²	£	22.84	to £	26.67

Item	Unit		Rate			Rate
Dry lining/sheet panelling; including battens; plugged to wall						
6.4 mm hardboard	m²	£	10.85	to	£	12.89
9.5 mm Gyproc Wallboard	m²	£	15.95	to	£	21.82
6 mm birch faced plywood	m²	£	16.84	to	£	20.93
6 mm WAM plywood	m²	£	20.93	to	£	25.78
15 mm chipboard	m²	£	160.14	to	£	17.86
15 mm melamine faced chipboard	m²	£	26.67	to	£	34.96
13.2 mm 'Formica' faced chipboard	m²	£	36.37	to	£	57.29
In situ wall finishes						
Extra over common brickwork for fair face and pointing both sides	m²	£	3.45	to	£	4.98
Comparative finishes						
one mist and two coats emulsion paint	m²	£	2.81	to	£	4.21
multi-coloured gloss paint	m²	£	4.98	to	£	6.38
two coats of lightweight plaster	m²	£	8.93	to	£	12.89
9.5 mm Gyproc Wallboard and skim coat	m²	£	11.87	to	£	16.08
12.5 mm Gyproc Wallboard and skim coat	m²	£	13.40	to	£	17.86
two coats of 'Thistle' plaster	m²	£	12.38	to	£	16.84
plaster and emulsion	m²	£	11.87	to	£	19.40
Extra for gloss paint in lieu of emulsion	m²	£	2.17	to	£	2.55
two coat render and emulsion	m²	£	20.93	to	£	29.60
Ceramic wall tiles; including backing						
economical	m²	£	22.84	to	£	43.64
medium quality	m²	£	43.64	to	£	77.20
high quality; to toilet blocks, kitchens and first aid rooms	m²	£	71.33	to	£	91.23
FLOOR FINISHES						
Chipboard flooring; t & g joints						
18 mm thick	m²	£	8.93	to	£	10.85
22 mm thick	m²	£	10.85	to	£	12.89
Wrought softwood flooring						
25 mm thick; butt joints; cleaned off and polished	m²	£	22.84	to	£	26.67
25 mm thick; t & g joints; cleaned off and polished	m²	£	24.75	to	£	30.62
Extra over concrete floor for						
power floating	m²	£	3.45	to	£	9.95
power floating; surface hardener	m²	£	7.53	to	£	14.42
Latex cement screeds						
3 mm thick; one coat	m²	£	4.98	to	£	5.74
5 mm thick; two coat	m²	£	6.64	to	£	7.53

SUPERSTRUCTURE - continued						
FLOOR FINISHES - continued						
Rubber latex non-slip solution and epoxy sealant	m²	£	8.42	to	£	19.91
Cement and sand (1:3) screeds						
25 mm thick	m²	£	9.95	to	£	10.85
50 mm thick	m²	£	11.87	to	£	13.91
75 mm thick	m²	£	16.84	to	£	18.88
Cement and sand (1:3) paving						
32 mm thick	m²	£	8.42	to	£	11.99
32 mm thick; surface hardener	m²	£	9.95	to	£	16.84
Screed only (for subsequent finish)	m²	£	12.89	to	£	19.91
Screed only (for subsequent finish); allowance for skirtings	m²	£	15.82	to	£	22.84
Mastic asphalt paving						
20 mm thick; BS 6925; black	m²	£	19.91	to	£	24.75
20 mm thick; BS 6925; red	m²	£	23.73	to	£	28.58
Granolithic						
20 mm thick	m²	£	10.85	to	£	17.74
25 mm thick	m²	£	14.80	to	£	19.91
25 mm thick; including screed	m²	£	24.75	to	£	28.58
38 mm thick; including screed	m²	£	34.96	to	£	41.73
Synthanite; on and including building paper						
25 mm thick	m²	£	22.84	to	£	28.58
50 mm thick	m²	£	31.52	to	£	38.79
75 mm thick	m²	£	38.79	to	£	46.57
Acrylic polymer floor finish						
10 mm thick	m²	£	22.84	to	£	28.58
Epoxy floor finish						
1.5 - 2 mm thick	m²	£	22.84	to	£	31.52
5 - 6 mm thick	m²	£	46.57	to	£	54.36
Polyester resin floor finish						
5 - 9 mm thick	m²	£	52.44	to	£	61.12
Quarry tile flooring						
150 x 150 x 12.5 mm thick; red	m²	£	28.58	to	£	32.54
150 x 150 x 12.5 mm thick; brown	m²	£	34.45	to	£	40.83
200 x 200 x 19 mm thick; brown	m²	£	42.75	to	£	48.49
average tiling	m²	£	28.58	to	£	48.49
tiling; including screed	m²	£	40.83	to	£	63.16
tiling; including screed and allowance for skirtings	m²	£	54.36	to	£	77.20

Item	Unit		Rate			Rate
Glazed ceramic tile flooring						
150 x 150 x 12 mm thick; black	m²	£	34.45	to	£	42.75
150 x 150 x 12 mm thick; antislip	m²	£	42.75	to	£	44.66
fully vitrified	m²	£	44.66	to	£	63.16
fully vitrified; including screed	m²	£	53.34	to	£	81.03
fully vitrified; including screed and allowance for skirtings	m²	£	61.12	to	£	97.10
Composition block flooring						
174 x 57 mm blocks	m²	£	71.33	to	£	79.11
Vinyl tile flooring						
2 mm thick; semi-flexible tiles	m²	£	8.93	to	£	12.89
2 mm thick; fully flexible tiles	m²	£	8.42	to	£	11.99
2.5 mm thick; semi-flexible tiles	m²	£	10.85	to	£	14.42
tiling; including screed	m²	£	22.84	to	£	28.58
tiling; including screed and allowance for skirtings	m²	£	24.75	to	£	33.43
tiling; antistatic	m²	£	41.73	to	£	47.59
tiling; antistatic; including screed	m²	£	51.42	to	£	63.54
Vinyl sheet flooring; heavy duty						
2 mm thick	m²	£	15.82	to	£	17.74
2.5 mm thick	m²	£	15.82	to	£	19.91
3 mm thick; needle felt backed	m²	£	10.85	to	£	14.42
3 mm thick; foam backed	m²	£	15.82	to	£	19.91
Sheeting; including screed and allowance for skirtings						
'Altro' safety flooring	m²	£	29.60	to	£	35.86
2 mm thick; Marine T20	m²	£	22.84	to	£	28.58
2.5 mm thick; Classic D25	m²	£	28.58	to	£	33.43
3.5 mm thick; stronghold	m²	£	35.86	to	£	41.73
flooring	m²	£	22.84	to	£	42.75
flooring; including screed	m²	£	33.43	to	£	57.29
Rubber tile flooring; smooth; ribbed or studded tiles						
2.5 mm thick	m²	£	28.58	to	£	33.43
5 mm thick	m²	£	33.43	to	£	38.79
5 mm thick; including screed	m²	£	45.04	to	£	57.29
Carpet tiles/Underlay						
Underlay	m²	£	4.98	to	£	6.89
nylon needlepunch (stick down)	m²	£	12.89	to	£	15.82
80% animal hair; 20% wool cord	m²	£	22.84	to	£	25.78
100% wool	m²	£	31.52	to	£	38.79
80% wool; 20% nylon antistatic	m²	£	33.43	to	£	46.57
economical; including screed and allowance for skirtings	m²	£	34.45	to	£	38.79
good quality	m²	£	31.52	to	£	46.57
good quality; including screed	m²	£	42.75	to	£	57.29
good quality; including screed and allowance for skirtings	m²	£	47.59	to	£	63.16

SUPERSTRUCTURE - continued					
FLOOR FINISHES - continued					
Access floors; excluding finish					
600 x 600 mm chipboard panels; faced both sides with galvanised steel sheet; on adjustable steel/aluminium pedestals; cavity height 100 - 300 mm high					
light grade duty	m²	£	46.57	to £	55.38
medium grade duty	m²	£	53.34	to £	62.14
heavy grade duty	m²	£	66.99	to £	84.47
extra heavy grade duty	m²	£	75.28	to £	84.47
600 x 600 mm chipboard panels; faced both sides with galvanised steel sheet; on adjustable steel/aluminium pedestals; cavity height 300 - 600 mm high					
medium grade duty	m²	£	61.25	to £	68.90
heavy grade duty	m²	£	68.90	to £	85.49
extra heavy grade duty	m²	£	76.56	to £	85.49
Common floor coverings bonded to access floor panels					
heavy-duty fully flexible vinyl to BS 3261, type A	m²	£	8.42	to £	23.73
anti-static grade sheet PVC to BS 3261	m²	£	15.31	to £	23.73
Comparative skirtings					
25 x 75 mm softwood skirting; painted; including grounds	m	£	9.44	to £	12.38
12.5 x 150 mm Quarry tile skirting; including backing	m	£	13.91	to £	18.37
13 x 75 mm granolithic skirting; including backing	m	£	19.91	to £	24.75
Entrance matting in aluminium-framed matwell	m²	£	335.59	to £	495.09

CEILING FINISHES					
Plastered ceilings					
Plaster to soffits					
lightweight plaster	m²	£	9.95	to £	13.40
plaster and emulsion	m²	£	12.89	to £	19.91
Extra for gloss paint in lieu of emulsion	m²	£	2.17	to £	2.55
Plasterboard to soffits					
9.5 mm Gyproc lath and skim coat	m²	£	13.40	to £	16.84
9.5 mm Gyproc insulating lath and skim coat	m²	£	14.42	to £	16.84
plasterboard, skim and emulsion	m²	£	15.31	to £	20.93
Extra for gloss paint in lieu of emulsion	m²	£	2.17	to £	2.55
plasterboard and Artex	m²	£	9.95	to £	14.42
plasterboard, Artex and emulsion	m²	£	13.40	to £	17.86
plaster and emulsion; including metal lathing	m²	£	21.82	to £	31.52

Description	Unit		Rate			Rate
Other board finishes; with fire-resisting properties; excluding decoration						
12.5 mm Gyproc Fireline; half hour	m^2	£	10.85	to	£	14.42
6 mm Supalux; half hour	m^2	£	14.42	to	£	16.84
two layers of 12.5 mm Gyproc Wallboard; half hour	m^2	£	15.95	to	£	19.40
two layers of 12.5 mm Gyproc Fireline; one hour	m^2	£	18.37	to	£	22.84
9 mm Supalux; one hour; on fillets	m^2	£	20.93	to	£	25.78
Specialist plasters; to soffits						
sprayed accoustic plaster; self-finished	m^2	£	28.58	to	£	39.81
rendering; 'Tyrolean' finish	m^2	£	29.60	to	£	42.75
Other ceiling finishes						
50 mm wood wool slabs as permanent lining	m^2	£	13.40	to	£	16.84
12 mm Pine tongued and grooved boarding	m^2	£	17.86	to	£	21.82
16 mm Softwood tongued and grooved boardings	m^2	£	21.82	to	£	25.78
Suspended and integrated ceilings						
Suspended ceiling						
economical; exposed grid	m^2	£	22.84	to	£	30.62
jointless; plasterboard	m^2	£	28.58	to	£	35.86
semi-concealed grid	m^2	£	30.62	to	£	38.79
medium quality; 'Minatone'; concealed grid	m^2	£	33.43	to	£	45.68
high quality; 'Travertone'; concealed grid	m^2	£	38.41	to	£	49.51
metal linear strip; 'Dampa'/'Luxalon'	m^2	£	42.11	to	£	53.59
metal tray	m^2	£	43.38	to	£	57.42
egg-crate	m^2	£	47.21	to	£	100.80
open grid; 'Formalux'/'Dimension'	m^2	£	86.77	to	£	107.18
Integrated ceilings						
coffered; with steel services	m^2	£	99.53	to	£	164.60
DECORATIONS						
Emulsion						
two coats	m^2	£	2.04	to	£	2.55
one mist and two coats	m^2	£	2.55	to	£	3.45
Artex plastic compound						
one coat; textured	m^2	£	3.32	to	£	4.47
Wall paper	m^2	£	4.47	to	£	7.78
Gloss						
primer and two coats	m^2	£	3.83	to	£	5.23
primer and three coats	m^2	£	5.23	to	£	6.51

SUPERSTRUCTURE - continued						
DECORATIONS - continued						
Comparative steel/metalwork finishes						
primer only	m²	£	0.77	to	£	1.53
grit blast and one coat zinc chromate primer	m²	£	1.66	to	£	2.55
touch up primer and one coat of two pack epoxy zinc phosphate	m²	£	2.30	to	£	2.93
primer gloss three coats	m²	£	5.23	to	£	6.38
sprayed mineral fibre; one hour	m²	£	9.95	to	£	15.31
sprayed mineral fibre; two hour	m²	£	16.59	to	£	19.91
sprayed vermiculite cement; one hour	m²	£	11.23	to	£	16.72
sprayed vermiculite cement; two hour	m²	£	13.53	to	£	19.91
intumescent coating with decorative top seal; half hour	m²	£	17.99	to	£	19.91
intumescent coating with decorative top seal; one hour	m²	£	28.58	to	£	34.96
Comparative woodwork finishes						
primer only	m²	£	1.40	to	£	1.53
two coats gloss; touch up primer	m²	£	2.93	to	£	3.32
three coats gloss; touch up primer	m²	£	3.83	to	£	4.98
primer and two coat gloss	m²	£	4.47	to	£	5.23
primer and three coat gloss	m²	£	5.87	to	£	6.38
polyurethene lacquer two coats	m²	£	2.93	to	£	3.57
polyurethene laquer three coats	m²	£	4.47	to	£	5.23
flame-retardent paint three coats	m²	£	6.64	to	£	8.42
FITTINGS AND FURNISHINGS						
Comparative wrought softwood shelving						
25 x 225 mm; including black japanned brackets	m	£	12.25	to	£	14.16
25 mm thick slatted shelving; including bearers	m²	£	47.59	to	£	56.27
25 mm thick cross-tongued shelving; including bearers	m²	£	64.06	to	£	71.33
Allowances per gross floor area for:-						
Reception desk, shelves and cupboards for general areas						
economical	m²	£	3.96	to	£	8.93
medium quality	m²	£	7.15	to	£	14.16
high quality	m²	£	11.36	to	£	21.82
Extra for						
high quality finishes to reception areas	m²	£	7.15	to	£	10.34
full kitchen equipment	m²	£	10.34	to	£	12.38
Furniture and fittings to general office areas						
economical	m²	£	8.42	to	£	11.36
medium quality	m²	£	11.36	to	£	17.86
high quality	m²	£	19.91	to	£	29.22

Description	Unit		Rate		
General fittings and equipment					
internal planting	m^2	£	22.84	to £	29.22
signs, notice-boards, shelving, fixed seating, curtains and blinds	m^2	£	12.38	to £	14.16
SANITARY AND DISPOSAL INSTALLATIONS					
Note: Material prices vary considerably, the following composite rates are based on average prices for mid priced fittings:					
Lavatory basins; vitreous china; chromium plated taps, waste, chain and plug, cantilever brackets					
white	nr	£	188.85	to £	247.54
coloured	nr	£	247.54	to £	349.62
Low level WC's; vitreous china pan and cistern; black plastic seat; low pressure ball valve; plastic flush pipe; fixing brackets					
on ground floor - white	nr	£	178.64	to £	218.20
- coloured	nr	£	237.34	to £	257.75
one of a range; on upper floors - white	nr	£	347.07	to £	408.32
- coloured	nr	£	405.77	to £	446.60
Bowl type wall urinal; white glazed vitreous china flushing cistern; chromium plated flush pipes and spreaders; fixing brackets					
white	nr	£	158.22	to £	218.20
Shower tray; glazed fireclay; chromium plated waste, chain and plug, riser pipe, rose and mixing valve					
white	nr	£	465.74	to £	572.92
coloured	nr	£	495.09	to £	611.20
Sink; glazed fireclay; chromium plated waste, chain and plug; fixing brackets					
white	nr	£	237.34	to £	613.76
Sink; stainless steel; chromium plated waste, chain and self coloured plug					
single drainer	nr	£	204.80	to £	257.37
double drainer	nr	£	237.34	to £	297.31
Soil waste stacks; 3.15 m storey height; branch and connection to drain					
110 mm PVC	nr	£	316.45	to £	357.28
Extra for additional floors	nr	£	158.22	to £	188.85
100 mm cast iron; decorated	nr	£	640.55	to £	680.11
Extra for additional floors	nr	£	316.45	to £	357.28
Industrial buildings Allowance per m^2 of floor area					
minimum provision	m^2	£	10.34	to £	14.42
high provision	m^2	£	14.42	to £	21.82

Description	Unit		Rate			Rate
SUPERSTRUCTURE - continued						
SANITARY AND DISPOSAL INSTALLATIONS - continued						
Hot and cold water installations - Allowance per m² of floor area						
Complete installations	m²	£	7.15	to	£	29.73
SERVICES INSTALLATIONS						
HEATING, AIR-CONDITIONING AND VENTILATING INSTALLATIONS						
Gas or oil-fired radiator heating						
Gas-fired hot water service and central heating for						
three radiators	nr	£	2378.46	to	£	3368.64
extra for additional radiator	nr	£	267.96	to	£	376.42
LPHW radiator system - Allowance per m² of floor area	m²	£	63.16	to	£	87.41
Ventilation systems						
Local ventilation to						
WC's	nr	£	237.85	to	£	316.45
toilet areas - Allowance per m² of floor area	m²	£	4.34	to	£	10.72
Air extract system - Allowance per m² of floor area	m²	£	39.81	to	£	53.34
Air supply and extract system - Allowance per m² of floor area	m²	£	57.29	to	£	87.41
Heating and ventilation systems						
Heating and ventilation - Allowance per m² of floor area	m²	£	63.16	to	£	79.62
Warm air heating and ventilation - Allowance per m² of floor area	m²	£	118.67	to	£	151.84
Comfort cooling systems						
Stand-alone air-conditioning unit systems						
air supply and extract - Allowance per m² of floor area	m²	£	148.02	to	£	218.20
Air-conditioning systems						
Full air-conditioning with dust and humidity control - Allowance per m² of floor area	m²	£	178.64	to	£	297.31
ELECTRICAL INSTALLATIONS						
Based upon gross internal area serviced						
Mains and sub-mains switchgear and distribution						
Mains intake only	m²	£	1.79	to	£	3.57
Mains switchgear only	m²	£	2.81	to	£	9.44
Mains and sub-mains distribution						
to floors only	m²	£	5.49	to	£	10.34
to floors; including small power and supplies to equipment	m²	£	15.82	to	£	18.88
to floors; including lighting and power and supplies to equipment	m²	£	10.34	to	£	24.75

Item	Unit		Rate			Rate
Lighting installation						
General lighting; including luminaries	m²	£	27.69	to	£	47.59
Emergency lighting	m²	£	10.34	to	£	15.82
standby generators only	m²	£	2.81	to	£	11.36
Lighting and power installations to buildings						
plant area	m²	£	49.51	to	£	79.62
plant area; high provision	m²	£	71.33	to	£	93.15
office area	m²	£	106.80	to	£	128.11
office area; high provision	m²	£	142.66	to	£	170.86
Comparative fittings/rates per each point						
Fittings; excluding lamps or light fittings						
lighting point; PVC cables	nr	£	56.27	to	£	63.16
lighting point; PVC cables in screwed conduits	nr	£	120.33	to	£	188.34
lighting point; MICC cables	nr	£	102.85	to	£	160.14
switch socket outlet; PVC cables						
single	nr	£	58.70	to	£	66.35
double	nr	£	71.46	to	£	81.66
switch socket outlet; PVC cables in screwed conduit						
single	nr	£	93.15	to	£	99.53
double	nr	£	96.98	to	£	108.46
Based upon gross internal area serviced						
Fittings; excluding lamps or light fittings						
switch socket outlet; MICC cables						
single	nr	£	93.15	to	£	99.53
double	nr	£	99.53	to	£	111.01
Immersion heater point (excluding heater)	nr	£	86.77	to	£	99.53
Cooker point; including control unit	nr	£	123.64	to	£	197.78
GAS INSTALLATION						
Connection charge	nr	£	640.55	to	£	815.36
LIFT AND CONVEYOR INSTALLATIONS						
Goods lifts						
Hoist	nr	£	8445.84	to	£	30093.18

Item	Unit	Rate
SERVICES INSTALLATIONS - continued		
LIFT AND CONVEYOR INSTALLATIONS - continued		
Electric heavy duty goods lifts		
500 kg; 2 - 3 levels	nr	£ 32035.26 to £ 45042.80
1000 kg; 2 -3 levels	nr	£ 41549.11 to £ 56498.73
1500 kg; 3 levels	nr	£ 98046.56 to £ 116879.05
2000 kg; 2 levels	nr	£ 98046.56 to £ 116879.05
2000 kg; 3 levels	nr	£ 116879.05 to £ 135712.81
3000 kg; 2 levels	nr	£ 105424.40 to £ 126198.95
3000 kg; 3 levels	nr	£ 131634.71 to £ 155516.33
Oil hydraulic heavy duty goods lifts		
500 kg; 3 levels	nr	£ 98046.56 to £ 116879.05
1000 kg; 3 levels	nr	£ 103496.36 to £ 122316.08
2000 kg; 3 levels	nr	£ 120374.01 to £ 139095.48
Dock levellers		
Dock levellers	nr	£ 10338.15 to £ 24754.40
Dock leveller and canopy	nr	£ 14852.64 to £ 34171.28

Item	Unit	Rate
PROTECTIVE, COMMUNICATION AND SPECIAL INSTALLATIONS		
Based upon gross internal area serviced		
Fire fighting/protective installations		
Fire alarms/appliances		
smoke detectors, alarms and controls	m²	£ 3.57 to £ 8.93
hosereels, dry risers and extinguishers	m²	£ 5.54 to £ 13.30
Sprinkler installations		
single level sprinkler systems, alarms and smoke detectors; low hazard	m²	£ 13.10 to £ 19.91
single level sprinkler systems; alarms and smoke detectors; ordinary hazard	m²	£ 18.83 to £ 26.21
double level sprinkler systems; alarms and smoke detectors; high hazard	m²	£ 28.15 to £ 35.92
Lightning protection	m²	£ 0.70 to £ 2.78
Security/communication installations		
Security alarm system	m²	£ 1.98 to £ 2.87
Telephone system	m²	£ 1.08 to £ 2.37
Public address, television aerial and clocks	m²	£ 2.78 to £ 4.76
Closed-circuit television and public address system	m²	£ 4.36 to £ 5.24

BUILDERS' WORK IN CONNECTION WITH SERVICES						
General builders work to						
mains supplies; lighting and power to landlords areas	m²	£	2.04	to	£	5.74
central heating and electrical installation	m²	£	4.08	to	£	15.18
central heating; electrical and lift installations	m²	£	5.87	to	£	16.84
air-conditioning	m²	£	15.06	to	£	23.35
air-conditioning and electrical installation	m²	£	16.84	to	£	24.75
air-conditioning; electrical and lift installations	m²	£	18.88	to	£	27.69
General builders work; including allowance for plant rooms; to						
central heating and electrical installations	m²	£	26.67	to	£	33.69
central heating, electrical and lift installations	m²	£	33.69	to	£	41.73
air-conditioning	m²	£	47.59	to	£	55.38
air-conditioning and electrical installation	m²	£	61.12	to	£	66.99
air-conditioning; electrical and lift installations	m²	£	75.28	to	£	75.28
SITE WORK						
LANDSCAPING AND EXTERNAL WORKS						
Seeded and planted areas						
Plant supply, planting, maintenance and 12 months guarantee						
seeded areas	m²	£	3.57	to	£	7.18
turfed areas	m²	£	4.66	to	£	9.31
Planted areas (per m² of surface area)						
herbaceous plants	m²	£	3.98	to	£	5.33
climbing plants	m²	£	5.33	to	£	9.31
general planting	m²	£	11.94	to	£	24.27
woodland	m²	£	18.06	to	£	36.11
shrubbed planting	m²	£	23.78	to	£	66.49
dense planting	m²	£	29.71	to	£	59.22
shrubbed area including allowance for small trees	m²	£	35.73	to	£	83.49
Trees						
advanced nursery stock trees (12 - 20 cm girth)	tree	£	146.74	to	£	178.64
semi-mature trees; 5 - 8 m high - coniferous	tree	£	475.95	to	£	1198.16
semi-mature trees; 5 - 8 m high - deciduous	tree	£	722.22	to	£	1980.35
Paved areas						
Precast concrete paving slabs						
50 mm thick	m²	£	10.34	to	£	21.82
50 mm thick 'Texitone' slabs	m²	£	14.80	to	£	20.42
slabs on sub-base; including excavation	m²	£	23.73	to	£	33.94

SITE WORK - continued						
LANDSCAPING AND EXTERNAL WORKS - continued						
Paved areas - continued						
Precast concrete block paviors						
65 mm 'Keyblok' grey paving	m²	£	19.91	to	£	24.75
65 mm 'Mount Sorrel' grey paving	m²	£	19.91	to	£	23.73
65 mm 'Intersett' paving	m²	£	20.93	to	£	24.75
60 mm 'Pedesta' paving	m²	£	15.06	to	£	24.75
paviors on sub-base; including excavation	m²	£	28.58	to	£	40.83
Brick paviors; 229 x 114 x 38 mm paving bricks						
laid flat	m²	£	32.54	to	£	40.83
laid to herring-bone pattern	m²	£	53.34	to	£	59.21
paviors on sub-base; including excavation	m²	£	61.22	to	£	71.33
Granite setts						
200 x 100 x 100 mm setts	m²	£	83.45	to	£	91.23
setts on sub-base; including excavation	m²	£	100.93	to	£	108.72
York stone slab paving						
paving on sub-base; including excavation	m²	£	108.72	to	£	129.13
Cobblestone paving						
50 - 75 mm	m²	£	61.12	to	£	75.28
cobblestones on sub-base; including excavation	m²	£	71.33	to	£	93.15
Car parking alternatives						
Surface level parking; including lighting and drainage						
tarmacadam on sub-base	car	£	1131.81	to	£	1556.72
concrete interlocking blocks	car	£	1273.45	to	£	1698.36
Grasscrete precast concrete units filled with top soil and grass seed	car	£	692.87	to	£	948.07
At ground level with deck or building over	car	£	5807.08	to	£	6719.42
Garages etc						
single car park	nr	£	760.50	to	£	1194.34
single; traditional construction; in a block	nr	£	2375.91	to	£	3394.16
single; traditional construction; pitched roof	nr	£	2345.29	to	£	6931.23
double; traditional construction; pitched roof	nr	£	7765.74	to	£	9948.97
External Furniture						
Guard rails and parking bollards etc.						
open metal post and rail fencing 1 m high	m	£	132.70	to	£	154.40
galvanised steel post and rail fencing 2 m high	m	£	148.02	to	£	193.95
steel guard rails and vehicle barriers	m	£	47.21	to	£	71.46

External Furniture					
Parking bollards					
precast concrete	nr	£	107.18	to £	126.32
steel	nr	£	172.26	to £	227.13
cast iron	nr	£	201.61	to £	278.17
Vehicle control barrier; manual pole	nr	£	931.48	to £	1000.38
Galvanised steel cycle stand	nr	£	39.56	to £	51.04
Galvanised steel flag staff	nr	£	1048.87	to £	1329.59
Benches - hardwood and precast concrete	nr	£	196.50	to £	267.96
Litter bins					
precast concrete	nr	£	196.50	to £	227.13
hardwood slatted	nr	£	79.11	to £	107.18
cast iron	nr	£	336.86		
large aluminium	nr	£	592.06		
Bus stops	nr	£	357.28		
Bus stops incl basic shelter	nr	£	825.57		
Pillar box	nr	£	297.31		
Telephone box	nr	£	3261.46		
Fencing and screen walls					
Chain link fencing; plastic coated					
1.2 m high	m	£	16.46	to £	19.91
1.8 m high	m	£	23.73	to £	26.67
Timber fencing					
1.2 m high chestnut pale fencing	m	£	18.50	to £	21.82
1.8 m high close-boarded fencing	m	£	47.59	to £	59.21
Screen walls; one brick thick; including foundations etc					
1.8 m high facing brick screen wall	m	£	237.34	to £	297.31
1.8 m high coloured masonry block boundary wall	m	£	266.68	to £	335.59
EXTERNAL SERVICES					
Service runs laid in trenches including excavation.					
Water main					
75 mm uPVC main in 225 mm ductile iron pipe as duct	m	£	49.51		
Electric main					
600/1000 volt cables. Two core 25 mm cable including 100 mm clayware duct	m	£	32.54		
Gas main					
150 mm ductile or cast iron gas pipe	m	£	49.51		
Telephone					
British Telecom installation in 100 mm uPVC duct	m	£	19.91		
External lighting (per m^2 of lighted area)	m^2	£	2.42	to £	3.83

SITE WORK - continued						
DRAINAGE						
Overall £ /m² allowances based on gross areas						
Site drainage (per m² of paved areas)	m²	£	7.15	to	£	18.37
Building drainage (per m² of gross floor area)	m²	£	7.15	to	£	16.08
Drainage work beyond the boundary of the site and final connection	nr	£	1588.62	to	£	10678.84

PART 8

Oncosts and Profit

In Part 1 of this book, it is stressed that the cost information given in Parts 2 to 7 leads to a cost estimate that requires further adjustment before it is submitted as a tender. This part deals with those adjustments and includes a worked example of a Tender Summary.

RISK/OPPORTUNITY

The factors to be taken into account when gauging the possibility of the Estimator's prediction of cost being inadequate or excessive are given in Part 2. Clearly it is considered in parallel with profit and it is not possible to give any indicative guidance on the level of adjustment that might result. For the purpose of a preliminary estimate, it is suggested that no adjustment is made to the costs generated by the other parts of this book.

At the same time as making a general appraisal of risk/opportunity, management will look at major quantities and may suggest amendments to the unit rates proposed.

HEAD OFFICE OVERHEADS

An addition needs to be made to the net estimate to cover all costs incurred in operating the central services provided by head office. Apart from general management and accountancy, this will normally include the departments dealing with:

estimating
planning and design
purchasing
surveying
insurance
wages and bonus
site safety.

The appropriate addition varies with the extent of services provided centrally, rather than on site, and with size of organisation, but a range of 4% to 8% on turnover would cover most circumstances.

Some organisations would include finance costs with head office overheads, as a general charge to the company, but for the purposes of this book finance costs are treated separately (see below).

PROFIT

Obviously, the level of profit is governed by the degree of competition applicable to the job - which is in turn a function of the industry's current workload. Again, the appropriate addition is highly variable, but for the purposes of a preliminary estimate an addition of 2% to 5% onto nett turnover is suggested.

FINANCE COSTS - ASSESSMENT OF CONTRACT INVESTMENT

The following procedure may be followed to give an indication of the average amount of capital investment required to finance the contract. It must be emphasised that this method will not give an accurate investment as this can only be done by preparing a detailed cash flow related to the programme for the contract. The example is based on the same theoretical contract used for the worked example in Part 2 and should be read in conjunction with the Tender Summary that follows.

The average monthly income must first be assessed. This is done by deducting from the Tender total the contingency items and those items for which immediate payment is necessary.

	£	£
Tender total (excluding finance charges)		7,908,249
Deduct		
Subcontractors	1,500,000	
Prime cost sums	75,000	
Employer's contingencies	195,000	1,770,000
Amount to be financed	£	6,138,249

FINANCE COSTS - ASSESSMENT OF CONTRACT INVESTMENT - continued

The average monthly income is this sum (£6,138,249) divided by the contract period (12 months), that is, £511,522.

The average contract investment may now be calculated as follows :

		£	£
Plant and equipment to be purchased			75,000
Non time related			
Contractor		170,238	
Employer		6,975	
Other services, charges and fees		NIL	
Subtotal	£	177,213	
Take 50% as an average **[1]**			88,606
Stores and unfixed materials on site			10,000
Work done but not paid for			
2½ months at £511,522 (see table above) **[2]**		1,278,805	
Less retention at 5% **[3]**		63,940	1,214,865
Retention (5% with limit of 3%)			
Average retention **[4]** 3% of £ 6,138,249 (see table above)			184,147
Subtotal		£	1,572,618
Deduct			
Advance payment by client		NIL	
Bill loading **[5]**		150,000	
Creditors (suppliers)		450,000	600,000
Average contract investment		£	972,618

The interest charges that must be added to the Tender price (or absorbed from profit if capital needs to be borrowed) are therefore :

£ 972,618 x say 6.5% x 1 year = say £ 63,220

Notes

[1] These non time related oncosts and services are incurred as lump sums during the contract and, therefore, only 50% of such costs are taken for investment purposes

[2] This period depends on the terms of payment set out in the contract.

[3] Retention is deducted as full retention is taken into account later.

[4] Average retention will depend on the retention condition set out in the contract, account any partial completion dates.

[5] The contractor assesses here any financial advantage he may obtain by varying his items.

VALUE ADDED TAX

All of the figures quoted in this book exclude value added tax, which the conditions of contract normally make the subject of a separate invoicing procedure between the contractor and the employer.

Value Added Tax will be chargeable at the standard rate, currently 17½%, on supplies of services in the course of:

1. The construction of a non-domestic building
2. The construction or demolition of a civil engineering work
3. The demolition of any building, and
4. The approved alteration of a non-domestic protected building

TENDER SUMMARY

This summary sets out a suggested method of collecting together the various costs and other items and sums which, together, make up the total Tender sum for the example contract.

	£	£
Preliminaries and General Items (from Part 2)		
Contractor's site oncosts - time related	736,407	
Contractor's site oncosts - non time related	170,238	
Employer's requirements - time related	129,220	
Employer's requirements - non time related	6975	
Other services, charges and fees	NIL	
Temporary works not included in unit costs	117,800	
Plant not included in unit costs	193,858	1,354,498
Estimated net cost of measured work, priced at unit costs		3,918,002
	£	5,272,500
Allowance for fixed price contract		
6% on labour (assumed to be £1,200,000)	72,000	
4% on materials (assumed to be £1,650 000)	66,000	
4% on plant (assumed to be £360,000)	14,400	
5% on staff, overheads etc. (assumed to be £230,000)	11,500	163,900
Sub-contractors (net)		1,500,000
Prime cost sums		75,000
Adjustments made at Management Appraisal		
price adjustments, add say	75,000	
risk evaluation, add say	50,000	125,000
	£	7,136,400
Head office overheads and profit at 6 %		428,184
Finance costs (from previous page)		63,220
Provisional Sums		75,000
Dayworks Bill		75,000
	£	7,777,804
Employer's contingencies		194,445
TENDER TOTAL	£	7,972,249

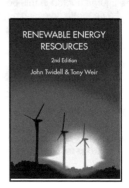

PART 9

Costs and Tender Prices Indices

The purpose of this part is to present historic changes in Civil Engineering costs and tender prices. It gives published and constructed indices and diagrammatic comparisons between building and Civil Engineering costs and tender prices and the retail price index, and will provide a basis for updating historical cost or tender price information.

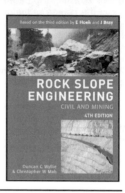

INTRODUCTION

It is important to distinguish between costs and tender prices; Civil Engineering costs are the costs incurred by a Contractor in the course of his business; Civil Engineering tender prices are the prices for which a Contractor undertakes work. Tender prices will be based on Contractor's costs but will also take into account market considerations such as the availability of labour and materials and the prevailing workload for Civil Engineering Contractors. This can mean that in a period when work is scarce tender prices may fall as costs are rising while when there is plenty of work prices will tend to increase at a faster rate than costs. This section comprises published Civil Engineering cost and tender indices, a constructed Civil Engineering cost index and comparisons of these with building cost and tender indices and with the retail price index.

PRICE ADJUSTMENT FORMULA INDICES

Cost indices for labour, plant and materials in Civil Engineering work are compiled and maintained by the Construction Industry Economics and Statistics Directorate (DTI) as Technical Secretariat to the Working Group on Building and Civil Engineering Indices. They are published in a monthly bulletin, formerly by HMSO, now by Tudorseed Construction, and are reproduced here with permission. Details of Tudorseed Construction, to whom enquiries concerning sales and subscriptions should be addressed, are given below. These indices were formerly known as NEDO or 'Baxter' indices. The original 1970 Series comprises 14 indices derived from Government sources. Two of them, DERV fuel (index 8) and Metal Sections (index 11B), are little used and a further two, indices 12 and 13, represent materials and labour specifically for structural steelwork. In December 1994 a new Series of indices was introduced, with 15 indices, including the new categories of ready Mixed Concrete (6), Plastic Products (8) and Sheet Steel Piling (15). The 1970 Series and the 1990 Series continue to be published in tandem. The reference numbers and titles are listed below. Quarterly values of the 1970 Series (excluding Indices 8 and 11B) are tabulated overleaf:

Index nr	1970 Series Title
1	Labour and supervision in Civil Engineering
2	Plant and road vehicles: provision and maintenance
3	Aggregates
4	Bricks and clay products
5	Cements
6	Cast iron products
7	Coated roadstone for road pavement & bituminous Products
8	DERV fuel
9	Gas oil fuel
10	Timber
11A	Steel for reinforcement
11B	Metal sections
12	Fabricated structural steel
13	Labour and supervision in fabricating and erecting Steelwork

Index nr	1990 Series Title
1	Labour and supervision
2	Plant and road vehicles
3	Aggregates
4	Bricks and clay products
5	Cements
6	Ready mixed concrete
7	Cast and spun iron products
8	Plastic products
9	Coated macadam and bituminous products
10	DERV fuel
11	Gas oil fuel
12	Timber
13	Steel for reinforcement
14	Metal sections
15	Sheet steel piling

Price Adjustment Formulae for Construction Contracts
Monthly Bulletin of Indices
are published by:

Tudorseed Construction
Unit 3, Ripon House,
35 Station Lane,
Hornchurch, Essex RM12 6JL
Tel: 01708 444678 Fax: 01708 443002

PRICE ADJUSTMENT FORMULA INDICES - continued

Quarterly values of price adjustment formula 1970 Series. Base: 1970 = 100 (except for index 11A which has a base date of July 1976).

Year	Q	1	2	3	4	5	6	7	9	10	11a
1998	1	974	852	1342	1494	1037	1654	2057	1351	841	283
	2	978	856	1358	1476	1045	1657	2079	1299	869	275
	3	1036	872	1326	1453	1056	1652	2077	1250	853	272
	4	1036	870	1309	1448	1048	1650	2044	1171	853	262
1999	1	1036	874	1376	1460	1063	1565	2083	1242	847	261
	2	1032	873	1358	1471	1077	1613	2064	1317	830	277
	3	1093	891	1370	1475	1087	1621	2090	1696	832	282
	4	1097	893	1363	1470	1081	1578	2086	1867	835	276
2000	1	1097	891	1399	1485	1107	1578	2220	1988	830	283
	2	1108	895	1400	1488	1092	1578	2283	2111	828	283
	3	1154	907	1405	1504	1086	1578	2276	2592	824	295
	4	1151	907	1389	1495	1068	1583	2277	2610	821	295
2001	1	1151	908	1408	1526	1079	1665	2363	2108	812	281
	2	1160	910	1419	1537	1099	1645	2414	2324	794	289
	3	1211	924	1422	1573	1091	1675	2429	2195	786	289
	4	1211	924	1423	1570	1080	1699	2429	1771	833	289
2002	1	1211	924	1437	1603	1119	1710	2440	1907	839	289
	2	1237	924	1702	1632	1127	1703	2702	1909	851	303
	3	1319	933	1676	1631	1120	1703	2813	2087	882	380
	4	1319	958	1620	1664	1114	1758	2744	2116	867	353
2003	1	1319	959	1668	1699	1134	1758	2824	2639	879	351
	2	1325	965	1729	1715	1123	1758	2935	2113	901	360
	3	1405	989	1729	1706	1117	1738	2816	2142	915	361
	4	1405	994	1642	1720	1117	1738	2819	2250	926	361
2004	1	1406	999	1697	1787	1116	1769	2906	2268	999	437
	2	1415	1006	1882	1754	1122	1777	2927	2510	1068	551
	3	1506	1038	1667	1780	1133	1777	2929	2952	1033	546
	4	1506	1042	1650	1783	1133	1820	2853	2872	975	540
2005	1	1507	1057	1657	1873	1171	1826	3022	3291	977	508
	2	1523	1063	1705	1919	1238	1815	3079	3573	986	459
	3	1639	1105	1661	1889	1234	1820	3081	4179	988	446
	4	1639	1106	1619	1909	1243	1839	3140	3868	1020	461
2006	1	1639*	1110*	1706*	1967*	1318*	1822*	3269*	4053*	1008*	463*

Provisional
Note: The indices shown are for the third month of each quarter

A CONSTRUCTED COST INDEX BASED ON THE PRICE ADJUSTMENT FORMULA INDICES

Although the above indices are prepared and published in order to provide a common basis for calculating reimbursement of increased costs during the course of a contract, they also present time series of cost indices for the main components of Civil Engineering work. They can therefore be used as the basis of an index for Civil Engineering work. The method used here is to construct a composite index by allocating weightings to each of the 10 indices, the weightings being established from an analysis of actual projects. The composite index is calculated by applying these weightings to the appropriate price adjustment formula indices and totalling the results; this index is again presented with a base date of 1970.

Constructed Civil Engineering Cost Index base: 1970 = 100

Year	First quarter	Second quarter	Third quarter	Fourth quarter	Annual Average
1987	701	709	728	730	717
1988	736	743	760	763	751
1989	774	793	820	837	806
1990	849	859	894	916	880
1991	913	912	928	928	920
1992	922	927	935	951	934
1993	954	968	968	977	967
1994	984	1001	1017	1026	1007
1995	1042	1064	1073	1075	1064
1996	1086	1103	1114	1125	1107
1997	1125	1130	1135	1144	1134
1998	1138	1152	1163	1156	1152
1999	1158	1166	1206	1212	1186
2000	1238	1256	1295	1307	1274
2001	1290	1308	1328	1314	1310
2002	1313	1385	1439	1440	1394
2003	1459	1477	1505	1495	1484
2004	1502	1559	1596	1597	1564
2005	1611	1652	1722	1718	1676
2006	1749*				

Note: * Provisional

Source: Davis Langdon based on DTI figures.

The figure on the following page illustrates graphically the movement of the Constructed Civil Engineering Index, the Davis Langdon Index of Building Costs and the Retail Price Index.

A CONSTRUCTED COST INDEX BASED ON THE PRICE ADJUSTMENT FORMULA INDICES

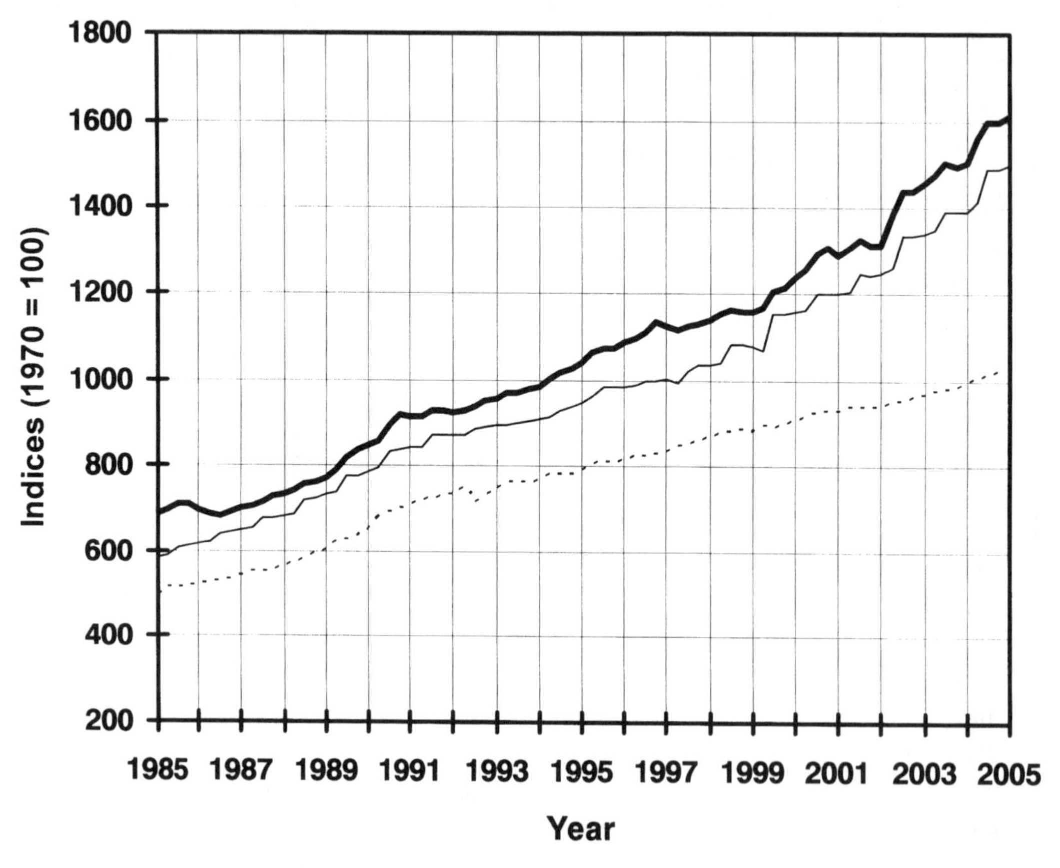

THE ROAD CONSTRUCTION TENDER PRICE INDEX

Civil Engineering work generally does not lend itself easily to the preparation of tender price indices in the same way as building work. There is, however, a published index for road construction tender and this is reproduced below with the permission of HMSO. The index is intended to indicate the movement in tender prices for road construction contracts. It is based on priced rates contained in accepted tenders for Road Construction, Motorway Widening and Major Maintenance Schemes. It is published with a base of 1995=100.

Tender Price Index of Road Construction Base: 1995 = 100

Year	First quarter	Second quarter	Third quarter	Fourth quarter	Annual Average
1990	82	82	79	79	81
1991	79	80	74	72	76
1992	71	70	70	68	70
1993	67	73	76	80	74
1994	83	87	92	93	89
1995	99	100	101	100	100
1996	96	94	100	101	98
1997	99	97	101	102	100
1998	99	102	97	96	99
1999	93	103	98	105	100
2000	111	114	114	115	114
2001	116	114	117	121	117
2002	122	123	121	119	121
2003	123	122	118	119	121
2004	120	120	120	127	122
2005	129	133	141*	142*	136*

Note: * Provisional

Source: DTI

The figure on the following page illustrates graphically the movement of the Tender Index of Road Construction, the Davis Langdon Building Tender Price Index and the Retail Price Index.

THE ROAD CONSTRUCTION PRICE INDEX

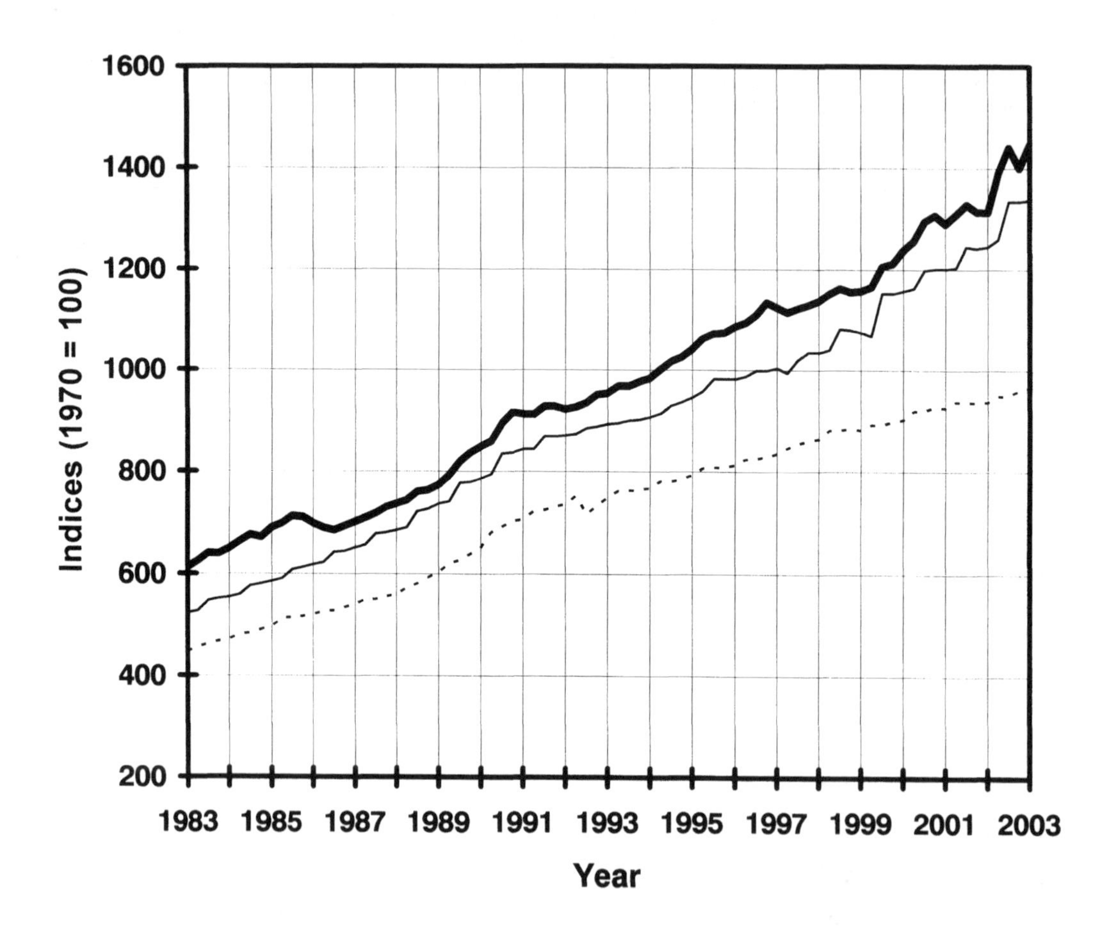

PART 10

Daywork

Taylor & Francis
Taylor & Francis Group plc

INTRODUCTION

"Dayworks" relates to work which is carried out incidental to a contract but where no other rates have been agreed. In the case of the ICE Conditions of Contract It is ordered by the Engineer pursuant to clause 52(5):

> "(5) The Engineer may if in his opinion it is necessary or desirable order in writing that any additional or substituted work shall be carried out on a daywork basis in accordance with the provisions of Clause 56(4)."

Clause 56(4) states that :

> "Where any work is carried out on a daywork basis the Contractor shall be paid for such work under the conditions and at the rates and prices set out in the daywork schedule included in the Contract or failing the inclusion of a daywork schedule he shall be paid at the rates and prices and under the conditions contained in the "Schedules of Dayworks carried out incidental to Contract Work" issued by The Civil Engineering Contractor's Association (formerly issued by The Federation of Civil Engineering Contractors), current at the date of carrying out of the daywork.
>
> The contractor shall furnish to the Engineer such records receipts and other documentation as may be necessary to prove amounts paid and/or costs incurred. Such returns shall be in the form and delivered at the times the Engineer shall direct and shall be agreed within a reasonable time.
>
> Before ordering materials the Contractor shall if so required submit to the Engineer quotations for the same for his approval."

The most recent Schedule is dated November 2002 (reprinted with amendments August 2003) and is published by the Civil Engineering Contractors Association. Copies may be obtained from :

CIVIL ENGINEERING CONTRACTORS ASSOCIATION
55 Tufton Street
London
SW1P 3QL
Tel: 020 7227 4620
Fax: 020 7227 4621

These schedules identify a number of items that are excluded from the rates and percentages quoted, but should be recovered by the Contractor when valuing his Daywork account. Under normal circumstances it is impractical to accurately value these items for each Daywork event, although certain items can be allowed for by means of a percentage addition. Suggested methods of calculating such additions are included in this section.

In Summary the Daywork schedule allows for an addition:-
- of 148% to wages paid to workmen
- of 88% to labour only sub-contractors and hired plant drivers
- of 12.5% to subsistence allowances and travel paid to workmen
- of 12.5% to materials used on dayworks
- of 12.5% to full cost of plant hired for dayworks
- of 12.5% to cost of operating welfare facilities
- of 12.5% to cost of additional insurance premiums for abnormal contract work or special conditions.

Part 3 of this book (Resources - Plant) includes detailed references to the plant section of the 2002 Daywork Schedule.

The text of the document is as follows :

SCHEDULES OF DAYWORKS CARRIED OUT INCIDENTAL TO CONTRACT WORK

The General clauses in the Schedule are repeated in full as follows:

1. Labour

"Add to the amount of wages paid to operatives .... 148%"

1.1 "Amount of wages" means :-

Actual wages and bonus paid, daily travelling allowances (fare and/or time), tool allowance and all prescribed payments including those in respect of time lost due to inclement weather paid to operatives at plain time rates and/or at overtime rates.

1.2 The percentage addition provides for all statutory charges at the date of publication and other charges including :-

National Insurance and Surcharge.
Normal Contract Works, Third Party & Employer's Liability Insurances.
Annual and Public Holidays with pay.
Statutory and industry sick pay.
Welfare benefits.
Industrial Training Levy.
Redundancy Payments.
Employment Rights Act 1996.
Employment Relations Act 1999.
Site Supervision and staff including foremen and walking gangers, but the time of the gangers or charge hands working with their gangs is to be paid for as for operatives.
Small tools – such as picks, shovels, barrows, trowels, hand saws, buckets, trestles, hammers, chisels and all items of a like nature.
Protective clothing.
Head Office charges and profit.

1.3 The time spent in training, mobilisation, demobilisation etc. for the Dayworks operation is chargeable.

1.4 All hired plant drivers and labour sub-contractor's accounts to be charged in full (without deduction of any cash discounts not exceeding 2.5%) plus 88%.

1.5 Subsistence or lodging allowances and periodic travel allowances (fare and/or time) paid to or incurred on behalf of operatives are chargeable at cost plus 12.5%.

2. Materials

"Add to the cost of materials ....12.5%"

2.1 The percentage addition provides for Head office charges and Profit.

2.2 The cost of materials means the invoiced price of materials including delivery to site without deduction of any cash discounts not exceeding 2.5%.

2.3 Unloading of materials :-

The percentage added to the cost of materials excludes the cost of handling which shall be charged in addition. An allowance for unloading into site stock or storage including wastage should be added where materials are taken from existing stock.

3. Supplementary Charges

3.1 Transport provided by contractors for operatives to, from, in and around the site to be charged at the appropriate Schedule rates.

3.2 Any other charges incurred in respect of any Dayworks operation including, tipping charges, professional fees, sub-contractor's accounts and the like shall be paid for in full plus 12.5% (without deduction of cash discounts not exceeding 2.5%). Labour subcontractors being dealt with in Section 1.

3.3 The cost of operating welfare facilities to be charged by the contractor at cost plus 12.5%.

3.4 The cost of additional insurance premiums for abnormal contract work or special site conditions to be charged at cost plus 12.5%.

3.5 The cost of watching and lighting specially necessitated by Dayworks is to be paid for separately at Schedule rates.

4. Plant

4.1 These rates apply **only** to plant already on site, exclusive of drivers and attendants, but inclusive of fuel and consumable stores unless stated to be charged in addition, repairs and maintenance, insurance of plant but excluding time spent on general servicing.

4.2 Where plant is hired specifically for Dayworks: plant hire (exclusive of drivers and attendants), fuel, oil and grease, insurance, transport etc., to be charged at full amount of invoice (without deduction of any cash discount not exceeding 2.5%) to which should be added consumables where supplied by the contractor, all plus 12.5%.

4.3 Fuel distribution, mobilisation and demobilisation are not included in the rates quoted which shall be an additional charge.

4.4 Metric capacities are adopted and these are not necessarily exact conversions from their imperial equivalents, but cater for the variations arising from comparison of plant manufacturing firms' ratings.

1.5 SAE rated capacities of plant means rated in accordance with the standards specified by the Society of Automotive Engineers.

4.6 Minimum hire charge will be for the period quoted.

4.7 Hire rates for plant not included below shall be settled at prices reasonably related to the rates quoted.

4.8 The rates provide for Head Office charges and Profit.

The Schedule then gives twenty-four pages of hire rates for a wide range of plant and equipment.

APPLICATION OF DAYWORKS

Generally

A check should be made on the accuracy of the recorded resources and times.

Tender documents generally allow the contractor to tender percentage variations to the figure calculated using the published percentage additions. These vary widely but a reasonable average indication can be along the lines of :

Labour .... 20% less
Materials .... 10% less
Plant .... 30% less

Labour

The Contractor should provide substantiation of the hourly rates he wishes to be paid for the various classes of labour and should demonstrate that the basic "amount of wages" does not include any of the items actually covered by the percentage addition.

The wage bill is intended to reflect the cost to the Contractor. The time involved is not restricted to the duration of the task, but also includes mobilisation and demobilisation, together with any training needed – which would

APPLICATION OF DAYWORKS - continued

Labour - continued

include induction courses required for Health & Safety requirements. The rate paid is the actual value of wages and bonuses paid (not simply the basic rate promulgated for the labour grade involved, and includes overtime rates if applicable, tool money, time lost due to inclement weather. In addition, daily travelling allowances are included, as are periodic travel allowances and also subsistence or lodging allowances.

Care should be taken that the matters deemed included in the percentage addition are not duplicated in the amount of wages. For example, it should be noted that foremen, gangers and other supervisory staff are covered by the percentage addition, unless they work in which case they are paid for at the correct rate for the task involved. The proportion of the time they spend working rather than supervising must be agreed. Refer to the amplification of labour categories in Part 3.

Hired or sub-contracted labour is paid at invoiced cost, adjusted only where any cash discount exceeded 2½%, in which case the excess percentage is deducted.

Materials

The cost of materials delivered to the site is simply the invoiced price of the materials plus any delivery charges.

Should the cash discount exceed 2½%, the excess percentage is deducted from the amount to be paid.

The percentage addition simply covers the cost of Head Office charges and profit. It does not include for unloading or temporarily storing the materials nor for distributing them on site to the work place. Such cost can be charged, even in cases where the materials may already be in the site stock.

Material waste should be added direct to the cost of materials used for each particular Daywork items as an appropriate percentage.

Handling and offloading materials

Schedule 2.3 states that an allowance for handling materials, and an allowance for unloading or storage including wastage should be an additional charge.

Example: For a 12 month, £8.0m Civil Engineering scheme where the total cost of materials that require handling (excluding Ready Mix concrete, imported fills, fuels and similar items) is £2,250,000.

The following gang is employed (part time) throughout the contract for offloading and handling of materials.

	Net cost
	£
2 labourers	24.22
Lorry (8T) with Hiab lift	19.85
	£ 44.07 /hour

Allow an average of 5 hours per week over 50 weeks

250 hours @ £44.07 /hour = £ 11,018

This cost as a percentage of the materials element of the contract

$$\frac{£11,018}{£2,250,000} \times 100 = 0.49\ \%$$

Supplementary charges

1. Transport

Schedule 3.1 states that transport provided by contractors to take operatives to and from the site as well as in and around the site shall be charged at the appropriate Schedule Rate. This would entail the driver and vehicle being included with the labour and plant parts of the Daywork calculation.

2. Any other charges

This relates to any other charges incurred in respect of any Dayworks operation and includes tipping charges, professional fees, sub-contractor's accounts and the like. Schedule 3.2 provides for full payment of such charges in full plus the addition of 12½% for Head Office charges and Profit.

Should the cash discount exceed 2½%, the excess percentage is deducted from the amount to be paid.

Welfare Facilities

Schedule 3.3 allows for the net cost of operating these facilities plus 12.5%.

Example: How the costs of operating welfare facilities may be charged to the Daywork account on a 12 month, £8.1m Civil Engineering scheme, where the total labour element is £1,200,000.

Facility	Weekly Cost £
Toilet unit (4 nr)	128.12
Jack leg hutments 24' (2 nr)	44.04
Jack leg hutment 12' (1 nr)	16.02
Labour to clean, maintain and make tea, etc. 1 man, 2 hours per day, 6 days x 6.86	133.32
Consumables (heat, light, soap, disinfectant, etc) say	50.00
Rates, insurance, taxes, etc (add 2%)	7.43
Total weekly cost	£ 378.93

Multiply by 50 weeks (construction period) plus 6 weeks (maintenance period)

Total cost to contract £ 378.93 x 56 = £ 21,220.00

Thus cost as a percentage of the Labour element of the contract =

$$\frac{£21,220}{£1,200,000} \times 100 = 1.76\ \%$$

Add, as schedule 3.3 12.5% 0.22 %

Percentage addition for facilities = 1.98 %

Insurances

Schedule 3.4 allows for the cost of additional insurance premiums for abnormal work or special site conditions to be charged at cost plus 12.5%.

Watching and lighting

Schedule 3.5 allows for all such costs necessitated by Dayworks to be paid for separately at Schedule rates.

Plant

The cost of the driver(s) and any required attendants such as banksmen should be covered in the labour section of the dayworks calculation.

The Schedule rates include fuel and consumable stores, repairs and maintenance (but not the time spent on general servicing) and insurance.

APPLICATION OF DAYWORKS – continued

Plant - continued

The Schedule rates only apply to machinery which is on site at the time of the work - where plant is specifically hired for the task, the Contractor is entitled to be paid the invoiced value. If the invoice excludes consumables used (fuel, oil and grease) then the Contractor is entitled to add the cost - together with insurance and any transport costs incurred in getting the equipment on site all subject to a 12½% addition for Head Office costs and profit.

Head Office charges and Profit allowances are included in the Schedule rates, 12½% being added to the charged value of hired plant.

1. General servicing of Plant

Schedule 4.1 specifically excludes time spent on general servicing from the Hire Rates.

General servicing in this context can be assumed to mean:

- Checking, replenishing (or changing, if applicable)
  i.e. engine lubrication
  transmission lubrications
  general greasing
  coolants
  hydraulic oils and brake systems
  filters
  tyres

- Inspecting special items
  e.g. buckets
  hoses/airlines
  shank protectors
  cables/ropes/hawsers
  rippers
  blades, steels, etc

Example : Assuming an 8 hour working day these operations could take a plant operator on average :

- large machine 20 minutes per day (equating to 1 hour for each 24 worked)
- medium machine 10 minutes per day (equating to 1 hour for each 48 worked)
- small machine 5 minutes per day (equating to 1 hour for each 96 worked)

The cost of the servicing labour would be as follows :

large machine	£17.68/hr ÷ 24 hrs	= £0.74
medium machine	£15.85/hr ÷ 48 hrs	= £0.33
small machine	£11.39/hr ÷ 96 hrs	= £0.12

These labour costs can be expressed as a percentage of the schedule hire rates :

D8 Dozer	£0.74 ÷ £82.19/hr x 100	= 0.9 %
JCB 3CX	£0.33 ÷ £15.51/hr x 100	= 2.1%
2 tonne dumper	£0.12 ÷ £3.74/hr x 100	= 3.2 %

Taking into account the range of these sizes of plant which are normally deployed on site, the following would provide a reasonable average percentage addition :

0.9 % x 2	=	1.8 %
2.1 % x 3	=	6.3 %
3.2 % x 6	=	19.2 %
Total	=	27.3 % for 11 items of plant - average = 2.5 %

2. Fuel distribution

Schedule 4.3 allows for charging for fuel distribution.

Assuming this is done with a towed fuel bowser behind a farm type tractor with driver/labourer in attendance, the operation cycle would involve visiting, service, and return or continue on to the next machine.

The attendance cost based on the hourly rate tractor/bowser/driver would be:
= £13.30 + £0.58 + £12.11 = £25.99

Example : A heavy item of plant, for example a Cat D8N Tractor Bulldozer with a 212 kW engine

- Fuel consumption is 30.4 litres/hr (38 litres/hr x 80% site utilisation factor)
- Fuel capacity is 200 litres
- Requires filling after 6.5 working hours operation (200 litres divided by 30.40 l/hr).
- Tractor/bowser service taking 30 minutes

The machine cost during this 6.5 hr period would be £82.19 x 6.5, i.e. £ 534.24
The attendance cost for the 30 min cycle would be £25.99 x 0.5, i.e. £13.00
The percentage addition for fuelling the machine would be :

$$\frac{£13.00}{£534.24} \times 100 = 2.4\%$$

Example : A medium sized item of plant, for example a JCB 3CX

- Fuel consumption is 5.6 litres/hr (7.5 litres/hr x 75% site utilisation factor)
- Fuel capacity is 90 litres
- Requires filling after 16 working hours operation (90 litres divided by 5.60 l/hr).
- Tractor/bowser service taking 10 minutes

The machine cost during this 16 hr period would be £15.51 x 16 hrs, i.e. £248.16
The attendance cost for the 10 min cycle would be £25.99 x 0.17 hrs, i.e. £4.42
The percentage addition for fuelling the machine would be :

$$\frac{£4.42}{£248.16} \times 100 = 1.8\%$$

Example : A medium sized item of plant, for example a 2 tonne dumper :

- Fuel consumption is 2.4 litres/hr (3 litres/hr x 80% site utilisation factor)
- Fuel capacity is 35 litres
- Requires filling after 14.5 working hours operation (35 litres divided by 2.40 l/hr).
- Tractor/bowser service taking 5 minutes

The machine cost during this 14.5 hr period would be £3.73 x 14.5 hrs, i.e. £54.09
The attendance cost for the 5 min cycle would be £25.99 x 0.08 hrs, i.e. £2.08
The percentage addition for fuelling the machine would be :

$$\frac{£2.08}{£54.09} \times 100 = 3.9\%$$

Considering the range of these categories of plant which are normally deployed on site, the following would provide a reasonable average percentage addition for the above :

2.4% x 2 = 4.8 %
1.8% x 3 = 5.4 %
3.9% x 6 = 23.4 %
Total = 33.6 % for 11 items of plant - average = 3.1 %

3. Mobilisation

Schedule 4.3 allows for charging for mobilisation and demobilisation.

PART 11

Professional Fees

CONSULTING ENGINEERS' FEES

Introduction

A scale of professional charges for consulting engineering services in connection with civil engineering works is published by the Association for Consultancy and Engineering (ACE)

Copies of the document can be obtained direct from : -

Association for Consultancy and Engineering
Alliance House
12 Caxton Street
London SW1H OQL
Tel 0207 222 6557
Fax 0207 222 0750

Comparisons

Instead of the previous arrangement of having different agreements designed for each major discipline of engineering, these new agreements have been developed primarily to suit the different roles that Consulting Engineers may be required to perform, with variants of some of them for different disciplines. The agreements have been standardised as far as possible whilst retaining essential differences.

Greater attention is required than with previous agreements to ensure the documents are completed properly. This is because of the perceived need to allow for a wider choice of arrangements, particularly of methods of payment.

The agreements are not intended to be used as unsigned reference material with the details of an engagement being covered in an exchange of letters, although much of their content could be used as a basis for drafting such correspondence.

Forms of Agreement

The initial agreements are for use where a Consulting Engineer is engaged as follows :

Agreement as a Lead Consultant

Agreement directly by the Client, but not as Lead Consultant

Agreement to provide design services for a design and construct Contractor

Short Form Agreement (Report and Advisory Services)

Agreement as a Project Manager

Agreement as Planning Supervisor in accordance with the Construction (Design and Management) Regulations 1994

Each of Agreements A, B and C are published in two variants

Variant 1 Civil and Structural Engineering

Variant 2 The Engineering of Electrical and Mechanical Services in Buildings

Each agreement comprises the following :-

Memorandum of Agreement

Conditions of Engagement

Appendix I - Services of the Consulting Engineer

Appendix II - Remuneration of the Consulting Engineer

CONSULTING ENGINEERS' FEES - continued

Memorandum of Agreement

There is a different memorandum for each agreement, reflecting in each instance the particular relationships between the parties. It is essential that the memorandum be fully completed. Spaces are provided for entry of important and specific details relevant to each commission, such as nominated individuals, limits of liability, requirements for professional indemnity insurance, the frequency of site visits and meetings, and requirements for collateral warranties. All the memoranda are arranged for execution under hand; some also have provision for execution as deeds.

Conditions of Engagement

These have been standardised as far as possible and thus contain much that is common between the agreements, but parts differ and are peculiar to individual agreements to reflect the responsibilities applying. The conditions can normally stand as drafted but clauses may be deleted and others be added should the circumstances so require for a particular commission.

Appendix I - Services

This appendix, which has significant differences between the agreements and variants, describes the services to be performed. These services include both standard Normal Services, the majority of which will usually be required, and standard Additional Services of which only some will be required. Standard Normal Services may be deleted if not required or not relevant to a particular commission; further Services, both Normal and Additional may be added in spaces provided. It may be agreed in advance, when known that certain of the Additional Services will clearly be required, that these will be treated and paid for as Normal Services for a particular commission.

Appendix II - Remuneration / Fees and Disbursements

This appendix provides alternate means of assessing the consulting engineer's fees and disbursements. It identifies, when completed, which of those services listed in Appendix I are to be performed within the overall fee applicable for Normal Services. Figures need to be entered on such details as time charge rates, fee percentages and interest rates on delayed payments. Alternatives which do not apply require deletion and those remaining completion, so that the appendix when incorporated within an engagement contract describes the exact arrangements applicable to that commission.

Collateral Warranties

The association is convinced that collateral warranties are generally unnecessary and should only be used in exceptional circumstances. The interests of clients, employers and others are better protected by taking out project or BUILD type latent defects insurance. Nevertheless, in response to observations raised when the pilot editions excluded any mention of warranties, references and arrangements have been included in the Memorandum and elsewhere by which Consulting Engineers may agree to enter into collateral warranty agreements; these should however only be given when the format and requirements thereof have been properly defined and recorded in advance of undertaking the commission.

Requirements for the provision of collateral warranties will be justified even less with commissions under Agreement D than with those under the other ACE agreements. Occasional calls may be made for them, such as when a client intends to dispose of property and needs evidence of a duty of care being owed to specific third parties, but these will be few and far between.

Remuneration

Guidance on appropriate levels of fees to be charged is given at the end of each agreement. Firms and their clients may use this or other sources, including their own records, to determine suitable fee arrangements.

Need for formal documentation

The Association of Consulting Engineers recommends that formal written documentation should be executed to record the details of each commission awarded to a Consulting Engineer. These Conditions are published as

model forms of agreement suitable for the purpose. However, even if these particular Conditions are not used, it is strongly recommended that, whenever a Consulting Engineer is appointed, there should be at least an exchange of letters defining the duties to be performed and the terms of payment.

Appointments outside the United Kingdom

These conditions of Engagement are designed for use within the UK. For work overseas it is impracticable to give definite recommendations; circumstances differ too widely between countries. There are added complications in documentation relating to local legislation, import customs, conditions of payment, insurance, freight, etc. Furthermore, it is often necessary to arrange for visits to be made by principals and senior staff whose absence abroad during such periods represents a serious reduction of their earning power. The additional duties, responsibilities and non-recoverable costs involved, and the extra work on general co-ordination, should be reflected in the levels of fees. Special arrangements are also necessary to cover travelling and other out-of-pocket expenses in excess of those normally incurred on similar work in the UK, including such matters as local cost-of-living allowances and the cost of providing home-leave facilities for expatriate staff.

CONDITIONS OF ENGAGEMENT

Obligations

The following is a brief summary of the conditions of engagement. It is recommended that reference should be made to the full document of the Association of Consulting Engineers Conditions of Engagement, 1995 before making an engagement.

Obligations of the Consulting Engineer

The responsibilities of the Consultant Engineer for the works are as set out in the actual agreement The various standard clauses in the Conditions relate to such matters as differentiating between Normal and Additional services, the duty to exercise skill and care, the need for Client's written consent to the assignment or transfer of any benefit or obligation of the agreement, the rendering of advice if requested on the appointment of other consultants and specialist sub- consultants, any recommendations for design of any part of the Works by Contractors or Sub-contractors (with the proviso that the Consulting Engineer is not responsible for detailed design of contractors or for defects or omissions in such design), the designation of a Project Leader, the need for timeliness in requests to the Client for information etc., freezing the design once it has been given Client approval and the specific exclusion of any duty to advise on the actual or possible presence of pollution or contamination or its consequences.

Obligations of the Client

The Consultant Engineer shall be supplied with all necessary data and information in good time. The Client shall designate a Representative authorised to make decisions on his behalf and ensure that all decisions, instructions, and approvals are given in time so as not to delay or disrupt the Consultant Engineer.

Site Staff

The Consulting Engineer may employ site staff he feels are required to perform the task, subject to the prior written agreement of the Client. The Client shall bear the cost of local office accommodation, equipment and running costs.

CONSULTING ENGINEERS' FEES – continued

CONDITIONS OF ENGAGEMENT – continued

Commencement, Determination, Postponement, Disruption and Delay

The Consulting Engineer's appointment commences at the date of the execution of the Memorandum of Agreement or such earlier date when the Consulting Engineer first commenced the performance of the Services, subject to the right of the Client to determine or postpone all or any of the Services at any time by Notice.

The Client or the Consulting Engineer may determine the appointment in the event of a breach of the Agreement by the other party after two weeks notice. In addition, the Consulting Engineer may determine his appointment after two weeks notice in the event of the Client failing to make proper payment.

The Consulting Engineer may suspend the performance of all or any of the Services for up to twenty-six weeks if he is prevented or significantly impeded from performance by circumstances outside his control. The appointment may be determined by either party in the event of insolvency subject to the issue of notice of determination.

Payments

The Client shall pay fees for the performance of the agreed service(s) together with all fees and charges to the local or other authorities for seeking and obtaining statutory permissions, for all site staff on a time basis, together with additional payments for any variation or the disruption of the Consulting Engineer's work due to the Client varying the task list or brief or to delay caused by the Client, others or unforeseeable events.

If any part of any invoice submitted by the Consulting Engineer is contested, payment shall be made in full of all that is not contested.

Payments shall be made within 28 days of the date of the Consulting Engineer's invoice; interest shall be added to all amounts remaining unpaid thereafter.

Ownership of Documents and Copyright

The Consulting Engineer retains the copyright in all drawings, reports, specifications, calculations etc. prepared in connection with the Task; with the agreement of the Consulting Engineer and subject to certain conditions, the Client may have a licence to copy and use such intellectual property solely for his own purpose on the Task in hand, subject to reservations.

The Consulting Engineer must obtain the client's permission before he publishes any articles, photographs or other illustrations relating to the Task, nor shall he disclose to any person any information provided by the Client as private and confidential unless so authorised by the Client.

Liability, Insurance and Warranties

The liability of the Consulting Engineer is defined, together with the duty of the Client to indemnify the Consulting Engineer against all claims etc. in excess of the agreed liability limit.

The Consulting Engineer shall maintain Professional Indemnity Insurance for an agreed amount and period at commercially reasonable rates, together with Public Liability Insurance and shall produce the brokers' certificates for inspection to show that the required cover is being maintained as and when requested by the Client.

The Consulting Engineer shall enter into and provide collateral warranties for the benefit of other parties if so agreed.

Disputes and Differences

Provision is made for mediation to solve disputes, subject to a time limit of six weeks of the appointment of the mediator at which point it should be referred to an independent adjudicator. Further action could be by referring the dispute to an arbitrator.

QUANTITY SURVEYORS' FEES

Introduction

Authors' Note:
The Royal Institution of Chartered Surveyors formally abolished standard Quantity Surveyor's fee scales with effect from 31st December 1998. However, in the absence of any alternative guidance and for the benefit of readers, extracts from relevant fee scales have been reproduced in part with the permission of the Royal Institution of Chartered Surveyors, which owns the copyright.

Summary of Scale of Professional Charges

Scale No 38. issued by The Royal Institution of Chartered Surveyors provides an itemised scale of professional charges for Quantity Surveying Services for Civil Engineering Works which is summarised as follows :-

1.0. Generally

1.1. The Scale of professional charges is applicable where the contract provides for the bills of quantities and final account to be based on measurements prepared in accordance with or based on the principles of the Standard Method of Measurement of Civil Engineering Quantities issued by the Institution of Civil Engineers.

1.2. The fees are in all cases exclusive of travelling and other expenses (for which the actual disbursement is recoverable unless there is some prior arrangement for such charges) and of the cost of reproduction of bills of quantities and other documents, which are chargeable in addition at net cost.

1.3. The fees are in all cases exclusive of services in connection with the allocation of the cost of the works for purposes of calculating value added tax for which there shall be an additional fee based on the time involved.

1.4. If any of the materials used in the works are supplied by the employer or charged at a preferential rate, then the actual or estimated market value thereof shall be included in the amounts upon which fees are to be calculated.

1.5. The fees are in all cases exclusive of preparing a specification of the materials to be used and the works to be done.

1.6. If the quantity surveyor incurs additional costs due to exceptional delays in construction operations or any other cause beyond the control of the quantity surveyor then the fees may be adjusted by agreement between the employer and the quantity surveyor.

1.7. If the works are substantially varied at any stage or if the quantity surveyor is involved in abortive work there shall be an additional fee based on the time involved.

1.8. The fees and charges are in all cases exclusive of value added tax which will be applied in accordance with legislation.

1.9. The scale is not intended to apply to works of a civil engineering nature which form a subsidiary part of a building contract or to buildings which are ancillary to a civil engineering contract. In these cases the fees to be charged for quantity surveying services shall be in accordance with the scales applicable to building works.

QUANTITY SURVEYORS' FEES - continued

Summary of Scale of Professional Charges - continued

1.10. When works of both categories I* and II** are included in one contract the fee to be charged shall be calculated by taking the total value of the sections of work in each of the categories and applying the appropriate scale from the beginning in each case. General items such as preliminaries (and in the case of post contract fees contract price fluctuations) or sections of works which cannot be specifically allocated to either category shall be apportioned pro-rata to the values of the other sections of the works and added thereto in order to ascertain the total value of works in each category.

1.11. When a project is the subject of a number of contracts then, for the purpose of calculating fees, the value of such contracts shall not be aggregated but each contract shall be taken separately and the scale of charges applied as appropriate.

1.12. Roads, railways, earthworks and dredging which are ancillary only to any Category II** work shall be regarded as Category II** work. Works or sections of works of Category I* which incorporate piled construction shall be regarded as Category II** works.

* Category I. Works or sections of works such as monolithic walls for quays, jetties, dams and reservoirs; caissons; tunnels; airport runways and tracks; roads; railways; and earthworks and dredging.

** Category II. Works or sections of works such as piled quay walls; suspended jetties and quays; bridges and their abutments; culverts; sewers; pipe-lines; electric mains; storage and treatment tanks; water cooling towers and structures for housing heavy industrial and public utility plant, e.g. furnace houses and rolling mills to steel works; and boiler houses, reactor blocks and turbine halls to electricity generating stations.

1.13 No addition to the fees given hereunder shall be made in respect of works of alteration or repair where such works are incidental to the new works. If the work covered by a single contract is mainly one of alteration or repair then an additional fee shall be negotiated.

1.14. In the absence of agreement to the contrary, payments to the quantity surveyor shall be made by instalments by arrangement between the employer and the quantity surveyor.

1.15. Copyright in bills of quantities and other documents prepared by the quantity surveyor is reserved to the quantity surveyor.

Scale of charges

For the full Scale of Fees for Professional Charges for Quantity Surveying Services together with a detailed description of the full service provided the appropriate RICS Fee Scale should be consulted.

PART 12

Approximate Estimates

1. INDUSTRIAL AND COMMERCIAL BUILDINGS AND CIVIL ENGINEERING FACILITIES

Prices given under this heading are average prices on a `fluctuating basis' for typical buildings based on the first quarter 2007 Building Works tender price level index of 500 (1976 = 100). Unless otherwise stated, prices INCLUDE overheads and profit but EXCLUDE preliminaries, loose or special equipment and fees for professional services.

Prices are based upon the total floor area of all storeys, measured between external walls and without deductior for internal walls, columns, stairwells, liftwells and the like.

As in previous editions it is emphasised that the prices must be treated with reserve in that they represent the average of prices from our records and cannot provide more than a rough guide to the probable cost of a building or structure.

In many instances normal commercial pressures together with a limited range of available specifications ensure that a single rate is sufficient to indicate the prevailing average price. However, where such restrictions do not apply a range has been given; this is not to suggest that figures outside this range will not be encountered, but simply that the calibre of such a type of building can itself vary significantly.

As elsewhere in this edition, prices do not include Value Added Tax, which should be applied at the current rate to all non-domestic building.

	Unit	Cost excluding VAT £
Surface car parking	m²	60 to 84
Multi-storey car parks		
split level	m²	269 to 367
split level with brick facades	m²	299 to 412
flat slab	m²	328 to 418
warped	m²	334 to 382
Underground car parks		
partially underground under buildings	m²	435 to 513
completely underground under buildings	m²	562 to 857
completely underground with landscaped roof	m²	930 to 1104
Railway stations	m²	1836 to 3112
Bus and coach stations	m²	909 to 1523
Bus garages	m²	872 to 955
Garage showrooms	m²	669 to 1076
Garages, domestic	m²	371 to 597
Airport facilities (excluding aprons)		
airport terminals	m²	1649 to 3608
airport piers/satellites	m²	2013 to 4498
apron/runway - varying infrastructure content	m²	96 to 180
Airport campus facilities		
cargo handling bases	m²	597 to 974
distribution centres	m²	299 to 597
hangars (type C and D aircraft)	m²	1267 to 1493
hangars (type E aircraft)	m²	1493 to 3745
TV, radio and video studios	m²	1105 to 1798
Telephone exchanges	m²	909 to 1452
Telephone engineering centres	m²	764 to 955
Branch Post Offices	m²	955 to 1296

INDUSTRIAL AND COMMERCIAL BUILDINGS AND CIVIL ENGINEERING FACILITIES - continued

	Unit	Cost excluding VAT £
Postal Delivery Offices/Sorting Offices	m²	764 to 1135
Mortuaries	m²	1738 to 2419
Sub - Stations	m²	1296 to 1948

INDUSTRIAL FACILITIES

Agricultural storage buildings	m²	460 to 616
Factories		
for letting (incoming services only)	m²	323 to 460
for letting (including lighting, power and heating)	m²	430 to 592
nursery units (including lighting, power and heating)	m²	520 to 783
workshops	m²	597 to 968
maintenance/motor transport workshops	m²	597 to 1033
owner occupation - for light industrial use	m²	538 to 783
owner occupation - for heavy industrial use	m²	1022 to 1176
Factory/office buildings - high technology production		
for letting (shell and core only)	m²	586 to 783
for letting (ground floor shell, first floor offices)	m²	955 to 1237
for owner occupation (controlled environment, fully finished)	m²	1237 to 1643
Light industrial/office building		
economic shell and core with heating only	m²	520 to 914
medium shell with heating and ventilation	m²	812 to 1207
high quality shell and core with air-conditioning	m²	1069 to 1948
Distribution centres		
low bay; speculative	m²	328 to 466
low bay; owner occupied	m²	490 to 861
low bay; owner occupied and chilled	m²	711 to 1426
high bay; owner occupied	m²	801 to 1159
Warehouses		
low bay (6-8 m high) for letting (no heating)	m²	323 to 401
low bay for owner occupation (including heating)	m²	382 to 681
high bay (9-18 m high) for owner occupation (including heating)	m²	579 to 824
Cold stores, refrigerated stores	m²	681 to 1469

WATER AND TREATMENT FACILITIES

		£
Reinforced Concrete tanks:		
including all excavation; fill; structural work; valves;penstocks; pipeworks per m³ of concrete in structure. treatment tanks	m³	515 to 580
fire ponds and lagoons	m³	452 to 524
reservoirs	m³	462 to 534
Reinforced Concrete river weirs, quay and wave walls:		
including temporary dams; caissons overpumping,anchorages, all structural works.	m²	351 to 540

	Unit	Cost excluding VAT £
Reinforced Concrete dams (upto 12 m high):		
arch dam, including excavation anchorages and structural work only. m^3 of structure	m^3	1173 to 1713
flat slab dam including excavation anchorages and structural work only. m^3 of structure	m^3	891 to 1396
Earth dams:		
rock fill with concrete core m^3 of completed structure	m^3	223 to 796
hydraulic fill embankment dam m^3 of completed structure	m^3	125 to 353

2. FOUNDATIONS FOR STRUCTURES

		£
In-situ concrete foundations complete with associated excavation and disposal:		
trench fill strip foundation; 450 x 1000 mm deep	m	66
trench fill strip foundation; 600 x 1200 mm deep	m	94
strip foundation; 900 mm deep	m	97
strip foundation; 1200 mm deep	m	118
column/stanchion base; unreinforced; 900 x 900 x 450 mm	nr	110
column/stanchion base; reinforced 50 kg/m^3; 1750 x 1750 x 500 mm	nr	319
column/stanchion base; reinforced 50 kg/m^3; 2200 x 2200 x 600 mm	nr	598
column/stanchion base; reinforced 75 kg/m^3; 2400 x 2400 x 600 mm	nr	717
pile cap; reinforced 75 kg/m^3; 900 x 900 x 1400 mm; for one pile	nr	318
pile cap; reinforced 75 kg/m^3; 2700 x 900 x 1400 mm; for two piles	nr	798
pile cap; reinforced 75 kg/m^3; 2700 x 2700 x 1400 mm; for four piles	nr	1845
pile cap; reinforced 75 kg/m^3; 3700 x 2700 x 1400 mm; for six piles	nr	2493
In situ concrete slabs comprising concrete C30 slab with tamped finish; reinforced with two layers of A393 fabric; laid in bays size 20 m x 10 mon sub-base of 250 mm granular material; including all associated excavation (250 mm depth), formwork and formed expansion joints with filler		
200 mm thick	m^2	26.60
250 mm thick	m^2	31.60
275 mm thick	m^2	34.00
300 mm thick	m^2	36.50
Extra over cost of slab		
for hand trowel finish	m^2	0.82
for power float finish	m^2	2.18
for wire brush finish	m^2	1.71
for additional sub base depth (including excavation and disposal)		
50 mm	m^2	1.33
100 mm	m^2	2.68
150 mm	m^2	4.01

2. FOUNDATIONS FOR STRUCTURES - continued

The prices following have been assembled from the relevant items in the unit costs section. They are intended to give a broad overall price or help in comparisons between a number of basic construction procedures. These approximate estimates are for construction only. They do not include for preliminaries, design/supervision costs, land purchase or OH&P etc.

Prices in this section are based on the same information and outputs as used in the unit costs section. Costs per m^2 or m^3 of completed structure.

3. EARTH RETENTION AND STABILISATION

	Unit	£
Reinforced in-situ concrete retaining wall: *(including excavation; reinforcement; formwork; expansion joints; granular backfill and 100 mm land drain; profiled formwork finish to one side typical retaining wall, allowing for profiling finishes)*		
1.0 m high	m^2	433
3.0 m high	m^2	339
6.0 m high	m^2	374
9.0 m high	m^2	421
Precast concrete block, earth retaining wall: *(including granular fill; earth anchors and proprietary units or concrete panels, strips, fixings and accessories (Reinforced Earth))*		
1.5 m high	m^2	350
3.0 m high	m^2	323
6.0 m high	m^2	350
Precast, reinforced concrete unit retaining wall (in-situ foundation): *(including excavation and fill; reinforced concrete foundation; pre-cast concrete units, joints)*		
1.00 high	m^2	427
2.00 high	m^2	433
3.00 high	m^2	456
Precast concrete crib wall *(including excavation; stabilisation and foundation work)*		
up to 1.0 m high	m^2	279
up to 1.5 m high	m^2	272
up to 2.5 m high	m^2	267
up to 4.0 m high	m^2	252
Timber crib walling		
up to 1.5 m high	m^2	135
up to 3.7 m high	m^2	169
up to 5.9 m high	m^2	209
up to 7.4 m high	m^2	243
Rock gabions *including preparation; excluding anchoring (see Page 159)*		
1 m thick	m^2	71

4. BRIDGEWORKS

The following prices are based on recovered data and information from approximately 50 separate structures completed as part of actual projects.

The prices include for the works described to the bridge decks and abutments, but exclude any approach works.

	Span	Unit			
Road Bridges					
Reinforced in-situ concrete viaduct *(including excavation; reinforcement; formwork; concrete; bearings; expansion joints; deck waterproofing; deck finishings; P1 parapet.)*					
per m² of deck maximum span between piers or abutments :	15 m	m²	1204	to	1356
	20 m	m²	1146	to	1204
	25 m	m²	1110	to	1146
Reinforced concrete bridge with precast beams *(including excavation; reinforcement; formwork; concrete; bearings; expansion joints; deck waterproofing; deck finishings; P1 parapet).*					
per m² of deck maximum span between piers or abutments :	12 m	m²	1286	to	1356
	17 m	m²	1204	to	1286
	22 m	m²	1146	to	1239
	27 m	m²	1099	to	1181
Reinforced concrete bridge with pre-fabricated steel beams *(including excavation; reinforcement; formwork; concrete; bearings; expansion joints; deck waterproofing; deck finishings; P1 parapet.)*					
per m² of deck maximum span between piers or abutments :	20 m	m²	1239	to	1321
	30 m	m²	1169	to	1239
	40 m	m²	1134	to	1204
Footbridges					
Reinforced in-situ concrete with precast beams *(including excavation; reinforcement; formwork; concrete; bearings; expansion joints; deck waterproofing; deck finishings; P6 parapet.)*					
widths up to 6 m wide; per m² of deck maximum span between piers or abutments :	5 m	m²	1110	to	1146
	10 m	m²	1029	to	1075
	20 m	m²	1075	to	1110
Structural steel bridge with concrete foundations					
width up to 4 m wide per m² of deck maximum span :	10 m	m²	1026	to	1086
	12 m	m²	1038	to	1086
	16 m	m²	1038	to	1098
	20 m	m²	1086	to	1110
Timber footbridge (stress graded with concrete piers)					
per m² of deck maximum span :	12 m	m²	923	to	947
	18 m	m²	947	to	982

5. HIGHWAY WORKS

The following prices are the approximate costs per metre run of roadway, and are based on information from a number of sources including engineers estimates, tenders, final account values etc on a large number of highways contracts.

Motorway and All Purpose Road prices include for earthworks, structures, drainage, pavements, line markings, reflective studs, footways signs, lighting, motorway communications, fencing and barrier works as well as allowance for accommodation works, statutory undertakings and landscaping as appropriate to the type and location of the carriageway. The earthworks elements can be adjusted by reference to factors detailed at the end of this sub-section.

Motorway and All Purpose Road prices do NOT include for the cost of associated features such as interchanges side roads, underbridges, overbridges, culverts, sub-ways, gantries and retaining walls. These are shown separately beside the cost range for each road type, based on statistical frequency norms.

	Unit	Range £			Feature £
MOTORWAYS					
The following costs are based on a 850 mm construction comprising 40 mm wearing course, 60 mm base course, 250 mm road base, sub-base and 350 mm capping layer; central reserve incorporating two 0.7 m wide hardstrips; no footpaths or cycle paths included					
Rural motorways					
grassed central reserve and outer berms; no kerbs or edgings; assumption that 30% of excavated material is unsuitable for filling; costs allow for forming embankments for 50% of highway length average 4.70 m high and 50% of length in cuttings average 3.90 m deep; accommodation fencing each side; allowance of 25% of length having crash barriers and 20% of length having lighting					
dual two lane (D2M_R); 25.60 m overall width; each carriageway 7.30 m with 3.30 m hard shoulder; 4.00 m central reserve	m	2,019	to	2,468	1,558
dual three lane (D3M_R); 32.60 m overall width; each carriageway 11.00 m with 3.30 m hard shoulder; 4.00 m central reserve	m	2,428	to	2,968	1,689
dual four lane (D4M_R); 39.80 m overall width; each carriageway 14.60 m with 3.30 m hard shoulder; 4.00 m central reserve	m	2,826	to	3,453	1,821
Urban motorways					
hard paved central reserve and outer berms; precast concrete kerbs; assumption that 30% of excavated material is unsuitable for filling; costs allow for forming embankments for 50% of highway length average 3.20 m high and 50% of length in cuttings average 1.60 m accommodation fencing each side; allowance of 25% of length having crash barriers and 20% of length having lighting					
dual two lane (D2M_U); 23.10 m overall width; each carriageway 7.30 m with 2.75m hard shoulder; 3.00 E:Em central reserve	m	1,946	to	2,379	2,802
dual three lane (D3M_U); 30.50 m overall width; each carriageway 11.00 m with 2.75 m hard shoulder; 3.00 m central reserve	m	2,329	to	2,846	2,949

dual four lane (D4M_U); 37.70 m overall width; each carriageway 14.60 m with 2.75 m hard shoulder; 3.00 m central reserve	m	2,701	to	3,301	3,096

ALL PURPOSE ROADS

The following costs are based on a 800 mm construction comprising 40 mm wearing course, 60 mm base course, 200 mm road base, 150 mm sub-base and 350 mm capping layer; no footpaths or cycle paths included

Rural all purpose roads

grassed central reserve; no kerbs or edgings; assumption that 30% of excavated material is unsuitable for filling; costs allow for forming embankments for 50% of highway length average 4.00m high and 50% of length in cuttings average 3.75m deep; allowance of 25% of length having crash barriers and 20% of length having lighting

dual two lane (D2AP_R); 25.60m overall width; each carriageway 7.30m ; 4.00m central reserve	m	1,580	to	1,932	1,196
dual three lane (D3AP_R); 32.60m overall width; each carriageway 11.00m ; 4.00m central reserve	m	1,953	to	2,387	1,291

Urban all purpose roads

hard paved central reserve; precast concrete kerbs; assumption that 30% of excavated material is unsuitable for filling; costs allow for forming embankments for 50% of highway length in average 2.2 m high and 50% of length in cuttings average 1.62 m deep; allowance of 25% of length having crash barriers and 20% of length having lighting

dual two lane (D2AP_U); 23.10m overall width; each carriageway 7.30 m with 2.75m hard shoulder; 3.00m central reserve	m	1,591	to	1,945	2,541
dual three lane (D3AP_U); 30.50m overall width; each carriageway 11.00 m C456with 2.75m hard shoulder; 3.00m central reserve	m	1,901	to	2,324	2,709

Other roads

Rural All-Purpose Roads					
Single carriageway all-purpose road (carriageway is 7.3 m wide)	m	1,075	to	1,311	555
Wide single carriageway all-purpose road (carriageway is 10.0 m wide)	m	1,181	to	1,441	555
Rural Link Roads					
Two lane link road (carriageway is 7.3 m wide)	m	1,205	to	1,477	543
Single lane link road (carriageway is 3.7 m wide)	m	0,780	to	0,945	343
Rural Motorway or Dual Carriageway Slip Roads					
Single carriageway slip road (carriageway is 5.0 m wide)	m	1,004	to	1,228	366
Urban Motorway or Dual Carriageway Sliproads					
Single carriageway slip road (carriageway is 6.0 m wide)	m	1,564	to	0,000	669
Wide single carriageway all purpose road with footway each side (carriageway is 10.0 m wide each footway is 3.0 m wide)	m	1,862	to	2,268	752
Cost of nominal 3.0 m cycle track to all-purpose roads (one side only)	m	155	to	179	

5. HIGHWAY WORKS - continued

Other roads - continued	Unit	Range £	Feature £
Urban Link Roads			
Two lane link road (carriageway is 7.3 m wide)	m	1,719 to 2,149	752
Single lane link road (carriageway is 3.7 m wide)	m	1,182 to 1,445	549

The following are approximate costs for the installation of roads and drains to serve as part of the development of infrastructure for Housing, Retail or Industrial development. The density (i.e. percentage of area used for roads, etc., is also given to enable adjustments). NB: excludes car parking.

Type A Construction = medium duty carriageway consisting of 100 mm surfacing, roadbase up to 115 mm subbase of 150 mm.

Type B Construction = heavy duty carriageway consisting of reinforced concrete slab 225 mm thick subbase 130 mm thick capping layer 280 mm thick.

Density of facility		Cost/unit £		Cost/hectare £		Density of road per hectare
per hectare	per acre	Type A	Type B	Type A	Type B	
5	(2)	3,740	4,300	18,700	21,510	2.00%
15	(6)	2,180	2,510	32,730	37,640	3.50%
20	(8)	1,780	2,040	35,540	40,870	3.80%
25	(10)	1,530	1,760	38,340	44,090	4.10%
30	(12)	1,400	1,610	42,080	48,390	4.50%
37	(15)	1,310	1,510	48,630	55,920	5.20%
50	(20)	1,220	1,400	60,780	69,900	6.50%

	Unit	£
Turning or passing bay = 35 m² overall area; suitable for cars, vans		
Type A construction	nr	3,280
Type B construction	nr	3,770
Turning or passing bay = 100 m² overall area; suitable for semi trailer		
Type A construction	nr	9,360
Type B construction	nr	10,760
Bus lay-by = 40 m² overall area		
Type B construction	nr	4,310
Parking lay-by = 200 m² overall area		
Type A construction	nr	18,710
Type B construction	nr	21,510
Vehicle crossing verge/footway/central reserve	m²	94
Footway construction (bit-mac plus edgings)	m²	23

6. CIVIL ENGINEERING WORKS SITE UTILITIES AND INFRASTRUCTURE

The following prices have been compiled for various services and provide average costs for site utilities and infrastructure works.

	Unit	To estate road	To major road
UNDERPASSES		£	£
Provision of underpasses to new roads, constructed as part of a road building programme			
pedestrian underpass 3 m wide x 2.5 m high	m	3,620	4,460
vehicle underpass 7 m wide x 5 m high	m	18,490	22,530
vehicle underpass 14 m wide x 5 m high	m	-	45,040
ROAD CROSSINGS			
Costs include road markings, beacons, lights, signs, advance danger signs, etc.			
4 way traffic signal installation	nr	40,980	57,860
zebra crossing	nr	4,630	5,180
pelican crossing	nr	17,970	19,330
pedestrian guard railing	m	117	141

STREET FURNITURE

There is an almost infinite variety of items with which local authorities and private developers can enhance the roadside and public areas, including statutory requirements such as lighting. It is therefore impossible to price all the different street furniture available. The following however gives a selection of the more common items:-

	Unit	£
reflectorised traffic signs 0.25-0.75 m² area on steel post	nr	99 to 174
internally illuminated traffic signs (dependent on area)	nr	205 to 281
externally illuminated traffic signs (up to 4 m²	nr	508 to 1,338
lighting to pedestrian areas and estates roads on 4-6 m columns with up to 70 W lamps	nr	228 to 338
lighting to main roads on 10-12 m columns with 250 W lamps	nr	526 to 643
lighting to main roads on 12-15 m columns with 400 W high pressure sodium lighting	nr	678 to 812
benches - hardwood and precast concrete	nr	199 to 269
litter bins: precast concrete	nr	199 to 228
hardwood slatted	nr	82 to 105
cast iron	nr	339
large aluminium	nr	596
concrete bollards	nr	64 to 123
steel bollards	nr	58 to 111
bus stops	nr	362
bus stops inc basic shelter	nr	830
pillar box on post	nr	298
telephone box	nr	3285

6. CIVIL ENGINEERING WORKS SITE UTILITIES AND INFRASTRUCTURE - continued

STREET FURNITURE - continued

	Unit	£
Hi mast radio/beacon - 60 m	nr	30,860
automatic barrier equipment	nr	21,859

FOOTPATHS AND PAVINGS

Costs include excavation, base course as necessary and precast concrete edgings on foundations to one or both sides

	Unit	£
bitumen macadam footpath	m^2	31
precast concrete paving flags	m^2	45
precast concrete block pavings	m^2	57
clay brick paving	m^2	70
granite setts	m^2	103
cobbled paving	m^2	82

SURFACE CAR PARKS

Surface car parking is the cheapest way of providing car parking. The cost of rooftop or basement car parking can be 10 to 15 times the cost of the surface car parking. The vogue in car parking for all except industrial schemes is for concrete block pavers, with different colours marking out parking bays and zones.

Guideline figures for parking requirements are shown below:-

	one car space per :	
offices	22 to 25	m^2 gross floor area
industrial - factories	45 to 55	m^2 gross floor area
- warehouses	200	m^2 gross floor area
shops	20 to 25	m^2 gross floor area
superstores	10	m^2 gross floor area
cinemas, theatres	3 to 5	seats
hospital	3	hospital beds
residential	1 to 2	dwellings

For surface level car parking the area required per car will not generally show much variation. Site shape, position of the building on the site, and parking configuration will be the main determinants of the area to be allowed per car. A fairly low range of 20-23 m^2 / car will usually suffice.

Typical costs for surface car parking including lighting and drainage are illustrated below:-

	£/m^2	£/car
Tarmacadam surfaced, marked out with thermoplastic road paint	47	1,110 to 1,282
Interlocking or herringbone concrete block paving, marked out with coloured blocks	51	1,264 to 1,453
Grasscrete precast concrete units filled with top soil and grass seeded	55	1,335 to 1,536

Note: Costs include forecourts, aprons and access areas but not approach roads.

SERVICES

Costs of services to a site are built up of connection charges and service runs. This can vary significantly depending upon the availability or otherwise of a suitably sized main in the neighbourhood of the site. However, typical service charges for an estate of 200 houses might be as follows:-

	Charge per house £	
Water	500 to 1,000	
Electric	300 if all electric	(plus cost of substation £ 12,000 - 17,000 total)
	500 if all gas	(plus cost of substation £ 12,000 - 17,000 total)
Gas	500 to 600	(plus cost of governing station £ 11,000 total)
Telephone	150	
Sewerage	300 to 500	

	Unit	£
Mains laid in trenches including excavation and filling		
Water main		
75 mm PVC-U main in 150 mm ductile iron pipe as duct	m	100
Electric main		
600/1000 volt cables. Two core 25 mm cable including 100 mm clayware duct	m	41
Gas main		
150 mm ductile or cast iron gas pipe	m	75
Telephone		
British Telecom installation in 110 mm PVC-U duct	m	31
Drainage		
100/150 mm clay or PVC-U pipes on granular bed and surround up to 3 m deep	m	51
100/150 mm cast iron pipes on concrete beds up to 3 m deep	m	110
300 mm clay drain pipe on granular bed and surround up to 3 m deep	m	110
450 mm concrete pipe on concrete beds up to 3 m deep	m	140
900 mm diameter concrete sewage pipe on granular bed up to 3.5 m deep	m	200
Brick manholes in commons, rendered internally, clay channel with three branches, concrete cover slab, cast iron cover and frame 900 - 1500 mm deep	nr	1080
Precast concrete manholes with precast concrete rings up to 1500 mm diameter, channel with three branches, concrete cover slab and frame 900-1500 mm deep	nr	1360
Vitrified clay or precast concrete road gully 450 x 900 mm deep with concrete surround, brick seating and cast iron grating	nr	230

7. LANDSCAPING

Landscaping as a subject matter is sufficiently large to fill a book (see Spons Landscaping and External Works Price Book). However, there are certain items that arise on a majority of projects and other discrete items for which indicative costs can be produced. The following rates include for normal surface excavation but exclude bulk excavation, levelling or earth shifting and land drain provision.

Soft landscaping

	Unit	£
cultivate ground, remove rubbish and plant with grass seed	m²	5.65
ditto and turf	m²	7.32
shrubbed areas	m²	47.10
shrubbed areas including allowance for small trees	m²	62.30
standard tree in tree pit including stake	nr	52
ditto but with tree guard and precast tree grid slabs	nr	185

Sports pitches

The provision of sports facilities will involve different techniques of earth shifting and cultivation and usually will be carried out by specialist landscaping contractors.

Costs include for cultivating ground, bringing to appropriate levels for the specified game, applying fertiliser, weedkiller, seeding and rolling and white line marking with nets, posts etc as required.

	Unit	£
football pitch (114 x 72 m)	nr	17,100
cricket outfield (160 x 142 m)	nr	57,800
cricket square (20 x 20 m) including imported marl or clay loam bringing to accurate levels, seeding with cricket square type grass	nr	5,300
bowling green (38 x 38 m) rink, including french drain and gravel path on four sides	nr	20,900
grass tennis courts 1 court (35 x 17 m) including bringing to accurate levels, chain link perimeter fence and gate, tennis posts and net	nr	21,300
two grass tennis courts (35 x 32 m) ditto	pair	36,200
artificial surface tennis courts (35 x 17 m) including chain link fencing, gate, tennis posts and net	nr	16,800
two courts (45 x 32 m) ditto	pair	31,500
artificial football pitch including sub-base, bitumen macadam open textured base and heavy duty astro-turf carpet	nr	332,100
golf putting green	hole	1,730
pitch and putt course	hole	5,300 to 8,100
full length golf course, full specifications inc watering system	hole	18,800 to 35,600
championship course	hole	up to 132,000

Parklands

As with all sports pitches, parklands will involve different techniques of earth shifting and cultivation.
The following rates include for normal surface excavation.

	Unit	£
parklands, including cultivating ground applying fertiliser etc. and seeding with parks type grass	ha	16,250
general sportsfield	ha	19,340
lakes including excavation up to 10 m deep, laying 1.5 mm thick butyl rubber sheet and spreading top soil evenly on top to depth of 300 mm		
under 1 ha in area	ha	355,400
between 1 and 5 ha in area	ha	332,100
extra for planting aquatic plants in lake top soil	m²	55

Playground Equipment

Modern swings with flat rubber safety seats:		
four seats, two bays	nr	1,300
Stainless steel slide, 3.40 m long	nr	1,440
Climbing frame - igloo type 3.20 x 3.75 m on plan x 2.00 m high	nr	1,500
Seesaw comprising timber plank on sealed ball bearings 3960 x 230 x 70 mm thick	nr	1,040
Wicksteed tumbleguard type safety surfacing around play equipment	m²	88
Bark particles type safety surfacing 150 mm thick on hardcore bed	m²	11

Land Drainage

The above rates exclude provision of any land drainage. If land drainage is required on a project, the propensity of the land to flood will decide the spacing of the land drains. However, some indicative figures can be given for land drainage.

Costs include for excavation and backfilling of trenches and laying agricultural clay drain pipes with 75 mm diameter lateral runs average 600 mm deep and 100 mm diameter main runs average 750 mm deep.

land drainage to parkland with laterals at 30 m centres and main runs at 100 m centres	ha	3,250
land drainage to sportsfields with laterals at 10 m centres and main runs at 33 mm centres	ha	9,430

8. TEMPORARY WORKS

Bailey bridges

	Unit	£
Installation of temporary Bailey bridges:		
(including temporary concrete abutments; erection maintenance; dismantling)		
span up to 10 m :		
Hire costs (for 52 weeks)		13,020
Delivery/collect		1,120
Erect/dismantle		4,160
Concrete abutments		5,110
Demolish after dismantling		840
Allowance for maintenance, etc		2,500
	nr	26,750
span 15 m	nr	30,280
span 20 m	nr	38,810
span 25 m	nr	42,200

Cofferdams

Installation of cofferdams (based on driven steel sections with recovery value) :

(including all plant, for installation and dismantling; loss of materials; pumping and maintenance excluding excavation and disposal of material - backfilling on completion.)

Cost range based on 12 weeks installation

	Unit	£ range
Depth of drive up to 5 m, diameter or side length		
up to 2 m	nr	3,700 to 3,900
up to 10 m	nr	18,000 to 20,100
up to 20 m	nr	31,900 to 40,100
Depth of drive 5 - 10 m, diameter or side length		
up to 2 m	nr	5,300 to 7,500
up to 10 m	nr	23,500 to 37,500
up to 20 m	nr	43,200 to 74,900
Depth of drive 10 - 15 m, diameter or side length		
up to 2 m	nr	7,500 to 10,400
up to 10 m	nr	36,400 to 52,100
up to 20 m	nr	68,000 to 104,300
Depth of drive 15 - 20 m, diameter or side length		
up to 2 m	nr	12,800 to 16,000
up to 10 m	nr	36,300 to 64,200
up to 20 m	nr	91,000 to 128,200

Above based on soft-medium ground conditions

add for medium-hard	+ 20%
hard but not rock	+ 33%

Access Scaffolding

The following are guideline costs for the hire and erection of proprietary scaffold systems (tube and coupling). Costs are very general and are based on a minimum area of 360 m² at 1.80m deep.

	Unit	£
1. Approximate hire (supply and fix) of patent scaffold per 4 week (cost dependant upon length of hire, quality of system, number of toe boards, handrails etc.)	m²	4.7 to 6.2
Based on this a typical cost of a 60 x 10m (600 m²) area for 8 weeks would be about £4,200		
2. Approximate hire (supply and fix of mobile access towers per 4 week hire (refer also to plant hire in Section 3)	m²	15.9 to 23.8
3. Additional costs of Pole ladder access, per 4.0 m high, per 4 week	nr	6.0 to 6.8
4. Additional cost of stair towers extra overthe cost of scaffold system per 2 m rise, per 4 week	m²	66.90
5. Additional cost of hoarding around base perimeter (using multi use ply sheeting) per 4 week period	m²	0.94 to 1.64
6. Additional cost of polythene debris netting (no re-use)	m²	0.18
7. Additional cost of Monaflex 'T' plus weather-proof sheeting including anchors and straps (based on 3 uses)	m²	0.88

Erection of scaffolding system is based on 3 men erecting a 16 m² bay in about 1 hour and dismantling the same in 20 minutes (i.e. experienced scaffolders)

Note: Although scaffolding is essentially plant hire, allowance must be made for inevitable loss and damage to fittings, for consumables used and for maintenance during hire periods.

Earthwork Support

The following are comparative costs for earthwork and trench support based on the hire of modular systems (trench box hydraulic) with an allowance for consumable materials (maximum 5 day hire allowance).

	Unit	£
Earthwork support not exceeding 1m deep, distance between opposing faces not exceeding 2m	m²	4.2
Earthwork support not exceeding 1m deep distance between opposing faces 2-4m	m²	7.2
Earthwork support not exceeding 2m deep distance between opposing faces not exceeding 2m	m²	8.1

8. TEMPORARY WORKS - continued

Earthwork Support - continued

	Unit	£
Earthwork support not exceeding 2m deep distance between opposing faces 2-4m	m^2	14.4
Earthwork support not exceeding 4m deep distance between opposing faces not exceeding 2m	m^2	20.5
Earthwork support not exceeding 4m deep distance between opposing face 2-4m	m^2	29.6

For larger excavations requiring Earthwork support refer to cofferdam estimates; or sheet piling within unit cost sections.

The following are approximate weekly hire costs for a range of basic support equipment used on site (see also Plant Costs Section - Part 3).

	Unit	£/week
Steel Sheet piling		
AU or AZ series section	tonne	24.10 to 30.12
Trench Sheeting		
Standard overlapping sheets	m^2	1.12 to 1.42
Interlocking type	m^2	2.19 to 3.14
Heavy duty overlapping sheets		
6 mm thick	m^2	3.60 to 4.12
8 mm thick	m^2	3.72
driving cap	nr	10.04
extraction clamp	nr	20.08
Trench box with hydraulic walings - as above		
Trench struts		
No. 0 - 0.3 - 0.40 m	nr	0.47
No. 1 - 0.5 - 0.75 m	nr	0.53
No. 2 - 0.7 - 1.14 m	nr	0.53
No. 3 - 1.0 - 1.75 m	nr	0.65

PART 13

Outputs

This part lists a selection of OUTPUT CONSTANTS for use within various areas of Civil Engineering Work.

DISPOSAL OF EXCAVATED MATERIALS

Outputs Per Hundred Cubic Metres

Tipper Capacity and Length of Haul	Driver (hours)	Attendant Labour (hours)	Number of cycles	Average Speed (km/hr)	Cycle time (minutes)				
					Loading	Haul	Discharge	Return	Total
3 m³ tipper:									
1 km haul	8.8	0.0	43.3	10	0.70	6.0	0.7	5.5	12.2
5 km haul	26.3	0.0	43.3	16	0.70	18.8	0.7	17.0	36.5
9 km haul	41.9	0.0	43.3	18	0.70	30.0	0.7	27.3	58.0
12 km haul	52.7	0.0	43.3	19	0.70	37.9	0.7	34.4	73.0
15 km haul	65.8	0.0	43.3	19	0.70	47.4	0.7	43.1	91.2
8 m³ tipper:									
1 km haul	10.6	0.8	16.3	3	2.0	20.0	0.9	18.2	39.1
5 km haul	19.7	0.8	16.3	8	2.0	37.5	0.9	34.1	72.5
9 km haul	25.7	0.8	16.3	11	2.0	49.1	0.9	44.6	94.6
12 km haul	26.9	0.8	16.3	14	2.0	51.4	0.9	46.8	99.1
15 km haul	31.4	0.8	16.3	15	2.0	60.0	0.9	54.5	115.4
12 m³ tipper:									
1 km haul	7.1	0.7	10.8	3	2.90	20.0	1.1	18.2	39.3
5 km haul	13.1	0.7	10.8	8	2.90	37.5	1.1	34.1	72.7
9 km haul	17.1	0.7	10.8	11	2.90	49.1	1.1	44.6	94.8
12 km haul	17.9	0.7	10.8	14	2.90	51.4	1.1	46.8	99.3
15 km haul	20.8	0.7	10.8	15	2.90	60.0	1.1	54.5	115.6
15 m³ tipper:									
1 km haul	5.8	0.8	8.7	3	3.70	20.0	1.5	18.2	39.7
5 km haul	10.6	0.8	8.7	8	3.70	37.5	1.5	34.1	73.1
9 km haul	13.8	0.8	8.7	11	3.70	49.1	1.5	44.6	95.2
12 km haul	14.5	0.8	8.7	14	3.70	51.4	1.5	46.8	99.7
15 km haul	16.8	0.8	8.7	15	3.70	60.0	1.5	54.5	116.0

Man hours include round trip for tipper and driver together with attendant labour for positioning during loading and unloading.

The number of cycles are based on the stated heaped capacity of the tipper divided into the total volume of 100 m³ to be moved, being multiplied by a bulking factor of x 1.30 to the loose soil volume.

The average speeds are calculated assuming that the vehicles run on roads or on reasonably level firm surfaces and allow for acceleration and deceleration with the return empty journey being say 10% faster than the haul.

The cycle time shows in detail the time spent being loaded (calculated using a 1.5 m³ loader with a cycle time of 22 seconds), haul journey, discharge (turning / manoeuvring / tipping) and return journey.

BREAKING OUT OBSTRUCTIONS BY HAND

Breaking out pavements, brickwork, concrete and masonry by hand and pneumatic breaker

Description	Unit	By hand using picks, shovels & points	Using Compressor			
			7 m³ Compressor (2 Tool)		10 m³ Compressor (3 Tool)	
		Labour	Compressor	Labour	Compressor	Labour
Break out bitmac surfaces on sub-base or hardcore						
75 mm thick	m²/hr	1.00	25	13	50	17
100 mm thick	m²/hr	1.00	20	10	33	11
Break out asphalt roads on hardcore:						
150 mm thick	m²/hr	0.60	8	4	11	4
225 mm thick	m²/hr	0.50	6	3	8	3
300 mm thick	m²/hr	0.40	4	2	7	2
Remove existing set paving	m²/hr	0.80	10	5	17	6
Break out brickwork in cement mortar: 215 mm thick	m²/hr	0.20	3	2	5	2
Break out concrete in areas						
100 mm thick	m²/hr	0.90	9	5	14	5
150 mm thick	m²/hr	0.50	7	4	10	4
225 mm thick	m²/hr	0.30	4	2	6	2
300 mm thick	m²/hr	0.20	3	1	4	1
Break out reinforced concrete	m³/hr	0.02	0.40	0.20	0.60	0.20
Break out sandstone	m³/hr	0.03	0.50	0.30	0.80	0.30

Loading loose materials and items by hand

Material	Unit	Loading into Vehicles	
		Tonne	m^3
Bricks	hr	1.7	2.9
Concrete, batches	hr	1.4	1.2
Gulley grates and frames	hr	1.0	
Kerb	hr	1.1	
Paving slabs	hr	0.9	
Pipes, concrete and clayware	hr	0.9	
Precast concrete items	hr	0.9	
Soil	hr	1.4	2.0
Steel reinforcement	hr	0.8	
Steel sections, etc	hr	1.0	
Stone and aggregates:			
bedding material	hr	1.5	1.8
filter/subbase	hr	1.3	1.4
rock fill (6" down)	hr	1.3	1.3
Trench planking and shoring	hr	0.9	2.9

CONCRETE WORK

Placing ready mixed concrete in the works

Description	Labour Gang (m^3 per hour)
MASS CONCRETE	
Blinding	
150 mm thick	5.50
150 - 300 mm thick	6.25
300 - 500 mm thick	7.00
Bases and oversite concrete	
not exceeding 150 mm thick	5.00
not exceeding 300 mm thick	5.75
not exceeding 500 mm thick	6.75
exceeding 500 mm thick	7.00
REINFORCED CONCRETE	
Bases	
not exceeding 300 mm thick	5.50
not exceeding 500 mm thick	6.25
exceeding 500 mm thick	6.75
Suspended slabs (not exceeding 3m above pavement level)	
not exceeding 150 mm thick	3.75
not exceeding 300 mm thick	4.75
exceeding 300 mm thick	5.75
Walls and stems (not exceeding 3m above pavement):	
not exceeding 150 mm thick	3.50
not exceeding 300 mm thick	4.50
exceeding 300 mm thick	5.00
Beams, columns and piers (not exceeding 3m above pavement):	
sectional area not exceeding 0.03 m^2	2.00
sectional area not exceeding 0.03 - 1.0 m^2	2.50
sectional area exceeding 1.0 m^2	3.50

Fixing bar reinforcement

All bars delivered to site cut and bent and marked, including craneage and hoisting (maximum height 5 m).

Description		up to 6 mm		7 to 12 mm		13 to 19 mm		over 19 mm	
(fix only)	Unit	steelfixer	labourer	steelfixer	labourer	steelfixer	labourer	steelfixer	labourer
Straight round bars									
to beams, floors, roofs and walls	t / hr	0.03	0.03	0.04	0.04	0.06	0.06	0.08	0.08
to braces, columns, sloping roofs and battered walls	t / hr	0.01	0.01	0.02	0.02	0.03	0.03	0.05	0.05
Bent round bars									
to beams, floors, roofs and walls	t / hr	0.02	0.02	0.03	0.03	0.03	0.03	0.04	0.04
to braces, columns, sloping roofs and battered walls	t / hr	0.01	0.01	0.01	0.01	0.02	0.02	0.03	0.03
Straight, indented or square bars									
to beams, floors, roofs and walls	t / hr	0.02	0.02	0.04	0.04	0.05	0.05	0.07	0.07
to braces, columns, sloping roofs and battered walls	t / hr	0.01	0.01	0.02	0.02	0.02	0.02	0.04	0.04
Bent, indented or square bars									
to beams, floors, roofs and walls	t / hr	0.02	0.02	0.02	0.02	0.03	0.03	0.03	0.03
to braces, columns, sloping roofs and battered walls	t / hr	0.01	0.01	0.01	0.01	0.02	0.02	0.02	0.02
(Average based on Gang D)		0.125		0.148		0.163		0.225	

CONCRETE WORK - continued

Erecting formwork to beams and walls

Erect and strike formwork	**Unit**	**Joiner**	**Labourer**
Walls - vertical face:(first fix)			
up to 1.5 m	m^2/hr	1.7	0.8
1.5 to 3.0 m	m^2/hr	1.4	0.7
3.0 to 4.5 m	m^2/hr	1.2	0.6
4.5 to 6.0 m	m^2/hr	1.0	0.5

Erecting formwork to slabs

Erect and strike formwork	**Unit**	**Joiner**	**Labourer**
Horizontal flat formwork at heights:(first fix)			
up to 3.0 m	m^2/hr	1.11	1.11
3.0 to 3.6 m	m^2/hr	1.05	1.05
3.6 to 4.2 m	m^2/hr	1.00	1.00
4.2 to 4.8 m	m^2/hr	0.95	0.95
4.8 to 5.4 m	m^2/hr	0.90	0.90
5.4 to 6.0 m	m^2/hr	0.83	0.83

Multipliers for formwork

Description	**Multiplier**
Walls built to batter	1.20
Walls built circular to large radius	1.70
Walls built circular to small radius	2.10
Formwork used once	1.00
Formwork used twice, per use	0.85
Formwork used three times, per use	0.75
Formwork used four times, per use	0.72
Formwork used five times, per use	0.68
Formwork used six times, per use	0.66
Formwork used seven or more times, per use	0.63
Formwork to slope not exceeding 45 degrees from horizontal	1.25

DRAINAGE

Laying and jointing flexible-jointed clayware pipes

Diameter of pipe in mm	Drainage Gang	In trench not exceeding 1.5 m	In trench not exceeding 3 m	In trench 3 - 4.5 m
Pipework				
100 mm	m/hr	10	8	7
150 mm	m/hr	7	6	5
225 mm	m/hr	5	4	3
300 mm	m/hr	4	3	2
375 mm	m/hr	3	2	2
450 mm	m/hr	2	2	1
Bends				
100 mm	nr/hr	20	17	14
150 mm	nr/hr	17	14	13
225 mm	nr/hr	13	10	8
300 mm	nr/hr	10	8	7
375 mm	nr/hr	7	6	5
450 mm	nr/hr	3	3	2
Single junctions				
100 mm	nr/hr	13	10	8
150 mm	nr/hr	7	6	5
225 mm	nr/hr	6	5	4
300 mm	nr/hr	4	3	3
375 mm	nr/hr	3	2	2
450 mm	nr/hr	2	2	1

DRAINAGE - continued

Precast concrete manholes in sections

Description	Unit	Pipelayer	Labourer
Place 675 mm dia shaft rings	m/hr	1.00	0.30
Place 900 mm manhole rings	m/hr	0.60	0.20
Place 1200 mm manhole rings	m/hr	0.40	0.10
Place 1500 mm manhole rings	m/hr	0.30	0.10
Place 900/675 mm tapers	nr/hr	1.00	0.30
Place 1200/675 mm tapers	nr/hr	0.60	0.20
Place 1500/675 mm tapers	nr/hr	0.50	0.20
Place cover slabs to 675 mm rings	nr/hr	2.50	0.80
Place cover slabs to 900 mm rings	nr/hr	2.00	0.70
Place cover slabs to 1200 mm rings	nr/hr	1.40	0.50
Place cover slabs to 1500 mm rings	nr/hr	0.90	0.30
Build in pipes and make good base:			
150 mm diameter	nr/hr	4.00	-
300 mm diameter	nr/hr	2.00	-
450 mm diameter	nr/hr	1.00	-
Benching 150 mm thick	m^2/hr	1.20	1.20
Benching 300 mm thick	m^2/hr	0.60	0.60
Render benching 25 mm thick	m^2/hr	1.10	1.10
Fix manhole covers frame	nr/hr	1.30	1.30

PART 14

Tables and Memoranda

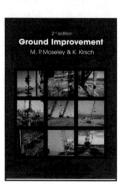

QUICK REFERENCE CONVERSION TABLES

	Imperial		Metric	
1. LINEAR				
0.039	in	1	mm	25.4
3.281	ft	1	metre	0.305
1.094	yd	1	metre	0.914
2. WEIGHT				
0.020	cwt	1	kg	50.802
0.984	ton	1	tonne	1.016
2.205	lb	1	kg	0.454
3. CAPACITY				
1.760	pint	1	litre	0.568
0.220	gal	1	litre	4.546
4. AREA				
0.002	in^2	1	mm^2	645.16
10.764	ft^2	1	m^2	0.093
1.196	yd^2	1	m^2	0.836
2.471	acre	1	ha	0.405
0.386	$mile^2$	1	km^2	2.59
5. VOLUME				
0.061	in^3	1	cm^3	16.387
35.315	ft^3	1	m^3	0.028
1.308	yd^3	1	m^3	0.765
6. POWER				
1.310	HP	1	kW	0.746

CONVERSION FACTORS - METRIC TO IMPERIAL

Multiply Metric	Unit	By	To obtain Imperial	Unit
Length				
kilometre	km	0.6214	statute mile	ml
metre	m	1.0936	yard	yd
centimetre	cm	0.0328	foot	ft
millimetre	mm	0.0394	inch	in
Area				
hectare	ha	2.471	acre	
square kilometre	km^2	0.3861	square mile	sq ml
square metre	m^2	10.764	square foot	sq ft
square metre	m^2	1550	square inch	sq in
square centimetre	cm^2	0.155	square inch	sq in
Volume				
cubic metre	m^3	35.336	cubic foot	cu ft
cubic metre	m^3	1.308	cubic yard	cu yd
cubic centimetre	cm^3	0.061	cubic inch	cu in
cubic centimetre	cm^3	0.0338	fluid ounce	fl oz
Liquid volume				
litre	l	0.0013	cubic yard	cu yd
litre	l	61.02	cubic inch	cu in
litre	l	0.22	Imperial gallon	gal
litre	l	0.2642	US gallon	US gal
litre	l	1.7596	pint	pt
Mass				
metric tonne	t	0.984	long ton	lg ton
metric tonne	t	1.102	short ton	sh ton
kilogram	kg	2.205	pound, avoirdupois	lb
gram	g or gr	0.0353	ounce, avoirdupois	oz

Unit mass

kilograms/cubic metre	kg/m³	0.062	pounds/cubic foot	lbs/cu ft
kilograms/cubic metre	kg/m³	1.686	pounds/cubic yard	lbs/cu yd
tonnes/cubic metre	t/m³	1692	pound/cubic yard	lbs/cu yd
kilograms/sq centimetre	kg/cm²	14.225	pounds/square inch	lbs/sq in
kilogram-metre	kg.m	7.233	foot-pound	ft-lb

Force

meganewton	MN	9.3197	tons force	tonf
kilonewton	kN	225	pounds force	lbf
newton	N	0.225	pounds force	lbf

Pressure and stress

meganewton per square metre	MN/m²	9.3197	tons force/square foot	tonf/ft²
kilopascal	kPa	0.145	pounds/square inch	psi
bar		14.5	pounds/square inch	psi
kilograme metre	kgm	7.2307	foot-pound	ft-lb

Energy

kilocalorie	kcal	3.968	British thermal unit	Btu
metric horsepower	CV	0.9863	horse power	hp
kilowatt	kW	1.341	horse power	hp

Speed

kilometres/hour	km/h	0.621	miles/hour	mph

CONVERSION FACTORS - IMPERIAL TO METRIC

Multiply Imperial	Unit	By	To obtain metric	Unit
Length				
statute mile	ml	1.609	kilometre	km
yard	yd	0.9144	metre	m
foot	ft	0.3048	metre	m
inch	in	25.4	millimetre	mm
Area				
acre	acre	0.4047	hectare	ha
square mile	sq ml	2.59	square kilometre	km^2
square foot	sq ft	0.0929	square metre	m^2
square inch	sq in	0.0006	square metre	m^2
square inch	sq in	6.4516	square centimetre	cm^2
Volume				
cubic foot	cu ft	0.0283	cubic metre	m^3
cubic yard	cu yd	0.7645	cubic metre	m^3
cubic inch	cu in	16.387	cubic centimetre	cm^3
fluid ounce	fl oz	29.57	cubic centimetre	cm^3
Liquid volume				
cubic yard	cu yd	764.55	litre	l
cubic inch	cu in	0.0164	litre	l
Imperial gallon	gal	4.5464	litre	l
US gallon	US gal	3.785	litre	l
US gallon	US gal	0.833	Imperial gallon	gal
pint	pt	0.5683	litre	l
Mass				
long ton	lg ton	1.016	metric tonne	tonne
short ton	sh ton	0.907	metric tonne	tonne
pound	lb	0.4536	kilogram	kg
ounce	oz	28.35	gram	g

CONVERSION TABLES

Unit mass				
pounds/cubic foot	lb/ cu ft	16.018	kilograms/cubic metre	kg/m^3
pounds/cubic yard	lb/cu yd	0.5933	kilograms /cubic metre	kg/m^3
pounds/cubic yard	lb/cu yd	0.0006	tonnes/cubic metre	t/m^3
foot-pound	ft-lb	0.1383	kilogram-metre	kg.m
Force				
tons force	tonf	0.1073	meganewton	MN
pounds force	lbf	0.0045	kilonewton	kN
pounds force	lbf	4.45	newton	N
Pressure and stress				
pounds/square inch	psi	0.1073	kilogram/sq. centimetre	kg/cm^2
pounds/square inch	psi	6.89	kilopascal	kPa
pounds/square inch	psi	0.0689	bar	
foot-pound	ft-lb	0.1383	kilogram metre	kgm
Energy				
British Thermal Unit	Btu	0.252	kilocalorie	kcal
horsepower (hp)	hp	1.014	metric horsepower	CV
horsepower (hp)	hp	0.7457	kilowatt	kW
Speed				
miles/hour	mph	1.61	kilometres/hour	km/h

CONVERSION TABLES – continued

Length

Millimetre	mm	1 in	=	25.4 cm	1 mm	=	0.0394 in
Centimetre	cm	1 in	=	2.54 cm	1 cm	=	0.3937 in
Metre	m	1 ft	=	0.3048 m	1 m	=	3.2808 ft
		1 yd	=	0.9144 m	1 m	=	1.0936 yd
Kilometre	km	1 mile	=	1.6093 km	1 km	=	0.6214 mile

Note:

1 cm	=	10 mm	1 ft	=	12 in
1 m	=	100 cm	1 yd	=	3 ft
1 km	=	1,000 m	1 mile	=	1,760 yd

Area

Square millimetre	mm^2	1 in^2	=	645.2 mm^2	1 mm^2	=	0.0016 in^2
Square centimetre	cm^2	1 in^2	=	6.4516 cm^2	1 cm^2	=	1.1550 in^2
Square metre	m^2	1 ft^2	=	0.0929 m^2	1 m^2	=	10.764 ft^2
		1 yd^2	=	0.8361 m^2	1 m^2	=	1.1960 yd^2
Square Kilometre	km^2	1 $mile^2$	=	2.590 km^2	1 km^2	=	0.3861mile2

Note:

1 cm^2	=	100 m^2	1 ft^2	=	144 in^2
1 m^2	=	10,000 cm^2	1 yd^2	=	9 ft^2
1 km^2	=	100 hectares	1 $mile^2$	=	640 acres
			1 acre	=	4, 840 yd^2

Volume

Cubic Centimetre	cm^3	1 cm^3	=	0.0610 in^3	1 in^3	=	16.387 cm^3
Cubic Decimetre	dm^3	1 dm^3	=	0.0353 ft^3	1 ft^3	=	28.329 dm^3
Cubic metre	m^3	1 m^3	=	35.3147 ft^3	1 ft^3	=	0.0283 m^3
		1 m^3	=	1.3080 yd^3	1 yd^3	=	0.7646 m^3
Litre	L	1 L	=	1.76 pint	1 pint	=	0.5683 L
		1 L	=	2.113 US pt	1 pint	=	0.4733 US L

Note:

1 dm^3	=	1,000 cm^3	1 ft^3	=	1.728 in^3
1 m^3	=	1,000 dm^3	1 yd^3	=	27 ft^3
1 L	=	1 dm^3	1 pint	=	20 fl oz
1 HL	=	100 L	1 gal	=	8 pints

Mass

Milligram	mg	1 mg	=	0.0154 grain	1 grain	=	64.935 mg
Gram	g	1 g	=	0.0353 oz	1 oz	=	28.35 g
Kilogram	kg	1 kg	=	2.2046 lb	1 lb	=	0.4536 kg
Tonne	t	1 t	=	0.9842 ton	1 ton	=	1.016 t

Note:	1 g	=	1,000 mg	1 oz	=	437.5 grains
	1 kg	=	1000 g	1 lb	=	16 oz
	1 t	=	1,000 kg	1 stone	=	14 lb
				1 cwt	=	112 lb
				1 ton	=	20 cwt

Force

Newton	N	1lbf	=	4.448 N	1 kgf	=	9.807 N
Kilonewton	kN	1lbf	=	0.00448 kN	1 ton f	=	9.964 kN
Meganewton	MN	100 tonf	=	0.9964 MN			

Pressure and stress

Kilonewton per square metre	kN/m^2	1 lbf/in^2	=	6.895 kN/m^2
		1 bar	=	100 kN/m^2
Meganewton per square metre	MN/m^2	1 $tonf/ft^2$	=	107.3 kN/m^2 = 0.1073 MN/m^2
		1 kgf/cm^2	=	98.07 kN/m^2
		1 lbf/ft^2	=	0.04788 kN/m^2

Temperature

Degree Celcius ^{o}C

$$^{o}C = \frac{5 \times (^{o}F - 32)}{9} \qquad ^{o}F = \frac{(9 \times {}^{o}C) + 32}{5}$$

CONVERSION TABLES – continued

Metric Equivalents

1 km	=	1000 m
1 m	=	100 cm
1 cm	=	10 mm
1 km^2	=	100 ha
1 ha	=	10,000 m^2
1 m^2	=	10,000 cm^2
1 cm^2	=	100 mm^2
1 m^3	=	1,000 litres
1 litre	=	1,000 cm^3
1 metric tonne	=	1,000 kg
1 quintal	=	100 kg
1 N	=	0.10197 kg
1 kg	=	1000 g
1 g	=	1000 mg
1 bar	=	14.504 psi
1 cal	=	427 kg.m
1 cal	=	0.0016 cv.h
torque unit	=	0.00116 kw.h
1 CV	=	75 kg.m/s
1 kg/cm^2	=	0.97 atmosphere

English Unit Equivalents

1 mile	=	1760 yd
1 yd	=	3 ft
1 ft	=	12 in
1 sq mile	=	640 acres
1 acre	=	43,560 sq ft
1 sq ft	=	144 sq in
1 cu ft	=	7.48 gal liq
1 gal	=	231 cu in
	=	4 quarts liq
1 quart	=	32 fl oz
1 fl oz	=	1.80 cu in
	=	437.5 grains
1 stone	=	14 lb
1 cwt	=	112 lb
1 sh ton	=	2000 lb
1 lg ton	=	2240 lb
	=	20 cwt
1 lb	=	16 oz, avdp
1 Btu	=	778 ft lb
	=	0.000393 hph
	=	0.000293 kwh
1 hp	=	550 ft-lb/sec
1 atmosph	=	14.7 lb/in^2

CONVERSION TABLES – continued

Power Units

kW	=	Kilowatt
HP	=	Horsepower
CV	=	Cheval Vapeur (Steam Horsepower)
	=	French designation for Metric Horsepower
PS	=	Pferderstarke (Horsepower)
	=	German designation of Metric Horsepower
1 HP	=	1.014 CV = 1.014 PS
	=	.7457 kW
1 PS	=	1 CV = .9863 HP
	=	.7355 kW
1 kW	=	1.341 HP
	=	1.359 CV
	=	1.359 PS

SPEED CONVERSION

km/h	m/min	Mph	fpm
1	16.7	0.6	54.7
2	33.3	1.2	109.4
3	50.0	1.9	164.0
4	66.7	2.5	218.7
5	83.3	3.1	273.4
6	100.0	3.7	328.1
7	116.7	4.3	382.8
8	133.3	5.0	437.4
9	150.0	5.6	492.1
10	166.7	6.2	546.8
11	183.3	6.8	601.5
12	200.0	7.5	656.2
13	216.7	8.1	710.8
14	233.3	8.7	765.5
15	250.0	9.3	820.2
16	266.7	9.9	874.9
17	283.3	10.6	929.6
18	300.0	11.2	984.3
19	316.7	11.8	1038.9
20	333.3	12.4	1093.6
21	350.0	13.0	1148.3
22	366.7	13.7	1203.0
23	383.3	14.3	1257.7
24	400.0	14.9	1312.3
25	416.7	15.5	1367.0
26	433.3	16.2	1421.7
27	450.0	16.8	1476.4
28	466.7	17.4	1531.1
29	483.3	18.0	1585.7
30	500.0	18.6	1640.4

SPEED CONVERSION – continued

km/h	m/min	Mph	fpm
31	516.7	19.3	1695.1
32	533.3	19.9	1749.8
33	550.0	20.5	1804.5
34	566.7	21.1	1859.1
35	583.3	21.7	1913.8
36	600.0	22.4	1968.5
37	616.7	23.0	2023.2
38	633.3	23.6	2077.9
39	650.0	24.2	2132.5
40	666.7	24.9	2187.2
41	683.3	25.5	2241.9
42	700.0	26.1	2296.6
43	716.7	26.7	2351.3
44	733.3	27.3	2405.9
45	750.0	28.0	2460.6
46	766.7	28.6	2515.3
47	783.3	29.2	2570.0
48	800.0	29.8	2624.7
49	816.7	30.4	2679.4
50	833.3	31.1	2734.0

FORMULAE

Two dimensional figures

Figure	Diagram of figure	Surface area	Perimeter
Square	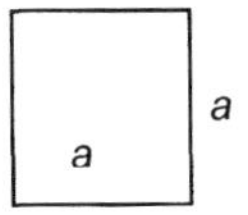	a^2	4a
Rectangle	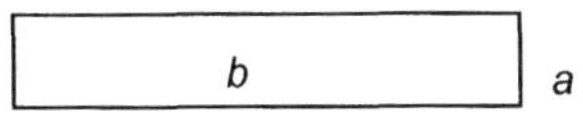	ab	2(a + b)
Triangle	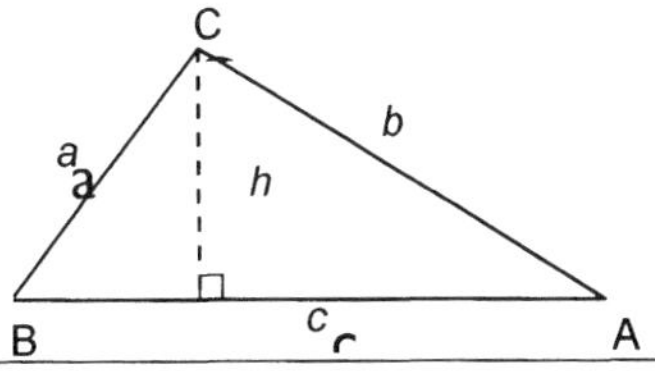	$\frac{1}{2}$ ch $\frac{1}{2}$ ab sin C $\sqrt{s(s-a)(s-b)(s-c)}$ where $s = \frac{1}{2}(a + b + c)$	a + b + c
Circle	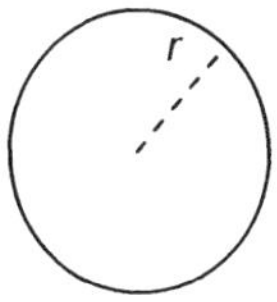	πr^2 $\frac{1}{4}\pi d^2$ where 2r = d	$2\pi r$ πd
Parallelogram	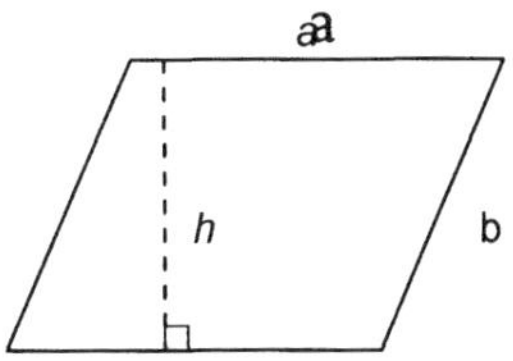	ah	2(a + b)
Trapezium	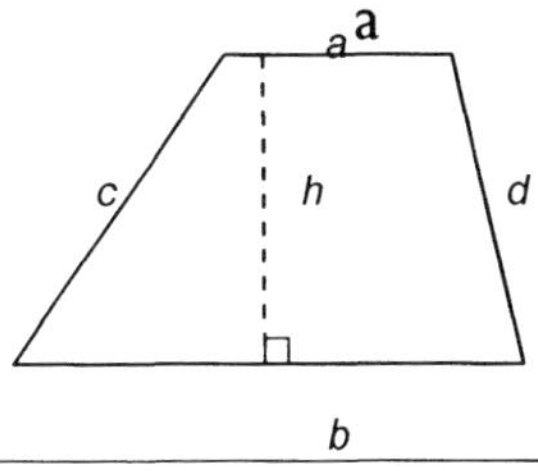	$\frac{1}{2}h(a + b)$	a + b + c + d

FORMULAE – continued

Figure	Diagram of figure	Surface area	Perimeter
Ellipse	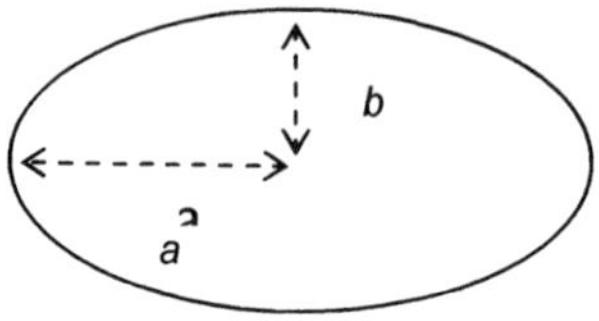 b, a	Approximately $\pi a b$	$\pi (a + b)$
Hexagon	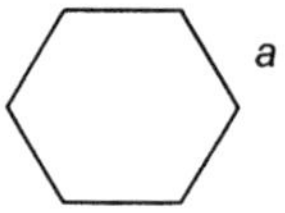 a	$2.6 \times a^2$	
Octagon	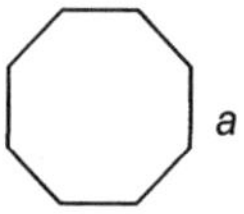 a	$4.83 \times a^2$	
Sector of circle	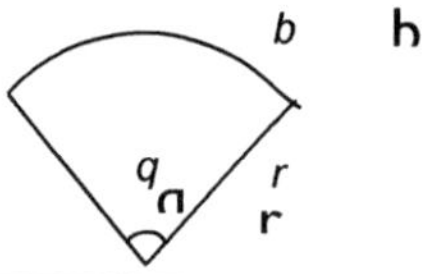 b, q, r	$\frac{1}{2}$ rb or $\frac{q}{360} \pi r^2$ note b= angle $\frac{q}{360} \times \pi 2r$	
Segment of A circle	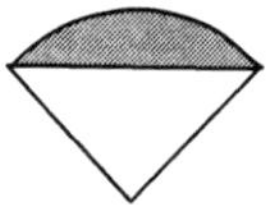	S - T where S = area of sector T = area of triangle	
Bellmouth	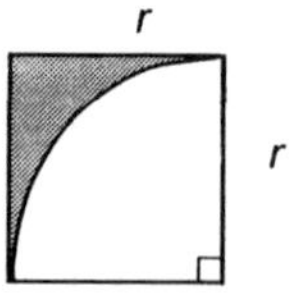 r, r	$\frac{3}{4} \times r^2$	
Three dimensional figures			
Cube	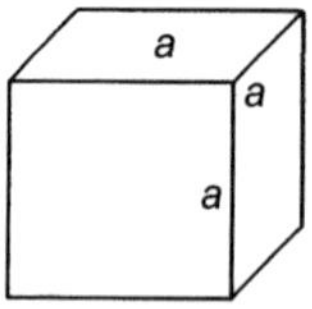 a, a, a	$6a^2$	a^3
Cuboid/rectangular block	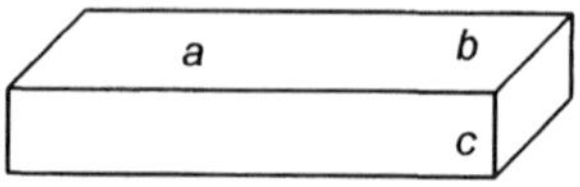 a, b, c	$2(ab + ac + bc)$	abc

Figure	Diagram of figure	Surface area	Perimeter
Prism / triangular block	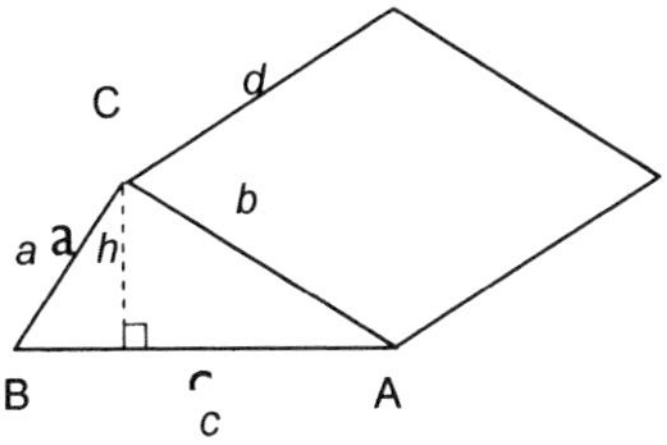	$bd + hc + dc + ad$	$\frac{1}{2} hcd$ $\frac{1}{2} ab \sin C\, d$ $d \sqrt{s(s-a)(s-b)(s-c)}$ where $s = \frac{1}{2}(a + b + c)$
Cylinder	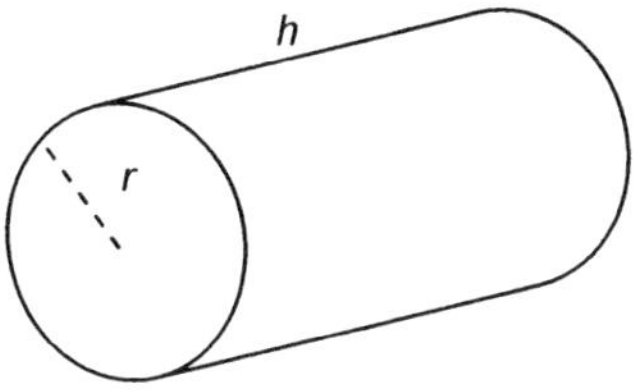	$2\pi rh + 2\pi^2$ $\pi dh + \frac{1}{2}\pi d^2$	$\pi r^2 h$ $\frac{1}{4}\pi d^2 h$
Sphere	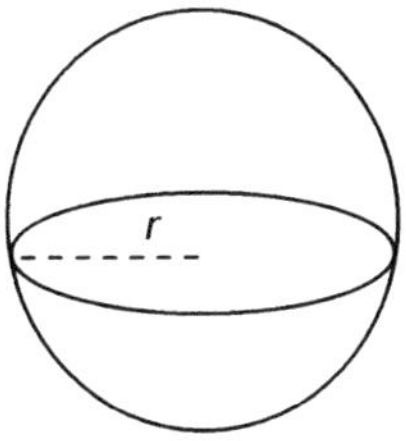	$4\pi r^2$	$4/3 r^3$
Segment of sphere	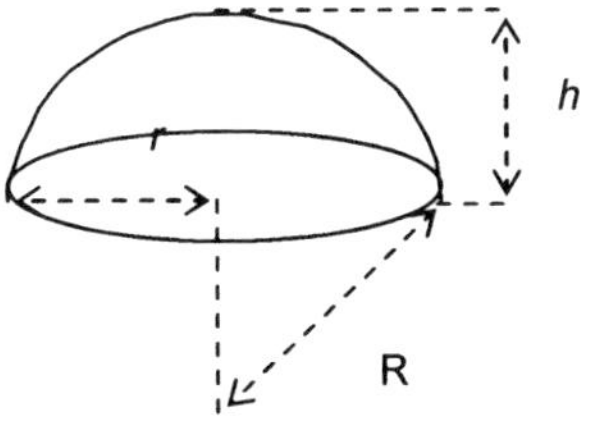	$2\pi Rh$	$1/6\ \pi h\,(3r^2 + h^2)$ $1/3\ \pi h^2\,(3R - H)$
Pyramid	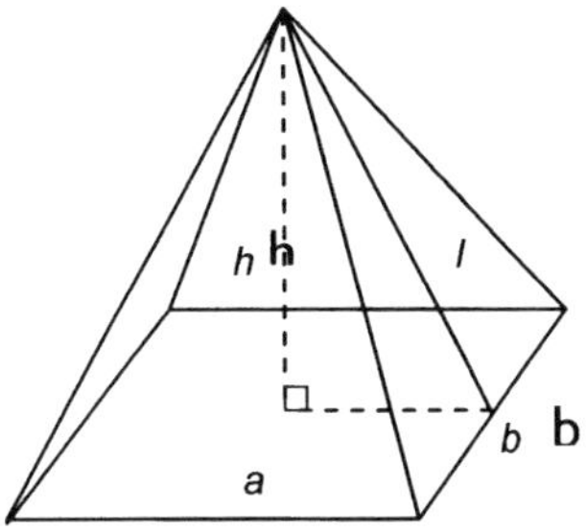	$(a + b)\, l + ab$	$1/3\ abh$

FORMULAE – continued

Figure	Diagram of figure	Surface area	Perimeter
Frustrum of a pyramid	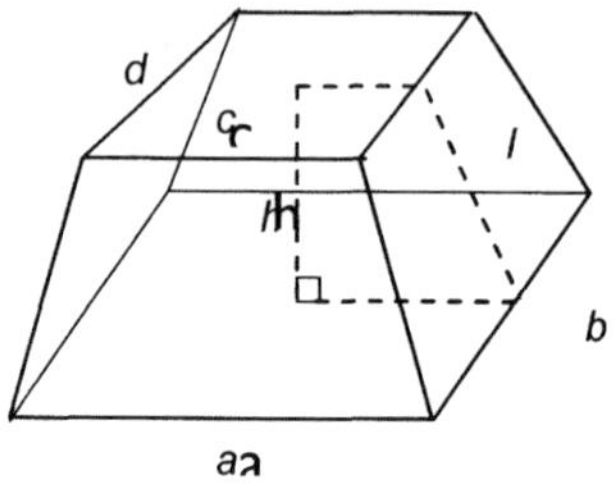 d c l h b a	$l\,(a+b+c+d) + \sqrt{(ab+cd)}$ [regular figure only]	$h/3(ab + cd + \sqrt{abcd})$
Cone	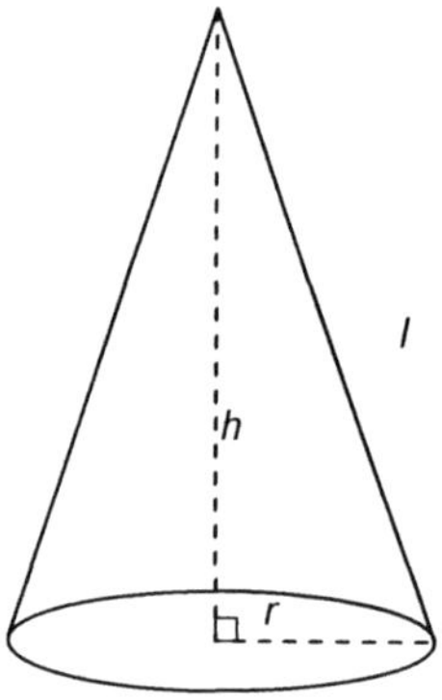 l h r	$\pi rl + \pi r^2$ $\frac{1}{2}\pi dh + \frac{1}{4}\pi d^2$	$1/3\ \pi r^2 h$ $1/12\ \pi d^2 h$
Frustrum of a cone	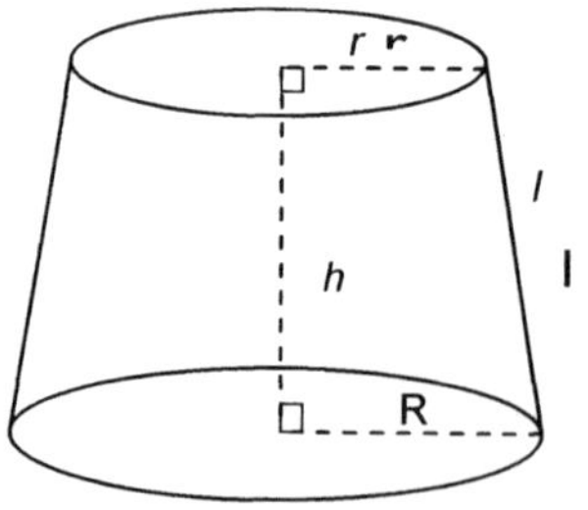 r l h R	$\pi r^2 + \pi R^2 + \pi h\,(R + r)$	$1/3\pi\,(R^2 + Rr + r^2)$

Formula	Description
Pythagoras theorem	$A^2 = B^2 + C^2$ where A is the hypotenuse of a right-angled triangle and B and C are the two adjacent sides
Simpsons Rule	The Area is divided into an even number of strips of equal width, and therefore has an odd number of ordinates at the division points area = $\frac{S\,(A + 2B + 4C)}{3}$ where S = common interval (strip width) A = sum of first and last ordinates B = sum of remaining odd ordinates C = sum of the even ordinates The Volume can be calculated by the same formula, but by substituting the area of each co-ordinate rather than its length.
Trapezoidal Rule	A given trench is divided into two equal sections, giving three ordinates, the first, the middle and the last. volume = $\frac{S}{2} \times (A + B + 2C)$ where S = width of the strips A = area of the first section B = area of the last section C = area of the rest of the sections
Prismoidal Rule	A given trench is divided into two equal sections, giving three ordinates, the first, the middle and the last. volume = $\frac{L \times (A + 4B + C)}{6}$ where L = total length of trench A = area of the first section B = area of the middle section C = area of the last section

EARTHWORK

Weights of typical materials handled by excavators

The weight of the material is that of the state in its natural bed and includes moisture. Adjustments should be made to allow for loose or compacted states.

Material	kg/m³	lb/cu yd
Adobe	1914	3230
Ashes	610	1030
Asphalt, rock	2400	4050
Asphalt, pavement	1920	3240
Basalt	2933	4950
Bauxite: alum ore	2619	4420
Bentonite	1600	2700
Borax	1730	2920
Caliche	1440	2430
Cement	1600	2700
Chalk (hard)	2406	4060
Cinders	759	1280
Clay: dry	1908	3220
Clay: damp	1985	3350
Coal: bituminous	1351	2280
Coke	510	860
Concrete, stone aggregate	2345	3960
Concrete, cinder aggregate	1760	2970
Conglomerate	2204	3720
Dolomite	2886	4870
Earth (loam): dry	1796	3030
Earth (loam): damp	1997	3370
Earth: wet, mud	1742	2940
Feldspar	2613	4410
Felsite	2495	4210
Fluorite	3093	5220
Gabbro	3093	5220
Gneiss	2696	4550
Granite	2690	4540
Gravel, dry	1790	3020
Gravel, wet	2092	3530

Material	kg/m³	lb/cu yd
Gypsum	2418	4080
Lignite	1244	2100
Limestone	2596	4380
Macadam, pavement	1685	2840
Masonry rubble	2325	3920
Magnesite, magnesium ore	2993	5050
Marble	2679	4520
Marl	2216	3740
Mud	1742	2940
Peat	700	1180
Potash	2193	3700
Pumice	640	1080
Quartz	2584	4360
Rhyolite	2400	4050
Rock : Earth mixture (75:25)	1955	3300
Rock : Earth mixture (75:25)	1720	2900
Rock : Earth mixture (75:25)	1575	2660
Sand: dry	1707	2880
Sand: wet	1831	3090
Sand and gravel - dry	1790	3020
- wet	2092	3530
Sandstone	2412	4070
Schist	2684	4530
Shale	2637	4450
Slag (blast)	2868	4840
Slate	2667	4500
Topsoil	1440	2430
Trachyte	2400	4050
Traprock	2791	4710

Snow - dry	130	220
- wet	509.61	860
Water	1000	1685
Quarry waste	1438	2425
Hardcore (consolidated)	1928	3250

EARTHWORK – continued

Transport Capacities

Type of vehicle	Capacity of vehicle	
	Payload t	Heaped capacity m^3
Wheelbarrow	150	0.10
1 tonne dumper	1250	1.00
2.5 tonne dumper	4000	2.50
Articulated dump truck (Volvo A20 6x4)	18500	11.00
Articulated dump truck (Volvo A35 6x6)	32000	19.00
Large capacity rear dumper (Euclid R35)	35000	22.00
Large capacity rear dumper (Euclid R85)	85000	50.00

Bulkage of soils (After excavation)

Type of soil	Approximate bulking of 1 m^3 after excavation
Vegetable soil and loam	25 - 30%
Soft clay	30 - 40%
Stiff clay	10 - 20%
Gravel	20 - 25%
Sand	40 - 50%
Chalk	40 - 50%
Rock, weathered	30 - 40%
Rock, unweathered	50 - 60%

Shrinkage of materials (On being deposited)

Type of soil	Approximate shrinking of 1 m^3 after excavation
Clay	10%
Gravel	8%
Gravel and sand	9%
Loam and light sandy soils	12%
Loose vegetable soils	15%

Voids in material used as sub bases or beddings

Material	m^3 of voids/m^3
Alluvium	0.37
River grit	0.29
Quarry sand	0.24
Shingle	0.37
Gravel	0.39
Broken stone	0.45
Broken bricks	0.42

Angles of repose

Type of soil		degrees
Clay	- dry	30
	- damp, well drained	45
	- wet	15 - 20
Earth	- dry	30
	- damp	45
Gravel	- moist	48
Sand	- dry or moist	35
	- wet	25
Loam		40

Slopes and angles

Ratio of base to height	Angle in degrees
5 : 1	11
4 : 1	14
3 : 1	18
2 : 1	27
1½ : 1	34
1 : 1	45
1 : 1½	56
1 : 2	63
1 : 3	72
1 : 4	76
1 : 5	79

EARTHWORK – continued

Grades (In Degrees and Percents)

Degrees	Percent	Degrees	Percent
1	1.8	24	44.5
2	3.5	25	46.6
3	5.2	26	48.8
4	7.0	27	51.0
5	8.8	28	53.2
6	10.5	29	55.4
7	12.3	30	57.7
8	14.0	31	60.0
9	15.8	32	62.5
10	17.6	33	64.9
11	19.4	34	67.4
12	21.3	35	70.0
13	23.1	36	72.7
14	24.9	37	75.4
15	26.8	38	78.1
16	28.7	39	81.0
17	30.6	40	83.9
18	32.5	41	86.9
19	34.4	42	90.0
20	36.4	43	93.3
21	38.4	44	96.6
22	40.4	45	100.0

Bearing powers

Ground conditions	Bearing Power		
	kN/m²	lb/in²	metric t/m²
Rock (broken)	483	70	50
Rock (solid)	2,415	350	240
Clay, dry or hard	380	55	40
medium dry	190	27	20
soft or wet	100	14	10
Gravel, cemented	760	110	80
Sand, compacted	380	55	40
clean dry	190	27	20
Swamp and alluvial soils	48	7	5

Earthwork support

Maximum depth of excavation in various soils without the use of earthwork support:

Ground conditions	Feet (ft)	Metres (m)
Compact soil	12	3.66
Drained loam	6	1.83
Dry sand	1	0.3
Gravelly earth	2	0.61
Ordinary earth	3	0.91
Stiff clay	10	3.05

It is important to note that the above table should only be used as a guide.
Each case must be taken on its merits and, as the limited distances given above are approached, careful watch must be kept for the slightest signs of caving in.

CONCRETE WORK

Weights of concrete and concrete elements

Type of Material	kg/m³	lb/cu ft
Ordinary concrete (dense aggregates)		
Non-reinforced plain or mass concrete		
Nominal weight	2305	144
Aggregate - limestone	2162 to 2407	135 to 150
- gravel	2244 to 2407	140 to 150
- broken brick	2000 (av)	125 (av)
- other crushed stone	2326 to 2489	145 to 155
Reinforced concrete		
Nominal weight	2407	150
Reinforcement - 1%	2305 to 2468	144 to 154
- 2%	2356 to 2519	147 to 157
- 4%	2448 to 2703	153 to 163
Special concretes		
Heavy concrete		
Aggregates - barytes, magnetite	3210 (min)	200 (min)
steel shot, punchings	5280	330
Lean mixes		
Dry-lean (gravel aggregate)	2244	140
Soil-cement (normal mix)	1601	100

Type of material		kg/m² per mm thick	lb/sq ft per inch thick
Ordinary concrete (dense aggregates)			
Solid slabs (floors, walls etc.)			
Thickness:	75 mm or 3 in	184	37.5
	100 mm or 4 in	245	50
	150 mm or 6 in	378	75
	250 mm or 10 in	612	125
	300 mm or 12 in	734	150
Ribbed slabs			
Thickness:	125 mm or 5 in	204	42
	150 mm or 6 in	219	45
	225 mm or 9 in	281	57
	300 mm or 12 in	342	70
Special concretes			
Finishes etc			
Rendering, screed etc Granolithic, terrazzo		1928 to 2401	10 to 12.5
Glass-block (hollow) concrete		1734 (approx)	9 (approx)
Prestressed concrete		Weights as for reinforced concrete (upper limits)	
Air-entrained concrete		Weights as for plain or reinforced concrete	

CONCRETE WORK – continued

Average weight of aggregates

Materials	**Voids %**	**Weight kg/m³**
Sand	39	1660
Gravel 10 - 20 mm	45	1440
Gravel 35 - 75 mm	42	1555
Crushed stone	50	1330
Crushed granite (over 15 mm)	50	1345
(n.e. 15 mm)	47	1440
'All-in' ballast	32	1800 - 2000

Material	**kg/m³**	**lb/cu yd**
Vermiculite (aggregate)	64-80	108-135
All-in aggregate	1999	125

Common mixes (per m³)

Recommended mix	Class of work suitable for: -	Cement (kg)	Sand (kg)	Coarse aggregate (kg)	No 50kg bags cement per m³ of combined aggregate
1:3:6	Roughest type of mass concrete such as footings, road haunching over 300mm thick	208	905	1509	4.00
1:2.½:5	Mass concrete of better class than 1:3:6 such as bases for machinery, walls below ground etc.	249	881	1474	5.00
1:2:4	Most ordinary uses of concrete, such as mass walls above ground, road slabs etc. and general reinforced concrete work	304	889	1431	6.00
1:1½:3	Watertight floors, pavements and walls, tanks, pits, steps, paths, surface of 2 course roads, reinforced concrete where where extra strength is required	371	801	1336	7.50
1:1:2	Work of thin section such as fence posts and small precast work	511	720	1206	10.50

CONCRETE WORK – continued

Prescribed mixes for ordinary structural concrete
Weights of cement and total dry aggregates in kg to produce approximately one cubic metre of fully compacted concrete together with the percentages by weight of fine aggregate in total dry aggregates.

Concrete grade	Nominal max. size of aggregate (mm)	40		20		14		10	
	Workability	Med.	High	Med.	High	Med.	High	Med.	High
	Limits to slump that may be expected (mm)	50-100	100-150	25-75	75-125	10-50	50-100	10-25	25-50
7	Cement (kg)	180	200	210	230	-	-	-	-
	Total aggregate (kg)	1950	1850	1900	1800	-	-	-	-
	Fine aggregate (%)	30-45	30-45	35-50	35-50	-	-	-	-
10	Cement (kg)	210	230	240	260	-	-	-	-
	Total aggregate (kg)	1900	1850	1850	1800	-	-	-	-
	Fine aggregate (%)	30-45	30-45	35-50	35-50	-	-	-	-
15	Cement (kg)	250	270	280	310	-	-	-	-
	Total aggregate (kg)	1850	1800	1800	1750	-	-	-	-
	Fine aggregate (%)	30-45	30-45	35-50	35-50	-	-	-	-
20	Cement (kg)	300	320	320	350	340	380	360	410
	Total aggregate (kg)	1850	1750	1800	1750	1750	1700	1750	1650
	Sand								
	Zone 1 (%)	35	40	40	45	45	50	50	55
	Zone 2 (%)	30	35	35	40	40	45	45	50
	Zone 3 (%)	30	30	30	35	35	40	40	45
25	Cement (kg)	340	360	360	390	380	420	400	450
	Total aggregate (kg)	1800	1750	1750	1700	1700	1650	1700	1600
	Sand								
	Zone 1 (%)	35	40	40	45	45	50	50	55
	Zone 2 (%)	30	35	35	40	40	45	45	50
	Zone 3 (%)	30	30	30	35	35	40	40	45
30	Cement (kg)	370	390	400	430	430	470	460	510
	Total aggregate (kg)	1750	1700	1700	1650	1700	1600	1650	1550
	Sand								
	Zone 1 (%)	35	40	40	45	45	50	50	55
	Zone 2 (%)	30	35	35	40	40	45	45	50
	Zone 3 (%)	30	30	30	35	35	40	40	45

Weights of bar reinforcement

Nominal sizes (mm)	Cross-sectional area (mm^2)	Mass kg/m	Length of bar m/tonne
6	28.27	0.222	4505
8	50.27	0.395	2534
10	78.54	0.617	1622
12	113.10	0.888	1126
16	201.06	1.578	634
20	314.16	2.466	405
25	490.87	3.853	260
32	804.25	6.313	158
40	1265.64	9.865	101
50	1963.50	15.413	65

Weights of bars (at specific spacings)

Weights of metric bars in kilogrammes per square metre

Size	Spacing of bars in millimetres									
(mm)	75	100	125	150	175	200	225	250	275	300
6	2.96	2.220	1.776	1.480	1.27	1.110	0.99	0.89	0.81	0.74
8	5.26	3.95	3.16	2.63	2.26	1.97	1.75	1.58	1.44	1.32
10	8.22	6.17	4.93	4.11	3.52	3.08	2.74	2.47	2.24	2.06
12	11.84	8.88	7.10	5.92	5.07	4.44	3.95	3.55	3.23	2.96
16	21.04	15.78	12.63	10.52	9.02	7.89	7.02	6.31	5.74	5.26
20	32.88	24.66	19.73	16.44	14.09	12.33	10.96	9.87	8.97	8.22
25	51.38	38.53	30.83	25.69	22.02	19.27	17.13	15.41	14.01	12.84
32	84.18	63.13	50.51	42.09	36.08	31.57	28.06	25.25	22.96	21.04
40	131.53	98.65	78.92	65.76	56.37	49.32	43.84	39.46	35.87	32.88
50	205.51	154.13	123.31	102.76	88.08	77.07	68.50	61.65	56.05	51.38

Basic weight of steelwork taken as 7850 kg/m^3
Basic weight of bar reinforcement per metre run = 0.00785 kg/mm^2
The value of PI has been taken as 3.141592654

CONCRETE WORK - continued

Fabric reinforcement

Preferred range of designated fabric types and stock sheet sizes

Fabric reference	Longitudinal wires			Cross wires			Mass
	Nominal wire size (mm)	Pitch (mm)	Area (mm/m²)	Nominal wire size (mm)	Pitch (mm)	Area (mm/m²)	(kg/m²)
Square mesh							
A393	10	200	393	10	200	393	6.16
A252	8	200	252	8	200	252	3.95
A193	7	200	193	7	200	193	3.02
A142	6	200	142	6	200	142	2.22
A98	5	200	98	5	200	98	1.54
Structural mesh							
B1131	12	100	1131	8	200	252	10.90
B785	10	100	785	8	200	252	8.14
B503	8	100	503	8	200	252	5.93
B385	7	100	385	7	200	193	4.53
B283	6	100	283	7	200	193	3.73
B196	5	100	196	7	200	193	3.05
Long mesh							
C785	10	100	785	6	400	70.8	6.72
C636	9	100	636	6	400	70.8	5.55
C503	8	100	503	5	400	49.00	4.34
C385	7	100	385	5	400	49.00	3.41
C283	6	100	283	5	400	49.00	2.61
Wrapping mesh							
D98	5	200	98	5	200	98	1.54
D49	2.5	100	49	2.5	100	49	0.77
Stock sheet size	Length 4.8 m			**Width 2.4 m**		**(Sheet area 11.52m²)**	

Wire

SWG	6g	5g	4g	3g	2g	1g	1/0g	2/0g	3/0g	4/0g	5/0g
diameter In mm	0.192 4.9	0.212 5.4	0.232 5.9	0.252 6.4	0.276 7.0	0.300 7.6	0.324 8.2	0.348 8.8	0.372 9.5	0.400 0.2	0.432 1.0
Area In mm²	0.029 19	0.035 23	0.042 27	0.050 32	0.060 39	0.071 46	0.082 53	0.095 61	0.109 70	0.126 81	0.146 95

Average weight kg/m³ of steelwork reinforcement in concrete for various building elements

	kg/m³ concrete
Substructure	
Pile caps	110 - 150
Tie beams	130 - 170
Ground beams	230 - 330
Bases	90 - 130
Footings	70 - 110
Retaining walls	110 - 150
Superstructure	
Slabs - one way	75 - 125
Slabs - two way	65 - 135
Plate slab	95 - 135
Cantilevered slab	90 - 130
Ribbed floors	80 - 120
Columns	200 - 300
Beams	250 - 350
Stairs	130 - 170
Walls - normal	30 - 70
Walls - wind	50 - 90

Note: **For exposed elements add the following % :**

Walls 50%, Beams 100%, Columns 15%

CONCRETE WORK – continued

Normal Curing Periods

Conditions under which concrete is maturing	**Minimum periods of protection for different types of cement**					
	Number of days (where the average surface temperature of the concrete exceeds 10°C during the whole period)			Equivalent maturity (degree hours) calculated as the age of the concrete in hours multiplied by the number of degrees Celsius by which the average surface temperature of the concrete exceeds -10°C		
	Other	SRPC	OPC or RHPC	Other	SRPC	OPC or RHPC
Hot weather or drying winds	7	4	3	3500	2000	1500
Conditions not covered by 1	4	3	2	2000	1500	1000

KEY

OPC - Ordinary Portland Cement

RHPC - Rapid-hardening Portland cement

SRPC - Sulphate-resisting Portland cement

Minimum period before striking formwork

	Minimum period before striking		
	Surface temperature of concrete		
	16 °C	**17 °C**	**t °C(0-25)**
Vertical formwork to columns, walls and large beams	12 hours	18 hours	$\frac{300}{t+10}$ hours
Soffit formwork to slabs	4 days	6 days	$\frac{100}{t+10}$ days
Props to slabs	10 days	15 days	$\frac{250}{t+10}$ days
Soffit formwork to beams	9 days	14 days	$\frac{230}{t+10}$ days
Props to beams	14 days	21 days	$\frac{360}{t+10}$ days

MASONRY

Weights of bricks and blocks

Walls and components of walls	kg/m² per mm thick	lb/sq ft per inch thick
Blockwork		
Hollow clay blocks Average)	1.15	6
Common clay blocks	1.90	10
Brickwork		
Engineering clay bricks	2.30	12
Refactory bricks	1.15	6
Sand-lime (and similar) bricks	2.02	10.5

Weights of stones

Type of stone		kg/m^3	lb/cu ft
Natural stone (solid)			
Granite		2560 to 2927	160 to 183
Limestone	- Bath stone	2081	130
	- Marble	2723	170
	- Portland stone	2244	140
Sandstone		2244 to 2407	140 to 150
Slate		2880	180
Stone rubble (packed)		2244	140

MASONRY - continued

Quantities of bricks and mortar

Materials per m² of wall:		
Thickness	**No. of Bricks**	**Mortar m³**
Half brick (112.5 mm)	58	0.022
One brick (225 mm)	116	0.055
Cavity, both skins (275 mm)	116	0.045
1.5 brick (337 mm)	174	0.074
Mass brickwork per m³	464	0.36

Mortar mixes: Quantities of Dry Materials

	Imperial cu yd			Metric m³		
Mix	Cement cwts	Lime cwts	Sand cu yds	Cement tonnes	Lime tonnes	Sand cu m
1:3	7.0	-	1.04	0.54	-	1.10
1:4	6.3	-	1.10	0.40	-	1.20
1:1:6	3.9	1.6	1.10	0.27	0.13	1.10
1:2:9	2.6	2.1	1.10	0.20	0.15	1.20
0:1:3	-	3.3	1.10	-	0.27	1.00

Mortar mixes for various uses

Mix	Use
1:3	Construction designed to withstand heavy loads in all seasons
1:1:6	Normal construction not designed for heavy loads. Sheltered and moderate conditions in spring and summer. Work above d.p.c - sand, lime bricks, clay blocks etc.
1:2:9	Internal partitions with blocks which have high drying shrinkage, pumice blocks, etc. any periods
0:1:3	Hydraulic lime only should be used in this mix and may be used for construction not designed for heavy loads and above d.p.c spring and summer.

Quantities of bricks and mortar required per m² of walling

Description	Unit	Nr of bricks required	Mortar required (m³) No frogs	Single frogs	Double frogs
Standard bricks					
Brick size 215 x 102.5 x 50 mm					
half brick wall (103 mm)	m²	72	0.022	0.027	0.032
2 x half brick cavity wall (270 mm)	m²	144	0.044	0.054	0.064
one brick wall (215 mm)	m²	144	0.052	0.064	0.076
one and a half brick wall (328 mm)	m²	216	0.073	0.091	0.108
mass brickwork	m³	576	0.347	0.413	0.480
Brick size 215 x 102.5 x 65 mm					
half brick wall (103 mm)	m²	58	0.019	0.022	0.026
2 x half brick cavity wall (270 mm)	m²	116	0.038	0.045	0.055
one brick wall (215 mm)	m²	116	0.046	0.055	0.064
one and a half brick wall (328 mm)	m²	174	0.063	0.074	0.088
mass brickwork	m³	464	0.307	0.360	0.413
Metric modular bricks	**Perforated**				
Brick co-ordinating size 200 x 100 x 75 mm					
90 mm thick	m²	67	0.016	0.019	
190 mm thick	m²	133	0.042	0.048	
290 mm thick	m²	200	0.068	0.078	
Brick co-ordinating size 200 x 100 x 100 mm					
90 mm thick	m²	50	0.013	0.016	
190 mm thick	m²	100	0.036	0.041	
290 mm thick	m²	150	0.059	0.067	
Brick co-ordinating size 300 x 100 x 75 mm					
90 mm thick	m²	33	-	0.015	
Brick co-ordinating size 00 x 100 x 100 mm					
90 mm thick	m²	44	0.015	0.018	

Note: Assuming 10 mm deep joints.

MASONRY – continued

Mortar required per m^2 blockwork (9.88 blocks/m^2)

Wall thickness	75	90	100	125	140	190	215
Mortar m^3/m^2	0.005	0.006	0.007	0.008	0.009	0.013	0.014

Mortar Group	Cement: lime: sand	Masonry cement: sand	Cement: sand with plasticiser
1	1 : 0-0.25:3		
2	1 : 0.5 :4-4.5	1 : 2.5-3.5	1 : 3-4
3	1 : 1:5-6	1 : 4-5	1 : 5-6
4	1 : 2:8-9	1 : 5.5-6.5	1 : 7-8
5	1 : 3:10-12	1 : 6.5-7	1 : 8

Group 1: strong inflexible mortar

Group 5: weak but flexible.

All mixes within a group are of approximately similar strength.
Frost resistance increases with the use of plasticisers.
Cement:lime:sand mixes give the strongest bond and greatest resistance to rain penetration.
Masonry cement equals ordinary Portland cement plus a fine neutral mineral filler and an air entraining agent.

Calcium Silicate Bricks

Type	Strength	Location
Class 2 crushing strength	14.0N/mm^2	not suitable for walls
Class 3	20.5N/mm^2	walls above dpc
Class 4	27.5N/mm^2	cappings and copings
Class 5	34.5N/mm^2	retaining walls
Class 6	41.5N/mm^2	walls below ground
Class 7	48.5N/mm^2	walls below ground

The Class 7 calcium silicate bricks are therefore equal in strength to Class B bricks.

Calcium silicate bricks are not suitable for DPCs

Durability of Bricks

FL	Frost resistant with low salt content
FN	Frost resistant with normal salt content
ML	Moderately frost resistant with low salt content
MN	Moderately frost resistant with normal salt content

Brickwork Dimensions

No. of Horizontal Bricks	Dimensions mm	No. of Vertical courses	No. of Vertical courses
1/2	112.5	1	75
1	225.0	2	150
1 1/2	337.5	3	225
2	450.0	4	300
2 1/2	562.5	5	375
3	675.0	6	450
3 1/2	787.5	7	525
4	900.0	8	600
4 1/2	1012.5	9	675
5	1125.0	10	750
5 1/2	1237.5	11	825
6	1350.0	12	900
6 1/2	1462.5	13	975
7	1575.0	14	1050
7 1/2	1687.5	15	1125
8	1800.0	16	1200
8 1/2	1912.5	17	1275
9	2025.0	18	1350
9 1/2	2137.5	19	1425
10	2250.0	20	1500
20	4500.0	24	1575
40	9000.0	28	2100
50	11250.0	32	2400
60	13500.0	36	2700
75	16875.0	40	3000

MASONRY – continued

Standard available block sizes

Block	Co-ordinating size Length x height	Work size	Thicknesses (work size)
A	400 x 100	390 x 90	) 75, 90, 100, 140 &
	400 x 200	440 x 190	) 190 mm
	450 x 225	440 x 215	) 75, 90, 100, 140,
			) 190 & 215 mm
B	400 x 100	390 x 90	) 75, 90, 100, 140 &
	400 x 200	390 x 190	) 190 mm
	450 x 200	440 x 190	)
	450 x 225	440 x 215	) 75, 90, 100, 140,
	450 x 300	440 x 290	) 190 & 215 mm
	600 x 200	590 x 190	)
	600 x 225	590 x 215	)
C	400 x 200	390 x 190	)
	450 x 200	440 x 190	)
	450 x 225	440 x 215	) 60 & 75 mm
	450 x 300	440 x 290	)
	600 x 200	590 x 190	)
	600 x 225	590 x 215	)

TIMBER

Weights of timber

Material	kg/m^3	lb/cu ft
General	806 (avg)	50 (avg)
Douglas fir	479	30
Yellow pine, spruce	479	30
Pitch pine	673	42
Larch, elm	561	35
Oak (English)	724 to 959	45 to 60
Teak	643 to 877	40 to 55
Jarrah	959	60
Greenheart	1040 to 1204	65 to 75
Quebracho	1285	80
Material	**kg/m^2 per mm thickness**	**lb/sq ft per inch thickness**
Wooden boarding and blocks		
Softwood	0.48	2.5
Hardwood	0.76	4
Hardboard	1.06	5.5
Chipboard	0.76	4
Plywood	0.62	3.25
Blockboard	0.48	2.5
Fibreboard	0.29	1.5
Wood-wool	0.58	3
Plasterboard	0.96	5
Weather boarding	0.35	1.8

TIMBER – continued

Conversion tables (for sawn timber only)

Inches >	Millimetres	Feet >	Metres
1	25	1	0.300
2	50	2	0.600
3	75	3	0.900
4	100	4	1.200
5	125	5	1.500
6	150	6	1.800
7	175	7	2.100
8	200	8	2.400
9	225	9	2.700
10	250	10	3.000
11	275	11	3.300
12	300	12	3.600
13	325	13	3.900
14	350	14	4.200
15	375	15	4.500
16	400	16	4.800
17	425	17	5.100
18	450	18	5.400
19	475	19	5.700
20	500	20	6.000
21	525	21	6.300
22	550	22	6.600
23	575	23	6.900
24	600	24	7.200

Planed softwood

The finished end section size of planed timber is usually 3/16" less than the original size from which it is produced. This however varies slightly dependant upon availability of material and origin of species used.

Standard (timber) to cubic metres and cubic metres to standards (timber)

m^3	m^3/Standards	Standard
4.672	1	0.214
9.344	2	0.428
14.017	3	0.642
18.689	4	0.856
23.361	5	1.070
28.033	6	1.284
32.706	7	1.498
37.378	8	1.712
42.05	9	1.926
46.722	10	2.140
93.445	20	4.281
140.167	30	6.421
186.890	40	8.561
233.612	50	10.702
280.335	60	12.842
327.057	70	14.982
373.779	80	17.122
420.502	90	19.263
467.224	100	21.403
1 cu metre = 35.3148 **1 cu ft= 0.028317 cu metres** **1 std = 4.67227 cu metres**		

Standards (timber) to cubic metres and cubic metres to standards (timber)

1 cu metre	=	35.3148 cu ft	=	0.21403 std
1 cu ft	=	0.028317 cu metres		
1 std	=	4.67227 cu metres		

TIMBER - continued

Basic sizes of sawn softwood available (cross sectional areas)

Thickness (mm)	Width (mm)								
	75	100	125	150	175	200	225	250	300
16	x	x	x	x					
19	x	x	x	x					
22	x	x	x	x					
25	x	x	x	x	x	x	x	x	x
32	x	x	x	x	x	x	x	x	x
36	x	x	x	x					
38	x	x	x	x	x	x	x		
44	x	x	x	x	x	x	x	x	x
47*	x	x	x	x	x	x	x	x	x
50	x	x	x	x	x	x	x	x	x
63	x	x	x	x	x	x	x		
75		x	x	x	x	x	x	x	x
100		x		x		x		x	x
150				x		x			x
200						x			
250								x	
300									x

* This range of widths for 47 mm thickness will usually be found to be available in construction quality only.

Note: The smaller sizes below 100 mm thick and 250 mm width are normally but not exclusively of European origin. Sizes beyond this are usually of North and South American origin.

Basic lengths of sawn softwood available (metres)

1.80	2.10	3.00	4.20	5.10	6.00	7.20
2.40	3.30	4.50	5.40	6.30		
2.70	3.60	4.80	5.70	6.60		
	3.90			6.90		

Note: Lengths of 6.00 m and over will generally only be available from North American species and may have to be recut from larger sizes.

Reductions from basic size to finished size of timber by planing of two opposed faces

Purpose	15 - 35 mm	36 - 100 mm	101 - 150 mm	over 150 mm
Constructional timber	3 mm	3 mm	5 mm	6 mm
Matching interlocking boards	4 mm	4 mm	6 mm	6 mm
Wood trim not specified in BS 584	5 mm	7 mm	7 mm	9 mm
Joinery and cabinet work	7 mm	9 mm	11 mm	13 mm

Note: The reduction of width or depth is overall the extreme size and is exclusive of any reduction of the face by the machining of a tongue or lap joints.

STRUCTURAL METALWORK

Weights of metalwork

Material	kg/m³	lb/cu ft
Metals, steel construction, etc		
Iron - cast	7207	450
- wrought	7687	480
- ore - general	2407	150
- (crushed) Swedish	3682	230
Steel	7854	490
Copper - cast	8731	545
- wrought	8945	558
Brass	8497	530
Bronze	8945	558
Aluminium	2774	173
Lead	11322	707
Zinc (rolled)	7140	446
	g/mm² per metre	**lb/sq ft per foot**
Steel bars	7.85	3.4
Structural steelwork	Net weight of member @ 7854 kg/m³	
rivetted	+ 10% for cleats, rivets, bolts, etc	
welded	+ 1.25% to 2.5% for welds, etc	
Rolled sections		
beams	+ 2.5%	
stanchions	+ 5% (extra for caps and bases)	
Plate		
web girders	+ 10% for rivets or welds, stiffeners, etc	
	kg/m	**lb/ft**
Steel stairs : industrial type		
1 m or 3ft wide	84	56
Steel tubes		
50 mm or 2 in bore	5 to 6	3 to 4
Gas piping		
20 mm or 3/4 in	2	1¼

Universal Beams BS 4: Part 1: 1993

Designation	Mass Kg/m	Depth of Section mm	Width of Section mm	Thickness		Surface Area m^2/m
				Web mm	Flange mm	
1016 x 305 x 487	487.0	1036.1	308.5	30.0	54.1	3.20
1016 x 305 x 438	438.0	1025.9	305.4	26.9	49.0	3.17
1016 x 305 x 393	393.0	1016.0	303.0	24.4	43.9	3.15
1016 x 305 x 349	349.0	1008.1	302.0	21.1	40.0	3.13
1016 x 305 x 314	314.0	1000.0	300.0	19.1	35.9	3.11
1016 x 305 x 272	272.0	990.1	300.0	16.5	31.0	3.10
1016 x 305 x 249	249.0	980.2	300.0	16.5	26.0	3.08
1016 x 305 x 222	222.0	970.3	300.0	16.0	21.1	3.06
914 x 419 x 388	388.0	921.0	420.5	21.4	36.6	3.44
914 x 419 x 343	343.3	911.8	418.5	19.4	32.0	3.42
914 x 305 x 289	289.1	926.6	307.7	19.5	32.0	3.01
914 x 305 x 253	253.4	918.4	305.5	17.3	27.9	2.99
914 x 305 x 224	224.2	910.4	304.1	15.9	23.9	2.97
914 x 305 x 201	200.9	903.0	303.3	15.1	20.2	2.96
838 x 292 x 226	226.5	850.9	293.8	16.1	26.8	2.81
838 x 292 x 194	193.8	840.7	292.4	14.7	21.7	2.79
838 x 292 x 176	175.9	834.9	291.7	14.0	18.8	2.78
762 x 267 x 197	196.8	769.8	268.0	15.6	25.4	2.55
762 x 267 x 173	173.0	762.2	266.7	14.3	21.6	2.53
762 x 267 x 147	146.9	754.0	265.2	12.8	17.5	2.51
762 x 267 x 134	133.9	750.0	264.4	12.0	15.5	2.51
686 x 254 x 170	170.2	692.9	255.8	14.5	23.7	2.35
686 x 254 x 152	152.4	687.5	254.5	13.2	21.0	2.34
686 x 254 x 140	140.1	383.5	253.7	12.4	19.0	2.33
686 x 254 x 125	125.2	677.9	253.0	11.7	16.2	2.32
610 x 305 x 238	238.1	635.8	311.4	18.4	31.4	2.45
610 x 305 x 179	179.0	620.2	307.1	14.1	23.6	2.41
610 x 305 x 149	149.1	612.4	304.8	11.8	19.7	2.39
610 x 229 x 140	139.9	617.2	230.2	13.1	22.1	2.11
610 x 229 x 125	125.1	612.2	229.0	11.9	19.6	2.09
610 x 229 x 113	113.0	607.6	228.2	11.1	17.3	2.08
610 x 229 x 101	101.2	602.6	227.6	10.5	14.8	2.07
533 x 210 x 122	122.0	544.5	211.9	12.7	21.3	1.89
533 x 210 x 109	109.0	539.5	210.8	11.6	18.8	1.88

STRUCTURAL METALWORK – continued

Universal Beams – continued

Designation	Mass Kg/m	Depth of Section mm	Width of Section mm	Thickness		Surface Area m²/m
				Web mm	Flange mm	
533 x 210 x 101	101.0	536.7	210.0	10.8	17.4	1.87
533 x 210 x 92	92.1	533.1	209.3	10.1	15.6	1.86
533 x 210 x 82	82.2	528.3	208.8	9.6	13.2	1.85
457 x 191 x 98	98.3	467.2	192.8	11.4	19.6	1.67
457 x 191 x 89	89.3	463.4	191.9	10.5	17.7	1.66
457 x 191 x 82	82.0	460.0	191.3	9.9	16.0	1.65
457 x 191 x 74	74.3	457.0	190.4	9.0	14.5	1.64
457 x 191 x 67	67.1	453.4	189.9	8.5	12.7	1.63
457 x 152 x 82	82.1	465.8	155.3	10.5	18.9	1.51
457 x 152 x 74	74.2	462.0	154.4	9.6	17.0	1.50
457 x 152 x 67	67.2	458.0	153.8	9.0	15.0	1.50
457 x 152 x 60	59.8	454.6	152.9	8.1	13.3	1.50
457 x 152 x 52	52.3	449.8	152.4	7.6	10.9	1.48
406 x 178 x 74	74.2	412.8	179.5	9.5	16.0	1.51
406 x 178 x 67	67.1	409.4	178.8	8.8	14.3	1.50
406 x 178 x 60	60.1	406.4	177.9	7.9	12.8	1.49
406 x 178 x 50	54.1	402.6	177.7	7.7	10.9	1.48
406 x 140 x 46	46.0	403.2	142.2	6.8	11.2	1.34
406 x 140 x 39	39.0	398.0	141.8	6.4	8.6	1.33
356 x 171 x 67	67.1	363.4	173.2	9.1	15 7	1.38
356 x 171 x 57	57.0	358.0	172.2	8.1	13.0	1.37
356 x 171 x 51	51.0	355.0	171.5	7.4	11.5	1.36
356 x 171 x 45	45.0	351.4	171.1	7.0	9.7	1.36
356 x 127 x 39	39.1	353.4	126.0	6.6	10.7	1.18
356 x 127 x 33	33.1	349.0	125.4	6.0	8.5	1.17
305 x 165 x 54	54.0	310.4	166.9	7.9	13.7	1.26
305 x 165 x 46	46.1	306.6	165.7	6.7	11.8	1.25
305 x 165 x 40	40.3	303.4	165.0	6.0	10.2	1.24
305 x 127 x 48	48.1	311.0	125.3	9.0	14.0	1.09
305 x 127 x 42	41.9	307.2	124.3	8.0	12.1	1.08
305 x 127 x 37	37.0	304.4	123.3	7.1	10.7	1.07
305 x 102 x 33	32.8	312.7	102.4	6.6	10.8	1.01
305 x 102 x 28	28.2	308.7	101.8	6.0	8.8	1.00
305 x 102 x 25	24.8	305.1	101.6	5.8	7.0	0.992

Designation	Mass Kg/m	Depth of Section mm	Width of Section mm	Thickness		Surface Area m^2/m
				Web mm	Flange mm	
254 x 146 x 43	43.0	259.6	147.3	7.2	12.7	1.08
254 x 146 x 37	37.0	256.0	146.4	6.3	10.9	1.07
254 x 146 x 31	31.1	251.4	146.1	6.0	8.6	1.06
254 x 102 x 28	28.3	260.4	102.2	6.3	10.0	0.904
254 x 102 x 25	25.2	257.2	101.9	6.0	8.4	0.897
254 x 102 x 22	22.0	254.0	101.6	5.7	6.8	0.890
203 x 133 x 30	30.0	206.8	133.9	6.4	9.6	0.923
203 x 133 x 25	25.1	203.2	133.2	5.7	7.8	0.915
203 x 102 x 23	23.1	203.2	101.8	5.4	9.3	0.790
178 x 102 x 19	19.0	177.8	101.2	4.8	7.9	0.738
152 x 89 x 16	16.0	152.4	88.7	4.5	7.7	0.638
127 x 76 x 13	13.0	127.0	76.0	4.0	7.6	0.537

STRUCTURAL METALWORK – continued

Universal Columns BS 4: Part 1: 1993

Designation	Mass Kg/m	Depth of Section mm	Width of Section mm	Thickness		Surface Area m²/m
				Web mm	Flange mm	
356 x 406 x 634	633.9	474.7	424.0	47.6	77.0	2.52
356 x 406 x 551	551.0	455.6	418.5	42.1	67.5	2.47
356 x 406 x 467	467.0	436.6	412.2	35.8	58.0	2.42
356 x 406 x 393	393.0	419.0	407.0	30.6	49.2	2.38
356 x 406 x 340	339.9	406.4	403.0	26.6	42.9	2.35
356 x 406 x 287	287.1	393.6	399.0	22.6	36.5	2.31
356 x 406 x 235	235.1	381.0	384.8	18.4	30.2	2.28
356 x 368 x 202	201.9	374.6	374.7	16.5	27.0	2.19
356 x 368 x 177	177.0	368.2	372.6	14.4	23.8	2.17
356 x 368 x 153	152.9	362.0	370.5	12.3	20.7	2.16
356 x 368 x 129	129.0	355.6	368.6	10.4	17.5	2.14
305 x 305 x 283	282.9	365.3	322.2	26.8	44.1	1.94
305 x 305 x 240	240.0	352.5	318.4	23.0	37.7	1.91
305 x 305 x 198	198.1	339.9	314.5	19.1	31.4	1.87
305 x 305 x 158	158.1	327.1	311.2	15.8	25.0	1.84
305 x 305 x 137	136.9	320.5	309.2	13.8	21.7	1.82
305 x 305 x 118	117.9	314.5	307.4	12.0	18.7	1.81
305 x 305 x 97	96.9	307.9	305.3	9.9	15.4	1.79
254 x 254 x 167	167.1	289.1	265.2	19.2	31.7	1.58
254 x 254 x 132	132.0	276.3	261.3	15.3	25.3	1.55
254 x 254 x 107	107.1	266.7	258.8	12.8	20.5	1.52
254 x 254 x 89	88.9	260.3	256.3	10.3	17.3	1.50
254 x 254 x 73	73.1	254.1	254.6	8.6	14.2	1.49
203 x 203 x 86	86.1	222.2	209.1	12.7	20.5	1.24
203 x 203 x 71	71.0	215.8	206.4	10.0	17.3	1.22
203 x 203 x 60	60.0	209.6	205.8	9.4	14.2	1.21
203 x 203 x 52	52.0	206.2	204.3	7.9	12.5	1.20
203 x 203 x 46	46.1	203.2	203.6	7.2	11.0	1.19
152 x 152 x 37	37.0	161.8	154.4	8.0	11.5	0.912
152 x 152 x 30	30.0	157.6	152.9	6.5	9.4	0.901
152 x 152 x 23	23.0	152.4	152.2	5.8	6.8	0.889

Joists BS 4: Part 1: 1993

Designation	Mass Kg/m	Depth of Section mm	Width of Section mm	Thickness		Surface Area m^2/m
				Web mm	Flange mm	
254 x 203 x 82	82.0	254.0	203.2	10.2	19.9	1.210
203 x 152 x 52	52.3	203.2	152.4	8.9	16.5	0.932
152 x 127 x 37	37.3	152.4	127.0	10.4	13.2	0.737
127 x 114 x 29	29.3	127.0	114.3	10.2	11.5	0.646
127 x 114 x 27	26.9	127.0	114.3	7.4	11.4	0.650
102 x 102 x 23	23.0	101.6	101.6	9.5	10.3	0.549
102 x 44 x 7	7.5	101.6	44.5	4.3	6.1	0.350
89 x 89 x 19	19.5	88.9	88.9	9.5	9.9	0.476
76 x 76 x 13	12.8	76.2	76.2	5.1	8.4	0.411

Parallel Flange Channels

Designation	Mass Kg/m	Depth of Section mm	Width of Section mm	Thickness		Surface Area m^2/m
				Web mm	Flange mm	
430 x 100 x 64	64.4	430	100	11.0	19.0	1.23
380 x 100 x 54	54.0	380	100	9.5	17.5	1.13
300 x 100 x 46	45.5	300	100	9.0	16.5	0.969
300 x 90 x 41	41.4	300	90	9.0	15.5	0.932
260 x 90 x 35	34.8	260	90	8.0	14.0	0.854
260 x 75 x 28	27.6	260	75	7.0	12.0	0.79
230 x 90 x 32	32.2	230	90	7.5	14.0	0.795
230 x 75 x 26	25.7	230	75	6.5	12.5	0.737
200 x 90 x 30	29.7	200	90	7.0	14.0	0.736
200 x 75 x 23	23.4	200	75	6.0	12.5	0.678
180 x 90 x 26	26.1	180	90	6.5	12.5	0.697
180 x 75 x 20	20.3	180	75	6.0	10.5	0.638
150 x 90 x 24	23.9	150	90	6.5	12.0	0.637
150 x 75 x 18	17.9	150	75	5.5	10.0	0.579
125 x 65 x 15	14.8	125	65	5.5	9.5	0.489
100 x 50 x 10	10.2	100	50	5.0	8.5	0.382

STRUCTURAL METALWORK – continued

Equal Angles BS 4848: Part 4: 1972

Designation	Mass kg/m	Surface area m^2/m
200 x 200 x 24	71.1	0.790
200 x 200 x 20	59.9	0.790
200 x 200 x 18	54.2	0.790
200 x 200 x 16	48.5	0.790
150 x 150 x 18	40.1	0.59
150 x 150 x 15	33.8	0.59
150 x 150 x 12	27.3	0.59
150 x 150 x 10	23.0	0.59
120 x 120 x 15	26.6	0.47
120 x 120 x 12	21.6	0.47
120 x 120 x 10	18.2	0.47
120 x 120 x 8	14.7	0.47
100 x 100 x 15	21.9	0.39
100 x 100 x 12	17.8	0.39
100 x 100 x 10	15.0	0.39
100 x 100 x 8	12.2	0.39
90 x 90 x 12	15.9	0.35
90 x 90 x 10	13.4	0.35
90 x 90 x 8	10.9	0.35
90 x 90 x 7	9.61	0.35
90 x 90 x 6	8.30	0.35

Unequal Angles BS 4848: Part 4: 1972

Designation	Mass kg/m	Surface area m^2/m
200 x 150 x 18	47.1	0.69
200 x 150 x 15	39.6	0.69
200 x 150 x 12	32.0	0.69
200 x 100 x 15	33.7	0.59
200 x 100 x 12	27.3	0.59
200 x 100 x 10	23.0	0.59
150 x 90 x 15	26.6	0.47
150 x 90 x 12	21.6	0.47
150 x 90 x 10	18.2	0.47
150 x 75 x 15	24.8	0.44
150 x 75 x 12	20.2	0.44
150 x 75 x 10	17.0	0.44
125 x 75 x 12	17.8	0.40
125 x 75 x 10	15.0	0.40
125 x 75 x 8	12.2	0.40
100 x 75 x 12	15.4	0.34
100 x 75 x 10	13.0	0.34
100 x 75 x 8	10.6	0.34
100 x 65 x 10	12.3	0.32
100 x 65 x 8	9.94	0.32
100 x 65 x 7	8.77	0.32

STRUCTURAL METALWORK – continued

Structural tees split from universal beams BS 4: Part 1: 1993

Designation	Mass kg/m	Surface area m^2/m
305 x 305 x 90	89.5	1.22
305 x 305 x 75	74.6	1.22
254 x 343 x 63	62.6	1.19
229 x 305 x 70	69.9	1.07
229 x 305 x 63	62.5	1.07
229 x 305 x 57	56.5	1.07
229 x 305 x 51	50.6	1.07
210 x 267 x 61	61.0	0.95
210 x 267 x 55	54.5	0.95
210 x 267 x 51	50.5	0.95
210 x 267 x 46	46.1	0.95
210 x 267 x 41	41.1	0.95
191 x 229 x 49	49.2	0.84
191 x 229 x 45	44.6	0.84
191 x 229 x 41	41.0	0.84
191 x 229 x 37	37.1	0.84
191 x 229 x 34	33.6	0.84
152 x 229 x 41	41.0	0.76
152 x 229 x 37	37.1	0.76
152 x 229 x 34	33.6	0.76
152 x 229 x 30	29.9	0.76
152 x 229 x 26	26.2	0.76

Universal Bearing Piles BS 4: Part 1: 1993

Designation	Mass	Depth of Section	Width of Section	Thickness		
				Web	Flange	
	Kg/m	mm	mm	mm	mm	
356 x 368 x 174	173.9	361.4	378.5	20.3	20.4	
356 x 368 x 152	152.0	356.4	376.0	17.8	17.9	
356 x 368 x 133	133.0	352.0	373.8	15.6	15.7	
356 x 368 x 109	108.9	346.4	371.0	12.8	12.9	
305 x 305 x 223	222.9	337.9	325.7	30.3	30.4	
305 x 305 x 186	186.0	328.3	320.9	25.5	25.6	
305 x 305 x 149	149.1	318.5	316.0	20.6	20.7	
305 x 305 x 126	126.1	312.3	312.9	17.5	17.6	
305 x 305 x 110	110.0	307.9	310.7	15.3	15.4	
305 x 305 x 95	94.9	303.7	308.7	13.3	13.3	
305 x 305 x 88	88.0	301.7	307.8	12.4	12.3	
305 x 305 x 79	78.9	299.3	306.4	11.0	11.1	
254 x 254 x 85	85.1	254.3	260.4	14.4	14.3	
254 x 254 x 71	71.0	249.7	258.0	12.0	12.0	
254 x 254 x 63	63.0	247.1	256.6	10.6	10.7	
203 x 203 x 54	53.9	204.0	207.7	11.3	11.4	
203 x 203 x 45	44.9	200.2	205.9	9.5	9.5	

STRUCTURAL METALWORK – continued

Hot Formed Square Hollow Sections EN 10210 S275J2H & S355J2H

Size (mm)	Wall thickness (mm)	Mass (kg/m)	Superficial area (m^2/m)
40 x 40	2.5	2.89	0.154
	3.0	3.41	0.152
	3.2	3.61	0.152
	3.6	4.01	0.151
	4.0	4.39	0.150
	5.0	5.28	0.147
50 x 50	2.5	3.68	0.194
	3.0	4.35	0.192
	3.2	4.62	0.192
	3.6	5.14	0.191
	4.0	5.64	0.190
	5.0	6.85	0.187
	6.0	7.99	0.185
	6.3	8.31	0.184
60 x 60	3.0	5.29	0.232
	3.2	5.62	0.232
	3.6	6.27	0.231
	4.0	6.90	0.230
	5.0	8.42	0.227
	6.0	9.87	0.225
	6.3	10.30	0.224
	8.0	12.50	0.219
70 x 70	3.0	6.24	0.272
	3.2	6.63	0.272
	3.6	7.40	0.271
	4.0	8.15	0.270
	5.0	9.99	0.267
	6.0	11.80	0.265
	6.3	12.30	0.264
	8.0	15.00	0.259

Hot Formed Square Hollow Sections - continued

Size (mm)	Wall thickness (mm)	Mass (kg/m)	Superficial area (m^2/m)
80 x 80	3.2	7.63	0.312
	3.6	8.53	0.311
	4.0	9.41	0.310
	5.0	11.60	0.307
	6.0	13.60	0.305
	6.3	14.20	0.304
	8.0	17.50	0.299
90 x 90	3.6	9.66	0.351
	4.0	10.70	0.350
	5.0	13.10	0.347
	6.0	15.50	0.345
	6.3	16.20	0.344
	8.0	20.10	0.339
100 x 100	3.6	10.80	0.391
	4.0	11.90	0.390
	5.0	14.70	0.387
	6.0	17.40	0.385
	6.3	18.20	0.384
	8.0	22.60	0.379
	10.0	27.40	0.374
120 x 120	4.0	14.40	0.470
	5.0	17.80	0.467
	6.0	21.20	0.465
	6.3	22.20	0.464
	8.0	27.60	0.459
	10.0	33.70	0.454
	12.0	39.50	0.449
	12.5	40.90	0.448

STRUCTURAL METALWORK – continued

Hot Formed Square Hollow Sections EN 10210 S275J2H & S355J2H - continued

Size (mm)	Wall thickness (mm)	Mass (kg/m)	Superficial area (m^2/m)
140 x 140	5.0	21.00	0.547
	6.0	24.90	0.545
	6.3	26.10	0.544
	8.0	32.60	0.539
	10.0	40.00	0.534
	12.0	47.00	0.529
	12.5	48.70	0.528
150 x 150	5.0	22.60	0.587
	6.0	26.80	0.585
	6.3	28.10	0.584
	8.0	35.10	0.579
	10.0	43.10	0.574
	12.0	50.80	0.569
	12.5	52.70	0.568
Hot formed from seamless hollow	16.0	65.2	0.559
160 x 160	5.0	24.10	0.627
	6.0	28.70	0.625
	6.3	30.10	0.624
	8.0	37.60	0.619
	10.0	46.30	0.614
	12.0	54.60	0.609
	12.5	56.60	0.608
	16.0	70.20	0.599
180 x 180	5.0	27.30	0.707
	6.0	32.50	0.705
	6.3	34.00	0.704
	8.0	42.70	0.699
	10.0	52.50	0.694
	12.0	62.10	0.689
	12.5	64.40	0.688
	16.0	80.20	0.679

Hot Formed Square Hollow Sections - continued

Size (mm)	Wall thickness (mm)	Mass (kg/m)	Superficial area (m^2/m)
200 x 200	5.0	30.40	0.787
	6.0	36.20	0.785
	6.3	38.00	0.784
	8.0	47.70	0.779
	10.0	58.80	0.774
	12.0	69.60	0.769
	12.5	72.30	0.768
	16.0	90.30	0.759
250 x 250	5.0	38.30	0.987
	6.0	45.70	0.985
	6.3	47.90	0.984
	8.0	60.30	0.979
	10.0	74.50	0.974
	12.0	88.50	0.969
	12.5	91.90	0.968
	16.0	115.00	0.959
300 x 300	6.0	55.10	1.18
	6.3	57.80	1.18
	8.0	72.80	1.18
	10.0	90.20	1.17
	12.0	107.00	1.17
	12.5	112.00	1.17
	16.0	141.00	1.16
350 x 350	8.0	85.40	1.38
	10.0	106.00	1.37
	12.0	126.00	1.37
	12.5	131.00	1.37
	16.0	166.00	1.36
400 x 400	8.0	97.90	1.58
	10.0	122.00	1.57
	12.0	145.00	1.57
	12.5	151.00	1.57
	16.0	191.00	1.56
(Grade S355J2H only)	20.00 *	235.00	1.55

*** SAW process**

STRUCTURAL METALWORK – continued

Hot Formed Square Hollow Sections JUMBO RHS : JIS G3136

Size (mm)	Wall thickness (mm)		Mass (kg/m)	Superficial area (m^2/m)
350 x 350	19.0		190.00	1.33
	22.0		217.00	1.32
	25.0		242.00	1.31
400 x 400	22.0		251.00	1.52
	25.0		282.00	1.51
450 x 450	12.0		162.00	1.76
	16.0		213.00	1.75
	19.0		250.00	1.73
	22.0		286.00	1.72
	25.0		321.00	1.71
	28.0	*	355.00	1.70
	32.0	*	399.00	1.69
500 x 500	12.0		181.00	1.96
	16.0		238.00	1.95
	19.0		280.00	1.93
	22.0		320.00	1.92
	25.0		360.00	1.91
	28.0	*	399.00	1.90
	32.0	*	450.00	1.89
	36.0	*	498.00	1.88
550 x 550	16.0		263.00	2.15
	19.0		309.00	2.13
	22.0		355.00	2.12
	25.0		399.00	2.11
	28.0	*	443.00	2.10
	32.0	*	500.00	2.09
	36.0	*	555.00	2.08
	40.0	*	608.00	2.06
600 x 600	25.0	*	439.00	2.31
	28.0	*	487.00	2.30
	32.0	*	550.00	2.29
	36.0	*	611.00	2.28
	40.0	*	671.00	2.26

Hot Formed Square Hollow Sections JUMBO RHS : JIS G3136

Size (mm)	Wall thickness (mm)		Mass (kg/m)	Superficial area (m^2/m)
700 x 700	25.0	*	517.00	2.71
	28.0	*	575.00	2.70
	32.0	*	651.00	2.69
	36.0	*	724.00	2.68
	40.0	*	797.00	2.68
Note:				
*** SAW process**				

Hot Formed Rectangular Hollow Sections: EN10210 S275J2h & S355J2H

Size (mm)	Wall thickness (mm)	Mass (kg/m)	Superficial area (m^2/m)
50 x 30	2.5	2.89	0.154
	3.0	3.41	0.152
	3.2	3.61	0.152
	3.6	4.01	0.151
	4.0	4.39	0.150
	5.0	5.28	0.147
60 x 40	2.5	3.68	0.194
	3.0	4.35	0.192
	3.2	4.62	0.192
	3.6	5.14	0.191
	4.0	5.64	0.190
	5.0	6.85	0.187
	6.0	7.99	0.185
	6.3	8.31	0.184
80 x 40	3.0	5.29	0.232
	3.2	5.62	0.232
	3.6	6.27	0.231
	4.0	6.90	0.230
	5.0	8.42	0.227
	6.0	9.87	0.225
	6.3	10.30	0.224
	8.0	12.50	0.219
76.2 x 50.8	3.0	5.62	0.246
	3.2	5.97	0.246
	3.6	6.66	0.245
	4.0	7.34	0.244

STRUCTURAL METALWORK – continued

Hot Formed Rectangular Hollow Sections - continued

Size (mm)	Wall thickness (mm)	Mass (kg/m)	Superficial area (m^2/m)
76.2 x 50.8	5.0	8.97	0.241
	6.0	10.50	0.239
	6.3	11.00	0.238
	8.0	13.40	0.233
90 x 50	3.0	6.24	0.272
	3.2	6.63	0.272
	3.6	7.40	0.271
	4.0	8.15	0.270
	5.0	9.99	0.267
	6.0	11.80	0.265
	6.3	12.30	0.264
	8.0	15.00	0.259
100 x 50	3.0	6.71	0.292
	3.2	7.13	0.292
	3.6	7.96	0.291
	4.0	8.78	0.290
	5.0	10.80	0.287
	6.0	12.70	0.285
	6.3	13.30	0.284
	8.0	16.30	0.279
100 x 60	3.0	7.18	0.312
	3.2	7.63	0.312
	3.6	8.53	0.311
	4.0	9.41	0.310
	5.0	11.60	0.307
	6.0	13.60	0.305
	6.3	14.20	0.304
	8.0	17.50	0.299
120 x 60	3.6	9.70	0.351
	4.0	10.70	0.350
	5.0	13.10	0.347
	6.0	15.50	0.345
	6.3	16.20	0344
	8.0	20.10	0.339

Hot Formed Rectangular Hollow Sections - continued

Size (mm)	Wall thickness (mm)	Mass (kg/m)	Superficial area (m^2/m)
120 x 80	3.6	10.80	0.391
	4.0	11.90	0.390
	5.0	14.70	0.387
	6.0	17.40	0.385
	6.3	18.20	0.384
	8.0	22.60	0.379
	10.0	27.40	0.374
150 x 100	4.0	15.10	0.490
	5.0	18.60	0.487
	6.0	22.10	0.485
	6.3	23.10	0.484
	8.0	28.90	0.479
	10.0	35.30	0.474
	12.0	41.40	0.469
	12.5	42.80	0.468
160 x 80	4.0	14.40	0.470
	5.0	17.80	0.467
	6.0	21.20	0.465
	6.3	22.20	0.464
	8.0	27.60	0.459
	10.0	33.70	0.454
	12.0	39.50	0.449
	12.5	40.90	0.448
200 x 100	5.0	22.60	0.587
	6.0	26.80	0.585
	6.3	28.10	0.584
	8.0	35.10	0.579
	10.0	43.10	0.574
	12.0	50.80	0.569
	12.5	52.70	0.568
	16.0	65.20	0.559
250 x 150	5.0	30.40	0.787
	6.0	36.20	0.785
	6.3	38.00	0.784

STRUCTURAL METALWORK – continued

Hot Formed Rectangular Hollow Sections – continued

Size (mm)	Wall thickness (mm)	Mass (kg/m)	Superficial area (m^2/m)
250 x 150	8.0	47.70	0.779
	10.0	58.80	0.774
	12.0	69.60	0.769
	12.5	72.30	0.768
	16.0	90.30	0.759
300 x 200	5.0	38.30	0.987
	6.0	45.70	0.985
	6.3	47.90	0.984
	8.0	60.30	0.979
	10.0	74.50	0.974
	12.0	88.50	0.969
	12.5	91.90	0.968
	16.0	115.00	0.959
400 x 200	6.0	55.10	1.18
	6.3	57.80	1.18
	8.0	72.80	1.18
	10.0	90.20	1.17
	12.0	107.00	1.17
	12.5	112.00	1.17
	16.0	141.00	1.16
450 x 250	8.0	85.40	1.38
	10.0	106.00	1.37
	12.0	126.00	1.37
	12.5	131.00	1.37
	16.0	166.00	1.36
500 x 300	8.0	98.00	1.58
	10.0	122.00	1.57
	12.0	145.00	1.57
	12.5	151.00	1.57
	16.0	191.00	1.56
	20.0	235.00	1.55

Hot Formed Circular Hollow Sections EN 10210 S275J2H & S355J2H

Outside Diameter (mm)	Wall thickness (mm)	Mass (kg/m)	Superficial area (m^2/m)
21.3	3.2	1.43	0.067
26.9	3.2	1.87	0.085
33.7	3.0	2.27	0.106
	3.2	2.41	0.106
	3.6	2.67	0.106
	4.0	2.93	0.106
42.4	3.0	2.91	0.133
	3.2	3.09	0.133
	3.6	3.44	0.133
	4.0	3.79	0.133
48.3	2.5	2.82	0.152
	3.0	3.35	0.152
	3.2	3.56	0.152
	3.6	3.97	0.152
	4.0	4.37	0.152
	5.0	5.34	0.152
60.3	2.5	3.56	0.189
	3.0	4.24	0.189
	3.2	4.51	0.189
	3.6	5.03	0.189
	4.0	5.55	0.189
	5.0	6.82	0.189
76.1	2.5	4.54	0.239
	3.0	5.41	0.239
	3.2	5.75	0.239
	3.6	6.44	0.239
	4.0	7.11	0.239
	5.0	8.77	0.239
	6.0	10.40	0.239
	6.3	10.80	0.239
88.9	2.5	5.33	0.279
	3.0	6.36	0.279
	3.2	6.76	0.27
	3.6	7.57	0.279

STRUCTURAL METALWORK – continued

Hot Formed Circular Hollow Sections - continued

Outside Diameter (mm)	Wall thickness (mm)	Mass (kg/m)	Superficial area (m^2/m)
88.9	4.0	8.38	0.279
	5.0	10.30	0.279
	6.0	12.30	0.279
	6.3	12.80	0.279
114.3	3.0	8.23	0.359
	3.2	8.77	0.359
	3.6	9.83	0.359
	4.0	10.09	0.359
	5.0	13.50	0.359
	6.0	16.00	0.359
	6.3	16.80	0.359
139.7	3.2	10.80	0.439
	3.6	12.10	0.439
	4.0	13.40	0.439
	5.0	16.60	0.439
	6.0	19.80	0.439
	6.3	20.70	0.439
	8.0	26.00	0.439
	10.0	32.00	0.439
168.3	3.2	13.00	0.529
	3.6	14.60	0.529
	4.0	16.20	0.529
	5.0	20.10	0.529
	6.0	24.00	0.529
	6.3	25.20	0.529
	8.0	31.60	0.529
	10.0	39.00	0.529
	12.0	46.30	0.529
	12.5	48.00	0.529
193.7	5.0	23.30	0.609
	6.0	27.80	0.609
	6.3	29.10	0.609
	8.0	36.60	0.609
	10.0	45.30	0.609

Hot Formed Circular Hollow Sections - continued

Outside Diameter (mm)	Wall thickness (mm)	Mass (kg/m)	Superficial area (m^2/m)
193.7	12.0	53.80	0.609
	12.5	55.90	0.609
219.1	5.0	26.40	0.688
	6.0	31.50	0.688
	6.3	33.10	0.688
	8.0	41.60	0.688
	10.0	51.60	0.688
	12.0	61.30	0.688
	12.5	63.70	0.688
	16.0	80.10	0.688
244.5	5.0	29.50	0.768
	6.0	35.30	0.768
	6.3	37.00	0.768
	8.0	46.70	0.768
	10.0	57.80	0.768
	12.0	68.80	0.768
	12.5	71.50	0.768
	16.0	90.20	0.768
273.0	5.0	33.00	0.858
	6.0	39.50	0.858
	6.3	41.40	0.858
	8.0	52.30	0.858
	10.0	64.90	0.858
	12.0	77.20	0.858
	12.5	80.30	0.858
	16.0	101.00	0.858
323.9	5.0	39.30	1.02
	6.0	47.00	1.02
	6.3	49.30	1.02
	8.0	62.30	1.02
	10.0	77.40	1.02
	12.0	92.30	1.02
	12.5	96.00	1.02
	16.0	121.00	1.02

STRUCTURAL METALWORK – continued

Hot Formed Circular Hollow Sections - continued

Outside Diameter (mm)	Wall thickness (mm)	Mass (kg/m)	Superficial area (m^2/m)
355.6	6.3	54.30	1.12
	8.0	68.60	1.12
	10.0	85.30	1.12
	12.0	102.00	1.12
	12.5	106.00	1.12
	16.0	134.00	1.12
406.4	6.3	62.20	1.28
	8.0	79.60	1.28
	10.0	97.80	1.28
	12.0	117.00	1.28
	12.5	121.00	1.28
	16.0	154.00	1.28
457.0	6.3	70.00	1.44
	8.0	88.60	1.44
	10.0	110.00	1.44
	12.0	132.00	1.44
	12.5	137.00	1.44
	16.0	174.00	1.44
508.0	6.3	77.90	1.60
	8.0	98.60	1.60
	10.0	123.00	1.60
	12.0	147.00	1.60
	12.5	153.00	1.60
	16.0	194.00	1.60

SPACING OF HOLES IN ANGLES

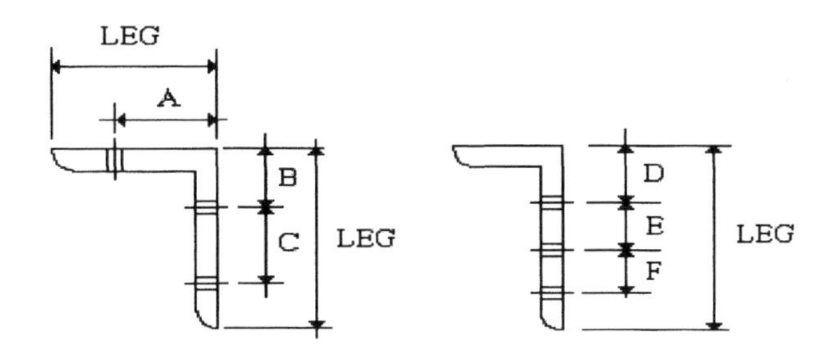

Nominal Leg Length mm	Spacing of holes						Maximum diameter of bolt or rivet		
	A	B	C	D	E	F	A	B and C	D, E and F
200		75	75	55	55	55		30	20
150		55	55					20	
125		45	60					20	
120									
100	55						24		
90	50						24		
80	45						20		
75	45						20		
70	40						20		
65	35						20		
60	35						16		
50	28						12		
45	25								
40	23								
30	20								
25	15								

PAVING AND SURFACING

Precast Concrete Kerbs to BS 7263

Straight kerb units: length from 450 to 915 mm

150mm high x 125mm thick			
bullnosed		type BN	
half battered		type HB3	
255mm high x 125mm thick			
45 degree splayed		type SP	
half battered		type HB2	
305mm high x 150mm thick			
half battered		type HB1	
Quadrant kerb units			
150 mm high x 305 and 455 mm radius to match		type BN	type QBN
150 mm high x 305 and 455 mm radius to match		type HB2, HB3	type QHB
150 mm high x 305 and 455 mm radius to match		type SP	type QSP
255 mm high x 305 and 455 mm radius to match		type BN	type QBN
255 mm high x 305 and 455 mm radius to match		type HB2, HB3	type QHB
225 mm high x 305 and 455 mm radius to match		type SP	type QSP
Angle kerb units			
305 x 305 x 225 mm high x 125 mm thick			
bullnosed external angle		type XA	
splayed external angle to match type SP		type XA	
bullnosed internal angle		type IA	
splayed internal angle to match type SP		type IA	
Channels			
255 mm wide x 125 mm high flat		type CS1	
150 mm wide x 125 mm high flat type		CS2	
255 mm wide x 125 mm high dished		type CD	
Transition kerb units			
from kerb type SP to HB	left handed	type TL	
	right handed	type TR	
from kerb type BN to HB	left handed	type DL1	
	right handed	type DR1	
from kerb type BN to SP	left handed	type DL2	
	right handed	type DR2	

Radial kerbs and channels

All profiles of kerbs and channels	
External radius	**Internal radius**
1000 mm	3000
2000	4500
3000	6000
4500	7500
6000	9000
7500	1050
9000	1200
1050	
1200	

Precast Concrete Edgings to BS 7263

Round top type ER	Flat top type EF	Bullnosed top type EBN
150 x 50 mm	150 x 50 mm	150 x 50 mm
200 x 50	200 x 50	200 x 50
250 x 50	250 x 50	250 x 50

Bases

Cement Bound Material for Bases and Sub-bases	
CBM1:	very carefully graded aggregate from 37.5 - 75ym, with a 7-day strength of 4.5N/mm^2
CBM2:	same range of aggregate as CBM1 but with more tolerance in each size of aggregate with a 7-day strength of 7.0N/mm^2
CBM3:	crushed natural aggregate or blastfurnace slag, graded from 37.5mm - 150ym for 40mm aggregate, and from 20 – 75 ym for 20mm aggregate, with a 7-day strength of 10N/mm^2
CBM4:	crushed natural aggregate or blastfurnace slag, graded from 37.5mm - 150ym for 40mm aggregate, and from 20 - 75ym for 20mm aggregate, with a 7-day strength of 15N/mm^2

PAVING AND SURFACING – continued

Interlocking block and other pavings

Sizes of Precast Concrete Paving Blocks to BS 6717: Part 1

Type R blocks	Type S
200 x 100 x 60 mm	Any shape within a 295 mm space
200 x 100 x 65	
200 x 100 x 80	
200 x 100 x 100	

Sizes of Clay Brick Pavers to BS 6677: Part 1

200 x 100 x 50 mm thick

200 x 100 x 65

210 x 105 x 50

210 x 105 x 65

215 x 102.5 x 50

215 x 102.5 x 65

Type PA: 3 kN

Footpaths and pedestrian areas, private driveways, car parks, light vehicle traffic and over-run.

Type PB: 7 kN

Residential roads, lorry parks, factory yards, docks, petrol station forecourts, hardstandings, bus stations.

Weights and sizes of paving and surfacing

Description of Item		Quantity per tonne
Paving 50 mm thick	900 x 600 mm	15
Paving 50 mm thick	750 x 600 mm	18
Paving 50 mm thick	600 x 600 mm	23
Paving 50 mm thick	450 x 600 mm	30
Paving 38 mm thick	600 x 600 mm	30
Path edging	914 x 50 x 150 mm	60
Kerb (including radius and tapers)	125 x 254 x 914 mm	15
Kerb (including radius and tapers)	125 x 150 x 914 mm	25
Square channel	125 x 254 x 914 mm	15
Dished channel	125 x 254 x 914 mm	15
Quadrants	300 x 300 x 254 mm	19
Quadrants	450 x 450 x 254 mm	12
Quadrants	300 x 300 x 150 mm	30
Internal angles	300 x 300 x 254 mm	30
Fluted pavement channel	255 x 75 x 914 mm	25
Corner stones	300 x 300 mm	80
Corner stones	360 x 360 mm	60
Cable covers	914 x 175 mm	55
Gulley kerbs	220 x 220 x 150 mm	60
Gulley kerbs	220 x 200 x 75 mm	120

PAVING AND SURFACING – continued

Weights and sizes of paving and surfacing - continued

Material	**kg/m³**	**lb/cu yd**
Tarmacadam	2306	3891
Macadam (waterbound)	2563	4325
Vermiculite (aggregate)	64-80	108-135
Terracotta	2114	3568
Cork - compressed	388	24
	kg/m²	**lb/sq ft**
Clay floor tiles, 12.7 mm	27.3	5.6
Pavement lights	122	25
Damp proof course)	5	1
	kg/m² per mm thickness	**lb/sq ft per inch thickness**
Paving Slabs (stone)	2.3	12
Granite setts	2.88	15
Asphalt	2.30	12
Rubber flooring	1.68	9
Poly-vinylchloride	1.94 (avg)	10 (avg)

Coverage (m²) per cubic metre of materials used as sub bases or capping layers

Consolidated thickness laid in (mm)	**Square metre coverage**		
	Gravel	**Sand**	**Hardcore**
50	15.80	16.50	-
75	10.50	11.00	-
100	7.92	8.20	7.42
125	6.34	6.60	5.90
150	5.28	5.50	4.95
175	-	-	4.23
200	-	-	3.71
225	-	-	3.30
300	-	-	2.47

Approximate rate of spreads

Average thickness of course mm	Description	Approximate rate of spread			
		Open Textured		Dense, Medium & Fine Textured	
		kg/m²	m²/t	kg/m²	m²/t
35	14 mm open textured or dense wearing course	60-75	13-17	70-85	12-14
40	20 mm open textured or dense base course	70-85	12-14	80-100	10-12
45	20 mm open textured or dense base course	80-100	10-12	95-100	9-10
50	20 mm open textured or dense, or 28 mm dense base course	85-110	9-12	110-120	8-9
60	28 mm dense base course, 40 mm open textured of dense base course or 40 mm single course as base course		8-10	130-150	7-8
65	28 mm dense base course, 40 mm open textured or dense base course or 40 mm single course	100-135	7-10	140-160	6-7
75	40 mm single course, 40 mm open textured or dense base course, 40 mm dense roadbase	120-150	7-8	165-185	5-6
100	40 mm dense base course or roadbase	-	-	220-240	4-4.5

Surface Dressing Roads: Coverage (m²) per tonne of Material

Size in mm	Sand	Granite chips	Gravel	Limestone Chips
Sand	168	-	-	-
3	-	148	152	165
6	-	130	133	144
9	-	111	114	123
13	-	85	87	95
19	-	68	71	78

PAVING AND SURFACING - continued

Sizes of Flags to BS 7263

Reference	Nominal Size	Thickness
A	600 x 450 mm	50 and 63 mm
B	600 x 600	50 and 63
C	600 x 750	50 and 63
D	600 x 900	50 and 63
E	450 x 450	50 and 70 chamfered top surface
F	400 x 400	50 and 65 chamfered top surface
G	300 x 300	50 and 60 chamfered top surface

Sizes of Natural Stone Setts to BS 435

Width		Length		Depth
100 mm	x	100 mm	x	100 mm
75	x	150 to 250	x	125
75	x	150 to 250	x	150
100	x	150 to 250	x	100
100	x	150 to 250	x	150

SEEDING/TURFING AND PLANTING

BS 3882: 1994 Topsoil Quality

Topsoil grade	Properties
Premium	natural topsoil, high fertility, loamy texture, good soil structure, suitable for intensive cultivation
General Purpose	natural or manufactured topsoil of lesser quality than Premium, suitable for agriculture or amenity landscape, may need fertilizer or soil structure improvement.
Economy	selected subsoil, natural mineral deposit such as river silt or greensand. The grade comprises two subgrades; "Low clay" and "High clay" which is more liable to compaction in handling. This grade is suitable for low production agricultural land and amenity woodland or conservation planting areas.

Forms of Trees to BS 3936: 1992

Standards:	shall be clear with substantially straight stems. Grafted and budded trees shall have no more than a slight bend at the union. Standards shall be designated as Half, Extra light, Light, Standard, Selected standard, Heavy, and Extra heavy.
Sizes of Standards	
Heavy standard	12-14 cm girth x 3.50 to 5.00 m high
Extra Heavy standard	14-16 cm girth x 4.25 to 5.00 m high
Extra Heavy standard	16-18 cm girth x 4.25 to 6.00 m high
Extra Heavy standard	18-20 cm girth x 5.00 to 6.00 m high
Semi-mature trees:	between 6.0m and 12.0 m tall with a girth of 20 to 75 cm at 1.0 m above ground.
Feathered trees:	shall have a defined upright central leader, with stem furnished with evenly spread and balanced lateral shoots down to or near the ground.
Whips:	shall be without significant feather growth as determined by visual inspection.
Multi-stemmed trees:	shall have two or more main stems at, near, above or below ground.

Seedlings grown from seed and not transplanted shall be specified when ordered for sale as:

1+0 one year old seedling		
2+0 two year old seedling		
1+1 one year seed bed,	one year transplanted	= two year old seedling
1+2 one year seed bed,	two years transplanted	= three year old seedling
2+1 two year seed bed,	one year transplanted	= three year old seedling
1u1 two years seed bed,	undercut after 1 year	= two year old seedling
2u2 four years seed bed,	undercut after 2 years	= four year old seedling

Cuttings: The age of cuttings (plants grown from shoots, stems, or roots of the mother plant) shall be specified when ordered for sale. The height of transplants and undercut seedlings/cuttings (which have been transplanted or undercut at least once) shall be stated in centimetres. The number of growing seasons before and after transplanting or undercutting shall be stated.

0+1	one year cutting
0+2	two year cutting
0+1+1	one year cutting bed, one year transplanted = two year old seedling
0+1+2	one year cutting bed, two years transplanted = three year old seedling

SEEDING/TURFING AND PLANTING - continued

Grass Cutting Capacities in m“² per hour

Speed mph	Width Of Cut in metres												
	0.5	0.7	1.0	1.2	1.5	1.7	2.0	2.0	2.1	2.5	2.8	3.0	3.4
1.0	724	1127	1529	1931	2334	2736	3138	3219	3380	4023	4506	4828	5472
1.5	1086	1690	2293	2897	3500	4104	4707	4828	5069	6035	6759	7242	8208
2.0	1448	2253	3058	3862	4667	5472	6276	6437	6759	8047	9012	9656	10944
2.5	1811	2816	3822	4828	5834	6840	7846	8047	8449	10058	11265	12070	13679
3.0	2173	3380	4587	5794	7001	8208	9415	9656	10139	12070	13518	14484	16415
3.5	2535	3943	5351	6759	8167	9576	10984	11265	11829	14082	15772	16898	19151
4.0	2897	4506	6115	7725	9334	10944	12553	12875	13518	16093	18025	19312	21887
4.5	3259	5069	6880	8690	10501	12311	14122	14484	15208	18105	20278	21726	24623
5.0	3621	5633	7644	9656	11668	13679	15691	16093	16898	20117	22531	24140	27359
5.5	3983	6196	8409	10622	12834	15047	17260	17703	18588	22128	24784	26554	30095
6.0	4345	6759	9173	11587	14001	16415	18829	19312	20278	24140	27037	28968	32831
6.5	4707	7322	9938	12553	15168	17783	20398	20921	21967	26152	29290	31382	35566
7.0	5069	7886	10702	13518	16335	19151	21967	22531	23657	28163	31543	33796	38302

Number of plants per m² for various planting distances

mm	0.10	0.15	0.20	0.25	0.35	0.40	0.45	0.50	0.60	0.75	0.90	1.00	1.20	1.50
0.10	100.00	66.67	50.00	40.00	28.57	25.00	22.22	20.00	16.67	13.33	11.11	10.00	8.33	6.67
0.15	66.67	44.44	33.33	26.67	19.05	16.67	14.81	13.33	11.11	8.89	7.41	6.67	5.56	4.44
0.20	50.00	33.33	25.00	20.00	14.29	12.50	11.11	10.00	8.33	6.67	5.56	5.00	4.17	3.33
0.25	40.00	26.67	20.00	16.00	11.43	10.00	8.89	8.00	6.67	5.33	4.44	4.00	3.33	2.67
0.35	28.57	19.05	14.29	11.43	8.16	7.14	6.35	5.71	4.76	3.81	3.17	2.86	2.38	1.90
0.40	25.00	16.67	12.50	10.00	7.14	6.25	5.56	5.00	4.17	3.33	2.78	2.50	2.08	1.67
0.45	22.22	14.81	11.11	8.89	6.35	5.56	4.94	4.44	3.70	2.96	2.47	2.22	1.85	1.48
0.50	20.00	13.33	10.00	8.00	5.71	5.00	4.44	4.00	3.33	2.67	2.22	2.00	1.67	1.33
0.60	16.67	11.11	8.33	6.67	4.76	4.17	3.70	3.33	2.78	2.22	1.85	1.67	1.39	1.11
0.75	13.33	8.89	6.67	5.33	3.81	3.33	2.96	2.67	2.22	1.78	1.48	1.33	1.11	0.89
0.90	11.11	7.41	5.56	4.44	3.17	2.78	2.47	2.22	1.85	1.48	1.23	1.11	0.93	0.74
1.00	10.00	6.67	5.00	4.00	2.86	2.50	2.22	2.00	1.67	1.33	1.11	1.00	0.83	0.67
1.20	8.33	5.56	4.17	3.33	2.38	2.08	1.85	1.67	1.39	1.11	0.93	0.83	0.69	0.56
1.50	6.67	4.44	3.33	2.67	1.90	1.67	1.48	1.33	1.11	0.89	0.74	0.67	0.56	0.44

Grass Clippings Wet: Based on 3.5 m³ /ton

Annual Kg/100 m²	Average 20 cuts Kg/100m²	m² /tonne	m² /m³
32.0	1.6	61162.1	214067.3

Nr of Cuts	22	20	18	16	12	4
Kg/cut	1.45	1.60	1.78	2.00	2.67	8.00
	Area capacity in m² of 3 tonne vehicle per load					
	206250	187500	168750	150000	112500	37500
Load m³	**100 m² units / m³ of vehicle space**					
1	196.4	178.6	160.7	142.9	107.1	35.7
2	392.9	357.1	321.4	285.7	214.3	71.4
3	589.3	535.7	482.1	428.6	321.4	107.1
4	785.7	714.3	642.9	571.4	428.6	142.9
5	982.1	892.9	803.6	714.3	535.7	178.6

FENCING AND GATES

Types of Preservative to BS 5589:1989

Creosote (tar oil) can be "factory" applied	by pressure to BS 144: pts 1&2
	by immersion to BS 144: pt 1
	by hot and cold open tank to BS 144: pts 1&2
Copper/chromium/arsenic (CCA)	by full cell process to BS 4072 pts 1&2
Organic solvent (OS)	by double vacuum (vac vac) to BS 5707 pts 1&3
	by immersion to BS 5057 pts 1&3
Pentachlorophenol (PCP)	by heavy oil double vacuum to BS 5705 pts 2&3

Boron diffusion process (treated with disodium octaborate to BWPA Manual 1986.

Note: Boron is used on green timber at source and the timber is suppled dry.

Cleft Chestnut Pale Fences to BS 1722:Part 4:1986

Pales	Pale spacing	Wire lines	
900 mm long	75 mm	2	temporary protection
1050	75 or 100	2	light protective fences
1200	75	3	perimeter fences
1350	75	3	perimeter fences
1500	50	3	narrow perimeter fences
1800	50	3	light security fences

FENCING AND GATES – continued

Close-boarded Fences to BS 1722 :Pt 5: 1986

Close-boarded fences 1.05 to 1.8m high

Type BCR (recessed) or BCM (morticed) with concrete posts 140 x 115 mm tapered and Type BW with timber posts:

Palisade Fences to BS 1722:pt 6:1986.

Wooden palisade fences

Type WPC with concrete posts 140 x 115 mm tapered and Type WPW with timber posts.

For both types of fence:

Height of fence 1050 mm: two rails:

Height of fence 1200 mm: two rails:

Height of fence 1500 mm: three rails:

Height of fence 1650 mm: three rails:

Height of fence 1800 mm: three rails:

Post and Rail Fences to BS 1722:part 7

Wooden post and rail fences

Type MPR 11/3 morticed rails and Type SPR 11/3 nailed rails

Height to top of rail 1100 mm

Rails: three rails 87 mm 38 mm

Type MPR 11/4 morticed rails and Type SPR 11/4 nailed rails

Height to top of rail 1100 mm

Rails: four rails 87 mm 38 mm.

Type MPR 13/4 morticed rails and Type SPR 13/4 nailed rails

Height to top of rail 1300 mm

Rail spacing 250 mm, 250 mm, and 225 mm from top

Rails: four rails 87 mm 38 mm.

Steel Fences to BS 1722 :Pt 9: 1992

Mild steel fences: round or square verticals; flat standards and horizontals.

Tops of vertical bars may be bow-top, blunt, or pointed

Round or square bar railings.

	Fence height	Top/bottom rails and flat posts	Vertical bars
Light	1000 mm	40 x 10 mm 450 mm in ground	12 mm diameter at 115 mm centres
	1200 mm	40 x 10 mm 550 mm in ground	12 mm diameter at 115 mm centres
	1400 mm	40 x 10 mm 550 mm in ground	12 mm diameter at 115 mm centres
Light	1000 mm	40 x 10 mm 450 mm in ground	16 mm diameter at 120 mm centres
	1200 mm	40 x 10 mm 550 mm in ground	16 mm diameter at 120 mm centres
	1400 mm	40 x 10 mm 550 mm in ground	16 mm diameter at 120 mm centres
Medium	1200 mm	50 x 10 mm 550 mm in ground	20 mm diameter at 125 mm centres
	1400 mm	50 x 10 mm 550 mm in ground	20 mm diameter at 125 mm centres
	1600 mm	50 x 10 mm 600 mm in ground	22 mm diameter at 145 mm centres
	1800 mm	50 x 10 mm 600 mm in ground	22 mm diameter at 145 mm centres
Heavy	1600 mm	50 x 10 mm 600 mm in ground	22 mm diameter at 145 mm centres
	1800 mm	50 x 10 mm 600 mm in ground	22 mm diameter at 145 mm centres
	2000 mm	50 x 10 mm 600 mm in ground	22 mm diameter at 145 mm centres
	2200 mm	50 x 10 mm 600 mm in ground	22 mm diameter at 145 mm centres

Gates

Timber Field Gates to BS 3470: 1975

Gates made to this standard are designed to open one way only.

All timber gates are 1100 mm high.

Width over stiles 2400, 2700, 3000, 3300, 3600, and 4200 mm.

Gates over 4200 mm should be made in two leaves.

Steel Field Gates to BS 3470: 1975

Heavy duty:	width over stiles 2400, 3000, 3600 and 4500 mm
Light duty:	width over stiles 2400, 3000, and 3600 mm

All steel gates are 1100 mm high.

FENCING AND GATES – continued

Domestic Front Entrance Gates to BS 4092:part 1: 1966

Metal gates: Single gates are 900 mm high minimum, 900 mm, 1000 mm and 1100 mm wide

Domestic Front Entrance Gates to BS 4092:part 2: 1966.

Wooden gates: All rails shall be tenoned into the stiles

Single gates are 840 mm high minimum, 801 mm and 1020 mm wide

Double gates are 840 mm high minimum, 2130, 2340 and 2640 mm wide

Timber Bridle Gates to BS 5709:1979 (Horse Or Hunting Gates)

Gates open one way only

Minimum width between posts	1525 mm
Minimum height	1100 mm

Timber Kissing Gates to BS 5709:1979

Minimum width	700 mm
Minimum height	1000 mm
Minimum distance between shutting posts	600 mm
Minimum clearance at mid-point	600 mm

Metal Kissing Gates to BS 5709:1979

Sizes are the same as those for timber kissing gates.

Maximum gaps between rails 120 mm

Categories of Pedestrian Guard Rail to BS 3049:1976

Class A for normal use;

Class B where vandalism is expected;

Class C where crowd pressure is likely;

Squared timber for general fences

Posts	Fence height	Strut	Straining post
1300 x 75 x 75 mm	600 mm	1200 x 75 x 75 mm	1450 x 100 x 100 mm
1500 x 75 x 75 mm	800 mm	1400 x 75 x 75 mm	1650 x 100 x 100 mm
1600 x 75 x 75 mm	900 mm	1500 x 75 x 75 mm	1750 x 100 x 100 mm
1700 x 75 x 75 mm	1050 mm	1600 x 75 x 75 mm	1850 x 100 x 100 mm
1800 x 75 x 75 mm	1150 mm	1750 x 75 x 75 mm	2000 x 125 x 100 mm

Timber Posts to BS 1722: Part 2

Cleft Chestnut Pale Fences to BS 1722:part 4:1986

Timber posts for wire mesh and hexagonal wire netting fences.
Round timber for general fences

Posts	Fence height	Strut	Straining post
1300 x 65 mm dia.	600 mm	1200 x 80 mm dia	1450 x 100 mm dia
1500 x 65 mm dia	800 mm	1400 x 80 mm dia	1650 x 100 mm dia
1600 x 65 mm dia.	900 mm	1500 x 80 mm dia	1750 x 100 mm dia
1700 x 65 mm dia.	1050 mm	1600 x 80 mm dia	1850 x 100 mm dia
1800 x 65 mm dia.	1150 mm	1750 x 80 mm dia	2000 x 120 mm dia

Concrete Posts to BS 1722: Part 1

Concrete posts for chain link fencing:

Posts and straining posts	Fence height	Strut
1570 mm 100 x 100 mm	900 mm	1500 mm x 75 x 75 mm
1870 mm 125 x 125 mm	1200 mm	1830 mm x 100 x 75 mm
2070 mm 125 x 125 mm	1400 mm	1980 mm x 100 x 75 mm
2620 mm 125 x 125 mm	1800 mm	2590 mm x 100 x 85 mm
3040 mm 125 x 125 mm	1800 mm	2590 mm x 100 x 85 mm (with arms)

Concrete Posts to BS 1722: Part 2

Concrete posts for rectangular wire mesh (field) fencing

Posts	Fence height	Strut	Straining post
1270 x 100 x 100 mm	600 mm	1200 x 75 x 75 mm	1420 x 100 x 100 mm
1470 x 100 x 100 mm	800 mm	1350 x 75 x 75 mm	1620 x 100 x 100 mm
1570 x 100 x 100 mm	900 mm	1500 x 75 x 75 mm	1720 x 100 x 100 mm
1670 x 100 x 100 mm	600 mm	1650 x 75 x 75 mm	1820 x 100 x 100 mm
1820 x 125 x 125 mm	1150 mm	1830 x 75 x 100 mm	1970 x 125 x 125 mm

FENCING AND GATES – continued

Steel Posts to BS 1722: Part 1

Rolled steel angle iron posts for chain link fencing:

Posts	Fence height	Strut	Straining post
1500 x 40 x 40 x 5 mm	900 mm	1500 x 40 x 40 x 5 mm	1500 x 50 x 50 x 6 mm
1800 x 40 x 40 x 5 mm	1200 mm	1800 x 40 x 40 x 5 mm	1800 x 50 x 50 x 6 mm
2000 x 45 x 45 x 5 mm	1400 mm	2000 x 45 x 45 x 5 mm	2000 x 60 x 60 x 6 mm
2600 x 45 x 45 x 5 mm	1800 mm	2600 x 45 x 45 x 5 mm	2600 x 60 x 60 x 6 mm
3000 x 50 x 50 x 6 mm with arms	1800 mm	2600 x 45 x 45 x 5 mm	3000 x 60 x 60 x 6 mm

Rolled Steel Angle Posts to BS 1722: Part 2

Rolled steel angle posts for rectangular wire mesh (field) fencing

Posts	Fence height	Strut	Straining post
1200 x 40 x 40 x 5 mm	600 mm	1200 x 75 x 75 mm	1350 x 100 x 100 mm
1400 x 40 x 40 x 5 mm	800 mm	1400 x 75 x 75 mm	1550 x 100 x 100 mm
1500 x 40 x 40 x 5 mm	900 mm	1500 x 75 x 75 mm	1650 x 100 x 100 mm
1600 x 40 x 40 x 5 mm	1000 mm	1600 x 75 x 75 mm	1750 x 100 x 100 mm
1750 x 40 x 40 x 5 mm	1150 mm	1750 x 75 x 100 mm	1900 x 125 x 125 mm

DRAINAGE

Weights and dimensions - vitrified clay pipes

Product	Nominal diameter (mm)	Effective length (mm)	BS 65 limits of tolerance min (mm)	max (mm)	Crushing Strength (kN/m)	Weight kg/pipe	kg/m
Supersleve	100	1600	96	105	35.00	14.71	9.19
	150	1750	146	158	35.00	29.24	16.71
Hepsleve	225	1850	221	236	28.00	84.03	45.42
	300	2500	295	313	34.00	193.05	77.22
	150	1500	146	158	22.00	37.04	24.69

Weights and dimensions - vitrified clay pipes – continued

Hepseal	225	1750	221	236	28.00	85.47	48.84
	300	2500	295	313	34.00	204.08	81.63
	400	2500	394	414	44.00	357.14	142.86
	450	2500	444	464	44.00	454.55	181.63
	500	2500	494	514	48.00	555.56	222.22
	600	2500	591	615	57.00	796.23	307.69
	700	3000	689	719	67.00	1111.11	370.45
	800	3000	788	822	72.00	1351.35	450.45
Hepline	100	1600	95	107	22.00	14.71	9.19
	150	1750	145	160	22.00	29.24	16.71
	225	1850	219	239	28.00	84.03	45.42
	300	1850	292	317	34.00	142.86	77.22
Hepduct	90	1500	-	-	28.00	12.05	8.03
(Conduit)	100	1600	-	-	28.00	14.71	9.19
	125	1750	-	-	28.00	20.73	11.84
	150	1750	-	-	28.00	29.24	16.71
	225	1850	-	-	28.00	84.03	45.42
	300	1850	-	-	34.00	142.86	77.22

Nominal internal diameter (mm)	Nominal wall thickness (mm)	Approximate weight kg/m
150	25	45
225	29	71
300	32	122
375	35	162
450	38	191
600	48	317
750	54	454
900	60	616
1200	76	912
1500	89	1458
1800	102	1884
2100	127	2619

DRAINAGE - continued

Wall thickness, weights and pipe lengths vary, depending on type of pipe required.

The particulars shown above represent a selection of available diameters and are applicable to strength class 1 pipes with flexible rubber ring joints.

Tubes with Ogee joints are also available.

Weights and dimensions - PVC-U pipes

	Nominal size	Mean outside diameter (mm)		Wall thickness	Weight
		min	max	(mm)	kg/m
Standard pipes					
	82.4	82.4	82.7	3.2	1.2
	110.0	110.0	110.4	3.2	1.6
	160.0	160.0	160.6	4.1	3.0
	200.0	200.0	200.6	4.9	4.6
	250.0	250.0	250.7	6.1	7.2
Perforated pipes					
- heavy grade	As above	As above	As above	As above	As above
- thin wall	82.4	82.4	82.7	1.7	-
	110.0	110.0	110.4	2.2	-
	160.0	160.0	160.6	3.2	-

Width of trenches required for various diameters of pipes

Pipe diameter (mm)	Trench n.e. 1.5 m deep (mm)	Trench over 1.5 m deep (mm)
n.e.100	450	600
100-150	500	650
150-225	600	750
225-300	650	800
300-400	750	900
400-450	900	1050
450-600	1100	1300

Flow of water which can be carried by various sizes of pipe

Clay or concrete pipes

		Gradient of pipeline							
		1:10	1:20	1:30	1:40	1:50	1:60	1:80	1:100
Pipe size		**Flow in litres per second**							
DN 100	15.0	8.5	6.8	5.8	5.2	4.7	4.0	3.5	
DN 150	28.0	19.0	16.0	14.0	12.0	11.0	9.1	8.0	
DN 225	140.0	95.0	76.0	66.0	58.0	53.0	46.0	40.0	

Plastic pipes

	Gradient of pipeline							
	1:10	1:20	1:30	1:40	1:50	1:60	1:80	1:100
Pipe size	**Flow in litres per second**							
82.4mm i/dia	12.0	8.5	6.8	5.8	5.2	4.7	4.0	3.5
110mm i/dia	28.0	19.0	16.0	14.0	12.0	11.0	9.1	8.0
160mm i/dia	76.0	53.0	43.0	37.0	33.0	29.0	25.0	22.0
200mm i/dia	140.0	95.0	76.0	66.0	58.0	53.0	46.0	40.0

Vitrified (Perforated) Clay Pipes and Fittings to BS EN 295-5 1994

Length not specified

75 mm bore	**250 mm bore**	**600 mm bore**
100	300	700
125	350	800
150	400	1000
200	450	1200
225	500	

Pre-cast Concrete Pipes: Pre-stressed Non-pressure Pipes and Fittings: Flexible Joints to BS 5911 :Pt.103: 1994

Rationalized metric nominal sizes: 450, 500

Length:	500 -	1000 by 100 increments
	1000 -	2200 by 200 increments
	2200 -	2800 by 300 increments
Angles: length:	450 - 600 angles 45, 22.5,11.25 °	
	600 or more angles 22.5, 11.25 °	

DRAINAGE - continued

Pre-cast concrete pipes: un-reinforced and circular Manholes and soakaways to BS 5911 Pt.200: 1994

Nominal Sizes:

Shafts:	675, 900 mm
Chambers:	900, 1050, 1200, 1350, 1500, 1800, 2100, 2400, 2700, 3000 mm.
Large chambers:	To have either tapered reducing rings or a flat reducing slab in order to accept the standard cover.
Ring depths:	1. 300 - 1200 mm by 300 mm increments except for bottom slab and rings below cover slab, these are by 150 mm increments.
	2. 250 - 1000 mm by 250 mm increments except for bottom slab and rings below cover slab, these are by 125 mm increments.
Access hole:	750 x 750 mm for DN 1050 chamber 1200 x 675 mm for DN 1350 chamber

Pre-cast concrete Inspection Chambers and Gullies to BS 5911 :Pt.230: 1994

Nominal sizes:	375 diameter, 750, 900 mm deep 450 diameter, 750, 900, 1050, 1200 mm deep
Depths:	from the top for trapped or un-trapped units: centre of outlet 300 mm invert (bottom) of the outlet pipe 400 mm
Depth of water seal for trapped gullies:	85 mm, rodding eye int. diam. 100 mm
Cover slab:	65 mm min.

Ductile Iron Pipes to BS EN 598 : 1995

Type K9 with flexible joints should be used for surface water drainage. 5500 mm or 8000 mm long

80 mm bore	400 mm bore	1000 mm bore
100	450	1100
150	500	1200
200	600	1400
250	700	1600
300	800	
350	900	

Bedding flexible pipes: Pvc-u or Ductile Iron

Type 1	100mm fill below pipe, 300mm above pipe: single size material
Type 2	100mm fill below pipe, 300mm above pipe: single size or graded material
Type 3	100mm fill below pipe, 75mm above pipe with concrete protective slab over
Type 4	100mm fill below pipe, fill laid level with top of pipe
Type 5	200mm fill below pipe, fill laid level with top of pipe
Concrete	25mm sand blinding to bottom of trench, pipe supported on chocks, 100mm concrete under the pipe, 150mm concrete over the pipe.

Bedding Rigid Pipes: Clay Or Concrete

(for vitrified clay pipes the manufacturer should be consulted)

Class D:	Pipe laid on natural ground with cut-outs for joints, soil screened to remove stones over 40mm and returned over pipe to 150mm min depth. Suitable for firm ground with trenches trimmed by hand.
Class N:	Pipe laid on 50mm granular material of graded aggregate to Table 4 of BS 882, or 10 mm aggregate to Table 6 of BS 882, or as dug light soil (not clay) screened to remove stones over 10mm. Suitable for machine dug trenches.
Class B:	As Class N, but with granular bedding extending half way up the pipe diameter.
Class F:	Pipe laid on 100mm granular fill to BS 882 below pipe, minimum 150mm granular fill above pipe: single size material. Suitable for machine dug trenches.
Class A:	Concrete 100mm thick under the pipe extending half way up the pipe, backfilled with the appropriate class of fill. Used where there is only a very shallow fall to the drain. Class A bedding allows the pipes to be laid to an exact gradient.

DRAINAGE – continued

Concrete surround	25 mm sand blinding to bottom of trench, pipe surround on chocks, 100 mm concrete under pipe, 150 mm concrete over pipe. It is preferable to bed pipes under slabs or wall in granular material.

PIPED SUPPLY SYSTEMS

Identification of Service Tubes From Utility to Dwellings

Utility	Colour	Size	Depth
British Telecom	grey	54 mm outside diameter	450 mm
Electricity	black	38 mm outside diameter	450 mm
Gas	yellow	42 mm outside diameter rigid	450 mm
Water	may be blue	(normally untubed)	750 mm

ELECTRICAL SUPPLY/POWER/LIGHTING SYSTEMS

Electrical Insulation Class En 60.598 BS 4533

Class 1:	luminaires comply with class 1 (I) earthed electrical requirements
Class 2:	luminaires comply with class 2 (II) double insulated electrical requirements
Class 3:	luminaires comply with class 3 (III) electrical requirements

Protection to Light Fittings

BS EN 60529:1992 Classification for degrees of protection provided by enclosures.
(IP Code - International or ingress Protection)

1st characteristic:		against ingress of solid foreign objects.
The figure	2	indicates that fingers cannot enter
	3	that a 2.5 mm diameter probe cannot enter
	4	that a 1.0 mm diameter probe cannot enter
	5	the fitting is dust proof (no dust around live parts)
	6	the fitting is dust tight (no dust entry)
2nd characteristic:		ingress of water with harmful effects:
The figure	0	indicates unprotected
	1	vertically dripping water cannot enter
	2	water dripping 15° (tilt) cannot enter
	3	spraying water cannot enter
	4	splashing water cannot enter
	5	jetting water cannot enter
	6	powerful jetting water cannot enter
	7	proof against temporary immersion
	8	proof against continuous immersion
Optional additional codes:		A-D protects against access to hazardous parts;
	H	High voltage apparatus
	M	fitting was in motion during water test
	S	fitting was static during water test
	W	protects against weather
Marking code arrangement:		(example) IPX5S = IP (International or Ingress Protection); X (denotes omission of first characteristic); 5 = jetting; S = static during water test.

RAIL TRACKS

	kg/m of track	lb/ft of track
Standard guage		
Bull-head rails, chairs, transverse timber (softwood) sleepers etc.	245	165
Main lines		
Flat-bottom rails, transverse prestressed concrete sleepers, etc.	418	280
Add for electric third rail	51	35
Add for crushed stone ballast	2600	1750
	kg/m²	lb/sq ft
Overall average weight – rails connections, Sleepers, ballast, etc.	733	150
	kg/m of track	lb/ft of track
Bridge rails, longitudinal timber sleepers, etc.	112	75

Heavy Rails

British Standard Section No.	Rail height mm	Foot width mm	Head width mm	Min web thickness mm	Section weight kg/m
Flat Bottom Rails					
60A	114.30	109.54	57.15	11.11	30.62
70A	123.82	111.12	60.32	12.30	34.81
75A	128.59	114.30	61.91	12.70	37.45
80A	133.35	117.47	63.50	13.10	39.76
90A	142.88	127.00	66.67	13.89	45.10
95A	147.64	130.17	69.85	14.68	47.31
100A	152.40	133.35	69.85	15.08	50.18
110A	158.75	139.70	69.85	15.87	54.52
113A	158.75	139.70	69.85	20.00	56.22
50 'O'	100.01	100.01	52.39	10.32	24.82
80 'O'	127.00	127.00	63.50	13.89	39.74
60R	114.30	109.54	57.15	11.11	29.85
75R	128.59	122.24	61.91	13.10	37.09
80R	133.35	127.00	63.50	13.49	39.72
90R	142.88	136.53	66.67	13.89	44.58
95R	147.64	141.29	68.26	14.29	47.21
100R	152.40	146.05	69.85	14.29	49.60
95N	147.64	139.70	69.85	13.89	47.27
Bull Head Rails					
95R BH	145.26	69.85	69.85	19.05	47.07

Light Rails

British Standard Section No.	Rail height mm	Foot width mm	Head width mm	Min web thickness mm	Section weight kg/m
Flat Bottom Rails					
20M	65.09	55.56	30.96	6.75	9.88
30M	75.41	69.85	38.10	9.13	14.79
35M	80.96	76.20	42.86	9.13	17.39
35R	85.73	82.55	44.45	8.33	17.40
40	88.11	80.57	45.64	12.3	19.89

RAIL TRACKS - continued

Bridge and Crane Rails

British steel No.	Rail height mm	Foot width mm	Head width mm	Head / web thickness mm	Section weight kg/m
Bridge Rails					
13	48.00	92	36.00	18.0	13.31
16	54.00	108	44.50	16.0	16.06
20	55.50	127	50.00	20.5	19.86
28	67.00	152	50.00	31.0	28.62
35	76.00	160	58.00	34.5	35.38
50	76.00	165	58.50	-	50.18
Crane Rails					
56	101.50	171.00	76.00	35.0	56.81
89	114.00	178.00	102.00	51.0	89.81
101	155.00	165.00	100.00	45.0	100.38
164	150.00	230.00	140.00	75.0	166.83
175CR	152.40	152.40	107.95	38.1	86.92

Fish Plates

British Standard Section No.	Overall plate length		Hole diameter	Finished weight per pair	
	4 Hole	6 Hole		4 Hole	6 Hole
	mm	mm	mm	kg/pair	kg/pair
For British Standard Heavy Rails: Flat Bottom Rails					
60A	406.40	609.60	20.64	9.87	14.76
70A	406.40	609.60	22.22	11.15	16.65
75A	406.40	-	23.81	11.82	17.73
80A	406.40	609.60	23.81	13.15	19.72
90A	457.20	685.80	25.40	17.49	26.23
100A	508.00	-	pear	25.02	-
110A (shallow)	507.00	-	27.00	30.11	54.64
113A (heavy)	507.00	-	27.00	30.11	54.64
50 'O' (shallow)	406.40	-	-	6.68	10.14
80 'O' (shallow)	495.30	-	23.81	14.72	22.69
60R (shallow)	406.40	609.60	20.64	8.76	13.13
60R (angled)	406.40	609.60	20.64	11.27	16.90
75R (shallow)	406.40	-	23.81	10.94	16.42
75R (angled)	406.40	-	23.81	13.67	-
80R (shallow)	406.40	609.60	23.81	11.93	17.89
80R (angled)	406.40	609.60	23.81	14.90	22.33
For British Standard Heavy Rails: Bull head rails					
95R BH (shallow)	-	457.20	27.00	14.59	14.61
For British Standard Light Rails: Flat Bottom Rails					
30M	355.6	-	-	-	2.72
35M	355.6	-	-	-	2.83
40	355.6	-	-	3.76	-

THE AGGREGATES TAX

The Tax

The Aggregates Tax came into operation on 1 April 2002 in the UK, except for Northern Ireland where it is being phased in over five years from 2003.

It is currently levied at the rate of £1.60 per tonne on anyone considered to be responsible for commercially exploiting 'virgin' Aggregates in the UK and should naturally be passed by price increase to the ultimate user.

All material falling within the definition of 'Aggregates' is liable for the levy unless it is specifically exempted.

It does not apply to clay, soil, vegetable or other organic matter.

The hope is that this will :-

- Encourage the use of alternative materials that would otherwise be disposed of to landfill sites.
- promote development of new recycling processes, such as using waste tyres and glass
- promote greater efficiency in the use of virgin aggregates
- reduce noise + vibration, dust + other emissions to air, visual intrusion, loss of amenity + damage to wildlife habitats

The intention is for part of the revenue from the levy to be recycled to business and communities affected by Aggregates extraction, through :-

- a 0.1 percentage point cut in employers' national insurance contributions
- a new £35 million per annum 'Sustainability Fund' to reduce the need for virgin materials and to limit the effects of extraction on the environmental where it takes place. A number of key priorities were identified with a promise to give effect to them through existing programmes. Following a mid-term review in December 2003, this fund was extended for a further three years.

Definition

'Aggregates' means any rock, gravel or sand which is extracted or dredged in the UK for aggregates use. It includes whatever substances are for the time being incorporated in it or naturally occur mixed with it.

'Exploitation' is defined as involving any one or a combination of any of the following :-

- being removed from its original site
- becoming subject to a contract or other agreement to supply to any person
- being used for construction purposes
- being mixed with any material or substance other than water, except in permitted circumstances

Incidence

It is a tax on primary Aggregates production – i.e. 'virgin' Aggregates won from a source and used in a location within the UK territorial boundaries (land or sea). The tax is not levied on aggregates which are exported nor on Aggregates imported from outside the UK territorial boundaries.

It is levied at the point of sale.

Exemption from tax

An 'aggregate' is exempt from the levy if it is :-

- material which has previously been used for construction purposes
- aggregate that has already been subject to a charge to the Aggregates levy
- aggregate which was previously removed from its originating site before the commencement date of the levy
- aggregate which is being returned to the land from which it was won
- aggregate won from a farm land or forest where used on that farm or forest
- rock which has not been subjected to an industrial crushing process

THE AGGREGATES TAX – continued

Exemption from tax – continued

- aggregate won by being removed from the ground on the site of any building or proposed building in the course of excavations carried out in connection with the modification or erection of the building and exclusively for the purpose of laying foundations or of laying any pipe or cable
- aggregate won by being removed from the bed of any river, canal or watercourse or channel in or approach to any port or harbour (natural or artificial), in the course of carrying out any dredging exclusively for the purpose of creating, restoring, improving or maintaining that body of water
- aggregate won by being removed from the ground along the line of any highway or proposed highway in the course of excavations for improving, maintaining or constructing the highway + otherwise than purely to extract the aggregate
- drill cuttings from petroleum operations on land and on the seabed
- aggregate resulting from works carried out in exercise of powers under the New Road and Street Works Act 1991, the Roads (Northern Ireland) Order 1993 or the Street Works (Northern Ireland) Order 1995
- aggregate removed for the purpose of cutting of rock to produce dimension stone, or the production of lime or cement from limestone.
- aggregate arising as a waste material during the processing of the following industrial minerals:
  - ball clay
  - barytes
  - calcite
  - china clay
  - coal, lignite, slate or shale
  - feldspar
  - flint
  - fluorspar
  - fuller's earth
  - gems and semi-precious stones
  - gypsum
  - any metal or the ore of any metal
  - muscovite
  - perlite
  - potash
  - pumice
  - rock phosphates
  - sodium chloride
  - talc
  - vermiculite

However, the levy is still chargeable on any aggregates arising as the spoil or waste from or the by-products of the above exempt processes. This includes quarry overburden.

Anything that consists 'wholly or mainly' of the following is exempt from the levy (note that 'wholly' is defined as 100% but 'mainly' as more than 50%, thus exempting any contained aggregates amounting to less than 50% of the original volumes :

- clay, soil, vegetable or other organic matter
- coal, slate or shale
- china clay waste and ball clay waste

Relief from the levy either in the form of credit or repayment is obtainable where :

- it is subsequently exported from the UK in the form of aggregate
- it is used in an exempt process
- where it is used in a prescribed industrial or agricultural process
- it is waste aggregate disposed of by dumping or otherwise, e.g. sent to landfill or returned to the originating site

Environment Act 1995

Provides for planning permissions for minerals extraction (which includes aggregates) to be reviewed and updated on a regular basis by the mineral planning authorities.

Minerals planning guidance notes were to be updated to ensure that all mineral operators provide local benefits that clearly outweigh the likely impacts of mineral extraction.

The Rural White Paper published in November 2000

This set out the government's 'vision of a countryside that is sustainable economically, socially and environmentally' apart from economic support, action was proposed to deal with a number of environmental threats, one is to reduce the environmental impact (such as damage to biodiversity and visual intrusion) of aggregate extraction by imposing an Aggregates levy.

Finance Act 2001

This provided for a levy being charged on aggregate which is subject to commercial exploitation. It placed the levy under the care and management of HM Revenue & Customs.

Impact

The British Aggregates Association suggests that the additional cost imposed by quarries is likely to be in the order of £2.65 per tonne on mainstream products, applying an above average rate on these in order that by-products and low grade waste products can be held at competitive rates, as well as making some allowance for administration and increased finance charges.

With many gravel aggregates costing in the region of £7.50 to £9.50 per tonne, there is a significant impact on construction costs.

Avoidance

An alternative to using new aggregates in filling operations is to crush and screen rubble which may become available during the process of demolition and site clearance as well as removal of obstacles during the excavation processes.

Example : Assuming that the material would be suitable for fill material under buildings or roads, a simple cost comparison would be as follows (note that for the purpose of the exercise, the material is taken to be 1.80 t/m³ and the total quantity involved less than 1,000 m³) :-

Importing fill material :

		£/m³	£/tonne
Cost of 'new' aggregates delivered to site		17.10	9.50
Addition for Aggregates Tax		2.88	1.60
Total cost of importing fill materials	£	19.98	11.10

Disposing of site material :

		£/m³	£/tonne
Cost of removing materials from site	£	9.50	5.28

THE AGGREGATES TAX – continued

Avoidance – continued

Crushing site materials :

		£/m³	£/tonne
Transportation of material from excavations or demolition places to temporary stockpiles		1.25	.69
Transportation of material from temporary stockpiles to the crushing plant		1.00	.56
Establishing plant and equipment on site; removing on completion		.75	.42
Maintain and operate plant		3.75	2.08
Crushing hard materials on site		5.50	3.05
Screening material on site		.75	.42
Total cost of crushing site materials	£	13.00	7.22

From the above it can be seen that potentially there is a great benefit in crushing site materials for filling rather than importing fill materials.

Setting the cost of crushing against the import price would produce a saving of £6.98 per m³. If the site materials were otherwise intended to be removed from the site, then the cost benefit increases by the saved disposal cost to £16.48 per m³.

Even if there is no call for any or all of the crushed material on site, it ought to be regarded as a useful asset and either sold on in crushed form or else sold with the prospects of crushing elsewhere.

Specimen Unit rates

Establishing plant and equipment on site; removing on completion		
Crushing plant	trip	£500
Screening plant	trip	£250
Maintain and operate plant		
Crushing plant	week	£3,000
Screening plant	week	£750
Transportation of material from excavations or demolition places to temporary stockpiles	m³	£1.25
Transportation of material from temporary stockpiles to the crushing plant	m³	£1.00
Breaking up material on site using impact breakers		
mass concrete	m³	£5.50
reinforced concrete	m³	£6.50
brickwork	m³	£2.50
Crushing material on site		
mass concrete ne 1000m³	m³	£5.50

mass concrete 1000 - 5000m^3	m^3	£5.00
mass concrete over 5000m^3	m^3	£4.50
reinforced concrete ne 1000m^3	m^3	£6.50
reinforced concrete 1000 - 5000m^3	m^3	£6.00
reinforced concrete over 5000m^3	m^3	£5.50
brickwork ne 1000m^3	m^3	£5.00
brickwork 1000 - 5000m^3	m^3	£4.50
brickwork over 5000m^3	m^3	£4.00
Screening material on site	m^3	£0.75

CAPITAL ALLOWANCES

Introduction

Capital Allowances provide tax relief by prescribing a statutory rate of depreciation for tax purposes in place of that used for accounting purposes. They are utilised by government to provide an incentive to invest in capital equipment, including commercial property, by allowing the majority of taxpayers a deduction from taxable profits for certain types of capital expenditure, thereby deferring tax liabilities.

The capital allowances most commonly applicable to real estate are those given for capital expenditure on both new and existing industrial buildings, and plant and machinery in all commercial buildings.

Other types of allowances particularly relevant to property are hotel and enterprise zone allowances, which are in fact variants to industrial buildings allowances code. Enhanced rates of allowances are available to small and medium sized enterprises and on certain types of energy saving plant and machinery, whilst reduced rates apply to items with an expected economic life of more than 25 years.

The Act

The primary legislation is now contained in the Capital Allowances Act 2001. Amendments to the Act have been made in each subsequent Finance Act.

The Act is arranged in 12 Parts and was published with an accompanying set of Explanatory Notes.

Plant and Machinery

The Finance Act 1994 introduced major changes to the availability of Capital Allowances on real estate. A definition was introduced which precludes expenditure on the provision of a building from qualifying for plant and machinery, with prescribed exceptions.

List A in Section 21 of the 2001 Act sets out those assets treated as parts of buildings:-

- *Walls, floors, ceilings, doors, gates, shutters, windows and stairs.*
- *Mains services, and systems, for water, electricity and gas.*
- *Waste disposal systems.*
- *Sewerage and drainage systems.*
- *Shafts or other structures in which lifts, hoists, escalators and moving walkways are installed.*
- *Fire safety systems.*

Similarly, List B in Section 22 identifies excluded structures and other assets.

Both sections are, however, subject to Section 23. This section sets out expenditure, which although being part of a building, may still be expenditure on the provision of Plant and Machinery.

List C in Section 23 is reproduced below:

CAPITAL ALLOWANCES – continued

LIST C EXPENDITURE UNAFFECTED BY SECTIONS 21 AND 22

Sections 21 and 22 do not affect the question whether expenditure on any item in List C is expenditure on the provision of Plant or Machinery

1. Machinery (including devices for providing motive power) not within any other item in this list.
2. Electrical systems (including lighting systems) and cold water, gas and sewerage systems provided mainly –
   (a) to meet the particular requirements of the qualifying activity, or
   (b) to serve particular Plant or Machinery used for the purposes of the qualifying activity.
3. Space or water heating systems; powered systems of ventilation, air cooling or air purification; and any floor or ceiling comprised in such systems.
4. Manufacturing or processing equipment; storage equipment (including cold rooms); display equipment; and counters, checkouts and similar equipment.
5. Cookers, washing machines, dishwashers, refrigerators and similar equipment; washbasins, sinks, baths, showers, sanitary ware and similar equipment; and furniture and furnishings.
6. Lifts, hoists, escalators and moving walkways.
7. Sound insulation provided mainly to meet the particular requirements of the qualifying activity.
8. Computer, telecommunication and surveillance systems (including their wiring or other links).
9. Refrigeration or cooling equipment.
10. Fire alarm systems; sprinkler and other equipment for extinguishing or containing fires.
11. Burglar alarm systems.
12. Strong rooms in bank or building society premises; safes.
13. Partition walls, where moveable and intended to be moved in the course of the qualifying activity.
14. Decorative assets provided for the enjoyment of the public in hotel, restaurant or similar trades.
15. Advertising hoardings; signs, displays and similar assets.
16. Swimming pools (including diving boards, slides and structures on which such boards or slides are mounted).
17. Any glasshouse constructed so that the required environment (namely, air, heat, light, irrigation and temperature) for the growing of plants is provided automatically by means of devices forming an integral part of its structure.
18. Cold stores.
19. Caravans provided mainly for holiday lettings.
20. Buildings provided for testing aircraft engines run within the buildings.
21. Moveable buildings intended to be moved in the course of the qualifying activity.
22. The alteration of land for the purpose only of installing Plant or Machinery.
23. The provision of dry docks.
24. The provision of any jetty or similar structure provided mainly to carry Plant or Machinery.
25. The provision of pipelines or underground ducts or tunnels with a primary purpose of carrying utility conduits.
26. The provision of towers to support floodlights.
27. The provision of –
   (a) any reservoir incorporated into a water treatment works, or
   (b) any service reservoir of treated water for supply within any housing estate or other particular locality.
28. The provision of –
   (a) silos provided for temporary storage, or
   (b) storage tanks.
29. The provision of slurry pits or silage clamps.
30. The provision of fish tanks or fish ponds.
31. The provision of rails, sleepers and ballast for a railway or tramway.

32. The provision of structures and other assets for providing the setting for any ride at an amusement park or exhibition.
33. The provision of fixed zoo cages.

Case Law

The fact that an item appears in List C does not automatically mean that it will qualify for capital allowances. It only means that it may potentially qualify.

Guidance about the meaning of plant has to be found in case law. The cases go back a long way, beginning in 1887. The current state of the law on the meaning of plant derives from the decision in the case of Wimpy International Ltd v Warland and Associated Restaurants Ltd v Warland in the late 1980s. The Judge in that case said that there were three tests to be applied when considering whether or not an item is plant.

1. Is the item stock in trade? If the answer yes, then the item is not plant.
2. Is the item used for carrying on the business? In order to pass the business use test the item must be employed in carrying on the business; it is not enough for the asset to be simply used in the business. For example, product display lighting in a retail store may be plant but general lighting in a warehouse would fail the test.
3. Is the item the business premises or part of the business premises? An item cannot be plant if it fails the premises test, i.e. if the business use is as the premises (or part of the premises) or place on which the business is conducted. The meaning of part of the premises in this context should not be confused with the law of real property. The Inland Revenue's internal manuals suggest there are four general factors to be considered, each of which is a question of fact and degree:
   - Does the item appear visually to retain a separate identity
   - With what degree of permanence has it been attached to the building
   - To what extent is the structure complete without it
   - To what extent is it intended to be permanent or alternatively is it likely to be replaced within a short period

There is obviously a core list of items that will usually qualify in the majority of cases. However, many other still need to be looked at on a case-by-case basis. For example, decorative assets in a hotel restaurant may be plant but similar assets in an office reception area would almost certainly not be.

Refurbishment Schemes

Building refurbishment projects will typically be a mixture of capital costs and revenue expenses, unless the works are so extensive that they are more appropriately classified a redevelopment. A straightforward repair or a "like for like" replacement of part of an asset would be a revenue expense, meaning that the entire amount can be deducted from taxable profits in the same year.

Where capital expenditure is incurred that is incidental to the installation of plant or machinery then Section 25 of the 2001 Act allows it to be treated as part of the expenditure on the qualifying item. Incidental expenditure will often include parts of the building that would be otherwise disallowed, as shown in the Lists reproduced above. For example, the cost of forming a lift shaft inside an existing building would be deemed to be part of the expenditure on the provision of the lift.

CAPITAL ALLOWANCES – continued

Rate of Allowances

Capital Allowances on plant and machinery are given in the form of writing down allowances at the rate of 25% per annum on a reducing balance basis. For every £100 of qualifying expenditure £25 is claimable in year 1, £18.75 in year 2 and so on until either the all the allowances have been claimed or the asset is sold. In addition to the basic rate, enhanced rates are available to certain types of business and on certain classes of asset.

Capital expenditure on plant and machinery by a small and medium sized enterprise attracts a 40% first year allowance. In subsequent years allowances are available at the 25% rate mentioned above. The definition of small and medium sized enterprise is broadly one that meets at least 2 of the following criteria:

- Turnover not more than £22.8m
- Assets not more than £11.4m
- Not more than 250 employees

First year allowances are not generally available on the provision of plant and machinery for leasing.

The Enhanced Capital Allowances Scheme

The scheme is one of a series of measures introduced to ensure that the UK meets its target for reducing greenhouse gases under the Kyoto Protocol. 100% first year allowances are available on products included on the Energy Technology List published on the DETR website at www.eca.gov.uk and other technologies supported by the scheme. All businesses will be able to claim the enhanced allowances, but only investments in new and unused Machinery and Plant can qualify. Leased assets only qualify from 17 April 2002.

There are currently sixteen technologies covered by the scheme.

- Air-to-air energy recovery.
- Automatic monitoring and targeting.
- Boilers.
- Combined heat and power e.g. utilisation of the heat produced as a waste by-product in power stations.
- Compact heat exchangers.
- Compressor air equipment.
- Heat pumps for space heating.
- HVAC zone controls.
- Lighting.
- Motors.
- Pipe work Insulation.
- Refrigeration equipment.
- Solar thermal systems.
- Thermal screens.
- Variable Speed Drives.
- Warm air and radiant heaters.
- Radiant and warm air heaters
- Solar heaters
- Energy-efficient refrigeration equipment

The Finance Act 2003 introduced a new category of environmentally beneficial plant and machinery qualifying for 100% first-year allowances. The Water Technology List includes five categories of sustainable water use products, namely water meters, flow controllers, leak detection equipment, low flush toilets, efficient taps and rainwater-harvesting equipment.

Buildings and structures as defined above and long life assets as discussed below cannot qualify.

Long Life Assets

A reduced writing down allowance of 6% per annum is available on plant and machinery that is a long-life asset. A long-life asset is defined as plant and machinery that can reasonably be expected to have a useful economic life of at least 25 years. The useful economic life is taken as the period from first use until it is likely to cease to be used as a fixed asset of any business. It is important to note that this likely to be a shorter period than an item's physical life.

Plant and machinery provided for use in a building used wholly or mainly as dwelling house, showroom, hotel, office or retail shop or similar premises, or for purposes ancillary to such use, cannot be long-life assets.

In contrast plant and machinery assets in buildings such as factories, cinemas, hospitals and so on are all potentially long-life assets.

Industrial Building Allowances

An industrial building (or structure) is defined in Sections 271 and 274 of the 2001 Act and includes buildings used for the following qualifying purposes:

- Manufacturing
- Processing
- Storage
- Agricultural contracting
- Working foreign plantations
- Fishing
- Mineral extraction

The following undertakings are also qualifying trades:

- Electricity
- Water
- Hydraulic power
- Sewerage
- Transport
- Highway undertakings
- Tunnels
- Bridges
- Inland navigation
- Docks

The definition is extended to include buildings provided for the welfare of workers in a qualifying trade and sports pavilions provided and used for the welfare of workers in any trade. Vehicle repair workshops and roads on industrial estates may also form part of the qualifying expenditure.

Retail shops, showrooms, offices, dwelling houses and buildings used ancillary to a retail purpose are specifically excluded.

CAPITAL ALLOWANCES – continued

Writing Down Allowances

Allowances are given on qualifying expenditure at the rate of 4% per annum on a straight-line basis over 25 years. The allowance is given if the building is being used for a qualifying purpose on the last day of the accounting period. Where the building is used for a non-qualifying purpose that year's allowance is lost.

A purchaser of used industrial building will be entitled to allowances equal to, in very simple terms, the original construction cost after adjustment for any periods of non-qualifying use. The allowances will be spread equally over the remaining period to the date twenty-five years after first use. The only exception to this rule is where the purchaser acquires the building unused from a property developer when the allowances will be based on the purchase price.

Hotel Allowances

Industrial Building Allowances are also available on capital expenditure incurred on constructing a "qualifying hotel". The building must not only be a "hotel" in the normal sense of the word, but must also be a "qualifying hotel" as defined in Section 279 of the 2001 Act, which means satisfying the following conditions:

- The accommodation is in buildings of a permanent nature
- It is open for at least 4 months in the season (April to October)
- It has 10 or more letting bedrooms
- The sleeping accommodation consists wholly or mainly of letting bedrooms
- The services that it provides include breakfast and an evening meal (i.e. there must be a restaurant), the making of beds and cleaning of rooms.

A hotel may be in more than one building and swimming pools, car parks and similar amenities are included in the definition.

Enterprise Zones

A 100% first year allowance is available on capital expenditure incurred on the construction (or the purchase within two years of first use) of any commercial building within a designated enterprise zone, within ten years of the site being so designated. Like other allowances given under the industrial buildings code the building has a life of twenty-five years for tax purposes.

The majority of enterprise zones had reached the end of their ten-year life by 1993. However, in certain very limited circumstances it may still be possible to claim these allowances up to twenty years after the site was first designated.

Flats Over Shops

Tax relief is available on capital expenditure incurred on or after 11 May 2001 on the renovation or conversion of vacant or underused space above shops and other commercial premises to provide flats for rent.

In order to qualify the property must have been built before 1980 and the expenditure incurred on or in connection with:

- Converting part of a qualifying building into a qualifying flat.
- Renovating an existing flat in a qualifying building if the flat is, or will be a qualifying flat.
- Repairs incidental to conversion or renovation of a qualifying flat, and
- The cost of providing access to the flat(s).

The property must not have more than 4 storeys above the ground floor and it must appear that, when the property was constructed, the floors above the ground floor were primarily for residential use. The ground floor must be authorised for business use at the time of the conversion work and for the period during which the flat is held for letting. Each new flat must be a self-contained dwelling, with external access separate from the ground-floor premises. It must have no more than 4 rooms, excluding kitchen and bathroom. None flats can be "high value" flats, as defined in the legislation. The new flats must be available for letting as a dwelling for a period of not more than 5 years.

An initial allowance of 100 per cent is available or, alternatively, a lower amount may be claimed, in which case the balance may be claimed at a rate of 25 per cent per annum in subsequent a years. The allowances may be recovered if the flat is sold or ceases to be let within 7 years.

Business premises renovation allowance

The Business Premises Renovation Allowance (BPRA) was first announced in December 2003. The idea behind the scheme is to bring long-term vacant properties back into productive use by providing 100 per cent capital allowances for the cost of renovating and converting unused premises in disadvantaged areas. The legislation is included in Finance Act 2005 and subject to EU state aid approval BPRA will be introduced later in 2005 and run for five years.

The legislation is identical in many respects to that for flat conversion allowances. The scheme will apply to properties within one of the 1,997 currently designated Enterprise Areas and will apply to the same buildings that previously qualified for SDLT relief. A list of areas and a postcode search tool are available on the Inland Revenue's stamp duty website.

BPRA will be available to both individuals and companies who own or lease business property that has been unused for 12 months or more. Allowances will be available to a person who incurs qualifying capital expenditure on the renovation of business premises.

Agricultural Buildings

Allowances are available on capital expenditure incurred on the construction of buildings and works for the purposes of husbandry on land in the UK. Agricultural building means a building such as a farmhouse or farm building, a fence or other works. A maximum of only one-third of the expenditure on a farmhouse may qualify.

Husbandry includes any method of intensive rearing of livestock or fish on a commercial basis for the production of food for human consumption, and the cultivation of short rotation coppice. Over the years the Courts have held that sheep grazing and poultry farming are husbandry, and that a dairy business and the rearing of pheasants for sport are not. Where the use is partly for other purposes the expenditure can be apportioned.

The rate of allowances available and the way in which the system operates is very similar to that described above for industrial buildings. However, no allowance is ever given if the first use of the building is not for husbandry. A different treatment is also applied following acquisition of a used building unless the parties to the transaction elect otherwise.

Other Capital Allowances

Other types of allowances include those available for capital expenditure on Mineral Extraction, Research and Development, Know-How, Patents, Dredging and Assured Tenancy.

VAT AND CONSTRUCTION

Introduction

Value Added Tax (VAT) is a tax on the consumption of goods and services. The UK adopted VAT when it joined the European Community in 1973. The principal source of European law in relation to VAT is the EC Sixth Directive 77/388 which is currently restated and consolidated in the UK through the VAT Act 1994 and various Statutory Instruments as amended by subsequent Finance Acts. The tax is administered in the UK by HM Customs & Excise.

VAT Notice 708: Buildings and construction (July 2002) gives an interpretation of the law in connection with construction works from the point of view of Customs and Excise. VAT tribunals and court decisions since the date of this publication will affect the application of the law in certain instances. The Notice is available from any VAT Business Advice Centre. The telephone and address is in local telephone books under "Customs and Excise". It is also available on Customs and Excise website at www.hmce.gov.uk.

The scope of VAT

VAT is payable on:

- Supplies of goods and services made in the UK
- By a taxable person
- In the course or furtherance of business; and
- Which are not specifically exempted or zero-rated.

Rates of VAT

There are three rates of VAT:

- A standard rate, currently 17.5%
- A reduced rate, currently 5%; and
- A zero rate.

Additionally some supplies are exempt from VAT and others are outside the scope of VAT.

Recovery of VAT

When a taxpayer makes taxable supplies he must account for VAT at the appropriate rate of either 17.5% or 5%. This VAT then has to be passed to Customs & Excise. This VAT will normally be charged to the taxpayer's customers.

As a VAT registered person, the taxpayer can reclaim from Customs and Excise as much of the VAT incurred on their purchases as relates to the standard-rated, reduced-rated and zero-rated onward supplies they make. A person cannot however reclaim VAT that relates to any non-business activity or to any exempt supplies they make.

At predetermined intervals the taxpayer will pay to Customs and Excise the excess of VAT collected over the VAT they can reclaim. However if the VAT reclaimed is more than the VAT collected, the taxpayer can reclaim the difference from Customs and Excise.

Example

X Ltd constructs a block of flats. It sells long leases to buyers for a premium. X Ltd has constructed a new building designed as a dwelling and will have granted a long lease. This sale of a long lease is VAT zero-rated. This means any VAT incurred in connection with the development that which X Ltd will have paid (e.g. payments for consultants and certain preliminary services) will be reclaimable. For reasons detailed below the builder employed by X Ltd will not have charged VAT on his construction services.

VAT AND CONSTRUCTION - continued

Taxable Persons

A taxable person is an individual, firm, company etc who is required to be registered for VAT. A person who makes taxable supplies above certain value limits is required to be registered. The current registration limit is £58,000 for 2004-05. The threshold is exceeded if at the end of any month the value of taxable supplies in the period of one year then ending is over the limit, or at any time, if there are reasonable grounds for believing that the value of the taxable supplies in the period of 30 days than beginning will exceed £58,000.
A person who makes taxable supplies below these limits is entitled to be registered on a voluntary basis if they wish, for example in order to recover VAT incurred in relation to those taxable supplies.

In addition, a person who is not registered for VAT in the UK but acquires goods from another EC member state, or make distance sales in the UK, above certain value limits may be required to register for VAT in the UK.

VAT Exempt Supplies

If a supply is exempt from VAT this means that no tax is payable – but equally the person making the exempt supply cannot normally recover any of the VAT on their own costs relating to that supply.

Generally property transactions such as leasing of land and buildings are exempt unless a landlord chooses to standard-rate its supplies by a process known as electing to waive exemption – more commonly known as opting to tax. This means that VAT is added to rental income and also that VAT incurred, on say, an expensive refurbishment, is recoverable.

Supplies outside the scope of VAT

Supplies are outside the scope of VAT if they are:

- Made by someone who is not a taxable person
- Made outside the UK; or
- Not made in the course or furtherance of business

In course or furtherance of business

VAT must be accounted for on all taxable supplies made in the course or furtherance of business with the corresponding recovery of VAT on expenditure incurred.

If a taxpayer also carries out non-business activities then VAT incurred in relation to such supplies is not recoverable.

In VAT terms, business means any activity continuously performed which is mainly concerned with making supplies for a consideration. This includes:

- Any one carrying on a trade, vocation or profession;
- The provision of membership benefits by clubs, associations and similar bodies in return for a subscription or other consideration; and
- Admission to premises for a charge.

It may also include the activities of other bodies including charities and non-profit making organisations.

Examples of non-business activities are:

- Providing free services or information;
- Maintaining museums, or particular historic sites;
- Publishing religious or political views.

Construction Services

In general the provision of construction services by a contractor will be VAT standard rated at 17.5%, however, there are a number of exceptions for construction services provided in relation to certain residential and charitable use buildings.

The supply of building materials is VAT standard rated at 17.5%, however, where these materials are supplied as part of the construction services the VAT liability of those materials follows that of the construction services supplied.

Zero-rated construction services

The following construction services are VAT zero-rated including the supply of related building materials.

The construction of new dwellings

The supply of services in the course of the construction of a building designed for use as a dwelling or number of dwellings is zero-rated other than the services of an architect, surveyor or any other person acting as a consultant or in a supervisory capacity.

The following conditions must be satisfied in order for the works to qualify for zero-rating:

1. the work must not amount to the conversion, reconstruction or alteration of an existing building;
2. the work must not be an enlargement of, or extension to, an existing building except to the extent that the enlargement or extension creates an additional dwelling or dwellings;
3. the building must be designed as a dwelling or number of dwellings. Each dwelling must consist of self-contained living accommodation with no provision for direct internal access from the dwelling to any other dwelling or part of a dwelling;
4. statutory planning consent must have been granted for the construction of the dwelling, and construction carried out in accordance with that consent;
5. separate use or disposal of the dwelling must not be prohibited by the terms of any covenant, statutory planning consent or similar provision.

The construction of a garage at the same time as the dwelling can also be zero-rated as can the demolition of any existing building on the site of the new dwelling

A building only ceases to be an existing building (see points 1. and 2. above) when it is:

1. demolished completely to ground level; or when
2. the part remaining above ground level consists of no more than a single façade (or a double façade on a corner site) the retention of which is a condition or requirement of statutory planning consent or similar permission.

The construction of a new building for 'relevant residential or charitable' use

The supply of services in the course of the construction of a building designed for use as a relevant residential or charitable building is zero-rated other than the services of an architect, surveyor or any other person acting as a consultant or in a supervisory capacity.

A 'relevant residential' use building means:

1. a home or other institution providing residential accommodation for children;
2. a home or other institution providing residential accommodation with personal care for persons in need of personal care by reason of old age, disablement, past or present dependence on alcohol or drugs or past or present mental disorder;
3. a hospice;
4. residential accommodation for students or school pupils
5. residential accommodation for members of any of the armed forces;
6. a monastery, nunnery, or similar establishment; or
7. an institution which is the sole or main residence of at least 90% of its residents.

VAT AND CONSTRUCTION – continued

The construction of a new building for 'relevant residential or charitable' use – continued

A 'relevant residential' purpose building does not include use as a hospital, a prison or similar institution or as an hotel, inn or similar establishment.

A 'relevant charitable' use means use by a charity:

1. otherwise than in the course or furtherance of a business; or
2. as a village hall or similarly in providing social or recreational facilities for a local community.

Non qualifying use which is not expected to exceed 10% of the time the building is normally available for use can be ignored. The calculation of business use can be based on time, floor area or head count subject to approval being acquired from Customs and Excise.

The construction services can only be zero-rated if a certificate is given by the end user to the contractor carrying out the works confirming that the building is to be used for a qualifying purpose i.e. for a 'relevant residential or charitable' purpose. It follows that such services can only be zero-rated when supplied to the end user and, unlike supplies relating to dwellings, supplies by sub contractors cannot be zero-rated.

The construction of an annex used for a 'relevant charitable' purpose

Construction services provided in the course of construction of an annexe for use entirely or partly for a 'relevant charitable' purpose can be zero-rated.

In order to qualify the annexe must:

1. be capable of functioning independently from the existing building;
2. have its own main entrance; and
3. be covered by a qualifying use certificate.

The conversion of a non-residential building into dwellings or the conversion of a building from non-residential use to 'relevant residential' use where the supply is to a 'relevant' housing association

The supply to a 'relevant' housing association in the course of conversion of a non-residential building or non-residential part of a building into:

1. a building or part of a building designed as a dwelling or number of dwellings; or
2. a building or part of a building for use solely for a relevant residential purpose,

of any services related to the conversion other than the services of an architect, surveyor or any person acting as a consultant or in a supervisory capacity are zero-rated.

A 'relevant' housing association is defined as:

1. a registered social landlord within the meaning of Part I of the Housing Act 1996
2. a registered housing association within the meaning of the Housing Associations Act 1985 (Scottish registered housing associations), or
3. a registered housing association within the meaning of Part II of the Housing (Northern Ireland) Order 1992 (Northern Irish registered housing associations).

If the building is to be used for a 'relevant residential' purpose the housing association should issue a qualifying use certificate to the contractor completing the works.

The construction of a permanent park for residential caravans

The supply in the course of the construction of any civil engineering work 'necessary for' the development of a permanent park for residential caravans of any services related to the construction can be VAT zero-rated. This includes access roads, paths, drainage, sewerage and the installation of mains water, power and gas supplies.

Certain building alterations for "handicapped" persons

Certain goods and services supplied to a "handicapped" person, or a charity making these items and services available to "handicapped persons" can be zero-rated. The recipient of these goods or services needs to give the supplier an appropriate written declaration that they are entitled to benefit from zero rating.

The following services (amongst others) are zero-rated:

1. the installation of specialist lifts and hoists and their repair and maintenance
2. the construction of ramps, widening doorways or passageways including any preparatory work and making good work
3. the provision, extension and adaptation of a bathroom, washroom or lavatory; and
4. emergency alarm call systems

Approved alterations to protected buildings

A supply in the course of an 'approved alteration' to a 'protected building' of any services other than the services of an architect, surveyor or any person acting as consultant or in a supervisory capacity can be zero-rated.

A 'protected building' is defined as a building that is:

1. designed to remain as or become a dwelling or number of dwellings after the alterations; or
2. is intended for use for a 'relevant residential or charitable purpose' after the alterations; and which is;
3. a listed building or scheduled ancient monument.

A listed building does not include buildings that are in conservation areas but not on the statutory list, or buildings included in non-statutory local lists.

An 'approved alteration' is an alteration to a 'protected building' that requires and has obtained listed building consent or scheduled monument consent. This consent is necessary for any works that affect the character of a building of special architectural or historic interest.

It is important to note that 'approved alterations' do not include any works of repair or maintenance or any incidental alteration to the fabric of a building that results from the carrying out of repairs or maintenance work.

A 'protected building' that is intended for use for a 'relevant residential or charitable purpose' will require the production of a qualifying use certificate by the end user to the contractor providing the alteration services.

Listed Churches are 'relevant charitable' use buildings and where 'approved alterations' are being carried out zero-rate VAT can be applied. Additionally since April 1 2001,

listed places of worship can apply for a grant for repair and maintenance works equal to the difference between the VAT paid at 17.5% on the repair and maintenance works and the amount that would have been charged if VAT had been 5%.

With effect from 1 April 2004, it will be possible to reclaim the full amount of VAT paid on eligible works carried out on or after 1 April 2004. Information relating to the scheme can be obtained from the website at www.lpwscheme.org.uk.

DIY Builders and Converters

Private individuals who decide to construct their own home are able to reclaim VAT they pay on goods they use to construct their home by use of a special refund mechanism made by way of an application to Customs & Excise. This also applies to services provided in the conversion of an existing non-residential building to form a new dwelling.

The scheme is meant to ensure that private individuals do not suffer the burden of VAT if they decide to construct their own home.

VAT AND CONSTRUCTION - continued

DIY Builders and Converters – continued

Charities may also qualify for a refund on the purchase of materials incorporated into a building used for non-business purposes where they provide their own free labour for the construction of a 'relevant charitable' use building.

Reduced-rated construction services

The following construction services are subject to the reduced rate of VAT of 5%, including the supply of related building materials.

A changed number of dwellings conversion

In order to qualify for the 5% rate there must be a different number of 'single household dwellings' within a building than there were before commencement of the conversion works. A 'single household dwelling' is defined as a dwelling that is designed for occupation by a single household.

These conversions can be from 'relevant residential' purpose buildings, non-residential buildings and houses in multiple occupation.

A house in multiple occupation conversion

This relates to construction services provided in the course of converting a 'single household dwelling', a number of 'single household dwellings', a non-residential building or a 'relevant residential' purpose building into a house for multiple occupation such as a bedsit accommodation.

A special residential conversion

A special residential conversion involves the conversion of a 'single household dwelling', a house in multiple occupation or a non-residential builidng into a 'relevant residential' purpose building such as student accommodation or a care home.

Renovation of derelict dwellings

The provision of renovation services in connection with a dwelling or 'relevant residential' purpose building that has been empty for three or more years prior to the date of commencement of construction works.can be carried out at a reduced rate of VAT of 5%.

Installation of energy saving materials

The supply and installation of certain energy saving materials including insulation, draught stripping, central heating and hot water controls and solar panels in a residential building or a building used for a relevant charitable purpose..

Grant-funded of heating equipment or connection of a gas supply

The grant funded supply and installation of heating appliances, connection of a mains gas supply, supply, installation, maintenance and repair of central heating systems, and supply and installation of renewable source heating systems, to qualifying persons. A qualifying person is someone aged 60 or over or is in receipt of various specified benefits.

Installation of security goods

The grant funded supply and installation of security goods to a qualifying person.

Building Contracts

Design and build contracts

If a contractor provides a design and build service relating to works to which the reduced or zero rate of VAT is applicable then any design costs incurred by the contractor will follow the VAT liability of the principal supply of construction services.

Management contracts

A management contractor acts as a main contractor for VAT purposes and the VAT liability of his services will follow that of the construction services provided. If the management contractor only provides advice without engaging trade contractors his services will be VAT standard rated.

Construction Management and Project Management

The project manger or construction management is appointed by the client to plan, manage and co-ordinate a construction project. This will involve establishing competitive bids for all the elements of the work and the appointment of trade contractors. The trade contractors are engaged directly by the client for their services. The VAT liability of the trade contractors will be determined by the nature of the construction services they provide and the building being constructed.

The fees of the construction manager or project manager will be VAT standard rated. If the construction manager also provides some construction services these works may be zero or reduced rated if the works qualify.

Liquidated and Ascertained Damages

Liquidated damages are outside of the scope of VAT as compensation. The employer should not reduce the VAT amount due on a payment under a building contract on account of a deduction of damages. In contrast an agreed reduction in the contract price will reduce the VAT amount.

Davis Langdon Crosher & James

ACOUSTICAL INVESTIGATION AND RESEARCH ORGANISATION LTD
Duxon's Turn, Maylands Avenue,
Hemel Hempstead, Herts HP2 4SB
Tel: 01442 247 146
Fax: 01442 256 749
Website: www.airo.co.uk

AGGREGATE CONCRETE BLOCK ASSOCIATION
See BRITISH PRECAST CONCRETE FEDERATION

ALUMINIUM FEDERATION LTD
Broadway House, Calthorpe Road, Five Ways,
Birmingham B15 1TN
Tel: 0121 456 1103
Fax: 0121 456 2274
Website: www.alfed.org.uk

AMERICAN HARDWOOD EXPORT COUNCIL
3 St Michael's Alley, London EC3V 9DS
Tel: 0207 626 4111
Fax: 0207 626 4222
Website: www.ahec-europe.org

ARBORICULTURAL ADVISORY & INFORMATION SERVICE
Tel: 0897 161 147

ARBORICULTURAL ASSOCIATION
Ampfield House, Ampfield, Nr Romsey, Hants
SO51 9PA
Tel: 01794 368 717
Fax: 01794 368 978
Website: www.trees.org.uk

ARCHITECTURAL ASSOCIATION SCHOOL OF ARCHITECTURE
34 - 36 Bedford Square, London WC1B 3ES
Tel: 002 7887 4000
Fax: 020 7414 0782
Website: www.aaschool.ac.uk

ARCHITECTS AND SURVEYORS INSTITUTE
St Mary House, 15 St Mary Street, Chippenham,
Wilts SN15 3WD
Tel: 01249 444 505
Fax: 01249 443 602
Website: www.ciob.org.uk

ASBESTOS INFORMATION CENTRE LTD
ATSS House, Station Road East
Stowmarket, Suffolk IP14 1RQ
Tel: 01449 676900
Fax: 01449 770028
Website: www.aic.org.uk

ASBESTOS REMOVAL CONTRACTORS ASSOCIATION
ARCA House, 237 Branston Road,
Burton upon Trent, Staffordshire, DE14 3BT
Tel:01283 513 126
Fax: 01283 568 228
Website: www.arca.org.uk

ASSOCIATION OF BUILDERS HARDWARE MANUFACTURERS
42 Heath Street, Tamworth, Staffs B79 7JH
Tel: 01827 52337
Fax: 01827 310 827
Website: www.abhm.org.uk

ASSOCIATION OF CONSULTING ENGINEERS
Alliance House, 12 Caxton Street, London
SW1H 0QL
Tel: 0207 222 6557
Fax: 0207 222 0750
Website: www.acenet.co.uk

ASSOCIATION OF LOADING AND ELEVATING EQUIPMENT MANUFACTURERS
Orbital House, 85 Croydon Road, Caterham,
Surrey CR3 6PD
Tel: 01883 334494
Fax: 01883 334490
Website: www.alem.org.uk

ASSOCIATION OF PROJECT MANAGEMENT
150 West Wycombe Road, High Wycombe,
Bucks HP12 3AE
Tel: 0845 458 1944
Fax: 01494 528 937
Website: www.apm.org.uk

BRITISH AGGREGATE CONSTRUCTION MATERIALS INDUSTRIES LTD
See QUARRY PRODUCTS ASSOC

BRITISH AIRPORTS AUTHORITY PLC
Corporate Office, 130 Wilton Road, London
SW1V 1LQ
Tel: 0207 834 9449
Fax: 0207 932 6699

BRITISH ANODISING ASSOCIATION
See ALUMINIUM FEDERATION

BRITISH ARCHITECTURAL LIBRARY
RIBA, 66 Portland Place, London W1B 1AD
Tel: 0207 580 5533
Fax: 0207 251 1541
Website: www.architecture.com

BRITISH ASSOCIATION OF LANDSCAPE INDUSTRIES
Landscape House, Stoneleigh Park, Warwickshire CV8 2LG
Tel: 0870 770 4971
Fax: 0870 770 4972
Website: www.bali.org.uk

BRITISH BOARD OF AGRÉMENT
PO Box 195, Bucknalls Lane, Garston, Watford, Herts WA25 9BA
Tel: 01923 665 300
Fax: 01923 665 301
Website: www.bbacerts.co.uk

BRITISH CABLE ASSOCIATION
37a Walton Rd, East Molesey, Surrey KT8 9DW
Tel: 0208 941 4079
Fax: 0208 783 0104

BRITISH CEMENT ASSOCIATION
Riverside House, 4 Meadows Business Park, Station Approach, Blackwater, Camberley, Surrey GU17 9AB
Tel: 01276 608700
Fax: 01276 608701
Website: www.bca.org.uk

BRITISH CERAMIC RESEARCH
Queens Road, Penkhull, Stoke-on-Trent, Staffs ST4 7LQ
Tel: 01782 764 444
Fax: 01782 412 331
Website: www.ceram.com

BRITISH CONSTRUCTIONAL STEELWORK ASSOCIATION LTD
4 Whitehall Court,Westminster, London SW1A 2ES
Tel: 0207 839 8566
Fax: 0207 976 1634
Website: www.bcsa.org.uk

BRITISH ELECTROTECHNICAL AND ALLIED MANUFACTURERS ASSOCIATION
Westminster Tower, 3 Albert Embankment, London SE1 7SL
Tel: 0207 793 3000
Fax: 0207 793 3003
Website: www.beama.org.uk

BRITISH FIRE PROTECTION SYSTEMS ASSOCIATION LTD
55 Eden Street, Kingston-upon-Thames, Surrey KT1 1BW
Tel: 0208 549 5855
Fax: 0208 547 1564
Website: www.bfpsa.org.uk

BRITISH FIRE SERVICES ASSOCIATION
86 London Road, Leicester LR2 0QR
Tel + Fax: 0116 254 2879

BRITISH FLUE & CHIMNEY MANUFACTURERS ASSOCIATION
See FEDERATION OF ENVIRONMENTAL TRADE ASSOCIATIONS

BRITISH GEOLOGICAL SURVEY
Kingsley Dunham Centre
Keyworth, Nottingham NG12 5GG
Tel: 0115 936 3100
Fax: 0115 936 3200
Website: www.bgs.ac.uk

BRITISH INSTITUTE OF ARCHITECTURAL TECHNOLOGISTS
397 City Road, London EC1V 1NH
Tel: 0207 278 2206
Fax: 0207 8373194
Website: www.biat.org.uk

BRITISH LIBRARY LENDING DIVISION
Thorpe Arch, Boston Spa, Wetherby, West Yorks LS23 7BQ
Tel: 01937 546 000

BRITISH LIBRARY, SCIENCE REFERENCE AND INFORMATION LIBRARY
The British Library, St Pancras, 96 Euston Road, London NW1 2DB
Tel: 0870 444 1500
Website: www.bl.uk

BRITISH NON-FERROUS METALS FEDERATION
Broadway House, 60 Calthorpe Road, Edgbaston, Birmingham B15 1TN
Tel: 0121 456 6110
Fax: 0121 456 2274

BRITISH PLASTICS FEDERATION
6 Bath Place, Rivington Street, London EC2A 3JE
Tel: 0207 457 5000
Fax: 0207 457 5045
Website: www.bpf.co.uk

BRITISH PRECAST CONCRETE FEDERATION
60 Charles St,
Leicester LE1 1FB
Tel: 0116 253 6161
Fax: 0116 251 4568
Website: www.britishprecast.org.uk

BRITISH REINFORCEMENT MANUFACTURERS ASSOCIATION (BRMA)
See UK STEEL ASSOCIATION

BRITISH RUBBER MANUFACTURERS ASSOCIATION LTD
90 Tottenham Court Road, London W1P 0BR
Tel: 0207 457 5040
Fax: 0207 631 5471
Website: www.brma.co.uk

BRITISH STAINLESS STEEL ASSOCIATION
Broomgrove, 59 Clarkhouse Road, Sheffield S10 2LE
Tel: 0114 267 1260
Fax: 0114 266 1252
Website: www.bssa.org.uk

BRITISH STANDARDS INSTITUTION
389 Chiswick High Road, Chiswick W4 4AL
Tel: 0208 996 9000
Fax: 0208 996 7001
Website: www.bsi-global.com.uk

BRITISH WATER
1 Queen Anne's Gate, London SW1 9BT
Tel: 0207 957 4554
Fax: 0207 957 4565
Website: www.britishwater.co.uk

BRITISH WOOD PRESERVING & DAMP PROOFING ASSOCIATION
1 Gleneagles House, Vernon, Gate, South Street, Derby, DE1 1UP
Tel: 01332 225 100
Fax: 01332 225 101
Website: www.bwpda.co.uk

BRITISH WOODWORKING FEDERATION
55 Tufton Street, London SW1 3QL
Tel: 0870 458 6939
Fax: 0870 458 6949
Website: www.bwf.org.uk

THE BUILDING CENTRE GROUP
26 Store Street,
London WC1E 7BT
Tel: 0207 692 4000
Fax: 0207 580 9641
Website: www.buildingcentre.co.uk

BUILDING MAINTENANCE INFORMATION (BMI)
3 Cadogan gate, London SW1X 0AS
Tel: 0207 695 1500
Fax: 0207 695 1501

BUILDING RESEARCH ESTABLISHMENT (BRE)
Bucknalls Lane, Garston, Watford, Herts WD25 9XX
Tel: 01923 664 000
Fax: 01923 664 010
Website: www.bre.co.uk

BUILDING RESEARCH ESTABLISHMENT: SCOTLAND (BRE)
Kelvin Rd, East Kilbride, Glasgow G75 0RZ
Tel: 01355 576 200
Fax: 01355 576 210

BUILDING SERVICES RESEARCH AND INFORMATION ASSOCIATION
Old Bracknell Lane West, Bracknell, Berks RG12 7AH
Tel: 01344 465 600
Fax: 01344 465 626
Website: www.bsria.co.uk

CASTINGS TECHNOLOGY INTERNATIONAL
7 East Bank Road, Sheffield
Tel: 0114 272 8647
Fax: 0114 273 0854
Website: www.castingsdev.com

CEMENT ADMIXTURES ASSOCIATION
38 Tilehouse, Green Lane, Knowle,
West Midlands B93 9EY
Tel: + Fax: 01564 776 362
Website: www.admixtures.org.uk

CHARTERED INSTITUTE OF ARBITRATORS
International Arbitration Centre
12 Bloomsbury Square, London WC1A 2LP
Tel: 0207 421 7444
Fax: 0207 404 4023
Website: www.arbitrators.org.uk

CHARTERED INSTITUTE OF BUILDING (CIOB)
Englemere, Kings Ride, Ascot, Berks SL5 7TB
Tel: 01344 630 700
Fax: 01344 630 777
Website: www.ciob.org.uk

CHARTERED INSTITUTION OF WATER AND ENVIRONMENTAL MANAGEMENT
15 John Street, London WC1N 2EB
Tel: 0207 831 3110
Fax: 0207 405 4967
Website: www.ciwem.org.uk

CIVIL ENGINEERING CONTRACTORS ASSOCIATION
55 Tufton Street, Westminster, London SW1P 3QL
Tel: 020 7227 4620
Fax: 020 7227 4621
Website: www.ceca.co.uk

CLAY PIPE DEVELOPMENT ASSOCIATION
Copsham House, 53 Broad Street, Chesham, Bucks HP5 3EA
Tel: 01494 791 456
Fax: 01494 792 378
Website: www.cpda.co.uk

COLD ROLLED SECTIONS ASSOCIATIONS
National Metal Forming Centre, 47 Birmingham Road, West Bromwich, B70 6PY
Tel: 0121 601 6350
Fax: 0121 601 6373
Website: www.crsauk.com

CONCRETE PIPELINE SYSTEMS ASSOCIATION
60 Charles St, Leicester LE1 1FB
Tel: 0116 253 6161
Fax: 0116 251 4568
Website: www.concretepipes.co.uk

CONFEDERATION OF BRITISH INDUSTRY
Centre Point, 103 New Oxford St, London WC1A 1DU
Tel: 0207 379 7400
Fax: 0207 240 1578
Website: www.cbi.org.uk

CONSTRUCTION CONFEDERATION
55 Tufton Street, Westminster, London SW1P 3QL
Tel: 0870 8989090
Fax: 0207 8989095
Website: www.thecc.org.uk

CONSTRUCTION EMPLOYERS FEDERATION
143 Malone Rd, Belfast BT9 6SU
Tel: 02890 877 143
Fax: 02890 877 155
Website: www.cefni.co.uk

CONSTRUCTION INDUSTRY RESEARCH & INFORMATION ASSOCIATION (CIRIA)
Classic House, 174-180 Old Street, London EC1V 9BP
Tel: 020 7549 3000
Fax: 020 7253 0523
Website: www.ciria.org.uk

CORUS CONSTRUCTION & INDUSTRIAL
PO Box L, Brigg Road, Scunthorpe, North Lincolnshire, DN16 1BP
Tel: 01724 404040
Fax: 01724 402191
Website: www.corusgroup.com

DEPARTMENT OF TRADE AND INDUSTRY
1 Victoria Street, London SW1H 0ET
Tel: 0207 215 5000
Fax: 0207 828 3258

DEPARTMENT FOR TRANSPORT
Ashdown House, 123 Victoria Street, London SW1E 6DE
Tel: 020 7944 3000
Website: www.dft.gov.uk

ELECTRICAL CONTRACTORS ASSOCIATION (ECA)
ESCA House, 34 Palace Court, Bayswater, London W2 4HY
Tel: 0207 313 4800
Fax: 0207 221 7344
Website: www.eca.co.uk

ELECTRICAL CONTRACTORS ASSOCIATION OF SCOTLAND (SELECT)
The Walled Garden, Bush Estate, Midlothian EH26 0SB
Tel: 0131 445 5577
Fax: 0131 445 5548
Website: www.select.org.uk

FEDERATION OF ENVIRONMENTAL TRADE ASSOCIATIONS
2 Waltham Court, Milley Lane, Hare Hatch, Reading RG10 9TH
Tel: 0118 940 3416
Fax: 0118 940 6258
Website: www.feta.co.uk

FEDERATION OF MASTER BUILDERS
Gordon Fisher House, 14 – 15 Gt James St, London WC1N 3DP
Tel: 0207 242 7583
Fax: 0207 404 0296
Website: www.fmb.org.uk

FEDERATION OF PILING SPECIALISTS
Forum Court, 83 Copers Cope Road, Beckenham, Kent BR3 1NR
Tel: 0208 663 0947
Fax: 0208 663 0949
Website: www.fps.org.uk

FEDERATION OF PLASTERING AND DRYWALL CONTRACTORS
The Building Centre, 26 Store Street London WC1E 7BT
Tel: 020 7580 3545
Fax: 020 7580 3288
Website: www.fpdc.org.uk

FENCING CONTRACTORS ASSOCIATION
Warren Road, Trelleck, Monmouthshire NP25 4PQ
Tel: 07000 560 722
Fax: 01600 860 614
Website: www.fencingcontractors.org

FLAT ROOFING ALLIANCE
Fields House, Gower Road, Haywards Heath, West Sussex RH16 4PL
Tel: 01444 440 027
Fax: 01444 415 616
Website: www.fra.org.uk

HEALTH & SAFETY LABORATORY
Business Development Unit Health & Safety Laboratory, Harpur Hill, Buxton, Derbyshire SK17 9JN
Tel: 01298 218 218
Fax: 01298 218 822
Website: www.hsl.gov.uk

HEATING AND VENTILATORS CONTRACTORS ASSOCIATION
ESCA House, 34 Palace Court, London W2 4JG
Tel: 0207 313 4900
Fax: 0207 727 9268
Website: www.hvca.org.uk

HM LAND REGISTRY (HQ)
32 Lincolns Inn Fields, London WC2A 3PH
Tel: 0207 917 8888
Fax: 0207 955 0110
Website: www.landreg.gov.uk

ICOM ENERGY ASSOCIATION
6th Floor, The MacLaren Building, 35 Dale End, Birmingham BI4 7LN
Tel: 0121 200 2100
Fax: 0121 200 1306
Website: www.icomenergyassociation.org.uk

INSTITUTE OF CONCRETE TECHNOLOGY
4 Meadows Business Park, Station Approach, Blackwater, Camberley GU17 9AB
Tel/Fax: *01276 37831*
Website: www.ictech.org

INSTITUTE OF MATERIALS, MINERALS & MINING
1 Carlton House Terrace, London SW1Y 5DB
Tel: 0207 451 7300
Fax: 0207 839 1702
Website: www.materials.org.uk

INSTITUTE OF QUALITY ASSURANCE
12 Grosvenor Crescent, London SW1X 7EE
Tel: 0207 245 6722
Fax: 0207 245 6755
Website: www.iqa.org.uk

INSTITUTION OF BRITISH ENGINEERS
Clifford Hill Court, Clifford Chambers
Stratford Upon-Avon
Warwickshire CU37 8AA
Tel: 01789 298 739
Fax: 01789 294 442
Website: www.britishengineers.com

INSTITUTION OF CIVIL ENGINEERS
1 Great George St, London SW1P 3AA
Tel: 0207 222 7722
Fax: 0207 222 7500
Website: www.ice.org.uk

INSTITUTION OF ELECTRICAL ENGINEERS
Savoy Place, London WC2R 0BL
Tel: 0207 240 1871
Fax: 0207 240 7735
Website: www.iee.org.uk

INSTITUTION OF INCORPORATED ENGINEERS
Savoy Hill House, Savoy Hill, London WC2R 0BS
Tel: 0207 836 3357
Fax: 0207 497 9006
Website: www.iie.org.uk

INSTITUTION OF MECHANICAL ENGINEERS
1 Birdcage Walk, London SW1H 9JJ
Tel: 0207 222 7899
Fax: 0207 222 4557
Website: www.imeche.org.uk

INSTITUTION OF STRUCTURAL ENGINEERS
11 Upper Belgrave Street, London SW1X 8BH
Tel: 0207 235 4535
Fax: 0207 235 4294
Website: www.istructe.org.uk

INSTITUTION OF WATER AND ENVIRONMENTAL MANAGEMENT
See CHARTERED INSTITUTION OF WATER AND ENVIRONMENTAL MANAGEMENT

INSTITUTION OF WASTES MANAGEMENT
9 Saxon Court, St Peter's Gardens, Marefair, Northampton NN1 1SX
Tel: 01604 620 426
Fax: 01604 621 339
Website: www.ciwm.co.uk

INTERNATIONAL CONCRETE BRICK ASSOCIATION
See BRITISH PRECAST CONCRETE FEDERATION

INTERPAVE
See BRITISH PRECAST CONCRETE FEDERATION

THE JOINT CONTRACTS TRIBUNAL
9 Cavendish Place, London W1G 0QD
Tel: 0207 637 8650
Fax: 0207 637 8670
Website: www.jctltd.co.uk

LANDSCAPE INSTITUTE
33 Great Portland Street, London W1W 8QG
Tel: 0207 299 4500
Fax: 0207 299 4501
Website: www.l-i.org.uk

LEAD DEVELOPMENT ASSOCIATION
42 Weymouth Street, London W1G 6NP
Tel: 0207 499 8422
Fax: 0207 493 1555
Website: www.ldaint.org

MASTIC ASPHALT COUNCIL LTD
PO Box 77, Hastings,
Kent TN35 4WL
Tel: 01424 814 400
Fax: 01424 814 446
Website: www.masticasphaltcouncil.co.uk

MET OFFICE
Fitzroy Road, Exeter, Devon EX1 3PB
Tel: 0870 900 0100
Fax: 0870 900 5050
Website: www.met-office.gov.uk

NATIONAL ASSOCIATION OF STEEL STOCKHOLDERS
6th Floor, McLaren Building, 35 Dale End,
Birmingham B4 7LN
Tel: 0121 200 2288
Fax: 0121 236 7444
Website: www.nass.org.uk

ORDNANCE SURVEY
Romsey Road, Maybush, Southampton
SO16 4GU
Tel: 0845 605 0505
Fax: 02380 792 615
Website: www.ordsvy.gov.uk

PIPELINE INDUSTRIES GUILD
14/15 Belgrave Square, London SW1X 8PS
Tel: 0207 235 7938
Fax: 0207 235 0074
Website: www.pipeguild.co.uk

PLASTIC PIPE MANUFACTURERS SOCIETY
89 Cornwall Street, Birmingham B3 3BY
Tel: 0121 236 1866
Fax: 0121 200 1389

PUBLIC RECORDS OFFICE (THE NATIONAL ARCHIVES)
Ruskin Avenue, Q Richmond, Surrey TW9 4DU
Tel: 0208 876 3444
Fax: 0208 392 5286
Website: www.nationalarchives.gov.uk

QUARRY PRODUCTS ASSOCIATION
Gillingham House, 38-44 Gillingham Street,
London SW1V 1HU
Tel: 0207 963 8000
Fax: 0207 963 8001
Website: www.qpa.org

REINFORCED CONCRETE COUNCIL
Riverside House, 4 Meadows Business Park,
Station Approach, Camberley GU17 9AB
Tel: 01276 607140
Fax: 01276 607141
Website: www.rcc-info.org.uk

THE RESIN FLOORING FEDERATION
Association House, 99 West Street
Farnham, Surrey GU9 7EN
Tel: 01252 739 149
Fax: 01252 739 140
Website: www.ferfa.org.uk

ROYAL INSTITUTE OF BRITISH ARCHITECTS
66 Portland Place, London W1B 1AD
Tel: 0207 580 5533
Fax: 0207 255 1541
Website: www.riba.org.uk

ROYAL INSTITUTION OF CHARTERED SURVEYORS (RICS)
12 Great George Street, Parliament Square
London SW1P 3AD
Tel: 0870 333 1600
Fax: 0207 222 9430
Website: www.rics.org.uk

SCOTTISH BUILDING EMPLOYERS FEDERATION
Carron Grange, Carrongrange Avenue,
Stenhousemuir, FK5 3BQ
Tel: 01324 555 550
Fax: 01324 555 551
Website: www.scottish-building.co.uk

SCOTTISH ENTERPRISE
5 Atlantic Quay, 150 Broomielaw,
Glasgow G2 8LU
Tel: 0141 248 2700
Fax: 0141 221 3217
Website: www.scottish-enterprise.com

SOCIETY OF GLASS TECHNOLOGY
Don Valley House, Saville Street East,
Sheffield S4 7UQ
Tel: 0114 263 4455
Fax: 0114 263 4411
Website: www.societyofglasstechnology.org

SPECIALISED ACCESS ENGINEERING AND MAINTENANCE ASSOCIATION
Carthusian Court, 12 Carthusian Street,
London EC1M 6EZ
Tel: 020 7397 8122
Fax: 020 7397 8121
Website: www.saema.org

STEEL CONSTRUCTION INSTITUTE
Silwood Park, Ascot, Berks SL5 7QN
Tel: 01344 623 345
Fax: 01344 622 944
Website: www.steel-sci.org

SWIMMING POOL AND ALLIED TRADES ASSOCIATION (SPATA)
SPATA House, 1A Junction Road, Andover, Hants SP10 3QT
Tel: 01264 356 210
Fax: 01264 332 628
Website: www.spata.co.uk

THERMAL INSULATION CONTRACTORS ASSOCIATION
TICA House, Allington Way, Yarm Road Business Park, Darlington, County Durham DL1 4QB
Tel: 01325 466 704
Fax: 01325 487 691
Website: www.tica-acad.co.uk

THERMAL INSULATION MANUFACTURERS AND SUPPLIERS ASSOCIATION
Association House, 99 West Street, Farnham Surrey GU9 7EN
Tel: 01252 739 154
Fax: 01252 739 140
Website: www.timsa.org.uk

TIMBER TRADE FEDERATION LTD
Clareville House, 26-27 Oxenden Street, London SW1Y 4EL
Tel: 0207 839 1891
Fax: 0207 930 0094
Website: www.ttf.co.uk

TOWN AND COUNTRY PLANNING ASSOCIATION
17 Carlton House Terrace, London SW1Y 5AS
Tel: 0207 930 8903
Fax: 0207 930 3280
Website: www.tcpa.org.uk

TRADA TECHNOLOGY LTD
Stocking Lane, Hughenden Valley, High Wycombe, Bucks HP14 4ND
Tel: 01494 569 602
Fax: 01494 565 487
Website: www.trada.co.uk

TWI - THE WELDING INSTITUTE
Granta Park, Great Abington, Cambridge CB1 6AL
Tel: 01123 891 162
Fax: 01223 892 588
Website: www.twi.co.uk

UK STEEL ASSOCIATION
Broadway House, Tothill Street, London SW1H 9NQ
Tel: 0207 222 7777
Fax: 0207 222 2782
Website: www.uksteel.org.uk

PART 15

Updates

Taylor & Francis
Taylor & Francis Group plc

INTRODUCTION

This section has been compiled as an aid in making adjustments to the rates and prices contained in Parts 3 to 5, 7 and 12 of the Book.

Users are advised to mark up the relevant pro-forma in line with any adjustments to the rates price and costs which will be reported in the free quarterly updates. This will allow the reader to obtain maximum benefit of the cost research that has been used to compile the data contained in the Book.

LABOUR COSTS

Revisions to labour costs as detailed on page 33.

Categories of labour	General	Skill rate 4	Skill rate 3	Skill rate 2	Skill rate 1	Craft rate
Update 1						
Update 2						
Update 3						

MATERIAL PRICES

This section details market prices of materials incorporating any appropriate trade discounts and delivery to site (except where noted), although distribution after delivery is not included.

Material Prices	End Oct	End Dec	End Mar
Materials generally			
Aluminium cladding			
Bar and fabric reinforcement			
Brickwork/blockwork			
Cast iron pipes and fittings			
Clayware pipes and fittings			
Concrete and cement			
Concrete manholes			
Concrete pipes and fittings			
Consumable stores			
Contractors site equipment			
Concrete culverts			
Earth retention			
Fencing			
Geotextiles			

MATERIALS - continued

MATERIAL PRICES	End Oct	End Dec	End Mar
Gulley grates and frames			
Soil instrumentation			
Interlocking steel sheet piling and bearing piles			
Joint fillers and waterstops			
Landscaping			
Manhole covers and frames			
Paint/stains/protective coatings			
Polymer channels and fittings			
Precast concrete kerbs etc.			
Quarry products, aggregates etc.			
Scour and erosion protection			
Septic tanks, cesspools etc.			
Shuttering timber, nails etc.			
Steel pipes and fittings			
Structural steelwork			
Timber			
PVC-U drain pipes and fittings			
Miscellaneous and minor items			

PLANT PRICES

This section indicates general hire prices of plant transported to site; prices of fuel are indicated separately.

PLANT	End Oct	End Dec	End Mar
access platforms			
agriculture type tractors			
asphalt/road construction			
compaction			
cleaners sweepers			
compressors			

PLANT	End Oct	End Dec	End Mar
compressor tools			
concrete equipment			
concrete pumps			
concrete skips			
cranes			
diesel generators			
excavators			
excavator equipment			
hoists and lifting			
piling plant			
portable accommodation			
rollers			
scaffolding and staging			
tractors (dozers, loaders and scrapers)			
transport (tippers and dumpers)			
transport (vans etc.)			
trench sheets			
pumping and dewatering			
miscellaneous			
power tools			
site tools			

UNIT COSTS

This section indicates general movement inclusive of labour, plant and materials cost changes as well as specialists.

UNIT COSTS	End Oct	End Dec	End Mar
Ground investigation, Geotechnical and other specialist processes			
Demolition and Site Clearance			
Fencing, Hedges etc.			

UNIT COSTS - continued

UNIT COSTS	End Oct	End Dec	End Mar
Safety Fencing, Parapets			
Drainage Work			
Earthworks - Excavation			
- Filling (imported)			
- Landscaping			
Concrete and Reinforcement			
Pre-cast items			
Formwork			
Pavements - Surfacing flexible			
- Surfacing concrete			
Kerbs and footways			
Signs and lighting columns			
- Trenches			
- Electrical Work			
Piling Works			
Structural Steelwork			
Timber Work			
Painting			
Rail Track			
Tunnels			
Waterproofing			
Roofing and Cladding			
Bridge Bearings			
Expansion joints			
Brickwork, Blockwork, Stonework			

APPROXIMATE ESTIMATES

The following details the changes in approximate estimate costs given in Part 12.

APPROXIMATE ESTIMATES	End Oct	End Dec	End Mar
Industrial and commercial facilities			
Water & treatment facilities			
Foundations for structures			
Concrete slabs for structures			
Earth retention and stabilisation			
- Concrete retaining walls			
- Crib walls			
- Gabions			
Bridgeworks			
- Concrete			
- Steel			
- Timber			
Highway works			
Civil engineering utilities & infrastructure			
- Roadworks, etc.			
- Services			
- Landscaping			
Temporary works			
- Bailey bridges			
- Piled cofferdams			
- Trench supports			
- Scaffolding			

Index

CD-Rom Single-User Licence Agreement

We welcome you as a user of this Taylor & Francis CD-ROM and hope that you find it a useful and valuable tool. Please read this document carefully. **This is a legal agreement** between you (hereinafter referred to as the "Licensee") and Taylor and Francis Books Ltd. (the "Publisher"), which defines the terms under which you may use the Product. **By breaking the seal and opening the package containing the CD-ROM you agree to these terms and conditions outlined herein. If you do not agree to these terms you must return the Product to your supplier intact, with the seal on the CD case unbroken.**

1. Definition of the Product

The product which is the subject of this Agreement, *Spon's Civil Engineering and Highway Works Price Book 2007 on CD-ROM* (the "Product") consists of:

1.1 Underlying data comprised in the product (the "Data")
1.2 A compilation of the Data (the "Database")
1.3 Software (the "Software") for accessing and using the Database
1.4 A CD-ROM disk (the "CD-ROM")

2. Commencement and licence

2.1 This Agreement commences upon the breaking open of the package containing the CD-ROM by the Licensee (the "Commencement Date").
2.2 This is a licence agreement (the "Agreement") for the use of the Product by the Licensee, and not an agreement for sale.
2.3 The Publisher licenses the Licensee on a non-exclusive and non-transferable basis to use the Product on condition that the Licensee complies with this Agreement. The Licensee acknowledges that it is only permitted to use the Product in accordance with this Agreement.

3. Multiple use

For more than one user or for a wide area network or consortium, use is only permissible with the purchase from the Publisher of a multiple-user licence and adherence to the terms and conditions of that licence.

4. Installation and Use

4.1 The Licensee may provide access to the Product for individual study in the following manner: The Licensee may install the Product on a secure local area network on a single site for use by one user.
4.2 The Licensee shall be responsible for installing the Product and for the effectiveness of such installation.
4.3 Text from the Product may be incorporated in a coursepack. Such use is only permissible with the express permission of the Publisher in writing and requires the payment of the appropriate fee as specified by the Publisher and signature of a separate licence agreement.
4.4 The CD-ROM is a free addition to the book and no technical support will be provided.

5. Permitted Activities

5.1 The Licensee shall be entitled:
  5.1.1 to use the Product for its own internal purposes;
  5.1.2 to download onto electronic, magnetic, optical or similar storage medium reasonable portions of the Database provided that the purpose of the Licensee is to undertake internal research or study and provided that such storage is temporary;
  5.1.3 to make a copy of the Database and/or the Software for back-up/archival/disaster recovery purposes.
5.2 The Licensee acknowledges that its rights to use the Product are strictly set out in this Agreement, and all other uses (whether expressly mentioned in Clause 6 below or not) are prohibited.

6. Prohibited Activities

The following are prohibited without the express permission of the Publisher:

6.1 The commercial exploitation of any part of the Product.
6.2 The rental, loan, (free or for money or money's worth) or hire purchase of this product, save with the express consent of the Publisher.
6.3 Any activity which raises the reasonable prospect of impeding the Publisher's ability or opportunities to market the Product.
6.4 Any networking, physical or electronic distribution or dissemination of the product save as expressly permitted by this Agreement.
6.5 Any reverse engineering, decompilation, disassembly or other alteration of the Product save in accordance with applicable national laws.
6.6 The right to create any derivative product or service from the Product save as expressly provided for in this Agreement.
6.7 Any alteration, amendment, modification or deletion from the Product, whether for the purposes of error correction or otherwise.

7. General Responsibilities of the License

7.1 The Licensee will take all reasonable steps to ensure that the Product is used in accordance with the terms and conditions of this Agreement.

7.2 The Licensee acknowledges that damages may not be a sufficient remedy for the Publisher in the event of breach of this Agreement by the Licensee, and that an injunction may be appropriate.

7.3 The Licensee undertakes to keep the Product safe and to use its best endeavours to ensure that the product does not fall into the hands of third parties, whether as a result of theft or otherwise.

7.4 Where information of a confidential nature relating to the product of the business affairs of the Publisher comes into the possession of the Licensee pursuant to this Agreement (or otherwise), the Licensee agrees to use such information solely for the purposes of this Agreement, and under no circumstances to disclose any element of the information to any third party save strictly as permitted under this Agreement. For the avoidance of doubt, the Licensee's obligations under this sub-clause 7.4 shall survive the termination of this Agreement.

8. Warrant and Liability

8.1 The Publisher warrants that it has the authority to enter into this agreement and that it has secured all rights and permissions necessary to enable the Licensee to use the Product in accordance with this Agreement.

8.2 The Publisher warrants that the CD-ROM as supplied on the Commencement Date shall be free of defects in materials and workmanship, and undertakes to replace any defective CD-ROM within 28 days of notice of such defect being received provided such notice is received within 30 days of such supply. As an alternative to replacement, the Publisher agrees fully to refund the Licensee in such circumstances, if the Licensee so requests, provided that the Licensee returns the Product to the Publisher. The provisions of this sub-clause 8.2 do not apply where the defect results from an accident or from misuse of the product by the Licensee.

8.3 Sub-clause 8.2 sets out the sole and exclusive remedy of the Licensee in relation to defects in the CD-ROM.

8.4 The Publisher and the Licensee acknowledge that the Publisher supplies the Product on an "as is" basis. The Publisher gives no warranties:

8.4.1 that the Product satisfies the individual requirements of the Licensee; or

8.4.2 that the Product is otherwise fit for the Licensee's purpose; or

8.4.3 that the Data are accurate or complete of free of errors or omissions; or

8.4.4 that the Product is compatible with the Licensee's hardware equipment and software operating environment.

8.5 The Publisher hereby disclaims all warranties and conditions, express or implied, which are not stated above.

8.6 Nothing in this Clause 8 limits the Publisher's liability to the Licensee in the event of death or personal injury resulting from the Publisher's negligence.

8.7 The Publisher hereby excludes liability for loss of revenue, reputation, business, profits, or for indirect or consequential losses, irrespective of whether the Publisher was advised by the Licensee of the potential of such losses.

8.8 The Licensee acknowledges the merit of independently verifying Data prior to taking any decisions of material significance (commercial or otherwise) based on such data. It is agreed that the Publisher shall not be liable for any losses which result from the Licensee placing reliance on the Data or on the Database, under any circumstances.

8.9 Subject to sub-clause 8.6 above, the Publisher's liability under this Agreement shall be limited to the purchase price.

9. Intellectual Property Rights

9.1 Nothing in this Agreement affects the ownership of copyright or other intellectual property rights in the Data, the Database of the Software.

9.2 The Licensee agrees to display the Publishers' copyright notice in the manner described in the Product.

9.3 The Licensee hereby agrees to abide by copyright and similar notice requirements required by the Publisher, details of which are as follows:

"© 2007 Taylor & Francis. All rights reserved. All materials in *Spon's Civil Engineering and Highway Works Price Book 2007* are copyright protected. © 2006 Adobe Systems Incorporated. All rights reserved. No such materials may be used, displayed, modified, adapted, distributed, transmitted, transferred, published or otherwise reproduced in any form or by any means now or hereafter developed other than strictly in accordance with the terms of the licence agreement enclosed with the CD-ROM. However, text and images may be printed and copied for research and private study within the preset program limitations. Please note the copyright notice above, and that any text or images printed or copied must credit the source."

9.4 This Product contains material proprietary to and copyedited by the Publisher and others. Except for the licence granted herein, all rights, title and interest in the Product, in all languages, formats and media throughout the world, including copyrights therein, are and remain the property of the Publisher or other copyright holders identified in the Product.

10. Non-assignment

This Agreement and the licence contained within it may not be assigned to any other person or entity without the written consent of the Publisher.

11. Termination and Consequences of Termination.

11.1 The Publisher shall have the right to terminate this Agreement if:

11.1.1 the Licensee is in material breach of this Agreement and fails to remedy such breach (where capable of remedy) within 14 days of a written notice from the Publisher requiring it to do so; or

11.1.2 the Licensee becomes insolvent, becomes subject to receivership, liquidation or similar external administration; or

11.1.3 the Licensee ceases to operate in business.

11.2 The Licensee shall have the right to terminate this Agreement for any reason upon two month's written notice. The Licensee shall not be entitled to any refund for payments made under this Agreement prior to termination under this sub-clause 11.2.

11.3 Termination by either of the parties is without prejudice to any other rights or remedies under the general law to which they may be entitled, or which survive such termination (including rights of the Publisher under sub-clause 7.4 above).

11.4 Upon termination of this Agreement, or expiry of its terms, the Licensee must:

11.4.1 destroy all back up copies of the product; and

11.4.2 return the Product to the Publisher.

12. General

12.1 **Compliance with export provisions**

The Publisher hereby agrees to comply fully with all relevant export laws and regulations of the United Kingdom to ensure that the Product is not exported, directly or indirectly, in violation of English law.

12.2 **Force majeure**

The parties accept no responsibility for breaches of this Agreement occurring as a result of circumstances beyond their control.

12.3 **No waiver**

Any failure or delay by either party to exercise or enforce any right conferred by this Agreement shall not be deemed to be a waiver of such right.

12.4 **Entire agreement**

This Agreement represents the entire agreement between the Publisher and the Licensee concerning the Product. The terms of this Agreement supersede all prior purchase orders, written terms and conditions, written or verbal representations, advertising or statements relating in any way to the Product.

12.5 **Severability**

If any provision of this Agreement is found to be invalid or unenforceable by a court of law of competent jurisdiction, such a finding shall not affect the other provisions of this Agreement and all provisions of this Agreement unaffected by such a finding shall remain in full force and effect.

12.6 **Variations**

This agreement may only be varied in writing by means of variation signed in writing by both parties.

12.7 **Notices**

All notices to be delivered to: Spon's Price Books, Taylor & Francis Books Ltd., 2 Park Square, Milton Park, Abingdon, Oxfordshire, OX14 4RN, UK.

12.8 **Governing law**

This Agreement is governed by English law and the parties hereby agree that any dispute arising under this Agreement shall be subject to the jurisdiction of the English courts.

If you have any queries about the terms of this licence, please contact:

Spon's Price Books
Taylor & Francis Books Ltd.
2 Park Square, Milton Park, Abingdon, Oxfordshire, OX14 4RN
Tel: +44 (0) 20 7017 6672
Fax: +44 (0) 20 7017 6702
www.sponpress.com

CD-ROM Installation Instructions

System requirements

Minimum

- Pentium processor
- 32 MB of RAM
- 10 MB available hard disk space
- CD-ROM drive
- Microsoft Windows 95/98/2000/NT/ME/XP
- SVGA screen
- Internet connection

Recommended

- Pentium 266 MHz processor
- 256 MB of RAM
- 100 MB available hard disk space
- CD-ROM drive
- Microsoft Windows 2000/NT/XP
- XVGA screen
- Internet connection

Microsoft ® is a registered trademark and Windows™ is a trademark of the Microsoft Corporation.

How to install *Spon's Civil Engineering and Highway Works Price Book 2007 CD-ROM*

Windows 95/98/2000/NT

Spon's Civil Engineering and Highway Works Price Book 2007 CD-ROM should run automatically when inserted into the CD-ROM drive. If it fails to run, follow the instructions below.

- Click the **Start** button and choose **Run.**
- Click the **Browse** button.
- Select your CD-ROM drive.
- Select the Setup file (setup.exe) then click **Open.**
- Click the OK button.
- Follow the instructions on screen.
- The installation process will create a folder containing an icon for *Spon's Civil Engineering and Highway Works Price Book 2007 CD-ROM* and also an icon on your desktop.

How to run the *Spon's Civil Engineering and Highway Works Price Book 2007 CD-ROM*

- Double click the icon (from the folder or desktop) installed by the Setup program.
- Follow the instructions on screen.

Installation

The CD-ROM is a free addition to the book and no technical support will be provided. For help with the use of the CD-ROM please visit www.pricebooks.co.uk

Multiple-user use of the Spon Press CD–ROM

To buy a licence to install your Spon Press Price Book CD–ROM on a secure local area network or a wide area network, and for the supply of network key files, for an agreed number of users please contact:

Spon's Price Books
Taylor & Francis Books Ltd.
2 Park Square, Milton Park, Abingdon, Oxfordshire, OX14 4RN
Tel: +44 (0) 207 7017 6672
Fax: +44 (0) 207 7017 6072
www.sponpress.com

Number of users	Licence cost
2–5	£390
6–10	£780
11–20	£1170
21–30	£1750
31–50	£3500
51–75	£5000
76–100	£6000
Over 100	Please contact Spon for details